PRINCIPLES OF
MICROBIOLOGY

PRINCIPLES OF
MICROBIOLOGY

Alice Lorraine Smith
A.B., M.D., F.C.A.P., F.A.C.P.

Professor of Pathology, The University of Texas Health Science
Center at Dallas, Texas; formerly Assistant Professor of Microbiology,
Department of Nursing, Dominican College and St. Joseph's Hospital,
Houston, Texas

WITH 352 ILLUSTRATIONS

Eighth edition

The C. V. Mosby Company
Saint Louis   1977

Eighth edition

Copyright © 1977 by The C. V. Mosby Company

All rights reserved. No part of this book may be reproduced in any manner without written permission of the publisher.

Previous editions copyrighted 1951, 1954, 1957, 1961, 1965, 1969, 1973

Printed in the United States of America

The C. V. Mosby Company
11830 Westline Industrial Drive, St. Louis, Missouri 63141

Library of Congress Cataloging in Publication Data

Smith, Alice Lorraine, 1920-
   Principles of microbiology.

   Includes bibliographical references and index.
   1. Medical microbiology. I. Title. [DNLM:
1. Microbiology. QW4 S642p]
QR46.C35 1977   616.01   76-30332
ISBN 0-8016-4681-2

TS/CB/B 9 8 7 6 5 4 3 2

To
Ned II and **Frederick IV**

PREFACE

"The quick harvest of applied science is the usable process,
the medicine, the machine. The shy fruit of pure science is Understanding."

LINCOLN BARNETT

The quick harvest is indeed apparent, for scientific advances come thick and fast, as it were, plowing the furrows of knowledge and sowing the seeds of discovery. Microbiology, never a static science, is rapidly caught up. Again it is time for a revision, time to take stock, time to rework, "time to plant and . . . time to pluck up that which is planted."

Here then is the text reworked to cover the contemporary scene and designed to be assimilated readily by students in health science training programs. The basic pattern of prior editions continues, comprising six units in orbit around microorganisms. It begins with basic concepts in the first unit, includes laboratory methods for the study of microbes in the second, and, in subsequent units, develops the events of microbial injury, indicts culprits, and emphasizes restraints; it also includes benefits that microbes confer.

What follows from the contact of microbes with living cells of the human body is a theme permeating this book. One full unit (Unit III) is devoted to it, concentrating on defenses inherent in the body. It discusses phagocytes, the immune system, kinds of immunity, the dual nature of the immune response, allergy, and key laboratory reactions in immunology.

A small and fairly compact unit (Unit IV) categorizes various agents destroying or impeding microbes. The action of antimicrobial agents is noted, and unfavorable side effects with antimicrobial drugs are stressed. Admittedly, with the increasing use and availability of commercial, prepackaged, sterile disposable units of all kinds, drastic changes are occurring in our concepts of sterilization. However, certain standard, long-reliable measures for practical sterilization in the health field still merit consideration. Hands, for instance, are not as yet disposable.

The largest unit of the book (Unit V) makes up the roster of significant pathogens and parasites, stressing their identity and nature of injury. Infections are matched to agents.

Preface

An unusual unit and the last one, Unit VI relates the student to the microbial life of our environment. A survey, and yet a practical unit, it accommodates such items as measures to safeguard food and fluoridation of water. Two chapters are paired to focus on the best available information on immunization from the United States Public Health Service, the American Academy of Pediatrics, the United States Armed Forces, the World Health Organization (WHO), and the Center for Disease Control (CDC). The companion chapters sort out modern biologic products, outline technics in passive immunization, tabulate latest schedules for active immunization, and give crucial guidelines for administration of biologic products.

This book in frame and substance must remain relevant to the here and now in health careers. It has been edited and updated to keep it so. New topics include the serologic diagnosis of protozoan and metazoan diseases, experimental production of dental caries, practical definition of cell-mediated immunity, the scanning electron microscope, Dane particle, and laboratory tests to detect allergy. New emphases come from expansion of such topics as the biologic classification of microbes, the lymphoid system's role in immunity, T and B cells, immunoglobulin E and allergy, viral hepatitis, influenza and the influenza virus, animals as sources of infection, BCG vaccination, and health information for the traveler. The historical survey of the second chapter, "Milestones of Progress," has been augmented.

The classification of bacteria throughout this text is that of *Bergey's Manual of Determinative Bacteriology* (1974), with the notable exception of the arrangement of enteric bacteria in the scheme of W. H. Ewing. The revisions in the new *Bergey's Manual* necessitated changes in terminology and reorganization of material with a new format for the chapters. Please note that rickettsias, chlamydiae (grouped with the rickettsias), and actinomycetes are now classed as bacteria. Bedsonia is chlamydia, and pneumococcus is streptococcus. New names such as *Branhamella catarrhalis* (formerly *Neisseria catarrhalis*), *Yersinia pestis* (old name, *Pasteurella pestis*), and *Francisella tularensis* (old name, *Pasteurella tularensis*) are the standard ones. For viruses, the most modern classification remains that based on their properties. One change is made; arboviruses are preferably designated togaviruses.

Projects for the laboratory survey at the end of each unit are adaptable to the needs of students and suitable for use with varied, and sometimes limited, facilities. Five chapters of the text support the laboratory sections with discussions of both conventional and newly emerging methods for study of microbes. *For maximum laboratory safety—at the outset, during, and after the laboratory session—please heed the warnings!*

"What is the use of a book," thought Alice, "without pictures or conversations?" (Lewis Carroll, *Alice's Adventures in Wonderland*). *That* Alice is partly right on both counts, but *this* Alice can only comply with the first and hope that the "pictures" herein are good and forceful. In this edition the illustrations have been carefully selected to enrich the prose. Every teacher knows that tables dramatize information. Immunization schedules, sterilization maneuvers, serologic testing, incubation periods, comparative sizes, metric equivalents, Arthropoda and diseases, differential characteristics, and biologic properties can thus be arranged effectively.

Current references are gathered at the end of a unit or after related chapters. In addition to thought-provoking questions for review at the end of a chapter, a cluster of exercises arranged at the end of each unit allows for evaluation of the student's progress. Sources for the glossary are found in the text, standard medical dictionaries, and Webster's unabridged dictionary.

This revision would not have been possible without the counsel, technical know-how, and cooperation of certain talents at The University of Texas Health Science Center at Dallas. In the Department of Pathology, I gratefully acknowledge the kindness of Dr. V. A. Stembridge, Chairman; Drs. R. C. Reynolds, B. D. Fallis, Frank Vellios, and C. S. Petty, Professors; Mrs. Phyllis Kitterman, Secretary, Mrs. Linda Bolding, Laboratory Technical Assistant, and Mr. Donald Calhoun, Photographer for the Medical Examiner of Dallas County; in the Department of Medical Illustration Services, Dr. W. R. Christensen, Chairman; R. J. Castanie, Medical Illustrator, and Miss Jean Gionas, Medical Photographer; and in the library, Mrs. Elinor Reinmiller, Reference Librarian, and other able members of Dr. Donald Hendricks' staff. I am indebted to Mrs. Earline Kutscher, Chief Technologist, and her staff at the Bacteriology Laboratory of Parkland Memorial Hospital for invaluable assistance.

Now a special word of appreciation to teachers and students whose ever-welcome criticisms and comments have guided me; may I voice a heartfelt thanks to the many of you who have used this text and who carefully consider this new edition.

Alice Lorraine Smith

CONTENTS

Contents

Contents

UNIT ONE
MICROBIOLOGY
PRELUDE AND PRIMER

1 Definition and dimension

Take interest, I implore you, in those sacred dwellings which one designates by the expressive term: laboratories. Demand that they be multiplied, that they be adorned. These are the temples of the future —temples of well-being and of happiness. There it is that humanity grows greater, stronger, better.

LOUIS PASTEUR

DEFINITION

Biology is the science that treats of living organisms. The branch of biology dealing with *microbes*—organisms structured as one cell and studied with the microscope— is *microbiology* ("microbe-biology"). Microbiology considers the occurrence in nature of the microscopic forms of life, their reproduction and physiology, their participation in the processes of nature, their helpful or harmful relationships with other living things, and their significance in science and industry.

Within the province of microbiology lies the study of bacteria, viruses, fungi, and protozoa. Subordinate sciences are *bacteriology*, the study of bacteria; *virology*, the study of viruses; *mycology*, the study of fungi (unicellular and multicellular plants); and *protozoology*, the study of protozoa (unicellular animals). Many microbes are parasitic. The science dealing with organisms dependent on living things for their sustenance is *parasitology*. So closely associated with microbiology as to be considered a part of it is the science of *immunology*, the study of those mechanisms whereby one organism deals with the harmful effects of another.

Although man has lived with microorganisms from time immemorial and has used certain of their activities such as fermentation to his advantage, the science of microbiology is a product of only the last 100 years or so. The studies of Antonj van Leeuwenhoek in the seventeenth century had shown the existence of microscopic forms of life, but it was not until the work of Louis Pasteur toward the end of the nineteenth century (some 200 years later) that the science of microbiology really took

3

shape. The new science stated the germ theory of disease, demonstrated patterns of communicable disease, and gave man a measure of protection he had not known in his struggle against the injurious forces in the biologic environment.* In its time this very young science has influenced practically every phase of human endeavor.

DIMENSION
Biologic classification

All living things are classified in a scheme wherein categories represent successively dependent and related groups. The highest possible levels are designated kingdoms, of which there may be two or three. For years, the traditional two were the plant and animal kingdoms. Today there are forces for change in this approach. The following basic terms in classification are listed in ascending order:

1. Species—organisms sharing a set of biologic traits and reproducing only their exact kind
   a. Strain—organisms within the species varying in a given quality
   b. Type—organisms within the species varying immunologically
2. Genus (*pl.*, genera)—closely related species
3. Family—closely related genera
4. Order—closely related families
5. Class—closely related orders
6. Phylum (*pl.*, phyla)†—related classes

The lower forms of life incorporate variable features of both plants and animals and do not show the dramatic differences of the higher forms. It is difficult to define many microbes as either plant or animal, and as bacteria and other microbes long classified as plants have been more closely studied, the inconsistencies appear even greater. Because of this fact, a third biologic compartment with equivalent rank to the plant and animal kingdoms has been advocated to sift out the simpler units, designating them as *protists*. Basically most protists are one-cell units and remain so throughout their life history. Even if they pile cells up in large plantlike masses, their component cells remain the same and do not differentiate.

There is an alternate scheme of classification of living things, again based on complexity of structure but this time focused directly on the nucleus. It is apparent that "higher" organisms possess a true nucleus; "lower" ones do not. Thus two distinct categories emerge. The obvious nucleus in the higher forms is complete, with the expected number of chromosomes and mitotic apparatus. These organisms are termed "eucaryotic" (using the Greek word that means true nucleus). This category (or kingdom) would be one to contain some protists and the plants and animals.

In the lower forms of life, nuclear function is carried out by only a single chromosome devoid of any membrane. Lower forms are small and less complex in other

---

*For scientific knowledge to bring results, such as in the organization of public health programs, it must be disseminated. Such is the aim of *health education*. To the individual it explains the mechanisms by which he can protect himself against microbial hazards. To the social group it designates the available community resources.

†In plant biology, the term *division* is used instead of phylum. In Table 1-1 (p. 6) note use of the term *division* for either of the two major classifications in the kingdom to include all procaryotes.

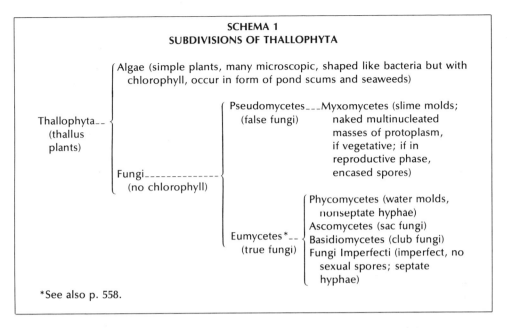

SCHEMA 1
SUBDIVISIONS OF THALLOPHYTA

Thallophyta__ (thallus plants)

Algae (simple plants, many microscopic, shaped like bacteria but with chlorophyll, occur in form of pond scums and seaweeds)

Fungi_____ (no chlorophyll)

Pseudomycetes___Myxomycetes (slime molds; (false fungi) naked multinucleated masses of protoplasm, if vegetative; if in reproductive phase, encased spores)

Eumycetes*__ (true fungi)

Phycomycetes (water molds, nonseptate hyphae)
Ascomycetes (sac fungi)
Basidiomycetes (club fungi)
Fungi Imperfecti (imperfect, no sexual spores; septate hyphae)

*See also p. 558.

ways. For instance they do not contain such membrane-bound organelles as mitochondria. They are designated "procaryotic," the second category to encompass all bacteria and a small group of blue-green algae (algae demonstrate plantlike photosynthesis).

Procaryotes, as a group, share distinctive properties. They possess certain unique components in their cell walls, and they display remarkable capabilities with regard to carbon storage, nitrogen fixation, obligate anaerobiosis, and derivation of energy from oxidation of inorganic compounds.

Scope. Microorganisms usually surveyed in a treatise of microbiology include bacteria (unicellular procaryotes as indicated previously), fungi, and lower forms of animal life.

Fungi are best known as plants and are classified in Thallophyta, one of the four divisions of the plant kingdom. Thallophytes, or thallus plants (Greek, *thallos*, young shoot or branch), are defined as simple forms of plant life that do not differentiate into true roots, stems, or leaves. The term *fungi* as ordinarily used refers to molds, yeasts, and certain related microorganisms. Schema 1 gives the subdivisions of thallus plants.

Some of the most important disease-producing agents known to man are lower forms of animal life. They include the unicellular protozoa, the simplest ones, and a restricted number of the more complex multicellular or metazoan animals as well (Chapters 31 and 32).

Naming of microbes

The scientific name of a living organism is usually made up of two words that are Latin or Greek in form. The first name begins with a capital letter and denotes the genus. The second name begins with a small letter and denotes the species. Either the genus or the species name may be derived from the proper name of a person or place or from a term describing some feature of the organism. The proper name may be that of the scientific investigator or that of the related geographic area. Biologic characteristics indicated include color, location in nature, disease produced, and presence of certain enzymes. For example, *Staphylococcus aureus* is the scientific name for bacteria of genus *Staphylococcus* (Greek *staphylē*, bunch of grapes + *kokkos*, berry) and species *aureus* (Latin *aureus*, golden). It indicates that the bacteria grow in typical clusters and produce a golden pigment. Honoring Sir David Bruce who discovered it, *Brucella melitensis* (pertaining to the island of Malta) by its name indicates its disease—Malta fever, or undulant fever—and the geographic area where first recognized.

Whether a term indicates a class, order, or family may be determined from its ending. Classes end in *etes*, orders end in *ales*, and families end in *aceae*. For instance, Phycomycetes is the name of a class, Spirochaetales is the name of an order, and Bacillaceae is the name of a family.

Classification of bacteria

The classification of bacteria is difficult. This applies both to the separation of bacteria into groups and to the placing of certain organisms into the proper group. Biologic classification is based largely on morphology, but the morphology of bacteria as a whole is so uniform that it is useful only in dividing bacteria into comparatively large groups. Shape has been an important factor in general classification, but for more exact identification such criteria as staining reactions, cultural characteristics, biochemical behavior, and immunologic differences must be used.

In this book we adhere to the scientific classification embodied in *Bergey's Manual of Determinative Bacteriology*, with certain important exceptions to be noted

Text continued on p. 20.

TABLE 1-1. ABBREVIATED CLASSIFICATION OF MICROBES FROM *BERGEY'S MANUAL OF DETERMINATIVE BACTERIOLOGY* (1974)*

Kingdom Procaryotae† (highest level taxon encompassing microbes wherein nucleoplasm lacks basic protein and is not bounded by nuclear membrane)

    Division I. **The Cyanobacteria** (blue-green algae with gliding motility, producing oxygen in light; photosynthetic procaryotes as single cells or simple or branched chains of cells; photopigments include chlorophyll a)

    Division II. **The Bacteria** (unicellular procaryotes multiplying by growth and division, usually binary; if cells remain together, arrangement classical; true branching may be seen; motility from flagella or by gliding, twitching, snapping, or darting motions; majority encased in rigid cell wall [constancy of form], most of which contain peptidoglycans; photosynthesis, if carried out, is anaerobic and bacteriochlorophylls used; chemosynthesis requires aerobic or anaerobic conditions, with some microbes facultative; endospores formed in some species, arthrospores and cysts in others, but no heterocysts)

*Based on data from Buchanan, R. E., and Gibbons, N. E., co-editors: *Bergey's manual of determinative bacteriology*, ed. 8, Baltimore, 1974, The Williams & Wilkins Co.

†*Eucaryotae,* the corresponding taxon at the same level, includes other protists and plants and animals.

‡Photosynthetic.

PART 1   PHOTOTROPHIC‡ BACTERIA
  Order I. **Rhodospirillales** (mostly water bacteria; gram negative, variably shaped, all with bacteriochlorophylls and carotenoid pigments; purple-violet, purple, red, or orange-brown, brown, or green colors from photopigments in cell suspensions; some can fix nitrogen; purple [sulfur or nonsulfur] and green sulfur groups of bacteria here)
PART 2   GLIDING BACTERIA
  Order I. **Myxobacterales** (slime bacteria; gram negative, strict aerobes with slow gliding movements found on soil and decomposing plant and animal matter; no photosynthetic pigments; chemo-organotrophs; energy-yielding mechanism respiratory, never fermentative; fruiting bodies formed from cell aggregates often brightly colored and macroscopic; bacteriolytic and cellulytic [attacking cellulose] groups here)
  Order II. **Cytophagales** (rods or filaments, gram negative, with slow or rapid gliding; no fruiting bodies; chemolithotrophs, chemo-organotrophs, or mixotrophs)
PART 3   SHEATHED BACTERIA
  (sheath present, may be encrusted with iron or manganese oxides; single cells; flagella may be found)
    Genus *Leptothrix* (gram **negative**, strictly aerobic straight rods in chains within a sheath; also free-swimming as single cells, in pairs, or in motile short chains; sheaths often impregnated with hydrated ferric or manganic oxides; prevalent in iron-containing, uncontaminated, slow-running fresh waters)
    Genus *Streptothrix* (thin, gram-negative rods in chains; strictly aerobic; widely distributed in fresh water and in activated sludge; barely visible sheaths not encrusted)
PART 4   BUDDING AND/OR APPENDAGED BACTERIA
  (soil and water bacteria reproducing by budding; may have excreted appendages and hold-fasts; in some a semirigid appendage, the prostheca, proceeds out from the cell [stalk of *Caulobacter* genus] extending the length of the rod, a small bit of glue at its tip)
PART 5   SPIROCHETES
  Order I. **Spirochaetales**
    Family I. **Spirochaetaceae**
      Genus I. *Spirochaeta* (helical cells, motile, free living in H_2S-containing mud and sewage)
      Genus II. *Cristispira* (helical cells with 2 to 10 complete turns; commensal in mollusks)
      Genus III. *Treponema*
        Treponema pallidum (type species; syphilis)
        Treponema pertenue (yaws)
        Treponema carateum (pinta, a chronic skin disease of children endemic in South and Central America)
        Treponema macrodentium (oral microflora)
        Treponema denticola [*microdentium*] (oral microflora)
        Treponema [*Borrelia*] *vincentii* (oral microflora)
        Species incertae sedis: Treponema buccale [*Borrelia buccalis*] (large oral treponemes)
      Genus IV. *Borrelia*
        Borrelia recurrentis (louse-borne epidemic relapsing fever)
        Borrelia species (tick-borne endemic relapsing fever)
      Genus V. *Leptospira*
        Leptospira interrogans [*icterohaemorrhagiae*] (type species; leptospirosis)
PART 6   SPIRAL AND CURVED BACTERIA
    Family I. **Spirillaceae**
      Genus I. *Spirillum*
        Spirillum minor (one type of rat-bite fever in man)
      Genus II. *Campylobacter* (slender spiral rods found in reproductive and alimentary tracts of man and animals; some species pathogenic)
        Campylobacter [*Vibrio*] *fetus* (type species; abortion in sheep and cattle; infections in man)

Continued.

TABLE 1-1. ABBREVIATED CLASSIFICATION OF MICROBES FROM *BERGEY'S MANUAL OF DETERMINATIVE BACTERIOLOGY* (1974)—cont'd

PART 7   GRAM-NEGATIVE AEROBIC RODS AND COCCI
　　Family I. **Pseudomonadaceae**
　　　　Genus I. *Pseudomonas*
　　　　　　Pseudomonas aeruginosa (type species; wound, burn, and urinary tract infections)
　　　　　　Pseudomonas [Actinobacillus] mallei (glanders and farcy in horses and donkeys; infection transmissible to man)
　　　　　　Pseudomonas pseudomallei (human and animal melioidosis)
　　　　Genus II. *Xanthomonas* (plant pathogens)
　　　　Genus III. *Zoogloea* (motile gram-negative rods in natural waters and sewage)
　　　　Genus IV. *Gluconobacter* (ellipsoids or rods in flowers, souring fruits, vegetables, cider, wine, baker's yeast, garden soil; ropiness in beer and wort)
　　Family II. **Azotobacteraceae** (large, motile, aerobic gram-negative rods fixing atmospheric nitrogen; found in soil, water, and on leaf surfaces)
　　　　Genus I. *Azotobacter* (nitrogen fixation)
　　　　Genus II. *Azomonas* (nitrogen fixation)
　　　　Genus III. *Beijerinckia* (nitrogen fixation)
　　　　Genus IV. *Derxia* (found in tropical soils of Asia, Africa, South America; fixation of atmospheric nitrogen)
　　Family III. **Rhizobiaceae** (nitrogen fixation; symbionts in root nodules of legumes; cortical overgrowths in plants)
　　　　Genus I. *Rhizobium* (nitrogen fixation)
　　　　Genus II. *Agrobacterium* (plant pathogens [tumorigenic phytopathogens]; gall hypertrophies on stems of more than 40 plants; free nitrogen not fixed; found in soil)
　　Family IV. **Methylomonadaceae** (gram-negative bacteria using only one-carbon organic compounds such as methane and methanol as carbon source)
　　Family V. **Halobacteriaceae** (high concentration of sodium chloride for growth; found in salterns, salt lakes, Dead Sea, proteinaceous material preserved with solar salt [fish, sausage casings, and hides])
　　　　Genus I. *Halobacterium* (salt-loving rods)
　　　　Genus II. *Halococcus* (halophilic cocci; red colonies)
　　Genera of uncertain affiliation
　　　　Genus *Alcaligenes [Achromobacter]* (motile aerobic rods or cocci, common saprophytes in intestines of vertebrates, in dairy products, rotting eggs, fresh water, soil; important in decomposition and mineralization processes)
　　　　　　Alcaligenes faecalis (type species; some strains denitrify)
　　　　Genus *Acetobacter* (motile aerobic vinegar rods; oxidize ethanol to acetic acid; found on fruits and vegetables, in souring fruit juices, vinegar, alcoholic beverages)
　　　　　　Acetobacter aceti (acetic acid bacteria)
　　　　Genus *Brucella* (brucellosis)
　　　　　　Brucella melitensis (type species; Malta fever; infection in goats, sheep, and cattle)
　　　　　　Brucella abortus (abortion in cattle; disease in man)
　　　　　　Brucella suis (infection in pigs, other animals, and man)
　　　　Genus *Bordetella*
　　　　　　Bordetella pertussis (type species; whooping cough)
　　　　　　Bordetella parapertussis (whooping cough-like disease)
　　　　　　Bordetella bronchiseptica (found in respiratory tract of animals, sometimes man; rodent bronchopneumonia)
　　　　Genus *Francisella*
　　　　　　Francisella [Pasteurella] tularensis (type species; tularemia)
　　　　Genus *Thermus* (gram-negative nonmotile rods and filaments; often pigmented; common in hot springs, hot water tanks, and thermally polluted rivers; unusually thermostable enzymes, ribosomes, plasma membrane)

PART 8   GRAM-NEGATIVE FACULTATIVELY ANAEROBIC RODS
  Family I. **Enterobacteriaceae**
    Genus I. *Escherichia*
      Escherichia coli (type species; the colon bacillus; important opportunist pathogens)
    Genus II. *Edwardsiella*
      Edwardsiella tarda (type species; found in intestine of snakes, sometimes in man)
    Genus III. *Citrobacter*
      Citrobacter freundii (type species; found in water, food, and excreta of man)
    Genus IV. *Salmonella* (salmonellosis)
      Subgenus I
        Salmonella cholerae-suis (type species; salmonellosis)
        Salmonella hirschfeldii (paratyphoid C bacilli; enteritis)
        Salmonella typhi (typhoid fever)
        Salmonella paratyphi-A (paratyphoid [enteric] fever)
        Salmonella schottmuelleri (enteritis)
        Salmonella typhimurium (food poisoning in man)
        Salmonella enteritidis (enteritis)
        Salmonella gallinarum (fowl typhoid)
      Subgenus II
        Salmonella salamae (type species of subgenus)
      Subgenus III
        Salmonella arizonae (type species of subgenus; isolated from reptiles)
      Subgenus IV
        Salmonella houtenae (type species of subgenus)
    Genus V. *Shigella* (shigellosis)
      Shigella dysenteriae (type species; bacillary dysentery plus effects of diffusible neurotoxin)
      Shigella flexneri (bacillary dysentery)
      Shigella boydii (baccillary dysentery)
      Shigella sonnei (one cause of summer diarrhea in young children; milder form bacillary dysentery in adults)
    Genus VI. *Klebsiella*
      Klebsiella pneumoniae (type species; pneumonia, infections of respiratory and urinary tracts)
      Klebsiella ozaenae (found in ozena and chronic respiratory disease)
      Klebsiella rhinoscleromatis (found in rhinoscleroma, a granulomatous disorder of nose and pharynx associated with nodular induration of tissues)
    Genus VII. *Enterobacter*
      Enterobacter cloacae (type species; opportunist pathogens)
      Enterobacter [Aerobacter] aerogenes (opportunist pathogens)
    Genus VIII. *Hafnia* (Hafnia, the old name for Copenhagen; opportunist pathogens)
    Genus IX. *Serratia*
      Serratia marcescens (type species; opportunist pathogens)
    Genus X. *Proteus* (urinary tract infections, community and hospital acquired)
      Proteus vulgaris (type species; urinary tract and wound infections, rarely peritonitis, meningitis)
      Proteus mirabilis (most frequent species in medical specimens)
      Proteus morganii (one cause of summer diarrhea in infants)
      Proteus rettgeri (gastroenteritis)
      Proteus [Providencia] inconstans (urinary tract infections)
    Genus XI. *Yersinia*
      Yersinia [Pasteurella] pestis (type species; plague)
      Yersinia [Pasteurella] pseudotuberculosis (pseudotuberculosis in animals, usually mesenteric lymphadenitis; septicemia in man)

Continued.

TABLE 1-1. ABBREVIATED CLASSIFICATION OF MICROBES FROM *BERGEY'S MANUAL OF DETERMINATIVE BACTERIOLOGY* (1974)—cont'd

PART 8   GRAM-NEGATIVE FACULTATIVELY ANAEROBIC RODS
   Family I. **Enterobacteriaceae**—cont'd
      Yersinia enterocolitica (widespread; been found in sick and healthy animals and in material likely contaminated by their feces; enterocolitis in young children, mesenteric lymphadenitis, variety of other infections)
     Genus XII. *Erwinia* (plant pathogens)
   Family II. **Vibrionaceae**
    Genus I. *Vibrio*
      Vibrio cholerae [*comma*] (type species; cholera)
      Vibrio cholerae biotype *eltor* (El Tor vibrio; cholera)
      Vibrio parahaemolyticus (acute gastroenteritis)
      Vibrio [*Photobacterium*] *fischeri* (luminescent saltwater bacteria)
    Genus II. *Aeromonas* (motile gas-forming rods)
      Aeromonas hydrophila (type species; nonluminescent freshwater bacteria; infections of cold-blooded animals—red leg [bacteremia] in frogs and septicemia in snakes; rarely infections in compromised host)
    Genus III. *Plesiomonas* (motile rods growing mostly on mineral media containing ammonia as sole source of nitrogen and glucose as sole source of carbon; found in feces; infectious gastroenteritis reported in man)
    Genus IV. *Photobacterium* (luminescent saltwater bacteria)
    Genus V. *Lucibacterium* (light-emitting bacteria; luminescent saltwater bacilli; found on surfaces of dead fish)
  Genera of uncertain affiliation
    Genus *Chromobacterium* (soil and water bacteria producing violet pigment violacein; infections in animals; food spoilage)
    Genus *Flavobacterium* (proteolytic soil and water bacteria producing yellow, orange, or red pigments; found on vegetables and in dairy products; rare infection in man by unpigmented organism)
    Genus *Haemophilus*
      Haemophilus influenzae (type species; purulent meningitis in young children; acute respiratory infection; acute conjunctivitis)
      Haemophilus suis (with virus, causes swine influenza)
      Haemophilus haemolyticus (commensal in upper respiratory tract of man)
      Haemophilus parainfluenzae (found in upper respiratory tract of man and cats)
      Haemophilus parahemolyticus (found in upper respiratory tract of man; associated with acute pharyngitis; pleuropneumonia and septicemia in swine)
      Haemophilus aphrophilus (endocarditis and other infections in man)
      Haemophilus ducreyi (chancroid)
      Species incertae sedis, A: Haemophilus aegyptius (Koch-Weeks bacillus; acute infectious conjunctivitis)
      Species incertae sedis, B: Haemophilus vaginalis (gram-variable bacilli and coccobacilli showing metachromatic granules and arranged like *Corynebacterium*, found in human genital tract; nonspecific vaginitis and urethritis)
    Genus *Pasteurella*
      Pasteurella multocida (type species; chicken cholera; shipping fever of cattle; hemorrhagic septicemia in warm-blooded animals; cat- and dog-bite wound infections in man)
      Pasteurella pneumotropica (infections in animals; dog-bite wounds in man)
      Pasteurella haemolytica (enzootic pneumonia of sheep and cattle; septicemia of lambs)
    Genus *Actinobacillus* (actinobacillosis)

Actinobacillus lignieresii (type species; actinobacillosis of cattle [wooden tongue] and of sheep)

Actinobacillus equuli (actinobacillosis in horses and pigs)

Genus *Cardiobacterium* (found in human nose and throat)

Cardiobacterium hominis (type species; endocarditis in man)

Genus *Streptobacillus* [*Haverhillia*] (rods and filaments in chains with filaments showing bulbous swellings, like a string of beads; parasites and pathogens of rats and other mammals)

Streptobacillus moniliformis [*Actinomyces muris ratti*] (type species; necklace-shaped bacteria found in nasopharynx of rats; streptobacillary rat-bite fever)

Genus *Calymmatobacterium* (pleomorphic encapsulated rods like safety pins)

Calymmatobacterium [*Donovania*] *granulomatis* (type species; granuloma inguinale)

PART 9 GRAM-NEGATIVE ANAEROBIC BACTERIA

Family I. **Bacteroidaceae**

Genus I. *Bacteroides* (bacteroidosis)

Bacteroides fragilis (type species; opportunists in visceral and wound infections; most common anaerobe in soft tissue infections)

Bacteroides oralis (gingival crevice of man; oral, upper respiratory, and genital infections)

Bacteroides corrodens (part of normal flora of man and animals; opportunists in infections of respiratory and alimentary tracts)

Bacteroides melaninogenicus (opportunists in infections of mouth, soft tissue, and respiratory, alimentary, and urogenital tracts; brown to black pigment on blood agar)

Genus II. *Fusobacterium* (purulent or gangrenous infections)

Fusobacterium nucleatum (type species; wound and respiratory tract infections)

Fusobacterium varium (wound and serous cavity infections; intestinal contents of roaches, termites)

Fusobacterium necrophorum [*Sphaerophorus necrophorus*] (abscesses of man and animals)

Fusobacterium mortiferum (visceral abscesses and septicemia in man)

Genus III. *Leptotrichia* (oral cavity of man; not known as pathogens but found in clinical material)

Leptotrichia buccalis (type species)

Genera of uncertain affiliation

Genus *Butyrivibrio* (many members, biochemically versatile in rumen of most ruminants and in intestinal tract of other mammals)

Genus *Selenomonas* (gastrointestinal tract of mammals; dirty river water)

PART 10 GRAM-NEGATIVE COCCI AND COCCOBACILLI [AEROBES]

Family I. **Neisseriaceae**

Genus I. *Neisseria* (parasites of mucous membranes of mammals)

Neisseria gonorrhoeae (type species; gonorrhea)

Neisseria meningitidis (epidemic cerebrospinal fever)

Neisseria sicca (nasopharynx of man)

Neisseria subflava [*flava*] [*perflava*] (nasopharynx of man; yellowish green pigment)

Neisseria flavescens (xanthophil pigment)

Neisseria mucosa (rhinopharynx of man)

Genus II. *Branhamella* (parasites of mammalian mucous membranes)

Branhamella [*Neisseria*] *catarrhalis* (type species; venereal discharges; catarrhal inflammations)

Genus III. *Moraxella* (parasites of mucous membranes of man and warm-blooded animals)

Moraxella lacunata (type species; pink-eye—conjunctivitis)

Genus *Acinetobacter* (opportunist pathogens)

Acinetobacter calcoaceticus [var. *anitratus* or *Herellea vaginicola*, acid-forming strains; var. *lwoffi* or *Mima polymorpha*, oxidase-negative, nonacid-forming strains] (type species; soil and water bacteria; pathogenicity uncertain)

Continued.

TABLE 1-1. ABBREVIATED CLASSIFICATION OF MICROBES FROM *BERGEY'S MANUAL OF DETERMINATIVE BACTERIOLOGY* (1974)—cont'd

PART 11   GRAM-NEGATIVE ANAEROBIC COCCI
   Family I. **Veillonellaceae** (alimentary tract parasites of man, ruminants, rodents, and pigs)
      Genus I. *Veillonella* (oral microflora; intestinal and respiratory tracts of man and animals)
         Veillonella parvula (type species; oral microflora)
         Veillonella alcalescens (oral microflora)
      Genus II. *Acidaminococcus* (amino acid cocci—amino acids sole energy source; intestinal tract of man and pigs)
      Genus III. *Megasphaera* (large cocci; rumen of cattle and sheep)
PART 12   GRAM-NEGATIVE CHEMOLITHOTROPHIC BACTERIA
   a. Organisms oxidizing ammonia or nitrite
   Family I. **Nitrobacteraceae** (soil and water bacteria converting ammonia to nitrite, nitrite to nitrate, and fixing carbon dioxide; nitrifying bacteria—one group oxidizes ammonia, the other nitrite)
      Genus I. *Nitrobacter* (nitrate rods oxidizing nitrite to nitrate)
      Genus II. *Nitrospina* (nitrate spines oxidizing nitrite to nitrate; straight slender rods found in South Atlantic Ocean)
      Genus III. *Nitrococcus* (nitrate spheres, yellowish to red, oxidizing nitrite to nitrate; found in South Pacific Ocean)
      Genus IV. *Nitrosomonas* (ellipsoidal, yellowish to red bacteria oxidizing ammonia to nitrite)
      Genus V. *Nitrosospira* (nitrous spirals, yellowish to red, oxidizing ammonia to nitrite)
      Genus VI. *Nitrosococcus* (nitrous spheres oxidizing ammonia to nitrite)
      Genus VII. *Nitrosolobus* (nitrite-producing lobes; pleomorphic and lobate, yellowish to red bacteria oxidizing ammonia to nitrite; found in soils from South America, Southwest Africa, and Russia)
   b. Organisms metabolizing sulfur and sulfur compounds*
      Genus 1. *Thiobacillus* (sulfur rodlets, small rods in soil, seawater, fresh water, acid mine water, sewage, sulfur springs, and near sulfur deposits or where hydrogen sulfide produced)
      Genus 2. *Sulfolobus* (lobed sulfur-oxidizing organisms resembling mycoplasmas in solfatara areas containing hot acid environments, both soil and water)
      Genus 3. *Thiobacterium* (small sulfur rods near surface of sulfurous brackish and marine waters)
      Genus 4. *Macromonas* (large cylindric or bean-shaped cells in fresh waters)
      Genus 5. *Thiovulum* (small sulfur egglike cells, cytoplasm at one end of cell, a large vacuole at the other; found in fresh and seawaters where sulfide-containing waters contact oxygen-containing waters)
      Genus 6. *Thiospira* (sulfur spirals in waters overlaying sulfurous muds)
   c. Organisms depositing iron and/or manganese oxides
   Family I. **Siderocapsaceae** (iron bacteria of iron-bearing waters)
      Genus I. *Siderocapsa* (iron bacteria of fresh water; may be attached to surface of water plants)
      Genus II. *Naumanniella* (encapsulated rods widespread in iron-bearing waters; capsule encrusted with iron compounds and manganese oxide)
      Genus III. *Ochrobium* (bacteria yellowish from iron oxides; widespread in fresh waters bearing iron)
      Genus IV. *Siderococcus* (iron cocci in fresh water and mud)

*Since all the sulfur-metabolizing genera have not been grown in pure culture, *Bergey's Manual* considers them as of uncertain affiliation.

FIG. 1-1. Methane gas burning out of a hollow increment borer bit drilled 20 cm into a cottonwood growing on shore of Lake Wingra, Wisconsin. (Timed exposure at night.) (From Zeikus, J. G., and Ward, J. C.: Science **184:**1181, 1974. Copyright by the American Association for the Advancement of Science.)

PART 13   METHANE-PRODUCING BACTERIA
Family I. **Methanobacteriaceae** (gram-positive or gram-negative rods or cocci, very strict anaerobes, motile or nonmotile, in a highly specialized physiologic group; to gain energy for growth, they reduce carbon dioxide–forming methane (Fig. 1-1) or ferment such compounds as acetate and methanol; widespread in nature in anaerobic habitats —sediments of natural waters, soil, anaerobic sewage digestors, and gastrointestinal tract of animals and man)
Genus I. *Methanobacterium* (methane-producing rodlets)
Genus II. *Methanosarcina* (methane-producing sarcinae—large spherical cells in regular packets)
Genus III. *Methanococcus* (methane cocci)
PART 14   GRAM-POSITIVE COCCI
a. Aerobic and/or facultatively anaerobic
Family I. **Micrococcaceae** (saprophytes and important pathogens; normal occupants of skin and associated structures in man)
Genus I. *Micrococcus* (small aerobic cocci common in soil and fresh water and on skin of man and animals)
Micrococcus luteus [*Sarcina lutea*] (type species; golden yellow pigment produced; pattern of tetrads)
Genus II. *Staphylococcus* (primary relation to skin and mucous membranes of warm-blooded animals; wide host range; important pathogenic strains)
Staphylococcus aureus (type species; type lesion: the abscess)
Staphylococcus epidermidis (normal occupants of skin and mucosae of man; opportunists in stitch abscesses and other wound infections)

Continued.

TABLE 1-1. ABBREVIATED CLASSIFICATION OF MICROBES FROM *BERGEY'S MANUAL OF DETERMINATIVE BACTERIOLOGY* (1974)—cont'd

PART 14   GRAM-POSITIVE COCCI
  Family I. **Micrococcaceae** — cont'd
    Genus III. *Planococcus* (motile cocci in seawater)
      Planococcus citreus (type species; yellowish orange pigment produced)
  Family II. **Streptococcaceae** (saprophytic and pathogenic bacteria in chains)
    Genus I. *Streptococcus*
      Streptococcus pyogenes (type species; many types of infection, usually of diffuse or spreading nature [cellulitis])
      Streptococcus equisimilis (upper respiratory tract infections in man and animals; erysipelas and puerperal fever)
      Streptococcus zooepidemicus (septicemia of cows, rabbits, and swine, wound infections of horses)
      Streptococcus equi (strangles in horses)
      Streptococcus dysgalactiae (mastitis in cows, polyarthritis, or joint-ill, in lambs)
      Streptococcus [*Diplococcus*] *pneumoniae* (lobar pneumonia, bronchopneumonia)
      Streptococcus anginosus (varied infections; alpha- and gamma-hemolytic strains formerly known as *Streptococcus* MG relate to primary atypical pneumonia)
      Streptococcus agalactiae (mastitis of cows; human infections)
      Streptococcus salivarius (tongue, saliva, and feces of man)
      [*Streptococcus mutans*] (dental caries)
      Streptococcus mitis (human saliva, sputum, feces)
      Streptococcus bovis (alimentary canal of cows, sheep, and other ruminants)
      Streptococcus equinus (alimentary tract of horses)
      Streptococcus thermophilus (in milk and milk products; starter culture for Swiss cheese and yogurt)
      Streptococcus faecalis (enterococcus in intestine of man and warm-blooded animals; urinary tract infections, endocarditis)
      Streptococcus lactis (important in dairy industry, some strains starter cultures in manufacture of cheese and cultured milk drinks)
      Streptococcus cremoris (raw milk and milk products)
    Genus II. *Leuconostoc* (colorless nostoc; saprophytes found in slimy sugar solutions, on fruits and vegetables, in milk and dairy products)
      Leuconostoc mesenteroides (type species; used in industrial fermentation)
    Genus III. *Pediococcus* (microaerophilic saprophytes in fermenting plant material, especially spoiled beer)
    Genus IV. *Aerococcus* (microaerophilic saprophytes widely distributed in air, in meat brines, on raw and processed vegetables)
  b. Anaerobic
  Family III. **Peptococcaceae** (occupants of alimentary and respiratory tracts of man and animals and of normal human female genital tract; found in soil and on surface of cereal grains; lesions of human female genitalia)
    Genus I. *Peptococcus* (lesions of viscera and serous cavities; postpartum septicemia; black colonies seen among species)
      Peptococcus niger (type species)
    Genus II. *Peptostreptococcus* (puerperal fever; pyogenic infections, septic war wounds, osteomyelitis, pleurisy, gangrene, sinusitis, dental infection, vulvovaginitis)
    Genus III. *Ruminococcus* (in rumen and cecum and colon of animals; important there in fermentation of cellulose)
    Genus IV. *Sarcina* (nearly spherical cells in packets of eight or more; found in soil, mud, surface of cereal seeds, and in stomach contents of man and animals)

PART 15   ENDOSPORE-FORMING RODS AND COCCI
  Family I. **Bacillaceae** (saprophytic gram-positive rods, mostly; endospore formation dominant feature in defining genera)
    Genus I. *Bacillus* (genus of great diversity in properties of members; several species produce antibiotics, some strains more than one; because of pattern of spore formation in nature, few species have distinctive habitats)
      Bacillus subtilis (type species; endospores widespread, may even be in heat-treated surgical dressings and canned foods; some strains produce antibiotics)
      Bacillus licheniformis (source of bacitracin; spores with high heat tolerance)
      Bacillus cereus (found in foods)
      Bacillus anthracis (anthrax)
      Bacillus thuringiensis (microbial insecticide)
      Bacillus polymyxa (source of polymyxin; participates in retting of flax)
      Bacillus stearothermophilus (thermophilic; endospores highly resistant to heat)
      Bacillus brevis (source of tyrothricin)
    Genus II. *Sporolactobacillus* (bacteria-like *Lactobacillus* but sporeforming)
    Genus III. *Clostridium* (strict anaerobes, motile rods; soil and water bacteria; occupants of intestinal tract of man and animals)
      Clostridium butyricum (type species; soil, animal feces, cheese, naturally soured milk)
      Clostridium beijerinckii (wound infections)
      Clostridium sporogenes (wound infections; low-grade bacteremia in debilitated patients)
      Clostridium botulinum (potent exotoxin causes botulism)
      Clostridium histolyticum (wound infections)
      Clostridium novyi (gas gangrene)
      Clostridium perfringens [*Bacterium welchii*] (gas gangrene)
      Clostridium septicum (gas gangrene)
      Clostridium tertium (wound infections, low-grade bacteremia in debilitated patients)
      Clostridium tetani (potent exotoxin causes tetanus)
    Genus IV. *Desulfotomaculum* (sausage-shaped bacteria reducing sulfur; strict anaerobes; common soil and water saprophytes)
    Genus V. *Sporosarcina* (sporeforming spherical cells, strict aerobes)
PART 16   GRAM POSITIVE, ASPOROGENOUS, ROD-SHAPED BACTERIA
  Family I. **Lactobacillaceae** (rods, nonmotile, anaerobic or facultative, highly saccharoclastic; unusual as pathogens; found in fermenting animal and plant products)
    Genus I. *Lactobacillus* (found in dairy products and effluents, grain and meat products, water, sewage, beer, wine, fruits, fruit juices, pickled vegetables, sourdough, mash; some species in tooth decay)
      Lactobacillus delbrueckii (type species; fermenting grain and vegetable mashes)
      Lactobacillus leichmannii (compressed yeast, grain mash)
      Lactobacillus lactis (milk, cheese; starter cultures in manufacture of cheese)
      Lactobacillus bulgaricus (sour milks as yogurt)
      Lactobacillus acidophilus (feces of infants, mouth and vagina of young human adults)
      Lactobacillus casei (milk, cheese, dairy products, sourdough, cow dung, silage, human alimentary tract and vagina)
      Lactobacillus brevis (milk, kefir, cheese, sauerkraut, sourdough, soils, ensilage)
  Genera of uncertain affiliation
    Genus *Listeria* (listeriosis)
      Listeria monocytogenes (type species; listeriosis in man and animals)
    Genus *Erysipelothrix* (parasites of mammals, birds, fish; widespread in nature)
      Erysipelothrix rhusiopathiae [*insidiosa*] (type species; swine erysipelas, erysipeloid in man)

Continued.

TABLE 1-1. ABBREVIATED CLASSIFICATION OF MICROBES FROM *BERGEY'S MANUAL OF DETERMINATIVE BACTERIOLOGY* (1974)—cont'd

PART 17 ACTINOMYCETES AND RELATED ORGANISMS

Coryneform group of bacteria

Genus I. *Corynebacterium*

Section I. Human and animal parasites and pathogens (club bacteria, nonsporing, gram-positive, irregular rods; widespread in nature)

Corynebacterium diphtheriae (type species; diphtheria—the effect of its highly lethal exotoxin)

Corynebacterium pseudotuberculosis (pseudotuberculosis; chronic purulent infections in warm-blooded animals, rarely in man)

Corynebacterium xerosis (harmless occupant of conjunctiva)

Corynebacterium kutscheri (parasite and opportunist pathogen of mice and rats)

Corynebacterium pseudodiphtheriticum (normal in throat of man; nontoxigenic)

Corynebacterium equi (pneumonia of horses)

Section II. Plant pathogenic corynebacteria

Section III. Nonpathogenic corynebacteria (soil, water, air)

Genus II. *Arthrobacter* (jointed gram-positive rods, strict aerobes; among dominant soil bacteria)

Genera of uncertain affiliation

Genus A. *Brevibacterium* (coryneform bacteria found in water, soil, dairy products, insects; widespread in nature, sewage; many in industry)

Genus B. *Microbacterium* (small diphtheroid rods with rounded ends; found in dairy products)

Genus III. *Cellulomonas* (coryneform bacteria with ability to attack cellulose)

Genus IV. *Kurthia* (coryneform bacteria in fresh and putrefying meats, meat products, and on working surfaces in meat plants)

Family I. **Propionibacteriaceae** (gram-positive, nonsporeforming, anaerobic, pleomorphic rods; saccharolytic members; found in skin, respiratory and alimentary tracts of most animals; some members in soft tissue infections)

Genus I. *Propionibacterium* (nonmotile, anaerobic to aerotolerant bacteria producing propionic and acetic acids; some pathogens here)

Propionibacterium freudenreichii (type species; raw milk, Swiss cheese, and dairy products)

Propionibacterium [Corynebacterium] acnes (soft tissue abscesses, wound infections; common laboratory contaminant)

Genus II. *Eubacterium* (motile or nonmotile, obligately anaerobic rods; found in cavities of man and animals, in plant products, soil; infections of soft tissue)

Eubacterium foedans (type species; dental tartar; varied infections)

Order I. **Actinomycetales** (gram-positive soil bacteria, mostly aerobic, forming branching filaments, which in some families develop into mycelium; some members acid-alcohol fast; some species fix nitrogen as obligate symbionts in plant root nodule; some members pathogenic to man, animals, and plants)

Family I. **Actinomycetaceae** (diphtheroid bacteria; no mycelium, nonacid fast, usually facultative anaerobes; branching filaments)

Genus I. *Actinomyces* (actinomycosis)

Actinomyces bovis (type species; lumpy jaw in cattle)

Actinomyces israelii (human actinomycosis)

Actinomyces naeslundii (oral cavity of man—in tonsillar crypts and dental calculus)

Genus II. *Arachnia* (branched diphtheroid rods, gram-positive, nonacid-fast pathogens)

Arachnia propionica (type species; human actinomycosis)

Genus III. *Bifidobacterium* (bifid bacteria; highly variable rods, gram positive, non-acid fast)

Bifidobacterium bifidum [*Lactobacillus bifidus*] (type species; stools and alimentary tract of breast-fed and bottle-fed infants, and adults)

Genus IV. *Bacterionema* (thread-shaped bacteria, "whip handle" cells, nonacid fast with filamentous branching morphology; on teeth and in oral cavity, especially in calculus and plaque deposits)

Genus V. *Rothia* (gram-positive, nonacid-fast aerobes with filamentous branching; common in normal mouth and throat)

Family II. **Mycobacteriaceae** (important acid-fast bacteria, parasitic and saprophytic; branching inconspicuous)

Genus I. *Mycobacterium* (mycobacteriosis, chronic granulomatous disease of varied forms, tuberculosis)

Mycobacterium tuberculosis (type species, human tuberculosis)

Mycobacterium bovis (tuberculosis in cattle, man)

Mycobacterium kansasii (yellow bacillus, Group I photochromogen; chronic pulmonary disease similar to tuberculosis)

Mycobacterium marinum (tuberculosis of saltwater fish, swimming pool granulomas in man)

Mycobacterium scrofulaceum (scrofula scotochromogen; suppurative cervical lymphadenitis in children)

Mycobacterium intracellulare (Battey bacillus; severe chronic pulmonary disease in man)

Mycobacterium avium (tuberculosis in birds)

Mycobacterium ulcerans (skin ulcers in man)

Mycobacterium phlei (timothy bacillus, hay bacillus; widespread in nature)

Mycobacterium smegmatis (smegma bacillus)

Mycobacterium fortuitum (mycobacteriosis in animals and in man; found in soil)

Mycobacterium paratuberculosis (Johne's disease in cattle and sheep)

Mycobacterium leprae (leprosy)

Mycobacterium lepraemurium (rat leprosy)

Family III. **Frankiaceae** (symbiotic filamentous, mycelium-forming bacteria, which induce and live in root nodules of a wide variety of plants; nodules can fix molecular nitrogen)

Genus I. *Frankia* (characteristic root nodules formed and inhabited by bacteria; also free state in soil)

Family IV. **Actinoplanaceae** (soil bacteria; distinctive mycelium; shape of sporangia and spore structure make division into genera)

Genus I. *Actinoplanes* (globose spores)

Genus II. *Spirillospora* (spiral spores)

Genus III. *Streptosporangium* (spores coiled within sporangia)

Genus IV. *Amorphosporangium* (irregularly shaped sporangia)

Genus V. *Ampullariella* (bottle-shaped sporangia)

Genus VI. *Pilimelia* (rod-shaped spores, end-to-end in parallel chains; about 1000 per sporangium)

Genus VII. *Planomonospora* (single, large, motile spore in each sporangium)

Genus VIII. *Planobispora* (longitudinal pair of large spores in sporangium)

Genus IX. *Dactylosporangium* (finger-shaped sporangia)

Genus X. *Kitasatoa* (sporangia club-shaped; motile spores in single chains within sporangium)

Family V. **Dermatophilaceae** (nonacid-fast, gram-positive organisms; skin lesions of mammals)

Genus I. *Dermatophilus* (characteristic mycelium; skin pathogen)

Genus II. *Geodermatophilus* (rudimentary mycelium, tuber-shaped thallus; soil organisms)

Family VI. **Nocardiaceae** (nocardiosis; gram-positive aerobic actinomycetes; mycelium rudimentary or extensive; spore production variable)

Continued.

TABLE 1-1. ABBREVIATED CLASSIFICATION OF MICROBES FROM *BERGEY'S MANUAL OF DETERMINATIVE BACTERIOLOGY* (1974)—cont'd

PART 17 ACTINOMYCETES AND RELATED ORGANISMS
  Family VI. **Nocardiaceae**—cont'd
    Genus I. *Nocardia* (acid-fast to partially acid-fast, gram-positive, obligate aerobes; some strains pigmented; found in soil)
      Nocardia asteroides (pulmonary nocardiosis, chronic subcutaneous abscesses, mycetomas)
      Nocardia brasiliensis (mycetoma)
    Genus II. *Pseudonocardia* (false nocardia; soil organisms, nonacid fast)
  Family VII. **Streptomycetaceae** (gram-positive, aerobic, primarily soil forms; well-developed branched mycelium and special spores for multiplication; many species important in production of antibiotics)
    Genus I. *Streptomyces* (about 500 distinctive antibiotics come out of this genus; many members produce one or more antibacterial, antifungal, antialgal, antiviral, antiprotozoal, or antitumor antibiotics [see Table 16-2, p. 290, for examples]; well over 500 species known)
    Genus II. *Streptoverticillium* (whorled actinomycetes)
    Genus III. *Sporichthya* (spores motile in water)
    Genus IV. *Microellobosporia* (small spores in a pod)
  Family VIII. **Micromonosporaceae** (saprophytic soil forms)
    Genus I. *Micromonospora* (well-developed branched septate mycelium, small single spores; nonacid-fast proteolytic and cellulolytic organisms; some antibiotics found here)
      Micromonospora purpurea (source of gentamicin complex)
    Genus II. *Thermoactinomyces* (heat-loving ray "fungus")
    Genus III. *Actinobifida* (ray "fungus" with bifurcations)
    Genus IV. *Thermomonospora* (heat-loving, single-spored organisms)
    Genus V. *Microbispora* (small two-spored organisms)
    Genus VI. *Micropolyspora* (small many-spored organisms)
PART 18 RICKETTSIAS
  Order I. **Rickettsiales** (procaryotic microbes; parasitic forms relate to reticuloendothelial and vascular endothelial cells or erythrocytes of vertebrates; mutualistic forms in insects; arthropods are vectors or primary hosts; disease in vertebrates and invertebrates)
    Family I. **Rickettsiaceae** (intracellular parasites of tissue cells, not erythrocytes; arthropod vectors important)
    Tribe I. **Rickettsieae** (adaptation to existence in arthropods but can infect vertebrate hosts; man usually incidental host)
      Genus I. *Rickettsia* (human pathogens; growth in cytoplasm, sometimes in nucleus, but not in vacuoles of cells; not cultivated in cell-free media)
        Rickettsia prowazekii (type species; typhus fever—man the reservoir)
        Rickettsia typhi (murine typhus)
        Rickettsia rickettsii (Rocky Mountain spotted fever)
        Rickettsia sibirica (Siberian tick typhus)
        Rickettsia conorii (fièvre boutonneuse, tick-bite fever)
        Rickettsia australis (Queensland tick typhus)
        Rickettsia akari (rickettsialpox)
        Rickettsia tsutsugamushi (scrub typhus)
      Genus II. *Rochalimaea* (like *Rickettsia* genus but usually in extracellular position in arthropod host; can be cultured in host cell–free media)
        Rochalimaea quintana (type species; trench fever—man primary host)
      Genus III. *Coxiella* (growth in vacuoles of host cell; highly resistant outside cells)
        Coxiella burnetii (type species; Q fever)
    Tribe II. **Ehrlichieae** (minute rickettsia-like organisms; only few species; adapted to invertebrate existence; pathogens of mammals but not man)

Genus IV. *Ehrlichia* (tick-borne diseases of dogs, cattle, sheep, goats, and horses)
Genus V. *Cowdria* (heartwater diseases of domestic ruminants)
Genus VI. *Neorickettsia* (helminth-borne disease of dogs, wolves, jackals, and foxes)
Tribe III. **Wolbachieae** (symbionts in arthropods but not in vertebrates; miscellaneous group)
Genus VII. *Wolbachia* (associated with arthropods — seldom pathogenic for hosts)
Genus VIII. *Symbiotes* (symbiotic in insect host)
Genus IX. *Blattabacterium* (symbiotic in cockroaches)
Genus X. *Rickettsiella* (pathogenic for insect larva and other insect hosts)
Family II. **Bartonellaceae** (intracellular parasites, sometimes extracellular, in or on red blood cells of vertebrates; arthropod transmission; cultivatable on cell-free media)
Genus I. *Bartonella* (occur in or on erythrocytes and within fixed tissue cells; often have flagella; found in man and sandfly *Phlebotomus*)
Bartonella bacilliformis (type species; Oroya fever, verruga peruana)
Genus II. *Grahamella* (intraerythrocytic; no flagella; no multiplication in fixed tissue cells; not found in man; grahamellosis of rodents and other mammals)
Family III. **Anaplasmataceae** (very small viruslike particles within or on red blood cells of vertebrates; anemia main feature of disease; arthropod transmission)
Genus I. *Anaplasma* (parasites form inclusions in red cells — no appendages; anaplasmosis of ruminants)
Genus II. *Paranaplasma* (no appendages to parasitic inclusions; infections in cattle)
Genus III. *Aegyptianella* (inclusions in red cells; aegyptianellosis in birds)
Genus IV. *Haemobartonella* (parasites within or outside erythrocytes, ring forms rare; pathogens in rodents, dogs, cats)
Genus V. *Eperythrozoon* (parasites on red cells and in plasma; ring forms common; infections in various animals)
Order II. **Chlamydiales** (gram-negative parasites of vertebrates causing various diseases; obligately intracellular reproduction typical)
Family I. **Chlamydiaceae**
Genus I. *Chlamydia* [*Bedsonia*] (parasites of tissue cells of vertebrates; three ecologic niches: (a) man — oculourogenital and respiratory diseases; (b) birds — respiratory and generalized diseases; (c) mammals [not primates] — respiratory, placental, arthritic, and enteric diseases)
Chlamydia trachomatis (type species; trachoma, inclusion conjunctivitis, lymphopathia venereum, urethritis, proctitis in man)
Chlamydia psittaci (ornithosis and psittacosis in birds; pneumonitis in cattle, sheep, goats; polyarthritis in sheep, cattle, pigs; placentitis in cattle and sheep)
PART 19 MYCOPLASMAS
Class **Mollicutes** (procaryotes lacking true cell wall; sometimes ultramicroscopic; saprophytes, parasites, or pathogens for animals; possibly arthropod-borne plant pathogens)
Order I. **Mycoplasmatales**
Family I. **Mycoplasmataceae** (sterol required for growth)
Genus I. *Mycoplasma* (pleuropneumonia group; tiny, very pleomorphic, delicate saprophytes and pathogens)
Mycoplasma mycoides (type species; contagious pleuropneumonia of cattle)
Mycoplasma gallisepticum (diseases in poultry)
Mycoplasma neurolyticum (rolling disease in mice and rats, epidemic conjunctivitis in mice; common in healthy as well as in diseased mice)
Mycoplasma pulmonis (infectious catarrh and pneumonia of mice and rats)
Mycoplasma canis (inhabitants of upper respiratory and genital tracts of dogs)
Mycoplasma hyorhinis (arthritis in swine)
Mycoplasma pneumoniae (cold-hemagglutin–associated primary atypical pneumonia of man)
Mycoplasma agalactiae (mastitis of animals)

Continued.

TABLE 1-1. ABBREVIATED CLASSIFICATION OF MICROBES FROM *BERGEY'S MANUAL OF DETERMINATIVE BACTERIOLOGY* (1974)—cont'd

PART 19   MYCOPLASMAS
    Family I. **Mycoplasmataceae**—cont'd
        Mycoplasma arthritidis (purulent polyarthritis of rats)
        Mycoplasma orale [*pharyngis*] (common parasites of human oropharynx)
        Mycoplasma salivarium (gingival crevice of man; possible role in periodontal disease)
        Mycoplasma hominis (common parasites of lower genitourinary tract in humans;
           potential pathogens in postpartum fever and in pelvic inflammatory disease)
        Mycoplasma fermentans (possible role in rheumatoid arthritis)
        "T-mycoplasmas" (common mucous membrane parasites in urogenital tract of man
           and animals)
    Family II. **Acholeplasmataceae** (sterol not required for growth)
        Genus I. *Acholeplasma* (free-living saprophytes; mammalian and avian parasites;
           wide host range; possible pathogens)
        Acholeplasma laidlawii (type species; saprophytes from sewage as well as parasites
           in animals)

---

in the text. This is the one most generally accepted in the United States. Table 1-1 presents in abbreviated form an overall survey of Bergey's classification of the microorganisms designated as bacteria.*

Bacteria that live in or about the human body have been much more completely studied than those from animals or any other source in nature. Of these, the ones producing disease are most familiar. Discussions in subsequent chapters will necessarily be limited largely to microorganisms significantly related to human disease.

---

*The American Type Culture Collection, located outside Washington, D.C., in Rockville, Maryland, 20852, maintains and distributes for research or other purposes authentic cultures of practically all known microorganisms. This nonprofit, independent agency grew out of the Bacteriological Collection and Bureau for the Distribution of Bacterial Cultures, established in 1911, at the American Museum of Natural History in New York City. There are over 20,000 strains of bacteria, fungi, algae, protozoa, viruses, and animal cell lines in this collection. Each year it supplies thousands of cultures of bacteria, fungi, and viruses to industrial and educational institutions such as medical schools, breweries, wineries, oil companies, pharmaceutical houses, and food processors.

QUESTIONS FOR REVIEW

1. What is biology? Microbiology?
2. Give the divisions of microbiology. Briefly define each.
3. List basic units used to classify both animals and plants.
4. State the purpose of health education.
5. How is the scientific name of an animal or plant derived?
6. Give the word ending that indicates reference to a class, order, or family. Carefully spell out three examples.
7. What is the difference between thallophytes and other plants? Do all forms of plant life contain chlorophyll?
8. Comment regarding criteria for classification of bacteria.
9. What classification of bacteria is most generally used at present?
10. What is the American Type Culture Collection?

REFERENCES. See at end of Chapter 9.

2 Milestones of progress*

Microbes were probably the first living things to appear on the earth. The study of fossil remains indicates that microbial infections and infectious diseases existed thousands of years ago. Epidemics and pandemics have been witnessed at all periods of written history, but an exact diagnosis of many outbreaks incompletely described in their time is not possible today.

Microbes were not seen until some three centuries ago because lenses of sufficient power to render them visible were not perfected until that time. Even after microbes were discovered, almost 200 years elapsed without any great progress in their study being made.

1546: *Girolamo Fracastoro wrote a treatise (De Contagione et Contagiosis Morbis), in which he said disease was caused by minute "seed" and was spread from person to person.* The idea of communicability of disease by contact is an ancient one. Fracastoro (1483-1553), a sixteenth-century scholar of northern Italy, published a work, which contained his theory of contagion, accounts of contagious diseases, and comments regarding their cure. Using the word *contagion* in a sense different from its accepted meaning today, Fracastoro defined it as an "infection" passing from one individual to another by contact alone, by fomites, or at a distance. He spoke of "seeds" or "germs" of disease, but he compared contagion to exhalation of an onion causing shedding of tears. He did realize from his observations, however, that all contagions do not behave alike, some attacking one organ and some another. For example, he believed in the infectivity of consumption (tuberculosis) and knew that the lungs were mainly involved. His work represents a great, although isolated, landmark in the doctrine of infectious diseases and was the result of wide and practical study of infectious diseases prevalent in his day.

---

*To facilitate review and permit a general study of the history of microbiology, important contributions (milestones) are briefly set up in italics. In most cases the milestones, not necessarily in exact chronologic order, are followed by a short discussion. Also, in several parts of the book, historical items in footnotes are related to the text.

1658: Athanasius Kircher, a Jesuit priest, published a treatise on microscopy in which he recorded seven experiments on the nature of putrefaction. The magnifying glass was known to the ancients. Not quite 100 years after the birth of Christ, Seneca, the Roman philosopher-statesman, recorded the magnifying effects of even a glass bulb filled with water. In 1590 father and son spectacle makers in Holland by the names of Johannes and Zacharias accidentally discovered the magnification produced by two convex lenses in sequence. If the lenses were arranged in a metal tube with sliding barrels, they found the magnified image could be focused. In 1624 Galileo, the Italian astronomer who made several telescopes, also made a microscope. The first person to engage in what might be called medical microscopy, however, was Kircher (1602-1680), who was well informed in mathematics, physics, medicine, and music and first drew attention to the Egyptian hieroglyphics.

1675: Antonj van Leeuwenhoek of Holland first described microbes (bacteria and protozoa) under the microscope. He made his own microscopes, which were superior to any of that time. He is known as the father of bacteriology because it was he who first accurately described the different shapes of bacteria (coccal, bacillary, and spiral) and pictured their arrangement in infected material (1683). Leeuwenhoek (1632-1723), the bedellus of Delft, was a wealthy man who devoted the major portion of his time to scientific studies. He fashioned about 250 microscopes, grinding most of the lenses himself. Leeuwenhoek was obsessed with his microscopic findings and examined everything he could think of with his scopes. He is also known as the father of scientific microscopy. In addition to his work in microbiology, Leeuwenhoek made other contributions to medicine: He gave the first complete account of the red blood cell, discovered spermatozoa, demonstrated the capillary connections between arteries and veins, and made other important anatomic observations.

Kircher's microscope magnified only 32 times. The better scopes fashioned by Leeuwenhoek magnified about 270 times. The compound light microscope of today magnifies 1000 times, the electron microscope 100,000 times.

1762: Marc Antony von Plenciz, Viennese physician, published a thesis stating his belief that disease was caused by living organisms and that each disease had its own organism. Not only did von Plenciz (1705-1786) believe that there was an organism for each disease, but he also thought that the organisms multiplied within the body and suggested that they might be transferred from person to person through the air. This is reminiscent of the treatise *(De Contagione)* written by Fracastoro in 1546.

1748: John Needham, probably the greatest supporter of the theory of spontaneous generation, published material advocating that theory.

1765: Abbé Lazzaro Spallanzani introduced experiments to disprove the doctrine of spontaneous generation. One of the factors that accelerated the development of microbiology was the argument that had been going on for years concerning spontaneous generation. The proponents believed that living forms sprang from non-living matter, and the opponents believed that every living thing descended from parents like itself. The older proponents of the theory believed that eels originated from mud, and they gave formulas for producing mice from decaying rags and cheese.

As time went on, both sides conceded that spontaneous generation did not occur in the higher forms of life. Much experimentation was done on both sides, and the argument continued until Pasteur completely and finally disproved the theory of spontaneous generation and showed the beliefs of its opponents to be correct.

Spallanzani (1729-1799), spoken of as the great Italian master of experiment, performed several experiments to disprove the doctrine of spontaneous generation. He concluded that what he called "animalcules" carried by air into infusions of organic material were the explanation for spontaneous generation. Much of Spallanzani's work was criticized by a man of Welsh origin, John Needham (1713-1781), whose observations on infusions of organic material in bottles closed with cork stoppers led him to a firm belief in spontaneous generation. To answer Needham's criticisms, Spallanzani performed a series of experiments whereby he boiled the infusion, removed the air from the flask, and hermetically sealed the container. He observed that, after a flask of infusion had remained barren a long time, only a small crack in the neck was sufficient to allow for development of animalcules in the infusion. This carefully done work might have ended the discussion on spontaneous generation had the precautions taken by Spallanzani been observed by subsequent workers.

1796: *Edward Jenner, an English physician, introduced the modern method of vaccination to prevent smallpox.* One of the world's epochal contributions to preventive medicine was made on May 14, 1796, when Jenner (1749-1823), having been impressed by the countryside tradition in Gloucestershire, England, that milkmaids who contracted cowpox while milking were subsequently immune to smallpox, performed a vaccination against smallpox by transferring material from a cowpox pustule on the hand of a milkmaid (Sarah Nelmes) to the arm of a small boy named James Phipps. Six weeks later the boy was inoculated with smallpox and failed to develop the disease. In 1798 Jenner published his results in 23 cases. By 1800 about 6000 persons had been inoculated with cowpox to prevent smallpox, and the scientific basis of the method, which in a more refined form is used today, was firmly established. The word *vaccination* was first used by Pasteur out of deference to Jenner.

1837: *Theodor Schwann proved that yeasts were living things.* Schwann (1810-1882) may be regarded as the founder of the germ theory of putrefaction and fermentation. In 1837 he concluded that the processes of putrefaction and fermentation were related to living agents that obtained their sustenance from fermentable or putrescible material. In the course of these experiments Schwann discovered and gave an accurate account of yeasts and their mode of reproduction by budding. He anticipated Pasteur's work when he asserted that fermentation of sugar was a chemical decomposition brought about by yeast attacking sugar.

1838-1839: *The cell theory was advanced.* In 1665 the Englishman Robert Hooke (1635-1703) first referred to the cell in an early work on microscopy. With his simple and imperfect microscope, he studied thin slices of a cork bottle stopper, noting the "little empty rooms," or cells, each with its own distinct walls. The word *cell*, derived from Latin *cella*, meaning storeroom, has persisted to designate the basic unit for all living things. The acceptance of the biologic theory that the structural unit of all plant and animal life is the cell came after the conclusions of Matthias Jacob Schleiden

(1804-1881), German sea lawyer and botanist, and Theodor Schwann. Their conclusions were voiced in 1838 and 1839 and were based on their scientific studies of plants and animals.

1850: *Casimir Joseph Davaine observed minute infusoria (anthrax bacilli) in the blood of sheep dead of anthrax and transmitted the disease by inoculating this blood into other animals.* Davaine (1812-1882) was a French pathologist, parasitologist, and experimenter who, with Pierre Rayer (1793-1867), first saw the bacillus that causes anthrax. He was a doctor in general practice who had to keep his experimental animals in the garden of a friend. Yet he was one of the early writers on infectious diseases and published classical works on anthrax and septicemia.

1843: *Oliver W. Holmes, physician-poet, published an important article on the contagiousness of puerperal sepsis.*

1861: *Ignaz Philipp Semmelweis published his views on the cause of puerperal fever in a book in which he stated that the cause of this disease could be found in "blood poisoning."* Oliver Wendell Holmes (1809-1894) published an article "On the Contagiousness of Puerperal Fever," which appeared in an early New England medical journal in 1843. In this article he indicated that this disease process might be spread by the examining hand of the nurse or doctor. Some time later Semmelweis (1818-1865), a Hungarian obstetrician in Vienna, noted that the death rate seemed to be higher in certain clinics where the medical students and physicians came directly to the wards from the morgue or autopsy room. He witnessed the postmortem examination of a pathologist who had died of an infectious process that developed after a dissection wound, and he observed that the changes brought about by the infection in this man were similar to those present in women dead of puerperal fever. Semmelweis then rightly concluded as to the infectious nature of puerperal fever and instituted certain practices intended to prevent the spread of the infection. On the wards that he supervised, all hands had to be cleansed carefully before a patient was examined, and the rooms were scrupulously cleaned. The mortality on his service dropped immediately. He made a statement in 1847 to the Vienna Medical Society that the cause of puerperal fever could be found in "blood poisoning" and published a book expressing his views in 1861. However, these views as to the infectious nature of the disease met with strong opposition (as did those of Holmes earlier). They were not accepted for 20 years, long after the death of Semmelweis.

1868: *Without knowing its cause, Jean Antoine Villemin showed that tuberculosis could be transmitted from man to animal and from animal to animal by inoculation of infectious material.* The early physicians from the time of Hippocrates believed that tuberculosis was infectious or contagious. Villemin, an Alsatian physician of the nineteenth century (1827-1892), was the first to prove experimentally (before the bacteriologic era) that tuberculosis could be transmitted by the injection of tuberculous material from man into animals.

1861: *Louis Pasteur disproved the theory of spontaneous generation.*

1863-1865: *Pasteur devised the process of destroying bacteria known as pasteurization.*

1881: *Pasteur developed anthrax vaccine.*

1885: *The first preventive treatment for rabies (the Pasteur treatment) was used by Louis Pasteur.* The development of modern bacteriology began with the work of the great French scientist, Pasteur, who probably has had more influence on future generations than any other man who ever lived. He was born the son of a tanner at Dôle, France, in 1822, and was not a physician but a chemist. At about 30 years of age, having already distinguished himself as a chemist, Pasteur became interested in the process of fermentation, which, among other things, led to his disproving the theory of spontaneous generation.

By 1865 Pasteur had proved that the "diseases of wine" could be prevented without altering the flavor by heating the wine for a short time to a temperature (55° to 60° C) a little more than halfway between its freezing and boiling points. This process known as *pasteurization* is employed throughout the civilized world today to preserve milk and certain other perishable foods.

Pasteur proved that pébrine, a disease of silkworms that threatened to destroy the silk industry of France, was caused by an animal parasite and devised methods for its prevention. Later he confirmed Koch's studies on anthrax, which was the first animal disease proved to be caused by bacteria. He discovered the staphylococcus, described the streptococcus, and was codiscoverer of one of the organisms causing gas gangrene.

Pasteur's next contribution, the one for which he is most widely known, was the development of the Pasteur treatment for rabies. He gave the first treatment in 1885, and his method, modified slightly, is used throughout the world today. Pasteur died in 1895, and his body lies in the Pasteur Institute of France.

1865: *Joseph, Lord Lister, English surgeon, applied antiseptic treatment to the prevention and care of wound infections.* The famous English surgeon Lister (1827-1912) was so impressed with the similarity between wound infections and certain fermentative and putrefactive changes, which Pasteur had already proved to be caused by microorganisms, that he concluded that wound infections, too, were caused by microbes. With this idea in mind, he protected wounds with dressings saturated with a solution of carbolic acid (phenol) and devised operating room procedures calculated to destroy microorganisms. The establishment of these methods was so far reaching in effect that Joseph, Lord Lister will always be known as the father of antiseptic surgery.

1873: *William Budd recognized the role of milk and water in the transmission of typhoid fever.* His very accurate study of typhoid fever outbreaks in 1856 led Budd (1811-1880) to believe that the disease was infectious and that the causative agent was excreted in the feces of the patient. He suspected that contaminated milk and water played an important part in the spread of the disease. In 1872 he noted that an outbreak in Lausen, Switzerland, was related to contaminated water. In a treatise on typhoid fever in 1873 Budd summarized his views. The discovery of the causative organism of typhoid fever did not come until 1880 (Karl Joseph Eberth, 1835-1926); Georg Gaffky (1850-1918), one of Koch's co-workers, was the first to cultivate it (1884).

1876: *Robert Koch reported the isolation of the anthrax bacillus in pure culture and the proof of its infectiousness.*

1881: *Koch had perfected his technic of isolating bacteria in pure culture. He introduced the use of solid culture media.*

1882: *Koch discovered Mycobacterium tuberculosis, the organism causing tuberculosis.*

1890: *Koch discovered tuberculin.* The first contribution of the German bacteriologist Robert Koch (1843-1910) was the isolation of the anthrax bacillus. (That this agent alone caused anthrax had been confirmed by Pasteur.) Koch is remembered for both his discovery of important disease-producing organisms and his fundamental contributions to bacteriologic technic. He was the first to stain bacterial smears as is done today. He devised a liquefiable solid culture medium (gelatin) and worked out pure culture technics (triggered by chance observations on a boiled potato left out in his laboratory one day).

1880: *Alphonse Laveran discovered the parasite of malaria.*

1895-1898: *Sir Ronald Ross discovered that mosquitoes transmitted malaria.* In 1880 Charles Louis Alphonse Laveran (1845-1922), a French Army surgeon, discovered the parasite of malaria, and within the next 20 years its transmission by the mosquito was worked out by Ross (1857-1932), an English pathologist and parasitologist. Ross also devised suitable methods for the elimination of mosquitoes over wide areas.

1888: *Roux and Yersin discovered diphtheria toxin.* Émile Roux (1853-1933), research associate of Pasteur, was considered the greatest French bacteriologist after Pasteur. Roux and Élie Metchnikoff were two of Pasteur's most noted pupils. Roux, who adored Pasteur, said that in the hands of his master the scientific approach was one "which resolves each difficulty by an easily interpreted experiment, delightful to the mind, and at the same time so decisive that it is as satisfying as a geometrical demonstration."

Together with Alexandre Yersin (1863-1943), a Swiss bacteriologist, Roux made a number of important contributions to the bacteriology of diphtheria, one of which was the demonstration of a toxic substance in filtrates of broth cultures of the diphtheria organism.

1889: *Hans Buchner discovered the bactericidal effect of blood serum.* Buchner (1850-1902), a German bacteriologist and immunologist, by his experimental work demonstrated the bactericidal effect of blood serum. He showed that bactericidal substances occur in blood and also in cell-free serum.

1889: *Kitasato cultivated the tetanus bacillus and proved it to be the cause of lockjaw.* In 1889 Shibasaburo Kitasato (1852-1931), a pupil and colleague of Koch, cultivated the tetanus bacillus and proved it to be the cause of tetanus or lockjaw. Since he was unable to find this organism in animals that had died of the disease, he felt the disease process to be the result of chemical intoxication rather than of bacterial invasion.

1888: *Ludwig Brieger discovered tetanus toxin.*

1890: *Kitasato and von Behring discovered tetanus antitoxin (passive immunization). Von Behring discovered diphtheria antitoxin.*

1894: *Roux and Martin produced antitoxin by immunizing horses.*

1913: *Von Behring first used toxin-antitoxin to produce a permanent active immunity against diphtheria in human beings.*

1914: *Park used toxin-antitoxin mixtures for immunization in New York City.*

1923: *Ramon prepared diphtheria toxoid, which has supplanted toxin-antitoxin.*

In 1888 Brieger (1849-1909), a Berlin physician, discovered the toxin of tetanus and studied its action. In 1890 Kitasato and Emil von Behring (1854-1917), also an assistant in Koch's laboratory, found that experimental animals treated with tetanus toxin developed a transferable immunity. The cell-free serum from such animals rendered harmless the toxin elaborated by the bacillus. This protective serum could be given to other laboratory animals. Normal serum did not affect tetanus toxin. Kitasato and von Behring published their results, and in a footnote in their article the word *antitoxic* was used. From this word, *antitoxin* became firmly established. Only a week after the discovery of tetanus antitoxin was announced, von Behring published another article, this one on immunization against diphtheria. In 1894 Emile Roux and his co-worker Louis Martin began to immunize horses to produce antitoxin. Martin (1864-1946) also was noted for his studies on diphtheria and antidiphtheritic serum.

Before the end of the nineteenth century, William Hallock Park (1863-1939) organized the first municipal bacteriologic laboratory in the United States. As dynamic head of this laboratory in New York City, he was the first to apply the best known methods of the time to protect the children of the city against diphtheria. For his widespread immunization program he used the toxin-antitoxin combination (first used by von Behring) and deserves the credit for practical application of von Behring's work.

By 1923 Gaston Leon Ramon (1885-1963), French bacteriologist and veterinarian, had prepared diphtheria toxoid, an agent devoid of the dangers of toxin-antitoxin and with superior immunizing capacity.

1892: *George M. Sternberg, called by Koch the "father of American bacteriology," published A Manual of Bacteriology.* Among Americans who added to the science of microbiology in the nineteenth century, Sternberg (1838-1915), military surgeon, pioneer bacteriologist, epidemiologist, and Surgeon General of the United States Army for nearly a decade, contributed much to the epidemiology of cholera and yellow fever. In 1881 he announced simultaneously with Pasteur the discovery of the pneumococcus. His treatise on bacteriology, published the year before he became Surgeon General, was an early landmark, being the first extensive presentation on the subject of microbiology published in the United States. During his term of office as Surgeon General, the Army Medical School was established (in 1893), probably the first school of modern hygiene in America; the Nurse and Dental Corps were organized; and in 1900, Major Walter Reed (1851-1902) was detailed to lead the famous Yellow Fever Commission, which incriminated the mosquito in the spread of yellow fever.

1896: *Gruber and Durham described the phenomenon of agglutination.* In 1896 Max von Gruber (1853-1927) and Herbert Edward Durham (1866-1945) discovered the phenomenon of agglutination. Practical application of this has resulted in agglutination tests for such diseases as typhoid fever, typhus fever, undulant fever, and tularemia.

1884: *Élie Metchnikoff announced the phagocytic theory of immunity.*

1898: *Paul Ehrlich expounded his "side-chain" theory of immunity.*

In 1884 Metchnikoff (1845-1916), zoologist and native of the Ukraine, proposed the phagocytic or cellular theory of immunity. (In some of his earliest experiments on phagocytosis Metchnikoff studied the ingestion of thorns from a rosebush by the larval form of a starfish.) With the development of the science of immunology in the 1800s two schools arose. The first, the French school, headed by Metchnikoff, believed the whole process of immunity to be centered in the ingestion and destruction of invading disease-producing agents by the leukocytes and certain of the body cells.

The second school of immunologists, the German school, was led by Paul Ehrlich (1854-1915), an outstanding German scientist who did important research in immunology from 1890 to 1900. Although he had drawn the outlines of his side-chain theory of immunity more than 10 years before, it was not until the turn of the century that he made his first exposition of it. In his studies Ehrlich conceived the structure of protein molecules to be similar to a benzene ring with unstable side chains. These could act as chemoreceptors that would combine with harmful materials to neutralize them and be released in excess after the need had been met. The German school believed that the establishment of immunity stemmed from the development in body fluids, especially in the blood, of certain substances known as antibodies or immune bodies with the capacity to destroy invading disease-producing agents. This was the humoral or chemical theory of immunity.

1909: *Ehrlich introduced salvarsan ("the great sterilizing dose") as a treatment for syphilis.* Paul Ehrlich was a genius of extraordinary activity who added a great mass to our knowledge of medical science, and a large number of technical laboratory methods he proposed are now in daily use. To him we are indebted for the basic procedures in staining, for he was one of the first to apply aniline dye solutions to cells and tissues. He founded and was the greatest exponent of the science of hematology and created the new branch of medicine known as *chemotherapy.* He began to synthesize many new drugs of slightly varying chemical formulas and to test each experimentally until a maximum therapeutic effect was obtained. Collaborating with a Japanese physician, Sahachiro Hata (1873-1938), Ehrlich found in the six hundred and sixth compound of a series tested, salvarsan (arsphenamine), a drug capable of controlling the manifestations of syphilis over a long period of time.

1898: *Beijerinck recognized the first virus (virus of tobacco mosaic disease).* In 1892 Dmitri Iwanowski reported the first experiments that pointed to the presence of an infectious agent much smaller than any known up to that time. The subject of these experiments was the virus of tobacco mosaic disease. This work was repeated and extended by Martinus Willem Beijerinck (1851-1931), Dutch botanist, in 1898. Iwanowski was doubtful of his own findings, but Beijerinck realized that a new type of infectious agent had been discovered. In addition to being the first virus discovered, the virus of tobacco mosaic was the first to be purified in a crystalline form. (About 1 ton of infected tobacco must be processed for 1 tablespoon of virus crystals.) This task was accomplished in 1935 by Wendell Meredith Stanley (1904-1971), biochemist and virologist. In 1955, in Stanley's laboratory at the University of California, the poliomyelitis virus was crystallized, the first virus affecting animals or man to be so purified.

1901: *The process of complement fixation was developed.* In 1901 Jules Jean Baptiste Bordet (1870-1961) and Octave Gengou (1875-1957) described the phenomenon of complement fixation, and in 1906 August von Wassermann (1866-1925) applied it in the Wassermann test, the first reliable serologic test for syphilis. In 1907, Leonor Michaelis (1875-1949), German physical chemist, described the phenomenon of precipitation, which is now used in such tests as the Kline, Kahn, Mazzini, VDRL, and Hinton tests for syphilis.

1901: *Karl Landsteiner discovered the basic human blood groups.*

1940: *The Rh factor was discovered by Landsteiner and Wiener.* Karl Landsteiner (1868-1943), American pathologist, had been associated with Max von Gruber in Vienna. When he came to America, he held an important post with the Rockefeller University of New York. He found that normal human blood falls into four main groups, the blood of some of these groups destroying blood belonging to other groups. This discovery laid the foundation for modern blood transfusion. In 1940 Landsteiner and Alexander S. Wiener (1907-   ) identified the Rh factor of blood. Landsteiner thought blood to be as unique a characteristic of an individual as fingerprints.

1906: *Novy and Knapp discovered the spirochete of the American variety of relapsing fever.* Frederick George Novy (1864-1957), American bacteriologist, together with R. E. Knapp, discovered the spirochete causing the American variety of relapsing fever. The first teaching course in bacteriology in the United States was established by Novy at the University of Michigan in 1889. Three years before, in 1886, the first department of bacteriology in any United States medical school had been instituted at Columbia University.

1894 to present: *Women microbiologists have made important contributions.*

1913: *Edna Steinhardt first used tissue culture technics to grow virus.*

1917-1923: *Alice C. Evans investigated undulant fever.*

1921-1927: *Gladys Dick assisted her husband George Dick in work on the bacteriology and serology of scarlet fever.*

1931: *Alice Woodruff grew a virus in a fertile egg.*

1928-1933: *Rebecca Lancefield published methods of classification of streptococci.*

1953 to present: *Sarah Stewart has carried on significant research in tumors in animals induced by virus.*

These milestones are by way of special mention of the distaff side who have played no inconsiderable part in the development of modern microbiology. One of the first such women was Anna Wessels Williams (1863-1954). In 1894 she was appointed assistant bacteriologist of a New York Board of Health laboratory just organized by W. H. Park. For 40 years she was closely associated with this outstanding health officer during a period of remarkable achievement in the field of preventive medicine in New York City. During the first part of this century the Chicago bacteriologist Ruth Tunnicliff (1876-1946) studied streptococci, anaerobic organisms, measles, and meningitis. Pearl L. Kendrick (1890-   ) has added much to our knowledge of whooping cough. Eleanor A. Bliss (1899-   ) is known for her work on streptococci and the sulfonamide drugs.

Senior bacteriologist at the National Institutes of Health for 27 years, Alice Catherine Evans (1881-1975) was a notable predecessor of women working in investigative laboratories. As a result of the work done by the Dicks, a husband-wife team in Chicago, important information was gained about the bacteriology of scarlet fever. In 1923 they causally linked the disease to hemolytic streptococci and in 1924 devised the Dick test. Rebecca C. Lancefield (1895-   ) developed the serologic classification of human and other groups of streptococci in wide use today.

Edna Steinhardt and her associates at Columbia University, who propagated cowpox virus in bits of cornea from the eye of a guinea pig, have been credited the first to cultivate virus by technics of tissue culture. The news in 1931 that viruses could be grown in embryonated hen's eggs made a major impact on the entire microbiologic world. A physiologist, another member of a husband-wife team, Alice Miles Woodruff, working with fowlpox virus, was first able to do this. The work was reported from the Pathology Laboratory at Vanderbilt University School of Medicine under the direction of E. W. Goodpasture. Since virus cultivation in fertile eggs and in tissue cultures produces an ample supply of virus for study or for vaccine production, the introduction and standardization of these two technics have in large measure contributed to the explosive surge of knowledge that has occurred in the field of virology in recent years. In fact, this period of the last several decades has been called the "golden age" of virology.

Feminine scientists figure prominently in the field of cancer research. A noted one with a background as a microbiologist is Sarah Stewart of the National Institutes of Health, who is working on tumors induced by viruses. In 1953 her first report came out on the salivary gland tumors in mice induced by the polyoma virus, a fairly common virus readily infecting many laboratory animals. With the help of Bernice Eddy and co-workers, she subsequently propagated the virus and showed that it can produce a great variety of cancers in the animals studied.

1928: *Penicillin was discovered by Sir Alexander Fleming of England.*

1940: *The value of penicillin as a therapeutic agent was determined by Florey and Chain in England.* In 1928, on return from a week's vacation, Fleming (1881-1955), Scottish bacteriologist, discovered a patch of green mold that had fallen haphazardly on one of his cultures, and he became interested in its antibacterial action. In 1929 he reported that this mold, *Penicillium notatum,* elaborated during its growth a substance that inhibited the development of certain bacteria. He called this substance penicillin. Fleming's own life at one time was saved from an overwhelming pneumonia by the very agent he had discovered. To his great delight, he made a speedy and dramatic recovery.

Fleming's monumental work went practically unnoticed until 1938, when Sir Howard Florey (1898-1968) and Ernst Boris Chain (1906-   ) at Oxford University began a systematic study of Fleming's findings. Not only did they develop suitable methods for producing penicillin and for demonstrating its usefulness, but they were also responsible for placing it before the American medical world and for gaining the interest of American industry. American industrial organizations have made possible the production and availability of penicillin on a wide scale.

1935-1936: *Domagk, Colebrook, and Kenny established the remarkable value of Prontosil (the forerunner of the sulfonamide compounds) in streptococcal infections.*

1938 to present: *Many important sulfonamide compounds have been developed.* Ehrlich, from his work in chemotherapy, had the idea that dyes might act as selective poisons, and other workers had found that certain azo dyes had some bactericidal power. In 1927 the pathologist Gerhard Domagk (1895-1964) began an eventful collaboration with two organic chemists, Mietzsch and Klarer, at the Elberfeld Laboratories of the German firm of I. G. Farbenindustrie. These two chemists prepared a number of azo dyes, and in 1932 Domagk found that one of them, when injected into mice infected with streptococci, had a definite curative power. The molecule of this effective agent contained a grouping of atoms that chemists call the *sulfonamide* group. This discovery was a considerable stimulus to further research, and subsequently more than 1000 new dyes were made. In 1932 Mietzsch and Klarer prepared a red dye that proved to be the most effective of the compounds, and it was given the trade name of Prontosil. In 1935 Domagk published his experimental data (in German), demonstrating the antibacterial powers of Prontosil, and about the same time several clinical reports confirming the therapeutic value of this compound in streptococcal infections appeared in the German literature. Shortly after, a group of French investigators showed that the compound Prontosil was broken down within the body to form sulfanilamide and demonstrated the antibacterial properties of this compound (first prepared by Paul Gelmo in 1908). Leonard Colebrook and Méavc Kenny were responsible for arousing the interest of the English-speaking world in bacterial chemotherapy and in these new sulfonamide compounds. They demonstrated the clinical effectiveness of sulfanilamide in puerperal fever, a streptococcal infection.

Since the introduction of Prontosil, more than 5000 sulfonamides have been prepared. Of these, only a few show antibacterial activity comparable to or greater than that of the parent drug.

1939: *René Jules Dubos discovered tyrothricin.*

1943: *Streptomycin was discovered by Waksman.*

1947 to present: *Many new antibiotic agents are constantly being developed and tested.* In 1939 Dubos (1901-   ), now of the Rockefeller University, discovered tyrothricin and demonstrated that microorganisms isolated from the soil can produce, under certain conditions, substances possessing antibacterial properties. Tyrothricin was the first member of the group of agents known as antibiotics to be given extensive study and was the first to be produced in a pure form. This information is not so new as it might seem because Pasteur in 1877 had demonstrated that the growth of anthrax bacilli was retarded by the presence of common soil organisms, and several antibiotic-like substances (example, pyocyanase) were prepared before 1900.*

---

*The ancient Egyptians painstakingly listed on papyri concoctions with activity corresponding to that of the modern day antibiotics. One item they felt to be most effective was "moldy wheaten loaf."

The development of streptomycin came as the result of a planned, careful search for an antibiotic agent effective for gram-negative bacteria and of minimum toxicity. Soils, composts, and other natural substrates harboring large numbers of organisms capable of producing antibiotic substances were carefully studied. Within a period of 5 years nearly 10,000 cultures were examined before streptomycin was obtained by Selman Abraham Waksman (1888-1973) and his co-workers at Rutgers University. This antibiotic agent was isolated from two strains of *Streptomyces* identified as *Streptomyces griseus*. Less than a year after the isolation of streptomycin, its effectiveness against the tubercle bacillus was demonstrated. For the first time in history an effective drug for the treatment of tuberculosis had been found.

Practical application of the observations of the soil microbiologist, the plant pathologist, and the medical bacteriologist has resulted in the development of many important antibiotics since World War II. Many diseases that formerly progressed unchecked are now harmless and easily controlled.

1941: *Albert Coons developed the fluorescent antibody test.* The fluorescent antibody technic represents an important example of the modern advances being made in the field of immunology. It was developed by an immunologist, Albert H. Coons (1912-   ), and his associates at Harvard Medical School, who demonstrated that fluorescein, a fluorescent dye, could be bound to an antibody without changing the behavior of the antibody. The attachment of the fluorescent dye would, however, cause the antibody to become luminous if it were viewed under ultraviolet light. These workers were the first to mark antibodies for future identification. Greer Williams refers to this test as one "equipped with neon lights."

1944: *Avery, MacLeod, and McCarty discovered the genetic role of DNA.*

1953: *Watson and Crick demonstrated its structure.* The story of deoxyribonucleic acid (DNA) goes back nearly a hundred years, but the dramatic chapters in the elucidation of its biologic significance did not unfold until well into the twentieth century. An acid substance referred to as nucleic acid, something as to its chemical composition, its relation to a carbohydrate (deoxyribose-deoxyribonucleic acid), and its association with chromosomes of the nucleus were known before Frederick Griffith, a British pathologist, made a significant discovery in 1928. When he inoculated mice (animals peculiarly susceptible to pneumococci) with a mixture of pneumococci, alive but *weakened*, and pneumococci, heat-killed but *virulent*, the mice developed a pneumonia produced by *virulent* organisms. A mysterious transformation had occurred whereby harmless microorganisms had changed their character. This work constituted a prologue to what has been called a revolution in biology.

In the years following, researchers at the Rockefeller University, because of their interest in the disease-producing properties of pneumococci, turned their attention to Griffith's heat-killed pneumococci and repeated his work. Over a period of 10 years they slowly eliminated, one by one, all genetic factors possibly involved in the transformation of the pneumococci, including protein of the nucleus (long thought to be the carrier of genetic information), the capsule of the microorganisms, and so on, until nothing was left except deoxyribonucleic acid (DNA), the nucleic acid of the nucleus. When DNA was removed from the heat-killed virulent pneumococci, it became apparent that it was indeed the transforming principle. By 1944, three workers, Oswald T. Avery, Colin MacLeod, and Maclyn McCarty, reported that they had for the first time the proof that DNA is the stuff of which heredity is made. The work of three biophysicists, one American, two British, is responsible for the elucidation of the molecular structure of DNA. In 1953 J. D. Watson and F. H. C. Crick reported their construction of the double helix model of the DNA molecule for which M. H. F. Wilkins had laid the groundwork by his x-ray diffraction studies.

1949-1963. *The dramatic development of poliomyelitis vaccine took place.* By the early 1930s research on poliomyelitis was well under way, and a vaccine had been prepared from the brain and spinal cord of monkeys. This first experimental vaccine was crude, only partially effective, and very impractical to make. Therefore, when the National Foundation for Infantile Paralysis (the National Foundation of today) began its support of scientific research in 1938, it could be rightly said that not a great deal was known about poliomyelitis. Before a successful program of immunization could be launched, more basic facts had to be obtained about the disease and its causative agent.

One of the most significant events in the history of the poliomyelitis vaccine was the report of John F. Enders and associates (Thomas H. Weller and Frederick C. Robbins) at Harvard that poliomyelitis virus could be successfully grown in cultures of living cells of nonnervous tissue. To this point poliomyelitis research had been slow and costly because of its dependence on monkeys as experimental laboratory animals, animals often hard to obtain and expensive to keep. The introduction of the tissue culture method marked the end of the monkey era in poliomyelitis research. Through tissue cultures extremely large amounts of virus, free of the presence of injurious impurities in nervous tissue, could be produced quite inexpensively and hundreds of preliminary and pertinent investigations as to the nature of the virus could be easily carried out.

In 1954 a field trial using the Salk vaccine was launched by the National Foundation for Infantile Paralysis. The vaccine was an experimental one that Jonas E. Salk (1914-   ) and associates at the University of Pittsburgh had prepared by formaldehyde inactivation of poliovirus. The events of this field trial were unique in medical history. It was a project of universal interest that reached out to involve the talents of research teams, the methods of commerce and industry, the facilities of universities and government, and the resources of the publicly subscribed National Foundation, meanwhile engrossing the press of the entire world. The success of the field trial was formally announced by the Poliomyelitis Evaluation Center of the University of Michigan on April 12, 1955, the anniversary of the death of Franklin D. Roosevelt.

The development of a live poliovirus vaccine started with a number of experimental, clinical, and epidemiologic studies on a series of small well-controlled trial runs in the United States. It then jumped to a mass immunization program of global extent and astronomic proportions. The trial vaccines for the final large-scale tests were experimental vaccines (comprising the three strains of poliomyelitis virus) developed by the major workers in the field: Hilary Koprowski, Herald A. Cox, and Albert B. Sabin (1906-   ). At the time that the Salk vaccine was introduced Sabin was busy searching for nonvirulent strains among the polioviruses growing in monkey tissue cultures in his laboratory. In 1956 he developed his first poliovirus for live virus (oral) vaccine.

From 1957 to 1959, beginning in the Belgian Congo, a program of mass immunization was carried out all over the world. In 1960 a committee set up by the United States Public Health Service in 1958 to follow developments in the laboratory and in the field trials reported that more than 60 million persons had been fed attenuated live poliovirus vaccines with no ill effects and that of the experimental vaccines available, the Sabin strains seemed to be more highly attenuated and therefore safer but were as immunogenic as the Cox and Koprowski vaccines. On the recommendation of its committee, the U.S. Public Health Service gave its approval in 1960 for the commercial manufacture of the Sabin vaccine under specified regulations. Although it had been used extensively abroad, the Sabin oral vaccine was first used in the United States in 1962. After the three strains of the Sabin oral vaccine had been licensed, immunization of the American public was carried out through community programs set up to dispense oral vaccine on a mass scale.

In the history of infectious disease, the conquest of poliomyelitis is a dramatic chapter. Once a major threat, this crippling disease has almost disappeared.

1954-1963: *Another viral vaccine, the measles vaccine, was developed.* In 1954 John F. Enders of Harvard University and Thomas Peebles, an associate, adapted measles virus to tissue cultures of human cells. The virus used was isolated from the blood of an 11-year-old boy named David Edmonston in Bethesda, Maryland. Designated the Edmonston strain, it was weakened by successive growth in cell cultures of human and animal tissues and in the chick embryo. The Edmonston strain has provided the source for the different measles vaccines subsequently used for research purposes and field trials. In initial tests, approximately 25,000 children in the United States were vaccinated with live virus vaccine.

Safe and effective viral vaccines have also been developed for rubella and mumps.

1964: *Baruch S. Blumberg discovered the Australia antigen.* In 1964 at the Institute for Cancer Research in Philadelphia, Blumberg, a geneticist, noted a foreign substance in the blood of an Australian aborigine that he labeled the Australia antigen. In 1968 Alfred M. Prince, a virologist, of New York City Blood Center found the same substance in the blood of patients with viral hepatitis, type B, and subsequently the Australia antigen was specifically linked to the disease. The isolation of this substance has stimulated a considerable amount of research in viral hepatitis.

1942: *Dr. Norman Gregg, an Australian ophthalmologist, published his observations relating congenital cataracts in infants to a severe epidemic of rubella (German measles) in Australia.* The epidemic had followed a 17-year period of minimal incidence of disease in Australia, and consequently many young adults were infected. Gregg noted that practically all mothers of affected infants had had German measles early in gestation, some even before they realized they were pregnant. His correlations showed that malformations may result from extrinsic agents, even infectious ones such as viruses, rather than being exclusively bound to inherent genetic or developmental mechanisms, a concept that has had wide-ranging effects.

1910: Peyton Rous (1879-1970) demonstrated experimentally a significant relation between cancer and viruses in chickens.
The last several years in biologic sciences have been remarkable ones as to the many and diverse discoveries of far-reaching significance. It is not possible in this chapter to mention the many implications and ramifications. We choose to conclude therefore with only a hint as to coming events in an amazing story—that of viruses and cancer.

At the turn of the century the suggestion was made that viruses might cause cancer. In 1910 Peyton Rous, pathologist and medical researcher of the Rockefeller University, used material from a large tumor on the breast of a Plymouth Rock hen to show that cell-free filtrates of it induced the same tumor in hens, that the induced tumor behaved as did the original one, and that the properties of the agent in his experiments were those of a filterable virus. This important landmark in cancer research was years ahead of the times. To the early fifties, nearly half a century, the viral etiology of cancer made slow headway although substantiated by a number of significant observations (Table 28-5), and for many years it took considerable courage to hold to such a theory.

1958: Denis Burkitt defined Burkitt's lymphoma epidemiologically and postulated a transmissible viral agent. In 1958 Burkitt defined the "jaw tumor in central African children" that now bears his name, and mapped out its geographic distribution. He described it as a cancer but much about it impressed him as an infectious disease. Partly because it is confined sharply to the yellow fever and mosquito belt of central Africa, he thought at first that it might be caused by a virus carried in a mosquito; later he dismissed the idea of an insect vector. This time the climate of the scientific world is right. The disease is a dramatic one, and the tools and technics for study are highly developed. Burkitt's lymphoma is thought by all to be caused by a virus, currently the Epstein-Barr virus.

1970: Howard Temin and Satoshi Mitzutani of the University of Wisconsin and David Baltimore at the Massachusetts Institute of Technology separately but almost simultaneously discovered reverse transcriptase. Reverse transcriptase is a specific viral enzyme in RNA viruses such as the Rous sarcoma virus that uses viral RNA as template to produce DNA, which, as genetic information, is inserted into the genes of the host cells. When it was found in 1940 that growth and heredity were controlled by DNA, it was thought that the genetic sequence could only go from DNA to RNA. The dramatic nature of the discovery of an RNA-directed enzyme lies in the fact that it showed that the genetic sequence could be reversed—information could go from RNA to DNA. The discovery has been most significant and productive for viral cancer research, and among many things has led to the identification of RNA tumor virus-like particles in a variety of human tumors, including human leukemias.

The relation of human neoplasms to viruses is one of the most challenging of research problems today. Scientifically we are living in exciting times!

REFERENCES. See at end of Chapter 9.

3 The basic unit

THE CELL

Every living thing, plant or animal, is made up of one or many cells. In the lowest forms of life the whole organism is only a single cell, one in which are carried on all the processes associated with life. Such is a *unicellular* organism. In the higher forms of life, multiplied millions of cells arrange themselves into groups of different type cells in order to meet the varied physiologic needs of that organism. Such an organism with a division of labor is a *multicellular* one.

Regardless of how simple or complex the organism, the unit of structure is the *cell*. Cells are the bricks with which living organisms are built. Not only is the cell the structural unit, but it is also the operational one. One living cell, be it ever so small, is a dynamic biologic unit, and the performance of the organism or of any of its parts is the sum of the functions of the constituent cells, acting singly or in groups. The science of the cell is *cytology*.

Anatomy of the cell

Protoplasm. A cell is composed of a colorless, translucent viscid substance of colloidal nature known as *protoplasm* (Greek, first formed), the physicochemical basis of life. Protoplasm is a complex but dynamic system incorporating units of varying size, physical nature, and chemical composition. Although adapted to cells of specific types, protoplasm presents basic features in all cells. It is made up principally of water (up to 90%). Contained therein are inorganic mineral ions and organic chemical compounds of varying complexity and physiologic significance—proteins (including enzymes), carbohydrates, fats, nucleic acids (Table 3-1), hormones (usually either protein or lipid or derivatives), and vitamins (chemically heterogeneous compounds).

Certain organic compounds in protoplasm have special biochemical significance in determining the form and biologic properties of the cell. Designated *macro-*

36

TABLE 3-1. OUTLINE OF CHIEF PROTOPLASMIC COMPOUNDS

	Biochemical compounds			
	Carbohydrates	Lipids	Proteins	Nucleic acids
Composition (building blocks)	Simple sugars, some with amino groups	Triglycerides of fatty acids; phosphorus in phospholipids	Amino acids (at least five elements here); most complex; can be very large	Pentose nucleotides; complex combinations with protein are nucleoproteins
Role in cell	Synthesis of cell wall Essential fuel for all living organisms Mostly found as glycogen (storage form of glucose)	Phospholipids important to cell membrane Large part of energy needs met by these fuels	Most are enzymes Synthesis of organelles and other parts of cell	DNA—primary function in heredity; the genetic code for cell structure and function RNA—protein synthesis

molecules, they are composed of organic chemical building blocks linked together in characteristic fashion. Important examples of protoplasmic macromolecules are select carbohydrates (polysaccharides), proteins, enzymes, and nucleic acids. Most of the unique qualities of protoplasm stem from its contained macromolecules.

Shape. In keeping with their location inside the body of an organism and the work that they do, cells vary in shape, being spherical, oblong, columnar, six sided, or spindle shaped.

Size. Cells are of microscopic size, except eggs of various animals (an ostrich egg is 4 to 6 inches in diameter and weighs about 4 pounds), striated muscle cells under certain conditions, and the cells of the pith, stem, and flesh of some pulpy fruits. Some cells are so small as to be barely visible with the highest magnification of the ordinary compound microscope.

For a living object of such small dimensions as the cell, a special unit of measurement is needed. For a long time this was the micromillimeter, or *micron,* designated by the Greek letter mu (μ). A micron is 1/1000 of a millimeter or 1/25,000 of an inch in length. (The paper on which this book is printed is about 100 μ thick.) A *millimicron* (mμ), a term also well known, is 1/1000 of a micron. Today many scientists prefer and use the equivalent terms *micrometer* (μm) for micron and *nanometer* (nm) for millimicron. (See Table 3-2 for metric equivalents.) Cells show considerable variation in size, but most range in diameter from 10 to 100 μm. The average diameter of the human red blood cell is 7.5 μm. Individual variations in the size of multicellular organisms of the same species depend on the number, not the size, of their cells.

TABLE 3-2. EQUIVALENTS IN THE METRIC SYSTEM

Unit symbol →	Meter (m)	Centimeter (cm)	Millimeter (mm)	Micron (μ)	Millimicron (mμ)	Angstrom (Å)
Meter (m)	1.0	100.0 10^2	1,000.0 10^3	1,000,000.0 10^6	1,000,000,000.0 10^9	10,000,000,000.0 10^{10}
Centimeter (cm)	0.01 10^{-2}	1.0	10.0	10,000.0 10^4	10,000,000.0 10^7	100,000,000.0 10^8
Millimeter (mm)	0.001 10^{-3}	0.1 10^{-1}	1.0	1,000.0 10^3	1,000,000.0 10^6	10,000,000.0 10^7
Micrometer (μm)	0.000001 10^{-6}	0.0001 10^{-4}	0.001 10^{-3}	1.0	1,000.0 10^3	10,000.0 10^4
Nanometer (nm)	0.000000001 10^{-9}	0.0000001 10^{-7}	0.000001 10^{-6}	0.001 10^{-3}	1.0	10.0
Tenth nanometer (0.1 nm)	0.0000000001 10^{-10}	0.00000001 10^{-8}	0.0000001 10^{-7}	0.0001 10^{-4}	0.1 10^{-1}	1.0

FIG. 3-1. Electron microscope. In this instrument particles $^1/_{10}$ millionth inch in diameter may be visualized. (Courtesy Forgflo Corporation, Sunbury, Pa.)

The information gained through the study of cells by the compound light microscope in over 100 years has contributed significantly to the development of modern medicine. However, the nature of light is such that it is not possible to get a clear image in the light microscope when the magnification of the object for study is greater than 2000 times. Hence many minute (submicroscopic) structures within the cell were not visualized until the electron microscope was directed to cytologic study.

ELECTRON MICROSCOPE. The electron microscope (Fig. 3-1), one of our most powerful research tools, differs from the compound microscope. In the latter, rays of ordinary light pass from the object being examined to the eye to be focused there, whereas in the electron microscope electrons pass through the specimen to be focused on a viewing screen from which a photograph is made—an *electron micrograph*. With the most modern electron microscopes the dimensions of the object under observation may be magnified 1,000,000 times. If careful photographic technics are used, a further enlargement can be accomplished so that today a potential magnification of 2 *million* or more times is possible. (With this magnification, a human hair viewed in its entirety in the electron microscope would appear twice the size of a California redwood.) The *scanning* electron microscope (Fig. 3-2) provides a three-dimensional quality to the study of specimens at magnifications up to 100,000

FIG. 3-2. Scanning electron microscope. (Courtesy Coates & Welter Instrument Corp., Sunnyvale, Calif.)

TABLE 3-3. COMPARATIVE SIZES OF BIOLOGIC OBJECTS

	Biologic object	Diameter Micrometers (μm)	Nanometers (nm)	Angstroms (Å)
Human eye	Ostrich egg	200,000	200,000,000	2,000,000,000
	Mature human ovum	120	120,000	1,200,000
Light microscope	Erythrocyte (red blood cell)	7.5	7,500	75,000
	Serratia marcescens (bacterium)	0.75	750	7,500
	Rickettsia	0.475	475	4,750
	Chlamydia psittaci	0.27	270	2,700
Electron microscope	*Mycoplasma*	0.15	150	1,500
	Influenza virus	0.085	85	850
	Genetic unit (Muller's	0.02 × 0.125‡	20 × 125‡	200 × 1,250‡
	estimation of largest	0.027	27	270
	size of a gene)	0.016	16	160
	Poliovirus			
	Tobacco necrosis (plant virus)			
	Egg albumin molecule	0.0025 × 0.01‡	2.5 × 10‡	25 × 100‡
	(protein molecule)	0.0001	0.1	1
	Hydrogen molecule			

Left margin vertical labels: "Limit of resolution", "Human eye", "Light microscope", "Electron microscope"

*Limit of resolution of human eye, 0.1 mm or 100 μm.
†Limit of resolution of compound light microscope, 0.2 μm.
‡Width × length.

times actual size. It sweeps a very narrow beam of electrons back and forth across a specimen, revealing its surface features rather than its internal structure.

A unit even smaller than the micrometer must be used in designating the size of objects too small to be visualized in the light microscope. This unit is given as a *tenth of a nanometer* or an *angstrom* (Å). In Table 3-3 the sizes of *microscopic* and *ultramicroscopic* objects are compared. An ostrich egg and mature human ovum are added to illustrate the magnitude of the size discrepancy.

Structure. If we magnify and examine a typical cell (Fig. 3-3), we see two main parts. The central portion is occupied by a more densely arranged and usually compact structure known as the *nucleus*. The *nuclear membrane* separates the *nucleoplasm* (protoplasm of the nucleus) from a zone of less dense spongy protoplasm surrounding and suspending the nucleus known as the cytoplasm. Generally, the cytoplasm of the cell (1) aids in cell multiplication, (2) assimilates and stores food, (3) eliminates waste products, and (4) secretes enzymes* and other products of physi-

---

*Enzymes enable metabolites in the cell to be transformed at a rate and temperature unattainable in research laboratories.

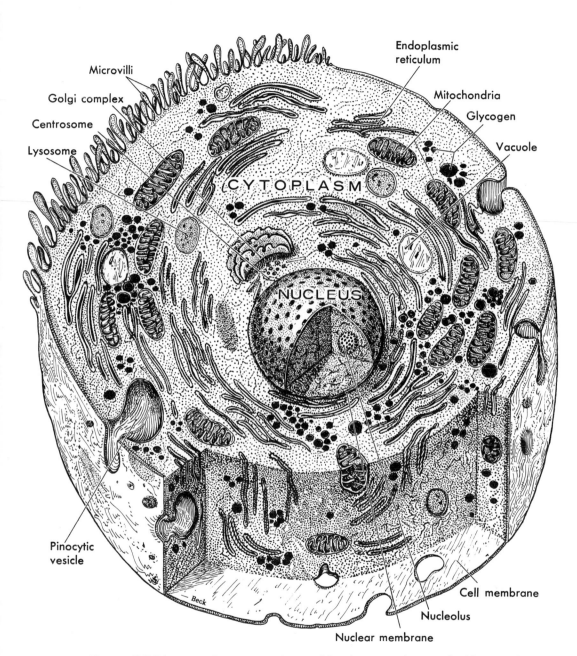

FIG. 3-3. The "cell." Diagram of structure observed in electron micrograph. (See text.) Note pinocytic vesicle (left) representing droplet of watery fluid entering cytoplasm. *Pinocytosis* (cell drinking) indicates uptake of certain substances in solution by the cell. (From Anthony, C. P., and Kolthoff, N. J.: Textbook of anatomy and physiology, ed. 9, St. Louis, 1975, The C. V. Mosby Co.)

ologic significance. The cytoplasm varies in different cells, for enzymes and enzyme systems may be possessed by only one specific kind of cell.

The nucleus (1) rules over the growth and development of the cell, (2) controls the metabolic processes that go on inside the cell, (3) transfers the hereditary characteristics of the cell, and (4) controls reproduction. In short, the nucleus is the center or initiating point for all vital activities of the cell, and a cell without a nucleus is dead. * One particularly important activity is protein synthesis. It is carried on in the cytoplasm but regulated by the nucleus.

The pattern of intracellular organization is not the same in all cells, and many cells do not show all the typical structures to be given.

CYTOPLASM. Within the ground substance of the cytoplasm are (1) numerous small purposeful configurations of living substance called *organelles* and (2) stores of lifeless materials, sometimes transient, referred to as *inclusions.* Organelles include the plasma membrane (plasmalemma), centrosome, centrioles, mitochondria, endoplasmic reticulum, ribosomes, Golgi apparatus, lysosomes, microtubules, and several kinds of filaments and fibrils. Inclusions may be starch granules, glycogen, fat globules, proteins, pigment granules, secretory products, and crystals. Certain organelles and inclusions were well known to cytologists, but with the electron microscope to reveal the complexities of cell architecture others were seen for the first time.

Encasing the cytoplasm is the *plasma membrane* (Fig. 3-4). This is a delicate film limiting the cytoplasm and presiding as a physiologic gatekeeper over the ex-

---

*The human red blood cell, with no nucleus, cannot divide and is limited in its metabolic behavior.

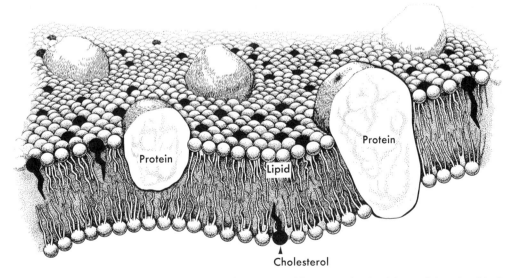

FIG. 3-4. Plasma membrane. Note globular (potato-like) proteins in this model embedded in biomolecular layer of lipids (including cholesterol). Proteins make up membrane's "active sites." (From Smith, A. L., Microbiology and pathology, ed. 11, St. Louis, 1976, The C. V. Mosby Co.)

change of food materials, metabolic waste products, and various other chemicals. Between the cell and its surroundings, numerous biologic events take place on or in the plasma membrane. A three-layered membrane 80 to 100 Å thick, it is formed chemically by a precise association of protein, fat, and carbohydrate molecules. Carbohydrate makes up less than 10%, lipids about 40%, and the balance is protein. About the plasma membrane in certain vegetable cells and adherent to it is the *cell wall*, made of cellulose; this wall is absent or indistinct in animal cells. The many thin, filamentous projections of cytoplasm seen along the free border of certain cells are referred to as *microvilli*.

The *cell center* or *centrosome* is a cytoplasmic region of altered texture usually adjacent to the nucleus. It is occupied by a pair or more of rod-shaped *centrioles*. In electron micrographs, centrioles are seen as hollow cylinders 300 to 500 nm long and 150 nm wide, placed at right angles to each other. Although the function of the centrosome is poorly understood, it is thought to play an important role in cell division.

The powerhouse of the cell is the *mitochondrion* (*pl.*, mitochondria), the site of many important biochemical reactions. A cell may contain a thousand or more. Barely visible through the light microscope as a threadlike form (0.2 to 3 μm), the mitochondrion as visualized through the electron microscope is an intricate oval or elongated structure consisting of a double membrane, the inner layer of which is a system of folds called *cristae* (Fig. 3-5). The mitochondria house the bulk of chemicals (enzymes) that provide energy for cell activities; they are the principal sites at which the energy in food materials is released to the cell. (Biochemically, mitochondria can generate this energy for the cell because they contain cytochrome chains that are linked by the Krebs cycle.) With special technics in isolated mitochondria, the necessary enzymes and coenzymes required for the complex chemical reactions can

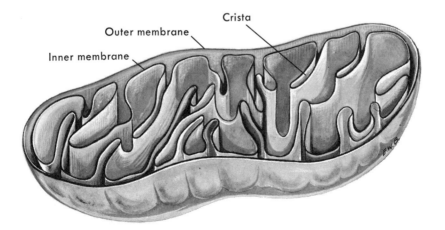

Crista

Outer membrane

Inner membrane

FIG. 3-5. Mitochondrion. Cutaway sketch to show inner and outer membranes and cristae. (From Schottelius, B. A., and Schottelius, D. D.: Textbook of physiology, ed. 17, St. Louis, 1973, The C. V. Mosby Co.)

be located primarily on the membranes. A mitochondrion may contain anywhere from 50 to 50,000 such enzymes, depending on the type of cell and its functional state.

The *endoplasmic reticulum* (the "cytoskeleton"), seen only by electron microscopy, is an elaborate pattern of channels with a secretory function. Its membrane-bound tubules measure 400 to 700 Å in diameter. The endoplasmic reticulum may be *granular (rough surfaced)* or *agranular (smooth surfaced)*. Many tiny, uniform, beadlike granules called *ribosomes*, also found free in the cytoplasmic matrix, are adherent to the outer surface of the membranes of the granular reticulum. Ribosomes are rich in ribonucleic acid (RNA) and represent the places in the cell where proteins are formed. A single cell may contain billions of them. Granular reticulum is highly developed in glandular cells that elaborate protein secretions. If the secretory product is an export from the cell, the ribosomes must be closely associated with the endoplasmic reticulum, for it is through the pathways of the reticulum that the product is transported to the Golgi region for packaging. The secretory product stored as droplets or granules for a while in the cytoplasm is released in time from the cell surface. Agranular endoplasmic reticulum is related to other metabolic functions of cells; in the liver cell it is involved in lipid and cholesterol metabolism, and in the cells of the gonads it is related to the production of steroid hormones.

Placed near one pole of the nucleus, the *Golgi apparatus* is made up of aggregates of flat saccules in parallel fashion with vesicles and vacuoles clustered about. Although it still remains somewhat a mystery, it does play an important role in cell secretion. The polysaccharide fraction of certain complex secretions is known to be formed in the Golgi complex and there to be bound to protein supplied from other organelles of the cell.

Greatly variable in different cells are small dense particles, *lysosomes*, limited by a membrane. Lysosomes are rich in lytic enzymes, including deoxyribonuclease and ribonuclease. They are also known as "suicide bags," for their potent enzymes could digest the cell containing them if they were released from the organelle. A prime function of lysosomes is tied in with the engulfment of foreign material in the body, and they are abundant in phagocytic cells (p. 180). Close by and related to the centrioles are *microtubules*, cytoskeletal elements helping to maintain cell shape. These are straight tubules 200 to 270 Å in diameter with a dense filamentous wall 50 to 70 Å thick. *Filaments* and *fibrils* are commonly seen in the cytoplasmic matrix of many cells. Filaments enable muscle cells to contract, but for many cytoplasmic filaments the biologic significance is undetermined. A newly described cytoplasmic organelle is the *peroxisome*. It is thought to play a role in carbohydrate metabolism. It contains a heavy concentration of peroxidase. Peroxisomes (protoplasmic microbodies) are called the "microkitchens" of the cell.

Cellular inclusions may serve as stores of energy for the cell, for example, deposits of glycogen (the polysaccharide storage form of carbohydrate) and fat (triglycerides of fatty acids). The enzymatic breakdown of the glucose derived from glycogen supplies energy to the cell and short-chain carbon skeletons to be reused in cell protoplasm. *Melanin,* the brown pigment of the skin and hair, is an example of an intracellular pigment. Certain *crystalline* inclusions visible by light and electron

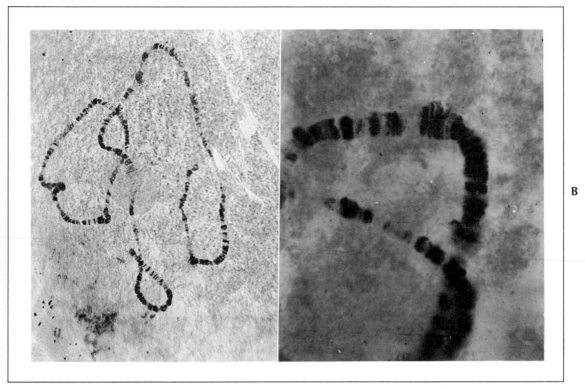

FIG. 3-6. Chromosomes from salivary gland of fly (species unidentified) prepared by "squash" technic. Note structure in photomicrographs **A,** low-power, and **B,** higher magnification. (From Brown, W. V., and Bertke, E. M.: Textbook of cytology, ed. 2, St. Louis, 1974, The C. V. Mosby Co.)

microscopy are assumed to be protein but are poorly understood. Tears, hydrochloric acid, mucus, milk, and digestive enzymes are well-known secretory products of cells. Mucus is one such product that may be readily visible microscopically as a droplet in the cytoplasm of the cell elaborating it.

NUCLEUS. The executive nucleus, usually found near the center of the cell, presides over the functional activities of that cell. Commonly a spherical body, it is made up of a framework of fibrils or threads on which are located the *chromatin granules,* which unite to form the *chromosomes* during cell division. For the time that the cell is not dividing, the chromosomes are in an extended state and cannot be counted. Early in cell division the chromosomes begin to contract and by metaphase (Fig. 3-6) the small rod-shaped bodies appear distinctly. The number and size of the chromosomes depend on the species and are constant for each species.

Structurally, a chromosome consists of coiled threads, each made up of a large number of strands of the giant molecule deoxyribonucleic acid, DNA. A chromosome is the unit of organization for DNA and represents a binding together of DNA, ribonucleic acid, histone (low molecular weight basic protein), and a more complex acidic residual protein.

45

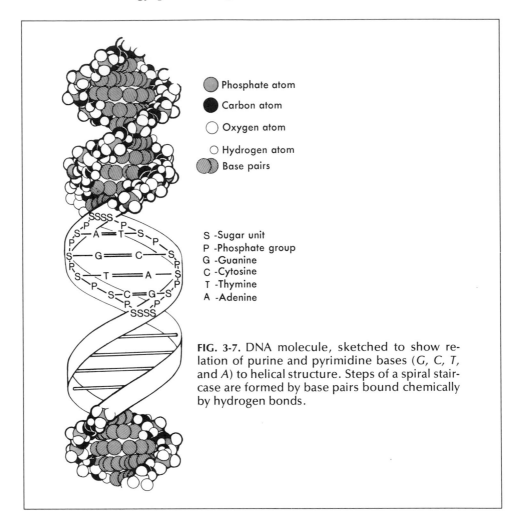

Phosphate atom

Carbon atom

Oxygen atom

Hydrogen atom

Base pairs

S -Sugar unit
P -Phosphate group
G -Guanine
C -Cytosine
T -Thymine
A -Adenine

FIG. 3-7. DNA molecule, sketched to show relation of purine and pyrimidine bases (*G, C, T,* and *A*) to helical structure. Steps of a spiral staircase are formed by base pairs bound chemically by hydrogen bonds.

In DNA three smaller chemical molecules—organic nucleotide bases, a pentose sugar (deoxyribose), and phosphoric acid—are fastened together in a characteristic spiral pattern. Two sugar-phosphate ladders with ever so many paired nucleotide rungs—that is, two identical, unbranched, rigid, intertwining, spiral chains—are thus formed. These are at least 1500 times as long as they are wide and are twisted in opposite directions about a central shaft (Fig. 3-7).

Every living organism must possess somewhere in its makeup a master plan that formulates all aspects of its appearance and behavior. For even simple organisms such a master plan encompasses vast amounts of biologic information that to be stored efficiently must be converted to some sort of code. Because of its vantage point in the nucleus of the cell, deoxyribonucleic acid is just right to do this biochemically. Its building blocks are linked in such a way that the four nucleotide bases (adenine, thymine, cytosine, and guanine) can serve as letters of a four-character code alphabet.

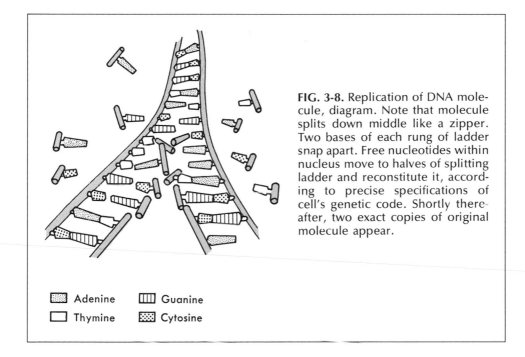

FIG. 3-8. Replication of DNA molecule, diagram. Note that molecule splits down middle like a zipper. Two bases of each rung of ladder snap apart. Free nucleotides within nucleus move to halves of splitting ladder and reconstitute it, according to precise specifications of cell's genetic code. Shortly thereafter, two exact copies of original molecule appear.

▦ Adenine	▥ Guanine	
☐ Thymine	▦ Cytosine	

Words can be formed in a biochemical and genetic language.* In the language of life the four characters are expressed in linkages of three, called triplet codes. There are 64 such triplet codes with so many possible arrangements that they can describe easily the 4 billion human beings populating the world. For a given organism the specifications for its every structure and life process are contained in specific chemical sequences in its DNA.

The chromosome, then, is a biologic document. From it can be transcribed biologic messages in code, over and over again, even the same message. The chromosomes carry the particular hereditary pattern for the given organism. The hereditary pattern for each of the 100 million million tiny cells of the human body is carried in each cell in its chromosomes.

An unusual property of the DNA molecule is its ability to reproduce itself, that is, to replicate into two exact copies (Fig. 3-8). (Replication occurs before cell division.) Because of this, the hereditary biochemical pattern, or *genetic code*, may be passed from one generation to the next. (Think of the thousands of billions of times the DNA of the fertilized human ovum† replicates itself to form the myriads of cells of the mature human being.)

If the chromosomes of the nucleus are compared to two measuring tapes side by side when the spirals are straightened, each point or locus marks the position of

*The Morse or International telegraphic code, composed of the dot and dash, is a two-character code.
†The DNA of the ova resulting in the entire earth's human population would fit into a ⅛-inch cube. One tenth of a trillionth of an ounce of DNA is present in one fertilized human egg, although the number of DNA molecules it contains is astronomically large and virtually incalculable.

a *gene*, which is paired with a matching gene on the corresponding tape. The majority of the life processes within the cell are carried on by enzymes, which are mostly protein. Since a gene physiologically is responsible for the development of a single protein,* collectively the influence of genes on the sum total of enzymatic action makes unicellular and multicellular organisms what they are biologically.

The other nucleic acid of physiologic significance, *ribonucleic acid*, or *RNA*, is similar to DNA chemically except that its pentose sugar is ribose, and it is laid out as a *single* coil, not a double one. About 10% of the cell's quota of RNA is found in the nucleolus, the rest in the cytoplasm. Between cell divisions the nucleus transfers information coded in the chromosomes into specific sequences of amino acids in *messenger* RNA, which then passes out of the nucleus to transmit instructions to the

---

*A given protein (such as an enzyme) is made up of hundreds or thousands of amino acid building blocks. The biologic property of the protein molecule is completely dependent on genes acting to align the amino acids precisely in the protein molecule.

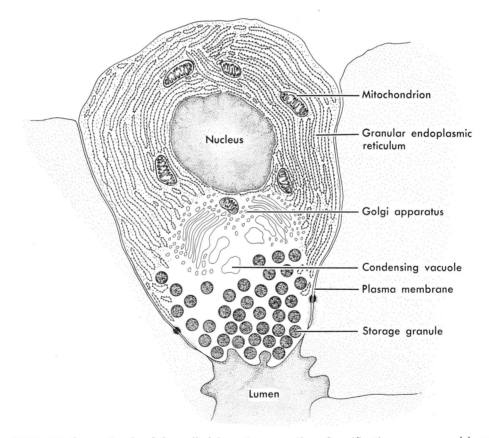

FIG. 3-9. Pathways in glandular cell elaborating secretion. Specifications are passed from nucleus to granular endoplasmic reticulum where product is synthesized. From there it moves to Golgi apparatus for packaging into condensing vacuoles. These ripen to storage granules that at the appropriate time empty contents into lumen. (From Smith, A. L.: Microbiology and pathology, ed. 11, St. Louis, 1976, The C. V. Mosby Co.)

ribosomes. These structures can decipher a sequence of several thousand words (in biochemical language) to map out complex protein patterns. (Cells of bacteria, plants, and animals contain both nucleic acids; viruses contain either, but only one of the two.)

Within the nucleus is usually found one spherical *nucleolus* (little nucleus), sometimes more, which is incorporated with the chromosomes into the background

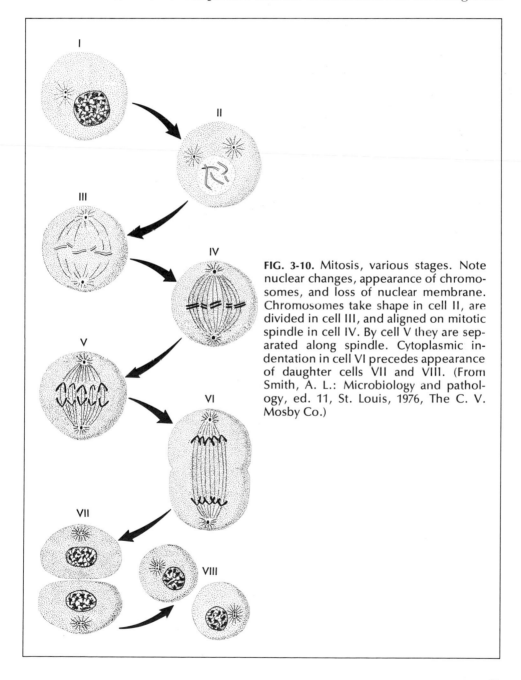

FIG. 3-10. Mitosis, various stages. Note nuclear changes, appearance of chromosomes, and loss of nuclear membrane. Chromosomes take shape in cell II, are divided in cell III, and aligned on mitotic spindle in cell IV. By cell V they are separated along spindle. Cytoplasmic indentation in cell VI precedes appearance of daughter cells VII and VIII. (From Smith, A. L.: Microbiology and pathology, ed. 11, St. Louis, 1976, The C. V. Mosby Co.)

material of the nucleus, the *nuclear sap*. There is no membrane about it. The nucleolus disappears during cell division but reforms in the daughter cells. It plays a key role in nucleic acid metabolism and protein synthesis and is itself composed largely of RNA. It is prominent in rapidly growing embryonic cells.

The outer boundary of the nucleus, the *nuclear membrane,* is seen in the electron micrograph to be made up of two membranes, each 75 Å thick, separated by a space 400 to 700 Å wide. The nuclear membrane is similar in its makeup to the endoplasmic reticulum and thought to be derived from it. The canals of the endoplasmic reticulum appear to open into small pores along the course of the nuclear envelope, and there may be numerous ribosomes on its cytoplasmic surface.

Interesting lines of communication are set up between the inner part of the nucleus and the outer portions of the cytoplasm, extending through to the outside of the cell. These pathways go from the pores of the nuclear membrane along the channels of the endoplasmic reticulum, incorporate the membranes of the Golgi apparatus, and end with the plasma membrane (Fig. 3-9).

Division of the cell

Every cell owes its existence to division of a preexisting one. Cells divide by the process of *mitosis* (Fig. 3-10), or indirect cell division, in which cleavage of the cell is preceded by a series of complicated nuclear and cytoplasmic changes. Mitotic division is characteristic of the higher animals and plants and may be seen in the lower forms as well.

QUESTIONS FOR REVIEW

1. Define protoplasm, cytology, mitosis, nanometer, macromolecule.
2. Outline the parts of a typical cell and give the function of each.
3. In your own words define the cell.
4. What structure controls and regulates vital activities in the living cell?
5. What instruments are used to study cells and unicellular organisms?
6. Give the unit of measurement for cells and unicellular organisms. State why it is used.
7. How are the hereditary characters of a cell transmitted?
8. What cells may be seen with the unaided eye?
9. What is DNA? Why is it important? Where does it occur?
10. State the function of a gene.
11. What is RNA? Where is it found in the cell? What is its physiologic significance?
12. Briefly characterize the important biologic molecules in protoplasm.

REFERENCES. See at end of Chapter 9.

4 The bacterial cell

Definition

Bacteria (*sing.*, bacterium) are minute unicellular microorganisms that ordinarily do not contain chlorophyll and may be able to move independently in their environment.

Classification (p. 4). In the early days of microbiology it was thought that bacteria belonged to the animal kingdom, but then for a period of years, it seemed more suitable to classify them with plants. Today we classify them as procaryotes. They are the smallest microorganisms having all the necessary protoplasmic equipment for growth and self-multiplication at the expense of available foodstuffs. They can start with rather simple substances and synthesize them into complicated organic moieties. They use food materials only in solution and excrete waste products in fluids that must diffuse outward. There is no special structure for intake of solids for digestion or for release of solid particles to the environment. Bacteria are morphologically simpler than the cells of the higher organisms, and as is true for procaryotes, they lack an organized nucleus. In spite of their relative simplicity, they have an elaborate and complicated life history.

Distribution. Bacteria are widely distributed in nature. They have adapted to every conceivable habitat. They are found within and on our bodies, in the food we eat, the water we drink, and the air we breathe. They are plentiful in the upper layers of the soil,* and no place on earth, except possibly the peaks of snowcapped mountains, is free of them. They are found in frozen Antarctica and in the hot water of the geysers in Yellowstone. They can be found in the stratosphere at a height of 20 kilometers (km) and in ocean sediments at a depth of 11,000 meters (m); they are able to grow from 5° to −2° C, the temperature of 90% of the world's oceans.

---

*Take a pinch of dirt. Hold it between your thumb and forefinger. You may be holding as many as 200 million bacteria.

Our skin has a large bacterial population, and bacteria make up a generous portion of the contents of the alimentary tract. There are several thousand species of bacteria; of this number, about 100 produce disease in man. The ratio is given as 30,000 nondisease-producing bacteria to one disease producer. Some of the bacteria that produce disease in man also produce disease in the lower animals. Others produce disease in lower animals only, and still others attack only plants. The majority, however, do not attack man, lower animals, or living plants, and either do not affect animals and plants at all or are actually helpful to them. In fact, if the activities of bacteria were to cease, all plant and animal life would soon become extinct.

Bacteria that cause disease are spoken of as *pathogenic;* those that do not cause disease are *nonpathogenic.*

Morphology

Shape. Bacteria assume three well-known shapes (Figs. 4-1 and 4-2)*:
1. Spherical—coccus
2. Rod-shaped—bacterium or bacillus
3. Spiral-shaped—vibrio, spirillum, spirochete

Cocci are not necessarily round but may be elongated, oval, or flattened on one side. Some bacilli are long and slender, whereas others are so short and plump as to be mistaken for cocci. These short, thick, oval bacilli are known as *coccobacilli.* The ends of bacilli are usually round but may be square or concave. A vibrio is a curved microbe shaped like a comma. A spirillum is a spiral organism whose long axis remains rigid when it is in motion. A spirochete is a spiral organism whose long axis bends when it is in motion.

When bacteria, especially cocci, divide, the manner in which they do so and their tendency to cling together often give them a distinct arrangement. Cocci that divide so as to form pairs are known as *diplococci.* The opposing sides of diplococci may be flattened (examples, gonococci and meningococci). Cocci that divide and cling end to end to form chains are known as *streptococci.* Those that divide in an irregular manner to form grapelike clusters or broad sheets are known as *staphylococci.* Other patterns for cocci are in groups of four *(tetrads)* and cubical packets of eight *(sarcinae).* No pathogenic cocci are found in the latter group. Bacilli that occur in pairs are known as *diplobacilli* and those in chains as *streptobacilli.* The diplobacillus and streptobacillus formations are unusual. When some bacilli divide, they bend at the point of division to give two organisms arranged in the form of a V. This is known as *snapping.* In other cases they tend to place themselves side by side. This is known as *slipping.*

---

*The following are the singular and plural forms:

Singular	*Plural*
Coccus	Cocci
Bacillus	Bacilli
Spirillum	Spirilla
Bacterium	Bacteria
Medium	Media

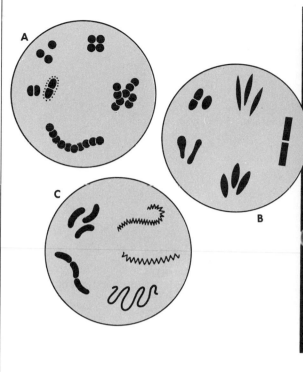

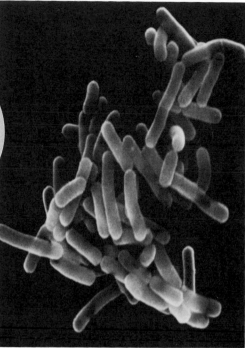

FIG. 4-1. Shapes of bacteria. **A,** Spherical—coccus (pair, diplococci; four, tetrad; chain, streptococci; cluster, staphylococci). **B,** Rod shaped—bacillus (including coccobacillus). **C,** Spiral shaped—vibrio, spirillum, and spirochete.

FIG. 4-2. Bacilli (*Lactobacillus* species), scanning electron micrograph. Note three-dimensional effect with this instrument. (Courtesy Dr. R. C. Reynolds, University of Texas Health Science Center at Dallas.)

Size. Bacteria are so small (no larger than 1/50,000 of an inch*) that the highest magnification of the ordinary compound microscope must be used to study them. Cocci range from 0.4 to 2 μm (μ) in diameter. The smallest bacillus is about 0.5 μm in length and 0.2 μm in diameter. The largest pathogenic bacilli are seldom greater than 1 μm in diameter and 3 μm in length; the average diameter and length of pathogenic bacilli are 0.5 and 2 μm, respectively. Nonpathogenic bacilli may be larger, reaching a diameter of 4 μm and a length of 20 μm. The spirilla are usually narrow organisms from 1 to 14 μm in length. Different species of bacteria show marked variation in size, and there is some variation within a species, but as a rule, the size of each species is fairly constant.

Structure. Bacteria, always unicellular, are so tiny and transparent (about the density of water) and so slightly refractile that, unless stained with dyes, they are dif-

---

*A cubic inch would hold 10 trillion medium-sized bacteria, or as many as there are stars in 100,000 galaxies.

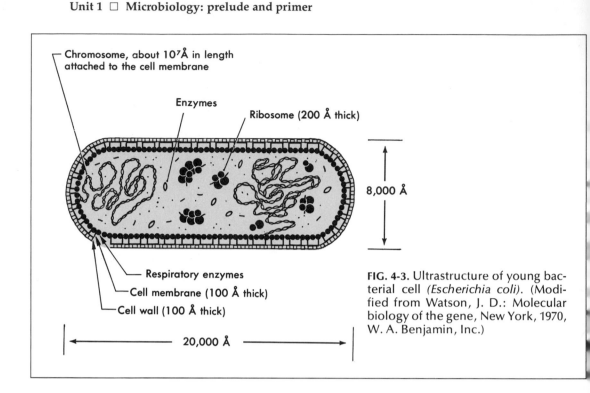

Chromosome, about 10^7Å in length attached to the cell membrane

Enzymes

Ribosome (200 Å thick)

8,000 Å

Respiratory enzymes

Cell membrane (100 Å thick)

Cell wall (100 Å thick)

20,000 Å

FIG. 4-3. Ultrastructure of young bacterial cell *(Escherichia coli)*. (Modified from Watson, J. D.: Molecular biology of the gene, New York, 1970, W. A. Benjamin, Inc.)

ficult to see even with the compound light microscope. When stained, they appear homogeneous or slightly granular. With the electron microscope, however, microbiologists can visualize minute details of bacterial structure (Fig. 4-3).

The shape of the bacterial cell is maintained by a rigid *cell wall*. The protoplasmic substance of bacteria exerts such a high osmotic pressure, equivalent to that of a 10% to 20% solution of sucrose, that in ordinary environments the cell wall is necessary to prevent the cell from bursting. If the bacterial cell is placed in a suitable hypertonic medium and the cell wall dissolved, the remainder of the bacterium is converted into a spherical *protoplast*. In an isotonic environment protoplasts remain viable and grow. The stability of the cell wall is derived from its chemical makeup; this varies in the two major groups of bacteria (p. 73). In gram-positive bacteria the chief component is mucopeptide—a polymer of the amino sugars N-acetylglucosoamine and N-acetylmuramic acid and short peptide linkages of amino acids. Sometimes teichoic acids or a mucopolysaccharide complex of amino sugars and simple monosaccharides may be present. In gram-negative bacteria the mucopeptide inner layer is chemically bound to two outer layers of lipopolysaccharide and lipoprotein. There are no teichoic acids.

The cell wall is so narrow that it cannot be seen with the ordinary compound light microscope. In ultrathin sections it is revealed by electron microscopy as a well-defined structure surrounding a distinct layer, the cell membrane or *plasma membrane* (Fig. 4-3), which separates it from the cytoplasm of the bacterial cell. The plasma membrane is the site of important enzyme systems, including the respiratory enzyme system (cytochrome enzymes). In fact, in bacteria it corresponds to the mito-

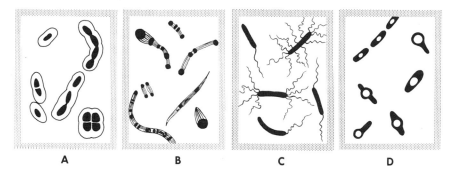

FIG. 4-4. Special features of bacteria. **A,** Capsules. **B,** Metachromatic granules. (Note pleomorphism in **B.**) **C,** Flagella. **D,** Spores.

chondria of higher organisms.* In regulating the passage of food materials and metabolic by-products between the interior of the cell (where metabolic activities are carried on) and the surroundings, it functions osmotically both as a barrier and as a link. It blocks the entry of certain substances and catalyzes the transport of others into the cell.

Surrounding many bacteria is a mucilaginous envelope or *capsule* (Fig. 4-4, *A*). Indistinct in most bacteria, it is well developed in a few (examples, *Streptococcus pneumoniae*, *Clostridium perfringens*, and *Klebsiella pneumoniae*). The capsule is formed by an accumulation of slime excreted by the bacterium. This material is usually a complex polysaccharide. If it is present about the cell in only small amounts, a distinct capsule does not appear.

Capsule formation is most prominent in organisms taken directly from the animal body, for when grown on artificial media, the same organisms often lose their ability to form capsules. A capsule does not stain with the ordinary bacteriologic dyes but may appear as a clear halo around the bacterium even two or three times as broad as the bacterium. It is stained by special methods. The presence of a capsule appears to enhance the virulence of an organism by protecting it against phagocytosis, and in some cases the capsule gives the organism its specific immunologic nature. For instance, relative to the nature of their capsules, pneumococci are divided into at least 82 types. The specific antigenic nature of a capsule depends on its carbohydrate content.

Within some bacteria (example, *Corynebacterium diphtheriae*) at the ordinary magnification of the light microscope are seen granules that stain more deeply than the remainder of the cell. Known as *metachromatic granules* (Fig. 4-4, *B*), they are enzymatically active and thought to be reserves of inorganic phosphate stored as polymerized metaphosphate (volutin). Sulfur-oxidizing bacteria convert excess hydrogen sulfide from the environment into intracellular granules of elemental sulfur. In some species granules are arranged irregularly within the cell, whereas in others they are located in one or both ends of the cell, where they are known as *polar bodies*.

---

*Bacteria do not contain mitochondria.

Electron microscopy reveals a dense packing of ribosomes in bacterial cytoplasm and the presence of spherules and various submicroscopic granules. The ribosomes, as would be expected, play an important role in protein synthesis. The submicroscopic granules are known to be biochemically complex and active; they may represent stores of food materials, even elimination products. There seems to be no relation of these granules to the ability of the bacteria containing them to produce disease.

As special staining methods designed to visualize the chemical substances making up the nucleus in the higher forms of life were applied to the study of bacteria, it became evident that bacteria contained nuclear material. This is seen in electron micrographs as a distinct and relatively transparent structure of rounded proportions with no detectable (nuclear) membrane. The nuclear material in bacteria consists essentially of deoxyribonucleic acid (DNA) present as a single chromosome, which, if unfolded, would stretch approximately 1 mm. It is hooked at one point to an infolding of the plasma membrane known as a *mesosome*. Many bacteria possess one chromosome, but in young, actively dividing cells, two or more may be seen. As it is for all cells, the nuclear material is precisely the governing force for the bacterial cell, and vital activities cannot be carried on in a bacterial cell without it.

TABLE 4-1. CHEMICAL MAKEUP OF A SINGLE, YOUNG, ACTIVELY DIVIDING COLON BACILLUS (ESCHERICHIA COLI)*

Component	Average molecular weight	Estimated number of molecules	Number of different kinds of molecules	% Total cell weight
Proteins	40,000	1,000,000	2000 to 3000	15
Carbohydrates and precursors	150	200,000,000	200	3
Lipids and precursors	750	25,000,000	50	2
Nucleic acids				
DNA	2,500,000,000	4	1	1
RNA	25,000 to 1,000,000	461,000	More than 1000	6
Amino acids and precursors	120	30,000,000	100	0.4
Nucleotides and precursors	300	12,000,000	200	0.4
Other small molecules (breakdown products of food molecules)	150	15,000,000	200	0.2
Inorganic ions (Na$^+$, K$^+$, Mg^{++}, Ca^{++}, Fe^{++}, Cl$^-$, PO$_4^-$, SO$_4^{--}$)	40	250,000,000	20	1
Water	18	40,000,000,000	1	70 to 75

*From Watson, J. D.: Molecular biology of the gene, New York, 1970, W. A. Benjamin, Inc.

Chemical composition. Even in a unit such as the bacterial cell, 500 times smaller than the average plant or animal cell, the chemical composition is exceedingly complex. To gain an idea as to the chemical makeup of any bacterial cell, let us look at one that, since it is easily grown and manipulated in the laboratory, is undergoing almost as intensive study today as man. This is the colon bacillus, *Escherichia coli* (Figs. 4-3 and 4-8).

Biochemically this microbe contains perhaps 3000 to 6000 different types of molecules (Table 4-1). Of these, specific proteins account for 2000 to 3000 kinds. The amount of DNA present is postulated to be that required to code for the necessary amino acid sequences in these proteins. About the DNA are 20,000 to 30,000 spherical ribosomes composed of protein (40%) and RNA (60%). Water, water-soluble enzymes, and a large number of various small and less complex molecules are associated with these nucleic acids.

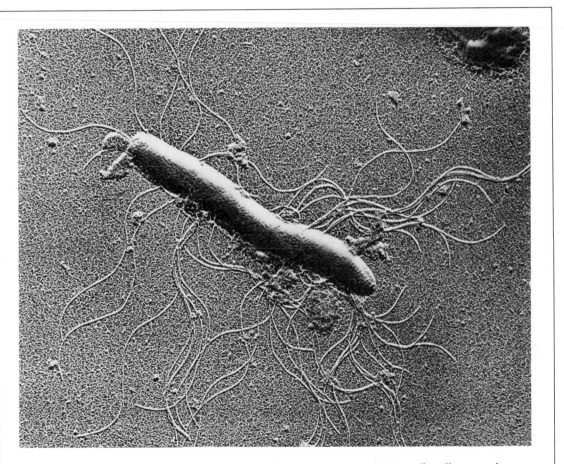

FIG. 4-5. Peritrichous bacillus from mouth, electron micrograph. Note flagella—number and arrangement. In preparation specimen shadowed with gold. (×28,000.) (Courtesy Dr. J. Swafford, Arizona State University, Tempe, Arizona; from Brown, W. V., and Bertke, E. M.: Textbook of cytology, ed. 2, St. Louis, 1974, The C. V. Mosby Co.)

Motility. Many bacilli and all spirilla are motile when suspended in a suitable liquid at the proper temperature.* True motility is seldom observed in cocci. The organs of bacterial locomotion are fine hairlike appendages known as *flagella* (little

*True motility, in which the organism changes its position in relation to its neighbors, should not be confused with *brownian motion*, a peculiar dancing motion possessed by all finely divided particles suspended in a liquid.

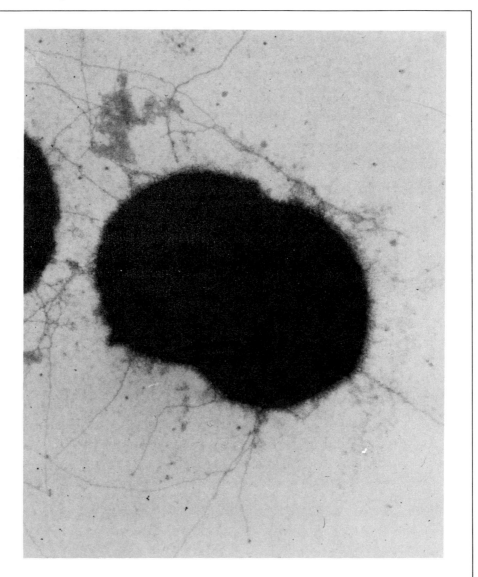

FIG. 4-6. Pili in electron micrograph of gram-negative gonococcus *(Neisseria gonorrhoeae)*. Negative staining with 2% uranyl acetate (p. 75). (× 88,020.) (Courtesy J. C. Bullard and Dr. S. J. Kraus of the Venereal Disease Research Unit, Center for Disease Control, Health Services and Mental Health Administration, Department of Health, Education and Welfare, Atlanta, Ga.)

whips) that spring from the bacterial cell and cause it to move along by their wavelike, rhythmic contractions (Figs. 4-4, *C*, and 4-5). Some spirochetes aid the action of their flagella with a sinuous motion of the cell body. A bacterium may have one flagellum (monotrichous), a few, or many flagella in a tuft (lophotrichous), and the flagella may be attached to one end, both ends, or all around the organism (peritrichous) (Fig. 4-5). Flagella, which chemically are elastic proteins, do not take the ordinary bacteriologic dyes but have to be stained by special methods.

Bacteria may be motile when grown on one medium and nonmotile when grown on another. Also they may be motile at one temperature and nonmotile at another. Different organisms travel at different rates of speed. The rod-shaped organism that causes typhoid fever *(Salmonella typhi)* is able to progress at a rate of 2000 times its own length per hour.

Pili (Latin, hairs) are surface projections like flagella found in gram-negative bacteria (Fig. 4-6). However, they are shorter and finer and do not propel the cell. Also called *fimbriae*, they may be part of the attachment of cells in conjugation; but otherwise their purpose is unknown.

Endospores. Under certain poorly understood conditions, some species of bacteria (examples, species in genera *Bacillus* and *Clostridium*) form within their cytoplasm bodies that are resistant to influences adverse to bacterial growth known as *spores (endospores)* (Fig. 4-4, *D*). Spore formation seems to be a trait of bacilli, being exceedingly rare in cocci and spirilla. Among bacilli about 150 species, most of which are nonpathogenic, form spores. The important pathogenic, sporeforming bacteria are those that cause tetanus, gas gangrene, botulism, and anthrax.

Since spores seem to form when conditions for bacterial growth are unfavorable, spore formation may be a protective mechanism, but in the life of certain species of bacteria it seems to be a normal phase. Bacteria that do not form spores and sporebearing bacteria in which spores are not forming are known as *vegetative* bacteria. Those in which spores are forming are *sporulating* bacteria. When conditions suitable for bacterial growth are established, the spore converts itself back to the actively multiplying form (germinates). The germinating spore becomes the vegetative form of the bacterium.

There is a temperature for each species at which spore formation is most active, and spore formation is preceded by a period of active vegetative reproduction. Some species form spores only in the presence of oxygen; others form spores only in its absence. Spore formation is not a reproductive phenomenon because a spore forms only one bacterium and a bacterium forms only one spore.

Sporulation means that the bacterial cell forms within its substance a round or oval, highly refractile body surrounded by a capsule. This body increases in size until it is broad or broader than the cell. The portion of the cell that remains gradually disintegrates, leaving only the spore. Spores from which the remainder of the cell has disappeared are *free* spores. In the electron microscope a spore has a structural pattern: a coat, a cortex, and a nuclear core of chromatinic material. Spores do not take the ordinary bacteriologic dyes but may be stained by special methods. Ordinary stains of sporulating organisms may show the spore as a clear, unstained area situated in the end of the bacterium (terminal), near its center (central), or in an intermediate position (subterminal). In shape, spores may be spherical, ellipsoidal, or cylindric. In anaerobic bacteria the spore is broader than the remainder of the bacterial cell, causing the bacterium to assume a spindle shape if the spore is central and a club shape if it is terminal or subterminal. The shape and position of the spores help to identify certain bacteria.

Although spores are especially resistant to heat, chemicals, and drying, the vegetative forms are no more resistant to these or other adverse influences than non-sporebearing bacteria. An idea of the protective value of spores can be drawn from the observation that some spores withstand boiling for hours, whereas the temperature of boiling water kills vegetative bacteria within 5 minutes. On the one hand, spores resist tremendous pressures, but on the other hand, they can persist in a vacuum approaching the emptiness of space. Spores have also been known to resist the temperature of liquid air ($-190°$ C) for 6 months. Their resistance is probably the result of the membrane about them and of the concentrated, water-free nature of their substance. Although only a few species of bacteria produce spores, their spores are everywhere. This means that *spore-killing* methods in bacteriologic and surgical sterilization and in the canning industry are absolutely necessary.

Reproduction

Cell division. The typical mode of bacterial reproduction, an asexual process, is by simple transverse division (binary fission) (Figs. 4-7 and 4-8). Bacteria do not divide by mitosis—there is no mitotic spindle. In preparation for division the nuclear

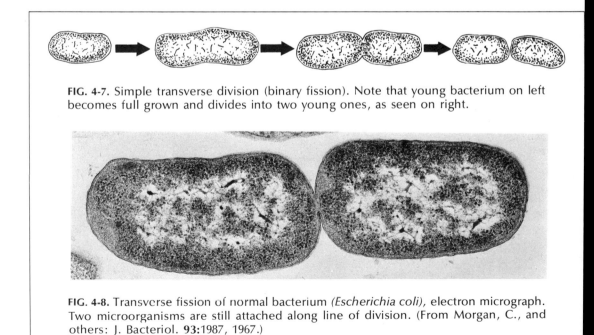

FIG. 4-7. Simple transverse division (binary fission). Note that young bacterium on left becomes full grown and divides into two young ones, as seen on right.

FIG. 4-8. Transverse fission of normal bacterium *(Escherichia coli)*, electron micrograph. Two microorganisms are still attached along line of division. (From Morgan, C., and others: J. Bacteriol. **93**:1987, 1967.)

chromosome of the dividing cell replicates to produce equal division of nuclear material into two sister chromosomes. Replication initiates active membrane synthesis at the periphery, and a transverse membrane is formed that moves into the bacterium. The membrane, along with a new cell wall, constricts the bacterium along its short axis and, partly because of the presence of the mesosome, pushes the sister chromosomes apart and into each of two daughter cells formed by the deepening constriction. Each newly separated cell soon elongates to full size and in turn divides. A newborn bacterium requires from 15 to 30 minutes to reach adult size and divide. Reproduction is always specific; for example, staphylococci always reproduce staphylococci.

Production of L-forms. In certain species of bacteria (examples, *Proteus, Bacteroides, Pseudomonas,* and some of the coliforms) a normal cell may swell to a large entity only to disintegrate into numerous particles, some 0.2 μm in diameter, known as *L-forms.* This kind of change may occur without known stimulus or with a well-defined one such as a drug. Although these forms possess distinguishing features, they are not only akin to the parent cell but can revert to it.

Variation

Living organisms, or the aggregate of living organisms, are seldom if ever exactly the same. Microbes are no exception. The deviation from the parent form in bacteria of the same species growing under different or identical conditions is known as *variation.*

Variation may be caused by external or internal influences. Environmental factors include the kind of culture medium, the temperature of growth, the length of time grown artificially, and the event of exposure to chemicals or to radiant energy (x rays, for example). Variation may also result from factors inherent in the bacteria themselves.

Observed variations. Variation may affect all the biologic properties—size, shape, biochemical nature, colonial characteristics, and physiologic activities—of bacteria and may be temporary or permanent.

DISSOCIATION. A change in the kind of colony formed on a semisolid culture medium as an example of bacterial variation is termed bacterial (or microbial) *dissociation.* Although a pure culture of an organism is used to streak the surface of a culture medium, the resultant colonies may present contrasting appearances. There may be *smooth* or *mucoid* colonies, regular in outline, round, moist, and glistening—the *S-type* colonies. At the other extreme are *rough* colonies, larger, irregular in outline, indented, and wrinkled—the *R-type* colonies. Between the two are the intermediate forms. By proper laboratory manipulation organisms forming S-type colonies can be made to form R-type colonies and vice versa. Usually organisms forming S-type colonies are more vigorous pathogens than those forming R-type colonies.

MORPHOLOGIC VARIANTS. Bacteria of the same species, growing under ideal conditions, may vary in size, shape, and appearance. This is *pleomorphism.* Growing under unfavorable conditions (for example, in old cultures), the members of some species assume irregular, bizarre shapes, and stain irregularly. These swollen, shrunken, or granular, aberrant (or abnormal) variants are known as *involution* or *degenerative forms.* Here variation in morphology probably reflects the presence of many dying as well as dead cells in the culture medium and the injurious effect of the accumulation therein of metabolic by-products.

Organisms that are surrounded by a capsule when grown in the animal body often lose that capsule when grown artificially. In some cases capsule formation may be restored by their return to a susceptible animal. Capsule formation is pronounced in anthrax bacilli infecting a susceptible animal, but from the artificial media of the laboratory there is hardly a sign of a capsule for the very same organism. Bacteria with capsules tend to form smooth colonies and those with none tend to produce rough colonies.

Adaptation. One of the attributes of living cells is their power to adapt themselves to their surroundings. This is probably truer of bacterial cells than of many other types of cells. For instance, certain bacteria that require specially prepared media to sustain their continued growth, when first isolated from the animal body, gradually acquire the ability to grow on media devoid of the growth-promoting and enriching materials necessary for their early growth. They may also be grown artificially in a gradually changing environment until at last they are able to grow under conditions of food supply, temperature, moisture, and oxygen supply far different from those in which they originally grew best.

Variations (changes in bacterial makeup) that represent physiologic adjustment to the environment are designated by the term *adaptation.*

Plasmids are extrachromosomal genetic elements widely seen among bacteria and not essential to viability. They determine bacterial traits crucial to adaptation.

ATTENUATION. An important form of adaptation is *attenuation* (an important concept in immunology), which indicates a loss in disease-producing ability of a given organism. An organism whose virulence is decreased is attenuated. A highly pathogenic organism may be rendered temporarily or permanently nonpathogenic if repeatedly subcultured on artificial laboratory media. For example, by cultivation of a strain of bovine tubercle bacilli on media containing bile until the strain had lost its ability to cause disease, a suitable preparation (BCG vaccine) was developed for vaccination against tuberculosis. Virulence, although artificially eliminated, may often be restored by animal passage, that is, the serial injection of microorganisms into and their recovery from susceptible animals.

Genetic factors. Certain internal factors operative in bacterial variations pertain to changes in the genetic makeup of the bacteria.

MUTATIONS. Mutations in bacteria are analogous to those in the higher forms of life. Since the specific intracellular enzyme (protein) is regulated biochemically by a specific gene positioned on the chromosome of the cell, the related structure (and function) of the cell must depend on the integrity of that gene. Within the gene the significant component is the sequence of the nucleotide bases, with any change in this pattern projecting an effect on the cell. Such a change constitutes a *mutation* and is inheritable.

Rearrangement of the nucleotide sequence of a gene can result from an error in replication or from breakage in the sugar-phosphate backbone of the DNA molecule.

Mutations occur spontaneously or are induced by certain mutagenic agents that can alter the nucleotide bases in such a way as to promote replication errors. The most effective mutagens are certain alkylating agents and forms of radiant energy.

INTERMICROBIAL TRANSFER. Variations resulting from the passage of nuclear material from one bacterium to another are easily demonstrated. If two bacterial strains, variants in a single species, are incubated together under special circumstances, such a transfer is borne out by the appearance in the culture of new organisms displaying qualities of both original strains. When genetic material (nuclear DNA) is transferred from one cell to another, the receiving cell does not gain the full complement of chromosome but only a portion. Immediately after the event of transfer this portion must be matched to the corresponding segments of the chromosome in the cell and genetic material exchanged and eliminated. From this rearrangement a newly formed chromosome emerges, containing DNA from two bacterial cells. This is the *recombinant chromosome,* shaped by the process of *recombination.*

Conjugation is a process effecting transfer of genetic material in a special way. On occasion a type of sexual reproduction occurs in bacteria wherein hereditary material is passed from one organism to the other on a transient physical contact (p. 294). In fact, in an electron micrograph (Fig. 4-9) two individual bacterial cells are seen to unite by means of a cytoplasmic bridge extending between them. Even male and female mating types have been defined for the closely related groups of

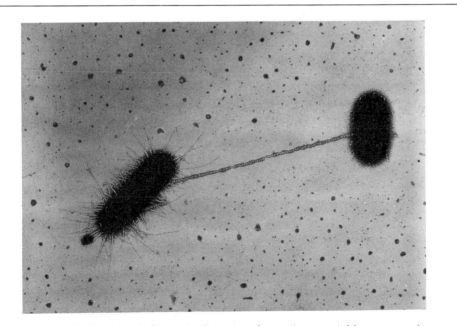

FIG. 4-9. Conjugation in bacteria. Passage of genetic material by means of a long thin filament between bacterium of one species and that of different species (both in family Enterobacteriaceae). A factor inducing resistance to certain antibiotics is known to be so transferred (see also p. 294). Note pili on surface of microbe on left. (Courtesy Dr. C. C. Brinton, Pittsburgh, Pa.)

63

enteric (gram-negative) bacteria in the family Enterobacteriaceae that have been studied. Conjugation can occur between any two members of the family, not just within a given genus. However, there is no new cell as a consequence. The progeny of the cell receiving the nuclear material are produced in the usual way, by binary fission of the parent.

Transformation is a process of direct transfer of nucleic acid. Certain species of bacteria release DNA into the medium in which they are growing, such as certain species of *Neisseria* that release it in their extracellular slime. A small number of bacteria in a given population, referred to as *transformable* bacteria, can pick up this DNA. How they do it is not known. Within the bacterial cell the DNA is integrated with the DNA already present, and what is left over is degraded. The net effect is the replacement of a *short* region of the chromosome with a new portion of genetic material.

Transduction is a special form of indirect transfer of genetic material. A bacterial virus or bacteriophage carries a small fragment of the chromosome from the bacterial cell in which it was produced to the bacterium that it invades. Transduction occurs in coliforms, enteric pathogens, and staphylococci.

These above-mentioned internal factors may pertain to changes in the genetic substance of bacteria. When nuclear material (DNA) is transferred from one bacterium to another (transformation), there is alteration of some property in the recipient.

Drug fastness. A microbial variation with considerable therapeutic importance is one that arises from a mutation or other genetic mechanism giving the organism increased tolerance for an antimicrobial drug. This tolerance for the drug is termed *drug resistance* or *drug fastness*. An organism that becomes resistant to a drug used in the treatment of disease is said to be drug fast. Strains of bacteria are always emerging resistant to one or more antibiotics, a resistance usually acquired when clinical infectious disease has been inadequately treated with the given antibiotic.

Bacteria may become tolerant of more than one antimicrobial drug. This *cross-resistance* is noted especially in the case of closely related antibiotics.

QUESTIONS FOR REVIEW

1. Define bacteria. Indicate their distribution.
2. Discuss the function of the cell wall. Of what is it composed?
3. Name and describe the three forms of bacteria.
4. What is the function of the plasma membrane?
5. Give two characteristics of bacteria at least partly dependent on the capsule.
6. How do bacteria move about in a fluid medium?
7. By what processes do bacteria reproduce?
8. What are spores?
9. Give examples of bacterial variation.
10. Discuss adaptation.
11. State the consequences of drug fastness and cross-resistance.
12. Briefly discuss the bacterial chromosome.
13. List significant contributions of the electron microscope to our knowledge of the bacterial cell.
14. Briefly define L-forms, conjugation, mutation, transformation, transduction, attenuation, recombination, pili.
15. Give implications of recombination. (Consult outside sources.)

REFERENCES. See at end of Chapter 9.

5 Visualization of microbes

PLAN OF ACTION

The microbiologist studies microbes in many ways. To visualize them, he must use an instrument of precision, the microscope. The test specimen from a contaminated site or diseased area that he inspects with the microscope may be unstained, or, as is often the case, it may be stained by one of several methods. He may make a culture from the suspect material. After the microbes have multiplied sufficiently to form visible growth, he may note the physical pattern of this growth, study the microorganisms in stained or unstained microscopic preparations made from the growth, and use some of it to determine their biochemical and biologic properties. He may inject microorganisms recovered from the test specimen into a suitable laboratory animal and observe their effect on the animal. Finally, he may apply well-known immunologic tests.

Note: In summary, the methods used to study microbes include (1) direct (microscopic) examination, (2) culture, (3) biochemical tests, (4) animal inoculation, and (5) immunologic reactions. They will be presented in a sequence of five chapters distributed through three units.*

Microbes must be handled with extreme caution and in accord with well-known principles of conduct in microbiologic laboratories. *Bacteria or other disease-producing agents can be very dangerous, and accidental laboratory infections can be fatal.* In one survey of 1300 infections among laboratory workers, 39 ended fatally. In many cases the infections occurred in research workers and highly trained technologists.

TOOLS FOR STUDY

The microbiologist has many instruments of precision at his command; some are in constant use; others are needed only in special investigations.

---

*Discussions of laboratory methods for identification and diagnosis of microbes are also included in material specifically related to the different microbes throughout the book.

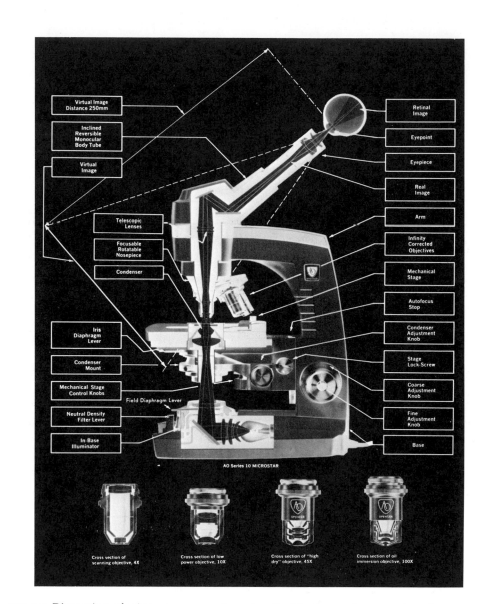

FIG. 5-1. Dissection of microscope—parts and optical features of monocular microscope. Note how light rays are reflected from mirror in built-in base illuminator through the microscope to eye of observer. This modern microscope is fitted with movable stage whereby specimen may be focused to objective, making it unnecessary for observer to change eye level. (Courtesy American Optical Corp., Scientific Instrument Division, Buffalo, N.Y.)

The microscope. The instrument most often used by the microbiologist and one to be handled with utmost care is the compound light microscope (Fig. 5-1). Needless to say, its workmanship should be of the highest quality.

GENERAL DESCRIPTION. Microscopes are of two kinds, simple and compound. A *simple* microscope is little more than a magnifying lens. A *compound* microscope incorporates two or more lens systems so that the magnification of one system is increased by the other. Practically, it consists of two parts, the supporting stand and the optical system. The supporting stand includes (1) a base and pillar, (2) an arm to support the optical system and house the fine adjustment, (3) a platform (stage) on which the object to be examined rests, and (4) a condenser and mirror fitted beneath the stage. The condenser and mirror focus the light from either an external source such as a special kind of microscope lamp or from an illuminating system fitted into the base of the scope. Where present, the built-in base illuminator houses the light source, a collecting lens system, mirror, and a condenser lens (Fig. 5-1). An iris field diaphragm, a variable transformer attached to the system, and properly placed filters permit adjustments of the light.

The optical system consists of a body tube, which supports the *ocular* lenses (eyepiece) at the top end and the *objective* lenses attached to a revolving *nosepiece* at the other end. The optical system is connected to the arm of the supporting stand by an *intermediate slide*, which moves up and down on the arm in response to movement of the *fine adjustment*. The intermediate slide contains the rack and pinion for the *coarse adjustment*, which acts directly on the tube of the optical system. The platform of the microscope is usually equipped with a mechanical device to hold the microslide firmly. The object is mounted on the microslide so that it can be moved from place to place by set screws. The advantages of this device are that the specimen can be examined systematically and, unless moved, the specimen remains in a fixed position.

The magnification of an objective is usually designated by its equivalent focal distance in inches or millimeters. By *equivalent focal distance* is meant the focal distance of a lens having the same magnification as the objective. The higher the number of the objective, the less is its magnification. American microscopes are usually fitted with 16, 4, and 1.8 mm objectives. The last is known as an *immersion objective* because for best results there must be a liquid (oil or water) between the objective and the object being examined. Usually this is immersion oil. The 16 mm objective magnifies 10 times; most 4 mm objectives magnify 43 times, and most 1.8 mm objectives magnify 97 times.

The oculars of a microscope are given 6×, 10×, and similar designations to indicate that they increase the magnification of the objective 6, 10, or more times, respectively. To obtain the magnification of any combination of ocular and objective lenses, multiply the magnification of objective by that of the ocular (Table 5-1). Remember that magnification refers to both the length and the width of an object; that is, a magnification of 100 means that the object is made to appear 100 times as long and 100 times as wide.

Ocular micrometer. Microbes can be measured microscopically by means of a device known as a micrometer. The simplest type is the ocular micrometer, which

TABLE 5-1. MAGNIFICATION WITH LENS COMBINATIONS OF THE COMPOUND MICROSCOPE

Oculars	Objectives		
	16 mm 10×	4 mm 43×	1.8 mm 97×
6×	60×	258×	582×
10×	100×	430×	970×
15×	150×	645×	1455×

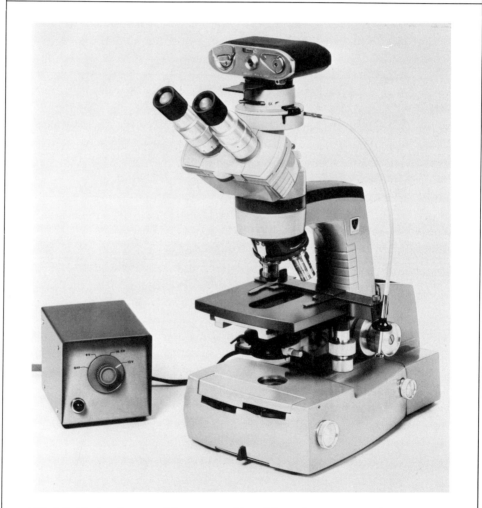

FIG. 5-2. Photomicrographic camera. Note binocular compound microscope fitted with camera for making photomicrographs (photographs of microscopic objects). Camera is fitted onto adaptor. (Courtesy American Optical Corp., Scientific Instrument Division, Buffalo, N.Y.)

consists of a scale on a glass disk that fits (scale side down) between the lenses of the eyepiece. The spaces between the lines on this scale do not represent true measurements. Real values are obtained by calibrating the ocular micrometer against a stage micrometer. This is a glass slide with a true measurement scale on it—the lines on the scale are either exactly 10 or 100 μm (μ) apart. By placing the stage micrometer in the position occupied by a smear of cells and by looking through the ocular of the microscope, one may superimpose the scale of the ocular on that of the stage micrometer. It is then easy to determine the actual unit length that the distance between the lines of the ocular micrometer represents.

Photomicrographic camera. Another instrument of great value to microbiologists is the photomicrographic camera, used to make pictures of objects seen under the microscope (Fig. 5-2). This gives an easily preserved visual record of microscopic findings and renders a wealth of material available for future study.

EXAMINATION OF UNSTAINED BACTERIA

It may be desirable to examine unstained bacteria to determine their biologic grouping, motility, and reaction to chemicals or specific serums. These features may be determined in a *hanging drop* preparation. A few species of bacteria that cannot be stained by the methods to be discussed are often examined by dark-field illumination.

Hanging drop preparations. To examine bacteria microscopically in a hanging drop, one must use (1) a platinum loop for transferring the material to be examined to (2) a cover glass to fit over (3) a hanging drop slide.

The platinum loop is a piece of fine platinum wire about 3 inches in length. One end is fastened in a handle and the other end is fashioned into a loop about 1/16 inch in diameter (Fig. 5-3). Platinum is used for making wire loops because heating this metal in a flame repeatedly to sterilize it does not destroy it; the wire cools quickly after being heated.

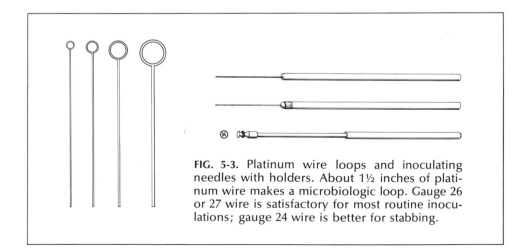

FIG. 5-3. Platinum wire loops and inoculating needles with holders. About 1½ inches of platinum wire makes a microbiologic loop. Gauge 26 or 27 wire is satisfactory for most routine inoculations; gauge 24 wire is better for stabbing.

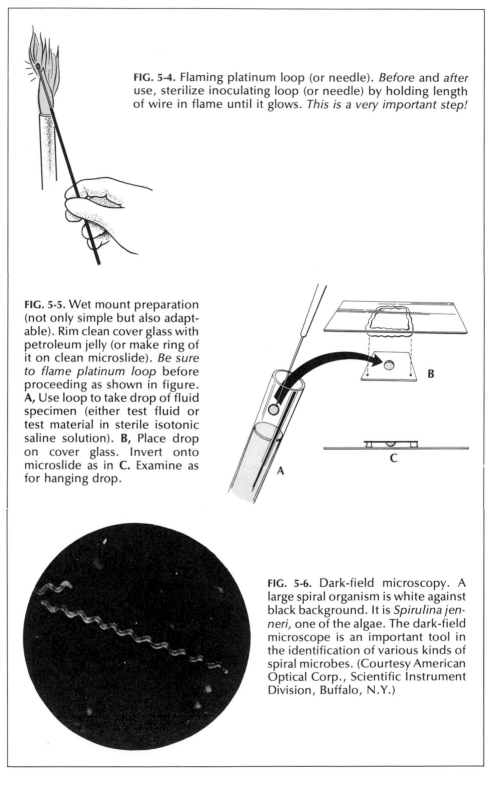

FIG. 5-4. Flaming platinum loop (or needle). *Before* and *after* use, sterilize inoculating loop (or needle) by holding length of wire in flame until it glows. *This is a very important step!*

FIG. 5-5. Wet mount preparation (not only simple but also adaptable). Rim clean cover glass with petroleum jelly (or make ring of it on clean microslide). *Be sure to flame platinum loop* before proceeding as shown in figure. **A,** Use loop to take drop of fluid specimen (either test fluid or test material in sterile isotonic saline solution). **B,** Place drop on cover glass. Invert onto microslide as in **C.** Examine as for hanging drop.

FIG. 5-6. Dark-field microscopy. A large spiral organism is white against black background. It is *Spirulina jenneri,* one of the algae. The dark-field microscope is an important tool in the identification of various kinds of spiral microbes. (Courtesy American Optical Corp., Scientific Instrument Division, Buffalo, N.Y.)

A hanging drop slide is a thick glass slide with a circular concavity or depression at its center. A cover glass is a piece of very thin glass about ⅞ inch square.

Make the preparation as follows: spread a small amount of petroleum jelly around the concavity of the slide. If the specimen to be examined is a culture growing on a solid medium or material such as thick pus, take up a loopful of specimen with the platinum loop and mix thoroughly with a drop of sterile isotonic salt solution placed in the center of the cover glass. If bacteria growing in a liquid medium are to be examined, transfer a drop of the fluid to the cover glass by means of the wire loop. Place the hanging drop slide over the cover glass in such a way that the center of the depression lies over the drop. The petroleum jelly seals the cover glass to the slide, holds it in place, and prevents evaporation. Invert the slide now so that the drop to be examined hangs from the bottom of the cover glass but does not touch the surface of the concavity at any point. The preparation is ready for microscopic examination. Examine with the 4 mm (high dry) lens, and reduce the amount of light passing through it by partly closing the diaphragm of the substage condenser of the microscope. When hanging drop preparations are observed, brownian motion and flowing of organisms in currents should not be mistaken for true motility.

The platinum loop should be sterilized immediately before and after each transfer of material containing bacteria(Fig. 5-4). Since hanging drop preparations contain living bacteria, discard the slide and cover glass into a suitable container of disinfectant after the examination is finished.

The *wet mount* (Fig. 5-5) is similar to the hanging drop preparation except that an ordinary microslide is used instead of the thick hanging drop slide with its central depression. Many of the applications are the same (p. 558).

Dark-field illumination. Dark-field illumination is used to examine certain delicate bacteria that are invisible in the living state in the light microscope, that cannot be stained by standard methods, or that are so distorted by staining as to lose their identifying characteristics. Its greatest usefulness is in the demonstration of *Treponema pallidum* in chancres and other syphilitic lesions, but it is of value in the examination of many other organisms as well (Fig. 5-6).

The material to be examined is placed on an ordinary slide and covered with a cover glass. Sealing the cover glass to the slide with a ring of melted paraffin prevents the cover glass from slipping and accidental infection of the fingers. Dark-field illumination depends on the use of a substage condenser so constructed that the light rays do not pass directly through the object being examined, as is the case with an ordinary condenser, but strike it from the sides at almost a right angle to the objective of the microscope. The microscopic field becomes a dark background against which bacteria or other particles appear as bright silvery objects (Fig. 5-6). A similar effect is seen when a beam of light enters a darkened room and renders visible particles of dust that cannot be seen in a better-lighted room.

EXAMINATION OF STAINED BACTERIA

Staining. The bodies of bacteria are so small that, when examined in hanging drop preparations, little of their finer structure can be made out; to be studied more closely,

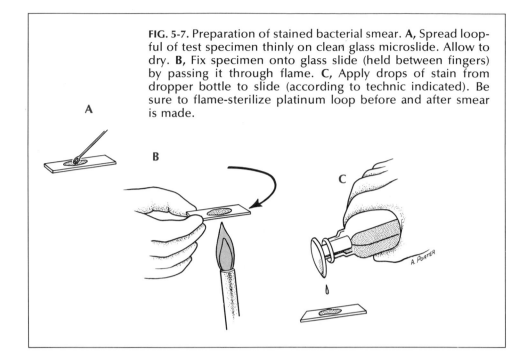

FIG. 5-7. Preparation of stained bacterial smear. **A,** Spread loopful of test specimen thinly on clean glass microslide. Allow to dry. **B,** Fix specimen onto glass slide (held between fingers) by passing it through flame. **C,** Apply drops of stain from dropper bottle to slide (according to technic indicated). Be sure to flame-sterilize platinum loop before and after smear is made.

they must be colored with some dye. This process is called *staining*. The dyes most often used are aniline dyes, derivatives of the coal tar product aniline. (Dye-impregnated paper strips for staining bacteria are commercially available.)

For a stained preparation, place a small amount of the material to be examined on a perfectly clean glass slide and spread out into a thin film by means of a platinum loop or swab (Fig. 5-7, *A*). The film is known as a *smear*. After it is allowed to dry in the air, slowly pass the slide, smear side up, through a flame two or three times (Fig. 5-7, *B*). *Flaming* kills the bacteria in the smear and causes them to stick to the slide. This is known as *fixing*. Other methods of fixing, such as immersion in methyl alcohol or in Zenker's solution, are sometimes used, but heat is the most suitable for routine work. Apply the stain to the fixed smear and wash off with water (Fig. 5-7, *C*); blot the slide dry between sheets of absorbent paper. (Flame sterilization of the platinum loop is sketched in Fig. 5-4.)

Three classes of stains are used in bacteriology: (1) simple stains, (2) differential stains, and (3) special stains. In addition, there is the process of negative staining.

SIMPLE STAINS. A simple stain is usually an aqueous or alcoholic solution of a dye. It is applied to the fixed smear from 1 to 5 minutes and washed off. The stained preparation is then ready for microscopic examination. Widely used simple stains are Löffler's alkaline methylene blue, carbolfuchsin, gentian violet, and safranine. The length of time that the stain remains on the smear depends on the avidity with which it acts. Sometimes added to the solution is a chemical that makes it stain more intensely. Such a chemical is a *mordant*.

Most bacteria stain easily and quickly with simple stains, some do not stain so readily, and a few do not stain at all. Capsules and spores are not stained with these simple stains but may give contrast as clear, *unstained* structures. Flagella cannot be stained in this way and are not seen as contrasting structures.

DIFFERENTIAL STAINS. More complex staining methods divide bacteria into groups, depending on their reaction to the chemicals used for staining. Of these, the *Gram stain* and the *acid-fast stain* are most often used.

The method of staining introduced by Hans Christian Gram* divides bacteria into two great groups: those that are gram positive and those that are gram negative. This method depends on the fact that when bacteria are stained with either crystal violet or gentian violet and the smear is then treated with a weak solution of iodine (mordant), the bodies of some bacteria combine with the dye and iodine to produce a color that cannot be removed by alcohol, acetone, or aniline, whereas the color is readily removed from certain other bacteria by these solvents. Bacteria from which the color cannot be removed are spoken of as being *gram positive*, and those from which it can be removed are spoken of as being *gram negative*. A few bacteria that sometimes keep the stain and that at other times do not are *gram variable*. Certain physiologic differences are generally correlated with Gram staining. Gram-positive bacteria tend to be more resistant to the action of oxidizing agents, alkalis, and proteolytic enzymes than are gram-negative ones. They are more susceptible to acids, detergents, sulfonamides, and antibiotics, such as penicillin, than are gram-negative ones. Many modifications have been devised for the original Gram's method. The method outlined below is often used. The explanations appended will apply to any technic used.

1. Make a thin smear of material for study and fix in a flame.
2. Stain with crystal violet or gentian violet (Gram I†) for 1 minute.
3. Blot thoroughly to take up excess stain.
4. Cover the smear with Gram's iodine solution (Gram II‡), the mordant, for 1 minute.
5. Drain and blot dry. Both gram-positive and gram-negative bacteria are now stained a dark violet or purple color.
6. Drop acetone on the smear (Gram III) until no more color flows off (about 5 to 10 seconds). Blot dry. All gram-negative bacteria are completely decolorized. The gram-positive ones are not affected.
7. Cover the smear with a stain (Gram IV§) that gives a contrast in color (counterstain) for 1 minute.
8. Wash with water, blot dry, and examine.

---

*Hans Christian Gram (1853-1938), a Danish physician working in Berlin, introduced this important method of differential staining in 1884. It remains essentially unaltered in use today.

†Gram I is usually a mixture of a 10% solution of crystal violet in 95% ethyl alcohol (solution A) with a 1% solution of ammonium oxalate in distilled water (solution B) in the ratio of 1 part of solution A to 4 parts of solution B (Hepler's ratio 1:8).

‡Gram II is prepared by dissolving 2 g of potassium iodide and 1 g of iodine crystals in 300 ml of distilled water; it should be stored in a brown glass bottle.

§Gram IV is usually made by mixing 10 ml of a 2.5% alcoholic solution of safranine with 90 ml of distilled water.

Stains used to give contrast in color are known as *counterstains*. The ones most often used in the Gram stain are safranine and dilute carbolfuchsin, both of which give a red color, and Bismarck brown, which, as its name implies, gives a brown color. Gram-negative bacteria are stained with the counterstain. Gram-positive bacteria do not stain with the counterstain because they are completely stained with the stain-iodine–bacterial cell combination that gives them their gram-positive microscopic appearance.

Table 5-2 gives the reaction to the Gram stain of important pathogenic bacteria.

A method of staining known as the acid-fast or Ziehl-Neelsen stain* is discussed next. When most bacteria and related forms are stained with carbolfuchsin, they stain easily, but when the smear is treated with acid, they are completely decolorized. It is difficult to stain certain other microbes with carbolfuchsin, but once stained, they retain the dye even when treated with an acid. Those that retain the stain are spoken of as being *acid fast*. The property of being acid fast probably derives from the presence of complex fatty substances within the bacterial cell.

The technic of acid-fast staining is as follows:†

1. Make a smear on the slide and fix.
2. Flood the slide with carbolfuchsin ‡ (red) and gently steam over a flame (do not boil) for 3 to 5 minutes. Keep the slide covered with dye.

---

*The credit for staining tubercle bacilli belongs to Paul Ehrlich (1854-1915), whose method slightly modified is used today. The modifications were made by Franz Ziehl (1857-1926) and Friederich Neelsen (1854-1894), whose names are attached to the staining method.
†There are several standard modifications of this method.
‡The carbolfuchsin solution for the acid-fast stain is made by combining 1 part of a saturated solution of basic fuchsin in 95% ethyl alcohol with 9 parts of a 5% aqueous solution of phenol.

TABLE 5-2. GRAM-STAIN REACTIONS FOR SOME PATHOGENIC BACTERIA

Gram positive (color: dark violet or purple)	Gram negative (color: that of counterstain – red if safranine)
Cocci	
Staphylococcus aureus	Neisseria gonorrhoeae
Streptococcus pneumoniae	Neisseria meningitidis
Streptococcus pyogenes	
Bacilli	
Bacillus anthracis	Bordetella pertussis
Clostridium species	Brucella species
Corynebacterium diphtheriae	Escherichia coli
Mycobacterium leprae	Francisella tularensis
Mycobacterium tuberculosis	Haemophilus influenzae
	Klebsiella pneumoniae
	Proteus vulgaris
	Pseudomonas aeruginosa
	Salmonella species
	Shigella species
	Yersinia pestis

3. Allow the slide to cool. Wash off the excess stain with water (all bacteria are now red).
4. Dip slide repeatedly in acid-alcohol* until all red color is removed from smear. Wash with water. (At this step the acid has removed the red color from all bacteria that are not acid fast. Acid-fast organisms are unaffected and remain stained bright red.)
5. Apply a counterstain for 1 minute to give contrast. Löffler's alkaline methylene blue † is frequently used. Since acid-fast organisms are completely saturated with the red carbolfuchsin, they do not take any of the counterstain. Nonacid-fast organisms, having had all their stain removed by the acid, stain a deep blue.)

Important examples of acid-fast organisms encountered in medicine are *Mycobacterium tuberculosis*, *Mycobacterium leprae*, the anonymous, or atypical, mycobacteria, and the actinomycete *Nocardia asteroides*.

SPECIAL STAINS. Important special stains are those for capsules, spores, flagella, and metachromatic granules. Stains primarily designed to demonstrate metachromatic granules are especially valuable in identifying *Corynebacterium diphtheriae* and in differentiating it from related organisms. Important stains of this type are Albert's stain, using toluidine blue and malachite green, and Neisser's stain, using methylene blue.

Negative (relief) staining. Microorganisms such as *Treponema pallidum*, not stained by ordinary dyes, may be made visible by the process known as negative or relief staining, in which the background, but *not* the microorganisms, is stained. The

---

*Acid-alcohol for decolorizing is 3% hydrochloric acid (concentrated) in 95% ethyl alcohol.
†Löffler's alkaline methylene blue is made by mixing 30 ml of a saturated solution of the dye methylene blue chloride in 95% alcohol (1.5% filtered) with 100 ml of a weak (0.01%) aqueous solution of potassium hydroxide. Modern samples of the dye have been considerably purified, and the addition of alkali may not be necessary, distilled water alone giving a satisfactory preparation for staining.

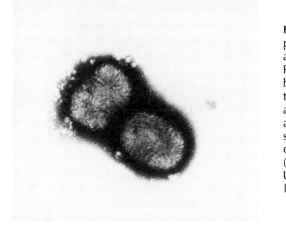

FIG. 5-8. Negative staining applied to study of viruses. Example here is virus of smallpox. Phosphotungstic acid provides background opaque to electrons when specimen is examined in electron microscope and one against which surface structures of viral particle are defined by contrast. (×120,000.) (Courtesy Dr. Derrick Baxby, University of Liverpool, England.)

microorganisms are mixed with India ink or 10% nigrosin (both of which are black); the mixture is spread out into a thin smear and allowed to dry. The microbes appear as colorless objects against a black background. Negative staining may be done with the dye Congo red; this has been used to display spiral organisms and chlamydiae (bedsoniae). Viruses are prepared by negative staining for visualization in the electron microscope (Fig. 5-8).

QUESTIONS FOR REVIEW
1. Name the procedures used in the study of microbes.
2. What are the purposes of a hanging drop preparation? How is one made? What is a wet mount?
3. Briefly define fixing, flaming, mordant, counterstain, gram variable, equivalent focal distance.
4. What are the three classes of stains used in microbiology? Describe each one and give examples.
5. How is a simple stain made? Do many bacteria stain with simple stains?
6. Name two very important differential stains. Give the underlying principles of each.
7. Name (a) two important gram-positive cocci, (b) three important gram-positive bacilli, (c) two important gram-negative cocci, (d) three important gram-negative bacilli.
8. Name two important acid-fast bacilli and one acid-fast actinomycete.
9. What is negative or relief staining?
10. Briefly describe the microscope.
11. Give the steps in the preparation of a bacterial smear.
12. When is dark-field illumination used to study microbes?

REFERENCES. See at end of Chapter 9.

6 Biologic attributes of bacteria

BIOLOGIC NEEDS
Environmental factors

For bacteria to grow and multiply most rapidly, certain requirements must be met: (1) Sufficient food of the proper kind must be present, (2) moisture must be available, (3) the temperature must be most suitable for the species, (4) the proper degree of alkalinity or acidity must be present, (5) the oxygen requirements of the species must be met, (6) light must be partially or completely excluded, and (7) by-products of bacterial growth must not accumulate in great amounts. Significant departure from any of these modifies bacterial growth although bacteria generally possess a greater degree of resistance to unfavorable conditions in their environment than do plants and animals.

Food materials prepared for the growth of bacteria in the laboratory are known as *culture media*. Some bacteria will grow on practically any properly prepared culture medium. Others grow only on specially nutritious ones, and a few will not grow on any artificial medium.

Nutrition. The protoplasm of the bacterial cell is composed of numerous organic compounds, including proteins, fats, and carbohydrates, as well as various inorganic components containing sulfur, phosphorus, calcium, magnesium, potassium, and iron. Proteins comprise about 50% of the dry weight of the cell (each species has a type of protein peculiar to itself), and bacterial nitrogen makes up 10%. In some species carbohydrates are plentiful, and important traits of the species depend on these compounds.

Nutrition is the provision of food materials (that is, chemical substances) to bacteria so that they can grow, maintain their constituents, and multiply. For their nourishment bacteria require sources of carbon and nitrogen, growth factors, certain mineral salts, and sources of energy. With the exception of some saprophytic species, all bacteria derive their carbon and nitrogen from organic matter. A number of minerals are required, the most important of the salts being those of calcium, phosphorus, iron, magnesium, potassium, and sodium. Certain minerals are prerequisite to the activation of enzymes.

Many microorganisms can synthesize all the organic compounds of their complex makeup if supplied with the basic nutrients. Many cannot, however, since they require vitamins and certain organic growth factors (a growth factor is utilized as the intact substance) for their activities. In this respect they resemble rather closely higher forms of life. In fact, such vitamins as nicotinic acid, pantothenic acid, para-aminobenzoic acid, biotin, and folic acid—requirements for animal nutrition—were first studied and identified as substances necessary for the growth of microorganisms.

KINDS OF ORGANISMS. Organisms that obtain their nourishment from nonliving organic material are known as *saprophytes*. Those that depend on living matter for their sustenance are *parasites*. *Facultative saprophytes* usually obtain nourishment from living matter but may obtain it from dead organic matter. *Facultative parasites* usually obtain nourishment from dead organic matter but may obtain it from living matter. Some pathogenic bacteria can exist only on living material (example, the spirochete of syphilis cannot be grown outside a living organism). Most, however, can lead either a parasitic or a saprophytic existence. A few pathogenic bacteria, usually saprophytic, may adapt themselves to a parasitic existence (example, bacteria that cause gas gangrene). The organism on which a parasite lives is known as a *host*.

Organisms that obtain their nutriments by breaking down organic matter into simpler chemical substances are *heterotrophic (organotrophic)*. Those that obtain them by building the organic compounds in protoplasm from the simpler inorganic substances are *autotrophic (lithotrophic)*. All pathogenic bacteria and many nonpathogenic ones are heterotrophic.

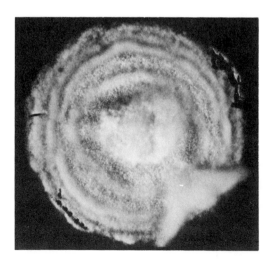

FIG. 6-1. Variation in growth rate from changes in temperature. Note rings of rapid and slow growth in giant colony of fungus *(Histoplasma capsulatum)*. One sector of mutant growth is seen in lower right-hand corner. (Courtesy Dr. R. H. Musgnug, Haddonfield, N.J.)

Moisture. Water is necessary for the growth and multiplication of bacteria. Not only is water a major component of the cytoplasm (on an average, 75% to 80% of the bacterial cell is made up of water), but it also dissolves the food materials in the environment of the bacterial cell so that they can be absorbed.

Drying is highly detrimental to bacterial growth. Delicate bacteria such as the gonococcus resist drying only a few hours, and even highly resistant bacteria such as the tubercle bacillus succumb to drying within a few days. Spores, however, may resist drying for years. As a rule, bacteria with capsules are more resistant in this respect than those with none.

Temperature. For each species of bacteria there is a minimum, optimum, and maximum temperature, meaning, respectively, the lowest temperature at which the species will grow, the temperature at which it grows best, and the highest temperature at which growth is possible (Fig. 6-1). The optimum temperature for a species corresponds to the average temperature of its usual habitat. For instance, bacteria that naturally live in or attack the human body live best at 37° C * (normal body temperature). The lowest temperature at which they will continue to multiply is around 20° C and the highest is from 42° to 45° C. In this temperature range, bacteria are *mesophiles*.

Many bacteria will not grow at a temperature more than a few degrees above or below their optimum. Some pathogens die off rapidly at only 38° C. Most saprophytic bacteria (mesophiles) grow best between 25° and 40° C, but some *thermophiles*, or heat-loving species, grow at a temperature above 45° and as high as 65° C. One species, *Sulfolobus acidocaldarius*, obtains its maximum growth at 85° C. A few *psychrophiles* or *cryophiles*, cold-loving species, grow at temperatures just above the freezing point (20° C or less). They proliferate slowly in the refrigerator. Many psychrophilic microbes have red pigments. Where they grow on the surface of ice and snow, they color it to give red snow.

Cold retards or stops bacterial growth, but when the bacteria are later exposed to a temperature favorable for their growth, multiplication is resumed. Refrigeration (4° to 6° C) is one of the best methods of preserving bacterial cultures, since bacteria are generally resistant to low temperatures and even to freezing. Prolonged freezing, however, destroys them.

High temperatures are much more injurious to bacteria than low ones and are used effectively in practical situations where bacteria and their spores must be destroyed (Chapter 15).

Reaction. For each species of bacteria there is a certain degree of alkalinity or acidity (a certain pH) at which growth is most rapid. The reaction of culture media must be carefully adjusted to the desired hydrogen ion concentration. The best growth of most microorganisms is found in a narrow pH range of not less than 6 nor more than

---

*The temperature scale called centigrade in America has been known in many other countries as Celsius after Anders Celsius (1701-1744), the Swede who originated it in 1742. According to the Eleventh General Conference on Weights and Measures (1960), the term *centigrade* is inexact, and the name Celsius replaces it, with the capital C retained.

8. Most pathogens grow best in a neutral or slightly alkaline medium. Regardless of the influence of the environment, the reaction of the interior of the cell is just at the neutral point (pH 7).

Oxygen. Organisms that grow in the presence of free atmospheric oxygen are known as *aerobes*. If an organism cannot develop at all in the absence of free oxygen, it is an *obligate aerobe*. Those that cannot grow in the presence of free oxygen but must obtain it from oxygen-containing compounds (inorganic sulfates, nitrates, and carbonates or certain organic compounds) are *anaerobes*. *Obligate anaerobes* are vulnerable to free oxygen; their enzyme systems are inactivated by atmospheric oxygen. Few pathogens are anaerobic. Organisms adaptable either to the presence of atmospheric oxygen or to its absence are *facultative*. Those growing best in an amount of oxygen less than that contained in the air are *microaerophiles*. The ones designated *capnophiles* need a 3% to 10% increase in carbon dioxide in the environment to initiate growth.

Light. Violet, ultraviolet,* and blue lights are highly destructive to bacteria, green light is much less so, and red and yellow lights have little bactericidal action. Because of its content of ultraviolet light, direct sunlight kills most bacteria within a few hours. Bright daylight has an effect similar to that of sunlight but one of less potency.

A few species of saprophytic bacteria containing chlorophyll can utilize sunlight to build up the compounds of which they are composed. This bacterial chlorophyll is scattered throughout the cytoplasm, unlike the chlorophyll of plant cells, which is contained in their chloroplasts.

By-products of bacterial growth. If it were not for inhibitory influences, bacteria could completely submerge the world. It is estimated that the progeny from the unrestricted growth of a single bacterium would be 280 trillion at the end of only 24 hours. In cultures, bacterial reproduction is so rapid that bacteria soon exhaust their food supply and release products that inhibit further bacterial growth. Notable are organic acids from carbohydrate metabolism that inhibit growth by changing the reaction of the medium. The practical application of this is found in the pickling industry, where the acid medium is used to prevent bacterial contamination.

Electricity and radiant energy. By itself electricity does not destroy bacteria, but it causes heat and changes that may be lethal in the medium in which the bacteria are growing. Electric light inhibits bacterial growth, but the inhibition is an effect of light, not of electricity.

The effects of roentgen rays are generally harmful to most bacteria, being about 100 times more effective in their action against bacteria than ultraviolet light. There are, however, notable exceptions. Microbial species exist that can tolerate an amount of radiation 10,000 times greater than the amount fatal for man.

Chemicals. Certain chemicals destroy bacteria, and others inhibit their growth (p. 275). Some attract bacteria *(positive chemotaxis)*, whereas others repel them *(negative chemotaxis)*.

Osmotic pressure. The bacterial cell is encased in a membrane said to be *semi-*

---

*The use of ultraviolet light in sterilization is discussed on p. 273.

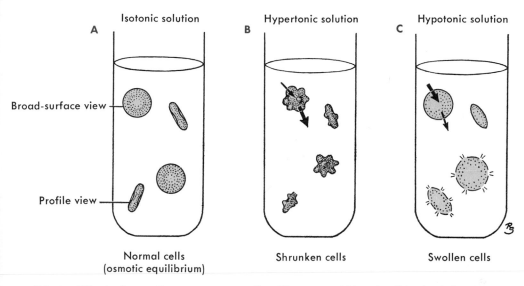

FIG. 6-2. Effect of osmotic pressure on cells. Observe red blood cell in the following. **A,** In isotonic solution, red cell does not change size or shape. **B,** In hypertonic solution, it shrinks. **C,** In hypotonic solution, red cell swells and bursts.

permeable because it allows water to pass freely in and out of the cell but gives a varying degree of resistance to dissolved substances in the fluid medium in which the cell is suspended. This makes the cell a small osmotic unit responsive to changes in its fluid environment (Fig. 6-2).

Under normal conditions there is a higher concentration of dissolved substances within the cell than without it. The greater osmotic pressure inside the cell keeps the protoplasm of the cell firmly against the cell wall, and the cell is said to be *turgid*. If a bacterial cell is placed in solutions having varying concentrations of dissolved substances, changes take place. In a solution with a high concentration of dissolved substances (*hypertonic* solution), water leaves the interior of the cell and the cell begins to shrink. If the difference in concentration between the interior and the exterior of the cell is not too great, the cell may be able to adjust to the hypertonic solution, regain its turgor, and continue its growth. If not, the cell continues to shrink and finally dies (*plasmolysis*).

If a cell is placed in a solution with a low concentration of dissolved substances or in distilled water (*hypotonic* solution), water passes into the cell, the cell swells, and it may burst (*plasmoptysis*). A solution containing that concentration of dissolved substances in which the cell neither swells nor shrinks is said to be *isotonic*.

Most bacteria resist small changes in osmotic pressure but are killed or inhibited by high concentrations either of salt (as used in brines) or of sugar. This fact is utilized in the preservation of foods such as syrups and jellies and in the preservation of meats in brine. That some microorganisms can adjust to a high concentration of sugar is seen in molds on jellies. To preserve foods safely, higher concentrations of sugar must be used than of salt.

That a few species of bacteria *(osmophiles)* do well in a hypertonic solution is seen from the bacterial life of the oceans. Even the Dead Sea with its high salt content supports a bacterial population *(halophiles)*. A small amount of various salts in the fluid medium for bacteria is beneficial to bacterial growth. Traces of such salts are furnished by natural foods such as meat extracts.

Interrelations of microbes

Symbiosis. Certain species of bacteria grow well together, and the associated species accomplish harmful or beneficial results that neither does alone. For instance, staphylococci and influenza bacilli multiply more rapidly when grown together than either does when grown alone. This is known as *synergism*.

Symbiosis refers to the relation of mutual benefit existing between two organisms. For example, there is the beneficial relation between the leguminous plants and the nitrogen-fixing bacteria living in the root nodules of these plants (p. 646). *Commensalism* (mutual tolerance) is the term applied when two organisms live together without benefit or injury to each other.

Antagonism. Sometimes the presence of certain species of microbes inhibits the growth of others. For instance, growth of the gonococcus is inhibited by the presence of almost any other species of bacteria. This is *antagonism*. Theories advanced to explain antagonism are (1) that one organism secretes a substance toxic to the growth of the other and (2) that one organism promotes a defense mechanism of the animal body against the other. The appearance of certain infections after the administration of antibiotics may be explained in terms of an antagonistic relationship between two organisms, only one of which succumbs to the action of the antibiotic. Released by the antibiotic, the other microorganism now becomes aggressive.

BIOLOGIC ACTIVITIES

Bacteria engage in a complex of biologic activities of varying importance. Of prime consideration are those relating to the growth and integrity of the microorganism. Those resulting in the elaboration of toxins and substances harmful to other living cells have special significance in the production or aggravation of disease. Other events in the life of the bacterial cell, although striking in their manifestations, are not vital and are of secondary importance.

Major metabolic events

Bacteria first drew attention to their very existence by the dramatic changes resulting from their metabolic activities. Today their biochemical capabilities are obvious when one considers the phenomena of fermentation (p. 86), putrefaction (p. 124), decay (p. 124), soil fertility (p. 646), and infectious disease (p. 161).

Much of the general body of information concerning metabolism in creatures more complex than bacteria has come from studies of comparable processes in bacteria and other microorganisms. Because bacteria multiply rapidly, we can learn much about their activities in 2 days. To obtain comparable information about man would require 200 years or more.

The bacterial cell makes two biologic demands of its environment: it must obtain therein the chemical ingredients with which it can build and maintain itself, and at the same time it must derive therefrom the energy necessary to do its work. *Metabolism* encompasses all biochemical reactions occurring within the bacterial cell by which these two requirements are met.

Enzymes. Enzymes are essential, very efficient ingredients in the complex maze of metabolic activity. Life is not possible without them.* The term *enzyme* is of Greek origin and means leavened (*en*, in; *zymē*, leaven).

CHARACTERISTICS. Enzymes are catalysts† and, being proteins, are therefore organic catalysts. Although produced by a living cell, an enzyme operates independently of that cell. Its activity is not lessened when it is separated from the cell that produced it. Over a thousand enzymes have been identified among living cells. Bacteria produce enzymes that are remarkably like those produced by the organs of higher forms of life.

Of the multitude of proteins built by living cells, the majority function as enzymes. A bacterium contains, on an average, about 2000 enzymes. Elaboration of proteins and therefore enzymes within it is under the direction of the DNA of its chromosome. For 2000 enzymes there would be a corresponding 2000 control positions (genes) in the bacterial chromosome (the one gene–one enzyme hypothesis).

Enzymatic action (Fig. 6-3) is specific; that is, each enzyme causes only its own peculiar type of chemical change on its specific substrate.‡ For the activation of an enzyme, a small molecule, usually an inorganic metallic ion such as magnesium, manganese, zinc, or iron, may be needed. The molecular *activator* does not participate in the process. Very small amounts of enzyme can induce most extensive chemical changes. One molecule of enzyme can catalyze the reaction for 10,000 to 1,000,000 moles§ of substrate per minute. Although the enzyme is not destroyed during the chemical reaction, its activity may be inhibited by the end products formed. The accumulation of end products may also block the formation of more enzyme. This is spoken of as a *negative feedback control.*

Enzymes are most active at temperatures from 35° to 40° C and within a limited pH range. They are delicate structures; a temperature of 60° C destroys them within 10 to 30 minutes (heat denaturation). Freezing retards their action but does not destroy them. A medium that is too acid or too alkaline inhibits their action or destroys them.

A *coenzyme* is a nonprotein organic compound that may be necessary for enzymatic activity. It serves to position the substrate to the enzyme. Although it is changed during the catalytic reaction, the coenzyme reappears at the end of the reaction in its original form. Many coenzymes come from dietary vitamins. It is possible for enzymes

---

*In space exploration the enzyme phosphatase, common to most forms of life on this planet, is considered a good indicator for the presence of life or prelife if detected in even trace amounts in outer space.

†A catalyst is a substance that accelerates a chemical reaction that otherwise would proceed either very slowly and ineffectively or not at all. In the reaction the catalyst is not chemically altered. It contributes nothing to the end product and does not furnish energy.

‡*Substrate* is the term designating the material acted on by the enzyme.

§One mole is one gram-molecular weight of substrate in a one-liter solution.

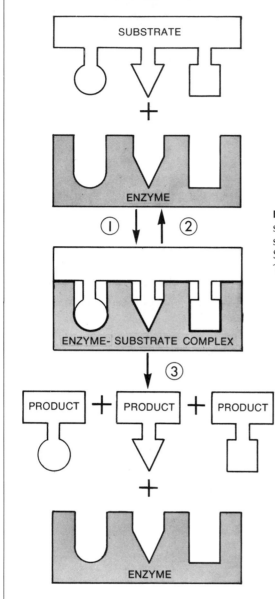

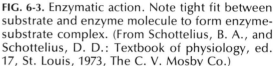

FIG. 6-3. Enzymatic action. Note tight fit between substrate and enzyme molecule to form enzyme-substrate complex. (From Schottelius, B. A., and Schottelius, D. D.: Textbook of physiology, ed. 17, St. Louis, 1973, The C. V. Mosby Co.)

performing the same chemical function to exist with different molecular structures. The physiologic equivalent for a given enzyme is an *isoenzyme*. An enzyme may have one or more isoenzymes. The enzyme lactic dehydrogenase has five.

Since the biochemical reactions occurring in bacteria are complicated and inter-related ones, bacterial enzymes may function in a group or as an *enzyme system*, with several enzymes being responsible for closely related chemical changes.

CLASSIFICATION. From the standpoint of a given cell, enzymes may be *exoenzymes* or *endoenzymes*. The former are elaborated within the cell and are diffused through the plasma membrane into the surrounding medium, where they are active. Endoenzymes are liberated only by the disintegration of the cell that produced them. Enzymes are also related to the chemical reactions catalyzed or to their substrates. They are named by adding the suffix *-ase* to the name of the substrate or to the name of the reaction.

In 1964 the Commission on Enzymes of the International Union of Biochemists adopted a classification separating enzymes into six major categories according to their type reaction:

1. *Oxidoreductases* catalyze oxidation-reduction reactions (includes oxidases and dehydrogenases).
2. *Transferases* catalyze the transfer of a specific group from one substrate to another (as in transamination or transmethylation). Examples are transaminases, transacetylases, and kinases.
3. *Hydrolases* catalyze hydrolytic reactions. Here are found digestive enzymes (proteinases, lipases, amylases) that break down large molecules to smaller, readily assimilated ones. Others include peptidases, esterases, and phosphatases.
4. *Lyases* (nonhydrolytic enzymes) catalyze the removal of specific groups from given substrates, such as when bonds are split between carbon atoms (decarboxylases) and between carbon and oxygen atoms (carbonic anhydrase). Others are deaminases and aldolases.
5. *Isomerases* catalyze the interconversion of two compounds of the same atomic composition but with different structure and properties. An isomerase is an important component of the glycolytic pathway.
6. *Ligases* catalyze the joining together of two molecules, coupled with the breakdown of adenosine triphosphate (ATP) or a similar energy-yielding triphosphate. Ligases are also known as synthetases and as "activating" enzymes.

Chemosynthesis. The plant cell, possessing chlorophyll obtains energy from the environment in the form of light (photosynthesis). Microorganisms with no chlorophyll must gain their energy from the chemical alteration of the substances at hand, a process termed *chemosynthesis.* Because of the complexity of their metabolic enzymes, most bacteria can break down the proteins, fats, carbohydrates, and other organic compounds that they contact. A series of enzymatic steps characterizes the chemical changes involved in chemosynthesis. The starting point is bacterial *digestion;* the end point is *biologic oxidation.*

DIGESTION. The principal sources of energy for bacteria in their environment are mostly complex molecules too large for the bacterial cell to deal with directly. They must be reduced to workable particle size. To accomplish this, bacteria rely on certain enzymes (hydrolases) that they release into the medium to split the large molecules chemically by a process that includes the addition of water. This is *hydrolysis.* The hydrolytic enzymes are examples of exoenzymes. Briefly, digestion is the exoenzyme-catalyzed hydrolysis of a large molecule to secure fragments small enough for passage into the bacterial cell.

ABSORPTION. After the bacterial cell has reduced the large molecules into smaller particles, it must next absorb them. There are two ways of doing this. Because of the osmotic pressure gradient, the cell can passively allow the molecules to move in by diffusion. This is slow and inefficient at best. On the other hand, the cell can actively participate to speed the movement. The manner in which this is brought about is referred to as *active transport*. As would be expected, the mechanisms of active transport, including the enzymes required, are located on the cell membrane, and their activity constitutes a "pump."

OXIDATION. Some of the molecules that the cell pulls into itself from without are ready for the sequences of biologic oxidation. Most are not, and these must be prepared, that is, changed into a form that can be oxidized. If a phosphate group is attached to the molecule (phosphorylation), a very significant chemical bonding is made. When this phosphorylated compound is oxidized, the chemical bond of the phosphate group traps energy not only in a form suitable for later use but also in large amounts. Phosphorylation is the most important preliminary sequence to oxidation, but other chemical reactions may be encountered at this step.

Biologic oxidation is the chemical setup whereby a cell is provided with energy that it can use biologically. The complicated reactions contributing to it, many of them chain reactions, are carried out notably by enzymes (oxidases, dehydrogenases, or oxidoreductases), coenzymes, and hydrogen acceptors or carriers. *Oxidases* act on a substrate to cause it to undergo oxidation, that is, a reaction wherein electrons are removed from a substrate with accompanying hydrogen ions. Enzymes making possible the removal of hydrogen (or electrons) are called *dehydrogenases*. As the hydrogen ions are removed, there must be a suitable compound to receive them, *a hydrogen acceptor* or *carrier*, which in so accepting becomes reduced. Hydrogen acceptors or carriers, which can be oxidized and reduced, function in the transport of hydrogen (or electrons) from tissue metabolites to oxygen or some other end product. Transfer of electrons constitutes oxidation and reduction, and the end results must include an oxidized product, a reduced one, and energy liberated or trapped.

STORAGE OF ENERGY. There is no dissipation of heat in biologic oxidation. The energy released is held in the chemical bond between an organic molecule and a phosphate group. Bacteria store this energy-rich bond solely on the organic molecule adenosine diphosphate, ADP. When a phosphate group is transferred to ADP, there is formed adenosine triphosphate, ATP, from which energy-rich reserves are available.

Classes of biologic oxidation. If the ultimate and final hydrogen acceptor is molecular oxygen, the events (aerobic) of biologic oxidation to generate energy are those of *respiration*. If it is an inorganic nitrate, sulfate, or carbonate, the metabolic processes are those of *anaerobic respiration*. The sulfate- and carbonate-reducing organisms are obligate anaerobes relying exclusively on these compounds for energy. The denitrifying bacteria are facultative aerobes using nitrate only in the absence of oxygen. If the final hydrogen acceptor is an organic compound, the processes of biologic oxidation are those of *fermentation*.

Fermentation, then, is metabolism utilizing organic compounds as both electron donors and electron acceptors. Many organic compounds can be fermented, and many microorganisms, including disease producers, can live anaerobically to satisfy their energy needs through fermentation reactions. Seven different biochemical types have been elucidated.

Using microbial metabolism of glucose, let us compare respiration (aerobic) with fermentation (anaerobic). In the presence of molecular oxygen, respiration takes place in all animal cells and in microbes, and glucose is changed in a biochemical sequence to carbon dioxide and water. In the absence of oxygen, microbes ferment the glucose but with different end products. Among bacteria generally, incomplete oxidation of this nature is the rule, with a wide variety of end products accumulating. The most commonly occurring sequence for glucose degradation is the Embden-Meyerhof glycolytic pathway in which glucose is changed to pyruvate.

Respiration is a much more efficient process than fermentation. Many more intermediate steps mean much more energy generated. The potential from complete oxidation of glucose is far greater than that from its conversion to alcohol and carbon dioxide; for instance, respiration yields over 30 molecules of ATP (the energy source) as compared to the net of only two for each mole of sugar fermented.

Anaerobic metabolism is restricted in the amount of energy that can be derived from the limited number of substrates serving as energy sources, but it can provide energy very quickly and continuously at a high rate as long as the fermentable substance is present.

Microbial fermentation reactions are of great value to man because of the many by-products of practical worth to home and industry. They are also useful in the laboratory study of microbes.

Macromolecular synthesis. In a microbial cell, basic physiologic phenomena are geared primarily for growth, that is, an orderly increase in mass or number of its constituents, most of which are macromolecules. Macromolecules are found in the protein-synthesizing groups of the cytoplasm, on membranes of organelles, in chromosomes, in the cell wall, and in the various enzyme systems. The metabolic activities of the cell revolve around these large molecules, and one of the crucial functions of the cell is their biosynthesis. To put together a macromolecule, two things are required: (1) the subunits for linkage and (2) the energy to do the work. The energy to induce linkage comes from the cellular reserves of ATP. The subunits must come from either the environment or the metabolic activities of the cell and depend on the chemical nature of the unit complex. For proteins these are amino acids; for carbohydrates, simple sugars; for lipids, glycerol (or other alcohols) and fatty acids; for nucleic acids, nucleotides; and for phospholipids, such substances as choline. The alignment of subunits in the macromolecule is determined in one of two ways. In nucleic acids and proteins, it is template directed: DNA serves as template for its own synthesis and for the synthesis of the types of RNA. Messenger RNA serves as template for the synthesis of proteins. In carbohydrates and lipids the arrangement of subunits is entirely enzymatic.

Although many metabolic activities are universally the same, some are unique to bacteria, for example the biochemical pathway by which the mucopeptide of the cell wall is structured. This provides a chemical basis for the design of an antimicrobial compound with a selective effect on one of the steps in the biosynthesis of the cell wall. Since bacterial protoplasm is under greatly increased osmotic pressure in relation to the isotonicity of the host cells, weakening the cell wall jacketing the bacterium could easily explode it.

Medically related activities

Toxins. Many species of bacteria produce poisonous substances, some of which are well known as *toxins*. The capacity to elaborate toxins is *toxigenicity*, and the resultant toxin may be almost entirely responsible for the specific action of the toxigenic bacteria. Toxins are of two types: *exotoxins*, diffused by the bacterial cell into the surrounding medium, and *endotoxins*, liberated only when the bacterial cell is destroyed. Some bacterial exotoxins are more deadly than any mineral or vegetable poison. Among the comparatively few bacteria (mostly gram positive) noted for their exotoxin production are *Corynebacterium diphtheriae*, *Clostridium tetani*, and *Clostridium botulinum*. In disease caused by *Corynebacterium diphtheriae* (diphtheria) and *Clostridium tetani* (tetanus), the bacteria grow restricted to a superficial area in the body. Within themselves these organisms produce little effect, but the soluble exotoxins elaborated at the site of growth are absorbed into the body to cause serious and often fatal illness. Such illness is better thought of as chemical poisoning (toxemia) rather than as disease induced by growth of bacteria.

It is interesting to note that some bacteria are harmful only when they are dis-

TABLE 6-1. DISTINGUISHING FEATURES OF TOXINS

Differential point	Exotoxins	Endotoxins
Location	Gram-positive bacteria	Gram-negative bacteria
Relation to cell	Extracellular	Closely bound to cell wall (release with destruction of cell)
Toxicity	Great	Weak
Tissue affinity	Specific (examples, nerves—tetanus toxin; adrenals, heart muscle—diphtheria toxin)	Nonspecific (local reaction—injection site; systemic reaction—fever, shock)
Chemical nature	Protein	Lipopolysaccharide complex
Stability	Unstable (denatured with heat, ultraviolet light)	Stable
Antigenicity	High (neutralizing antibodies)	Weak
Conversion to toxoid	Yes	No
Diseases where action important	Diphtheria Tetanus Gas gangrene Staphylococcal food poisoning Botulism	Typhoid fever and certain salmonelloses Asiatic cholera Brucellosis

eased. Evidence is accumulating that the above-mentioned organisms make exotoxin only because they are infected with bacterial virus (bacteriophage). The virus initiates (codes for) the production of toxin (p. 427).

Exotoxins are protein in composition and are antigenic and specific. When a small amount of a toxin is injected into an animal, it stimulates the production of *antitoxin*. The *specificity* of exotoxins refers to the fact that diphtheria bacilli elaborate a toxin that causes diphtheria and nothing else and tetanus bacilli elaborate a toxin that causes tetanus and nothing else. When an exotoxin is inactivated with formaldehyde, it no longer causes the disease but still can produce an immunity to the disease. Such modified toxins are known as *anatoxins* or *toxoids*. Toxoids are used to produce permanent immunity to diphtheria and tetanus.

Substances contained in the seeds of certain plants and some of the secretions of animals resemble exotoxins. Among these are crotin, derived from the seed of the croton plant; ricin, from the castor bean; and the venoms of snakes, spiders, and scorpions.

Endotoxins are complex lipopolysaccharides that are an integral part of the cell wall of most gram-negative bacteria. Endotoxins are not found in the surrounding medium when bacteria are grown in a liquid. They do not effectively promote the formation of antitoxins, do not possess the specificity of exotoxins, and cannot be converted into toxoids. The lipid fraction of the complex is related to the numerous biologic manifestations of endotoxin—fever, circulatory disturbances, certain immunologic effects, and so on (Table 6-1). Noteworthy for their production of endotoxins are *Salmonella typhi*, the cause of typhoid fever, and *Shigella flexneri*, one cause of bacillary dysentery.

The biochemical mechanisms by which toxins injure cells and produce disease are poorly understood.

Harmful metabolic products. Bacteria produce a number of enzymatic substances not directly toxic but related significantly to disease. *Hemolysins* are substances that cause the lysis (dissolution) of red blood cells. There are many types, among which are immune hemolysins, hemolysins of certain vegetables, hemolysins contained in venoms, and bacterial hemolysins. The bacterial hemolysins are of two types: (1) those that may be separated from the bacterial cells by filtration (filtrable hemolysins) and (2) those that are demonstrated about the bacterial colony on a culture medium containing red blood cells. The filtrable hemolysins are named after the bacteria that give rise to them (examples, *staphylolysin*, derived from staphylococci; *streptolysin*, derived from streptococci). The filtrable hemolysins have a certain degree of specificity; that is, a given hemolysin may act on the red blood cells of one species of animal but not on those of another species. They are protein in nature and inactivated by exposure to a temperature of 55° C for 30 minutes. In the body of an animal they induce the formation of antibodies against themselves.

Staphylococci, streptococci, pneumococci, and *Clostridium perfringens* are important producers of these hemolysins. An organism may have more than one filtrable hemolysin; for instance, streptococci produce two, known respectively as streptolysin O and streptolysin S.

The hemolysins detected when bacteria are grown on a culture medium containing blood are of two types, *alpha* and *beta*. Beta hemolysins give rise to a clear, colorless zone of hemolysis around the bacterial colony. Alpha hemolysins give rise to a greenish zone of partial hemolysis. The relation existing between filtrable hemolysins and those detected by growing the bacteria on a culture medium containing blood is not known. The relation between hemolysin production and virulence is obscure. As a rule, the hemolytic strains of pathogenic bacteria have greater capacities to cause disease than the nonhemolytic ones. On the other hand, certain nonpathogenic bacteria produce hemolysins. Hemolysins probably contribute to the invasive capacity of some bacteria.

Leukocidins destroy polymorphonuclear neutrophilic leukocytes. The leukocytes, or white blood cells, take an active part in the battle of the body against infection. Leukocidins are formed by pneumococci, streptococci, and staphylococci. Their relation to immunity is not known. It is likely that they enhance virulence. Certain hemolysins and leukocidins appear to be identical.

Coagulase is elaborated by bacteria to accelerate the coagulation of blood and, under suitable laboratory conditions, causes oxalated or citrated blood to clot. Coagulase formation appears to be confined to the staphylococci. The *coagulase test*, performed by mixing a culture of staphylococci with a suitable amount of oxalated or citrated blood under specified conditions, is used to differentiate pathogenic and nonpathogenic staphylococci. Because the former cause the blood to coagulate, they are said to be *coagulase positive*. The role of coagulase in immunity is not known. Possibly it protects organisms against the destructive action of leukocytes (phagocytosis). The coagulum from plasma that has leaked into the tissues forms a barrier between the bacteria and the white blood cells.

The *bacterial kinases* act on certain components of the blood to liquefy fibrin. Kinases interfere with coagulation of blood, since a blood clot is made up of red blood cells enmeshed in interlacing strands of fibrin. Kinases liquefy clots already formed. There seems to be a relation between the kinases and virulence because kinases destroy blood or fibrin clots that form around a site of infection to wall it off. The most important kinase is *streptokinase*, also known as *fibrinolysin*, elaborated by many hemolytic streptococci. Staphylococci and certain other bacteria also produce kinases. Strepotokinase has been used in certain locations in the body to dissolve clots and to prevent the formation of adhesions that would be laid down on the fibrin precipitated in the body cavities.

Hyaluronidase is secreted by certain bacteria to make the tissues more permeable to the bacteria elaborating it. It hydrolyzes hyaluronic acid, a constituent of the intercellular ground substance of many tissues that helps to hold the cells of the tissue together. Hyaluronidase was formerly spoken of as the "spreading factor." It is produced by pneumococci and streptococci.

Bacteriocins are bacterial protein or polypeptide substances produced by strains of a family of microbes and active only on certain other strains in the given family. *Colicins* are those produced by strains in the family Enterobacteriaceae. They are highly specific and seem to be directed chiefly against the bacterial membrane.

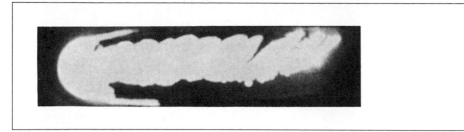

FIG. 6-4. Luminescent bacteria *(Vibrio fischeri)*. Photograph made in total darkness to show light produced by culture. (Courtesy Carolina Biological Supply Co., Burlington, N.C.; from Beaver, W. C., and Noland, G. B.: General biology, ed. 8, St. Louis, 1970, The C. V. Mosby Co.)

Nearly 20 different ones have been described and grouped by their bactericidal specificity.

Other effects

Pigment production. So far as is known, pigment production has no relation to disease and no significance other than in identification of pigment-producing organisms. Yellow pigments are most prevalent, but pigments of almost any color may occur. The red and yellow pigments probably belong to the same chemical group as those in turnips, egg yolk, and fruits. Pigments are produced by both parasitic and saprophytic bacteria; but most species of bacteria do not produce any pigment. Common pigment producers are *Staphylococcus aureus*, with a golden pigment, and *Pseudomonas aeruginosa*, with a bluish green pigment. *Serratia marcescens* produces a red pigment. Pigment-producing or *chromogenic* bacteria lose that property when grown under unfavorable conditions.

Heat. During the growth of all bacteria, heat is generated but in such small amounts that it can be detected in ordinary cultures only by most delicate methods. The heating of damp hay is in part caused by bacterial action.

Light. A few species of bacteria have the ability to produce light without emitting heat (bioluminescence) (Fig. 6-4). Most of them live in salt water, and some live a parasitic existence on the bodies of saltwater fish. Fox fire, seen on decaying organic material especially in the woods, is from luminescent bacteria. None of the light-producing bacteria is pathogenic.

Odors. Certain odors are characteristic of some species of bacteria; some arise from the decomposition of the material on which the bacteria are growing.

QUESTIONS FOR REVIEW

1. Name at least five requirements for bacteria to grow and multiply.
2. Classify bacteria with regard to food requirements, to oxygen requirements.
3. How are bacteria affected by extremes of temperature? By light?
4. What is the difference between negative and positive chemotaxis? Plasmoptysis and plasmolysis? Endotoxin and exotoxin? Alpha and beta hemolysis? Exoenzyme and endoenzyme? Hypertonic and hypotonic? Autotrophic and heterotrophic? Aerobe and anaerobe?
5. Briefly discuss bacterial metabolism. What are the two main functions of metabolism?
6. Define enzymes. Give salient features, including their classification.

7. What are toxoids? For what are they used?
8. Explain chemosynthesis, digestion, phosphorylation, oxidation-reduction, fermentation.
9. Define hemolysin, leukocidin, coagulase, streptokinase. Name a microbe producing each. What is the importance of the coagulase test?
10. What is hyaluronidase? Why is it sometimes called the "spreading factor?" In what infections is it important?
11. Briefly tell what is meant by catalyst, substrate, macromolecule, colicins, psychrophiles, halophiles, mesophiles, osmophiles, commensalism, symbiosis, antagonism, obligate, facultative, capnophiles, negative feedback control.

REFERENCES. See at end of Chapter 9.

LABORATORY SURVEY OF UNIT ONE

Information's pretty thin stuff unless mixed with experience.

CLARENCE DAY

Design

A laboratory course in microbiology is planned to (1) present the student with the scientific method, (2) impress the fact of microbes as living agents, and (3) demonstrate characteristics of microbes and their effects on the human body.

The following exercises may be varied to meet the needs of the individual student and fit the teaching facilities of the school. Details of technic have been omitted in this type of abbreviated survey because it is felt that the student can be taught technics to which the instructor is accustomed. In all instances the exercises are designed to be carried out with a minimum of equipment and a maximum of safety for all.

The student needs a notebook in which to record accurately each experiment: (1) its purpose, (2) how it is performed, (3) results obtained, and (4) conclusions reached. The notebook should be neat and be corrected frequently by the instructor.

SAFETY FIRST!

Rules for your own personal protection

Avoid any exposure of yourself or others to unnecessary danger. Do not take chances! Use common sense in an emergency. For safety in the laboratory, observe the following precautions:

1. Wash your hands thoroughly with soap and water at the beginning and end of each laboratory period. Disinfect your hands when necessary. Use hand lotion regularly to prevent dryness and cracking of the skin.
2. Wear a laboratory coat or jacket made of material easily cleaned and sterilized. Coat or jacket should be laundered at least once a week (more frequently when indicated).
3. Scrub with a cleansing agent; then disinfect the working surface of your desk at the beginning and end of each laboratory period. A suitable detergent-germicide may be used.
4. Do not put anything in your mouth during the laboratory period except as specifically indicated in the laboratory exercise (pipetting, etc.). Do not under any circumstances lick gummed labels or put pencils in your mouth. Do not eat, drink, or smoke in the laboratory during laboratory time.
5. At all times, keep your hands away from your face and mouth. Fingers and the areas under the nails can become contaminated quite easily.
6. Dispose of infectious material, cultures, and contaminated materials carefully and *only* in the way prescribed by the instructor. Note location of the special containers of disinfectant and use regularly.
7. Always flame-sterilize platinum loops and needles carefully, faithfully, and thoroughly before laying them aside.

8. Be very careful to avoid spilling any kind of contaminated material. If infectious material contacts the desk, hands, clothing, or floor, notify the instructor at once.
9. Report even a minor accident to the laboratory instructor *immediately!*
10. Arrange cultures and infectious material in a secure place on the desk. Do not allow unused equipment to accumulate in your work area. Avoid clutter.
11. Keep glassware and other equipment clean and in its proper place. Be clean and orderly at all times.
12. Keep personal items in the places designated and away from the work area.

The compound light microscope
Care of the microscope

"Faith" is a fine invention
When Gentlemen can *see*
But *Microscopes* are prudent
In an Emergency.

EMILY DICKINSON

The microscope, representing one of our most valued scientific instruments, is obviously "prudent" in a wide variety of situations. That its proper care is "prudent" is likewise evident. As you use the microscope, follow these instructions:

1. Keep both eyes open. You can do so with very little practice.
2. Many modern microscopes have built-in base illuminators. If yours does not and you must work with the mirror, avoid direct sunlight. North light is advantageous. Good results are obtained with daylight.
3. When you place a slide on the stage, see that it lies flat against the platform. Adjust the light so that the object is evenly illuminated.
4. Learn to focus with the low-power objectives (such as the 16 mm objective). Use the coarse adjustment to move the low-power objective until it nearly (but *not quite*) touches the cover glass or upper surface of the mounted specimen. Then focus *up* until the object comes plainly into view. Complete focusing with the fine adjustment. When the immersion objective is used, first place a drop of immersion oil on the object so that it can be clearly brought into view.
5. Keep the microscope clean and handle all parts with care. Do not touch the glass parts of the scope with the fingers. Do not allow chemicals to contact the microscope, as they may injure it. Clean the mechanical parts with an application of olive oil on gauze. Wipe the oil off with chamois or lens paper. Remove immersion oil from the optical glass parts by wiping with lens paper moistened with xylol. Do this as rapidly as possible to prevent injury to the optical settings.
6. Clean the microscope thoroughly when you are finished. Leave objectives with the lowest power in the working position. This precaution ensures that the least expensive objective would be injured should the optical system be jammed down accidentally. Keep the microscope covered when you are not using it.

PROJECT
Use of the compound light microscope

1. *Parts of the microscope*
   a. Place the microscope on a table in the proper working position at a convenient height. Be sure you are comfortably seated and that you do not have to stretch or that you are not cramped.
   b. Locate all parts shown in Fig. 5-1. Refer to the discussion of the microscope on pp. 66 and 67.
   c. By means of their numbers, locate the low-power, high dry, and oil immersion objectives. Explain the use of each in your notebook.

2. *Resolution*
   a. Place a prepared microslide on the stage of the microscope. (A section of tissue may be used.) Be sure that the specimen slide lies flat on the stage.
   b. Elevate the top of the condenser until it is flush with the surface of the stage.
   c. With the aid of the instructor, focus the low-power objective on the specimen.
   d. Observe the changes when the coarse and fine adjustments are manipulated. Note how an image comes in and out of focus.

3. *Illumination*
   a. Observe the changes in lighting brought about with a change in the position of the mirror. (*Note:* a microscope with a built-in lamp base does not have a mirror.) When you can fill the field of observation with light, have the instructor check to see if the lighting can be improved.
   b. Adjust the illumination so that it is even throughout the microscopic field of view.

4. *Contrast*
   a. Observe the changes in the specimen when the iris diaphragm is opened or closed.
   b. Note changes in contrast of specimen image in the microscope. Note lesser changes in focus.

5. *Magnification*
   a. Focus on a specimen image with the low-power objective. (Complete focusing with the fine adjustment.) Center specimen image.
   b. Move high-power objective into position. Complete focusing with fine adjustment. Center specimen image. Note change in magnification.
   Note: The constant movement of the fine adjustment is essential in producing a three-dimensional effect on the specimen image in microscopy. Rotate the fine adjustment slightly clockwise and then slightly counterclockwise.
   c. Under the direct supervision of the instructor, carefully focus with the oil immersion objective: first place a drop of immersion oil on the cover glass of the microslide (or on the upper surface of the specimen preparation). Lower the oil immersion objective with the coarse adjustment into the

drop of oil until the objective engages the drop and spreads it somewhat. The objective does not quite touch the surface of the microslide. Complete focus with the fine adjustment. Note change in magnification.

d. Under the supervision of the instructor, clean the microscope and return it to the locker.

6. *Problem*

a. Compare and calculate the magnification of the different objectives of your microscope with each eyepiece and complete the following table:

Focal length	Magnification objective lens	Final magnification with 8× ocular	Final magnification with 12.5× ocular
16 mm	10×		
4 mm	43×		
1.8 mm	97×		

PROJECT
Study of cells

1. *Cells from inside of cheek*
   a. Unstained
      (1) Scrape inner surface of the cheek with wooden tongue blade.
      (2) Mount scrapings in a drop of water on glass microslide and apply cover glass.
      (3) Examine with low power of the microscope.
      (4) Using the high dry objective for observation, sketch a few epithelial cells. You can recognize them by their flattened, pavement appearance. They may occur singly or in clusters. Locate any blood cells or bacteria that may be present. (If necessary close down the iris diaphragm to enhance contrast.)
   b. Stained
      (1) Mount scrapings from the inside of the cheek as outlined above, spreading the material out into a thin even smear with a toothpick applied to the microslide. Do *not* cover with a cover glass. Let dry.
      (2) Pass the slide rapidly through a flame two or three times to fix the smear.
      (3) Cover the smear with methylene blue stain for 3 minutes.
      (4) Wash stain off with distilled water. Blot between sheets of blotting or bibulous paper.
      (5) Examine with oil immersion objective. Check for proper illumination.
      (6) Compare stained preparation with the unstained one. In your notebook sketch three or four representative cells from each. (Use colored pencils for stained cells.)

3. *Cells in the urine*
    a. Place a 10 ml sample of urine in a conical centrifuge tube and centrifuge for 5 minutes at low speed (about 1500 rpm). Decant supernatant.
    b. Place a drop of the urinary sediment on a microslide and mount with cover glass.
    c. Examine with low-power and high-power objectives of microscope. Look for epithelial cells and leukocytes (white blood cells). Make a sketch showing the different objects seen. With the aid of the instructor, identify them. *Note:* Close down the iris diaphragm of the scope to enhance contrast in the cellular specimen.
4. *Cells of the blood*
    Note: Students may work in pairs to obtain finger puncture specimens of blood.
    a. Obtaining the specimen of blood
        (1) Cleanse side of the tip of the fourth finger of either hand with 70% alcohol, removing alcohol with dry sterile cotton.
        (2) With sterile lancet or needle, prick the finger with enough force to get a large drop of blood. Discard the *first* drop.
        (3) Gently squeeze finger until a second drop of blood forms.
    b. Making the blood smear
        (1) Touch a clean slide to a drop of blood on the finger. The drop of blood should be about 1.5 cm from one end of the microslide.
        (2) Place the slide on a level surface with the drop of blood up. With one hand, steady the slide, and with the other hand, place the end of another slide flat against the surface of the blood slide so that an angle of about 45 degrees is formed. The second slide acts as the spreader or pusher.
        (3) Draw the spreader slide back until its edge contacts the drop of blood. Let the blood spread along the junction of the two slides. Push spreader slide forward, keeping the 45 degrees of contact to form a film of blood. *Note:* The thickness or thinness of the blood smear depends on the acuteness of the angle and the rapidity with which the slide is pushed along. A wider angle and increased speed of spreading tend to thicken the blood film greatly. Since thin films are more easily studied, the angle of contact should be kept small and the speed of spread fairly slow.
    c. Staining the blood smear
        (1) Allow the blood film to dry in air.
        (2) Cover with Wright's stain* and let stand 4 to 6 minutes.

---

*Wright's stain and practically all reagents used in the laboratory exercises may be obtained in ready-to-use form—either as single reagents or in kit combinations—from any of several commercial sources. To prepare Wright's stain working solution in the laboratory, grind 0.3 g of commercial powder with a pestle in 3 ml of glycerin and successive small portions of 100 ml of acetone-free methyl alcohol in a mortar. Allow the solution so made to stand, with occasional shaking, for at least 1 week. Filter before use.

(3) Do not remove any stain but add an equal volume of distilled water or preferably phosphate buffer.

(4) Let remain until a metallic scum forms (usually 2 to 3 minutes).

(5) Wash with distilled water and dry between sheets of blotting paper. *Note:* A phosphate buffer* used for 5 to 10 minutes after the application of Wright's stain (instead of distilled water) gives nicely stained smears. The staining times for Wright's stain and phosphate buffer are variable and the actual time for a given stain must be determined by trial intervals of time.

d. Examining the smear

(1) Use the low-power objective of the microscope first to examine the blood film.

(2) Place a drop of immersion oil on a thinner part of the smear.

(3) Use the oil immersion lens to study the different kinds of cells present. *Note:* Red cells appear reddish orange in a well-stained smear. Those from normal adult human blood do not have nuclei. White cells have dark purple nuclei with light blue or lavender cytoplasm. Some white blood cells have granules in the cytoplasm.

(4) Sketch a few red cells and a few white cells. Note the proportion of red blood cells to white blood cells.

(5) From your study of this slide, do you think that there are different types of red blood cells or are all essentially alike? How many kinds of white blood cells do you see?

5. *Cells in pus*

a. Examine microscopically a prepared and stained smear of pus. Use the oil immersion objective.

b. Note the predominant cells—the pus cells. Sketch a few.

c. Do pus cells resemble certain white blood cells that you have just visualized? Does this mean that they are the same cells?

PROJECT
Microscopic appearance of bacteria

1. *Study of prepared slides*

a. Using the oil immersion objective of the microscope, study the appearance of bacteria in the prepared stained slides carefully. Draw representative fields.

b. Note the different shapes of bacteria. Draw cocci, bacilli, and spirilla.

c. Note the arrangements of bacteria. Draw bacteria arranged as staphylococci, diplococci, streptococci, and streptobacilli.

d. Note the following structures related to bacteria and draw: (1) spores, (2) capsules, (3) metachromatic granules, and (4) flagella.

---

*Potassium phosphate, monobasic, 6.63 g, and sodium phosphate, dibasic, 2.56 g, in 1 liter of distilled water give a buffer with a final pH of 6.4.

2. *Demonstration of motility of bacteria*
   a. Observe motility in hanging drop preparations set up as demonstrations.
   b. Contrast true motility with brownian motion. Define brownian motion.
   c. Note the steps in making a *hanging drop preparation* (demonstration and discussion by instructor).
3. *Use of the hanging drop*
   a. *Motile organisms*
      (1) Consult p. 69 for the method.
      (2) Use a broth culture of a motile organism to make a hanging drop preparation.
         Note: Be careful to sterilize the platinum loop in the flame each time *before* and *after* it is used.
      (3) Examine the preparation with the high dry objective of the microscope. The amount of light passing through the substage condenser must be somewhat reduced by partly closing the diaphragm. A properly made preparation shows the cells standing out distinctly against a dimly lighted background.
      (4) Discard carefully the preparation according to the method given by your instructor. *Why is this important?*
   b. *Nonmotile organisms*
      (1) Make hanging drop preparation of a suspension of carmine—note brownian motion.
      (2) Demonstrate similarly nonmotile organisms—note that they do not change their position in relation to each other. Note brownian motion.
4. *Demonstration of the use of the dark-field microscope*
   The instructor demonstrates the use of the dark-field microscope and briefly discusses its application.
5. *Application of negative or relief staining*
   a. Place a loopful or two of a broth culture or watery suspension of bacteria on a slide.
   b. Add an equal amount of commercial India ink that has been diluted with 2 parts of water.
   c. Mix thoroughly and spread out slightly.
   d. Allow to dry and examine with the oil immersion objective. In satisfactory portions of the smear the bacteria stand out as colorless bodies against a gray-brown or black background. With *relief* staining, the background is stained, *not* the bacteria.

PROJECT
Staining of bacteria

1. *Simple stains*
   a. Make three smears of organisms furnished by the instructor according to

directions given previously (p. 72). *Do not neglect proper sterilization of the platinum loop.*

   b. Fix the smears; then pour a few drops of methylene blue on the first, a few drops of gentian violet on the second, and a few drops of carbolfuchsin on the third. Let stains act for 1 minute.

   c. Wash the stains off with distilled water and drain the slides. Blot with blotting paper. Examine with the oil immersion lens. Make drawings. (Use colored pencils.) What is the color given to the organisms by the dye used in each smear?

2. *Gram stain*

   a. Obtain fixed smears already prepared by the instructor, or prepare such smears from suitable organisms furnished by the instructor.

   b. Refer to the technic of staining outlined previously (p. 73).

   c. Examine smears with the oil immersion objective of the microscope. Are the organisms furnished you gram positive or gram negative? How do you know?

3. *Acid-fast stain*

   a. Obtain prepared and fixed slides of sputum from the instructor.

   b. Refer to the technic of acid-fast staining outlined previously (p. 74).

   c. Examine smears with the oil immersion objective. What color is *Mycobacterium tuberculosis* when stained by this method? What is the color of nonacid-fast organisms? Sketch acid-fast bacilli. (Use colored pencils.)

4. *Special stains*

   a. Make a methylene blue stain of a sporeforming organism. Note spores, which appear as unstained areas in the cell body. Make a drawing.

   b. Make a Gram stain of pneumococci. Note the capsule, which stands out as an unstained halo around the cell. Make a drawing.

EVALUATION FOR UNIT ONE

Part I

In the following statements or questions, encircle the number in the column on the right that correctly completes the statement or answers the question.

1. The sciences that include the study of animals are:
   (a) Botany
   (b) Biology
   (c) Microbiology
   (d) Zoology
   (e) Bacteriology

   1. a and e
   2. b only
   3. b, c, and d
   4. c only
   5. e only

2. Sciences that include *only* unicellular organisms are:
   (a) Bacteriology
   (b) Botany
   (c) Zoology
   (d) Protozoology
   (e) Microbiology

   1. a, b, and c
   2. a only
   3. b and c
   4. a and d
   5. none of the above

3. In what order do we arrange the following groups of living organisms beginning with the largest or most general classification?
   (a) Species
   (b) Family
   (c) Phylum
   (d) Order
   (e) Genus
   (f) Class

   1. c, d, f, b, e, a
   2. b, c, a, f, e, d
   3. d, a, f, c, b, e
   4. c, f, d, b, e, a
   5. none of above is correct

4. Which of the following possess chlorophyll?
   (a) Bacteria
   (b) Fungi
   (c) Pond scums
   (d) Seaweeds
   (e) Slime molds

   1. a only
   2. b only
   3. a and b
   4. c and d
   5. c, d, and e

5. Which of the following are associated with the discovery of the genetic role of DNA?
   (a) MacLeod
   (b) Watson
   (c) Avery
   (d) Coons
   (e) Crick

   1. b only
   2. a, b, c, d, e
   3. a, b, c, e
   4. b, c, d, e
   5. e only

6. Which of the following are associated with the development of the poliomyelitis vaccine?
   (a) Salk
   (b) Edmonston
   (c) Enders
   (d) Dubos
   (e) Sabin

   1. a and b
   2. a, c, and e
   3. b, c, and e
   4. e only
   5. a only

7. In which of the following parts of the cell are the hereditary characteristics transmitted?
   (a) Cytoplasm
   (b) Nucleolus
   (c) Chromosomes
   (d) Mitochondrion
   (e) Endoplasmic reticulum

   1. a
   2. b
   3. c
   4. d
   5. e

8. Which of the following functions belong to the nucleus of a cell?
   - (a) Storage of food
   - (b) Control of metabolism
   - (c) Elimination of waste products
   - (d) Control of reproduction
   - (e) Transfer of hereditary characteristics
   1. a only
   2. a, b, and c
   3. b only
   4. b, d, and e
   5. a, b, d, and e
9. Which of the following functions belong to the cytoplasm of a cell?
   - (a) Storage of food
   - (b) Control of metabolism
   - (c) Enzyme secretion
   - (d) Assimilation of food
   - (e) Transfer of hereditary characteristics
   1. a only
   2. a and b
   3. a, b, and c
   4. a, c, d, and e
   5. a, c, d
10. Which of the following bacteria are spiral in shape?
    - (a) Diplococci
    - (b) Bacilli
    - (c) Vibrios
    - (d) Spirochetes
    - (e) Streptococci
    1. a and e
    2. a
    3. b
    4. c and d
    5. e
11. Which of the structures listed below are resistant to conditions adverse to bacterial growth?
    - (a) Capsules
    - (b) Flagella
    - (c) Metachromatic granules
    - (d) Spores
    - (e) Polar bodies
    1. a only
    2. d only
    3. a and d
    4. a, b, c, and d
    5. none of these
12. The method by which bacteria reproduce is:
    - (a) Budding
    - (b) Transverse fission
    - (c) Spore formation
    - (d) Binary fission
    - (e) Capsule formation
    1. a only
    2. b and d
    3. c only
    4. d only
    5. c and e
13. Which of the following maintains the shape of the bacterial cell?
    - (a) Spore
    - (b) Flagella
    - (c) Plasma membrane
    - (d) Capsule
    - (e) Cell wall
    1. e only
    2. a and e
    3. c and e
    4. d and e
    5. c only
14. A structure in the bacterial cell rich in enzymatic activity is the:
    - (a) Plasma membrane
    - (b) Cell wall
    - (c) Metachromatic granule
    - (d) Capsule
    - (e) Flagellum
    1. a and b
    2. d and e
    3. b and c
    4. b only
    5. a only
15. A protoplast is not:
    - (a) A bacterial cell with a spore
    - (b) A bacterial cell without a spore
    - (c) A bacterial nucleus
    - (d) A bacterial spore without a coat
    - (e) A bacterial cell without a cell wall
    1. a and e
    2. a, c, and e
    3. a, b, and c
    4. d only
    5. a, b, c, and d
16. The unit of measurement for most purposes in the study of cells with the compound light microscope is the:

(a) Angstrom
(b) Micrometer
(c) Millimicron
(d) Nanometer
(e) Meter

1. a
2. b
3. c
4. d
5. e

Part II

1. Match the discovery or contribution to the science of microbiology listed in column B to the name of the person responsible in column A. Give the *one* best answer.

COLUMN A

_____ von Behring
_____ Beijerinck
_____ Budd
_____ Eberth
_____ Ehrlich
_____ Fleming
_____ Holmes
_____ Jenner
_____ Koch
_____ Landsteiner
_____ Leeuwenhoek
_____ Lister
_____ Metchnikoff
_____ Pasteur
_____ Ross
_____ Schwann
_____ Semmelweis
_____ Snow
_____ Spallanzani
_____ Waksman
_____ von Wassermann

COLUMN B

1. Discovery of penicillin
2. Phagocytic theory of immunity
3. Humoral theory of immunity
4. Technic of pure culture
5. Preventive treatment for rabies
6. Observations on epidemiology of typhoid fever
7. Observations on epidemiology of cholera
8. Discovery of diphtheria antitoxin
9. Proof that yeasts are living things
10. Recognized first virus
11. Transmission of malaria
12. Discovery of basic human blood groups
13. Early experiments to disprove spontaneous generation
14. Laid foundation for aseptic surgery
15. Observations on transmission of puerperal sepsis
16. Discovery of streptomycin
17. Established serologic diagnosis of syphilis
18. Discovery of typhoid bacillus
19. Introduced smallpox vaccination
20. First drawing of bacteria

2. Consider the characteristics as listed in column on right. Indicate by a check mark in the appropriate column the relation to exotoxin or endotoxin.

Exotoxin	Endotoxin	
		Found in gram-positive bacteria
		Found in gram-negative bacteria
		Great toxicity
		Protein
		Lipopolysaccharide complex
		Important in diphtheria
		Weakly antigenic
		Conversion to toxoid

3. Match the constituent of the cell as given in column B to the item in column A that best indicates its structure or function. Give the *one* best answer.

COLUMN A
_____ Hydrolytic enzymes
_____ Protein synthesis
_____ Protein excretion
_____ Cell shape
_____ Glycogen
_____ Cellulose
_____ Pigment granules
_____ Cytochrome enzymes
_____ Cytoskeleton
_____ Exchange of food material
_____ Crystals
_____ Powerhouse of cell
_____ "Suicide bags"
_____ Nucleic acid metabolism
_____ Role in carbohydrate metabolism
_____ Unit of organization for DNA
_____ Genetic code

COLUMN B
1. Golgi apparatus
2. Microtubules
3. Chromosomes
4. Cell wall
5. Mitochondria
6. Endoplasmic reticulum
7. Ribosomes
8. Lysosomes
9. Plasma membrane
10. Inclusions
11. Gene
12. Nucleolus
13. Peroxisome
14. Nuclear sap

4. Identify the organisms from the following list as being either gram positive or gram negative by placing the number before the name in the appropriate column.

GRAM POSITIVE	GRAM NEGATIVE	GENUS OF ORGANISMS TO BE IDENTIFIED
_____	_____	1. *Neisseria*
_____	_____	2. *Staphylococcus*
_____	_____	3. *Streptococcus*
_____	_____	4. *Shigella*
_____	_____	5. *Escherichia*
_____	_____	6. *Salmonella*
_____	_____	7. *Mycobacterium*
_____	_____	8. *Clostridium*
_____	_____	9. *Haemophilus*
_____	_____	10. *Brucella*
_____	_____	11. *Pasteurella*
_____	_____	12. *Bacillus*
_____	_____	13. *Corynebacterium*
_____	_____	14. *Pseudomonas*
_____	_____	15. *Proteus*

5. Identify the organisms from the following list as being either acid fast or nonacid fast by placing the number before the genus name in the appropriate column.

ACID FAST	NONACID FAST	GENUS OF ORGANISMS TO BE IDENTIFIED
_____	_____	1. *Escherichia*
_____	_____	2. *Proteus*
_____	_____	3. *Mycobacterium*
_____	_____	4. *Nocardia*
_____	_____	5. *Actinomyces*
_____	_____	6. *Staphylococcus*

UNIT TWO
MICROBES
PROCEDURES FOR STUDY

7 Cultivation

The small size and similarity in appearance and staining reactions of microbes often make the identification by microscopic methods alone impossible. One of the most important ways to identify them is to observe their growth on artificial food substances prepared in the laboratory. This is cultivation of the microorganisms, or *culturing*. The food material on which they are grown is a *culture medium* (*pl.*, media), and the growth itself is a *culture*. Some 10,000 different kinds of culture media have been prepared.

Cultural methods help to identify microbes and also to pick up organisms in test material of low microbial content. When such material is placed on a suitable culture medium, each organism present multiplies many times. Practically speaking, the smear may contain few or no organisms, whereas the culture soon produces hundreds.

CULTURE MEDIA

General considerations. For the most satisfactory growth of bacteria and related forms on artificial culture media, the proper temperature, the right amount of moisture, the required oxygen tension, and the proper degree of alkalinity or acidity (pH) must be provided. The culture medium itself must contain the necessary nutrients and growth-promoting factors and, of course, be free from contaminating microorganisms; that is, *it must be sterile*.

Most disease-producing bacteria require complex foods similar in composition to the fluids of the animal body. For this reason the base of many culture media is an infusion of meat of neutral or slightly alkaline pH containing meat extractives, salts, and peptone to which various other ingredients may be added.*

---

*The basic infusion can be prepared by soaking 500 g of fresh lean ground beef in 1 liter of distilled water in the refrigerator. After 24 hours, the surface layer of fat is removed with absorbent cotton, the mixture is passed through muslin or gauze, and the meat discarded. After ingredients related to the organism under study are added and the reaction of the medium adjusted to neutral or slightly alkaline, the infusion is heated to 100° C for 20 minutes to remove coagulated tissue proteins and then filtered through coarse paper. At each step the volume is reconstituted to 1000 ml.

Culture media are of three types: (1) a liquid; (2) a solid that can be liquefied by heating but, when cooled, returns to the solid state; and (3) a solid that cannot be liquefied. Many times it is desirable that the culture medium be a solid, since bacteria and related forms can be viewed on the surface. Agar (Japanese seaweed) is a solidifying agent widely used. It melts completely at the temperature of boiling water and solidifies when cooled to about 40° C. With minor exceptions, it has no effect on bacterial growth and is not attacked by bacteria growing on it. The low temperature at which agar solidifies is crucial if test material must be inoculated directly into melted media before it solidifies. If the agar began to solidify at a temperature high enough to kill bacteria or fungi, this could not be done. Gelatin is a well-known but less frequently used solidifying agent.

Numerous enriching materials are found in different culture media—carbohydrates, serum, whole blood, bile, ascitic fluid, and hydrocele fluid. Carbohydrates are added (1) to increase the nutritive value of the medium and (2) to indicate the fermentation reactions of microbes being studied. Serum, whole blood, and ascitic fluid are added to promote the growth of the less hardy organisms. Dyes added to culture media act as indicators to detect the formation of acid when fermentation reactions are being tested or act as inhibitors of the growth of certain bacteria but not that of others. An example of an indicator dye is phenol red, which is red in an alkaline or neutral medium but yellow in an acid one. An example of an inhibitory dye is gentian violet, which inhibits the growth of most gram-positive bacteria.

In actual practice, media making is greatly simplified by the availability of commercially prepared mixtures containing the necessary and any special ingredients for growth of bacteria or fungi in just the right proportions. Such preparations are sold in the dehydrated state as a powder or tablet, either of which may be reconstituted in water.*

Before sterilization, hydrated culture media are poured into suitable test tubes or flasks that are closed with cotton plugs. After sterilization, some of the tubes of hot liquid media containing agar remain vertical for the agar to solidify, but some are laid on a flat surface with their mouths raised so that when the medium cools and solidifies there is a slanting surface, an *agar slant*. This gives a larger surface area for growth. Cotton plugs allow the access of moisture and oxygen but block the entrance of contaminating microorganisms; therefore the medium remains sterile until used, and microbes with which it is inoculated are not contaminated by those from the outside.

A large surface area for bacterial growth is provided when a solid culture medium partly fills a Petri dish. A *Petri dish* is a circular glass (or plastic) dish about 3 inches in diameter with perpendicular sides about ½ inch high. Inverted over it is a glass, metal, or plastic cover exactly like it except for a slightly greater diameter. The edges of the dishes are smooth, and a container is formed that prevents both the entrance and exit of bacteria. The Petri dish may be filled with a sterile agar medium still hot from the sterilization process, or a tube of solidified agar may be melted and poured into the

---

*Dehydrated culture medium (powder) is weighed out and dispensed into a given volume of distilled water. The mixture is heated carefully over a water bath to effect solution.

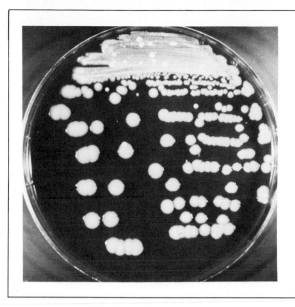

FIG. 7-1. Use of blood agar in Petri dish. Bacterial colonies are those of *Escherichia coli,* the colon bacillus.

dish. Until the agar has completely cooled, the lid is slightly raised on the Petri dish. When the agar is solid, the lid is lowered.

Most media are sterilized by autoclaving (pp. 270 and 303). Those that contain carbohydrates may have to be sterilized by the fractional method because many carbohydrates will not withstand the high temperature of the autoclave. Enrichment materials such as serum and whole blood that are injured by even moderate heat are collected in such a manner as to be kept sterile and are mixed with the medium only after it has been sterilized and allowed to cool. With agar media, these substances are added when the medium reaches 40° to 42° C because when the temperature falls slightly lower than this, the agar begins to solidify.

Almost every species of microbe has some medium on which it grows best. A few will grow only on media specially prepared for them, but most pathogenic ones will grow more or less luxuriantly on certain routine or standard media (Fig. 7-1). Table 7-1 lists media suitable for the isolation and growth of important pathogenic microorganisms.

Selective and differential media. Media that promote the growth of one organism and retard the growth of other organisms are *selective* media. Examples are bismuth sulfite agar and Petragnani medium. Bismuth sulfite agar, used to isolate typhoid bacilli from feces, promotes the growth of typhoid bacilli but retards the growth of bacteria normally resident in the feces. Petragnani medium promotes the growth of tubercle bacilli but retards the growth of any other organism present.

Media that distinguish organisms growing together are *differential* media. Examples are eosin–methylene blue (EMB) agar and MacConkey agar used in the differentiation of the gram-negative bacteria of the intestinal tract. The incorporation of lactose into such differential media makes possible a sharp separation between the organisms that ferment this sugar and those that do not. The colonies of lactose fer-

TABLE 7-1. MICROBES RELATED TO CULTURE MEDIA

Organism	Media most suitable for growth
Actinomyces bovis	Brain-heart infusion glucose broth and agar; thioglycollate broth medium (Brewer modified) (anaerobic)
Bacillus anthracis	Growth on almost all media
Bacteroides	Blood agar (anaerobic system)
Blastomyces dermatitidis	Blood agar; beef infusion glucose agar; Sabouraud glucose agar (antibiotics added to media used)
Bordetella pertussis	Glycerin-potato-blood agar (Bordet-Gengou agar)
Brucella abortus, melitensis, and *suis*	Trypticase soy broth and agar
Candida albicans	Sabouraud glucose agar; brain-heart infusion blood agar; corn meal agar; rice infusion agar
Chlamydia (Bedsonia)	Chick embryo; cell culture
Clostridia of gas gangrene	Thioglycollate broth medium (Brewer modified); anaerobic blood agar; cooked meat medium under petroleum; Clostrisel agar
Clostridium tetani	Same as for clostridia of gas gangrene
Coccidioides immitis	Sabouraud glucose agar
Corynebacterium diphtheriae	Löffler blood serum medium; cystine-tellurite-blood agar
Cryptococcus neoformans	Blood agar; beef infusion glucose agar; Sabouraud glucose agar
Entamoeba histolytica	Entamoeba medium (Difco) with dilute horse serum and rice powder
Escherichia coli and coliforms	Eosin–methylene blue (EMB) agar; Endo agar; MacConkey agar; desoxycholate agar; blood infusion agar (growth on almost any medium)
Francisella tularensis	Cystine glucose blood agar
Fungi, including dermatophytes	Sabouraud glucose agar with antibiotics; Littman oxgall agar
Haemophilus influenzae	Chocolate agar with yeast extract; rabbit blood agar with yeast extract
Histoplasma capsulatum	Brain-heart infusion glucose blood agar with antibiotics; Sabouraud glucose agar with antibiotics; brain-heart infusion glucose broth
Mycobacterium tuberculosis	Slow growth on special media such as Petragnani medium, Löwenstein-Jensen medium, or Dubos oleic agar; Dorset egg medium
Neisseria gonorrhoeae	Chocolate agar
Neisseria meningitidis	Chocolate agar
Nocardia asteroides	Beef infusion blood agar; beef infusion glucose agar; Sabouraud glucose agar; Czapek agar; Dorset egg medium

TABLE 7-1. MICROBES RELATED TO CULTURE MEDIA—cont'd

Organism	Media most suitable for growth
Proteus vulgaris	EMB agar; Endo agar; MacConkey agar; desoxycholate agar; blood infusion agar (growth on almost any medium)
Pseudomonas aeruginosa	Blood agar; EMB agar; Endo agar; MacConkey agar; desoxycholate agar (growth on almost any medium)
Rickettsia	Chick embryo; cell culture
Salmonella	Desoxycholate agar; desoxycholate-citrate agar; *Salmonella-Shigella* (SS) agar; MacConkey agar; tetrathionate broth; selenite-F enrichment medium
Salmonella typhi	Bismuth sulfite agar; SS agar; desoxycholate agar; desoxycholate-citrate agar; MacConkey agar; tetrathionate broth; selenite-F enrichment medium
Shigella	SS agar; desoxycholate agar; desoxycholate-citrate agar; MacConkey agar; tetrathionate broth; selenite-F enrichment medium
Staphylococcus	Growth on almost any medium
Streptococcus	Blood infusion agar; trypticase soy broth; tryptose phosphate broth; brain-heart infusion media
Viruses	Chick embryo; cell culture

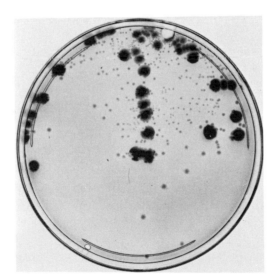

FIG. 7-2. Use of differential culture medium. On MacConkey agar dark (red purple) colonies of lactose-fermenting *Escherichia coli* contrast with colorless colonies of nonlactose-fermenting *Shigella* species.

menters are deeply colored; colonies of nonfermenters are colorless. This point is important because generally the pathogens in the intestinal tract do *not* ferment lactose, whereas the normal inhabitants, the coliforms, do (Fig. 7-2).

Selective and differential media are of great help in the diagnosis of infections in such areas of the body as the respiratory and intestinal tracts, where normally a variety of organisms resides. The presence of pathogenic bacteria may not disrupt the normal bacterial pattern for the area (the normal flora); such pathogens tend to be mixed with the other organisms in specimens taken from the site.

Synthetic media. The culture media just discussed are made up of components of variable composition such as meat infusions, serum, or other body fluids. Such media are *nonsynthetic media. Synthetic media,* on the other hand, contain ingredients of definite chemical composition. Used primarily in research work, such media have the advantage that whenever prepared their composition is the same. Therefore the results of microbial action on these media in one laboratory are strictly comparable with those in other laboratories thousands of miles away.

CULTURE METHODS*

Inoculation. A culture is made or inoculated (Fig. 7-3) when the specimen to be cultured, such as sputum, urine, or pus, is placed into a fluid medium or is rubbed gently over the surface of a solid medium with either a sterile swab or a flame-sterilized platinum wire loop (Fig. 5-4). The inoculated medium is then incubated for a period of time, routinely 24 to 48 hours.

Bacteriologic incubator. The bacteriologic incubator consists of an insulated cabinet (Fig. 7-4) fitted with an electrical heating element and a thermoregulator to maintain the temperature at a set point. A good thermoregulator in a properly constructed incubator maintains the temperature constant from day to day to within 0.5° C.

The incubator must be properly ventilated. It may be constructed so that circulation of air is brought about by the combined effects of gravity and the difference in weight of warm and cold air. Incubators are also ventilated with blowers or fans. All are fitted with perforated shelves and most with sets of double doors. While keeping out cold air, the inner glass doors allow the contents of the incubator to be viewed.

Inspection of cultures. Each microbe in the medium inoculated and incubated multiplies rapidly, and within a few hours there are many more microorganisms in the culture than there were in an equal amount of the material from which the culture was made. Consequently, bacteria or fungi are readily identified in cultures; in smears of the same test material they may be found with difficulty or not at all. Diphtheria bacilli can be found in throat cultures twice as often as in throat smears.

When a culture is made on a solid medium, all bacteria that grow from each bacterium deposited on the medium cling together to form a mass visible to the naked eye. This is called a *colony.* The colony has characteristics such as texture, size, shape, color, and adherence to the medium (Fig. 7-5). These are fairly constant for each

---

*Standard aseptic bacteriologic technics are implied in the discussion to follow.

FIG. 7-3. Inoculating a liquid culture. Note slanting position of test tubes during maneuver so that contaminating organisms from air are less likely to drop into the open tubes. Fingers of right hand are lifted away from handle of platinum loop to show how cotton plugs are held.

FIG. 7-4. Bacteriologic incubator, electrically heated, with temperature kept constant by delicate thermoregulator. This one also provides measured carbon dioxide atmosphere, a feature promoting growth of certain microbes. (Courtesy Lab-Line Instruments, Inc. Melrose Park, Ill.)

species and are valuable in differentiating one species from another. Theoretically, each organism should give rise to one colony, but two or more may cling together and, when planted on a medium, give rise to only one colony. If organisms that cling together are different, the colony will contain the different kinds. As a rule, with *good laboratory technic* a colony contains only one kind of microorganism.

Petri dish cultures permit good observation of colonies. The large surface area favors separation of individual microorganisms. Petri dish cultures are usually spoken of as plates and the process of making them as *plating* or *streaking*.

113

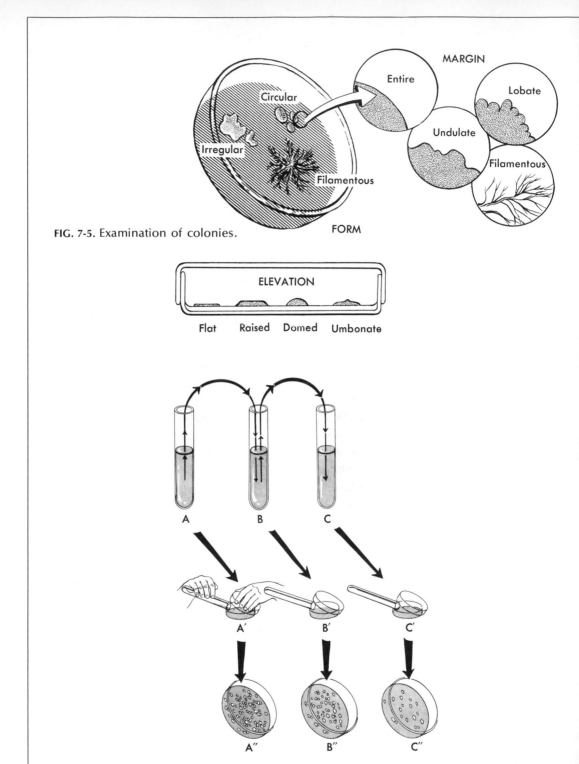

FIG. 7-5. Examination of colonies.

FIG. 7-6. Pure culture, pour plate method (see text). *Step I:* Inoculate tubes **A**, **B**, and **C**, as indicated by arrows. *Step II.* Transfer contents of each tube to Petri dishes **A'**, **B'**, and **C'**, respectively. Incubate plates. *Step III.* Inspect plates (**A''**, **B''**, and **C''**) for number and distribution of colonies. Fish representative colony for further study.

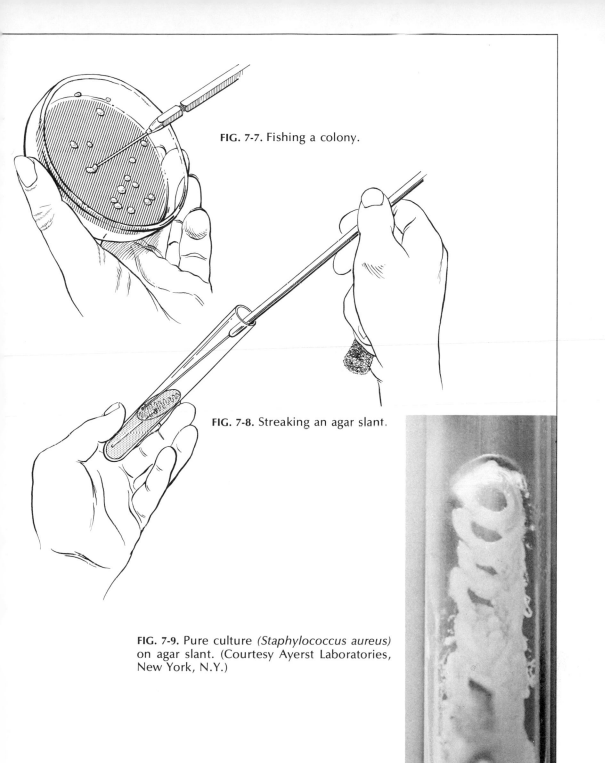

FIG. 7-7. Fishing a colony.

FIG. 7-8. Streaking an agar slant.

FIG. 7-9. Pure culture *(Staphylococcus aureus)* on agar slant. (Courtesy Ayerst Laboratories, New York, N.Y.)

Pure cultures. A *pure culture* contains only one kind of bacteria; a *mixed culture* contains two or more. As a rule, infectious material contains more than one kind of bacteria. So that one kind alone can be studied, it must be separated from all others and grown alone as a pure culture. Pure cultures are usually obtained by the pour plate or streak plate method. The pour plate method is as follows (Fig. 7-6):

1. Melt three tubes of agar in boiling water and then allow to cool to 40° or 42° C. If an enriching material that is injured by heat, such as serum or whole blood, is needed, add it to the medium at this time.
2. Transfer to one tube a loopful of test material from which you expect to obtain a pure culture. Replace the cotton plug and roll the tube between the palms of the hands to distribute the bacteria throughout the medium.
3. Flame-sterilize the loop thoroughly. Transfer 3 loopfuls of the contents of the inoculated tube to the second tube.
4. Mix the contents of the second tube with the inoculum. Sterilize the platinum loop. Transfer 5 loopfuls of the mixture to the third tube and mix.
5. Pour the contents of each tube of medium into a Petri dish and allow to solidify.
6. Incubate the Petri dishes for 24 to 48 hours. Observe the colonies carefully.

Successive dilution of the specimen in the three tubes reduces the number of bacteria and disperses them in the medium so that the colonies in the Petri dish cultures are more likely to be clearly separated from each other. Those that are to be studied further are removed from the Petri dish with a straight platinum wire and rubbed over the surface of one or more slants of suitable media. This is known as *fishing* (Figs. 7-7 and 7-8). If the colonies are not separated, it is impossible to fish one colony without touching others.

The inoculated slants are allowed to incubate for 24 to 48 hours, and these cultures are studied further (Fig. 7-9). In the majority of cases all bacteria growing on a slant will be alike because they all grew from the members of a single colony on the plate, and these in turn grew from a single bacterium in the original material. If, as may sometimes happen, the original colony on the plate contains two or more kinds of bacteria, the same two or more kinds of bacteria will grow on the slant. Separation is made by suspending some of the bacteria from the slant in sterile salt solution and replating.

Streak plates. To prepare a streak plate, a single loopful of infectious material is streaked over the surface of the solid medium (agar) in a Petri dish. There are several ways to do this. The patterns illustrated in Figs. 7-10 and 7-11 are widely used and give good separation of colonies.

Bacterial plate count. The accuracy of the bacterial plate count (Fig. 7-12) rests on the premise that when material containing bacteria is cultured, every bacterium present develops into a colony. This statement is not strictly true because some bacteria may fail to multiply, and two or more bacteria may cling together to form a single colony. Nevertheless, the method is of decided value in the examination of water and milk (pp. 656 and 665).

A bacterial plate count is carried out as follows. Four tubes containing 9 ml of sterile distilled water are needed. To the first tube, add 1 ml of the material to be ex-

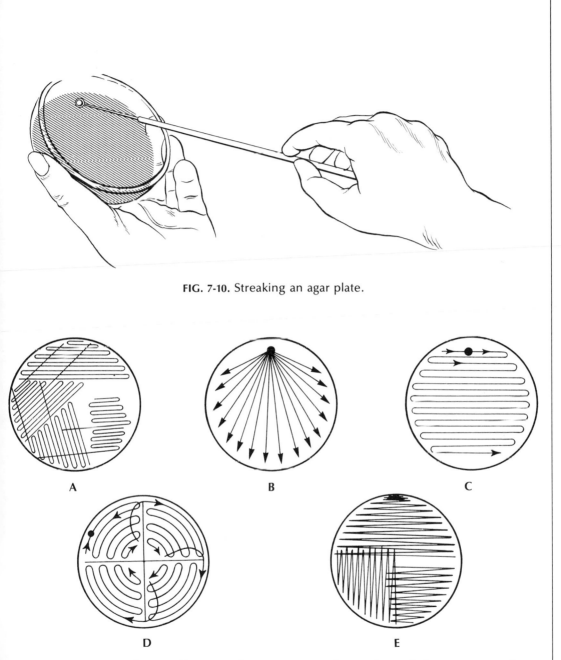

FIG. 7-10. Streaking an agar plate.

A

B

C

D

E

FIG. 7-11. Patterns for streaking agar plates. In **A, D,** and **E,** sections of plate are streaked successively. After each section is streaked, platinum loop is flame sterilized. A bit of inoculum for next section in turn is obtained when loop is passed back into or through section just inoculated. (From Smith, A. L.: Microbiology: laboratory manual and workbook, ed. 4, St. Louis, 1977, The C. V. Mosby Co.)

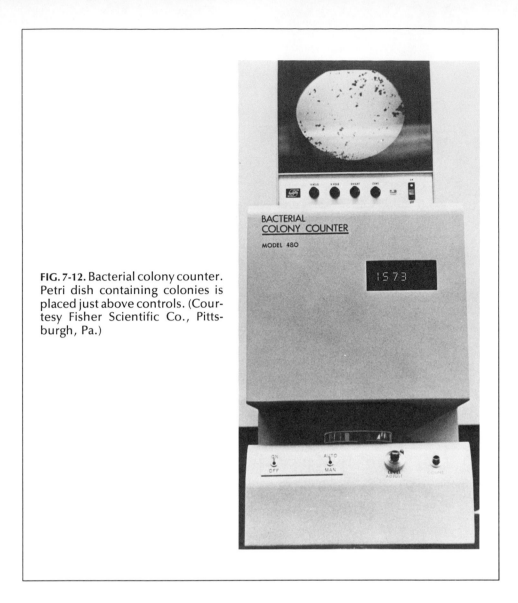

FIG. 7-12. Bacterial colony counter. Petri dish containing colonies is placed just above controls. (Courtesy Fisher Scientific Co., Pittsburgh, Pa.)

amined and mix. From this tube, transfer 1 ml to the second tube and mix. From the second tube, transfer 1 ml to the third tube and mix, and from the third tube, transfer 1 ml to the fourth tube and mix. This gives 1:10, 1:100, 1:1000, and 1:10,000 dilutions of the original material. With a sterile pipette transfer exactly 1 ml from each tube to a Petri dish and add sufficient melted agar (cooled to 42° C). Mix the contents of each dish by rotating and allow to solidify. After incubation for 24 to 48 hours, count the colonies that have developed. Let us say that the second plate shows 200 colonies. This means that at least 200 bacteria were present in the 1 ml of material placed in the tube, and since this was a 1:100 dilution, the original test material must have contained at least 20,000 bacteria/ml. More bacteria may have been introduced into the dish, since some may have failed to grow and some may have clung together in groups of two or more that developed into single colonies. In other words, the number of bacteria indicated by the count is certainly present, and more may be also.

The bacterial plate count is an important part of the microbiologic evaluation of urine from a patient with possible urinary tract infection. A measured amount of urine is handled as indicated above. A colony count is made from the bacterial growth on solid media. Quantitation helps to indicate whether bacteria recovered from urine are disease producers or contaminants. Generally, with less than 1000 to 10,000 colonies from 1 ml of specimen the microorganisms isolated represent normal flora of the urethra or contaminants.

Culture of anaerobic bacteria. To grow some anaerobic bacteria in a routine laboratory, a tube of suitable culture medium about 8 inches in length and about half full of medium may be heated in a boiling water bath for several minutes to expel oxygen present in the medium. The tube is quickly cooled and inoculated; then sterile melted petroleum jelly to make a layer about ¾ inch thick is poured onto the top of the culture, thus sealing it off from the air. If agar is used, inoculation is made when the heated medium cools to a temperature of 42° C. Thioglycollate broth, a special medium containing thioglycollic acid, supports the growth of anaerobes within the depths of a partly filled tube without special seal. Methylene blue indicator colors the upper layers of the fluid medium as oxidation occurs.

Cultures may be made in the usual way and may be placed in a specially constructed glass or metal chamber from which the oxygen either is removed by some chemical reaction that is made to take place in the chamber (Figs. 19-3 and 22-2) or replaced by a gas such as hydrogen that does not affect the growth of the bacteria. Better results are obtained with the more sophisticated methods indicated on p. 402.

Slide cultures. Slide cultures (p. 559), used to study fungi, may be placed on the stage of the microscope and actual growth observed from time to time. The depression of a sterile culture slide is filled with suitable medium, inoculated, and covered with a sterile cover glass. Liquid cultures may be made by inoculating a drop of medium on a sterile cover glass and proceeding as for a hanging drop preparation.

Cultures in embryonated hen's egg. Important sites of growth for viruses and rickettsias, including chlamydiae, are the yolk sac and embryonic membranes of the developing chick embryo (p. 499). Bacteria other than rickettsias are occasionally grown in this way.

Cultivation of microbes in cell cultures. Cell (tissue) cultures are composed of animal cells in a suitable substrate in which they multiply and grow. The cells may be supported on a solid or semisolid substrate such as fibrin, agar, and cellulose, or they may be suspended in a liquid. The cells used are taken from mammalian or fowl embryos or selected adult tissues such as the cornea, kidney, and lung, and minced tissue of the cell source is inoculated. Pathogenic agents such as rickettsias (also chlamydiae) and viruses that do not grow in lifeless media multiply with ease when incorporated into cell (tissue) cultures.

ANTIBIOTIC SUSCEPTIBILITY TESTING

The susceptibility (or resistance) of bacteria to different antibiotics may be determined by suitable laboratory procedures usually designated as susceptibility tests.

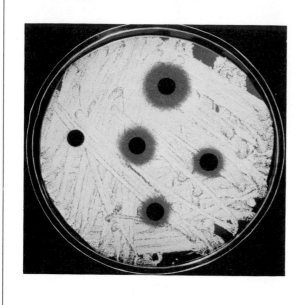

FIG. 7-13. Antibiotic susceptibility testing: single paper disk diffusion method. Petri dish of agar is heavily inoculated with test organisms. Disks of filter paper impregnated with different antibiotics are dropped onto freshly seeded surface. Note heavy white bacterial growth with zones of inhibition about four disks and none about one. A zone of inhibition indicates that growth of organisms would probably be limited in a patient receiving that antibiotic.

Two tests are used: (1) the disk susceptibility test (Fig. 7-13) and (2) the tube-dilution method. The first employs disks of filter paper impregnated with precise amounts of different antibiotics. A Petri dish is heavily inoculated with test organisms. The disks are dropped onto the freshly inoculated surface, and the plate is incubated for 24 hours. Meanwhile, the antibiotic diffuses out in the culture medium. If the organism is susceptible, a zone of inhibition (no growth) encircles the disk. If the microorganism is unaffected by the drug, growth will cover the area around the disk. Several preparations can be tested on one plate, and the method is simple and rapid. Results correlate well with the effectiveness of clinical treatment.

The United States Food and Drug Administration has set its stamp of approval on the standardized disk susceptibility test as precisely defined step by step in the *Federal Register* (**37**:20527-20529, Sept. 30, 1972), which is a modification of the well-known Kirby-Bauer method introduced in 1966. In the official test, Mueller-Hinton agar (beef infusion, peptone, starch, and agar at a pH of around 7.4) is used for culture of test microbes since it is low in interfering substances and it supports the growth of most pathogens for which evaluation is needed.

The second method of susceptibility testing utilizes a tube of culture medium containing both test organisms and a specific amount of given antibiotic. The tube is incubated and growth quantitated. Since only one drug can be tested at a time, this method is laborious.

With this sort of testing, microbes are reported as being susceptible or resistant to a given antibiotic. There are also technics whereby the amount of certain antibiotics can be determined in the blood. Automated methods of susceptibility testing are recently available with claims of speeding up the process to approximately 3 hours while maintaining good correlation with the Kirby-Bauer method.

QUESTIONS FOR REVIEW

1. Name important ingredients of culture media.
2. Why is agar the most commonly used solidifying agent?
3. What purposes do dyes serve in culture media?
4. What advantages are there to culture by comparison with other methods for studying microbes?
5. Explain differential media and selective media. Give examples.
6. Give the practicality of the Petri dish in microbiologic culture.
7. Outline the steps in making a pure culture. What is meant by a pure culture? What purpose does it serve?
8. Briefly describe how and under what circumstances a bacterial count is made.
9. Briefly explain slide culture, culture on synthetic media, culture of anaerobic bacteria, cell culture, fishing a colony, streaking, plating, inoculation of a culture.
10. How is antibiotic susceptibility testing carried out? What is its purpose?

REFERENCES. See at end of Chapter 9.

8 Laboratory identification

BIOCHEMICAL REACTIONS

Biochemical tests demonstrate the presence of enzyme systems within the microbial cell, such as those responsible for fermentation of carbohydrates and decomposition of proteins.

Miniaturization of microbiologic technics refers to the use of commercially prepackaged test units wherein basic reagents for a given biochemical reaction are premeasured, standardized, and compacted into a tablet or onto a paper disk or filter paper strip. The test unit is applied to cultures in liquid or solid media and is chemically designed so that enzymatic action of the test organism usually results in a color change or end point, a quick and easy observation. There are many varieties of such tests available.

Fermentation of sugars. Microbes ferment many organic compounds, including carbohydrates, generating in the process energy-rich chemical bonds. Through fermentation simple sugars can serve as the main source of energy for many different kinds of microbes, and in the laboratory it is practical to deal with these relatively simple test solutions. The end products of fermentation depend on the substrate, enzymes present, and conditions under which the reaction proceeds. For instance, the yeast enzymes that break down glucose into carbon dioxide and alcohol are without effect on sucrose (cane sugar), and certain enzymes produced by pneumococci and other streptococci change glucose to lactic acid. Common products of bacterial fermentation are lactic acid, formic acid, acetic acid, butyric acid, butyl alcohol, acetone, ethyl alcohol, and the gases carbon dioxide and hydrogen.

Fermentation reactions, though varying among species of microorganisms, are constant and therefore of great value in differentiating the species. From their specific action on a given sugar in the laboratory, bacteria (and other microbes) are classified as (1) those that do not ferment it, (2) those that do with the production of acid only, and (3) those that ferment it with the production of both acid and gas. Sugar-containing media are inoculated with test microbes and observed for gas production and acid formation. Gas production in liquid media is detected by an accumulation of gas that displaces the fluid medium contained in the closed arm of a special tube. It may be detected in a small tube (one end of which is sealed off) placed in an inverted position within a larger tube of liquid, the sealed end projecting slightly above the level of the fluid. As gas is formed in the inoculated liquid, it collects in the small tube within the depths of the culture and rises toward the sealed end (Fig. 8-1).

Solid media for fermentation studies are inoculated by plunging a straight needle carrying some of the test organisms deep into the medium. (This plunging maneuver

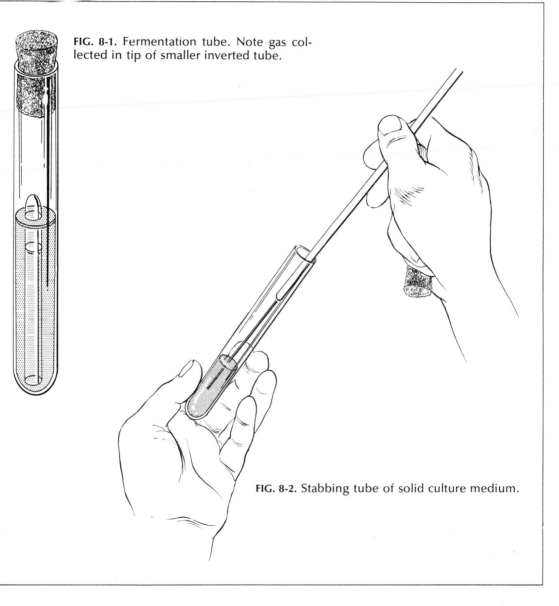

FIG. 8-1. Fermentation tube. Note gas collected in tip of smaller inverted tube.

FIG. 8-2. Stabbing tube of solid culture medium.

is known as "stabbing" [Fig. 8-2].) The solid medium may also be melted, cooled to 40° C, and then inoculated. Bubbles of gas produced disrupt the medium. Acid formation in both liquid and solid media is measured by color changes in indicators incorporated therein.

Many carbohydrates are used in routine fermentation studies. Important sugars are glucose, lactose, maltose, sucrose, xylose, mannitol, and salicin. The fermentation of lactose is crucial in the identification of bacteria recovered from the intestinal tract since it makes a division between the nonpathogenic coliforms and the pathogenic *Salmonella* and *Shigella*, the organisms that cannot ferment lactose (with a few exceptions) being the pathogens.

Hydrolysis of starch. When microbes are grown on starch agar, a clear zone develops around the colonies of those that digest starch. If the agar is covered with a weak solution of iodine, the *undigested* starch assumes a blue color.

Liquefaction of gelatin.* A gelatin medium is stabbed with a wire having some of the bacteria on it, incubated at 20° C, and observed for liquefaction. It may be incubated at 37° C (gelatin remains liquefied at this temperature) and then placed in the refrigerator, where the unliquefied gelatin solidifies. Incubation should be continued for at least 2 weeks unless liquefaction occurs before then, and a tube of uninoculated gelatin should be incubated as a control.

Indole production. Production of indole (from the amino acid tryptophan) is determined by culture of test organisms in a medium containing tryptophan. A strip of filter paper soaked with a saturated solution of oxalic acid may be hung over the culture and held in place either by the cotton plug or the screw cap of the culture tube. A pink color on the paper indicates indole production. Notable indole producers are *Escherichia coli* and *Proteus*, which avidly decompose proteins.

Nitrate reduction. Nitrate reduction means the removal of oxygen from the nitrate radical (NO_3) to convert it into nitrite (NO_2). Test organisms are incubated in broth containing 0.1% potassium nitrate, and the broth is tested for nitrite with sulfanilic acid and alpha-naphthylamine reagents. A red color means a positive test.

Deoxyribonuclease elaboration. Formation of DNase (deoxyribonuclease) is demonstrated by culture of test microbes on the surface of an agar plate into which deoxyribonucleic acid has been incorporated. After 24 hours' incubation, the surface is flooded with either 1N hydrochloric acid to highlight the clear zones about the DNase-positive growth or with 0.1% toluidine blue to display a bright rose-pink end point about that growth. Staphylococci produce this enzyme.

Hydrogen sulfide production. A stab culture in an agar that contains basic lead acetate is incubated for 1 to 4 days. The production of hydrogen sulfide from the sulfur-containing amino acids of the medium is indicated by the appearance of a black com-

---

*The proteolytic enzymes (protein-splitting enzymes, proteinases), produced by bacteria, split complex proteins into proteases, peptones, polypeptides, amino acids, ammonia, and free nitrogen. Protein decomposition is known as *putrefaction.* Some authorities restrict the term *putrefaction* to the decomposition of proteins by anaerobic bacteria, which results in the formation of hydrogen sulfide and other foul-smelling decomposition products, and they use the term *decay* for the decomposition of proteins by aerobic bacteria. The latter does not result in the malodorous decomposition products.

pound, lead sulfide, formed from the combination of the hydrogen sulfide with the lead acetate. Instead of being incorporated into the medium, the basic lead acetate may be impregnated on sterile filter paper suspended over the medium as in the test for indole production. Iron salts also may be used to detect hydrogen sulfide formation. The production of hydrogen sulfide facilitates the identification of *Brucella* species and enteric bacilli.

Splitting of urea. Certain microorganisms when cultivated on media containing urea convert it to ammonia. With phenol red as the indicator, the presence of urease is marked by a red color. The enzyme *urease*, responsible for the conversion, is produced by species of the genus *Proteus*. This test separates them from urease-negative *Salmonella* and *Shigella*.

Digestion of milk. Bacterial growth in sterile milk may be alkaline or acid and with or without curdling. Curdling may or may not be followed by liquefaction of the curd. Excessive gas production in milk produced by *Clostridium perfringens* is "stormy fermentation."

Oxidase reaction. The enzyme *oxidase* is produced by *Neisseria* species, and its detection is of great value in the identification of *Neisseria gonorrhoeae*. The oxidase reagent colors the positive colonies pink to red to black.

Niacin test. The niacin test is used to distinguish *Mycobacterium tuberculosis* from other species of mycobacteria. A 4% alcoholic aniline solution and a watery solution of cyanogen bromide are added to 1 or 2 ml of emulsified bacterial growth. A complex yellow compound is formed when niacin or nicotinic acid, formed by the tubercle bacillus but not by other mycobacteria, reacts with the cyanogen bromide and a primary amine.

Demonstration of specific enzymes. Reagent systems for the identification of diagnostic enzymes are commercially impregnated onto easy-to-use strips of paper suitable for application to bacterial cultures. Three enzymes of practical importance are so packaged: phenylalanine deaminase, cytochrome oxidase, and lysine decarboxylase.

Phenylalanine deaminase catalyzes the metabolism of phenylalanine to phenylpyruvic acid, which in turn reacts with ferric ions present to give a brownish or gray-black color. The presence of this enzyme in cultures of *Proteus* species is responsible for the positive reaction obtained with them.

$$\text{Phenylalanine} \xrightarrow{\text{Enzyme}} \text{Phenylpyruvic acid} + \text{Ferric ions} \longrightarrow \text{Color}$$

Cytochrome oxidase is produced by *Pseudomonas, Alcaligenes, Neisseria, Vibrio, Brucella,* and some *Halobacterium* species. This enzyme catalyzes a coupling reaction to give a blue color (the positive reaction) as follows:

$$\text{Dimethyl-}p\text{-phenylenediamine} + \text{Alpha naphthol} + \text{Oxygen} \xrightarrow{\text{Enzyme}} \text{Indophenol blue}$$

Lysine decarboxylase, which is formed by most of the *Salmonella* species, catalyzes the conversion of lysine to cadaverine, a more alkaline compound than lysine. In one commercial prepackaged test kit, the Prussian blue reaction is used to indicate the positive test.

$$\text{Lysine} \xrightarrow{\text{Enzyme}} \text{Cadaverine} + CO_2 \longrightarrow \text{Bromthymol blue (yellow) changing to blue}$$
$$\text{as pH rises}$$

Gas chromatography. Gas chromatography can be used to advantage in microbiology because bacteria (as a dry pellet) subjected to *pyrolysis,* that is, to heat-induced chemical degradation, break down into specific products. Conditions are rigidly controlled, and specially designed equipment is used. The gas chromatograph sorts and separates the pyrolytic fragments, then generates and records on a graph the precise pattern for the test organisms. Closely related bacteria may also be separated in a comparable way by gas-*liquid* chromatography; this time the diagnostic data are generated from *hydrolysis* of the sample.

Limulus test for endotoxemia. Serum or cerebrospinal fluid from a patient (or suitable test material) is incubated with lysates derived from the blood cells (amebocytes) of the horseshoe crab*(Limulus polyphemus).* In the presence of even minute amounts of bacterial endotoxin there is a greatly increased viscosity of the test medium (the gelation reaction), which can progress to a solid gel.

ANIMAL INOCULATION

The inoculation of suitable laboratory animals is an essential part of the study of many microbes. After an animal has been inoculated, it is observed for effects produced by the microbes. In some cases it is killed after a certain length of time and examined for evidence of disease. Smears and cultures are made, and gross and microscopic changes in the organs noted. In other cases the animal is not killed, but blood and body fluids are examined at intervals.

The well-known laboratory animals are guinea pigs, white mice, white rats, hamsters, and rabbits (Figs. 8-3 and 8-4). The inoculations may be given with syringe and needle either subcutaneously (beneath the skin), intradermally (between the layers of the skin), intravenously (into a vein), intraperitoneally (into the peritoneal cavity), subdurally (beneath the dura of the brain), or intracerebrally (into the brain).

Advantages. The advantages of animal inoculation in recovery and identification of microbes are as follows:

1. Some microbes are most easily detected by animal inoculation (examples, *Francisella tularensis* and, under some conditions, *Mycobacterium tuberculosis*).
2. The virulence of microbes can be exhibited (such as *Corynebacterium diphtheriae*).
3. It is sometimes the easiest way of obtaining the pure culture *(Streptococcus pneumoniae).*
4. It is often necessary in order to determine the action of drugs on microbes.
5. Microorganisms that cannot be cultured at all on artificial media can be readily recovered from suitable laboratory animals (spiral bacteria, rickettsias, and most viruses). A very few microbes can be propagated in no other way.

---

*The horseshoe crab is not a crab at all and not even closely related to any other living thing in the sea. Its blood is blue from the copper-containing molecule therein that carries oxygen.

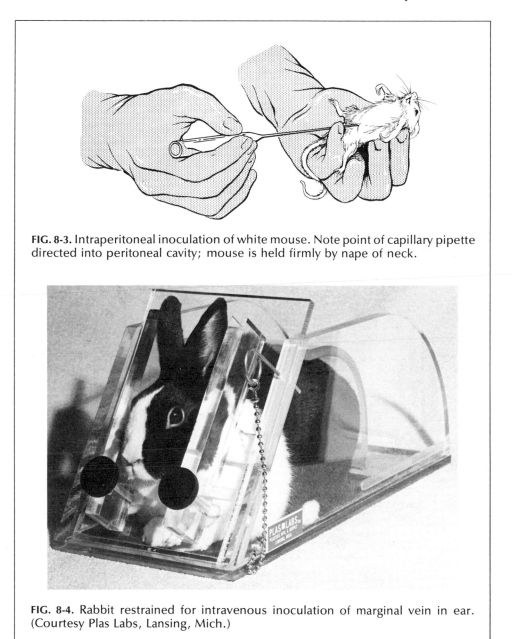

FIG. 8-3. Intraperitoneal inoculation of white mouse. Note point of capillary pipette directed into peritoneal cavity; mouse is held firmly by nape of neck.

FIG. 8-4. Rabbit restrained for intravenous inoculation of marginal vein in ear. (Courtesy Plas Labs, Lansing, Mich.)

6. The action of pathogenic and nonpathogenic agents on the animal body can be determined experimentally.
7. Animal inoculation is basic to the manufacture of antitoxins and other antiserums.

A good example of animal inoculation as the best method of detecting microbes is the ease with which tuberculosis of the kidney is proved in a patient whose urine shows no bacteria on direct microscopic examination but whose urine produces classic disease in a guinea pig. A good example of animal inoculation being used to obtain a pure culture is seen in pneumococcal typing. Some of the sputum in which the type of pneumococcus is to be determined is injected into the peritoneal cavity of a white mouse, and within only a few hours the growth of pneumococci has outstripped that of all other organisms to such an extent that the peritoneal contents consist of practically a pure culture of pneumococci ready to be typed.

SUMMARY: IDENTIFICATION OF BACTERIA

When the various methods are applied to the study of bacteria, it will be seen that classification into comparatively large groups may be made from several standpoints, as follows:

I. Direct microscopic examination
   A. Staining reactions
      1. Reaction to Gram stain
         a. Gram positive
         b. Gram negative
      2. Reaction to acid-fast stain
         a. Acid fast
         b. Nonacid fast
   B. Size
   C. Shape
      1. Coccus—spherical
      2. Bacillus—rod shaped
      3. Vibrio, spirillum, spirochete— spiral shaped
   D. Presence of spores
      1. Sporeformers
      2. Nonsporeformers
   E. Capsule formation
      1. Encapsulated
      2. Nonencapsulated
   F. Motility
      1. Motile
      2. Nonmotile
II. Culture
   A. Food requirements
      1. Saprophytes—growth on dead organic matter
      2. Parasites—growth on living matter
   B. Media most suitable for growth
      1. Growth on simple media
      2. Growth only on special media
      3. No growth on any media
   C. Appearance of growth on different media
   D. Oxygen requirements
      1. Aerobes—growth only in the presence of free oxygen
      2. Anaerobes—growth only in the absence of free oxygen
      3. Microaerophiles—growth in the presence of small amounts of free oxygen
   E. Optimum temperature for growth
   F. Characteristics of the pure culture (Fig. 7-5)
      1. Size, shape, texture of colonies
      2. Pigment production
   G. Production of hemolysins on blood agar
      1. Beta hemolytic—complete hemolysis of red blood cells
      2. Alpha hemolytic—partial hemolysis of red blood cells
      3. Nonhemolytic
III. Biochemical reactions
   A. Fermentation of sugars
      1. Fermentation with acid and gas
      2. Fermentation with acid only; no gas
      3. No fermentation
   B. Fermentation of lactose
      1. Lactose fermenters
      2. Nonlactose fermenters
   C. Splitting of urea
      1. Urease positive
      2. Urease negative
IV. Animal inoculation
   A. Disease production (virulence tests)
      1. Pathogenic—disease produced
      2. Nonpathogenic
   B. Toxin production
      1. Exotoxin producers
      2. Endotoxin producers

QUESTIONS FOR REVIEW

1. How are fermentation reactions used in the study of microbes?
2. List five sugars important in routine fermentation studies.
3. What two observable changes may occur in the culture medium when carbohydrates are fermented?
4. What is the importance of the fermentation of lactose?
5. List the routine laboratory animals.
6. State the advantages of animal inoculation.
7. What is meant by putrefaction? Decay?
8. Name two organisms that produce indole.
9. What genus produces urease?
10. Briefly outline the niacin test.
11. How may specific enzymes be demonstrated? Name three of diagnostic value.
12. Summarize methods used to classify and categorize bacteria.

REFERENCES. See at end of Chapter 9.

9 Specimen collection

In that it defines the diagnosis and nature of disease, the microbiologic investigation determines the mode of treatment and outlook for that patient affected. In the practice of scientific medicine, specimen collection must be a crucial first step on which rests the validity of each succeeding step.

GROUND RULES

In the collection of any specimen for microbiologic examination, remember the following:

1. Collect the specimen from the actual site of disease and do not contaminate it with microbes from nearby areas. For example, in making smears and cultures from ulcers in the throat, be very careful to take the specimen from the actual site of ulceration and not contaminate it unduly with secretions of the mouth.

2. Always use sterile paraphernalia to collect the specimen. Place it in a sterile container.

3. Sterilize material used in collecting the specimen as soon as possible after proper disposal of the specimen.

4. Collect adequate amounts of material. For instance, when pus is to be examined, collect several milliliters, if possible. If it is necessary to collect the material on swabs, use more than one swab.

5. Insofar as possible, do not send specimens to the laboratory on swabs (a swab can take up a small amount of material, dry quickly, and enmesh in its fibers important cells and organisms that fail to be transferred to smears or culture media).

   Note: If swabs must be used, specially devised transport media are available.

6. Collect the specimen in such a manner so as not to endanger others. Sputum or other excreta must not soil the outside of the container.

7. Take great care in handling specimens collected in cotton-plugged tubes. Cotton plugs can soak up a small specimen. Microbes from the environment pass through wet plugs and contaminate the specimen.
8. Whenever possible, make smears from the original material.
9. Do not add any preservative or antiseptic to the specimen. If possible, take the specimen *before* the patient has received any antimicrobial drug or *before* his wound has had local treatment. If the patient has already received such a drug, notify the laboratory.
10. Label the specimen properly with the patient's name, the source of the specimen, and the tentative diagnosis.
11. Deliver the specimen to the laboratory promptly.

In certain instances, the person delivering the specimen to the laboratory does so at a time when laboratory personnel are off duty or the laboratory is unmanned. If so, he may have to care for that specimen himself after he arrives. In this event, specimens already inoculated onto culture media are placed in the incubator and those not on culture media are placed in the refrigerator. When cultures must be made without delay, the medium chosen should be that most likely to grow the suspected organisms, as well as any other that might be present. Since it supports the growth of microorganisms better than any other culture medium found in the usual microbiologic laboratory, blood agar is best to use. Blood agar plates are superior to the tubed media, except in those instances in which the cultures must be shipped a distance to the laboratory. Plates are not easily transported.

CLINICAL SPECIMENS

Urine for microbiologic examination may be carefully collected as a clean-voided specimen. The collection of a urine specimen with a sterile catheter has long been recommended, especially in the female, but many physicians now feel that catheterization is unnecessary. No matter how carefully done, the maneuver extends infectious organisms up into the urinary tract. For most purposes a distinction can be made between infectious and contaminating microorganisms in a *carefully collected,* clean-voided specimen. The meatus (opening) of the urethra is gently cleaned with soap (or detergent) and water. In both voided and catheterized specimens the first portion should be discarded, and the last portion should be received in a sterile container. A "clean catch" urine specimen is one obtained during the midpart or toward the end of the act of voiding; it is received into a sterile receptacle. The sample should be promptly refrigerated or a suitable preservative added if laboratory examination is delayed. The urine in health is free of microorganisms.

Blood for cultures must be collected with special care because microbes, especially staphylococci, are present on the surface of the skin and within its superficial layers. A blood culture is taken by venipuncture (Figs. 9-1 and 9-2) as follows:

1. Paint the skin over the veins in the bend of the elbow with tincture of iodine or other suitable disinfectant (70% alcohol may be used).
   Note: Some persons are allergic to iodine!

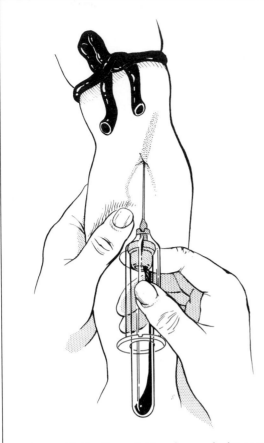

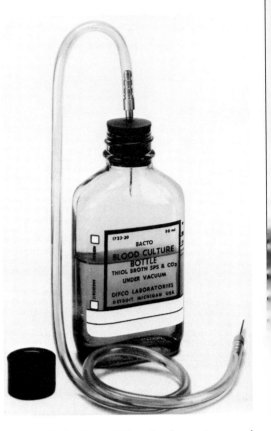

FIG. 9-1. Collection of blood sample by venipuncture. Note use of evacuated blood collection tube in plastic holder. (Syringe can also be used.)

FIG. 9-2. Collection of blood culture. Prepared medium with sterile connections for venipuncture under vacuum is used. (Courtesy Difco Laboratories, Detroit, Mich.)

2. Remove disinfectant with 70% alcohol (or make a second application of 70% alcohol).

3. Place a 70% alcohol compress on the area.

4. Secure a tourniquet, not too tightly, around the arm just above the elbow, and ask the patient to close and open his hand several times.

5. Puncture a prominent vein with a 20- or 21-gauge needle to which may be attached either a 20 ml syringe or Vacutainer, a specially designed holder for a vacuum tube (Fig. 9-1). Remove 10 to 15 ml of blood (or whatever amount is indicated for the particular laboratory test).

6. Add the blood directly to the culture medium or place it in a sterile bottle containing sodium citrate to prevent coagulation and carry it to the laboratory where it can be distributed to suitable culture media. To prevent contamination, remove the needle from the syringe before expelling the blood.

Bacteremia is often of short duration. If the blood culture is not collected at the proper time, the etiologic organism of the disease may not be found. Since the advent of the antimicrobial drugs, positive blood cultures are fewer. Even one dose of an antimicrobial drug before the culture is drawn may mask the infection. The bloodstream in health is free of microorganisms.

Blood for serologic examinations (such as the precipitation tests for syphilis and various agglutination tests for the continued fevers) may be collected from a vein in the bend of the elbow. The technic is the same as that for collecting blood for culture, except that (1) the preliminary sterilization consists of scrubbing the area with 70% alcohol only and (2) the blood is placed in a chemically clean test tube and allowed to clot. Five milliliters is usually sufficient.

Note: Hepatitis can be conveyed to the person from whom the blood is drawn by the use of a clean *but unsterile* syringe. Use sterile *disposable needles* and preferably sterile *disposable syringes* to prevent hepatitis. (Otherwise syringes and needles must be carefully autoclaved.)

Sputum is often collected improperly, and samples tend to be too small. Many specimens consist of secretions from the nose, mouth, and throat and contain no sputum at all. Sputum should represent a true pulmonary secretion and should be expelled after deep coughing. Instruct the patient, if need be, as to how to bring the specimen up from the lower part of the respiratory tract. If necessary, nebulized aerosols can be used. The teeth should be scrubbed with toothpaste and the mouth rinsed with sterile water before the specimen is taken. Usually more sputum is raised in the morning, and this sputum may contain *Mycobacterium tuberculosis*, when specimens taken later in the day do not. Generally, sputum specimens for bacteriologic study are collected daily for at least 3 separate days. The ideal container for sputum is a 6-ounce widemouthed bottle (or plastic receptacle) with a tight-fitting stopper or cover.

Children swallow their sputum, and adults do also in their sleep. In children the only way to collect sputum is to aspirate stomach contents. When it is impossible to obtain a satisfactory cough specimen of sputum in adults, the stomach contents may be examined. The stomach should be aspirated early in the morning before any food or water is taken.

Bronchial secretions are secured by means of the bronchoscope and may be subjected to microbiologic study. After bronchoscopy, sputum specimens may be collected again for a 3-day period.

Cultures from the nose and throat are made many times, since sore throat and nasal discharge are common disorders and often part of systemic illness. To take the culture just after an antiseptic has been used is to defeat its purpose because the antiseptic retards the growth of bacteria present. (This applies to other cultures as well as to throat cultures.) When a throat culture is taken, use a good light, and insofar as possible allow the swab to contact only the diseased area. Use at least two swabs. When cultures are to be taken from the nose, pass a small tightly wound swab directly back through the nose. Avoid large loose swabs because they may slip off the applicator and lodge in the nose.

When diphtheria is suspected, cultures made from both nose and throat, instead of from the throat alone, improve the detection rate. The microbes causing diphtheria are often found abundantly in cultures when they have not been seen on direct smears. Although this is true, smears should always be made when diphtheria-like lesions of the throat are encountered, since the exudate in lesions of Vincent's angina, the etiologic agents of which are detected only in smears, often closely resembles the membrane of diphtheria.

Cultures from the nasopharynx (the upper part of the throat behind the soft palate) are needed for the detection of meningococcal carriers and are used in infants and children suspected of having whooping cough. To collect specimens from this area, pass a thin swab (made of Dacron, cotton, or calcium alginate) through the nose into the nasopharyngeal area, gently rotate it, remove, and place in a suitable transport medium. Expedite transport to the laboratory. A satisfactory swab may be made by wrapping cotton around the end of a piece of wire bent at a right angle about 1 inch from the end. When specimens from the throat and nasopharynx are to be obtained, take care to avoid contamination with saliva.

Feces for routine microbiologic examination may be collected directly by the patient into a sterilized cardboard carton or other suitable container. At times it may be desirable for the patient to pass a stool into a larger, previously sterilized container, with a sterile tongue blade or spoon being used to remove a small amount to a sterile widemouthed bottle. The stool in the latter case should be examined superficially. Portions of the fecal material with mucus are preferable for microbiologic study.

Rectal swabs are very satisfactory for most microbiologic purposes and are easily obtained from both adults and children. If a disease process involves the lining of the terminal portion of the rectum, such swabs are likely to obtain material from within the focus of disease and are therefore more likely to contain disease-producing agents than the stool of that person. Swabs are of value in testing for carriers of bacteria causing typhoid fever, dysentery, or cholera. To obtain the rectal swab, clean the skin about the anus with soap, water, and 70% alcohol. Introduce a sterile cotton swab moistened with either sterile isotonic solution or sterile broth through the anus and gently rotate it about the circumference of the lower rectum to contact a large area of the mucosal lining. A fecal specimen may also be obtained from the physician's glove after digital examination of the rectum has been made.

In all cases the stool specimen should be sent to the laboratory at once because nonpathogenic intestinal bacteria quickly overgrow the pathogenic ones, and pathogenic amebas rapidly lose their motility if the fecal material containing them cools. If acute amebiasis is suspected, the warm feces should be examined immediately for pathogenic amebas showing their typical and diagnostic movement. If some delay is anticipated, receive the specimen into a vessel that has been warmed. A portion may then be placed in a small tightly corked bottle, which may be placed in a fruit jar that has been filled with water just a few degrees above body temperature. This will keep the specimen warm for a considerable time so that it can be speeded to the nearest laboratory. In the hospital the specimen may be rushed to the laboratory without this special preparation. Examination for typhoid bacilli is facilitated by collecting the feces in a special brilliant green bile medium because bile facilitates the growth of typhoid bacilli, and brilliant green retards the growth of many other intestinal organisms.

Pus from abscesses and boils may be obtained after drainage. The abscess or boil is painted with suitable disinfectant, the area is allowed to dry, and an incision is made with a sterile scalpel. Some of the contents are obtained by means of sterile swabs. If the lesion is opened widely, as much of the superficial portion should be removed as possible and the specimen taken from the deeper part. When this is not done, the specimen is usually contaminated with surface microorganisms. Protect all specimens taken on swabs from drying. This is best done by placing the swabs in a sterile test tube containing either a drop or two of sterile physiologic salt solution or a nutrient broth that may be directly inoculated with the swabs. The swabs, which are longer than the test tube, should be placed in the test tube in such a manner that the cotton pledget is just above the salt solution but does not touch it. A cotton plug is inserted to hold the swabs in place.

Smears for gonococci (see also p. 355) in the pubertal or adult female should be taken from the meatus of the urethra, the cervix uteri, and the rectum. In prepubertal and younger girls suspected of having gonorrheal vulvovaginitis, the smears and cultures are made from the vagina because it is the vagina that is primarily attacked. In acute gonorrhea in the male the smears are obtained from the urethra. In chronic cases in the male a specimen may be obtained from the physician's massage of the prostate gland and seminal vesicles.

Cerebrospinal fluid is obtained by a procedure known as *lumbar puncture* (Fig. 9-3). It is carried out by the physician as follows. After the overlying skin of the lower back has been disinfected and anesthetized, a special needle with stylet is introduced slightly to one side of the midline between the third and fourth lumbar vertebrae and passed into the spinal subarachnoid space. The strictest asepsis must be observed and a sterile dressing placed over the site of puncture. An infant or child seated with head and arms resting on a folded sheet, blanket, or pillow set against the abdomen may be held securely while the physician performs the lumbar puncture.

Cerebrospinal fluid may be obtained from the cisterna magna at the base of the brain by the skilled neurosurgeon. However, *cisternal puncture* is infrequent because of the danger at this location that the penetrating needle might impale the brainstem or cervical cord. The cerebrospinal fluid is normally sterile.

Pleural and peritoneal fluids are obtained through the lumen of a trochar or specially designed needle inserted through the wall of the chest or abdomen. With the trochar in place, fluid is aspirated with a syringe or freely drained into a sterile container. The strictest asepsis should be observed during this procedure, which is done by the physician.

Smears and cultures from the conjunctiva are made with swabs wet with sterile physiologic solution. The swabs should be handled carefully to prevent injury and spread of infection to adjacent parts of the eye or to the other eye.

Specimens for anaerobic culture are collected as carefully as possible: first, to avoid contamination with the abundance of anaerobes usually found in the normal

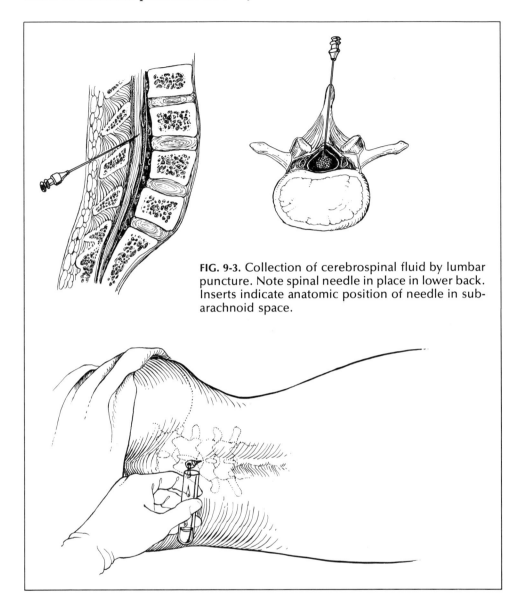

FIG. 9-3. Collection of cerebrospinal fluid by lumbar puncture. Note spinal needle in place in lower back. Inserts indicate anatomic position of needle in subarachnoid space.

flora of a given area, and second, to minimize the exposure of the specimen to the oxygen of the air. Direct aspiration (with needle and syringe) of a given focus of disease gives a workable preparation. For example, in pulmonary infection, direct lung puncture or transtracheal aspiration is done to obtain pulmonary secretions or tissue, and in urinary tract infection, percutaneous suprapubic bladder puncture is done to obtain urine.

Specimens for the viral and rickettsial laboratory, such as blood, cerebrospinal fluid, respiratory secretions, feces, and urine, should be collected with the strictest aseptic technic and placed in sterile containers. Further contamination of throat

TABLE 9-1. THE SPECIMEN AND ITS PATHOGENS

Specimen	Important pathogens
Blood	Anaerobic cocci
	Bacteroides species and related anaerobes
	Brucella
	Chlamydia psittaci
	Coxiella burnetii
	Escherichia coli and other coliforms
	Francisella tularensis
	Haemophilus influenzae
	Hepatitis viruses
	Leptospira
	Neisseria gonorrhoeae
	Neisseria meningitidis
	Opportunistic fungi
	Plasmodium
	Proteus
	Pseudomonas
	Rickettsias of spotted fevers
	Rubeola and rubella viruses
	Salmonella species
	Spirillum minor
	Staphylococcus aureus
	Streptobacillus moniliformis
	Streptococcus pneumoniae
	Streptococcus pyogenes and other streptococci
Urine	*Brucella*
	Candida albicans
	Escherichia coli and other coliforms
	Measles viruses
	Mumps virus
	Mycobacterium tuberculosis
	Proteus
	Pseudomonas aeruginosa
	Salmonella
	Staphylococcus aureus
	Streptococcus pyogenes and other streptococci
Nose and throat secretions	Adenoviruses
	Bordetella pertussis
	Chlamydia psittaci
	Coliforms
	Corynebacterium diphtheriae
	Enteroviruses
	Fungi
	Measles viruses
	Mumps virus (saliva)
	Mycobacteria
	Mycoplasma pneumoniae
	Myxoviruses
	Neisseria species

Continued.

TABLE 9-1. THE SPECIMEN AND ITS PATHOGENS — cont'd

Specimen	Important pathogens
Nose and throat secretions — cont'd	Predominance of *Haemophilus influenzae,* *Staphylococcus aureus,* or *Candida albicans* Respiratory viruses *Staphylococcus aureus* *Streptococcus pyogenes* and other streptococci
Sputum	*Actinomyces israelii* Anaerobic streptococci *Blastomyces dermatitidis* *Candida albicans* *Coccidioides immitis* *Cryptococcus neoformans* *Histoplasma capsulatum* *Klebsiella pneumoniae* *Mycobacterium tuberculosis* and other mycobacteria *Nocardia asteroides* *Staphylococcus aureus* *Streptococcus pneumoniae* *Streptococcus pyogenes*
Stool (feces)	*Candida albicans* Enteroviruses Hepatitis virus Metazoa (*Taenia,* etc.) *Proteus* Protozoa (*Entamoeba histolytica,* etc.) *Pseudomonas* *Salmonella* species *Shigella* species *Staphylococcus aureus* *Vibrio cholerae*
Cerebrospinal fluid	*Bacteroides* species Coliforms *Cryptococcus neoformans* Enteroviruses *Haemophilus influenzae,* type b *Listeria monocytogenes* *Mycobacterium tuberculosis* *Neisseria meningitidis* *Staphylococcus aureus* *Streptococcus pneumoniae* *Streptococcus pyogenes* Togaviruses (arboviruses)
Fluid from conjunctiva of eye	Adenoviruses *Chlamydia trachomatis* *Haemophilus* *Moraxella lacunata* *Neisseria gonorrhoeae* *Staphylococcus aureus* *Streptococcus pneumoniae* *Streptococcus pyogenes* and other streptococci

---

TABLE 9-1. THE SPECIMEN AND ITS PATHOGENS—cont'd

Specimen	Important pathogens
Pleural fluid	*Haemophilus influenzae* *Mycobacterium tuberculosis* *Staphylococcus aureus* *Streptococcus pneumoniae*
Peritoneal fluid	Coliforms Enterococci *Mycobacterium tuberculosis*
Pus, exudates, wound drainages, and the like	*Actinomyces* Anaerobic cocci (*Peptostreptococcus, Peptococcus*) *Bacillus anthracis* *Bacteroides* species and related anaerobic rods *Blastomyces* and other systemic fungi *Clostridium* species Coliforms *Corynebacterium diphtheriae* Enterococci *Francisella tularensis* Mycobacteria *Mycobacterium tuberculosis* *Nocardia* *Proteus* species *Pseudomonas* species *Staphylococcus aureus* *Streptococcus pyogenes* and other streptococci

---

swabs, nasal washings, stools, and specimens already containing bacteria is to be avoided. Swabs may be received into tubes holding a small amount of sterile nutrient broth.

Rush all specimens to the viral diagnostic laboratory immediately after collection! Often this means that they must be transported a distance to a medical center, a large hospital, or the laboratory of a state health department. When this is the case, freeze the specimen (with the exception of blood) immediately and keep it so until shipment can be made. Refrigerate blood samples and let clot. It is desirable to separate serum from the clot. Transit is best made with specimen container packed in dry ice and posted airmail special delivery. A frozen shipment is likely to be a satisfactory one provided that the dry or wet ice has not thawed out by the time the destination is reached. (Small packages on dry ice are likely to thaw in 8 to 12 hours or less.) To send blood samples on dry ice, place the clot and serum in separate tubes. Although cold shipment of fresh material is preferable most of the time, specimens may be preserved in sterile 50% glycerin solution or in buffered glycerol transport medium (but *never* in formalin).

The paired samples of blood for serologic examinations in viral and rickettsial diseases (p. 500) are to be collected aseptically, the blood allowed to clot, and the serum separated from the clot. The serum is sent at once to the out-of-town laboratory by airmail.

Remember that proper labels on the outside of the package help to prevent accidental infection of laboratory personnel receiving the specimen, and the diagnostic specimens discussed here should always be packaged so that they do not leak in transit.

RELATED PATHOGENS

In this chapter we have considered body fluids and anatomic sites in the body from which specimens for microbiologic examination are regularly taken. Table 9-1 lists certain routine specimens with the important pathogenic microorganisms encountered therein.

QUESTIONS FOR REVIEW

1. Why is careful specimen collection stressed?
2. State the ground rules for the collection of any specimen for microbiologic examination.
3. What precautions should be taken in collecting urine specimens for microbiologic examination? Is it always necessary to catheterize the female?
4. How should the skin be prepared for the collection of a blood culture? Why is this necessary?
5. What vein is usually used as a collection site for blood specimens?
6. What are the precautions to be considered in collecting sputum specimens? In collecting stool specimens?
7. How would you handle a specimen of pus that is to be cultured?
8. Briefly discuss the collection of cerebrospinal fluid.
9. How are specimens handled that must be shipped to an out-of-town laboratory for diagnosis of viral disease?

REFERENCES FOR UNITS ONE AND TWO

Alpert, N. L.: Gas chromatographs in the clinical laboratory, (Instrument series, Report No. 31) Lab. World 25:29, May, 1975.
American Type Culture Collection, ASM News 38:590, 1972.
Asimov, I.: Basic research—of what use? ASM News 42:1, 1976.
Automated antibiotic susceptibility testing, Lab. Management 13:25, Mar., 1975.
Bailey, W. R., and Scott, E. G.: Diagnostic microbiology, ed. 4, St. Louis, 1974, The C. V. Mosby Co.
Balows, A., and others, editors: Current techniques for antibiotic susceptibility testing, Springfield, Ill., 1974, Charles C Thomas, Publisher.
Barnes, I. J., and others: Microbes and man. A laboratory manual for students in the health sciences, San Francisco, 1974, W. H. Freeman & Co., Publishers.
Barry, A. L., and others: The antimicrobic susceptibility test: principles and practices, Philadelphia, 1976, Lea & Febiger.
Bartlett, R. C., and others: Blood cultures (Cumitech 1), Washington, D.C., 1974, American Society for Microbiology.
Bauer, J. D., and others: Clinical laboratory methods, ed. 8, St. Louis, 1974, The C. V. Mosby Co.
Beaver, W. C., and Noland, G. B.: General biology—the science of biology, ed. 8, St. Louis, 1970, The C. V. Mosby Co.
Blazevic, D. J., and Ederer, G. M., editors: Principles of biochemical tests in diagnostic microbiology, New York, 1975, John Wiley & Sons, Inc.
Bodily, H. L., and others, editors: Diagnostic procedures for bacterial, mycotic and parasitic infections, Washington, D. C., 1970, American Public Health Association, Inc.
Bologna, C. V.: Understanding laboratory medicine, St. Louis, 1971, The C. V. Mosby Co.
Bonner, J. T.: The size of life, Natural History 78:40, Jan., 1969.

Boyd, W.: An introduction to the study of disease, Philadelphia, 1971, Lea & Febiger.

Branson, D.: Microbiology for the small laboratory, Springfield, Ill., 1972, Charles C Thomas, Publisher.

Brock, T. D., editor: Milestones in microbiology, Washington D.C., 1961, American Society for Microbiology.

Brock, T. D.: Biology of microorganisms, Englewood Cliffs, N.J., 1974, Prentice-Hall, Inc.

Brown, W. V., and Bertke, E. M.: Textbook of cytology, ed. 2, St. Louis, 1974, The C. V. Mosby Co.

Buchanan, R. E., and Gibbons, N. E., co-editors: Bergey's manual of determinative bacteriology, ed. 8, Baltimore, 1974, The Williams & Wilkins Co.

Burrows, W.: Textbook of microbiology, Philadelphia, 1973, W. B. Saunders Co.

Busch, H.: The cell nucleus, New York, 1974, Academic Press, Inc.

Bush, C. L.: A simple solution to media processing, Lab. Management **4**:31, Apr., 1975.

Chedd, G.: The new biology, New York, 1972, Basic Books, Inc., Publishers.

Clark, G.: Staining procedures, Baltimore, 1973, The Williams & Wilkins Co.

Claude, A.: The coming of age of the cell, Science **189**:433, 1975.

Codino, R. J.: Hazards of viral hepatitis in clinical laboratories, Lab. Med. **6**:46, June, 1975.

Culling, C. J.: Modern microscopy: elementary theory and practice, London, 1974, Butterworth & Co. (Publishers), Ltd.

Culliton, B. J.: Cell membranes: a new look at how they work, Science **175**:1348, 1972.

Cunningham, C. H.: A laboratory guide in virology, Minneapolis, 1973, Burgess Publishing Co.

Danielli, J. F.: The bilayer hypothesis of membrane structure, Hosp. Pract. **8**:63, Jan., 1973.

Davidson, M., and others: Bacteriologic diagnosis of acute pneumonia; comparison of sputum, transtracheal aspirates and lung aspirates, J.A.M.A. **235**:158, 1976.

Davis, B. D., and others: Microbiology including immunology and molecular genetics, New York, 1973, Harper & Row, Publishers.

Davis, J. G.: Aspects of laboratory hygiene, Lab. Pract. **21**:101, 1972.

de Duve, C.: Exploring cells with a centrifuge, Science **189**:186, 1975.

Dickerson, R. E.: The structure and history of an ancient protein, Sci. Am. **226**:58, Apr., 1972.

Dubos, R. J.: Louis Pasteur, free lance of science, Boston, 1976, Little, Brown & Co.

Dubos, R. J.: Pasteur's dilemma—the road not taken, ASM News **40**:703, 1974.

Dunlop, R.: Leeuwenhock's lost years, Prism **1**:27, Aug., 1973.

Ederer, G. M., and others: Motility-indole-lysine-sulfide medium, J. Clin. Microbiol. **2**:266, 1975.

Editorial: G. Armauer Hansen, J.A.M.A. **224**:1757, 1973.

Elbrink, J., and Bihler, I.: Membrane transport: its relation to cellular metabolic rates, Science **188**:1177, 1975.

Fisher, K. D., and Nixon, A. U.: The science of life: contributions of biology to human welfare, New York, 1975, Plenum Publishing Corp.

Fox, C. F.: The structure of cell membranes, Sci. Am. **226**:30, Feb., 1972.

Frankel, S., and others, editors: Gradwohl's clinical laboratory methods and diagnosis, ed. 7, St. Louis, 1970, The C. V. Mosby Co.

Freeman, J. A., and Beeler, M. F.: Laboratory medicine—clinical microscopy, Philadelphia, 1974, Lea & Febiger.

Friedman, R. B., and others: Computer-assisted identification of bacteria, Am. J. Clin. Pathol. **60**:395, 1973.

Gasser, W.: Converting to the metric system: the international system of units, Cadence Clin. Lab. **7**:5, Mar.-Apr., 1976.

Gebhardt, L. P., and Nicholes, P. S.: Microbiology, ed. 5, St. Louis, 1975, The C. V. Mosby Co.

Gillies, R. R., and Dodds, T. C.: Bacteriology illustrated, ed. 3, Baltimore, 1973, The Williams & Wilkins Co.

Gray, P., editor: The encyclopedia of microscopy and microtechnique, New York, 1973, Van Nostrand Reinhold Co.

Green, M. H.: International and metric units of measurement, New York, 1973, Chemical Publishing Co., Inc.

Grimstone, A. V., and Skaer, R. J.: A guidebook to microscopical methods, New York, 1972, Cambridge University Press.

Heden, C-G, and Illeni, T., editors: New approaches to the identification of microorganisms, New York, 1975, John Wiley & Sons, Inc.

Holtzman, E.: The biogenesis of organelles, Hosp. Pract. **9:**75, Mar., 1974.

Horstmann, D. M.: Controlling rubella: problems and perspectives, Ann. Intern. Med. **83:**412, 1975.

Imshenetsky, A. A.: Microbiological research in space biology, ASM News **40:**518, 1974.

Jamieson, J. D.: Membranes and secretion, Hosp. Pract. **8:**71, Dec., 1973.

Jawetz, E., and others: Review of medical microbiology, Los Altos, Calif., 1976, Lange Medical Publications.

Jefferson, H., and others: Transportation delay and the microbiological quality of clinical specimens, Am. J. Clin. Pathol. **64:**689, 1975.

Johnson, N. E.: Coping with complications of intravenous therapy, Nursing '72, **2:**5, Feb., 1972.

Joklik, W. K., and Smith, D. T.: Zinsser microbiology, ed. 15, New York, 1972, Appleton-Century-Crofts.

Klainer, A. S., and Geis, I.: Agents of bacterial diseases, New York, 1973, Harper & Row, Publishers.

Klein, H. P.: Microbiology on Mars? ASM News **42:**207, 1976.

Kolata, G. B.: Ribosomes (I): Genetic studies with viruses, Science **190:**136, 1975.

Kolata, G. B.: Intracellular bacterial toxins: origins and effects, Science **190:**969, 1975.

Kolata, G. B.: DNA sequencing: a new era in molecular biology, Science **192:**645, 1976.

Konetzka, W. A.: On the design of objective examinations, ASM News **41:**171, 1975.

Kornberg, A.: DNA synthesis, San Francisco, 1974, W. H. Freeman & Co., Publishers.

Krieg, A. F., and others: Why are clinical laboratory tests performed? When are they valid? J.A.M.A. **233:**76, 1975.

Lamanna, C., and others: Basic bacteriology, Baltimore, 1973, The Williams & Wilkins Co.

Langford, T. L.: Nursing problem: bacteriuria and the indwelling catheter, Am. J. Nurs. **72:**113, Jan., 1972.

Lechevalier, H. A., and Solotorovsky, M.: Three centuries of microbiology, New York, 1974, Dover Publications, Inc.

Lee, L. W.: Elementary principles of laboratory instruments, St. Louis, 1974, The C. V. Mosby Co.

Lehmann, H. P.: Metrication of clinical laboratory data in SI units, Am. J. Clin. Pathol. **65:**2, 1976.

Leive, L., editor: Bacterial membranes and walls. Microbiology series, vol. 1, New York, 1973, Marcel Dekker, Inc.

Lennette, E. H., and others: Manual of clinical microbiology, Washington, D.C., 1974, American Society for Microbiology.

Lindberg, A. A., and others: Identification of gram-negative aerobic fermenters in a clinical bacteriological laboratory, Med. Microbiol. Immunol. **159:**201, 1974.

Locatcher-Khorazo, D., and Seegal, B. C.: Microbiology of the eye, St. Louis, 1972, The C. V. Mosby Co.

Lowis, M. J.: An identification key for some aerobic bacteria, Lab. Pract. **20:**331, 1971.

MacFaddin, J. F.: Manual of biochemical tests for the identification of medical bacteria, Baltimore, 1975, The Williams & Wilkins Co.

McFate, R. P.: Introduction to the clinical laboratory, ed. 3, Chicago, 1972, Year Book Medical Publishers, Inc.

Maisey, M. A.: Bacteriology, Nurs. Times **70:**142, Jan. 31, 1974.

Margileth, A. M., and Pedreira, F. A.: Office diagnosis of streptococcal upper respiratory tract disease, South Med. J. **68:**489, 1975.

Maugh, T. H., II: Ribosomes (II): a complicated structure begins to emerge, Science **190:**258, 1975.

Metric medicine for a metric America (editorial), Ann. Intern. Med. **76:**138, 1972.

Minckler, J., and others, editors: Pathobiology: an introduction, St. Louis, 1971, The C. V. Mosby Co.

Mitruka, B. M.: Gas chromatographic applications in microbiology and medicine, New York, 1975, John Wiley & Sons, Inc.

Moffet, H. L.: Clinical microbiology, 1975, Philadelphia, J. B. Lippincott Co.

Morgan, R. M., and Good, D. S.: Microbiology procedures: the selection and interpretation of current tests for physicians, nurses, and paramedical personnel, Springfield, Ill., 1973, Charles C Thomas, Publisher.

Morowitz, H. J.: On first looking into Bergey's Manual, Hosp. Pract. **10:**156, Apr., 1975.

Olby, R.: The path to the double helix, Seattle, 1974, University of Washington Press.

Olds, R. J.: Color atlas of microbiology, Chicago, 1975, Year Book Medical Publishers, Inc.

Palade, G.: Intracellular aspects of the process of protein synthesis, Science **189:**347, 1975.

Parish, H. J.: A history of immunization, Edinburgh, 1965, E. & S. Livingstone, Ltd.

Parish, H. J.: Victory with vaccines, Edinburgh, 1968, E. & S. Livingstone, Ltd.

Payne, J. E., and Kaplan, H. M.: Alternative techniques for venipuncture, Am. J. Nurs. **72:**702, 1972.

Porter, J. R.: Louis Pasteur Sesquiecentennial (1822-1972), Science **178:**1249, 1972.

Portnoy, B.: Diagnosis of viral disease in the clinical laboratory, Lab. Med. **1:**38, Sept., 1970.

Provine, H., and Gardner, P.: The gram-stained smear and its interpretation, Hosp. Pract. **9:**85, Oct., 1974.

Quinn, R. W.: The positive throat culture—what does it mean? South. Med. J. **67:**1009, 1974.

Racker, E.: Inner mitochondrial membranes: basic and applied aspects, Hosp. Pract. **9:**87, Feb., 1974.

Roller-Massar, A.: Discovering the basis of life: an introduction to molecular biology, New York, 1974, McGraw-Hill Book Co.

Rosenfield, R. E.: The past and future of immunohematology, Am. J. Clin. Pathol. **64:**569, 1975.

Russell, W. O.: Cell membranes and disease, Am. J. Clin. Pathol. **63:**618, 1975.

Schneierson, S. S.: Atlas of diagnostic microbiology, North Chicago, Ill., 1975, Abbott Laboratories.

Siegelman, A. M., and Friedman, B.: Microscopy in the clinical laboratory, Cadence Clin. Lab. **5:**6, Sept.-Oct., 1974.

Siekevitz, P.: Dynamics of intracellular membranes, Hosp. Pract. **8:**91, Nov., 1973.

Singer, S. J.: Architecture and topography of biologic membranes, Hosp. Pract. **8:**81, May, 1973.

Snyder, B.: Pitfalls in the gram stain, Lab. Med. **1:**41, July, 1970.

Sourkes, T. L.: Nobel Prize winners in medicine and physiology 1901-1965, New York, 1966, Abelard-Schuman.

Stanier, R. Y., and others: The microbial world, Englewood Cliffs, N.J., 1970, Prentice-Hall, Inc.

Steere, N. V., editor: CRC Handbook of laboratory safety, Cleveland, Ohio, 1971, CRC Press, Inc.

Stein, H. J.: Caution: biology may be hazardous to your health, Bioscience **21:**80, 1971.

Stewart, J. A.: Methods of media preparation for the biological sciences, Springfield, Ill., 1974, Charles C Thomas, Publisher.

Sussman, J. L., and Kim, S-H: Three-dimensional structure of a transfer RNA in two crystal forms, Science **192:**853, 1976.

Taaca, E. B.: Simulated patient "arm" helps in venipuncture instruction, Lab. Med. **6:**40, Dec., 1975.

The National Library of Medicine (editorial), J.A.M.A. **218:**1937, 1971.

Thomas, L.: The lives of a cell, New York, 1976, The Viking Press, Inc.

Tilton, R. C.: Methodologic advances in clinical microbiology, Lab. Med. **1:**54, Feb., 1970.

Toner, P. G., and Carr, K. E.: Cell structure, an introduction to biological electron microscopy, Edinburgh, 1971, E. & S. Livingstone, Ltd.

Tortora, G. J., and Becker, J. F.: Life Sciences, New York, 1972, Macmillan, Inc.

Trump, B. F.: The network of intracellular membranes, Hosp. Pract. **8:**111, Oct., 1973.

Ungvarski, P.: Mechanical stimulation of coughing, Am. J. Nurs. **71:**2358, 1971.

Walker, L. J., and Taub, H.: Fundamental skills in serology, Springfield, Ill., 1976, Charles C Thomas, Publisher.

Warren, L.: Isolation and properties of natural membranes, Hosp. Pract. **8:**127, Apr., 1973.

Washington, J. A., II, editor: Laboratory procedures in clinical microbiology, Boston, 1974, Little, Brown & Co.

Werner, M., editor: Microtechniques for the clinical laboratory, New York, 1976, John Wiley & Sons, Inc.

Wolfe, S. L.: Biology of the cell, Belmont, Calif., 1972, Wadsworth Publishing Co., Inc.

LABORATORY SURVEY OF UNIT TWO

PROJECT
Growth of bacteria in the laboratory—an overview
Part A—Demonstration by the instructor with discussion

1. Materials and equipment for culture
   a. *Petri dish*—what it is, how it is used, how manipulated with sterile technic, how culture medium is poured into the dish, why it is inverted in the incubator, how handled on the laboratory bench
   b. *Culture tubes*—how used, purpose of cotton plug or screw cap, care and handling of the cotton plug, how tubes are manipulated with sterile technic
   c. *Culture media*—solid and liquid, various kinds, various ways media dispensed, incorporation of specific substances for specific reasons as, for example, indicators, test reagents, and materials to enhance growth
   d. *Platinum loop and platinum needle*—manipulation, importance of flame sterilization
   e. *Microbiologic incubator*—its mechanism, how used
2. Precautions to avoid contamination, species variation from inconsistencies in technic, and spread of infection or undesired bacterial growth
3. Examination of bacterial growth—how to study it in the laboratory
   a. Enumerate and stress descriptive features. (Refer to Figs. 7-1, 7-2, 7-5, and 7-9.) List workable adjectives, useful terms, and the like.
   b. Comment on variations.
   c. Note how identification of bacteria is indicated from gross morphology.

Part B—Student participation: studying bacterial growth

Note: It is suggested that for the following exercise students work in groups of four (or more). Each group is furnished with a set of prepared cultures. Each member of the group carefully inspects each culture and records a description of its pattern of growth in the notebook. Twelve cultures have been selected. Although these make up a practical and workable set, they can easily be varied.

1. Study the following prepared cultures assigned to your group:
   a. Three Petri dish cultures (streak plates on trypticase soy agar) in each of which growth of one organism on one side is separate from growth of another organism on the opposite side
      (1) The first plate displays *Escherichia coli* and *Bacillus subtilis* (incubated 24 hours at 37° C).
      (2) The second plate shows *Micrococcus luteus* and *Serratia marcescens* (grown at room temperature for 24 hours).
      (3) The third plate shows *Streptococcus faecalis* and *Pseudomonas aeruginosa* (grown at room temperature for 24 hours).
   b. Two pour plates (trypticase soy agar incubated 24 hours at 37° C) showing growth of *Serratia marcescens* and *Pseudomonas aeruginosa*, respectively

144

c. Four agar slant cultures (trypticase soy agar incubated at room temper-
ature), one each for *Escherichia coli*, *Serratia marcescens*, *Streptococcus faecalis*, and *Pseudomonas aeruginosa*

d. Three broth cultures (trypticase soy broth grown at room temperature for 48 hours), one each for *Escherichia coli*, *Serratia marcescens*, and *Pseudomonas aeruginosa*

2. Note colonial morphology and gross features of bacterial growth in the laboratory setting.

a. Inspect growth found on the prepared Petri dishes (three), pour plates (two), agar slants (three), and broth cultures (four).

b. Record observations for each in your notebook.

(1) Note appearance of discrete colonies on agar plates. Wherever a bacterium contacts the surface of the medium, a colony develops. Count the number of colonies for each organism. Do colonies vary for a given organism? For different organisms? Are the edges of the colonies smooth or irregular? What might this mean? Is the growth on the agar slant discrete or confluent? Why?

(2) List terms and phrases that help to describe bacterial growth in colonies.

(3) Note location and nature of growth in broth culture. Why are both solid and liquid media used?

c. Compare the following:

(1) Growth patterns for the same microbe in different media

(2) Growth patterns of one microbe with those of another in different media, in the same medium

(3) Bacterial growth in solid media with that in fluid media

d. Keep the cultures in a place designated by the laboratory instructor so that you may observe any changes in them after a few days and after a week. Note differences with time.

e. Answer the following in your notebook:

(1) Is it possible to identify a bacterium by the nature of its growth under specified conditions in the laboratory?

(2) Must requirements for culture of bacteria in the laboratory be specific and constant if reproducible results are desired?

PROJECT
Conditions affecting growth of bacteria

1. *Influence of temperature on bacterial growth*

a. The instructor will furnish you three broth cultures just inoculated with bacteria. Note that all are clear because bacterial growth has not occurred yet.

b. Place one in the refrigerator, another at room temperature, and another in the microbiologic incubator.

c. Observe at end of 24 hours.

d. Record results. In which tube is the growth, as indicated by cloudiness, most abundant? What does this prove?

	Refrigerator temperature 4° C (24 hr)	Room temperature 20° C (24 hr)	Incubator temperature 37° C (24 hr)
Bacterial growth			

2. *Effect of reaction of medium (pH) on bacterial growth*

a. The instructor will furnish you five recently inoculated tubes of broth. Make additions of alkali and of acid to four of them as follows: (a) 0.2 ml 1N NaOH, (b) 0.5 ml 1N NaOH, (c) 0.2 ml 1N HCl, and (d) 0.5 ml 1N HCl. The fifth serves as a control.

b. Incubate at 37° C for 24 hours. Compare the multiplication of bacteria in the different tubes as indicated by clouding of the medium.

c. Record results.

	Alkaline pH		Acid pH		
	0.2 ml 1N NaOH	0.5 ml 1N NaOH	0.2 ml 1N HCl	0.5 ml 1N HCl	Control
Bacterial growth					

3. *Effects of sunlight on bacterial growth*

a. The instructor will furnish you a recently inoculated Petri dish. Invert and cover one half of the Petri dish in such a manner that direct sunlight is not allowed to strike the medium. A piece of black paper may be used.

b. Expose to sunlight for several hours.

c. Incubate at room temperature until colonies appear. On which half of the plate are the colonies more numerous? What does this prove?

PROJECT
Special activities of bacteria
Part A—Fermentation

1. *Fermentation of sugars*

a. The instructor will furnish each group of students three Smith fermentation tubes containing sterile sugar solutions. The first contains 10% dextrose (grape sugar) solution; the second contains 10% lactose (milk sugar) solution; and the third contains a 10% solution of sucrose (cane sugar).

b. Rub up a small piece of yeast cake in water and add a portion to each tube.

c. Set in a warm place and observe at the end of 24 hours.

d. Chart. What differences are noted? Why?

	Dextrose, 10% (grape sugar)	Lactose, 10% (milk sugar)	Sucrose, 10% (cane sugar)
Changes observed after inoculation with yeast			

2. *Prove that gas formed in no. 1 is carbon dioxide as follows:*
   a. Measure the length of the gas column in the arm of the fermentation tube.
   b. Remove a small amount of material from the bulb of the tube with a pipette and reserve for no. 3.
   c. Fill bulb of tube with 10% potassium hydroxide (KOH).
   d. Place thumb over mouth of tube and mix by inverting so that gas in arm of tube comes in contact with KOH.
   e. Collect all remaining gas in arm of tube. Reduction in amount of gas proves CO_2 was present and was absorbed according to the following chemical equation: $2KOH + CO_2 = K_2CO_3 + H_2O$.
3. *Prove that the fermentation in no. 1 produced alcohol as follows:*
   a. Filter the liquid removed in no. 2.
   b. To the filtrate add a few drops of weak iodine solution.
   c. Add enough 10% sodium hydroxide solution to change the color from a brown to a distinct yellow.
   d. Warm slightly. A distinct odor of iodoform will be obtained, and a yellow precipitate, appearing under the microscope as small hexagonal crystals, will be formed. If the amount of alcohol is small, the crystals may not appear until the solution has cooled or stood some time.
4. *Fermentation with acid and gas formation*
   a. The instructor will furnish you a phenol red dextrose broth fermentation tube recently inoculated with *Escherichia coli*.
   b. Incubate 24 to 48 hours.
   c. Note multiplication of bacteria as indicated by turbidity. What other two changes have taken place? What does each indicate? What is the purpose of the phenol red?
5. *Fermentation with reduction of pigment*
   a. Nearly fill a long narrow test tube with milk.
   b. Add enough weak aqueous solution of methylene blue to give the milk a distinct blue color.
   c. Add a little milk that has soured naturally.
   d. Mix and incubate at 37° C for several days.
   e. Observe. What change has taken place?

Part B—Pigment production by bacteria

1. The instructor will give you three agar slants. One has been inoculated with *Pseudomonas aeruginosa*, one with *Serratia marcescens*, and one with *Staphylococcus aureus*.

147

a. Incubate at 37° C until pigment production is well developed.
b. Observe. What color is produced by each? Is the color confined to the growth of bacteria or does it diffuse into the medium?
c. Record colors.

	Pseudomonas aeruginosa	Serratia marcescens	Staphylococcus aureus
Color produced			

2. Examine cultures of *Escherichia coli* and *Proteus vulgaris* prepared on eosin–methylene blue agar (EMB) or MacConkey agar (p. 109).
   a. Note pigmentation. Record findings. What color is produced by *Escherichia coli*? By *Proteus vulgaris*? What is the color of the culture medium?
   b. Explain and interpret the findings.

Color produced on:	Escherichia coli	Proteus vulgaris
EMB agar		
MacConkey agar		

PROJECT
Preparation of culture media
Part A—Demonstration of bacteriologic technics by the instructor with discussion

1. Flaming the mouth of a container to preserve sterility of contents
2. Holding tubes of culture media in a slanting position during inoculations
3. Removing and inserting cotton plugs or other stoppers during inoculations
4. Transferring bacterial growth
   a. Broth-to-agar slant cultures
   b. Agar slant-to-broth cultures
5. Pouring melted agar into a Petri dish

Part B—Preparation of nutrient broth

The composition of 1 liter of nutrient broth is as follows:

Meat extract	3 g
Peptone	5 g
Sodium chloride (may be omitted)	5 g
Distilled water	1 liter

Commercially, one can obtain Difco (or equivalent product*)—a dehydrated medium that contains the meat extract, peptone, and buffer substance in such propor-

---

*Commercial products are usually supplied with precise directions for use.

tions that the finished product has a pH of 6.8. It is not necessary to adjust the pH of this product. Using the commercial dehydrated medium to make nutrient broth, proceed as follows:

a. Dissolve 4 g of dehydrated nutrient broth (containing 1.5 g of meat extract and 2.5 g peptone) in 500 ml of distilled water contained in a liter Erlenmeyer flask. Add 2.5 g of sodium chloride if desired. (For the subsequent experiments, 500 ml amounts of nutrient broth will be sufficient.)

b. After solution is complete, pour 10 to 15 ml amounts in each of 25 culture tubes.

c. Plug tubes with cotton plugs and autoclave at 121° C for 15 minutes. During the process of plugging and when handling the culture medium, do not let the liquid contact the cotton plugs. Under no condition can the pressure of the autoclave be rapidly released when steam pressure sterilization is finished because rapid release of pressure causes the liquid in the tubes to boil over and wet the plugs. What are some of the objections to wetting of cotton plugs?

Part C—Preparation of nutrient agar

The composition of 1 liter of nutrient agar is as follows:

Meat extract	3 g
Peptone	5 g
Sodium chloride (may be omitted)	5 g
Agar	15 g
Distilled water	1 liter

The Difco dehydrated medium (or equivalent commercial product) contains the meat extract, peptone, and agar in a dehydrated condition buffered so that the final pH of the medium is 6.8. No further adjustment of pH is necessary.

To make nutrient agar, proceed as follows:

a. To 500 ml of cold distilled water contained in a liter flask, add 11.5 g dehydrated nutrient agar (Difco).

b. Place in pan of water or on boiling water bath and heat until solution is effected. The material may be boiled over a free flame, but there is danger of scorching.

c. In order to have sufficient culture medium for the experiments to follow in this and the next unit, put up the following:

(1) Tubes containing 5 ml of agar   6
(these are slanted to solidify after sterilization)
(2) Test tubes containing 25 ml agar   6
(3) Erlenmeyer flasks (150 ml) containing about 89 ml agar   4

d. Plug containers with cotton plugs and autoclave at 121° C for 15 minutes.

e. After pressure in autoclave has returned to normal, slant agar tubes to solidify as previously indicated.

PROJECT
Cultivation of bacteria

1. *Taking a throat culture*
   a. Demonstration by the instructor with discussion
      The instructor will show how a sterile cotton swab can be used to take a throat culture. The swab, after it has been rubbed over the back part of the throat of a given individual, is then rubbed over the surface of a suitable culture medium. No attempt in this exercise is made to separate bacterial organisms. Media such as Löffler serum medium, blood agar, or nutrient agar may be used.
   b. Student participation
      (1) Using a sterile swab, take a throat culture. (Students may work in pairs.) Apply swab to the surface of a suitable culture medium.
      (2) Incubate cultures and observe at the end of 24 hours.
      (3) Make gram-stained smears of the bacterial growth. Make drawings to show representative microorganisms.
2. *Transferring bacterial growth*
   a. Demonstration by the instructor with discussion
      The instructor reemphasizes the proper technics of sterile transfers, indicates the differences in ability of microorganisms to grow in various laboratory media, and explains the usefulness of cultural transfers.
   b. Student participation
      (1) Obtain a culture (tube or plate) showing growth of a nonpathogenic organism.
      (2) Make the transfer of organisms from your culture to a different kind of culture medium.
      (3) Incubate the culture for 24 to 48 hours. Observe and describe growth.

PROJECT
Isolation of pure cultures

1. *Pour plate method*
   Note: The students, working in small groups will use tubes of culture media that they have prepared, sterile Petri dishes, and a mixed culture furnished by the instructor.
   a. Refer to the pour plate method as outlined on p. 116.
   b. Use this method to obtain pure cultures of organisms in the mixed culture provided to you. Incubate the Petri dish cultures for 24 to 48 hours.
2. *Streak plate method*
   Note: The students are each given three Petri dishes of sterile culture medium, and mixed cultures are laid out in the laboratory.
   a. Refer to the paragraph on streak plates on p. 116 of the text and to the illus-

tration of the method in Fig. 7-11. Consult the instructor for assistance when you carry out this technic for the first time.

b. Make a streak plate using one of the mixed cultures available in the laboratory. Incubate for 24 to 48 hours.

c. Observe colonial growth as obtained by the two methods of isolation. How do the distributions of the colonies on the streak and pour plates differ?

d. Note the importance of making gram-stained smears for microscopic study of different types of colonies obtained. How can a gram-stained smear be used to check a pure culture? Make gram-stained smears for representative colonies that you have obtained from your cultures.

3. *Fishing of colonies*

a. Demonstration by the instructor with discussion

The instructor should demonstrate how given colonies are transferred from Petri dishes to agar slants. Why is it often necessary to do this?

PROJECT
Bacterial plate count

1. Work in small groups of two to four.
2. Obtain necessary materials for this experiment as follows:
   a. A small amount of milk
   b. Four tubes containing 9 ml sterile distilled water
   c. Five sterile 1 ml pipettes
   d. Four sterile Petri dishes
3. Place a flask of agar in a pan of water and heat water until agar melts. Let cool to about 40° C. This may be determined roughly by considering a temperature of 40° C as that temperature at which the flask may comfortably be held against the back of the hand. Keep at this temperature.
4. Proceed as follows:
   a. Transfer 1 ml of milk to a tube of 9 ml sterile distilled water and mix. With a fresh pipette transfer 1 ml of the mixture to a second tube of 9 ml sterile distilled water. Continue in this manner until the fourth tube is mixed, being careful to use a fresh sterile pipette to make each mixture. You now have 1:10, 1:100, 1:1000, and 1:10,000 dilutions of milk.
   b. With a sterile pipette transfer 1 ml of each dilution to a Petri dish. Begin with the highest dilution and go to the lowest when making transfers. The material is transferred to the Petri dishes by slightly raising the lid and depositing the material in the middle of the plate. Gently raise the lid and add about 15 ml of melted agar to each dish. Mix contents of each dish by gently rotating. Let agar solidify; invert and incubate 48 hours.
   c. Select a plate showing well-distributed, easily counted colonies and count the colonies present. Multiply the number of colonies counted by the times the milk in that plate was diluted. The result is the number of bacteria in 1 ml of the milk used. Why was a new pipette used for making each dilution

151

of the milk? Why can transfers from the tubes to Petri dishes be made with a single pipette by beginning with the highest dilution and proceeding to the lowest?

PROJECT
Animal inoculation

1. Demonstration by instructor with discussion
   a. Technic of animal inoculation
   b. Value of animal inoculation
      The organisms suspected and the results obtained should be explained.
2. Study of preserved organs from autopsies of guinea pigs inoculated with *Mycobacterium tuberculosis* and *Francisella tularensis*.
   Compare diseased animal organs with normal ones.

EVALUATION FOR UNIT TWO

Part I

In the following statements or questions, encircle the number in the column on the right that correctly completes the statement or answers the question.

1. Why is refrigeration a good method for preserving bacterial cultures?
   - (a) Cold is necessary for bacterial growth
   - (b) Cold retards bacterial growth
   - (c) Freezing may destroy bacteria
   - (d) Cold excludes oxygen
   - (e) Cold protects bacteria from sunlight

   1. a
   2. c
   3. a, c, and e
   4. b
   5. a, b, and d

2. Which of the following bacterial metabolic products are associated with disease production?
   - (a) Hemolysins
   - (b) Coagulases
   - (c) Leukocidins
   - (d) Kinases
   - (e) Spreading factor

   1. a and d
   2. a and b
   3. a, b, and c
   4. c, d, and e
   5. a, b, c, d, and e

3. Bacteria are capable of which of the following variations?
   - (a) A change from coccus to bacillus
   - (b) A change from encapsulated to a nonencapsulated organism
   - (c) A change from a nonvirulent to a virulent organism
   - (d) A change from a spirillum to a spirochete
   - (e) A change from an antibiotic-susceptible to an antibiotic-resistant organism

   1. all are true
   2. all but d are true
   3. all but a and d
   4. all but a, b, and e
   5. none is true

4. Which of the following organisms exist *only* on living material?
   - (a) Facultative saprophyte
   - (b) Strict parasite
   - (c) Facultative parasite
   - (d) Strict saprophyte

   1. b
   2. b and d
   3. c
   4. a

5. Which terms indicate that two organisms can live together?
   - (a) Commensalism
   - (b) Symbiosis
   - (c) Antagonism
   - (d) Positive chemotaxis
   - (e) Plasmoptysis

   1. d
   2. b and e
   3. a and b
   4. c
   5. a, b, and c

6. A spirochete can best be studied by which of the following preparations?
   - (a) Gram stain
   - (b) Hanging drop preparation
   - (c) Dark-field illumination
   - (d) Acid-fast stain
   - (e) Culture on an artificial medium

   1. a
   2. b
   3. c
   4. d
   5. e

7. In which groups would you *not* expect to find concentrated the organisms of greatest medical importance?
   - (a) Autotrophs
   - (b) Lithotrophs
   - (c) Psychrophiles
   - (d) Obligate halophiles
   - (e) Thermophiles

   1. a
   2. b
   3. all of these
   4. a, b, c, and d
   5. c and d

8. Strict anaerobes grow well:
   - (a) Under 5% to 10% increase of carbon dioxide tension
   - (b) In atmospheric oxygen
   - (c) In broth beneath layer of petroleum jelly
   - (d) In thioglycollate broth
   - (e) On routine blood agar

   1. a, b, and c
   2. a, b, c, and d
   3. b, c, d, and e
   4. c, d, and e
   5. c and d

9. The process by which the red cell gives up water to a hypertonic plasma is called:
   - (a) Hemolysis
   - (b) Crenation
   - (c) Osmosis
   - (d) Metabolism
   - (e) None of the above

   1. c
   2. b
   3. b and c
   4. a and c
   5. e

10. Testing the antibiotic susceptibility of a bacterial culture by the sensitivity disk method has:
    - (a) Become unnecessary since discovery of broad-spectrum drugs
    - (b) Little application to subsequent treatment of patient
    - (c) Been used only with penicillin
    - (d) Been too time consuming to be practical
    - (e) Interfered with routine blood cultures

    1. a and b
    2. c and d
    3. c and e
    4. a, b, c, and d
    5. none of these

11. The bacterial plate count is important in the microbiologic evaluation of urine. Generally the microbes recovered per milliliter of specimen are not considered significant until the count exceeds:
    - (a) 100/ml of urine
    - (b) 1000/ml of urine
    - (c) 10,000/ml of urine
    - (d) 100,000/ml of urine
    - (e) 1,000,000/ml of urine

    1. d
    2. e
    3. a
    4. b
    5. c

12. The usual complication that follows urethral catheterization is:
    - (a) Urinary tract infection
    - (b) Urinary tract closure
    - (c) Rupture of urethra
    - (d) Bleeding from the bladder
    - (e) Patient is unable to void naturally

    1. a and b
    2. a
    3. b
    4. c and d
    5. e

Part II

A. Before each principle, place the letter from the column on the right indicating the procedure based on that principle.

PRINCIPLES

_____ 1. Bacteria can be partially classified by their characteristic action on media containing different carbohydrates.

_____ 2. A single species of bacteria can be isolated from infectious material by making a dilute mixture of the medium and the infectious material. Then, when the inoculated medium is allowed to harden in a Petri dish, the colonies will be widely separated.

_____ 3. We can assume that for each colony appearing on a pour plate, there was only one bacterium originally present in the medium.

_____ 4. Oxygen can be removed from an airtight container by bringing about some chemical reaction in the container that uses up the oxygen.

PROCEDURES

(a) Pure culture, pour plate method
(b) Culture of anaerobic bacteria
(c) Pure culture, streak plate method
(d) Bacterial count, plate method
(e) Determination of fermentation reactions

B. Comparisons: Match the item in column B to the word or phrase in column A best describing it.

COLUMN A

Gram-positive with gram-negative bacteria:

_____ 1. Walls often contain teichoic acid
_____ 2. Walls contain mucopeptide
_____ 3. Walls rich in lipid
_____ 4. Usually more susceptible to antibiotics
_____ 5. Survive autoclaving
_____ 6. Usually more resistant to oxidizing agents

Capsules with flagella:

_____ 7. Rare in cocci
_____ 8. Complex polysaccharide
_____ 9. Elastic protein
_____ 10. Stained by special stains
_____ 11. Stained by routine Gram stain
_____ 12. Quellung reaction
_____ 13. Related to virulence

Respiration (aerobic oxidation) with fermentation:

_____ 14. Yields more energy per mole of substrate
_____ 15. End products carbon dioxide and water
_____ 16. End product ammonia
_____ 17. End product pyruvate
_____ 18. Carried out by facultative organisms
_____ 19. Carried out by strict anaerobes
_____ 20. Important laboratory application in identification of microbes

COLUMN B

(a) Gram-positive bacteria
(b) Gram-negative bacteria
(c) Both
(d) Neither

(a) Capsules
(b) Flagella
(c) Both
(d) Neither

(a) Aerobic oxidation of glucose
(b) Fermentation of glucose
(c) Both
(d) Neither

Differential culture media with selective media:

_____ 21. Retard growth of microbes in certain instances
_____ 22. Important in culture of feces
_____ 23. Petragnani for growth of tubercle bacilli, an example
_____ 24. MacConkey agar for growth of gram-negative bacilli, an example
_____ 25. Lactose may be incorporated
_____ 26. Dye added to indicate fermentation in certain instances
_____ 27. Dye added to suppress growth in certain instances
_____ 28. Selected embryonic tissues incorporated in certain instances

(a) Selective culture media
(b) Differential media
(c) Both
(d) Neither

D. Match the item in column B to the phrase best defining it in column A.

COLUMN A

_____ 1. Enhanced at times by radiant energy
_____ 2. Transfer of large segment of DNA
_____ 3. Smooth and rough colonies grown from pure culture
_____ 4. Variation in size, shape, and other features
_____ 5. Loss in disease-producing ability of microbe
_____ 6. Direct transfer of DNA into culture medium
_____ 7. Indirect transfer of genetic material by virus
_____ 8. Physiologic adjustment to environment

COLUMN B

(a) Adaptation
(b) Pleomorphism
(c) Dissociation
(d) Commensalism
(e) Attenuation
(f) Recombination
(g) Transformation
(h) Mutation
(i) Transduction
(j) Conjugation

E. Complete the following statements by filling in the blanks to the left. Be sure that the number of the answer corresponds with that in the statement. Note vocabulary for reference to the right.

1. _____ Certain bacteria can form a (1) when surrounding conditions are unfavorable for growth.

2. _____ A visible mass formed by rapid reproduction of an organism on culture media is called a (2).

3. _____ When the organism sought is found in a culture, we designate the culture a (3) one.

4. _____ Iodine is used as a (4) in the gram-staining procedure.

5. _____ When the presence of two species of microbes in the same environment favors the development of both, the condition is called (5).

6. _____ A growth of one kind of bacteria is called a (6) culture.

7. _____ Chromogenic bacteria produce (7).

8. _____ Microbes without chlorophyll must obtain their energy by (8).

9. _____ The (9) is an important structural constituent of cells.

10. _____ Excessive gas formation in milk produced by *Clostridium perfringens* is (10).

11. _____ The enzyme (11) is produced by *Neisseria* species.

12. _____ The (12) test is important in the differentiation of tubercle bacilli from atypical mycobacteria.

13. _____ If a culture contains two or more kinds of microbes, it is termed a (13) culture.

14. _____ Culture of media of known and definite chemical composition are (14) media.

15. _____ The enzyme (15) is responsible for conversion of urea to ammonia.

16. _____ A nonprotein compound necessary for enzyme action is (16).

17. _____ The temperature most suitable for growth of a given microbe is termed the (17) one.

18. _____ Organisms growing best in an amount of oxygen less than that of the air are (18).

19. _____ The organic catalyst of the body is the (19).

20. _____ A protein may be composed of several hundred (20) building blocks.

21. _____ The material acted on by a given enzyme is termed the (21).

22. _____ An enzyme released from a given cell into the extracellular medium is called (22).

23. _____ Microbes utilizing organic matter as a source of energy are termed (23).

24. _____ The process whereby a cell works actively to absorb molecules is (24).

25. _____ (25) destroys polymorphonuclear neutrophilic leukocytes.

26. _____ (26) accelerates clotting of blood.

157

UNIT THREE
MICROBES
PRODUCTION OF INFECTION

10 Role in disease

INFECTION

When microbes or certain other living agents enter the body of a human being or animal, multiply, and produce a reaction there, this is an *infection*. The host is *infected*, and the disease is an *infectious* one. The reaction of the body may or may not be accompanied by outward signs of disease. Infection is differentiated from *contamination*, which refers to the mere presence of infectious material (no reaction produced). Superficial wounds in skin and mucous membranes are often the site of a large microbial population. As long as these microorganisms do not invade the deeper tissues and induce a reaction there, they are contaminants instead of agents of infection.

RESIDENT POPULATION

Microbes normally present. *The mere presence of microbes in the body does not mean infection because microorganisms normally inhabit many parts of the body without invading the deeper tissues to cause disease.* Certain of the body fluids such as blood and urine are normally sterile, that is, free of the presence of microorganisms. In parts of the body other than the circulatory system and urinary tract, the secretions or excretions are normally in contact with a resident population of microorganisms. These microbes are consistently present in varying proportions. This pattern of growth, conspicuously bacterial, is associated with the well-being of the person and in an area such as the intestinal tract is even necessary to his health. These microorganisms constitute the *normal flora*. Table 10-1 gives their normal habitats in the human body.

INVASION OF BODY

Source of microbes causing infections. The microbes that cause infections fall into two classes: (1) those that can cause disease in healthy persons and reach such

161

TABLE 10-1. DISTRIBUTION OF RESIDENT POPULATION

Anatomic site	Important resident microbes
Skin	*Bacillus* species Coliforms Diphtheroids Enterococci Fungi (lipophilic) Mycobacteria Neisseriae *Peptostreptococcus* species *Propionibacterium acnes* *Proteus* species *Pseudomonas* species Staphylococci Streptococci
Mouth (including gingival crevice)	*Actinomyces* species Amebas *Bacteroides* species *Bifidobacterium* species Borreliae *Candida* species and other fungi Diphtheroids Hemophilic bacilli Lactobacilli *Leptotrichia buccalis* Mycoplasmas Neisseriae Pneumococci *Peptostreptococcus* species Spirochetes Staphylococci Streptococci *Trichomonas* species Veillonellae *Vibrio* species
Respiratory tract	*Acinetobacter* species *Actinomyces* species *Bacteroides* species *Candida albicans* and other fungi Diphtheroids Enterococci *Fusobacterium* species

*The stomach is normally free of microbial growth because of its acid content. Microorganisms are much more plentiful in the large than in the small intestine.

†A baby is born with a sterile intestinal tract, but before or with the first feeding, bacteria are introduced. If the child is breast fed, the predominant organism is said to be *Bifidobacterium bifidum* (formerly *Lactobacillus bifidus*). If the child is bottle fed, *Lactobacillus acidophilus* (bacteria of genus *Lactobacillus* convert carbohydrates to lactic acid) predominates. In addition, other bacteria are present.

TABLE 10-1. DISTRIBUTION OF RESIDENT POPULATION — cont'd

Anatomic site	Important resident microbes
	Hemophilic bacilli
	Mycobacteria
	Mycoplasmas
	Neisseriae
	Peptostreptococcus species
	Pneumococci
	Spirochetes
	Staphylococci
	Streptococci (aerobic and anaerobic)
	Veillonellae
Gastrointestinal tract*	*Actinomyces* species
(mostly in lower	*Aeromonas* species
ileum and colon)	*Alcaligenes faecalis*
	Bacteroides species
	Bifidobacterium species
	Candida species
	Clostridium species
	Coliforms
	Diphtheroids
	Entamoeba coli
	Enterococci
	Enteroviruses
	Eubacterium species
	Lactobacilli†
	Mycobacteria
	Mycoplasmas
	Penicillium species
	Peptococcus species
	Peptostreptococcus species
	Proteus species
	Pseudomonas aeruginosa
	Spirochetes
	Staphylococci
	Streptococci (aerobic and anaerobic)
	Yeasts
	Vibrio species
Genital tract	*Bacteroides* species
	Bifidobacterium species
	Coliforms
	Diphtheroids
	Enterococci
	Fusobacterium species
	Haemophilus vaginalis
	Lactobacilli
	Mycobacterium smegmatis
	Mycoplasmas
	Neisseriae
	Peptostreptococcus species
	Proteus species

Continued.

TABLE 10-1. DISTRIBUTION OF RESIDENT POPULATION—cont'd

Anatomic site	Important resident microbes
Genital tract—cont'd	Saprophytic yeasts
	Spirochetes
	Staphylococci
	Streptococci (aerobic and anaerobic)
	Veillonellae
Vagina	
Before puberty	Coliforms
	Diphtheroids
	Streptococci
During childberaring period	Döderlein's bacilli (*Lactobacillus* species)
After childbearing period	Coliforms
	Diphtheroids
	Streptococci
Organs of special sense	
Eye	Alpha streptococci
	Diphtheroids
	Hemophilic bacilli
	Pneumococci
	Staphylococci
External ear	*Bacillus* species
	Diphtheroids
	Nonpathogenic acid-fast organisms
	Staphylococci

persons directly or indirectly from animals, persons ill of the disease, or carriers (such microbes are pathogens and cause communicable diseases) and (2) those that attack the host at a time of injury or lowered resistance or when they themselves are increased in virulence (these are ordinarily saprophytes but can be opportunists). To the second group belong microorganisms that normally inhabit the body but produce disease only under given conditions and certain ones that are inadvertently introduced into the body by wounds, injuries, and such.

How microbes reach the body. According to the manner in which the causative agent reaches the body, infectious diseases may be classified as communicable and noncommunicable.

A *communicable disease* is one whose agent is directly or indirectly transmitted from host to host. Examples are diphtheria and tuberculosis. The host from which the infection is spread is usually of the same species as the recipient, but not necessarily so. For instance, cattle may transmit tuberculosis and undulant fever to man.

A *noncommunicable disease* is one whose agent either normally inhabits the body, only occasionally producing disease, or resides outside it, producing disease only when introduced into the body. For example, tetanus bacilli, inhabitants of the soil, produce disease only when introduced into abrasions or wounds. Although *not* communicable, tetanus is infectious. The term *contagious* is applied to diseases that are easily spread directly from person to person.

Infectious diseases are also classified as exogenous and endogenous. *Exogenous* infections are those in which the causative agent reaches the body from the outside and enters through one of the portals of entry. An *endogenous* infection is one from organisms normally present in the body. Endogenous infections occur when the defensive powers of the host are weakened or, for some reason, virulence of the microorganism is increased.

How microbes enter the body. Microorganisms invade the body by several avenues, and each species has its own favored one. The way of access to the body is known as the *portal of entry,* as indicated in the following areas.

SKIN. Most pathogens do not penetrate unbroken skin or mucous membrane, but some do. Staphylococci and some fungi can penetrate the hair follicles and cause disease in the deeper tissues of the skin. The organisms of tularemia pass through the unbroken skin, as do hookworm larvae. In their penetration of intact skin, malarial parasites have the help of a biting insect. Many bacteria are found in the superficial layers of the skin but under normal conditions are not able to penetrate further. As soon as the superficial barrier is broken, infection is easily accomplished.

RESPIRATORY APPARATUS. Pulmonary tuberculosis, pneumonia, and influenza are contracted by way of the respiratory tract, and the viruses causing measles, smallpox, and German measles enter the body this way.

ALIMENTARY TRACT. Some of the most important disease producers enter the body through the digestive tract, for example, the dysentery and typhoid bacilli, cholera vibrios, and amebas of dysentery. In many cases food and drink are the vehicles. The great majority of pathogenic microorganisms invade the body via the respiratory system or the digestive tract.

GENITOURINARY SYSTEM. Certain infections are acquired chiefly through the genitourinary system, notably the venereal or sexually transmitted diseases.

PLACENTA. Most microbes do not cross the placenta, but the spirochete of syphilis and the smallpox virus may.

EVENT OF INFECTION

Factors influencing the occurrence of infection. The fact that microbes have entered the body in no manner indicates that infection has occurred. Whether infection supervenes depends on (1) the portal of entry, (2) virulence of the organisms, (3) their number, and (4) defensive powers of the host (Chapter 12).

Most pathogenic bacteria have definite portals by which they enter the body, and they can fail to produce disease when introduced into the body by another route. For instance, typhoid bacilli produce typhoid fever when swallowed but only a slight local inflammation when rubbed on the abraded skin, whereas streptococci rubbed into the skin produce intense inflammation (cellulitis) but are generally without effect if swallowed. If streptococci are breathed into the lungs, they may cause pneumonia. A few bacteria such as *Francisella tularensis,* the cause of tularemia, can enter the body by several different routes and produce disease in each area. Usually the route of entry determines the disease process induced for a given organism.

By *virulence* or *pathogenicity* one means the ability of microbes to induce disease by overcoming the defensive powers of the host. There is a difference in virulence not only among species but also among members of the same species. As a rule, microbes are most virulent when freshly discharged from the person ill with the disease that they cause. Organisms harbored by carriers are less virulent.

Virulence may be increased by rapid transfer of organisms through a series of susceptible animals. As each animal becomes ill, the organisms are isolated from its excreta and transferred to a well animal. Herein are explained epidemics. The agent of the disease, by repeated passage from person to person, has become so virulent that everyone whom it contacts is made ill.

An organism that is highly virulent for one species of animals and less virulent for another may, with repeated passage through the animal for which it is less virulent, show a *transposal of virulence;* that is, it becomes less virulent for the animal for which it was originally highly virulent and highly virulent for the animal for which it was originally less so. Advantage is taken of this in the production of antirabies vaccine.

The *number* of microbes is crucial to infection. If only a few enter the body, they may well be overcome by the local defenses of the host even though they are highly virulent. Therefore, if infection is to occur, enough microorganisms to overcome the local defenses of the host must penetrate.

How microbes cause disease. When pathogens enter the body, two opposing forces are set in motion. The organisms strive to invade the tissues and colonize there. The body, utilizing its defensive powers, strives to block the invasion of the microbes, destroy them, and cast them off. If the body wins the contest, the microbes are destroyed, and the body suffers no ill effects. If the microbes prevail, infection occurs.

Microbes cause disease in various ways. In some cases the mechanical effects of microorganisms operate, for example, when the organisms of estivoautumnal malaria occlude the capillaries to the brain.

In most cases, however, biochemical effects are foremost. The ability of pathogenic microorganisms to produce disease is closely tied in with certain complex chemical substances that the organisms either elaborate and release or that are a vital part of their makeup. A soluble exotoxin (p. 88) is an example of a product released into body fluids and even into the bloodstream of the host. A constituent of the bacterial cell wall may be an injurious factor. A part of the microbial cell such as its capsule, though not harmful in itself, may so protect the microbe as to enhance its virulence. The carbohydrate (polysaccharide) capsules of the pneumococci invading the lower respiratory tract protect them from the defensive cells of the lungs, thereby facilitating the development of pneumonia.

Biochemical substances implicated in pathogenicity of microbes are many, and their action is not always well understood. Their major effects are summarized as follows:

    1. They interfere with mechanical blocks to the spread of infection set up in the body (bacterial kinases).

2. They slow or stop the ingestion of microbes by the phagocytic white blood cells (leukocidins).
3. They destroy body tissues (hemolysins, necrotizing exotoxins, lethal factor).
4. They cause generalized unfavorable reactions in the host, resulting in fever, discomfort, and aching (endotoxins). Such features are usually collected together under the term *toxicity*.

Most disease-producing microbes prefer a given part of the body; they have their favored site for involvement. This is *elective localization*. For instance, dysentery bacilli attack the intestine, pneumococci attack the lungs, and meningococci localize in the leptomeninges. Toxic products released by microbes also have a tissue affinity; the toxin of tetanus attacks the central nervous sytem, and the toxin of diphtheria affects not only the central nervous system but also the heart.

Local effects. By local effects one means the changes produced in the tissues in which given microbes are multiplying, summed up in the process of *inflammation*. Inflammation is the body's answer to injury. Its design is to halt the invasion and destroy the invaders. The features of the inflammatory process vary greatly with the different causative organisms and are significantly related to the disease-producing capacity of the microbe.

General effects. There are certain host reactions found in nearly all infectious diseases. In addition, many diseases have their own peculiar ones. Among the general effects are fever (p. 178), increased pulse rate (tachycardia), increased metabolic rate, and signs of toxicity. The degree of fever approximates the severity of the infection. Anemia is a result of prolonged and severe infections.

A very common effect of infection is a change in the total number of circulating leukocytes (white blood cells) and in the relative proportion of the different kinds. This is why total leukocyte and differential white cell counts are so important in the diagnosis of disease. In most infections the total number of leukocytes is increased *(leukocytosis)*. In some the number is decreased *(leukopenia)*, and in a few it is unchanged. If fever or leukocytosis fails to occur in infections where either ordinarily is present, a severe infection or decreased host resistance is indicated.

An important consequence of infection is the development of immunity (Chapter 12).

How disease-producing agents leave the body. Just as they have definite avenues of entry, pathogenic agents have definite routes of discharge from the body, known as *portals of exit*, which to a great extent depend on the part of the body that is diseased. The following list gives the important ones with examples:

1. *Feces*—bacteria of salmonellosis, bacillary dysentery, and cholera; protozoa of dysentery; viruses of poliomyelitis and infectious hepatitis
2. *Urine*—bacteria of typhoid fever, tuberculosis (when affecting the genitourinary tract), and undulant fever
3. *Discharges from the mouth, nose, and respiratory passages*—bacteria of tuberculosis, whooping cough, pneumonia, scarlet fever, and epidemic meningitis; viruses of measles, smallpox, mumps, poliomyelitis, influenza, and epidemic encephalitis

4. *Saliva*—virus of rabies
5. *Blood (removed by biting insects)*—protozoa of malaria; bacteria of tularemia; rickettsias of typhus fever and Rocky Mountain spotted fever; virus of yellow fever

PATTERN OF INFECTION

Course of infectious disease. The course of many infectious diseases extends over the following:

1. *Period of incubation*—interval between the time the infection is received and the appearance of disease (Table 10-2). In some diseases its length is constant. In others it varies greatly. The length of the incubation period depends on (a) the nature of the agent (for example, the incubation period of diphtheria is less than that of rabies), (b) its virulence, (c) resistance of the host, (d) distance from site of entrance to focus of action (for instance, the incubation period of rabies, which affects the brain, is shorter when the inoculation site is about the face), and (e) number of infectious agents invading the body.

2. *Period of prodromal symptoms*—short interval (prodrome) that sometimes follows the incubation period, described by such symptoms as headache and malaise.

3. *Period of invasion*—disease reaching its full development and maximum intensity. Invasion may be rapid (a few hours, as in pneumonia) or insidious (a few days, as in typhoid fever). At the onset of acute infectious disease, rigors and chills often precede the temperature rise. The sensation of cold is difficult to explain but probably results from a difference between the temperatures of the deep and superficial tissues of the body. To conserve heat, the superficial vessels of the skin constrict and sweating stops. The skin is pale and dry. As heat loss is decreased, the temperature rises rapidly.

4. *Fastigium or acme*—disease is at its height.

5. *Period of defervescence or decline*—stage during which manifestations subside. During this state profuse sweating occurs. Heat loss soon exceeds heat production. As the temperature falls, the normal hue of the body returns and sweating ceases. Defervescence may be by *crisis* (within 24 hours) or by *lysis* (within several days, with the temperature going down a little each day until it returns to normal). Fever that begins abruptly usually ends by crisis. During the stage of *convalescence* the patient regains his lost strength.

Many diseases are *self-limited*, which means that under ordinary conditions of host resistance and microbial virulence the disease will last a certain, rather definite length of time and recovery will take place.

Types of infection. In a *localized* infection the microbes remain confined to a particular anatomic spot (example, boils and abscesses). In a *generalized* infection, microorganisms or their products are spread generally over the body by the bloodstream or lymphatics. A *mixed* infection is one caused by two or more organisms. If a person infected with a given organism becomes infected with still another, a *primary* infection is complicated by a *secondary* one. Secondary infections of the skin and

TABLE 10-2. INCUBATION PERIODS OF IMPORTANT INFECTIOUS DISEASES*

Disease	Usual incubation period
Adenovirus infections	5 to 6 days
Amebiasis	2 weeks (varies)
Ascariasis	8 to 10 weeks
Brucellosis	5 to 30 days (varies)
Cat-scratch fever	3 to 10 days (varies)
Chickenpox	14 to 16 days
Cholera	1 to 3 days
Coccidioidomycosis, primary infection	1 to 3 weeks
Coxsackievirus infections	2 to 14 days
Diphtheria	2 to 6 days
Echovirus infections	3 to 5 days
Encephalitis	2 to 21 days (varies as to type)
Escherichia coli diarrhea	2 to 4 days
Gas gangrene	1 to 5 days
Gonorrhea	3 to 5 days
Hepatitis, viral, type A (infectious)	15 to 50 days
Hepatitis, viral, type B (serum)	6 weeks to 6 months
Herpesvirus infections	4 days
Histoplasmosis	5 to 18 days
Hookworm disease	6 weeks
Impetigo contagiosa	2 to 5 days
Infectious mononucleosis	2 to 6 weeks
Influenza	1 to 3 days
Leprosy	3 months to 20 years (?)
Leptospirosis	2 to 20 days
Measles (rubeola)	10 to 12 days
Meningitis, acute bacterial	1 to 7 days
Molluscum contagiosum	2 to 8 weeks
Mumps	14 to 21 days
Mycoplasmal pneumonia (primary atypical)	7 to 21 days
Pertussis	5 to 21 days
Pinworm infection	2 to 6 weeks
Plague	2 to 6 days
Poliomyelitis	7 to 14 days
Psittacosis	4 to 15 days
Rabies†	2 to 6 weeks (to 1 year)
Rocky Mountain spotted fever	3 to 12 days
Rubella	14 to 21 days
Salmonelloses	
Food poisoning (intraluminal)	6 to 72 hours
Enteric fever (extraluminal)	1 to 10 days
Typhoid fever	7 to 21 days
Scabies	4 to 6 weeks
Shigellosis (bacillary dysentery)	1 to 7 days
Smallpox	12 days
Streptococcal infections (scarlet fever, etc.)	2 to 5 days
Syphilis, primary lesion	10 to 90 days
Tetanus	3 days to 3 weeks
Trichinosis	2 to 28 days
Tuberculosis, primary lesion	2 to 10 weeks
Tularemia	1 to 10 days
Yellow fever	3 to 6 days

*In this table and throughout the book the length of the incubation period is usually that given in the "Red Book" of the American Academy of Pediatrics.
†Rabies incubation in dogs is 21 to 60 days.

respiratory tract are quite common, and in some cases the secondary infection is more dangerous than the primary, for example, the streptococcal bronchopneumonia that often follows measles, influenza, or whooping cough. Lowered body resistance resulting from the primary infection facilitates the development of the secondary infection.

A *focal* infection is one confined to a restricted area from which infectious material spreads to other parts of the body. Examples are infections of the teeth, sinuses, and prostate gland. An infection that does not cause any detectable manifestations is an *inapparent* or *subclinical* one. An infection held in check by the defensive forces of the body but which may spread when the body resistance is reduced is a *latent* infection. An infection from the accidental or surgical penetration of the skin or mucous membranes is sometimes spoken of as an *inoculation* infection.

When bacteria enter the bloodstream but do not multiply, the condition is spoken of as a *bacteremia*. If they enter the bloodstream and multiply, causing infection of the bloodstream itself, the condition is *septicemia*. Septicemia is the layman's "blood poisoning." When pyogenic bacteria (pus formers) in the bloodstream are spread to different parts of the body to lodge and set up new foci of disease, the condition is called *pyemia*. When toxins liberated by bacteria enter the bloodstream and cause disease, the condition is *toxemia*. Diphtheria is a good example of a toxemia. Saprophytic bacteria may grow on dead tissue such as a retained placenta or a gangrenous limb and produce poisons that cause disease when absorbed into the body. This is *sapremia*. Patients with chronic wasting diseases like cancer often die from the immediate effects of some bacterial infection, especially streptococcal and pneumococcal infections. These are *terminal* infections.

A *sporadic* disease occurs only as an occasional case in a community. An *endemic* disease is one that is constantly present to a greater or lesser degree in a community. When a disease attacks a larger number of persons in the community in a short time, an *epidemic* is said to exist. Endemic diseases may become epidemic. When a disease becomes epidemic in a great number of countries at the same time, it is said to be *pandemic*.

Epidemiology presents the pattern of disease in a given community and is the study of those factors influencing its presence or absence. The *incidence* of disease is the number of new cases per block of population in a specific time period. The *prevalence* of the disease is the number of cases in existence at any given time in that population.

SPREAD OF INFECTION

Transmission of communicable diseases. The etiologic agents of communicable diseases may be transmitted from the source of infection to the recipient by (1) direct contact, (2) indirect contact, or (3) insect carriers.

Direct contact is the term applied when an infection is spread more or less directly from person to person or from a lower animal to man. It does not necessarily mean actual bodily contact, except with venereal diseases (p. 359), but does indicate a rather close association. *Droplet infection*, infection by microbes cast off in the fine spray from the mouth and nose during coughing, talking, or laughing, is a form of direct contact, as is placental transmission. By direct contact, blood transfusion transmits malaria and viral hepatitis. Diseases spread directly from person to person include tuberculosis, diphtheria, measles, pneumonia, scarlet fever, the common cold, smallpox, syphilis, gonorrhea, and epidemic meningitis, and from lower animals to man, rabies and tularemia. (Some of these are transmitted also by indirect contact.)

Indirect contact refers to the spread of the agent of a disease by conveyers such as milk, other foods, water, air, contaminated hands, and inanimate objects. Infections arising from indirect contact are usually more widely separated in space and time than those arising by direct contact. The diseases ordinarily spread by indirect contact are those in which the infectious material enters the body via the mouth.

Food, including milk, may convey infection. Salmonellosis, bacillary dysentery, and cholera may be spread by contaminated food. Botulism is caused by the consumption of canned foods insufficiently heated. Trichinosis and tapeworm infection are contracted from improperly cooked pork.

Air is established in the spread of infection; particles of dried secretion from the mouth or nose *(droplet nuclei)* may float in the air for a considerable time and be carried a rather long distance. Naturally, airborne transfer operates in respiratory infections.

Contaminated fingers are important conveyors of infection. A person with contaminated fingers may pass infection to other people or to different parts of his own body. Fingers easily soil food and drink.

Fomites are inanimate objects that spread infection. The most important are items such as handkerchiefs, towels, blankets, bed sheets, diapers, pencils, and drinking cups. Money is grossly contaminated, especially the small unit coins and paper bills that constantly change hands. Potential pathogens can be cultured from 13% of coins and 42% of bills.

Filth is responsible for disease, and certain diseases are primarily ones of filth. Of these, typhus fever, as it is seen in parts of the world, is an example. However, filth plays little or no part in the spread of typhus fever in the United States; it is rare in this country. Persons who are unclean in their personal habits are more likely to contract an infectious disease than are those who are clean, and disease is more difficult to control in an unsanitary community.

Insects convey disease mechanically or biologically. In the *mechanical* transfer of disease, insects merely hold the pathogens on their feet or other parts of their body. Flies can so transfer the agents of typhoid fever and bacillary dysentery from the feces of patients to food to infect the consumer of that food. Flies act as more or less important carriers of 30 diseases. The distance that they may carry infectious agents is surprising—they have been known to travel miles in search of food.

In the *biologic* transfer of disease the insect bites a person or animal ill of a disease or a carrier and ingests some of the infected blood. The microbes therein undergo a cycle of development within the body of the insect within a specified time. The insect can then transfer the infection to a well person, usually by biting that person. Spread in

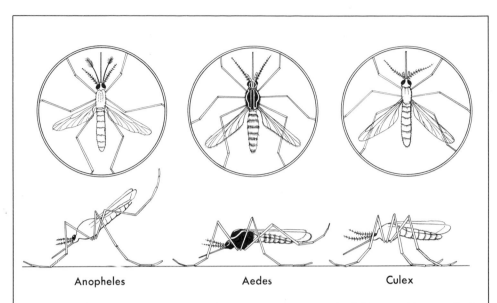

Anopheles   Aedes   Culex

FIG. 10-1. Three mosquitoes as biologic vectors in spread of disease. Note typical resting positions. Mosquitoes of genus *Anopheles* transmit malaria; members of genus *Aedes* transmit yellow fever and dengue fever; common house mosquito (genus *Culex*) is vector for togaviral (arboviral) encephalitis and filariasis.

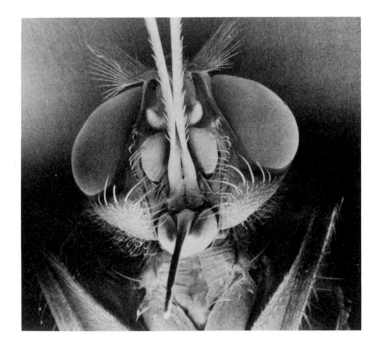

FIG. 10-2. Head of tsetse fly, scanning electron micrograph. (×60.) (Courtesy Eastman Kodak Co., Rochester, N.Y.)

this manner are malaria, yellow fever, and encephalitis by mosquitoes (Fig. 10-1); typhus fever by lice; Texas fever (a disease of cattle) by ticks; African sleeping sickness by tsetse flies (Fig. 10-2); and plague by fleas. For diseases that are biologically transferred to be prevalent in a community, both the transmitting insect and hosts to harbor the infection must be present. As a rule, a particular infection is spread by only one species of insect, and a given insect is able to spread only one type of infection. There are, however, important exceptions.

Sources of infection in communicable diseases. In only a few instances does the infectious agent live in lifeless surroundings outside the body long enough to maintain the source of infection. For the continuous existence of a disease there must be a *reservoir of infection*. Reservoirs of infection and the sources of most communicable diseases are found in practically all communities as (1) human beings or animals with typical disease, (2) human beings or animals with unrecognized disease, and (3) human or animal carriers. Plants may be the reservoir in some fungous infections.

Human carriers are persons who harbor pathogenic agents in their bodies but show no signs of illness. The carrier is infected but asymptomatic. *Convalescent carriers* harbor an organism during recovery from the related illness. *Active carriers* harbor an organism for a long time *after* recovery. *Passive carriers* shelter a pathogen without having had its disease. *Intestinal* and *urinary carriers* discharge it from the body via the feces and urine, respectively. *Oral carriers* discharge infectious material from the mouth. *Intermittent carriers* discharge organisms only at intervals. (Intestinal carriers are often intermittent carriers.) Human carriers play a significant role in the spread of diphtheria, epidemic meningitis, salmonellosis, amebic dysentery, bacillary dysentery, streptococcal infections, and pneumonia.

As a rule, an actual case of a disease is more likely to spread infection to others than a carrier. Carriers and persons with unrecognized disease keep epidemic diseases in existence during interepidemic periods; the number of carriers increases just before the epidemic (for example, diphtheria and acute bacterial meningitis). The prominence of carriers in the spread of a disease depends on the frequency with which people become carriers and the length of the carrier state.

Animals spread disease to man in a number of ways. Man may so acquire disease from (1) direct contact with an infected animal, (2) contamination of food by discharges of the animal, (3) insect or rodent vectors, (4) contaminated air or water, and (5) consumption of animal products such as milk or eggs.

Rabies is acquired by the bite of rabid dogs, cats, or other animals. Children may contract tuberculosis by drinking the raw milk of infected cows. Bubonic plague is essentially a disease of rats, transmitted from rat to man by the flea. Undulant fever may be transmitted to man by milk, or it may be acquired when man handles the meat of infected animals. Psittacosis is an infectious disease of parrots in which transmission may be airborne from the sick parrot to man. Those handling the hides of anthrax-infected animals may contract anthrax. Tetanus may be indirectly contracted from horses because the tetanus bacillus is a normal inhabitant of the intestinal canal of the horse. This is why wounds contaminated by barnyard dirt are likely to be followed by tetanus. Shellfish such as oysters may transmit typhoid fever, hepatitis,

TABLE 10-3. SOURCES OF INFECTION IN HOUSEHOLD PETS

Disease	Dogs	Cats	Farm animals	Caged birds (pigeons, parrakeets, parrots, myna birds, canaries, etc.)	Poultry	Rodents (mice, rabbits, rats, hamsters, etc.)	Reptiles (snakes, turtles, lizards, etc.)
Viral							
Encephalitides			✔	✔			
Rabies	✔	✔	✔				
Bacterial							
Anthrax	✔		✔				
Brucellosis	✔	✔	✔		✔	✔	
Chlamydial							
Cat-scratch disease		✔					
Psittacosis-ornithosis				✔	✔		
Leptospirosis	✔		✔	✔		✔	
Rickettsial							
Q fever			✔				
Salmonellosis	✔	✔	✔	✔	✔	✔	✔
Tuberculosis	✔	✔	✔				
Tularemia	✔	✔				✔	
Fungous							
Ringworm	✔	✔	✔			✔	
Protozoan							
Toxoplasmosis	✔	✔					
Metazoan							
Roundworm infection	✔	✔					
Scabies	✔	✔					
Tapeworm infection	✔	✔	✔				

and other enteric infections if human excreta contaminates the water in which they are grown.

All told, there are about 150 different infections that are spread from animals to man. Rats and mice carry more than 15 different diseases, and the casualties from rodent-borne diseases are greater than those from all of history's wars.

Animals found in the home as pets may be significant sources of infection (Table 10-3), since many persons live closely associated with them. The two most common household pets are cats and dogs. There are an estimated 110 million of them in the United States today. In North America the dog can transmit 24 diseases to man, and the cat can transmit twelve. Exotic animals that are becoming more popular as household pets may present real hazards. For example, monkeys in the home, pet shop, or even some zoos may spread such diseases to man as shigellosis, hepatitis, salmonellosis, and tuberculosis.

The most prevalent of the *zoonoses* (infections of animals secondarily transmissible to man) are cat-scratch disease, salmonellosis, and fungous infections. There are around 181 different diseases known to be transmitted naturally *from animals to man.*

Spread of disease in the jet age. This is an age of travel at high speeds. More and more people are and will be traveling over the world aboard the jumbo jets or supersonic airliners and consequently exposing themselves to the infections of the geographic localities they visit. The problems in public health are enormous. Some are as follows: First, disease is so easily and quickly spread as to jeopardize public health controls. A traveler can bring back smallpox, for example, into a country that had virtually eliminated it. Second, since no two points on the earth are separated by more than 48 hours' travel time, the tourist can return home during the incubation period (feeling well), to come down with the disease days or weeks later. By this time, the diagnosis may not be obvious. Third, the world sightseer may fail to take adequate precautions against a disease he inevitably contacts. The best example of this is malaria, the major risk. It is a widespread disease but nowhere is chemoprophylaxis required. The traveler must find out for himself what to do for protection. Even then, he may fail to take the suppressive drugs as required and may discontinue them after leaving the infected area, feeling they are no longer needed. Fourth, still another problem has to do with the establishment of international standards for sanitary facilities. Maintenance of these is required for the tourist in all areas, but mandatory aboard the big jet, in the airport terminal, and in the accommodations related to it.

KOCH'S POSTULATES

Absolute proof that an organism causes a given disease rests on the fulfillment of certain requirements known as Koch's postulates. * These are as follows:

1. The organism must be observed in every case of the disease.

---

*Robert Koch (1843-1910) stated certain principles related to the germ theory of disease with such clarity that they are known as Koch's postulates. They remain to this day the basis of the experimental investigation of infectious disease.

2. The organism must be isolated and grown in pure culture.
3. The organism must, when inoculated into a susceptible animal, cause the disease.
4. The organism must be recovered from the experimental animal and its identity confirmed.

In exceptional instances an organism has been accepted as the cause of a disease although all of these requirements have not been met.

QUESTIONS FOR REVIEW

1. Explain infection and contamination.
2. Give the routes by which microbes enter the body. Name two diseases whose causative agent enters the body by each route.
3. What is virulence? How do microbes produce disease?
4. What is one important local effect of bacterial invasion?
5. List some of the general effects of bacterial invasion.
6. What is the difference between bacteremia and septicemia?
7. Give the routes by which disease-producing agents leave the body. Name two agents eliminated by each route.
8. List five diseases spread by animals to man.
9. Define carrier. List the different kinds.
10. Name three diseases spread by carriers.
11. List important insect vectors and diseases spread.
12. State Koch's postulates.

REFERENCES. See at end of Chapter 14.

11 The body's defense

PROTECTIVE MECHANISMS

If infection occurred each time infectious and injurious agents entered our bodies, we would be constantly ill. When microorganisms attempt to invade the body, they must first overcome certain mechanical, physiologic, and chemical barriers existing on the body surface or in the body cavity at the site of entry. The body not only reacts to the invasion but is also endowed with a measure of protection against that event.

Anatomic barriers. During his agelong struggle for existence, man has developed certain mechanisms that enable him to overcome many agents of potential injury in his environment. In the first place, nature has provided him with special senses that act as watchdogs against danger. Man protects himself from major injuries by moving objects by batting his eyes and jumping aside. Although some of these acts are voluntary, they are executed as reflex movements so that they may be considered protective natural mechanisms.

The body's first line of defense is the epithelium that covers the exterior of the body and lines its internal surfaces. A condensation of resistant cells along the most superficial part of the skin offers a barrier against physical and chemical injuries or microbial invasion. Epithelium may become quite thick at a site of irritation (example, a callus or corn in the skin). The lining cells of the mucous membranes opening on body surfaces also afford a barrier, but these cells are softer and more vulnerable to injury than are the surface cells of the skin. The hairs in the anterior nares protect the respiratory tract by filtering bacteria and larger particles from the inspired air. The respiratory passages are lined with epithelial cells from the surface of which spring hairlike appendages, known as cilia, that sweep overlying material from the deeper portion of the tract to the upper portion from where it may be discharged to the environment.

Closure of the glottis during swallowing prevents food from entering the respiratory tract. Coughing and sneezing serve to expel mechanically any irritating materials from the respiratory tract. In a similar manner vomiting and diarrhea mechanically rid the intestinal tract of irritants.

Chemical factors. Body cavities opening on the surface are protected by secretions that wash away bacteria and foreign materials. Body fluids such as saliva, tears, gastric juice, and bile exert an antiseptic action that reduces microbial invasion. For instance, the gastric juice destroys bacteria and almost all important bacterial toxins except that of *Clostridium botulinum*. The antimicrobial action of mucus and tears is partly related to the presence of *lysozyme* (muramidase), a substance dissolving the cell walls of certain gram-positive bacteria. It is found in secretions coming from organs exposed to airborne bacteria. Lysozyme, an enzyme not an antibody, was discovered by Fleming (in 1922) before he discovered penicillin. Because of their acid reaction and content of fatty acids, sweat and sebaceous secretions on the skin have antimicrobial properties.

Another internal defense is a naturally occurring substance *interferon* (not an antibody). Its target action is against viruses to block viral infection. Many cells form it quickly if so threatened.

Physiologic reserves. The blood supply of many parts of the body is protected by a series of *anastomoses* (connections) among the branches of the supply vessels. If a vessel is occluded, the blood is detoured around the obstruction by way of the anastomoses. A *collateral* or *compensatory circulation* is set up.

When the body becomes chilled, the superficial vessels of the skin contract to prevent heat from being dissipated at the surface, and sweating stops to slow cooling caused by evaporation. To aid in this conservation of heat, the involuntary muscles of the skin contract (gooseflesh). In animals having hair or feathers, the hair or feathers are raised to enclose in their meshes a thick layer of air, which is a poor conductor of heat. The muscular activity associated with rigor and shivering increases heat production. When the body becomes too hot, the superficial vessels dilate, and sweating occurs. Heat is dissipated from the surface, and evaporation has a cooling effect.

The functional capacities of the vital organs (organs necessary for life) extend far beyong the normal demands of life. We have much more liver, pancreas, adrenal gland, and parathyroid gland tissue than we ordinarily use. That we are able to lead a normal life after one kidney has been removed or after one lung has collapsed illustrates the abundance of reserve possessed by our vital organs.

Vital organs (brain, heart, lungs) are enclosed within protective bony cases, and their surfaces are bathed with a watery fluid. The fluid surrounding the brain serves as a water bed to prevent undue jarring, and the fluid that bathes the surface of the heart and lungs acts as a lubricant.

Fever. Fever is a condition characterized by an increase in body temperature. It is usually thought to result because toxic substances arising from the disintegration of microbes and other cells in an injured (or inflamed) area gain access to the bloodstream. There are many causes, ranging from mechanical to microbial injury, but microbial infection is by far the most regular. Although the manifestations of fever may be disagreeable, it is primarily beneficial. If the temperature is not too high, fever accelerates the destruction of injurious agents by increasing phagocytosis and production of immune bodies (Chapter 12).

In health the body temperature of man remains remarkably constant and, as determined by a thermometer placed in the mouth, ranges from 96.7° F (35.94° C) to 99° F (37.22° C) with an average of about 98.6° F (37° C). The rectal temperature is about 1° F higher, and the temperature taken in the armpit is about 1° F lower than the oral temperature. The rectal temperature is the most dependable because both oral and axillary readings are influenced by many outside factors. The temperature of the internal organs is 2° or 3° F higher than that of the skin.

There is a close relation between pulse rate and temperature. Ordinarily an increase of around eight pulse beats per minute occurs for each 1.8° F increase in temperature. Certain exceptions to this rule are diagnostically significant. For instance, the pulse in typhoid fever, malaria, miliary tuberculosis, and yellow fever is comparatively slow.

Inflammation. *Inflammation* is the sum of the reactions in the body incited by an injury. Within itself, inflammation is not a pathologic condition (although usually discussed as one), but rather it is an exaggeration of physiologic processes set in motion by an irritant. The initial purpose of inflammation is to destroy the irritating and injurious agent and to remove it and its related by-products from the body. If this is not possible, the inflammatory process serves to limit its extension through and effects on the body. Finally, inflammation is the mechanism within the body for the repair and replacement of tissues damaged or destroyed by the offending agent.

RETICULOENDOTHELIAL SYSTEM (PHAGOCYTIC SYSTEM)

The reticuloendothelial (RE) system with its widely dispersed cells constitutes one of the main organs in the body designed primarily for man's defense against both the living and the nonliving agents in his environment that harm him. The RE cells not only can directly fight the invader, which they do by phagocytosis, but can also clean up the debris and eliminate the cellular and metabolic breakdown products incident to the encounter. In short, they are also scavengers.

The production and regulation of the cells in the peripheral blood and bone marrow are crucial functions of the RE system. Formation of the different blood cells (hemopoiesis) is possible because of the presence within the system of primitive nonphagocytic cells. Being multipotential, they give rise to the range of blood cell types found.

Phagocytosis. The ingestion of microbes or other particulate matter by cells is known as *phagocytosis*, and the cells that ingest such materials are phagocytes. Phagocytosis resembles closely the feeding process of unicellular organisms and is a universal response on the part of the body to penetration by microbes, alien cells, or other foreign particles. It is an essential protective mechanism against infection and probably plays an important part in natural immunity. Phagocytosis is a general process because the few kinds of body cells having this power can ingest many different kinds of particulate matter (bacteria, dead body cells, mineral particles, dusts, pigments).

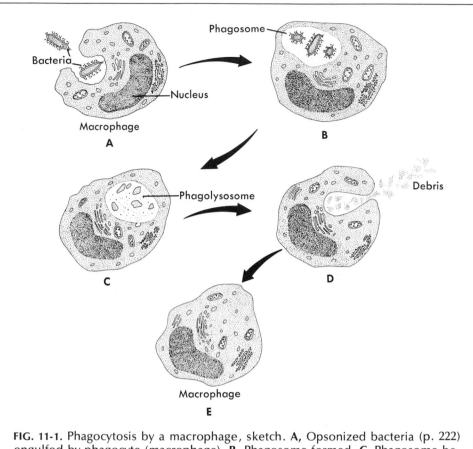

FIG. 11-1. Phagocytosis by a macrophage, sketch. **A,** Opsonized bacteria (p. 222) engulfed by phagocyte (macrophage). **B,** Phagosome formed. **C,** Phagosome becomes phagolysosome; bacteria digested. (To this point, process of phagocytosis is comparable in either a macrophage or neutrophil, not shown.) **D,** Debris is egested. (Neutrophil would succumb here.) **E,** Macrophage returns to resting state. (From Smith, A. L.: Microbiology and pathology, ed. 11, St. Louis, 1976, The C. V. Mosby Co.)

Study of the process has revealed that phagocytosis takes place in three phases. (Fig. 11-1). First, the particle (for example, a microbe) to be engulfed is isolated and incorporated into an invagination of the cell membrane of the phagocyte. The result is a *phagosome*, a vacuole (within the cell) surrounded by cytoplasm. Next, there is a burst of metabolic activity within the cell. The phagosome is moved into the interior of the cell contacting its lysosomes. When hydrolytic enzymes are released into the phagosome, it becomes a *phagolysosome*. A special kind of microbicidal system is activated in preparation for the second phase. This is the killing or inactivation of the microbe, so that in the third phase it can be digested.

If, after ingesting bacteria, phagocytes fail to destroy them, the phagocytes themselves may be destroyed with liberation of the ingested organisms. Relatively harmless microbes are usually completely destroyed. Sometimes the microbes persist and

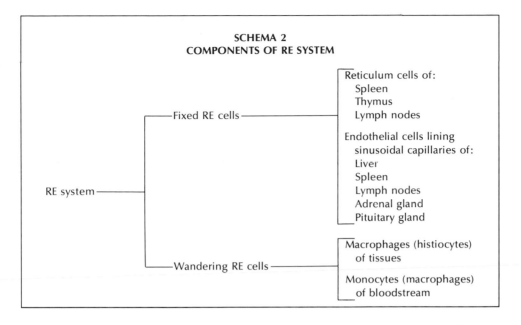

SCHEMA 2
COMPONENTS OF RE SYSTEM

RE system
— Fixed RE cells
  — Reticulum cells of:
    Spleen
    Thymus
    Lymph nodes
  — Endothelial cells lining
    sinusoidal capillaries of:
    Liver
    Spleen
    Lymph nodes
    Adrenal gland
    Pituitary gland
— Wandering RE cells
  — Macrophages (histiocytes)
    of tissues
  — Monocytes (macrophages)
    of bloodstream

even continue to multiply within the cytoplasm of the phagocyte. Virulent bacteria, especially those with capsules, are resistant to phagocytosis.

The student who wishes to see bacteria undergoing phagocytosis should stain a smear from an ordinary boil. Staphylococci can be found easily within the cytoplasm of the leukocytes. Different bacteria show differences in their susceptibility to phagocytosis. For instance, gonococci undergo phagocytosis easily, whereas tubercle bacilli are quite resistant.

Phagocytosis is not so great in the first 3 years of human life as it is later on.

Anatomy of RE system. The RE system is a group of cells in the body especially endowed with phagocytic powers—both the cells themselves and those to which they give rise. Although widespread throughout the body (Fig. 11-2), these cells are concentrated in lymph nodes, spleen, bone marrow, and liver, where they are arranged as the lining cells for the peculiar kind of sinusoidal arrangement common to these organs. Although scattered, these uniform cells are spoken of collectively as the RE system. There are two main cell types: (1) the stationary or fixed (littoral) cells and (2) the free or wandering cells. The wandering cells in tissues are known as *macrophages;* in the bloodstream, as *monocytes.*

The components of the RE system are shown in Schema 2.

Because the macrophage moves about in body fluids and because its observed properties of phagocytosis are dramatic, it receives a lot of attention. The details of its life history, including even its exact origin, are not completely worked out. It is known by a variety of names—RE cell, histiocyte, epithelioid cell, clasmatocyte, multinucleated giant cell (if confluence of several cells)—relating it (the same cell) to different circumstances and reflecting an active participation in its diverse roles.

In *granulomatous* inflammation the disease-producing agent is dealt with directly

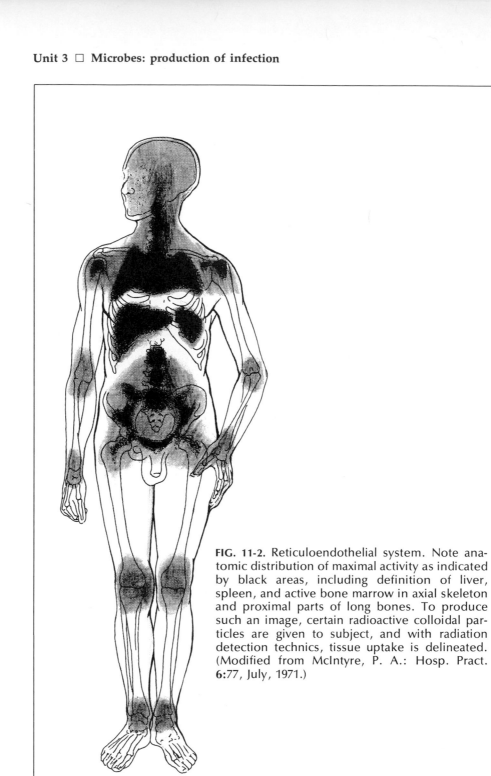

FIG. 11-2. Reticuloendothelial system. Note anatomic distribution of maximal activity as indicated by black areas, including definition of liver, spleen, and active bone marrow in axial skeleton and proximal parts of long bones. To produce such an image, certain radioactive colloidal particles are given to subject, and with radiation detection technics, tissue uptake is delineated. (Modified from McIntyre, P. A.: Hosp. Pract. **6:**77, July, 1971.)

by the RE system, and the main reacting cell is this wandering phagocyte or macrophage.

A most important derivative cell of the RE system, also phagocytic, is the *polymorphonuclear neutrophilic leukocyte*, familiarly known as the "poly," a cell especially designed to phagocytize bacteria (microphage *) and one of several kinds of white blood cells (leukocytes). With the majority of bacterial inflammations, there is an increase in the number of these cells at the site of injury and in the bloodstream *(leukocytosis)*. They have been produced in the bone marrow, where they are normally formed, in increased numbers in response to chemotaxins released from the inflammatory process into the bloodstream and circulated to the bone marrow. Neutrophils are rich in proteolytic enzymes, known as leukoproteases, contained within their lysosomes. Leukoproteases participate in intracellular digestion of phagocytized particles. In conditions that induce a leukocytosis, the phagocytic power of the leukocytes is enhanced in patients who are doing well.

The phagocytic cells especially related to processes of immunity (discussed in Chapter 12) are the polymorphonuclear neutrophils and the macrophages of the RE system.

LYMPHOID SYSTEM (IMMUNOLOGIC SYSTEM †)

Lymphoid tissue. Closely interwoven with the RE system (the phagocytic system) and encompassed by it is the lymphoid system (the immune system), which comprises the lymphoid tissues and organs of the body. Lymphoid tissue is widely distributed in the body; it is concentrated in lymph nodes, spleen, thymus, and Peyer's patches (in the ileum of the bowel) and scattered in the lining of the alimentary tract and bone marrow. Its major constituent cells are lymphocytes, which stand out in its mixed population of plasma cells, macrophages, and respective precursors (forebears). Lymphocytes are small round cells with a relatively large and darkly staining nucleus, and their constant companions, the plasma cells, are slightly larger cells with an eccentric nucleus of unique appearance.

In lymphoid tissue the component cells are packed more or less densely onto a spongelike support provided by a special type of stroma, and collections of lymphocytes, including precursors and progeny, in a well-defined, usually rounded aggregate constitute a lymphoid nodule. A nodule may be solitary, or several nodules may be grouped together, as in Peyer's patches. Lymphoid tissue organized into a small bean-shaped organ well enscribed by a connective tissue capsule becomes a *lymph node* (Fig. 11-3). The outer part is referred to as the cortex; the inner part is referred to as the medulla. In the cortex there is an orderly alignment of lymphoid follicles. Lymph nodes are found throughout the body, strategically placed in all major organ systems and body areas. Lymphocytes circulate in the bloodstream as single cells and account for one fifth of the white blood cells present.

---

*A *micro*phage ingests small particles, notably bacteria; a *macro*phage, larger ones.
†Immunologic function is an important property of lymphoid tissue. Mechanisms are discussed in the next chapter.

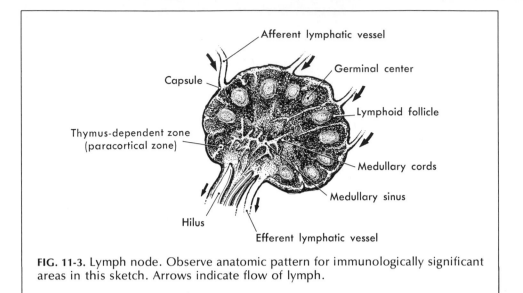

FIG. 11-3. Lymph node. Observe anatomic pattern for immunologically significant areas in this sketch. Arrows indicate flow of lymph.

Lymphatic circulation. In the exchange of nutrients for waste products that goes on continuously between the cells and the blood of the capillaries, excess fluid leaks into the tissue spaces. From here it moves into the lymphatic capillaries, the many small thin-walled tubes of the lymphatic circulation draining the vast intercellular region of the body. The fluid circulates as lymph into successively larger vessels and ultimately pours back into the bloodstream. Along the course of the lymphatic channels at definite intervals are placed the lymph nodes in chain formations (Fig. 11-4). All the lymph of the body is filtered and strained through these chains.

Spleen. The largest mass of lymphoid tissue in the body is the spleen, found in the upper left abdomen as a large, encapsulated lymphoid organ adapted to a peculiar circulation of blood. Like the lymph nodes it is a filter, but unlike them it strains the blood not the lymph stream. It functions as an important member of the RE system in the filtration of foreign particles from the blood, including living organisms. The fixed cells of the RE system line the lumens of all lymphatic channels, lymph sinuses, and splenic sinusoids.

Thymus. The thymus, a key organ in immunologic processes, is situated in the front part of the chest just behind the breast bone. It is a lobular, partly epithelial and partly lymphoid organ, which is present at birth and continues to enlarge until puberty. After puberty, it shrinks (atrophies) and is soon replaced by fatty tissue in the adult (Fig. 11-5). (This sequence of events has always suggested that the thymus plays a significant role in the individual's development.) At or about the time of birth it processes certain lymphocytes, which then migrate out to colonize the spleen, lymph nodes, and other areas of the body. These in turn give rise to cells that maintain cell-mediated immunity (Chapter 12).

Lymphatic vessels in infection. When an agent gains access to the tissues, it may sometimes be carried away from the site of entry in the lymphatic circulation and be

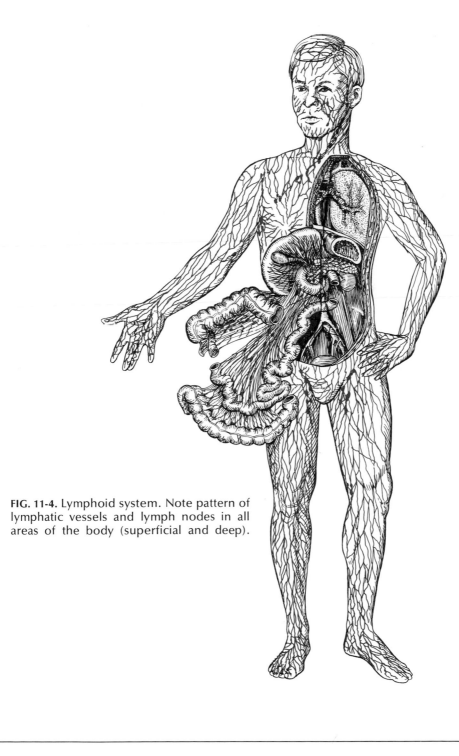

FIG. 11-4. Lymphoid system. Note pattern of lymphatic vessels and lymph nodes in all areas of the body (superficial and deep).

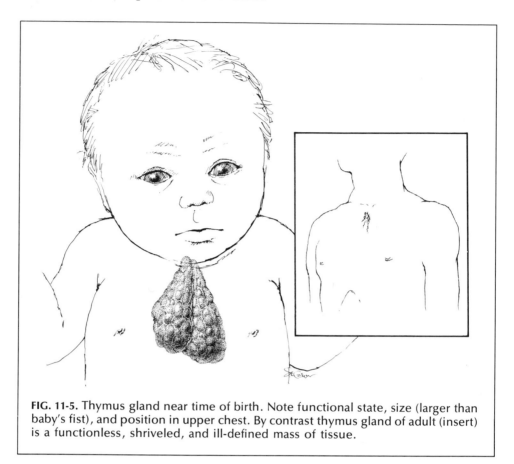

FIG. 11-5. Thymus gland near time of birth. Note functional state, size (larger than baby's fist), and position in upper chest. By contrast thymus gland of adult (insert) is a functionless, shriveled, and ill-defined mass of tissue.

deposited in the nearest lymph nodes. (In the lungs of persons residing in coal-burning regions, coal dust is thus deposited in drainage nodes.) Likewise, bacterial invaders may be borne from the site of disease to the regional nodes, where their presence stimulates phagocytic cells normally present to increased activity. Hopefully, these phagocytes dispose of the invaders. Should they fail to do so, the bacteria flow with the lymph stream through successive nodes in the chains into the thoracic duct, from there to be emptied into the bloodstream (bacteremia) and disseminated throughout the body. Infection of the bloodstream (septicemia) before the days of antibiotic therapy was a rapidly fatal condition.

RE responses in lymphoid organs. There is an intimate relationship between elements designated as lymphoid and those designated as reticuloendothelial. If an infection is localized, the RE cells of the regional lymph nodes proliferate and the nodes enlarge as a consequence (lymphadenopathy). Regional lymph nodes are an important second line of defense. Many injurious agents carried there are destroyed. If an infection becomes generalized, the RE system throughout the body responds. One notable feature of this is the enlargement of the spleen (splenomegaly) seen with such acute infectious processes as typhoid fever, malaria, and septicemia.

QUESTIONS FOR REVIEW

1. List mechanisms that serve to protect the body from disease or injury and explain how each functions.
2. What is lysozyme? Who discovered it?
3. What are cilia? What is their function in the respiratory passages?
4. What is a vital organ? Name three. How are the vital organs protected in the body?
5. Define fever.
6. What is considered normal body temperature?
7. What is the relation between pulse rate and temperature?
8. Briefly describe the RE system. What is its chief function?
9. Briefly discuss phagocytosis. What are the three phases? Why is this an important process?
10. Outline the anatomic distribution of the RE system.
11. Give the component cells and organs of the lymphoid system.
12. What is meant by lymphadenopathy? By splenomegaly?

REFERENCES. See at end of Chapter 14.

12 Immunologic concepts

MEANING OF IMMUNITY

Immunology is the division of biology concerned with the study of *immunity*. Today the concept of immunity is hard to define. In terms of the steady progression of knowledge in the field of immunology, the meaning of the word is rapidly expanding.

Basic to an understanding of the complexity of physiologic mechanisms encompassed by immunity are the concepts of "self" and "nonself," or "foreign." For a given individual to preserve his biologic integrity (self) he must be able to recognize and deal with factors in his environment threatening it. On the one hand, there is self, which must remain intact. On the other, there are the elements it contacts, not self, although perhaps of comparable makeup, and potentially harmful. Because of the many biologic parallels in nature, the distinction is a fine one. We require an arrangement whereby the cells and cell substances of our bodies are marked off as "ours" and the substances that appear on the scene are accosted as "intruders."

The scheme of things in the body that function in this discriminatory way is the machinery of immunity, and the elements making it up are collectively the *immune system*. The term *immunobiology* surveys the subject of immunity and its many ramifications in modern medicine.

Susceptibility is the reverse of immunity and the result of an absence of suppression of the factors that produce immunity.

IMMUNITY AND INFECTION

Infectious agents such as bacteria and other microorganisms are easily "foreign" to the body (nonself), their effect detrimental, and the cause-effect relationships clear cut. The applications of immunity with infectious processes have been dramatic. The classic definition that immunity is a highly developed state of resistance has referred directly to infectious disease.

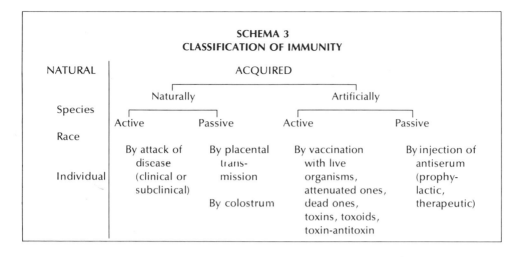

SCHEMA 3
CLASSIFICATION OF IMMUNITY

NATURAL	ACQUIRED			
	Naturally		Artificially	
Species	Active	Passive	Active	Passive
Race				
Individual	By attack of disease (clinical or subclinical)	By placental transmission	By vaccination with live organisms, attenuated ones, dead ones, toxins, toxoids, toxin-antitoxin	By injection of antiserum (prophylactic, therapeutic)
		By colostrum		

Kinds of immunity. Traditionally immunity has been classified as shown in Schema 3.

A *natural* immunity is a more or less permanent one with which a person or lower animal is born; that is, it is a natural heritage. It may be the heritage of a species, race, or individual; it is also known as *innate* or *genetic* immunity. *Species* immunity is that peculiar to a species (for example, man does not have distemper, nor do dogs have measles). A *racial* immunity is one possessed by a race (for example, ordinary sheep are susceptible to anthrax, whereas Algerian sheep seldom contract the disease). Species and racial immunities are more highly developed in plants than in animals. *Individual* immunity is a rare condition, and most so-called cases are from unrecognized infections.

An *acquired* immunity is one in which protection *must be obtained*. It is never the heritage of a species, race, or individual. It may be *naturally* acquired when a mother transmits antibodies (protective substances) to her fetus by way of the placenta or when a person has an attack of a disease. When a person has a disease such as measles, the disease is rather severe for a time, after which recovery occurs. The person will not then contract the disease again although repeatedly exposed; he has developed a permanent immunity.

An immunity may be *artificially* acquired by vaccination or the administration of an immune serum.

Depending on the part played by the body cells of the animal or person being immunized, acquired immunity (natural or artificial) is classified as *active* and *passive*. If a person has an attack of measles or is vaccinated against it, his body cells respond to the presence of the agent (or its products) by producing antibodies that destroy that agent should it again gain access to the body. When certain body cells react to the agent or its products in this way, the naturally acquired immunity is an *active* one. For example, if a horse is given frequent injections of diphtheria toxin, beginning with a small dose and with a gradually increased amount at each subsequent injection, the horse will eventually be able to withstand thousands of times the amount of toxin required to kill an untreated animal; it has become immune to the action of diphtheria toxin.

Immunity transferred by way of the placenta to the child from his mother is a *passive* immunity, which the child has naturally acquired. When the serum of an actively immunized animal is injected into a nonimmune animal, the latter becomes temporarily immune. This artificially acquired immunity is *passive* because the immunity-producing principle is introduced in the serum, and the cells of the recipient animal take no part in the process. For the first 4 to 6 months of his life an infant may have a passive immunity to measles, smallpox, diphtheria, mumps, tetanus, influenza, and certain staphylococcal and streptococcal infections because of the transfer of immune bodies from the blood of the mother to the child via the placenta. The immunity that the child receives in his mother's womb may be enhanced by the content of protective substances that he receives from his mother's milk (especially from colostrum). Naturally, the child will not receive any immunity if the mother is not herself immune. Except in newborn babies, all passive immunities are established by the administration of an immune serum (the serum of an animal that contains antibodies because the animal has been actively immunized).

Active immunity artificially acquired does not last so long as when it is naturally acquired, but active immunity always lasts longer than passive immunity. One should bear in mind that an active immunity can be established only with an attack of a given disease or with vaccination against it and that the immunity is slowly established (days or weeks) but of long duration (months or years). On the other hand, a passive immunity is ordinarily established at once by the injection of an immune serum but is of short duration (1 or 2 weeks). On these facts is based the principle of producing active immunity when there is no immediate danger from disease and establishing passive immunity when danger from disease is imminent or patient is already ill.

Level of immunity. It seems that when an infectious disease attacks a population that has never been exposed, the attack rate is very high and many succumb to its ravages. On the other hand, when a disease endemic in a population for years becomes epidemic, the number of persons attacked and the percentage of deaths are not so great. People develop a degree of inherent resistance to a given disease through exposure for generation after generation. There is little doubt that syphilis was at one time a far more virulent disease and more often fatal than it is now. The susceptibility of aboriginal people to tuberculosis is much greater than that of people in communities in which the disease has existed for a long time.

In 1951 measles was introduced into Greenland by a recently arrived visitor who attended a dance during the onset of symptoms. Within 3 months there were more than 4000 cases, with 72 deaths. The attack rate reached the unprecedented figure of 999 cases/1000 people. When measles was first introduced into the Fiji Islands in 1875, 30% of the population died.

IMMUNE SYSTEM

Immunity is important as a defense against infection, but in light of current achievements it is more. Broadly speaking, immunity comprises the things that help man to maintain his structural and functional integrity and to ward off certain kinds of injury. Some of the factors in immunity are hard to define. One cannot easily measure the basic mechanisms of resistance built into all living organisms and expressed as the inflammatory process, the presence of anatomic barriers, and the effect of naturally occurring antimicrobial substances. Therefore such factors are collected loosely into the concept of *innate* or *nonspecific immunity*.

Another approach to immunity emphasizes the study of the highly developed, precisely specialized physiologic mechanisms, the topics of *specific* or *adaptive immunity*. In adaptive immunity there are two expressions of fundamental processes, both related to cells in the lymphoid system. In one there is elaboration of chemicals* —this is *humoral immunity*. In the other there is focus on less well understood activities of certain cells—this is *cell-mediated* or *cellular immunity*.

Reflecting the dual nature of the immune response, our discussion will take the twofold approach with, first, a consideration of humoral immunity and, second, a note as to the cell-mediated kind.

Role of lymphoid tissue. In man the lymphoid tissue is responsible for the events of specific immunity. The cells (predominantly lymphocytes) in the different areas of lymphoid tissue look very much alike when viewed under the microscope, but in the immunologic setting they behave quite differently. At least two distinct cell populations have emerged. Both may have a common origin in bone marrow stem cells, but they diverge in development of function.

One is a so-called *thymus-dependent* (processed) *system* of cells because its early development is influenced and its destiny fixed by the thymus gland. The thymus gland may act directly on these cells or effect their maturation through secretion of a hormone (or hormones) referred to as *thymosin*. These are the *T cells*, also known as "helper," "suppressor," and "killer" cells. T cells are small lymphocytes that circulate in the blood and lymph; in fact, most of the circulating lymphocytes are T cells, although they also tend to congregate in certain foci within the lymphoid tissues. They have a very long life expectancy—many years, a decade or so—and regulate the cell-mediated immune responses (graft rejection; delayed hypersensitivity; immunologic surveillance against cancer; and those in infections from fungi, acid-fast bacteria, and viruses).

The other population of lymphoid cells is a *thymus-independent system* of cells (the *B cells*, Fig. 12-1). It owes its differentiation to a site not as yet identified in man (possibly in the bone marrow). The small lymphocytes of this cell population are short-lived—only 1 or 2 weeks. The thymus-independent system of cells, or B cells, is responsible for the production of humoral immunity, as seen in the body's main defense against bacterial infection (as with pneumococcal, streptococcal, or meningococcal infections).

The immune system can be compartmentalized into the B-cell system and the T-cell system, and the two systems can be sorted out for practical purposes in animals and man. Of course, there is overlapping of the two.

---

*Immunochemistry is the study of the complex chemical reactions in immunity.

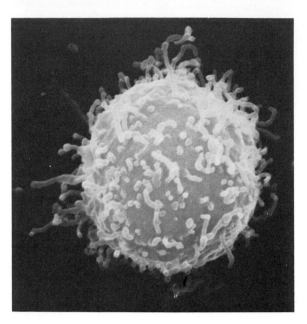

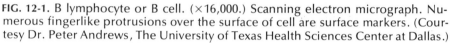

FIG. 12-1. B lymphocyte or B cell. (×16,000.) Scanning electron micrograph. Numerous fingerlike protrusions over the surface of cell are surface markers. (Courtesy Dr. Peter Andrews, The University of Texas Health Sciences Center at Dallas.)

If a lymph node is studied during the course of an immune response, definite changes may be seen within its cortex. With hormonally mediated responses, the follicles of the cortex become very active and greatly enlarged; numerous plasma cells appear within them. With cell-mediated immune responses the cortical follicles appear to be spared, but pronounced changes are found in the lymphoid tissue adjoining them. Within these *paracortical* areas numerous mononuclear cells containing increased amounts of RNA accumulate later to become small lymphocytes ready for release into the circulation.

Humoral immunity

Humoral immunity is that related to antibodies.

Antigens. When a person or animal becomes immune to a disease, the immunity is largely the result of the development within the body of substances capable of destroying or inactivating the causative agent of the disease should it gain access to the body. These substances, known as *antigens* or *immunogens*, are produced by the body in response to a specific stimulus. Microbes and their products may stimulate the body cells to antibody production, as may also certain vegetable poisons, snake venoms, and, from an animal of a different species, red blood cells, serum, and other proteins. Such a substance, eliciting the production of antibodies against itself when introduced into the animal body, is known as an *antigen*. To act as an antigen, it *must* *be* introduced into the body; that is, it must be a substance foreign to the body of a given individual. Antigens are usually high molecular weight substances of a protein nature, but in some cases complex carbohydrates (polysaccharides) may act as antigens. Enzymes and many hormones are antigenic. It is estimated that there are 1 million different antigens to which the human immune system can react directly.

When certain body proteins (examples, thyroglobulin and lens protein from the eye) of one animal are injected into another animal of the same species, antibodies against the proteins are produced. These antigens are *isoantibodies*. Naturally occurring isoantigens are those residing on the red blood cells and making up the blood groups (p. 209).

Comparatively simple chemical substances (certain low molecular weight lipids and carbohydrates) may be combined with the protein of an antigen to give it specificity. When separated from the protein of the antigen, these substances do not elicit the formation of antibodies but do combine with antibodies already formed against the antigen. These are *haptens* or *partial antigens*.

Certain chemicals that within themselves are not antigenic increase the potency of antigens. They are known as *adjuvants*. Among these are alum and aluminum hydroxide, used in preparation of toxoids. In addition to increasing the potency of the antigen, they concentrate it.

Antibodies. As you may recall, plasma is the liquid portion of the blood in which the red corpuscles and other formed elements are suspended. It is composed chiefly of water (90% to 92%), proteins (6% to 7%), and various mineral constituents. The plasma proteins are albumin, globulin, and fibrinogen.* By electrophoresis it is shown

---

Serum is the fluid plasma of the blood minus blood cells and fibrin precipitated from the plasma protein fibrinogen.

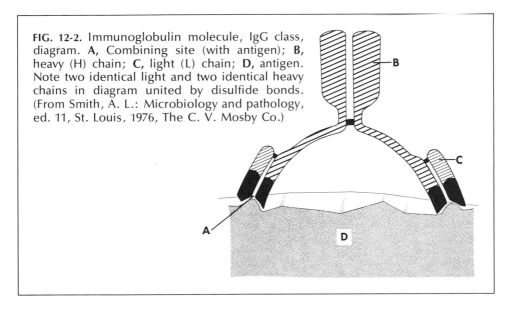

FIG. 12-2. Immunoglobulin molecule, IgG class, diagram. **A,** Combining site (with antigen); **B,** heavy (H) chain; **C,** light (L) chain; **D,** antigen. Note two identical light and two identical heavy chains in diagram united by disulfide bonds. (From Smith, A. L.: Microbiology and pathology, ed. 11, St. Louis, 1976, The C. V. Mosby Co.)

that the globulin is made up of these main fractions—alpha-1, alpha-2, beta, and gamma globulins. Antibodies migrate electrophoretically in the gamma and beta globulin regions. They are largely associated with gamma globulin, representing 1% to 2% of total serum protein. This is why this fraction (immune globulin) is used so extensively in the prevention of diseases such as infectious hepatitis, poliomyelitis, and measles.

NATURE. Much is being learned of the true nature of antibodies, the specialized globulins that react precisely with the antigen inducing their formation. Antibodies are better designated as *immunoglobulins* (Ig) and belong to a family of related proteins with distinctive properties when studied by special laboratory technics. They are very large and heterogeneous protein molecules and the most complex group known. Chemically, immunoglobulins adhere to a basic pattern on which significant

TABLE 12-1. ANALYSIS OF MAJOR CLASSES OF IMMUNOGLOBULINS (ANTIBODIES)

WHO term*	Where found in body	Sedimentation coefficient (ultracentrifugal analysis)†	Chemical composition known	Molecular weight
IgG	Serum (40% intravascular)	7S	4 polypeptide chains— 2 light, 2 heavy	160,000
IgA	Serum (40% intravascular) External secretions: saliva, parotid gland, gastrointestinal, respiratory tract, colostrum (secretory IgA)	9-15S	4 polypeptide chains— 2 light, 2 heavy	160,000 (monomer) to 900,000 (secretory IgA: 390,000)
IgM	Serum (80% intravascular)	19S	20 polypeptide molecules of five basic 4-chain units — 2 light, 2 heavy	1,000,000
IgD	Serum (75% intravascular)	7S		160,000
IgE	Serum	8S	4 polypeptide chains— 2 light, 2 heavy	196,000

*Designation recommended by the World Health Organization Committee on Nomenclature of
†The ultracentrifuge is an important tool in the identification of antibodies. By means of the exceed according to their molecular weights. Results are quantitated in Svedberg units (S).

variations are superimposed (Fig. 12-2). The basic subunit or building block is structured of four polypeptide chains in two pairs. Two of the paired chains are of greater molecular weight—heavy or H chains—and two of lower molecular weight (and shorter)—light or L chains. The light chains are linked to the heavy ones by single disulfide bonds.

Light chains are about 220 amino acids long; heavy chains are approximately 450 amino acids long. Light chains contain the same two kinds of protein, but in the heavy chains, proteins are found in at least five varieties. Therefore immunoglobulins (antibodies) have been classified according to the heavy chain, respectively, as immunoglobulins G, A, M, D, and E.

The tops of the heavy chains together with the light chains spread in the shape of a Y to form two "Fab" fragments ("Fragment, antigen-binding"), where indeed anti-

Estimated % Ig in population	Serum level (g/100 ml)	Crosses placenta	Binding of complement	% Carbohydrate	Electrophoretic mobility (principal)
75	1.2	Yes	+	2.5	γ
21	0.18	No	No	8	Slow β
7	0.12	No	+	10	Between γ and β
0.2	0.003	No	No	10	Between γ and β
0.5	0.0001	?	No	12	γ

Immunoglobulins.

ingly high gravitational forces possible in this instrument, serum protein fractions may be separated

TABLE 12-2. IMMUNOLOGIC ACTIVITIES OF MAJOR CLASSES OF IMMUNOGLOBULINS

Immuno-globulin	Participation in well-known antigen-antibody reactions						General remarks
	Agglu-tination	Precipi-tation	Comple-ment fixation	Lysis	Neutralization		
					Viruses	Toxins (and enzymes)	

IgG (gamma G, γG)	Weak	Strong	Strong	Weak	+	+	1. Ones best studied
							2. Responsible for passive immunity of newborn
							3. Identified here:
							  a. Certain Rh antibodies
							  b. Bacterial agglutinins
							  c. LE cell factor*
							  d. Antitoxins
							  e. Antiviral antibodies

IgA (gamma A, γA)	+	±	−	−	+	?	1. Known to be made by plasma cells
							2. Chief Ig in external secretions
							3. Protective function on body surfaces exposed to environment
							4. Identified here:
							  a. Diphtheria antitoxin
							  b. Blood group antibodies
							  c. Antibodies against *Brucella* and *Escherichia coli*
							  d. Antibodies against respiratory viruses
							  e. Antinuclear factors in collagen disease*
							  f. Certain other auto-antibodies* (against insulin in diabetes, thyroglobulin in chronic thyroiditis)

IgM (gamma M, γM)	Strong	Variable	Weak	Strong	+	−	1. Rapid protection here —first antibodies noted after antigen injection

*For discussion of autoimmunity and autoimmune diseases, see pp. 203 to 205.

TABLE 12-2. IMMUNOLOGIC ACTIVITIES OF MAJOR CLASSES OF IMMUNOGLOBULINS—cont'd

Immuno-globulin	Agglu-tination	Precipi-tation	Comple-ment fixation	Lysis	Viruses	Toxins (and enzymes)	General remarks
	Participation in well-known antigen-antibody reactions				**Neutralization**		
IgM (gamma M, γM) —cont'd							2. Powerful agglutinins and hemolysins here (700 to 1000 times stronger than those of IgG in agglutinating red cells or bacteria) 3. Identified here: a. Blood group antibodies (ABO) b. Antibodies against somatic O factors of gram negative bacteria c. Human anti-A isoantibody d. Isohemagglutinins, cold agglutinins, rheumatoid factor
IgD	?	?	?	?	?	?	Functions not identified
IgE (reagin)	−	−	−	+	−	−	1. Role in allergy— governs responses of immediate-type hypersensitivity (mast cell fixation) 2. Carries skin-sensitizing antibody

gen binding occurs. The tails of the heavy chains form the other main part of the molecule called the "Fc" fragment ("Fragment, crystalline"), which is involved in complement fixation and other biologic processes.

CLASSIFICATION. The major classes of immunoglobulins as defined by the World Health Organization Committee on Nomenclature of Immunoglobulins are given in Table 12-1 with their distinguishing features; Table 12-2 indicates their immunologic capacity.

FORMATION. Antibodies (immunoglobulins) are formed in the lymphoid tissues of the body in response to a particular antigen within the body. The main anatomic sites are the spleen, lymph nodes, and bone marrow, but antibody synthesis may be shown wherever there is lymphoid tissue—*with the exception of the thymus!*

The principal cells involved are lymphocytes (B cells primarily, but T cells are important), plasma cells (derivative cells of lymphocytes), and macrophages (mobile and sessile cells able to ingest fairly large particles of foreign matter). Although the macrophage participates in the induction of immunity, its role is still unsettled.

Although B cells and T cells look alike with the light microscope, significant differences on the cell surfaces can be demonstrated by special technics. Bound to the plasma membrane of the B cell are numerous (estimated 100,000) specific receptors shown to be molecules of immunoglobulin (sIg). Among these are also receptors for the third factor of complement (C3). By contrast, T lymphocytes show considerably fewer surface markers.

When an antigen (immunogen) gains access to an appropriate lymphoid area within the body, there follows an incompletely understood sequence of events wherein the antigen is recognized, processed, and bound to an antigen-sensitive lymphocyte. Mystery and speculation still cloud the earliest cellular changes. For antigens to attain the maximum immunologic response in production of immunoglobulin, they must induce an interaction between B and T cells with involvement of macrophages, although on occasion immunogens need affect B cells only.

On the surface of an antigen there is a special pattern of atoms designated as the *antigenic determinant,* or *marker.* The immunoglobulin resulting from the stimulus of a given antigen must be structured to make a neat fitting into the unique arrangement of the antigenic determinant.

T cells recognize the antigen in both cellular and humoral immunity and so affect and potentiate the processing of antigen by macrophages and reticulum cells of the lymphoid area. This results in changes in the antigen but not the loss of its antigenic marker. The altered immunogen is then presented to the B cells, which receive it at the specific receptor sites on their plasma membranes. The engagement of the immunogen with the membrane receptors is the signal for the lymphocyte to become immunologically active.

According to one theory of antibody formation, the given B cell, the immunologically responsive cell, responds to only one kind of antigen, a capacity in some way acquired before the encounter and most probably related to the specific nature of its surface immunoglobulin molecules. This is the "one cell–one antibody" theory. The immunogen attracted to the cell binds to a receptor on its surface because that marker is chemically built like an antibody. This starts the B cell on a course of immunologic activity, with formation of immunoglobulins as the net result. The B cell proliferates and soon generates a clone of differentiated and immunologically competent cells. Some of the derivative cells become plasma cells (with the influence of T cells as "helper cells"), known for a long time to synthesize and secrete antibody globulin. Others become lymphoid cells to disseminate through blood, lymph, and tissues. They establish a reservoir of antigen-sensitive cells, this time primed with a memory factor. When the same immunogen reappears on the scene, the immunologic response is enhanced and expedited.

Today we speak of lymphocytes as *immunologically competent* cells, meaning ones that can undertake an immunologic response when engaged by an antigen. During a lifetime, because of the many and varied antigens of our environments, each of us produces tens of thousands of different kinds of antibodies.

Antigen-antibody reaction. The antigen-antibody reaction is specific: a given antigen promotes the production of antibodies only against itself, and a given antibody acts only against the antigen promoting its development. A person vaccinated against smallpox is protected against smallpox only. The hemolysin in the serum of a rabbit that has been repeatedly injected with the red blood cells of a sheep dissolves the red blood cells of the sheep but not those of other animals. The amino acid sequence and the three-dimensional structure of an antibody determine its specificity. The difference of even one amino acid can be detected in a specific antigen-antibody combination.

Within the antibody molecule there is a pattern of small areas specially designed for a close-fitting union with the antigen (like a key fitted into a lock). Most immunoglobulins are bivalent; that is, their structure allows combination with the antigen molecule at two areas.

If the union of antigen and antibody occurs in such a way that the aggregation is

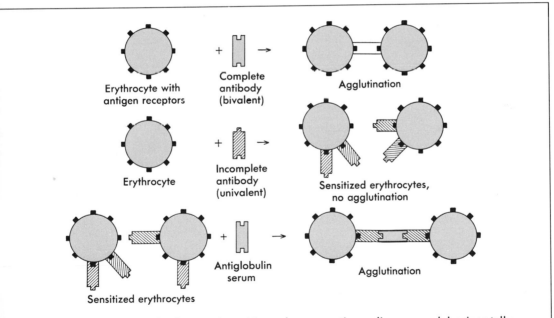

FIG. 12-3. Antigen-antibody reaction. Note three reactions diagrammed horizontally between antigens on erythrocytes and antibody. *Top,* Agglutination (visible clumping) results from bonding of antigens and antibody. *Middle,* Antigen-antibody complex, but without clumping. Red blood cells here are altered (sensitized). *Bottom,* In contact with antiglobulin serum (a special complete antibody, see also p. 213), sensitized red cells are clumped. (From Miale, J. B.: Laboratory medicine: hematology, ed. 4, St. Louis, 1972, The C. V. Mosby Co.)

manifest as a visible phenomenon, the antibody is said to be a *complete antibody* (Fig. 12-3). When union of an antibody occurs with specific antigen but the antibody lacks the capacity to change the surface properties of the antigen and therefore a visible reaction does not take place, that antibody is referred to as an *incomplete* or *blocking antibody*.

DETECTION OF ANTIBODIES AND THE ANTIGEN-ANTIBODY REACTION. Discussion of this subject is found in Chapter 13.

Complement. Although complement (alexin) is not an antibody, its presence is necessary for the complete action of certain *lytic* antibodies, notably hemolysins, bacteriolysins, and other cytolysins, and complement (the first four components) enhances the action of opsonins. Basically a complex enzyme system of some 11 proteins taking part in the antigen-antibody reaction, but only after antigen has been sensitized by its antibody, complement is present in fresh serum (10% of globulin fraction) and is not increased by immunization. Actually, the widely used term *complement* collects at least nine distinct complement components, numbered from one to nine, with inhibitors into the complement system. C1 is a complex of three proteins stabilized by calcium ions. C3, the third component, appears to play the most significant role in disease.

Complement is destroyed by exposure to room temperature for a few hours or to a temperature of 45° C for 30 minutes. The fresh serum of a guinea pig is regularly used as a source of complement because its content shows little variation in different animals and is comparatively high. Complement may be preserved by various chemical methods or by drying the fresh frozen guinea pig serum to a powder in a vacuum.

When certain antigens and antibodies combine to produce the immune complex, the complement system is activated. The components interact in a complicated sequence or cascade of events much like that in the blood-clotting system and thereby produce the immune phenomena such as cytolysis with which they are associated.

During the process of antigen-antibody-complement union, complement is destroyed or at least inactivated. This phenomenon is *complement fixation*. The activity of complement directed to cell membranes is responsible for the cytolytic effect so neatly demonstrated in the laboratory.

IMMUNE COMPLEX DISEASE. *Immune complex disease* is a recent term to indicate disease present because soluble antigen combines with circulating antibody in a given immunologic setting. Complement is fixed as a consequence, and the complex of antigen-antibody-complement is deposited on an endothelial wall of a blood vessel. Once deposited, the complex triggers a number of interrelated events that culminate in tissue injury and inflammation at the site (vasculitis). The best example is found in the glomerulus of the kidney in some forms of glomerulonephritis.

Cell-mediated immunity

In acute bacterial infections with gram-positive cocci the mechanisms of humoral immunity in the normal individual are efficient. Persons with agammaglobulinemia, who cannot make measurable amounts of antibody globulin, do not fare well with the cocci but adequately resist infections caused by viruses, fungi, protozoa, and certain other bacteria. Such resistance must stem from an immunologic mechanism other than the classic antigen-antibody combination.

To the host, infection with these latter agents presents a peculiar problem in defense, since these microbes colonize the interior of host cells. There they obtain shelter from antimicrobial elements of blood and tissue fluids. Some are in danger if they pass from cell to cell, but others are unaffected in transit. The patterns of cell-mediated immunity seem designed to meet this situation.

Although much less is known of immunity related to the activities of certain cells than is known of that related to antibodies, *cell-mediated (cellular) immunity*, because of its practical implications for tissue transplantation and rejection, is a subject of considerable interest. The notion that, without demonstrable antibody, cells can bring about immune responses was hard to accept for a long time.

Comparisons. Like humoral immunity, cell-mediated immunity is part of the body's defense. Unlike it, there is no demonstrable, clear-cut antibody. For this reason there is no equivalent of passive immunization with cellular immunity—it cannot be passed in serum. However, transfer can be made in a less predictable fashion with lymphoid cells from a sensitized host.

For the study of cellular immunity there are only a small number of immunologic procedures to match the extended list of well-standardized tests used to describe antibodies. The traditional skin test, as important as it is, depends on the subject's prior contacts with antigens and is influenced by local tissue factors. As illustrated by the skin test, the immune responses of cellular type emerge more slowly (several days) than those mediated by antibodies, which can evolve in a matter of hours.

In both kinds of immunity, contact with antigen triggers the sequence of events, but in cell-mediated immunity the antigen does not reach the lymphoid tissues; it seems blocked peripherally. With this type of immunity, concurrence with presence of antigen is crucial. If antigen is destroyed, the immune process subsides. In tuberculin testing, for example, the previously sensitized person has a reaction lasting until all injected tuberculoprotein is gone.

Principal cells. The mediating and primary reacting cells are small, *thymus-dependent* lymphocytes, T cells. Authorities think that these lymphocytes possess their unique specificity for reaction with a certain kind of antigen as a result of the prior influence of the thymus gland. When contact with that particular antigen is made, the cells become sensitized and respond immunologically.

Macrophages are secondary participants, important for their property of phagocytosis. They make up 80% to 90% of cells mobilized to the reaction site.

The antigen. Antigens that induce the sequence of events in cellular immunity are usually protein. The antigenic protein may be quite small or more complex, for example, the lipoprotein transplantation antigen in the mouse.

Events. The first step is the recognition of the antigen as "foreign" by the population of T cells. Then ensues *blastogenesis*, the transformation and enlargement of T cells into rapidly dividing blast cells, cells of an immature appearance. They elaborate and release certain factors called *lymphokines*, designed to summon macrophages to the scene and to program them for the attack. The fully equipped macrophages assemble in large numbers, primed to sequester and destroy that which does not belong. Although the activation of the macrophages by lymphocytes as an immunologic phenomenon is not well understood, it does mean that the microbicidal faculties of the macrophage are enhanced.

Some of the biologically active substances include *migration inhibitory factor,* blocking movement of macrophages; *chemotactic factor,* orienting migration of macrophages into the area; *blastogenic* or *mitogenic factor,* inducing blast transformation of lymphocytes; and *lymphotoxin,* damaging or killing cells. One important substance released is *transfer factor,* recoverable as a cell-free white blood cell extract that transmits immunologic information from one person to another; the passage of sensitivity in this way is measured by skin test results. Interferon is also elaborated by T cells (p. 503).

The cytotoxic effects of the activated lymphocytes coupled with the efforts of activated macrophages result in localization and destruction of antigen or antigen carrier. Some of the progeny of the lymphocytes resume the small lymphocyte status to recirculate as memory cells.

Manifestations. Cell-mediated immunity is a major component of the body's resistance to viruses, fungi, and certain bacteria. Classic expressions of it are found in delayed-type hypersensitivity (once known as bacterial hypersensitivity), graft rejection, and graft-versus-host reactions. It is also the capacity possessed by the body to resist most forms of cancer.

DISTURBANCES IN IMMUNITY

Unfortunately, the immunologic response is not always favorable in the body and may even be quite harmful. Considerable attention is being given today to the elucidation of both normal and abnormal mechanisms in immunity. As a result, immunologic disorders are better understood. Some have only recently been defined. *Immunopathology* is the study of the tissue lesions and alterations consequent to the immune reaction. In the following listing, the nature of the defect is indicated.

1. Immunologic accidents—best example, reaction following mismatched blood transfusion
2. Immunologic depression (immunosuppression)—administration of immunosuppressive agents materially interfering with or inhibiting the immune system (humoral or cellular); such are the steroid hormones in huge doses, irradiation, cytotoxic chemicals (antimetabolite drugs and alkylating agents), and antilymphocytic serum (ALS)
3. Immunologic deficiency—failure or impairment of normal development of immunologically competent cells of the lymphoid system of the body, both antibody deficiencies and defects of cell-mediated immunity included

    An absence of lymphocytes and plasma cells and hence of all immunoglobulins is termed *lymphoid aplasia.* Abnormal development of the thymus gland *(thymic dysplasia)* or a congenital absence of the thymus will result in impairment of immunologic mechanisms. Death comes early in infancy.

    An absence of plasma cells with very low levels of gamma globulin is either *hypogammaglobulinemia* (antibodies in greatly decreased amounts) or *agammaglobulinemia* (no demonstrable gamma globulin in the blood). An absence of or a defect in some of the immunoglobulin components of serum is *dysgammaglobulinemia*.

Certain patients with depressed cellular immunity are peculiarly vulnerable to infections caused by *Mycobacterium tuberculosis*, *Pseudomonas aeruginosa*, and *Pneumocystis carinii*. An infection such as vaccinia may progress even where there is a demonstrably high titer of circulating antibodies.

4. Immunologic effects of nonimmunologic diseases—infections with bacteria, viruses, and protozoa producing striking elevations in globulin levels of serum (*hyperglobulinemia*)

5. Allergic states—see Chapter 14

6. Autoimmunity and autoimmune diseases—defects of immunologic tolerance*

7. Immunologic aberrations secondary to malignancies of the lymphoid system— as would be expected, associated with striking abnormalities

The malignant prototype of the lymphoid cell can yield an abnormal type of globulin, often in large quantities. Examples of frankly abnormal globulins synthesized by malignant cells are the myeloma proteins, the macroglobulins of Waldenström, and the Bence Jones proteins. The malignancies include the leukemias, lymphomas, or other neoplastic cell proliferations arising in the reticuloendothelial tissues.

Autoimmunization. Generally the immunologic mechanisms in the body are protective. Microbes or substances that would enter and injure are attacked. Because the body recognizes its own cells and tissues, it differentiates between the protein of the foreign invader and that which belongs to itself. "Unfortunately," as William Boyd says, "the immune system is a two-edged sword which can be turned against the body in that biological paradox nicknamed autoimmunity. The same forces that normally reject foreign material act in reverse and reject the cells and tissues of the body itself with unpleasant consequences."†

Many potentially antigenic substances exist in or on the surface of a person's own cells, but before these can induce an immune response, it is thought that they must be altered in some way, possibly by the action of bacteria, viruses, chemicals, or drugs. Once that change has been made, these antigens, now *autoantigens*, stimulate the body of the individual in whom they occur, in certain instances, to form the corresponding antibodies, or *autoantibodies*. The process by which antibodies are made by the body against its own cellular constituents is referred to as *autoimmunization*. Its importance lies in the unfavorable and disease-producing effects of certain immune responses to the autoantigen. The category of human disease resulting is that of the *autoimmune diseases*.

Autoimmune diseases. Autoantibodies can exist without disease. Under certain conditions, however, the reaction of autoantibodies (or sensitized immunocytes of the

---

*Immunologic tolerance refers to the ability of the immune system to discriminate between what belongs to its host (which it tolerates) and what does not (which it attacks).

†From Boyd, W.: A textbook of pathology, Philadelphia, 1970, Lea & Febiger.

TABLE 12-3.   EXAMPLES OF DISEASE AND ANTIGEN IN AUTOIMMUNITY

Autoimmune disease	Autoantigen
Acquired hemolytic anemia	Antigens on surface of red blood cells
Idiopathic thrombocytopenic purpura	Platelet antigens
Rheumatoid arthritis	Denatured gamma globulin
Systemic lupus erythematosus	Nuclear material including DNA; certain cytoplasmic substances

FIG. 12-4. Lupus erythematosus (LE) cell, distinctive cell found in blood and bone marrow of patient with lupus. It is a mature neutrophil that has phagocytized a homogeneous mass of nuclear material (degraded DNA) in the presence of antinuclear autoantibody in serum.

lymphoid series) with the specific tissues (the autoantigens) does result in disease, the clinical and pathologic picture of which is determined by the particular tissue attacked immunologically, its distribution in the body, and the extent of the damage done by the reaction. Autoimmunity (autoallergy) is seen in disorders of the thyroid gland, brain, eye, skin, joints, kidneys, and blood vessels.

Six features are emphasized in discussions of autoimmune disease: (1) elevation of serum globulin, (2) presence of autoantibodies in the serum, (3) deposition of denatured protein in tissues, (4) infiltration of lymphocytes and plasma cells in damaged tissues, (5) benefit of treatment with corticosteroids, and (6) coexistence of multiple autoimmune disorders.

In some autoimmune disorders the autoantigens involved are known (Table 12-3), and circulating antoantibodies can be demonstrated in patient's serum, especially when there is an associated elevation of gamma globulin. As a group these diseases often give a history of preceding infection, tend to run in families, and are relieved by steroid hormones (ACTH and cortisone); the patient's serum gives a false positive test for syphilis. Virus particles have been associated with autoimmune diseases, but the relationship is unclear at this time.

DISSEMINATED (SYSTEMIC) LUPUS ERYTHEMATOSUS. One of the most interesting of the autoimmune diseases is disseminated lupus erythematosus. (Its name is derived from the Latin word for wolf.) A disease primarily of women, it runs an acute or subacute course interspersed with periods of improvement. It is postulated that immunologic injury derives from complexes formed when the autoantibodies, the antinuclear globulins, react with the breakdown products of nuclei in the normal wear and tear of cells (Fig. 12-4). Anatomically, connective tissue as prime target is damaged or destroyed. Since it is a tissue widely distributed over the body, lesions would be expected most anywhere but classically select the kidney, heart, spleen, and skin (butterfly rash on face). Also relative to the connective tissue injury are joint pains, fever, malaise, and anemia.

TISSUE TRANSPLANTATION

There are many times in medicine when it would be desirable to take a normal tissue (or organ) from a healthy person (donor) and transplant it into the body of a patient (recipient) whose corresponding tissue (or organ) is severely diseased. It is true that a piece of skin can be relocated as a graft from one site on the same body to another *(autograft);* that a transplant of living tissue from one identical, *not* fraternal, twin will take in the other *(isograft, isogeneric* transplant*); and that in plastic surgery pieces of nonviable bone, cartilage, and blood vessel from one individual can be used in another as a framework for new growth in certain areas *(homostatic graft).* But immunologically each of us is a rugged individual! In the natural course of things what would be lifesaving grafts of viable skin, bone marrow, kidney, and so on are cast off, rejected, because very subtle antigenic differences exist among even closely related members of the human race.

Graft rejection. A graft from one member to another within a noninbred species is a *homograft* or *allograft.* Without intervention of any kind this graft is characteristically rejected. For about the first week there is every indication that the graft has taken. After that time the situation changes. The circulation of blood established in the graft is cut down; the graft is infiltrated by mononuclear cells, loses its viability, and is soon cast off (Fig. 12-5).

What has happened immunologically is an expression of cell-mediated immunity. (Although antibodies may also play a part, they are not primarily implicated.) Small thymus-dependent lymphocytes detected the foreign antigen, became transformed, were sensitized in the regional lymph nodes, and released back into the circulation as "activated" lymphocytes. They returned to the graft site to set the rejection process in motion. Although they initiated it, the actual demolition of the graft was carried out by macrophages, with the help of a few polymorphonuclear leukocytes, that the lymphocytes secured to the site.

Transplantation antigens. *Histocompatibility* refers to the degree of compatibility, or lack of it, between two given tissues. Of some eight to ten genetically independent systems in the body affecting graft survival or rejection, there are two that are of practical importance.

The first, a very important histocompatibility factor in man, is the ABO blood group system, which is present on red blood cells and practically all tissues.

The second is determined by a complex genetic region known as the HL-A locus (histocompatibility locus-A) because it controls homograft immunity. It contains a group of antigens associated with cell membranes that are found in the majority of cells and tissues of the body, and especially in lymphoid and blood-forming tissues (not on red blood cells, however). This complex system of at least 31 antigens, of which a person may have as many as four, is most easily identified in leukocytes. Therefore

---

Heterografts or *xenografts* are those between members of different species.

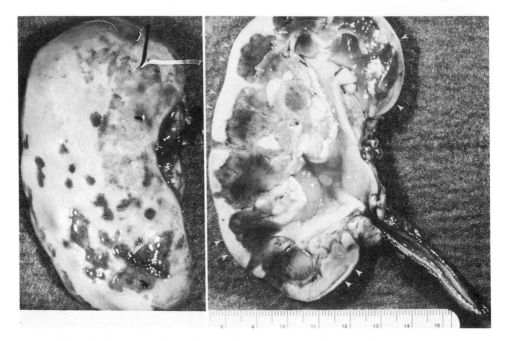

FIG. 12-5. Graft rejection of transplanted kidney (acute). Outer surface and cross section of kidney shown. Diffuse death and destruction of tissue seen in whitish areas of outer and sectioned surfaces. Mottling is produced by hemorrhage. (From Smith, A. L.: Microbiology and pathology, ed. 11, St. Louis, 1976, The C. V. Mosby Co.)

tissue typing is commonly done with white blood cells. Practically speaking, the antigens usually referred to as *transplantation antigens* are those of the HL-A system. Part of the graft, they are the ones that mark the graft as "trespasser" to the recipient host.

Testing for histocompatibility. When tissue transplantation is anticipated, typing and cross matching of blood specimens from the postulated donor and patient-recipient are done as though for blood transfusion. Any incompatibility in the ABO blood groups is an *absolute* contraindication to the use of the donor's tissue.

Tissue typing for transplantation, done with white blood cells or platelets, is partly analogous to red blood cell typing and cross matching for clinical blood transfusion. Of the various technics for tissue typing, lymphoagglutination and lymphocytotoxicity are widely used. *Lymphoagglutination* compares the clumping reactions of would-be donor and recipient lymphocytes when tested with a panel of selected antiserums. *Lymphocytotoxicity* measures the lethal effect of plasma membrane injury to presumed donor and recipient lymphocytes under similar conditions. Recipient serum can be mixed with the would-be donor's lymphocytes as a "cross match." Reaction in this indicates probability of immediate graft rejection. In evaluating the degrees of donor-recipient incompatibility for HL-A types determined from tissue typing, there are no absolute matches (except with identical twins). A "good" match still means differences, although small.

Immunosuppression. Various clinical and experimental technics have been designed to suppress or manipulate artificially the body's normal immune responses so that tissue transplantation would be possible. *Immunosuppression* (induction of tolerance) is the term used here. For instance, a potent antiproliferative drug such as 6-mercaptopurine (6-MP) that selectively interferes with either DNA or RNA synthesis may be used singly or combined with steroid hormones and irradiation. Depletion of lymphocytes has been attempted by drainage of the thoracic duct. Antilymphocyte globulin (ALG) has been administered to selectively eliminate the cellular population mediating the immune responses and perhaps to spare the sorely needed humoral ones.

The disadvantage of immunosuppressive agents is that they concurrently impair normal host resistance. Such an individual with either subnormal or practically nonexistent immune responses is spoken of as the *compromised host*. An effort is being made experimentally to induce tolerance to the graft under conditions that do not compromise the body's normal defenses when they are needed elsewhere in the body.

Graft-versus-host reaction. It has been observed in animals and a few humans that at times instead of the host rejecting the graft, the graft has rejected the host. This has been found especially with transplantation of bone marrow and lymphoid tissue. In the immunologically depressed host, the graft survives. The immunologically competent lymphoid cells of the graft recognize the "foreignness" of the environment, and the stage is set for rejection. The end of it—runt disease in animals, secondary disease in man—is characterized by wasting, diarrhea, dermatitis, infection, and death.

Present status. Much has been accomplished in tissue transplantation. Transplantation of the kidney is established; worldwide more than 6000 recipients of a kidney transplant have survived. Transplantation of other organs is still experimental. Fallout has been considerable in terms of new knowledge in basic disciplines, control of infection, and a breakthrough in cancer immunology. It was discovered that with immunosuppression there is also a greater likelihood for the development of certain cancers. Problems are formidable but still optimism and enthusiasm prevail.

IMMUNITY IN CANCER

The most significant advances in oncology (the study of cancerous growth) today focus on the immunologic aspects of neoplasms and on *immunotherapy*, their possible application to the treatment of cancer. Immunotherapy is based on the premise that the immune responses normally operating for prevention and control of cancer can be manipulated or adapted for the host to enhance his resistance to his cancer. It is still largely experimental.

In man and animals a relationship is known to exist between an individual's immunologic system and his cancer. Most of the evidence supporting present-day concepts of tumor immunology stems from an abundance of experimental work and many interesting observations made on patients with organ transplants. The incidence of cancer after organ transplant is about 100 times that observed in the same age group of the general population. (Immune mechanisms must be tampered with dramatically for organ transplantation.) Malignancies* complicate immunosuppression in many clinical settings, and in congenital immunologic deficiencies as well, there is a high incidence of cancer.

The immunology of cancer is complex, and a lot of work is being done on it. Most and maybe all neoplasms contain tumor-associated antigens (TAA) new and "foreign" to the host. Tumor antigens are tumor specific and provoke an antitumor response either in the host animal or on transplantation of the tumor into a comparable animal. The immune response in the host is a function of cell-mediated immunity, the T-cell system recognizing the antigens as foreign to the host and destroying them. The reaction is comparable to that of graft rejection. Tumor antigens arise because the cancer cell differs from the normal cell (not because of differences between donor and recipient animal). Both tumor and normal cells have identical histocompatibility or HL-A antigens; however, tumor antigens probably result from the initial transformation of the normal cell into a cancerous one. At this early time a neoplasm is most vulnerable to the immune response. Undoubtedly, many incipient ones are completely and quietly destroyed.

As would be expected, there are different kinds and degrees of tumor immunity. A special kind is found with tumors induced by viruses. Surface antigens appear as a direct effect of viral activity, and these are consistent for the same viral oncogen in other circumstances. Not so if the oncogenic mechanism is nonviral. That neoplasm has its own distinct antigen it is true, but even though the same carcinogen is applied in exactly the same way to other anatomic areas in the same animal, that same tumor antigen is not repeated. As a result of this effect, many antigenic types become possible with varying immunologic potential. If tumor cells are highly antigenic, they may immunize against themselves so efficiently that tumor growth is blocked.

In view of the existence of an immune mechanism, how does one explain the prevalence of clinical cancer? One can only surmise. It may be that the bulk of experimental cancers and spontaneous cancers in human beings arise at a time when immunologic defenses are down. It may be that the tumoricidal T cell cannot contact the tumor antigen bound to the membrane of the tumor cell; its action is in some way blocked. It may be that a small fraction of starter growths can make the clinical scene only because they combine the least in antigenicity with the highest in proliferative and aggressive activity.

IMMUNOLOGY OF RED BLOOD CELL

The red blood cell (Fig. 12-6) is composed of a compact stromal protein that supports molecules of lipid and hemoglobin. Antigens of the blood groups are contained in both the stroma and on the surface of the red cells. On the surface, exposed antigens are ready to react with specific antibodies. The spatial features of the relatively small antigenic groups out on the cell surface are such that each red cell can have a large number and variety of them.

---

*"Malignancies," "neoplasms," and "tumors" are scientific terms used for cancer. "Oncogenic" and "carcinogenic" refer to the ability to induce cancer, and an "oncogen" is such an agent.

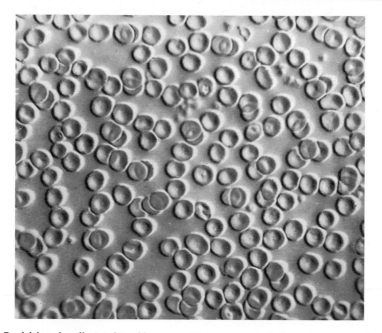

FIG. 12-6. Red blood cells in thin film on which vaporized chromium was shadow-cast at fixed angle. Note biconcave shape. (From Mountcastle, V. B., editor: Medical physiology, ed. 13, St. Louis, 1974, The C. V. Mosby Co.)

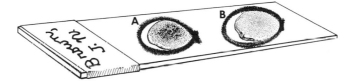

FIG. 12-7. Technic of blood grouping (typing). Make suspension in isotonic saline solution of red blood cells from blood to be typed. Mark out two circles (wells) on microslide; label **A** and **B.** Place drop of test suspension in each. In well labeled **B,** mix in drop of serum from person of blood group A (anti-B serum); in well labeled **A,** mix in drop of serum from person of blood group B (anti-A serum). Macroscopic agglutination of red blood cells in drop of typing serum seen in **A.** No agglutination in **B.** Well **A** contains anti-A typing serum and well **B,** anti-B serum. What type of blood do we have?

Agglutinins of the blood. The blood of every person falls into one of four ABO blood groups, depending on the ability of his serum to agglutinate the cells of persons whose blood is in another group (Fig. 12-7). This is the result of the distribution of two agglutinins in the serum and two agglutinogens (antigens) on the red blood cells. The agglutinins are known as *anti-A* and *anti-B*. The agglutinogens are designated A and B. Agglutinin *anti-A* causes cells containing agglutinogen A to clump. Agglutinin *anti-B* causes cells containing agglutinogen B to clump. The blood groups and their agglutinin and agglutinogen content are on p. 211 (Table 12-4).

TABLE 12-4. DETERMINATION OF BLOOD TYPE FROM AGGLUTINATION OF RED BLOOD CELLS IN SPECIFIC TYPING SERUM

If agglutination of red cells occurs in		Then blood type is	How many persons in the United States with it?	
Anti-A serum	Anti-B serum		If it is Rh positive	If it is Rh negative
−	−	O	1 in 3	1 in 15
+	−	A	1 in 3	1 in 16
−	+	B	1 in 12	1 in 67
+	+	AB	1 in 29	1 in 167

TABLE 12-5. OUTLINE OF TRANSFUSION REACTIONS

Kind	Cause	Clinical features
Hemolytic	Mismatched transfusion (example, blood type A given to patient, blood type O)	Severe chill; lumbar pain; nausea and vomiting; fever; suppression of urine (urine may be reddish brown); jaundice; death
Pyrogenic	Fever-producing substance in blood (sterile chemical contaminants, bacterial toxic products, antibodies to white blood cells)	Occurs 30 to 60 minutes after transfusion; flushing; nausea and vomiting; headache; muscular aches and pains; chills and fever
Contaminated blood	Break in aseptic technic: bacteria (usually gram-negative rods) introduced at time blood collected or with use of unsterile equipment	Chills, fever; generalized aching and pain; marked redness of skin; drop in blood pressure; shock
Allergic	Occurs in patients with history of allergy; cause unknown	Occurs within 1 or 2 hours of transfusion; itching of skin; hives or definite rash; swelling of face and lips; sometimes asthma
Sensitivity to donor white blood cells, platelets, or plasma	Multiple transfusions favor development of leukoagglutinins (those to white cells) and similar substances	Chill and fever; headache; malaise; serum may agglutinate white blood cell suspension from donor
Circulatory overload	Blood transfused too rapidly into person with failing circulation; likely to occur in elderly, debilitated, or cardiac patients, also in children	Difficulty in breathing; cyanosis; cough with frothy, blood-tinged sputum; heart failure
Embolic	Infusion of air: (1) transfusion under pressure, (2) tubing not completely filled before venipuncture	Sudden onset of cough, cyanosis, syncope, convulsions

Group AB   Serum contains no agglutinin; cells contain agglutinogens A and B
Group A   Serum contains agglutinin anti-B; cells contain agglutinogen A
Group B   Serum contains agglutinin anti-A; cells contain agglutinogen B
Group O   Serum contains agglutinins anti-A and anti-B; cells contain no
                  agglutinogen

If a person whose blood belongs to one group receives a transfusion of blood from a donor of another group, a hemolytic transfusion reaction is likely to occur. This is because the serum of the recipient may agglutinate the cells of the donor, or the serum of the donor may agglutinate the cells of the recipient. Reactions to transfusion of mismatched blood are associated with agglutination of red cells in the circulation, hemolysis of red cells, and liberation of free hemoglobin into the plasma; this leads to hemoglobinuria, fever, prostration, failure of kidney function, and in some cases death. A reaction is much more likely to occur if the recipient's serum agglutinates the donor's cells than if the donor's serum agglutinates the recipient's cells because the donor's serum is considerably diluted in the circulation of the recipient. This effect tends to minimize the agglutinating capacity of the donor's serum. For kinds of transfusion reactions, see Table 12-5.

Although not a transfusion reaction, a serious complication of blood transfusion is posttransfusion hepatitis,* a serious concern to blood banks. The link between the disease and the Australia antigen (Au) has led blood banks to screen routinely all would-be donors for its presence. The high-risk carriers of the agent (p. 545) are most likely to be the "commercial" donors from the skid-row, drug-addict, and prison-inmate populations.

THE RH FACTOR. An agglutinogen with no relation to the ABO blood groups is the *Rh factor*. When a guinea pig is given repeated injections of the red blood cells of a rhesus monkey, its serum acquires the ability to agglutinate the red cells of rhesus monkeys. It also agglutinates the red blood cells of a large majority of human beings (85% of the white race; more in others). The red blood cells that are agglutinated contain the Rh factor, and the person from whom the cells were removed is said to be *Rh positive*. Persons who do not have the Rh factor on their red blood cells are *Rh negative*. The plasma of Rh-negative persons does not contain agglutinins against the Rh factor, but such agglutinins may develop if blood transfusions of Rh-positive blood are given to these people. Such incompatible blood transfusions, if continued, may lead to serious reactions.

If an Rh-negative woman and Rh-positive man have children, one or more of the children will probably be Rh positive. The Rh-positive cells of her fetus can sometimes get into the bloodstream of the mother during gestation but usually in such small amounts that they do not induce antibody formation. It is thought that the Rh-negative mother has antibodies against the Rh factor during a given pregnancy because she was immunized in a previous one. When the Rh-positive cells of the baby enter the mother's circulation at time of delivery, they can be most effective in stimulating anti-

---

*Synonyms are hepatitis B, serum hepatitis, long-incubation hepatitis, tattoo hepatitis, postvaccinal hepatitis, HAA-positive hepatitis, homologous serum jaundice, syringe hepatitis.

body formation. Antibodies formed against the Rh-positive cells begin to appear about 6 weeks later.

When antibodies against the Rh factor are present in the plasma of the Rh-negative mother, they freely pass through the placenta to attack the red blood cells of the fetus and destroy them. As a consequence, the child may be born with *erythroblastosis fetalis* or *hemolytic disease of the newborn*. The disease attacks the fetus before it is born or within the first few days of postnatal life. Its features are anemia, jaundice, edema, and enlargement of the infant's spleen and liver. As would be expected, erythroblastosis is rare in the first pregnancy. Once it happens, the condition recurs in about 80% of subsequent pregnancies.

Designated the Rh "vaccine," an immunizing agent, Rh_0 (D) immune globulin (human)* is available to prevent a susceptible Rh-negative mother from developing anti-Rh antibodies. It is human gamma globulin containing anti-Rh antibodies and obtained from sensitized Rh-negative mothers who have given birth to erythroblastotic babies. Injected as a single dose within 72 hours of delivery, it cancels out the antigenic effect of the Rh-positive cells that have entered the circulation. Immune mechanisms for the production of Rh antibodies are blocked. In the event of future pregnancy with an Rh-positive fetus, the immunologic basis for hemolytic disease of the newborn has been eliminated.

Blood grouping (typing). Human blood is routinely grouped (or typed) in modern blood bank laboratories into one of the four major blood groups, and the presence or absence of the Rh factor is determined. This means that the blood is typed as O, A, B, or AB and as either Rh positive or Rh negative. However, the red blood cell is a very complex structure and contains many antigens or factors (at least 60), and new ones are constantly being discovered. As antigens are studied, an attempt is made to classify them into categories containing related antigens, designated *systems*. There are today at least 15 well-known systems of blood groups: (1) ABO, (2) MNS, (3) Rh, (4) P, (5) Lewis, (6) Kell, (7) Lutheran, (8) Duffy, (9) Kidd, (10) Sutter, (11) Vel, (12) Diego, (13) I, (14) Auberger, and (15) Xg (the only sex-linked blood group).

CROSS MATCHING. In routine blood grouping or typing for blood transfusion, the presence or absence of antigens in all of these systems is not determined. But it is important to know whether blood from a donor will be safe to give to a recipient, most often a patient with some medical or surgical condition. Certain technics have been devised to determine the relative safety of such a procedure and to prevent a blood transfusion reaction. The first step is called the *major cross match*, and in this the red blood cells of the donor are mixed with the serum of the person to receive the blood. The mixture is carefully observed for any sign of clumping of the red blood cells, any agglutination meaning that the donor blood is incompatible with the blood of the recipient. If the recipient were to receive this blood even in small amounts, a transfusion reaction could easily occur.

The second step is the *minor cross match*, in which red blood cells, this time from the recipient, are admixed with serum from the donor. Again the mixture is care-

---

*Trade name RhoGAM.

212

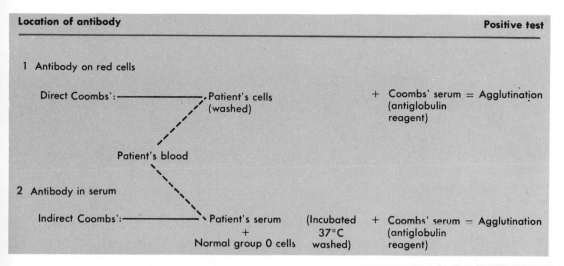

| Location of antibody | Positive test |

1 Antibody on red cells

Direct Coombs': ———————— Patient's cells (washed)   + Coombs' serum = Agglutination (antiglobulin reagent)

Patient's blood

2 Antibody in serum

Indirect Coombs': ———————— Patient's serum + Normal group 0 cells   (Incubated 37°C washed)   + Coombs' serum = Agglutination (antiglobulin reagent)

FIG. 12-8. Coombs' antiglobulin test (direct and indirect), outline. The antiglobulin test (patient's cells plus Coombs' serum) indicating presence of antibodies on red blood cells is *direct* Coombs' test. The *indirect* Coombs' test (patient's serum plus O cells plus Coombs' serum) identifies presence of antibodies in patient's serum. No agglutination is negative test. (Modified from Bove, J. R.: J.A.M.A. **200:**459, 1967.)

fully observed for any sign of clumping of the red blood cells, which would mean an incompatibility of the two bloods. Agglutination at this step indicates the possibility of a transfusion reaction should the donor blood be transfused into the recipient, but the reaction would probably not be so severe as an incompatibility picked up in the major cross match.

A third step is further indicated to detect unusual or uncommon antibodies that for some reason are not demonstrated in the major or minor cross match. In many blood bank laboratories this is accomplished by the use of *Coombs' test* (also known as the antiglobulin test). If we remember that antibodies are primarily related to the globulin fraction of the plasma proteins, and if we know that the protein globulin itself can function as an antigen to stimulate the production of antibodies, then it should be easy to understand that an antihuman globulin serum can be prepared by successively injecting human serum into a suitable animal such as the rabbit or goat. The serum of this immunized animal will then contain an antibody against the globulin of human serum. Since this fraction contains the antibodies, it means that this Coombs' serum will also have an action against antibodies, an action, as one can see, of a general or nonspecific nature. The Coombs test does not in any way identify antibodies; its chief use lies in the fact that it can detect them (Fig. 12-8). It helps to indicate the presence of antibodies even when an antigen-antibody combination has not resulted in an observable reaction. Such a test is of immeasurable value in blood typing and, on the whole, is easily carried out. If this test is negative when donor red blood cells and recipient serum are used in the procedure and if the major and minor cross matches show no incompatibility, it is then considered safe for the donor blood to be given as a transfusion to the recipient in question.

Unit 3 □ Microbes: production of infection

1. Fill out the table, indicating whether each example of acquired immunity (a) is obtained naturally or artificially and (b) is active or passive.

	Acquired			
Cause of immunity	Naturally	Artificially	Active	Passive
Attack of measles				
Attack of smallpox				
Vaccination against poliomyelitis				
Vaccination against smallpox				
Pasteur treatment				
Diphtheria antitoxin				
Diphtheria toxoid				
Tetanus antitoxin				
Tetanus toxoid				
Immune globulin				
Bacterial vaccine				
Placental transmission of antibodies				
Breast feeding				
Viral vaccine				

2. Classify immunity. Give examples.
3. What type of immunity is of the longest duration? The shortest? Explain.
4. What are immunoglobulins? How may they be classified?
5. Compare humoral immunity with cellular immunity.
6. What is the role of the lymphoid system in immunity? Role of thymus gland?
7. Comment on the immunologically competent cell.
8. Categorize in general terms the disturbances in immunity.
9. In the following diagram, place arrows pointing from the serum of each blood group to the cells agglutinated by that serum.

10. Why is a transfusion reaction more likely to occur when the donor's cells are agglutinated by the recipient's serum than when the recipient's cells are agglutinated by the donor's serum?

11. List the different kinds of transfusion reactions. Briefly explain.
12. Cite examples of immunosuppressive therapy.
13. What is meant by graft-versus-host reaction? How is it thought to come about?
14. State the importance of the transplantation antigens.
15. Explain the difference between autograft and heterograft; between isograft and homostatic graft.
16. Discuss briefly immunity in cancer.
17. Define or explain immunity, immunology, immunopathology, immunoglobulin, immunochemistry, immunobiology, immunosuppression, agglutination, antigenic determinant, isoantigen, adjuvant, immune system, susceptibility.
18. What practical application is made of the Coombs test?

REFERENCES. See at end of Chapter 14.

13 Immunologic reactions

REACTIONS FOR HUMORAL IMMUNITY

Because of the specificity of the antigen-antibody reaction, if either the antigen or antibody is known, it is possible to identify the other and, in many instances, to obtain reliable information as to how much of the other is present. Since antigen-antibody reactions are usually measured in serum,* they are called *serologic reactions*. In the clinical laboratory, *serology* is the study of the antigen-antibody reaction in a specimen of serum. With a positive serologic test, the unit that measures the number of antibodies present is the *titer*. Many technics are available. Some of the important ones will be discussed here.

Detection of antibodies

The type of antigen-antibody reaction that applies in any given situation depends largely on the physical state of available antigen. Practically speaking, antibodies (immunoglobulins) are detected by what they do to produce a *visible* reaction with a specific antigen under specified conditions (Table 13-1). In the hospital and clinical laboratory, antibodies are ordinarily classed as (1) complement-fixing antibodies, (2) antitoxins (neutralizing), (3) precipitins, (4) cytolysins, (5) agglutinins, (6) opsonins, (7) virus-neutralizing antibodies, and (8) fluorescent antibodies. In Table 13-1 this list of antibodies is tied in with antigens involved, and the nature of the antigen-antibody reaction is briefly indicated.

Kinds of reactions

Complement fixation. The complement-fixation test is designed to detect complement-fixing antibodies and is based on this fact: When serum containing the

---

*To collect serum: Allow a sample of blood to clot. The fibrin formed in the process entraps the blood cells. When the clot shrinks, the fluid expressed is the serum. Practically, serum is an important source of antibodies.

TABLE 13-1. DETECTION OF ANTIGEN-ANTIBODY COMBINATIONS

Positive reaction	Antigen	Antibody (in serum)	Nature of reaction
Complement fixation	Microbes	Complement-fixing antibodies	Inactivation of complement detected by hemolytic indicator system; *absence of hemolysis a positive test*
Flocculation	Exotoxins	Antitoxins	Flocculent precipitate
Precipitation	Microbes Animal proteins	Precipitins	Fine precipitate from clear solutions
Cytolysis Bacteriolysis Hemolysis	Intact cells Bacteria Red blood cells	Cytolysins Bacteriolysins Hemolysins	Cells dissolved
Agglutination	Agglutinogens Bacteria	Agglutinins	Gross clumping of cells
	Proteus bacilli	Heterophil antibodies	
	Human red blood cells	Isoantibodies	
	Sheep red blood cells	Heterophil antibodies	
Inhibition of viral hemagglutination	Human type O red cells coated with viral antigen	Inhibiting antibodies	Clumping of red cells *blocked*
Opsonization	Bacteria	Opsonins	Bacteria phagocytosed in greater numbers by leukocytes
Neutralization or protection	Viruses	Virus-neutralizing antibodies	Protection of an animal or tissue culture from harmful effects of agent
	Toxins	Antitoxins	
Fluorescence under ultraviolet light	Microbes	Fluorescein-tagged antibodies	Reaction tagged with marker

complement-fixing antibodies against the etiologic agent of a disease, the microbe itself, and complement are mixed in suitable proportions and incubated, the three enter into a combination whereby the complement is destroyed or inactivated (fixed). Since this combination is not accompanied by any visible change, an indicator system must be used to detect it. Because hemolysis is dramatic, a hemolytic system is a good one. In this system the combination of complement, *hemolysin (amboceptor)*, and sensitized red cells (of the species of animal against which hemolysin was prepared) effects *hemolysis* or dissolution of the red cells.

If a sample of serum from a patient is mixed with complement and a bacterial suspension and incubated, complement will be fixed when complement-fixing antibodies to the given bacteria are present in that serum. If suitable proportions of hemolysin and red blood cells are then added and the mixture incubated a second time,

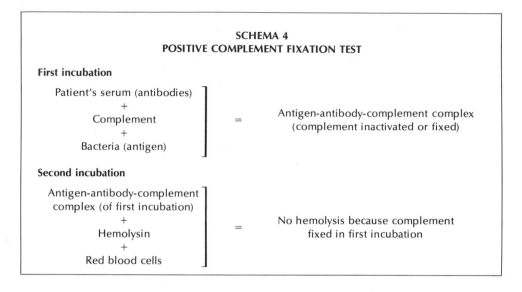

hemolysis *cannot* occur. This is a positive test, graphically illustrated in Schema 4. If, on the other hand, complement is *not* fixed during the first incubation, hemolysis can occur in the second incubation. This is a negative test, with the more colorful results.

WASSERMANN TEST. The Wassermann test for syphilis applies the principle of complement fixation. It has always been impossible to grow artificially the spirochetes of syphilis. When the Wassermann test first came into use, extracts of the livers of syphilitic fetuses were used as antigens because the liver of a syphilitic fetus was known to contain myriads of the spirochetes, which were the nearest approach to an extract of the organisms themselves. Today certain lipid extracts of normal organs are efficient antigens. A lipid extract of beef heart gives clinically reliable and consistent results. Of course, the term *antigen* applied to extracts of normal tissue used in the Wassermann test is scientifically incorrect. In other diseases the antigens used in complement fixation tests are derived directly from the causative organisms.

ASSAY OF COMPLEMENT. The complement system itself can be measured by the pattern of complement fixation outlined previously. The level of complement is altered in many disease states. The level of beta-1 complement (C3) is quite low in important diseases. Assay of complement with a hemolytic system can indicate both amount of complement (quantity) and its functional activity (quality).

Toxin-antitoxin flocculation. Antitoxins are antibodies that neutralize toxins, a combination demonstrable by flocculation. *Flocculation* refers to the presence of aggregates visible microscopically or macroscopically with certain antigen-antibody reactions. Toxin-antitoxin flocculations are similar to precipitin reactions (see below) but show a very sharp end point, the reaction blocked by either too much antigen or too much antibody.

Precipitation. When a suitable animal is immunized with certain bacteria, the animal's serum acquires the ability to cause a fine, powdery precipitate from the clear filtrate of the bacterial growth (*soluble* antigen of the bacterial extract is present). The antibodies formed are *precipitins*. The same phenomenon is noted when animals are immunized with soluble antigens of certain animal and vegetable proteins. Precipitins are specific; bacterial precipitins react only with the filtrates of the bacteria that produced them, and precipitins formed by immunizing an animal with the serum of an animal of a different species react only with the proteins of the species used for immunization. Precipitins do not form regularly in infectious disease, but where they are present they are nicely demonstrated by technics of immunodiffusion (p. 225).

The precipitin reaction has a wide medicolegal application in determining whether bloodstains are of human origin and in detecting the adulteration of one kind of meat with another. It is the test on which Lancefield's classification of streptococci is based. Flocculation tests for syphilis, such as the Kline, Kahn, and VDRL, are modified precipitation tests.

C-REACTIVE PROTEIN. A biologic coincidence occurs in a number of inflammatory conditions of either infectious or noninfectious nature. A peculiar protein (a globulin) that can form a precipitate when in contact with the somatic C-polysaccharide of pneumococci appears in the blood of the affected person. For this reason the protein is spoken of as C-reactive protein; it is *not* an antibody. The serum of an animal immunized to it is used in a precipitation test to detect C-reactive protein in serum of persons suspected of having one of the diseases in which the protein appears. Among such diseases are acute rheumatic fever, subacute bacterial endocarditis, staphylococcal infections, infections with enteric bacteria, and neoplasms.

Cytolysis. Antibodies that dissolve or lyse cells are known as *cytolysins* (or *amboceptors*). *Bacteriolysins*, cytolysins that lyse bacteria, are produced or at least increased by bacterial infection. Cytolysis requires both cytolytic antibodies and complement; if only one is present, cytolysis does not occur. A cytolysin of note is hemolysin, which dissolves red blood cells. It is prepared by giving an animal (most often a rabbit) a series of injections of the washed red blood cells from an animal of another species (most often man or sheep). Hemolysin develops in the serum of the recipient animal, specific *only* for the red blood cells of the species used.

Agglutination. If the serum of a person who has had a certain disease such as salmonellosis is mixed with a suspension of the bacteria that caused the disease, the bacteria will adhere to one another and form easily visible clumps that sink in the test tube. Likewise, if an animal receives several injections of a given microbe, its serum will acquire the ability to clump, or *agglutinate*, those microbes. This kind of clumping is *agglutination* (Figs. 13-1 and 13-2), and the antibodies that induce it are *agglutinins*. Particulate substances (antigens) that, when injected into an animal, induce the formation of agglutinins are known as *agglutinogens*. Agglutinogens may be microorganisms, red blood cells, or latex spheres coated with adsorbed antigen. In each instance there must be particles for clumping to occur.

Motile bacteria lost their motility before agglutination. Bacteria are not necessarily killed by agglutination, and dead bacteria are agglutinated as easily as are live ones. Agglutinated microorganisms are more readily ingested by phagocytic cells.

KINDS OF AGGLUTININS. Normally occurring agglutinins exist in the blood of

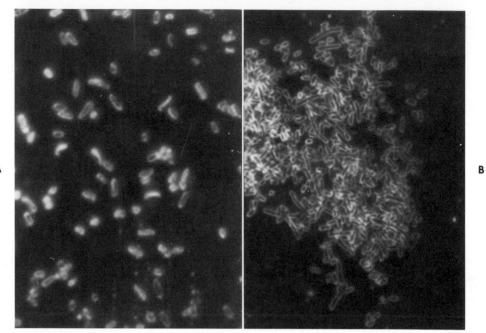

FIG. 13-1. Bacterial agglutination, dark-field illumination. **A,** Saline suspension of killed *Salmonella,* bacteria evenly spread with no clumping. **B,** Addition of positive antiserum; note large clump of organisms. (Courtesy Gwyn Hopkins and Dr. Gerard Noteboom, Dallas, Tex.)

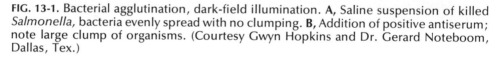

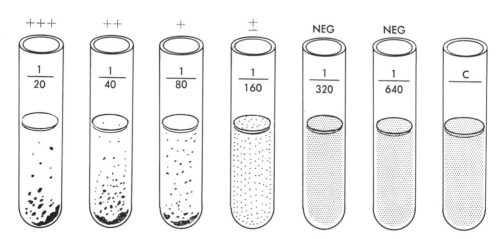

FIG. 13-2. Bacterial agglutination, test tube method. Dilute 0.2 ml of patient's serum serially in isotonic saline solution through first *six* tubes (left to right). Add 1 ml of suspension of bacteria to each tube through all seven. Final dilutions indicated by numbers on each tube (as 1 part serum in 20 parts, etc., suspension). Control tube on far right contains only bacterial suspension and saline solution. Note maximal agglutination of bacteria (large masses) in first three tubes on left, some in fourth, and none in fifth and sixth. Agglutination test positive; titer of serum 1:160.

certain persons. For instance, human serum sometimes shows a weak agglutinin content against typhoid bacilli although the person has never been knowingly infected. The agglutinin content is so low, however, that when the serum is diluted 20 or 40 times and then mixed with a suspension of organisms, agglutination does not occur.

Immune agglutinins are agglutinins brought about by infection or artificial immunization. Such agglutinins can occur plentifully in serum, and such serum will cause agglutination although diluted 500 to 1000 times. Agglutinin formation is usually specific; that is, when an organism is introduced into the body, the body forms agglutinins against that organism only or, in some cases, against very closely related organisms. Agglutinins for closely related organisms are known as *group agglutinins*. However, the agglutinin content of the serum is much higher for the organism directly inducing the production of the agglutinins than it is for the closely related ones.* Usually, in the first 7 to 10 days of an infection, the human (or animal) body has not manufactured enough agglutinins for positive identification.

Comparatively few organisms induce appreciable agglutinin formation with infection or artificial immunization. Important ones that do are *Salmonella typhi*, other *Salmonella* species, *Brucella* species, and *Francisella tularensis*. In infections with these (salmonellosis, undulant fever, tularemia) the agglutination test (Fig. 13-2) becomes a valuable diagnostic procedure. For example, serum from a patient is mixed with a suspension of *Salmonella typhi* in the *Widal test* for typhoid fever. If agglutination occurs when the serum is diluted enough to exclude the action of natural and group agglutinins, we know that the patient's serum contains immune agglutinins against *Salmonella typhi*. We may not know, however, whether the agglutinins result from a current attack of typhoid fever, a previous attack, or the recent administration of vaccine. (The diagnosis of typhoid fever is usually resolved by the case history and clinical manifestations.) In lower animals the agglutination test is used to detect Bang's disease (contagious abortion of cattle) and bacillary white dysentery in chickens caused by *Salmonella gallinarum*.

Sometimes *nonspecific* agglutinins are produced against organisms with apparently no relation to the disease. Nonspecific agglutination is of special diagnostic value in typhus fever and other rickettsial diseases. In typhus fever, nonspecific agglutinins develop against certain members of the *Proteus* genus of bacteria; typhus fever is *not* caused by *Proteus*. Many such agglutinins are probably formed as a result of the introduction into the body of *heterophil* antigens* (*hetero,* from an individual of a *different* species). Heterophil antigens cause heterophil antibodies to be formed not only against themselves but against other antigens as well. Such antigens are common in the lower orders of life. In typhus fever the causative agent of the disease, a rickettsia, seems to act as a heterophil antigen, and the antibodies formed react not only to the rickettsias but to certain types of *Proteus* bacilli as well. The agglutination of certain *Proteus* organisms by the serum of a patient with typhus fever constitutes the *Weil-Felix reaction*.

---

*The heterophil antigen is an exception to the tenet of antigen specificity.

The serum of human patients with infectious mononucleosis agglutinates the red blood cells of the sheep. This antibody is a *heterophil antibody* since it agglutinates an antigen (in the red cells) of a different species (the sheep). This diagnostic test is the *heterophil-antibody test*, or the *Paul-Bunnell test*.

Agglutinating serums are prepared by artificially immunizing separately several individual animals of the same species with known types of bacteria so that a battery of diagnostic typing serums is available. After an immunized animal is bled, its serum is separated and preserved for use. Such typing serums retain their potency for months. In testing, a suspension of unknown bacteria is mixed with the battery of specific serums. If one shows agglutination, the unknown bacteria must be the same as those against which that serum was prepared. In the application of the agglutination test to the identification of bacteria, the antibody is known and the antigen is unknown; in its application to the diagnosis of disease, the antigen is known but the antibody (in serum from a patient) is unknown.

Agglutination forms the basis of many useful laboratory tests and is thought to represent a physiologic mechanism for clearing the bloodstream of would-be invaders.

HEMAGGLUTINATION. Many types of antigens can be adsorbed onto the surface of a red blood cell. A red cell properly prepared with a test antigen can serve as a reagent for the detection of antibodies in the serum of a patient. This technic is *hemagglutination*. The titer of antibody is the dilution of serum in which no further agglutination of red cells takes place.

HEMAGGLUTINATION-INHIBITION. Red cell agglutination is used to identify certain viruses (p. 500). The action of a virus to clump red cells (viral hemagglutination) can be blocked by specific antibody to the virus. This is *hemagglutination-inhibition* and is a very valuable test in a number of viral infections.

Opsonization. Opsonins are substances that act on bacteria or other cells in such a manner (opsonization) as to render them more easily ingested by phagocytes and to favor their breakdown within the cytoplasm of the phagocyte. When the white blood cells of a person or animal suspected of having a disease are mixed with a suspension of the bacteria causing the disease, the degree of phagocytosis taking place is of diagnostic significance and can be evaluated in the *opsonocytophagic* test. This test has been used in the diagnosis of undulant fever and tularemia.

Phagocytosis (p. 179) is not an activity of the phagocytes alone but is dependent on the action of opsonins in the serum. Opsonins are to some degree present in normal serum but are present in an increased amount in immune serum, and their action is enhanced by complement.

NITROBLUE TETRAZOLIUM (NBT) TEST. When normal neutrophils (microphages) in an individual with a normally functioning immune system are confronted by bacterial pathogens, certain metabolic changes consequent to their phagocytic activity take place within them. (These changes probably relate to the cell membrane; opsonins are not pinpointed.) Because of these, the neutrophils can be shown to reduce in appreciable amounts a supravital dye, pale yellow nitroblue tetrazolium to blue-black formazan crystals. An in vitro test, the NBT test, has been devised to display precisely in human blood the ability of the neutrophils to do so. This test helps separate certain bacterial infections from nonbacterial illness in patients in whom clinical manifestations are confusing and monitor patients with increased vulnerability to infection (because of tissue transplant, immunosuppressive therapy, and the like).

Neutralization or protection tests. Neutralization tests are important in a number of infections caused by viruses and a few caused by bacteria. Antibodies that give immunity to viruses are known as *virus-neutralizing antibodies*. They can block infectivity of a given viral agent under test conditions in a susceptible host (p. 500).

Immunofluorescence. Antibodies can be chemically tied to the fluorescent dye fluorescein without changing their basic properties as antibodies, reacting with their specific antigens in the usual way. When such tagged antibodies, or fluorescent antibodies, are allowed to react with their specific antigens in cultures or smears containing certain microorganisms or in the tissue cells of the human or animal body (the *direct test*), the result is a precipitate formed from the combination of antigen and antibody that, because of the accompanying fluorescein, can be seen as a luminous area when viewed under ultraviolet light. A special type of microscope has been devised with which one can do just this. Specimens for testing with fluorescent antisera are taken from various body fluids and sites of disease.

The direct test is applicable to infections with the following agents:

Actinomyces   *Histoplasma capsulatum*
Blastomyces dermatitidis   Influenza virus
Bordetella pertussis   *Listeria monocytogenes*
Brucella   *Neisseria gonorrhoeae*
Candida albicans   *Neisseria meningitidis*
Cryptococcus neoformans   Parainfluenza viruses
Enteropathogenic *Escherichia coli*   Rabies virus
Enteroviruses   *Shigella*
Group A streptococci   *Staphylococcus aureus*
Haemophilus influenzae   *Streptococcus pneumoniae*
Herpesvirus

In the *indirect test*, nonfluorescent antibody is bound to its antigen as the first part of the test. In part two the combination is made visible after application of a second antibody (antihuman serum), this time a fluorescent one. The test is set up so that the unconjugated serum acts as antibody in the first part and as antigen when exposed to fluorescent serum in the second part. The indirect test has been used with infections caused by the following:

Brucella   *Leptospira*
Chlamydia trachomatis   *Mycoplasma pneumoniae*
Cryptococcus neoformans   *Plasmodium* of malaria
Cytomegalovirus   *Rickettsia prowazekii*
Epstein-Barr virus   Rubella virus
Haemophilus influenzae   *Toxoplasma gondii*
Herpesvirus   *Treponema pallidum*

Because of the rapid results with this technic, there is the wide variety of applications in identification of antigens related to bacteria, rickettsias, viruses, fungi, and protozoa. Pathogenic amebas can be quickly distinguished from nonpathogenic ones,

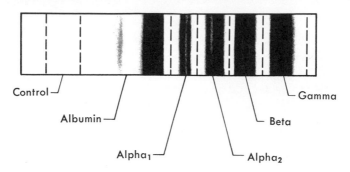

Control — Gamma
Albumin — Beta
Alpha₁ — Alpha₂

FIG. 13-3. Electrophoresis of normal human serum. Serum sample is placed near one end of cellulose acetate strip moistened with buffer. An electric potential is applied. Serum proteins migrate along strip at different speeds. Afterward, strip is dried and proteins are fixed and stained (bromphenol blue). Black bands correspond to separated and stained proteins, depth of color in each band proportional to amount of that protein present. Strip may be cut crosswise as indicated and protein fractions analyzed further. (From Bauer, J. D., and others: Clinical laboratory methods, ed. 8, St. Louis, 1974, The C. V. Mosby Co.)

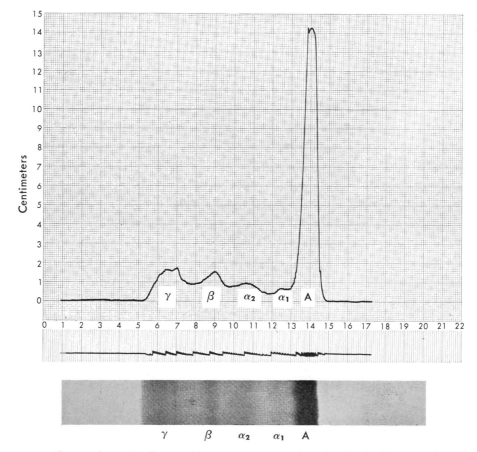

γ   β   α₂   α₁   A

FIG. 13-4. Electrophoresis of normal human serum. Color of individual protein fractions may be measured directly on cellulose acetate strip with recording densitometer and curve obtained as above. (From Bauer, J. D., and others: Clinical laboratory methods, ed. 8, St. Louis, 1974, The C. V. Mosby Co.)

streptococci may be detected quickly in throat swabs, rabies virus can be tagged in animals, and complement (especially C3) may also be assayed. In tissues, fluorescent staining helps to localize the precise anatomic sites of antigen-antibody combination in diseases such as lupus erythematosus, rheumatoid arthritis, and certain hypersensitivity reactions.

Special technics. There are several other laboratory tests contributing much to the study of the structure and nature of antibodies (immunoglobulins) and to the elucidation of the immunologic disorders.

ELECTROPHORESIS. When proteins are placed in an electrical field at a given pH, they will move in a distinct path according to their own specific electrical charges. The pattern of migration of different proteins can be visualized under prescribed test conditions on a suitable strip, such as paper, stained with dye (Figs. 13-3 and 13-4). *Zone electrophoresis* refers to the technic of using a stabilizing medium to trap migrating proteins into more or less separate areas or zones that can then be stained and identified later. Routinely used solid support media include paper, starch, agar, and cellulose acetate. Antibodies with zone electrophoresis tend to show up as an undifferentiated group. They can be sorted, however, by technics indicated below.

RADIAL IMMUNODIFFUSION. Under the appropriate conditions antigen-antibody combinations form a visible precipitate. *Gel diffusion* is a term used for the antigen-antibody precipitin reaction detected in semisolid media. If protein (antigen) is placed in a specially designed hole or well in an agarose gel medium (no electrical charge), it diffuses concentrically out of the well. Since specific antibody precipitates antigen in gels, this radial diffusion can be shown nicely in the clear gel if specific antibody has been impregnated into it. A ring of precipitate forms at the points of contact of antigen with specific antibody, and the reaction can be quantitated easily for amounts of protein (or antigen). Known also as *single gel diffusion*, this is an important method for analysis of immunoglobulins.

Electroimmunodiffusion is single gel diffusion in an electric field.

DOUBLE GEL DIFFUSION. In this method antibody is not incorporated throughout the agar as in single gel diffusion but is placed in a trench (or other reservoir) cut in the medium. The arrangement is such that soluble protein (antigen) and antibody diffuse out into the agar to form a band or line of precipitate at points of contact. If a mixture of proteins (or antigens) is analyzed, each specific antigen-antibody combination tends to present as a separate band at a definite position in the medium between the reservoir for the antigen and that for the antibody. This reaction is also a quantitative one.

In the *Ouchterlony technic* for double diffusion a microslide is covered with suitable diffusion medium. The center well is filled with serum (human or animal) containing known antibodies, and the test specimens containing soluble antigens are placed in peripheral wells opposite the center one. The microslide is incubated in a moist chamber for 24 to 48 hours, and the combination of specific antibody with antigen is read from the precipitation line. This particular precipitin-in-agar technic may be used at times as a simple, fairly rapid screening technic for the identification of various microorganisms.

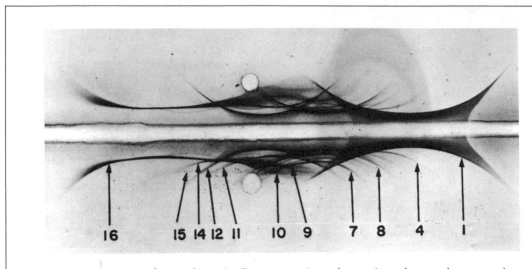

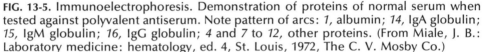

FIG. 13-5. Immunoelectrophoresis. Demonstration of proteins of normal serum when tested against polyvalent antiserum. Note pattern of arcs: *1*, albumin; *14*, IgA globulin; *15*, IgM globulin; *16*, IgG globulin; *4* and *7* to *12*, other proteins. (From Miale, J. B.: Laboratory medicine: hematology, ed. 4, St. Louis, 1972, The C. V. Mosby Co.)

IMMUNOELECTROPHORESIS. If the technics of electrophoresis and double gel diffusion are combined, we have the process of *immunoelectrophoresis*, a valuable method for separating complex mixtures of proteins (antigens). In the test specimen in an electric field the unknown proteins are separated and spread out in a series of differently charged masses. After electrophoresis, specific antibody is placed in a trough parallel to the line of migration of the proteins in the gel and allowed to diffuse inward in an incoming wave. As specific antibody meets antigen, precipitation occurs in a series of arcs (Fig. 13-5). The immunoprecipitin bands are readily seen, and the position and shape are consistent and stable for known proteins. Deviations represent abnormal constituents (Fig. 13-6).

COUNTERELECTROPHORESIS. This is a modification of immunodiffusion whereby both antigen and antibody are brought together in an electric field; that is, they are "driven" toward each other electrophoretically. It is only applicable under special circumstances.

RADIOIMMUNOASSAY (RIA) (COMPETITIVE BINDING ANALYSIS). This method is most sensitive for quantitating antigens that can be radioactively labeled. It is based on competition under test conditions for a limited number of binding sites in a fixed amount of specific antibody between (a) the known, a fixed level of purified antigen labeled with radioactive iodine, and (b) the unknown, an unlabeled test sample of antigen. The antigen and antibody complexes formed are separated, and the amount of radioactivity is determined. The concentration of the unknown, unlabeled antigen is calculated by comparison with reference standards.

226

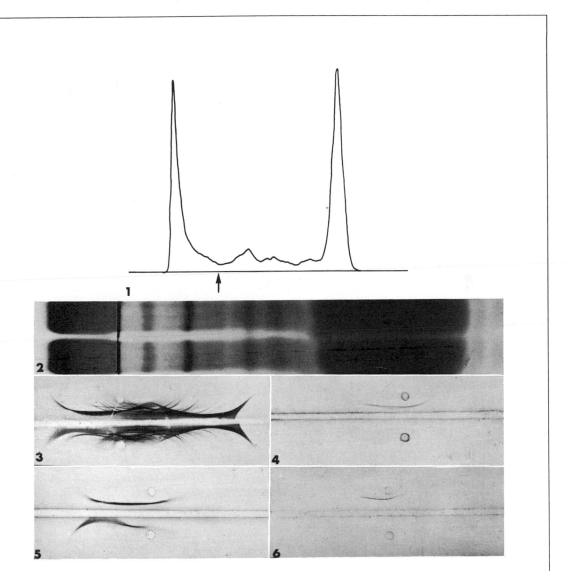

FIG. 13-6. Protein abnormalities, electrophoretic study. In several technics demonstrated, an example of a malignant proliferation of lymphoid system (multiple myeloma) is used, a disease associated with large quantities of abnormal globulins. **1,** Paper electrophoresis showing abnormal peak *(arrow).* **2,** Starch gel electrophoresis showing abnormal gamma globulin component. **3** to **6,** Immunoelectrophoresis (in each instance normal control sample is at top, patient specimen at bottom). **3,** Testing done with polyvalent antiserum. **4,** Testing with anti-IgA. **5,** With anti-IgG (which picks up, identifies, and locates the IgG) there is an increase of IgG. **6,** With anti-IgM there is a decrease of IgM. (From Miale, J. B.: Laboratory medicine: hematology, ed. 4, St. Louis, 1972, The C. V. Mosby Co.)

PROCEDURES FOR CELLULAR IMMUNITY

Cellular immunity may be measured by immunologic yardsticks although the technics are not so well established as those with which we evaluate humoral immunity.

Lymphocytes in peripheral blood. The presence, number, and kind of lymphocytes circulating in the peripheral blood can be determined. This is information of a general nature; nonetheless, it can be useful. If the total number of lymphocytes is decreased (lymphopenia, a significant finding), this is reflected in the total lymphocyte count (sometimes done sequentially). The critical level for lymphocytes is 1200/mm^3. The morphology of the lymphocytes may be studied in peripheral blood smears, and differential counts made. Thymus-derived T cells are easily recognized: they are small, mostly nucleus with little cytoplasm. Approximately 70% of peripheral blood lymphocytes are T cells; some 25% are B cells. The small percentage remaining are indeterminate.

Battery of skin tests. Skin testing is fairly straightforward for the presence or absence of cellular immunity. A battery of antigens is selected with the idea that most normal adults should have a positive skin test to at least one of them. The antigens used routinely include one for *Candida*, a mumps antigen, streptokinase-streptodornase, trichophytin, and purified protein derivative (PPD, tuberculin). Children, however, even under normal circumstances may show negative skin test results with all of these. In a child, therefore, sensitivity may be artificially induced with dinitrochlorobenzene and challenge made 2 weeks later.

Immunologic competency in vitro. There are tests based on the premise that cell-mediated immunity requires for its expression a direct interaction between the host lymphoid cell and the agent (microbe, foreign cell, and the like) that carries the antigen into the area of contact. When the natural situation is simulated, the immunologic behavior of the lymphocytes can be observed. Rather precise assays are being developed along these lines. One such assay starts with peripheral lymphocytes isolated in cell culture. A mitogen such as phytohemagglutinin (PHA) is added to stimulate increased DNA and RNA synthesis (in T cells only), a response of the lymphocytes measurable in two ways. One method quantitates uptake of radioactive thymidine. A second makes a differential count of the cells. Small T lymphocytes exposed to PHA become larger cells (blast transformation). The nucleolus swells and cytoplasmic organelles develop. After suitable staining, one can record the number of transformed blast cells.

Another assay starts with a mixed white cell culture, one in which test lymphocytes from a given person are cultured with lymphocytes from an unrelated donor. Competent lymphocytes can recognize the cells that do not belong, the nonself. This they do because of the histocompatibility antigens foreign to them on the donor cells. In severe immunologic deficiency disorders involving cell-mediated immunity the ability to recognize and respond to HL-A antigens is lost.

Migration inhibitory factor assay. Several methods have been devised to assay factors released after the interaction of sensitized lymphocytes with specific antigen. One such method is the assay of the migration inhibitory factor (MIF). For this test

TABLE 13-2. IMMUNOLOGIC APPROACH TO DISEASE

Test(s)	Application
Serologic	
1. Complement fixation	Diagnosis of syphilis, viral infections, and certain bacterial infections
2. Flocculation	Diagnosis of syphilis
3. Precipitation	Diagnosis of syphilis
Protein precipitation	Detection of origin of bloodstains
	Detection of meat adulteration
4. Agglutination	Identification (serologic typing) of bacteria; diagnosis of undulant fever, tularemia, and salmonelloses
a. Widal	Diagnosis of typhoid fever
b. Weil-Felix	Diagnosis of typhus fever and Rocky Mountain spotted fever
c. Paul-Bunnell (heterophil antibody)	Diagnosis of infectious mononucleosis
d. Viral hemagglutination	Diagnosis of viral infections—rubella, rubeola, influenza, parainfluenzal infection
5. Virus neutralization	Diagnosis of viral infections
6. Fluorescent antibody	Identification of protozoa, fungi, bacteria, rickettsias, chlamydiae, and viruses; diagnosis of rabies, syphilis, amebiasis
Intradermal (skin)	Diagnosis of infection, past or present, caused by microorganisms of
1. Cell-mediated immunity to infection with:	
a. *Bacteria*	
Brucellergen	Brucellosis
Lepromin	Leprosy
Mallein	Glanders
Tuberculin	Tuberculosis
b. Chlamydiae (also bacteria)	
Frei	Lymphopathia venereum
c. Fungi	
Blastomycin	Blastomycosis
Coccidioidin	Coccidioidomycosis
Cryptococcin	Cryptococcosis
Histoplasmin	Histoplasmosis
Oidiomycin	Candidiasis
Spherulin	Coccidioidomycosis
Sporotrichin	Sporotrichosis
Trichophytin	Epidermophytosis
d. Protozoa	
Leishmanin	Leishmaniasis
e. Viruses	
Mumps	Mumps
2. Susceptibility	
Schick	Susceptibility to diphtheria
Dick	Susceptibility to scarlet fever

cellular suspensions are prepared from peritoneal exudates of guinea pigs sensitized to antigen(s). In this material there are macrophages (70%), polymorphonuclear leukocytes (only a few), and lymphocytes, specifically sensitized. The exudative material is placed into capillary tubes, and the macrophages are observed as they migrate out of the tubes onto glass coverslips in culture chambers in the *absence of antigen*. When the antigen is added, migration is stopped. It takes only one sensitized lymphocyte to block movement of 99 macrophages. This reaction can also be quantitated. First, the patient is studied to determine whether sensitized lymphocytes are circulating and then as to what sensitivities are present to a given variety of tissue antigens.

Rosettes. T cells can be separated from B cells in a mixed population in a variety of ways. According to some observers, T cells may form rosettes with sheep red blood cells when incubated together. All agree that B cells form rosettes under comparable conditions, the antigen of the sheep red cell binding to the B cell membrane immunoglobulin. B cells, unlike T cells, can also be identified with fluorescent anti–immunoglobulin designed to indicate their surface receptors.

When red cells coated with anti–red cell antibody and complement are added to lymphocytes to demonstrate receptors for complement on the B cell surface, the red cells cluster forming rosettes about the B cells.

APPLICATION OF IMMUNOLOGIC METHODS

Some of the more frequently used immunologic methods for identifying and studying microbes and for defining their role in disease production are given in Table 13-2.

QUESTIONS FOR REVIEW

1. What is an agglutination test? Name diseases in which it is used as a diagnostic procedure.
2. Briefly define antitoxin, cytolysin, agglutinin, precipitation, precipitin, serology, heterophil antibody, opsonin, hemolysis, hemolysin, amboceptor, serum, hemagglutination, virus neutralization, flocculation, migration inhibitory factor, lymphopenia, blast transformation.
3. How may antibodies be detected?
4. What is meant by the titer of a serum?
5. What are the advantages of the immunofluorescent technics?
6. Explain the difference between the direct and indirect fluorescent antibody test.
7. Briefly discuss immunoelectrophoresis and its applications.
8. Diagram a negative complement fixation test.
9. State ways in which cell-mediated immunity may be evaluated.

REFERENCES. See at end of Chapter 14.

14 Allergy

ALLERGIC DISORDERS

General characteristics. Some persons exhibit unusual and heightened manifestations on contact with certain substances that are often, but not necessarily, protein and that do not affect the average or another person. For example, an individual may have asthma on contact with horse dander,* and many persons have hay fever on contact with plant pollens.

We have learned that immunity is a state wherein by prior contact a person obtains protection against an agent that would otherwise harm him. The reaction is specific for the particular agent, and a benefit comes from the immune setup. If we think of another situation, comparable immunologically but one in which the immune mechanisms injure or damage instead of protect, then we are referring to the state of *allergy* or *hypersensitivity*.† Allergy or hypersensitivity is a state of altered reactivity directly related to the operation of the immune system. The study of the effects in body tissues of the allergic reaction is included in *immunopathology*.

An antigenic substance that can trigger the allergic state is an *allergen*. It may be a protein or frequently a nonprotein of low molecular weight. Allergens reach the body by way of the respiratory tract or the digestive tract, by direct contact with skin or mucous membranes, and by mechanical injection. Of these routes, the respiratory pathway is most consistent. It is thought that unchanged protein must pass through the intestinal wall to the bloodstream for sensitization via the intestinal tract to occur. This can occur in children and in adults with digestive disturbances.

---

*Dander is animal dandruff—the composite of cast-off epithelial cells of the skin, sebum, and other body oils.

†Both terms are ordinarily used to mean conditions such as emotional upsets, nonallergic food reactions, toxic responses to drugs, and others that are nonimmunogenic. Since *hypersensitivity* is more often so used than *allergy*, some authorities believe that the term *allergy* has priority. Allergy is a term first used by von Pirquet.

231

TABLE 14-1. KINDS OF ALLERGIC DISORDERS

	Immediate-type reactions (anaphylactic type)	Delayed-type reactions (bacterial or tuberculin type)
Clinical state	Hay fever	Drug allergies
	Asthma	Infectious allergies
	Urticaria	Tuberculosis
	Allergic skin conditions	Tularemia
	Serum sickness	Brucellosis
	Anaphylactic shock	Rheumatic fever
		Smallpox
		Histoplasmosis
		Coccidioidomycosis
		Blastomycosis
		Trichinosis
		Contact dermatitis
Onset	Immediate	Delayed
Duration	Short—hours	Prolonged—days or longer
Allergens	Pollen	Drugs
	Molds	Antibiotics
	House dust	Microbes—bacteria, viruses, fungi, animal parasites
	Danders	Poison ivy and plant oils
	Drugs	Plastics and other chemicals
	Antibiotics	Fabrics, furs
	Horse serum	Cosmetics
	Soluble proteins and carbohydrates	
	Foods	
Passive transfer of sensitivity	With serum	With cells or cell fractions of lymphoid series

A passive sensitization may be transmitted from the mother to her child in utero via the placenta. This type of sensitization is short lived, as is true of passive immunity in the newborn infant.

An allergic person is often sensitive to several different allergens. At least 10% to 20% of persons in the United States develop allergic disorders under the natural conditions of life.

Kinds of allergy. As would be expected from our discussion of immunity, the reactions of allergy are separated into two main divisions (Table 14-1): (1) those in which the allergic responses result from the presence of humoral antibody, reactions of the *immediate* type, and (2) those in which the responses are not mediated by circulating antibodies but by specifically sensitized cells (that is, they are cell mediated), reactions of the *delayed* type.

Mechanisms. The mechanisms of allergy are complex and incompletely understood. Only a sketch (Fig. 14-1) is given of the background.

There is a sequence in the development of the immediate hypersensitivity reaction. It starts with the first exposure to an allergen, which initiates biosynthesis

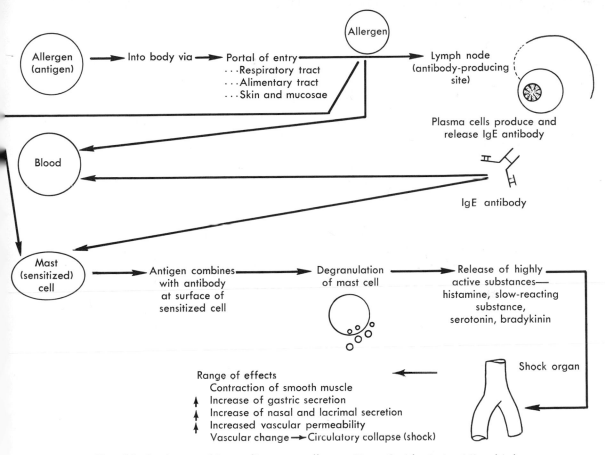

FIG. 14-1. Simplified schema of immediate-type allergy. (From Smith, A. L.: Microbiology and pathology, ed. 11, St. Louis, 1976, The C. V. Mosby Co.)

of the humoral antibody, immunoglobulin E (IgE). Of 5000 different immunoglobulin molecules found in normal persons, only one seems to be an IgE. Nearly everyone has some IgE in his serum, but in allergic individuals it is often present in increased amounts. Like all immunoglobulins, IgE is produced by cells in the lymphocyte–plasma cell series. It is formed mainly in tonsillar tissues, Peyer's patches, and generally near the mucosal surfaces of the gastrointestinal and respiratory passages (not to any considerable extent in the spleen and most lymph nodes, however). Individuals vary in their production of this antibody, but it is highly specific for the allergen inducing its formation.

From the first exposure the IgE is fixed to specific receptor sites on mast cells and basophils, sensitizing them. The mast cell (Fig. 14-2) is a tissue-based cell most numerous in the places where IgE is formed—in the lining areas of the respiratory and gastrointestinal tracts and under the skin. It is a cell about 1 μm wide known for its distinct appearance. Large, densely placed, metachromatically staining granules found in its cytoplasm are notable for their high concentration of pharmacologically potent compounds such as histamine, heparin, and proteolytic enzymes. Degranu-

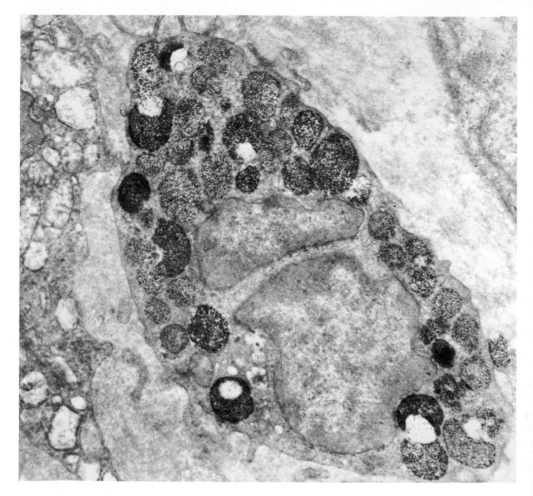

FIG. 14-2. Human mast cell (basophil of peripheral blood), electron micrograph. Note large dark granular masses (large dark blue cytoplasmic granules of Wright-stained blood smears). They store histamine and heparin. (From Barrett, J. T.: Textbook of immunology, ed. 2., St. Louis, 1974, The C. V. Mosby Co.)

lation or loss of these unique granules releases their contained substances into tissue fluids. Nonimmune as well as immune factors can cause degranulation. The basophil circulating in the bloodstream is a look-alike to the mast cell and mimics its responses in allergy. The two are the only cells to which IgE is affixed and the only two cells in the body containing histamine.

With the second and subsequent exposure to the allergen, the stage is set for the release of the mediators of allergy, that is, those compounds with the chemical potential to induce the manifestations of immediate-type allergy. The allergen reaches the sensitized cell, complexes there with its specific IgE, and is locked into the receptor site. For reasons not fully known, this combination at the surface of the sensitized cell sets in motion an active, energy-requiring process along its plasma membrane, which causes it to degranulate. In so doing, the cell releases histamine, the slow-reacting substance of anaphylaxis (SRS-A), and the other mediators that profoundly influence vascular permeability and smooth muscles. IgE bound to the sensitized cell has an extremely high order of biologic activity, and even very low concentrations of allergen can induce reactions ranging from a mild allergy to fatal anaphylactic shock.

The spectrum of effects in immediate-type allergy includes contraction of smooth muscle, increase of gastric secretion, increase of nasal and lacrimal secretions, increase of vascular permeability, and vascular changes that can lead to circulatory collapse (shock). Of the chemical mediators of allergy, histamine has been the most extensively studied, and the allergic reaction has even been said to constitute a response to too much histamine.

In immediate-type allergy the consequences of the antigen-antibody combination follow shortly thereafter, even within minutes, and the reaction reaches a maximum within a few hours. The responses occur in vascularized tissues and relate primarily to smooth muscle, blood vessels, and connective tissues. A specific shock tissue is important; that is, a target tissue or organ that, because of its anatomic makeup, shows maximal effects. Manifestations are acute and short lived. Epinephrine, antihistaminic drugs, and related compounds are of benefit. This type of reactivity can be transferred passively to a nonsensitized person by antibody-containing serum.

In delayed-type allergy, by contrast, there is no relation to conventional antibody. The phenomenon is closely associated with cells. The responses are set in motion by lymphocytes specifically modified so that they can respond to the specific allergen deposited at a given site. The main feature of the reaction is the direct destruction of target tissue containing the antigen. The condition may be passively transferred to a nonsensitized person if immunologically competent cells (lymphoid cells) are used.

In allergic reactions of this type a much longer period of time elapses between the presentation of the antigen and manifestations of the disorder. The period of incubation ranges from 1 to several days. Reactions do not favor any particular tissue but may occur anywhere. There is no specific shock organ or tissue as with immediate-type allergy. Many cells in the body are vulnerable. Manifestations tend to be slow in onset and prolonged. No histamine effect is demonstrated as in immediate-type reactions, and antihistamines are without effect. The steroid hormones seem to help in the treatment of some allergic disorders of this kind and in some of those of the immediate type as well. The exact mode of their action is not known.

Immediate-type allergic reactions

Anaphylaxis. In immediate-type allergy with sudden release of chemical mediators, the response is acute. The acute reaction is *anaphylaxis*, the prototype of immediate-type hypersensitivity. Systemic anaphylaxis or anaphylactic shock is a generalized reaction brought about in a sensitized animal on contact with adequate amounts of antigen administered so that rapid release and dissemination of mediators occurs (as intravascularly).

If a small amount of foreign protein (horse serum, egg white, and the like), in itself nonpoisonous, is injected into a suitable animal such as a guinea pig, the first and *sensitizing dose* will be without noticeable effect. When the second and larger injection, the *provocative dose*, is given after an interval of from 10 to 14 days, severe manifestations ensue. The guinea pig becomes restless and breathes with increasing difficulty within *1 or 2 minutes!* Frantic activity precedes its death in respiratory failure.

Anaphylaxis is specific; that is, for anaphylaxis to occur, the sensitizing and provocative doses must be of the same substance. If the serum of a sensitized animal is injected into a normal animal and if after an interval of 6 to 8 hours the normal animal is injected with some of the material to which the first animal is sensitive, signs of anaphylaxis appear in the normal animal. This is *passive anaphylaxis.* The passive transfer of the hypersensitive state assumes considerable importance when it is realized that the use of an allergic donor for blood transfusion may render the recipient hypersensitive. One interesting case is reported in which, after receiving blood from a donor sensitive to horse dander, the recipient had an attack of asthma on contact with a horse. If an actively sensitized animal survives an anaphylactic attack, it is desensitized for a few days but eventually becomes sensitive again.

ROLE OF SMOOTH MUSCLE. The dramatic manifestations of anaphylaxis result from the contraction of smooth muscle fibers, and the part of the body primarily attacked in a given animal depends on the distribution of smooth muscle fibers in the species. In the guinea pig, smooth muscle fibers are plentiful in the lungs; contrac-

---

TABLE 14-2. MAKEUP OF THE ANAPHYLACTIC SET*

Item	Number needed	Item	Number needed
Ampules of 1:1000 solution of epinephrine, 1 ml	2	Gauze sponges	
		Syringes, 2 ml	2
Hypodermic needles	2	Needles (1 and 4 inches long)	2
Ampules of aminophylline (0.24 g each)	2	Bottle of 5% solution of dextrose in distilled water, 1000 ml	1
Intravenous set	1	Ampule of diphenhydramine	1
Bottle of hydrocortisone (dilution to 2 ml gives 50 mg/ml)	1	hydrochloride, 10 ml (10 mg/ml)	
		Ampule of sterile water	1
Scalpel	1	Hemostat	1
Ampule of absorbable surgical suture (catgut) with needle	1	Syringe, 20 ml	1
Tongue depressors		Alcohol, 70% (disinfectant)	

*This kind of emergency must be anticipated. At the Mayo Clinic, Rochester, Minn., "anaphylactic sets" are placed in strategic areas. They contain the above items, quickly available, to be used in the event of such a reaction.

tion of these closes bronchioles and bronchi and leads to death from respiratory tract obstruction. In the rabbit, smooth muscle is plentiful in the pulmonary arteries. If the rabbit is used as test animal, these arteries contract, and a circulatory burden that leads to failure is thrown on the right side of the heart.

That anaphylactic shock causes contraction of smooth muscle may be proved in the following manner: Take a strip of fresh tissue containing an abundance of smooth muscle (for instance, the uterine muscle of the guinea pig) from an animal that has been sensitized and place it in a bath of Ringer's solution. Make one end of the strip stationary and attach the other end to a recording instrument (kymograph). When only a small amount of the sensitizing substance is added to the solution, vigorous contraction of the smooth muscle will occur and be registered on the kymograph. This is the *Schultz-Dale reaction.*

ANAPHYLAXIS IN MAN. In man, anaphylaxis can be an immediate, severe, and fatal reaction, running its course within seconds or minutes (Table 14-2). If the person is sensitive to the given allergen, a very small amount can precipitate the reaction.

When anaphylaxis was first described at the turn of the century, one of the chief causes then was a sting by a member of the insect order Hymenoptera. Included here are bees, yellow jackets, wasps, and hornets. The fact is still true. It is said today that far more deaths occur from insect stings than from snakebites.*

In all areas of our complex society the number of different substances to which man may become sensitized is astonishing. This is especially true in the field of medicine, where a variety of chemicals are used for many reasons so that most of the anaphylactic reactions today are the result of sensitivity to drugs that have been used in one way or another for diagnosis and treatment of disease. Reactions of this type are referred to as *iatrogenic* reactions. Table 14-3 lists chemical agents known to have caused systemic anaphylaxis in man.

Atopic allergy. Atopic allergy is the designation for a group of well-known human allergies to naturally occurring allergens with chronic manifestations of immediate-type allergy. Here are found hay fever and asthma. By contrast with systemic anaphylaxis, where larger doses of antigen are given intravenously, small doses of antigen repeatedly contact mucous membranes.

Allergic diseases are not inherited, but the special tendency for the individual to develop the state of altered reactivity in tissues is. Atopic allergy seems to be present more in some families than in others; however, all the members of an affected family do not have the same condition. One may have hay fever, another asthma, and still another some type of skin eruption. Seldom will all the members of a family manifest the hypersensitive state.

ASTHMA. Asthma is a recurrent type of breathlessness coming in acute episodes and described by certain types of respiratory movements and wheezing. It is an allergic condition most often caused by animal hair, feathers or dander, house dust, foods, microbes, and certain cosmetics. Important asthma-producing foods are milk,

---

*A person with no history of allergy can have a fatal reaction to an insect bite. If one has had a reaction, he should take every precaution in areas of contact, wear the necessary protective clothing, and, if need be, carry an emergency kit of appropriate medication. Insect-sting kits are available commercially.

TABLE 14-3. SYSTEMIC ANAPHYLAXIS IN MAN—DOCUMENTED AGENTS*

Chemical class	Trade name of drug	Route of inoculation	Deaths
Proteins			
Antiserum of horse (in passive immunization)		Parenteral	Yes
Antirabies serum			
Tetanus antitoxin			
Bivalent botulism antitoxin			
Hormones			
Insulin		Subcutaneous	
Corticotropin	Acthar, Cortrophin	Intravenous	
Enzymes			
Chymotrypsin		Intramuscular	
Trypsin	Parenzyme	Intramuscular	
Penicillinase	Neutrapen	Intramuscular	
Sting of Hymenoptera (bees, wasps, hornets, and yellow jackets)		Subcutaneous (insect bite)	Yes
Pollen			
Bermuda grass		Intradermal	Yes
Ragweed		Intradermal	Yes
Food			
Egg white		Intradermal	Yes
Buckwheat		Intradermal	Yes
Cotton seed		Intradermal	Yes
Glue		Intradermal	Yes
Polysaccharides			
Acacia (emulsifier)		Intravenous	
Dextran (plasma expander)	Expandex, Gentran	Intravenous	
Others: action as haptens			
Antibiotics			
Penicillin		Oral and parenteral	Yes
Demethylchlortetracycline hydrochloride	Declomycin	Oral	
Nitrofurantoin	Furadantin	Oral	
Streptomycin		Intramuscular	
Other medications			
Sodium dehydrocholate	Decholin	Intravenous	Yes
Thiamine		Subcutaneous and intravenous	Yes
Salicylates (aspirin)		Oral	
Procaine	Novocaine	Parenteral	Yes
Diagnostic agents			
Sulfobromophthalein	Bromsulphalein (BSP)	Intravenous	Yes
Iodinated organic contrast agents		Oral and intravenous	Yes

*After Austen, K. F.: J.A.M.A. **192:**108, 1965.

FIG. 14-3. Windborne pollen grain of ragweed looks like battered golf ball in scanning electron micrograph. (×2000.) Wind-pollinated plants produce pollen in huge quantities. A single ragweed plant may expel 1 million pollen grains in 1 day, 1 billion in one season. More than 250,000 tons of ragweed pollen may be blown hundreds of miles across this nation during a fall season.

milk products, eggs, meat, fish, and cereals. A person may become sensitive to the bacteria normally inhabiting the upper respiratory tract and thereby become a victim of asthmatic attacks. This type of asthma is spoken of as *endogenous* asthma. An estimated 4% of the population has asthma.

HAY FEVER. Hay fever, with its well-known nasal distress, results from sensitivity to pollens (Fig. 14-3), and the period of attack corresponds to the time of pollination of the offending plant or plants. To be important as a cause of hay fever, a plant must produce a light dry pollen easily carried a long distance by the wind. This excuses both goldenrod and roses, long accorded an unearned distinction as causes of hay fever. Early spring hay fever is usually caused by the pollen of trees. Late spring and early summer hay fever is most often from grass pollens, and more than 80% of the cases of fall hay fever result from ragweed pollen. Ragweed is found only in the United States, parts of Canada, and Mexico. The importance of individual trees, grasses, or other plants depends on geographic location. Ten percent of the population has hay fever.

Urticaria and allergic skin eruptions. Allergic skin conditions may come from foods, drugs, chemicals, and a wide variety of other allergens. Urticaria, known sometimes as nettle rash, is an allergic disorder of the skin depicted by the presence of wheals (whitish swellings) or hives (lesions of a strikingly transitory nature at times).

Serum sickness. After the administration of an immune serum a reaction known as serum sickness may appear in persons who have never had a previous injection of horse serum and, so far as is known, are not sensitive to horse proteins.

Serum sickness is common, and its manifestations are unpleasant but seldom life-threatening. It usually begins 8 to 12 days after the injection of an immune serum and is typified by skin eruption; swollen, painful, and stiff joints; enlargement of the lymph nodes; leukopenia; and decreased coagulability of the blood. A reaction may occur only around the site of injection (local serum disease). Serum sickness is thought to be caused by the combination of antibodies formed right after the injection with an excess of the serum. (Serum sickness is here classified as an immediate-type response although it is a disorder combining features of both major categories of hypersensitivity.)

An immune serum should be administered with extreme caution to asthmatic patients and to those who have had a previous injection of horse serum. Before an immune serum is given, tests (p. 692) to detect sensitivity to horse serum should always be done because the factor responsible for untoward reactions is the horse protein, the medium for many antitoxins and immune serums. The fall in blood pressure, the drop in body temperature, and respiratory difficulty resulting from the injection of an immune serum are most likely to appear when the second injection is given 2 or 3 weeks after the first, but severe reactions may occur when a second injection is given months or even years after the first. Anaphylaxis is most likely to occur in man after intravenous injection. Rarely, manifestations may lead to immediate collapse and death.

Arthus phenomenon. The Arthus (toxic complex) phenomenon shows us how immunologic mechanisms may damage tissues (Fig. 14-4). When material (an antigen) to which an individual is sensitized is *re*introduced into his body, the Arthus phenomenon may develop in a localized area of skin about the injection site. Although considered an immediate-type allergy, this reaction appears after a delay of several hours at least. Skin involvement results from an intricate sequence of events set in motion when the injected antigen combines with antibody present and precipitates as a complex in the subendothelial layer of blood vessels. The complex fixes complement and, in so doing, triggers an inflammatory response. The net effect is injury

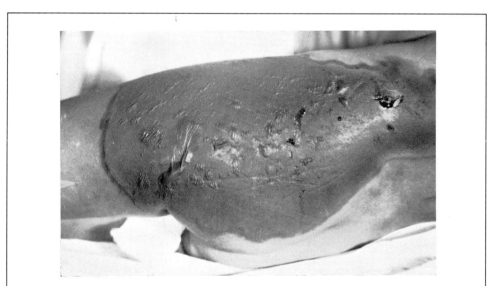

FIG. 14-4. Arthus phenomenon in sharply set-off area in skin of hip and thigh. Note discoloration associated with hemorrhage and death of tissue. (From Top, F. H., Sr.: Communicable and infectious diseases, ed. 6, St. Louis, 1968, The C. V. Mosby Co.)

to tissues and primarily to small blood vessels. The area of skin becomes swollen, reddened, hemorrhagic, and even necrotic (dead).

Delayed-type allergic reactions

Sensitivity to drugs.* It is not hard to find persons who exhibit drug allergy, that is, an untoward reaction to a certain drug or drugs. It is believed that when drugs are regularly taken into the body, a chemical combination may occur between the drug and certain body proteins, forming a complete antigen in a hapten-protein conjugate foreign to the body and one against which the mechanisms responsible for allergic manifestations are directed. True allergic drug reactions may be associated either with humoral antibodies or cell-mediated responses.

It has been estimated that about 500 drugs can bring about the allergic state. The ones most commonly responsible for a hypersensitive state in the person using them are opiates (morphine, codeine), salicylates (aspirin), barbiturates, iodides, bromides, arsenicals, sulfonamides, and antibiotics. Among the drugs inducing allergy, penicillin is the worst offender; an allergy to the drug occurs in 10% to 15% of the instances where it is given. For other drugs, fortunately the incidence is much lower. Because of the likelihood of sensitivity after topical use of sulfonamide drugs and penicillin as well, this type of therapy has almost completely disappeared.

Sensitivity to the antibiotics regularly appears in the form of a skin eruption. In a general way a reaction to penicillin and to certain other antibiotics often closely resembles serum sickness. The same is true of reactions to insulin and liver extract. If a previous reaction to penicillin has taken place or if there is reason to think that the patient is hypersensitive to the drug, it should *not* be given. Penicillin is said to have caused more deaths than any other drug.† No reliance should be placed on desensitization.

Infectious allergies (hypersensitivity to infection). Repeated or chronic infections may sensitize a patient to the etiologic microbes. This is well illustrated in the allergy to *Mycobacterium tuberculosis* and its products. When an extract of *Mycobacterium tuberculosis* is applied to the skin of the person who has or who has had tuberculosis, there is a reaction, and the tuberculin test is said to be positive. Patients may likewise be hypersensitive to other organisms such as *Brucella* (the causative agents of undulant fever). In fact, the manifestations of acute rheumatic fever are considered as an allergic reaction to streptococci. Other infections in which hypersensitivity to the causative agent is measured are leprosy, chancroid, lymphopathia venereum, coccidioidomycosis, blastomycosis, and histoplasmosis. In veterinary practice, hypersensitivity of infection is encountered in glanders, Bang's disease (undulant

---

Drug intolerance is the state in which a person receiving the drug reacts in a characteristic way to unusually small doses of the drug, doses with no physiologic effect. *Drug idiosyncrasy* refers to the situation in which the recipient of the drug reacts in an unusual way, such as when a dose of a barbiturate (a sedative) produces excitement.
†Penicillin as such does not induce hypersensitivity; its degradation products can become haptens. A skin test can be done with one of these, penicilloyl-polylysine, to indicate the probable risk of an allergic reaction.

fever), and Johne's disease. In some of these, diagnostic skin tests routinely detect sensitivity (Table 13-2).

Contact dermatitis. Contact dermatitis (an eczema or eczematous dermatitis) localizes in the skin, especially in areas exposed to direct physical contact with the varied irritant substances. As would be expected, this inflammatory condition is often seen on the hands. The first time that the skin contacts the offending substance there is no visible effect, but sensitization means that subsequent contacts will result in skin changes. The long list of offenders includes the well-known poisons of poison ivy and poison oak* and a vast host of substances related to trade and industry such as dyes, soaps, lacquers, plastics, woods, fabrics, furs, formalin, cosmetics, drugs, chemicals, metals, and explosives.

Autoallergies. See p. 203.

LABORATORY TESTS TO DETECT ALLERGY†

The clinical laboratory provides several routes to the diagnosis and evaluation of allergy.

Identification of the eosinophil in blood and other body fluids is a rather simple way of testing for the allergic state. The *eosinophil,* sometimes referred to as the cell of allergic inflammation, is one of the granular leukocytes and is normally seen in small numbers in the circulating blood. A cell just larger than the polymorphonuclear neutrophil, it is distinct for the large, round, bright red-orange granules within its cytoplasm demonstrated in Wright-stained blood smears. Its precise functions are not clear, but *eosinophilia,* an increase in its numbers, is typically associated with allergic reactions and also parasitic infections. In both these conditions there is usually a mild to moderate blood eosinophilia and at the site of involvement a fairly intense tissue eosinophilia. For instance, in allergic rhinitis from 30% to 90% of cells viewed microscopically in a stained smear of nasal secretions consistently are eosinophils, and they are also prominent in the sputum of asthmatics.

Basophil counts may used to study allergies, and there is an in vitro *basophil degranulation test.* A preparation of fresh basophils from either a human being or a rabbit (a good source) is mixed with patient's serum and stained supravitally. Specific antigen is added, and on the premise that in the presence of the specific antigen-antibody reaction degranulation occurs, the extent of degranulation in the preparation is measured by the reduction in the number of intact basophils. In a positive test over 30% of them have been degranulated.

The IgE level in serum can be measured by radioimmunoassay, and the increase with allergic and parasitic diseases can be demonstrated in this way. Normal values of this test are stated to be from 50 to 500 units/ml of serum.

The *RAST* (radio*a*llergo*s*orbent *t*est) test utilizes the technic of radioimmunoassay for detection of specific IgE against a variety of antigens in patients with atopic

---

*Poison ivy, oak, and sumac belong to the *Rhus* genus of trees and shrubs; the skin lesions may be called rhus dermatitis. The *Rhus* plants are responsible for more allergic contact dermatitis than all other allergens combined.
†See also p. 692.

allergy. Forty-five different allergens used include Bermuda grass, ragweed, oak, cedar, elm, house dust, and house mites.

The *skin test* is of prime value. If some of the antigen to which a person is sensitive is rubbed into a scratch on the skin or if dilute antigen is injected between the layers of the skin, a wheal and flare 0.5 to 2 cm in diameter will occur at the site within 20 to 30 minutes (cutaneous anaphylaxis). If the person is not sensitive to the antigen, no reaction appears.

The *ophthalmic test* helps to detect sensitization to horse serum. A drop of the *diluted* serum is instilled into the conjunctival sac; if the patient is sensitive to it, redness of the conjunctiva and a watery discharge will appear within 10 to 20 minutes. The reaction may be controlled by epinephrine.

The *patch test* is useful in detecting the cause of *contact dermatitis*. In this test suspect material is applied directly to the skin and held in place by means of adhesive tape from 1 to 4 days. A positive reaction reproduces the lesion from which the patient is suffering, with blister and papule formation.

DESENSITIZATION

In some cases desensitization may be accomplished by giving repeated injections of very small amounts of the antigen to which the patient is sensitive. A patient hypersensitive to horse serum may sometimes be given an immune serum if the administration is preceded by several injections of very small amounts at 30-minute intervals. The desensitizing doses are graded by the reaction of the patient, and the administration of an immune serum to a hypersensitive person should be undertaken only by those experienced in immune therapy. A similar but much prolonged method of desensitization is used in the treatment of hay fever and selected patients with asthma and urticaria. Desensitization, unlike sensitization, is of relatively short duration. In some forms of allergy it is accomplished with great difficulty.

QUESTIONS FOR REVIEW

1. Compare the two major categories of allergic disorders.
2. Briefly discuss anaphylaxis.
3. Describe briefly the most important allergic diseases.
4. What is serum sickness? Why is it less common than previously?
5. Comment on the chemical mediators of allergy.
6. Outline the laboratory diagnosis of allergy.
7. Explain the role in allergy of IgE.
8. Explain the relation of the basophil to allergic disorders.
9. How is desensitization accomplished?
10. List five agents known to have caused anaphylaxis in man.
11. What are the infectious allergies? How can they be detected?
12. Define or briefly explain hypersensitivity, allergy, sensitizing dose, provocative dose, allergen, immunopathology, Arthus phenomenon, Schultz-Dale reaction, iatrogenic, endogenous, insect-sting kit, drug intolerance, drug idiosyncrasy, passive anaphylaxis, cutaneous anaphylaxis.

REFERENCES FOR UNIT THREE

Abrams, B. L., and Waterman, N. G.: Dirty money, J.A.M.A. **219:**1202, 1972.
Anderson, C. L.: Community health, ed. 2, St. Louis, 1973, The C. V. Mosby Co.

Arber, W., and others, editors: Current topics in microbiology and immunology, vol. 67, New York, 1975, Springer-Verlag, New York, Inc.

Aronsohn, R. S., and Oberman, H. A.: Frozen red blood cells: current status of preservation and utilization, Univ. Mich. Med. Center J. **41**:24, Jan.-Dec., 1975.

Bahn, A. K.: Epidemiology as a field of practice, J. Am. Med. Wom. Assoc. **29**:387, 1974.

Barr, S. E.: Allergy to Hymenoptera stings, J.A.M.A. **228**:718, 1974.

Barrett, J. T.: Textbook of immunology, ed. 2, St. Louis, 1974, The C. V. Mosby Co.

Bellanti, J.: Immunology, Philadelphia, 1971, W. B. Saunders Co.

Benenson, A. S., editor: Control of communicable diseases in man, Washington, D.C., 1975, American Public Health Association, Inc.

Benzinger, T. H.: Clinical temperature, J.A.M.A. **209**:1200, 1969.

Bergman, N., and others: Antibacterial activity of human amniotic fluid, Am. J. Obstet. Gynecol. **114**:520, 1972.

Berland, T.: Mrs. Berland, your son is just plain lousy, Today's Health **52**:39, June, 1974.

Bigley, N. J.: Immunologic fundamentals, Chicago, 1975, Year Book Medical Publishers, Inc.

Billingham, R., and Silvers, W.: The immunobiology of transplantation, Englewood Cliffs, N.J., 1971, Prentice-Hall, Inc.

Black, F. L.: Infectious diseases in primitive societies, Science **187**:515, 1975.

Blackstock, R., and Hyde, R. M.: The role of thymus and bone marrow cells in immunologic competence: Part I, South. Med. J. **69**:97, 1976.

Blackstock, R., and others: The role of thymus and bone marrow cells in immunologic competence: Part II, South. Med. J. **69**:209, 1976.

Bodey, G. P.: Isolation for the compromised host, J.A.M.A. **233**:543, 1975.

Booth, B., and others: Modern concepts in clinical allergy, New York, 1973, Medcom Press.

Boyd, W.: Textbook of pathology, Philadelphia, 1970, Lea & Febiger.

Bruninga, G. L.: Complement—a review of the chemistry and reaction mechanisms, Am. J. Clin. Pathol. **55**:273, 1971.

Buckley, C. E., III: Immunologic evaluation of older patients, Postgrad. Med. **51**:235, 1972.

Burnet, F. M.: Auto-immunity and auto-immune disease, Philadelphia, 1972, F. A. Davis Co.

Cartwright, R. Y.: Commensal bacteria of the human respiratory tract, Nurs. Times **70**:418, Mar. 21, 1974.

Christie, A. B.: Infectious diseases: epidemiology and clinical practice, ed. 2, Edinburgh, 1974, E. & S. Livingstone, Ltd.

Civantos, F., and others: Protein immunoelectrophoresis, Lab. Med. **4**:37, Mar., 1973.

Cline, M. J., and others: UCLA Conference, Granulocytes in human disease, Ann. Intern. Med. **81**:801, 1974.

Clough, J. D., and Deodhar, S. D.: Clinical application of autoantibody testing in systemic lupus erythematosus and allied diseases, Lab. Med. **6**:26, June, 1975.

Cohen, E. P.: Must you have shots for hay fever? Today's Health **52**:54, July, 1974.

Crowle, A. J.: Radial immunodiffusion, Lab. Management **13**:40, Nov., 1975.

Culliton, B. J.: Immunology: two immune systems capture attention, Science **180**:45, 1973.

Culliton, B. J.: Restoring immunity: marrow and thymus transplants may do it, Science **180**:168, 1973.

Diamond, L. K.: The Rh problem through a retrospectroscope, Am. J. Clin. Pathol. **62**:311, 1974.

Diseases of pets, Today's Health **51**:26, May, 1973.

Dixon, F. J., and Kunkel, H. G., editors: Advances in immunology, vol. 19, New York, 1974, Academic Press, Inc.

Dodd, R. Y.: Transmissible disease and blood transfusion, Science **186**:1138, 1974.

Edelman, G. M.: Antibody structure and molecular immunology, Science **180**:830, 1973.

Erskine, A. G., and Wiener, A. S.: The principles and practices of blood grouping, St. Louis, 1973, The C. V. Mosby Co.

Evans, H. E., and others: Flora in newborn infants, Arch. Environ. Health **26**:275, 1973.

Fite, G. L.: Canine zoonoses (editorial), J.A.M.A. **231**:497, 1975.

Fluorescing antibodies, FDA papers **5**:19, Feb., 1971.

Force, D. C.: Ecology of insect host-parasitoid communities, Science **184**:624, 1974.

Freedman, S. O., and Gold, P.: Clinical immunology, New York, 1976, Harper & Row, Publishers.

Friedman, G. D.: Medical usage and abusage: "prevalence" and "incidence," Ann. Intern. Med. **84**:502, 1976.

Fudenberg, H. H., and others: Basic immunogenetics, New York, 1972, Oxford University Press.

Fudenberg, H. H., and others: The therapeutic uses of transfer factor, Hosp. Pract. **9:**95, Jan., 1974.

Giannella, R. A., and others: Influence of gastric acidity on bacterial and parasitic enteric infections, Ann. Intern. Med. **78:**271, 1973.

Gillett, J. D.: The mosquito, its life, activities, and impact on human affairs, New York, 1972, Doubleday & Co., Inc.

Gleich, G. J., and Jacob, G. L.: Immunoglobulin E antibodies to pollen allergens account for high percentage of total immunoglobulin E protein, Science **190:**1106, 1975.

Gonzalez-C, C. L., and Calia, F. W.: Bacteriologic flora of aspiration-induced pulmonary infections, Arch. Intern. Med. **135:**711, 1975.

Gottlieb, A. A., editor: Developments in lymphoid cell biology, Cleveland, Ohio, 1974, CRC Press, Inc.

Greenwalt, T. J., and Jamieson, G. A.: Transmissible diseases and blood transfusion, New York, 1975, Grune & Stratton, Inc.

Grindon, A. J.: The use of packed red blood cells, J.A.M.A. **235:**389, 1976.

Guttman, R. D., and others, guest editors: Immunology, Kalamazoo, Mich., 1972, The Upjohn Co.

Haesler, W. F., Jr.: Immunohematology, medical technology series, Philadelphia, 1972, Lea & Febiger.

Halliday, W. J.: Glossary of immunological terms, London, 1971, Butterworth & Co. (Publishers), Ltd.

Hegenauer, J., and Saltman, P.: Iron and susceptibility to infectious disease, Science **188:**1038, 1975.

Herbert, W. J., and Wilkinson, P. C., editors: A dictionary of immunology, Philadelphia, 1971, F. A. Davis Co.

Hersh, E. M., and others: Assessment of immunocompetence in the routine evaluation of the cancer patient, South. Med. J. **69:**356, 1976.

Hersh, W. M.: Role of BCG immunotherapy in malignant disease, South. Med. J. **68:**673, 1975.

Holborow, E. J.: An ABC of modern immunology, Boston, 1973, Little, Brown & Co.

Holmes, E. C., and others: Immunotherapy of malignancy in humans, J.A.M.A. **232:**1052, 1975.

Hubbert, W. T., and others, editors: Diseases transmitted from animals to man, Springfield, Ill., 1975, Charles C Thomas, Publisher.

Hussey, H. H.: All about antibodies, J.A.M.A. **235:**636, 1976.

Isler, C.: Blood, the age of components, RN **36:**31, June, 1973.

Isler, C.: Immunotherapy, RN **39:**35, Apr., 1976.

James, K. K.: Lymphocytes—a new dimension, Lab. Med. **7:**37, Apr., 1976.

Jenkins, G., and others: A review of transfusion complications, J. Am. Assoc. Nurse Anesthetists **43:**369, 1975.

Keller, S. E., and others: Decreased T-lymphocytes in patients with mammary cancer, Am. J. Clin. Pathol. **65:**445, 1976.

Kelly, J. F., and Patterson, R.: Anaphylaxis, course, mechanisms, and treatment, J.A.M.A. **227:**1431, 1974.

Kolata, G. B.: Antibody diversity: how many antibody genes? Science **186:**432, 1974.

Kountz, S. L.: Current status of kidney transplantation, New Physician **22:**550, Sept., 1973.

Laskin, A. I., and Lechevalier, H., editors: Macrophages and cellular immunity, Cleveland, Ohio, 1972, CRC Press, Inc.

Levin, A. S.: Transfer factor therapy: current status, South. Med. J. **68:**1465, 1975.

Lichtenstein, L. M.: Anaphylactic reactions to insect stings: a new approach, Hosp. Pract. **10:**67, Mar., 1975.

Luciano, J. R., and Tarpay, M.: Penicillin allergy, South. Med. J. **69:**118, 1976.

Marchalonis, J. J.: Lymphocyte surface immunoglobulins, Science **190:**20, 1975.

Marx, J. L.: Thymic hormones: inducers of T cell maturation, Science **187:**1183, 1975.

Marx, J. L.: Suppressor T cells: role in immune regulation. Science **188:**245, 1975.

Marx, J. L.: Antibody structure: now in three dimensions, Science **189:**1075, 1975.

Marx, J. L.: Immunology: role of immune response genes, Science **191:**277, 1976.

Maugh, T. H., II: Tissue cultures: transplantation without immune suppression, Science **181:**929, 1973.

Mausner, J. S., and Bahn, A. K.: Epidemiology—an introductory text, Philadelphia, 1974, W. B. Saunders Co.

McAllen, M. K.: Hay fever, Nurs. Mirror **138**:63, May 24, 1974.

McBride, G.: Antibodies yield their secrets and display therapeutic versatility, J.A.M.A. **235**:583, 1976.

McCluskey, R. T., and Cohen, S., editors: Mechanisms of cell-mediated immunity, New York, 1974, John Wiley & Sons, Inc.

McDevitt, H. O.: Genetic control of the antibody response, Hosp. Pract. **8**:61, Apr., 1973.

Medawar, P. B.: The new immunology, Hosp. Pract. **9**:48, Sept., 1974.

Mollison, P. L.: Blood transfusions in clinical medicine, Philadelphia, 1973, F. A. Davis Co.

Moody, L.: Asthma: physiology and patient care, Am. J. Nurs. **73**:1212, 1973.

Moore, F. D.: Transplant, the give and take of tissue transplantation, New York, 1972, Simon & Schuster, Inc.

Moser, R. H.: Ruminations. Host factors: an overview, J.A.M.A. **232**:516, 1975.

Movat, H. Z., editor: Inflammation, immunity and hypersensitivity, New York, 1971, Harper & Row, Publishers.

Nakamura, R. M.: Diagnostic laboratory tests for systemic lupus erythematosus and related disorders, Lab. Med. **6**:11, June, 1975.

Ness, P. M.: Plasma fractionation in the United States; a review for clinicians, J.A.M.A. **230**:247, 1974.

Nichols, A. L.: Radioimmunoassay, Lab. Management **13**:44, Nov., 1975.

Nichols, G. A., and Kucha, D. H.: Taking adult temperatures: oral measurements, Am. J. Nurs. **72**:1090, 1974.

Norman, P. S.: The clinical significance of IgE, Hosp. Pract. **10**:41, Aug., 1975.

Notkins, A. L.: Viral infections: mechanisms of immunologic defense and injury, Hosp. Pract. **9**:65, Sept., 1974.

Paine, R. T., and Zaret, T. M.: Ecological gambling, the high risks and rewards of species introductions, J.A.M.A. **231**:471, 1975.

Palmer, R. L.: Diagnostic aids for immunological disorders, Tex. Med. **67**:79, Mar., 1971.

Park, B. H.: Immunologic reconstitution in man—present status, South. Med. J. **68**:615, 1975.

Patterson, B. B.: Nitroblue tetrazolium reduction in neutrophils—a modification using the buffy coat, Lab. Med. **6**:50, Sept., 1975.

Race, R. R., and Sanger, R.: Blood groups in man, Philadelphia, 1975, J. B. Lippincott Co.

Reeves, W. C.: Can the war to contain infectious diseases be lost? Am. J. Trop. Med. **21**:251, 1972.

Reisfeld, R. A., and Kahan, B. D.: Markers of biological individuality, Sci. Am. **226**:28, June, 1972.

Remington, J. S.: The compromised host, Hosp. Pract. **7**:59, Apr., 1972.

Remmers, A. R., and others: Renal transplantation, Tex. Med. **71**:80, July, 1975.

Richards, F. F., and others: On the specificity of antibodies, Science **187**:130, 1975.

Richmond, R. S.: Donating blood, J.A.M.A. **232**:753, 1975.

Ringler, D. H., and Anver, M. R.: Fever and survival, Science **188**:166, 1975.

Ritchie, R. F.: Automated immunoanalysis of plasma proteins, Lab. Management **13**:32, Nov., 1975.

Roitt, I. M.: Essential immunology, Philadelphia, 1974, J. B. Lippincott Co.

Rose, N. R., and others, editors: Methods in immunodiagnosis, New York, 1973, John Wiley & Sons, Inc.

Rosenfield, R. E.: The past and future of immunohematology, Am. J. Clin. Pathol. **64**:569, 1975.

Ross, W. S.: 4000 kidneys later. . ., Today's Health **48**:35, Dec., 1970.

Samter, M., editor: Immunological diseases, Boston, 1971, Little, Brown & Co.

Schwartz, L. M., and Schwartz, P.: An exercise manual in immunology, Medcom Workbook/Manual Series, New York, 1975, Medcom Press.

Scott, B.: Asthma—the demon that thrives on myths, Today's Health **48**:42, June, 1970.

Sela, M., editor: The antigens, vol. II, New York, 1974, Academic Press, Inc.

Sell, S.: Immunology, immunopathology, and immunity, New York, 1975, Harper & Row, Publishers.

Sellars, W. A.: New frontiers in allergy and immunology, Dallas Med. J. **61**:412, 1975.

Shafer, A. W.: Use of blood and blood components, South. Med. J. **68**:631, 1975.

Shafer, A. W.: Adverse effects of transfusions, South. Med. J. **69**:476, 1976.

Sheffield, J.: The immunoglobulins, J. Am. Med. Technol. **35**:226, 1973.

Smith, G. P.: The variation and adaptive expression of antibodies, Cambridge, Mass., 1973, Harvard University Press.

Sokol, A. B., and Houser, R. G.: Dog bites: prevention and treatment, Clin. Pediatr. **10**:336, 1971.

Stevens, D. A., and others: Spherulin in clinical coccidioidomycosis, Chest **68**:697, 1975.

Stiehm, E. R., moderator: UCLA Conference—Diseases of cellular immunity, Ann. Intern. Med. **77**:101, 1972.

Stiehm, E. R., and Fulginiti, V. A.: Immunologic disorders in infants and children, Philadelphia, 1973, W. B. Saunders Co.

Stiller, C. R., and others: Autoimmunity: present concepts, Ann. Intern. Med. **82**:405, 1975.

Stone, M. J.: Immunotherapy of cancer, Dallas Med. J. **61**:458, 1975.

Stossel, T. P.: Phagocytosis, N. Engl. J. Med. **290**:717, 774, 833, 1974.

Sumida, S.: Transfusion of blood preserved by freezing, Philadelphia, 1973, J. B. Lippincott Co.

Taplin, D. and Mertz, P. M.: Flower vases in hospitals as reservoirs of pathogens, Lancet **2**:1279, 1973.

The control of lice and louse-borne diseases. Proceedings of a symposium, Sci. Pub. No. 263, Washington, D.C., 1973, Pan American Health Organization.

Thomas, E. D.: Progress in marrow transplantation, J.A.M.A. **235**:611, 1976.

Thompson, E., and others: Changes in antigenic nature of lymphocytes caused by common viruses, Br. Med. J. **4**:709, Dec. 22, 1972.

Thomson, D.: The ebb and flow of infection, J.A.M.A. **235**:269, 1976.

Vitetta, E. S., and Uhr, J. W.: Immunoglobulin receptors revisited, Science **189**:964, 1975.

Volpe, E. P., and Turpen, J. B.: Thymus: central role in the immune system of a frog, Science **190**:1101, 1975.

Vyas, G. N., and others, editors: Laboratory diagnosis of immunologic disorders, New York, 1975, Grune & Stratton, Inc.

Weiser, R. S., and others: Fundamentals of immunology, Philadelphia, 1969, Lea & Febiger.

Weiss, L.: The cells and tissues of the immune system: structure, functions, interactions, Englewood Cliffs, N.J., 1972, Prentice-Hall, Inc.

Wilson, W. R., and others: Incidence of bacteremia in adults without infection, J. Clin. Microbiol. **2**:94, 1975.

Wybran, J., and Fudenberg, H.: How clinically useful is T and B cell quantitation? (editorial), Ann. Intern. Med. **80**:765, 1974.

Zmijewski, C. A., and Fletcher, J. L.: Immunohematology, ed. 2, New York, 1972, Prentice-Hall, Inc.

LABORATORY SURVEY OF UNIT THREE*

PROJECT
Sources of infection
Part A—Bacteria on the human body

1. *Hands*
   a. Melt a tube of nutrient agar and allow it to cool until it feels just comfortable to the back of the hand.
   b. Wash your hands in a sterile dish containing 300 ml of sterile water. Do not use soap.
   c. With a sterile pipette place 1 ml of the wash water in a sterile Petri dish.
   d. Pour the cooled agar into the same Petri dish, replace cover, and set the dish on your desk.
   e. Slide the dish *gently* over the desk with a rotary motion until the melted medium and water are well mixed.
      Note: Do not allow the agar to come up the sides of the Petri dish and touch the cover.
   f. Let the medium harden, invert the dish, and incubate at 37° C for 24 to 48 hours.
   g. Observe the growth of colonies. Note the number and different kinds.
2. *Fingertips*
   a. Gently touch the ends of your fingers over the surface of a sterile agar plate.
   b. Incubate plate at 37° C for 24 hours and examine.
3. *Breathing passages*
   a. Normal breathing
      (1) Remove the lid from a sterile nutrient agar plate.
      (2) Hold the plate about 6 inches from your mouth and breathe normally but directly on the medium for about 1 minute.
      (3) Replace the cover, invert, and incubate at 37° C for 24 to 48 hours.
      (4) Observe the growth of colonies. Note different kinds.
   b. Violent coughing
      (1) Remove the lid from a sterile nutrient agar plate.
      (2) Hold the plate about 1 foot from your mouth and cough violently onto the surface of the agar.
      (3) Replace the cover, invert, and incubate at 37° C for 24 to 48 hours.
   c. Violent coughing—plate at arm's length from the mouth
      (1) Proceed as above.
      (2) Observe the growth of colonies on this plate. Note the different kinds.

---

*There is no end to the laboratory projects that may be carried out in connection with this unit. Some of them require such highly skilled technic and are so time-consuming that it would be unwise to include them. Others yield little information. Our selection is therefore limited to a few—simple to perform and yet applicable to the content of this unit.

(3) Which plate of the three prepared in a, b, and c of this exercise has the most colonies? Why?

4. *Other body sites*

   a. Select body sites such as the nose, throat, or ears. You can gently pass a sterile cotton swab moistened with sterile isotonic saline solution over or into the given area. Use the cotton swab to make a culture of the area chosen.

   b. Use tubes or plates of media that would be likely to grow all organisms present. Blood agar is suitable.

   c. Incubate cultures taken at 37° C for 24 to 48 hours.

   d. Observe the growth. Note the appearance of colonies.

   e. Make smears from colonies and stain by Gram's method. Note the shape of bacteria and whether gram positive or gram negative.

   f. Compare the bacterial growth obtained from the body sites selected for culture.

Part B—Bacteria on insects

1. Prepare a Petri dish of nutrient agar.
2. Allow an insect, such as a fly, to crawl for a time on the surface of the hardened agar under the cover of the dish.
3. Invert the plate after the fly has escaped and incubate at 37° C for 24 to 48 hours.
4. Note the colonies.
5. Make gram-stained smears and examine.

Part C—Bacteria in the air

1. Open a Petri dish and expose the surface of sterile nutrient agar to the air in the laboratory for 5 minutes. (Exposure is made by removing the lid from the dish but the lid is not inverted.)
2. Replace the lid, invert the dish, and place in the incubator.
3. Incubate for 24 to 48 hours.
4. Study the colonies obtained.
5. Make gram-stained smears, examine, and record findings.

Part D—Bacteria at specific sites in our surroundings

1. Sample areas of the laboratory and its environs—tabletops, doorknobs, cabinet handles, floors, water faucets, water from the faucets, walls, windowsills, and the like. Feel free to investigate your surroundings bacteriologically.
2. Expose the surface of the nutrient agar in a given Petri dish to the given site for 5 minutes. The agar surface may be applied directly to the specified site, or a sterile swab stick moistened with sterile saline solution may be rubbed over the given area and then used to inoculate the surface of the agar plate.
3. Cover the dish, invert, and incubate.
4. Note colonies. Describe them. Do you recognize any?

5. Make gram-stained smears, examine, and record findings. Do you encounter the same or different microbes as you sample the bacterial population of the above exercises?

PROJECT
Measure of immunity

Part A—Demonstration of serologic technic by the instructor with discussion

1. Technic of pipetting solutions for serologic tests
2. Technic of making serial dilutions for serologic testing
3. Technic of manipulating serologic tubes

Part B—Demonstration of complement fixation by the instructor with brief discussion

1. Principles of complement fixation
2. Explanation of results in positive and negative tests
3. Demonstration of positive and negative tests, with ample opportunity for student observation of the exhibits

Part C—Detection of antigen-antibody combinations

1. *Demonstration of phenomenon of agglutination—tube agglutination*
   a. Set up a row of seven serologic tubes in a test tube rack.
   b. To the first tube add 1.8 ml isotonic saline solution, and to each of the remaining six tubes add 1 ml isotonic saline solution.
   c. To the first tube add 0.2 ml agglutinating serum against *Salmonella typhi* (or suitable *Salmonella* species) and mix.
   d. From tube 1 remove 1 ml and transfer to tube 2 and mix. From tube 2 remove 1 ml, transfer to tube 3, and mix. Continue transferring and mixing until tube 6 has been mixed. Discard 1 ml. Tube 7 acts as a control.
   e. To all tubes add 1 ml of a suspension of *killed* typhoid bacilli. Calculate the dilution of agglutinating serum in each tube.
   f. Incubate tubes in the test tube rack at 37° C for 2 hours.
   g. Note clumping of bacteria (agglutination) in the tubes containing the lower dilutions of agglutinating serum.
   h. Relative to this experiment fill out the following table:

Tube	1	2	3	4	5	6	7
Dilution of serum							
Agglutination*							

*Complete, partial, or none.

Note: Suppose human serum had been used in this test instead of an artificial agglutinating serum and the results had been as indicated in the table. What would be the significance of the test? What would agglutination in tube 7 indicate? What is the name of the agglutination test for typhoid fever?

1. *Agglutination in blood typing (grouping)—slide agglutination*
   Note: Students should work in groups of two. The blood type of each student will be determined.
   a. Mark a microslide down the middle with a wax pencil. Mark the left end *A* and the right end *B*. Make two wax rings, one on each end of the slide (Fig. 12-7).
   b. Place 1 drop of anti-A serum in the wax ring on the *A* side of the slide and a drop of anti-B serum on the *B* side.
      Note: These serums may be obtained commercially. One drop of each is sufficient for the test.
   c. Puncture the finger or the earlobe of the person to be tested. Discard the first drop of blood. Transfer a minute drop of blood, by means of a clean applicator, to the drop of anti-A serum, mixing to make a smooth suspension of the cells. Discard applicator. With a fresh applicator, transfer a like drop to the anti-B serum and mix thoroughly. Do *not* use the same applicator for both serums. Why?
   d. Allow the slide to stand for 5 minutes, occasionally rolling or tilting it to ensure thorough mixing. Examine under the low power of the microscope.
      Note: It may be difficult to distinguish between true agglutination and formation of *rouleaux*. Stir with an applicator. Rouleaux will be broken up, but true agglutination will be unaffected.
   e. If there is no agglutination at the end of 5 minutes, cover each mixture with a cover glass and examine at 5- to 10-minute intervals, making a final reading at the end of 30 minutes.
      Note: Sometimes a final reading must be deferred for 60 minutes, especially with group A or AB blood; in this event it is advisable to ring the preparation with petroleum jelly to prevent evaporation.
   f. Determine the blood type as follows:
      (a) If the cells are not agglutinated on either end of the slide, the subject belongs in group O.
      (2) If the cells on both ends are agglutinated, the subject belongs in group AB.
      (3) If the cells on the A end are agglutinated but those on the B end are not, that person belongs in group A.
      (4) If the cells on the B end are agglutinated but those on the A end are not, that person belongs in group B.

3. *Demonstration of the phenomenon of hemolysis*
   a. Prepare serial dilutions of antisheep hemolysin.
      (1) Prepare a 1:100 dilution of antisheep hemolysin (having a titer of about 1:3000) by adding 0.1 ml of hemolysin to 9.9 ml of isotonic sodium chloride solution and mixing.
      (2) Set up a row of 10 clean serologic tubes in a test tube rack.

TABLE A. SERIAL DILUTION OF HEMOLYSIN

Tube No.	1	2	3	4	5	6	7	8*	9*	10*
Isotonic NaCl	—	0.5 ml	0.5 ml	0.5 ml	0.5 ml	0.5 ml	0.5 ml	0.5 ml	—	0.5 ml
Hemolysin	1.0 ml of 1:100 dilution	0.5 ml of 1:100 dilution No. 1	0.5 ml from mixture of No. 2	0.5 ml from mixture of No. 3	0.5 ml from mixture of No. 4	0.5 ml from mixture of No. 5	0.5 ml from mixture of No. 6 discard 0.5 ml.	—	0.5 ml of 1:100 dilution	—
Dilution of hemolysin	1:100							—	1:100	—
Total volume	0.5 ml	0.5 ml	0.5 ml	0.5 ml	0.5 ml	0.5 ml	0.5 ml	0.5 ml	0.5 ml	0.5 ml

*Controls.

TABLE B. HEMOLYSIS OF RED BLOOD CELLS

Tube No.	1	2	3	4	5	6	7	8*	9*	10*
Isotonic NaCl	1.7 ml	1.7 ml	1.7 ml	1.7 ml	1.7 ml	1.7 ml	1.7 ml	1.7 ml	2.0 ml	2.5 ml
Hemolysin serial dilution (from Table A)	0.5 ml 1:100	0.5 ml	0.5 ml	0.5 ml	0.5 ml	0.5 ml	0.5 ml	—	0.5 ml 1:100	—
Complement 1:30 dilution	0.3 ml	0.3 ml	0.3 ml	0.3 ml	0.3 ml	0.3 ml	0.3 ml	0.3 ml	—	—
Sheep red cell suspension, 2%	0.5 ml	0.5 ml	0.5 ml	0.5 ml	0.5 ml	0.5 ml	0.5 ml	0.5 ml	0.5 ml	0.5 ml
Presence of hemolysis?										
Final volume	3.0 ml	3.0 ml	3.0 ml	3.0 ml	3.0 ml	3.0 ml	3.0 ml	3.0 ml	3.0 ml	3.0 ml

*Controls.

(3) To the first tube add 1 ml of the 1:100 hemolysin and 0.5 ml of the 1:100 hemolysin to the ninth tube.

(4) To the remaining tubes, except the ninth, add 0.5 ml isotonic sodium chloride solution.

(5) To make serial dilutions of hemolysin, remove 0.5 ml from the first tube, add to the second tube, and mix. From the second tube remove 0.5 ml, transfer to the third, and mix. Continue in this manner until the contents of tube 7 have been mixed. Discard 0.5 ml of contents of this tube.

(6) Calculate the dilution of hemolysin so produced in each tube and chart the results in Table A, p. 252.

b. Add complement.

(1) To the first eight tubes add 0.3 ml of a 1:30 dilution of fresh guinea pig serum (complement).

(2) Note that complement is not added to tubes 9 and 10.

c. Add isotonic saline solution.

(1) Add 1.7 ml to the first eight tubes.

(2) Add 2 ml isotonic sodium chloride to tube 9 and 2.5 ml to tube 10.

(3) Note that by this part of the exercise all 10 tubes contain isotonic saline solution.

d. Add to all tubes 0.5 ml of a 2% suspension of sheep red blood cells.

e. Incubate tubes in the water bath at 37° C for 1 hour.

f. Observe. Note the solution of cells (hemolysis).

g. Chart the results using Table B, p. 252.

h. Answer the following questions:

(1) What is the highest dilution of hemolysin showing complete hemolysis?

(2) In what dilutions is hemolysis partial?

(3) In what dilution does hemolysis cease?

(4) Is hemolysis present or absent in the eighth tube? Why?

(5) Is it present in the ninth tube? Why?

(6) What would hemolysis in the tenth tube indicate?

TABLE C. SUGGESTED COMBINATIONS OF ANTIGEN AND ANTIBODY

	Antigen(s)		Antibody(ies)
Reference label:	A	B	Anti-AB
Test I	Human serum from "normal" volunteer	Human serum from "normal" volunteer	Rabbit antihuman immunoglobulin G
Test II	Human serum from "normal" volunteer	Cholera toxin	Mixture of rabbit anti-human immunoglobulin G and rabbit anticholera toxin
Test III	Goat immunoglobulin G	Human immunoglobulin G	Rabbit antihuman immunoglobulin G

4. *Demonstration of precipitin reaction—agar-gel diffusion (Ouchterlony technic)*

   Note: Students may participate in small groups, or the instructor may demonstrate and discuss the technic. The procedure for preparing the Ouchterlony plate (Petri dish) is outlined. For the three combinations of antigen and antibody given (Table C), three such plates are needed.

   a. Pour the first of two tubes of warm Noble agar (pH 7 with Merthiolate) into a flat-bottomed Petri dish to make a thin layer about 3 mm or so. Let harden.

   b. Use the diagram below as a guide. Place Petri dish over it. Using sterile forceps, place three templates (or cylinders) over small circles in the diagram, pressing them gently into the agar. (A ¼ inch or size 2 cork borer may be used.)

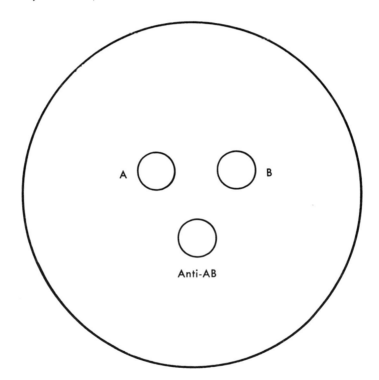

Diagram of Ouchterlony plate. *Anti-AB* is reference label for well to contain antiserum or antibody source; *A*, well to contain antigen designated A; and *B*, well to contain antigen B.

   c. Pour the second tube of warm agar onto the plate. Let harden. Remove templates leaving wells or reservoirs in the agar. Gingerly clear out excess agar from the wells. With wax pencil markings on the underside of the plate, identify reservoirs as indicated on diagram.

   d. Using a sterile Pasteur pipette for each test substance, fill wells slowly with 0.2 ml amounts.

e. Replace lid on dish. Leave plate at room temperature to be checked at regular intervals for 7 to 10 days.

f. Draw lines representing precipitin bands on diagram.
   Note: Antigen and specific antibody diffuse out in the agar. Where they meet in optimal proportions, white lines of precipitation form, the precipitin bands.

g. Interpret results.
   (1) If precipitin band forms a complete chevron, antigens A and B are serologically identical—reaction of identity.

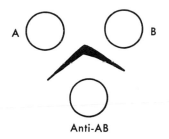

   (2) If the bands form a spur, antigens A and B are partly related serologically but one has reacted more fully with antibodies present than the other—reaction of partial identity.

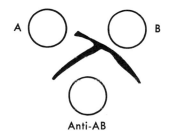

   (3) If lines completely cross, anti-AB (antiserum) contains antibodies against both A and B (antigens) but they are *not* related serologically—reaction of nonidentity.

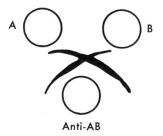

Part D—Phagocytosis

1. Examine stained smears of pus showing phagocytosis by pus cells.
2. Compare with smears of pus in which phagocytosis is absent.
3. What is the pus cell? What is its usual function?

255

PROJECT
Preparation of a bacterial vaccine
Part A—Taking the culture

Note: Students may work in groups of three or four. In the preparation of this vaccine no attempt will be made to procure one organism in a pure culture, but a mixed vaccine composed of all of the organisms in the throat will be made.

1. Prepare the following supplies and sterilize in autoclave at 121° F for 15 minutes:

Isotonic saline solution containing 0.5% phenol (in 200 ml flask plugged with cotton)	100 ml
Cotton swabs (swab ends down in 60 ml widemouthed bottle)	5
Small funnel with pledget of absorbent cotton to use as filter (set in mouth of flask to facilitate handling)	1
Vaccine bottle with diaphragm stopper (plug bottle with cotton and wrap stopper in paper)	1

2. Allow materials in the autoclave to cool.
3. Use a sterile swab to obtain material from the throat and inoculate four large tubes of nutrient agar.
4. Incubate cultures at 37° C for 24 hours.
5. Examine cultures by gross inspection and by making gram-stained smears. Are spores present?
   Note: If spores are found, discard the particular culture from which the gram-stained smear was made. Why?

Part B—Processing the culture

Note: During the preparation of the bacterial suspension you must use strict aseptic technic. Otherwise the vaccine will be contaminated by outside bacteria.

1. Use a sterile pipette to transfer 5 to 10 ml of sterile isotonic saline solution to each of the four tubes previously cultured.
2. Using a fresh swab for each tube, rub bacteria off the culture medium into the isotonic saline solution. Discard the swab and replace the cotton plug in the tube.
   Note: Swabs and contaminated material should be placed in a pan containing a small amount of water or disinfectant. After the preparation of the vaccine is completed, the pan can be boiled or its contents disinfected.
3. Remove the cotton plug from the vaccine bottle and place the funnel in the mouth of bottle. Filter contents of each culture tube into the vaccine bottle. Wash a small amount of isotonic sodium chloride through the funnel.
4. Remove the funnel and place in a waste pan.
5. Place the stopper in the vaccine bottle and secure in place with rubber bands.
   Note: The vaccine bottle must not become unstopped during the next step.
6. Place the vaccine bottle in a water bath at 60° C. Let remain 1 hour.

Part C—Testing the sterility of the vaccine

1. Remove the bottle from the water bath and sterilize the stopper with iodine.
2. Puncture the stopper with a sterile syringe and remove a small amount of the vaccine.
3. Plant the vaccine on a tube of nutrient agar. Incubate at 37° C.
4. Observe growth at the end of 24 and 48 hours. If the vaccine is satisfactory to give to a person, will bacterial growth be present or absent? What is a vaccine made from the patient's own organisms called?

Note: In the foregoing technic, anaerobic cultures and determination of the number of bacteria per milliliter have been purposely avoided because these procedures are too complicated to be attempted at this time.

PROJECT
Measure of allergy
Part A—Demonstration of anaphylaxis by the instructor with discussion*

Note: Prepare test guinea pigs 2 weeks beforehand.
1. Administer sensitizing dose—inject *intraperitoneally* 0.1 ml of horse serum (or of 5% egg albumin or other suitable foreign protein) in 2 ml of sterile isotonic saline solution.
2. Inject provocative (or shocking) dose—1.0 ml of horse serum (or of 5% egg albumin or other suitable foreign protein) *into the heart* of the test animal.
3. Observe and describe results. Encourage student comments.

Part B—Demonstration of skin testing for allergy by the instructor with discussion

---

*A 12-minute color and sound film, "Anaphylaxis in Guinea Pigs," no. 290, is available from the American Society for Microbiology, 1913 I St. N.W., Washington, D.C., 20006.

EVALUATION FOR UNIT THREE

Part I

Select the numbers on the right that represent completions of the statements or answers to the questions.

1. Which of these are infectious but *not* communicable diseases?
   - (a) Smallpox
   - (b) Tetanus
   - (c) Boils
   - (d) Typhoid fever
   - (e) Common cold

   1. c
   2. b and c
   3. c and d
   4. a, d, and e
   5. e

2. Which of the following articles from an active case of typhoid fever most likely carry the organisms?
   - (a) Bedpan
   - (b) Urinal
   - (c) Mouth wipes with nasal secretion
   - (d) Eating utensils
   - (e) Bloody bandage from arm wound

   1. c and d
   2. e
   3. c
   4. a and b
   5. none

3. In which of the following diseases should the nurse be careful to use a mask?
   - (a) Smallpox
   - (b) Gonorrhea
   - (c) Epidemic meningitis
   - (d) Malaria
   - (e) Diphtheria

   1. a, c, and e
   2. b
   3. a and d
   4. all but a
   5. all but b

4. The skin is the common portal of entry for the causative agents of all of the following diseases of man *except:*
   - (a) Malaria
   - (b) Tularemia
   - (c) Hookworm disease
   - (d) Measles
   - (e) Cholera

   1. a, b, and c
   2. a, b, c, and d
   3. a and c
   4. d and e
   5. e

5. In which of the following tests is the serum from a patient mixed with a suspension of bacteria to test its ability to clump bacterial cells?
   - (a) Agglutination test
   - (b) Complement fixation test
   - (c) Tuberculin test
   - (d) Precipitin test
   - (e) Schick test

   1. a
   2. b and c
   3. a and d
   4. c, d, and e
   5. all but e

6. A patient has type B blood. A person of which of the following blood types could be used in unusual circumstances as a donor?
   - (a) Type A
   - (b) Type AB
   - (c) Type B
   - (d) Type O

   1. b and c
   2. b and d
   3. d
   4. a
   5. all but a

   Which of these types would be the best type to give to this patient under ideal conditions?

7. A nurse beginning her communicable disease service is given an injection consisting of an antigen that has been modified. Which of the following might she be receiving?

258

(a) Gamma globulin	1. a
(b) Antitoxin	2. b and d
(c) Toxoid	3. c
(d) Immune serum	4. all but e
(e) Antivenin	5. none of above

8. Which of the following most nearly states your chances of developing an allergy during your lifetime?

(a) A tendency to become hypersensitive runs in families to a noticeable degree.	1. a
	2. b
(b) Allergy is always inherited.	3. c
(c) There is no connection between allergy and heredity.	4. d
	5. e
(d) Hypersensitivity may be inherited as a constitutional abnormality, or it may develop later in life.	
(e) If hypersensitivity runs in a family, all members will suffer the same type of allergic condition.	

9. A grown person has just received a stab wound. What facts should be ascertained before tetanus antitoxin is injected?

(a) The patient's temperature, pulse, and respiration are normal.	1. a and e
	2. b
(b) The patient has no allergic condition.	3. a, c, and d
(c) The patient has never reacted to horse serum and has not received a first dose within the last few months.	4. b, c, and d
	5. b, c, d, and e
(d) The patient gives no history of active immunization with toxoid.	
(e) The results of a skin or ophthalmic test for sensitivity to horse serum are negative.	

10. The cell directly related to antibody formation is:

(a) Polymorphonuclear neutrophilic leukocyte.	1. a
	2. b
(b) Red blood cell	3. c
(c) Plasma cell	4. d
(d) Squamous cell	5. a and c
(e) Eosinophil	

11. Normal body temperature is usually given as:

(a) 37° C	1. a
(b) 35° C	2. a and d
(c) 99° F	3. a and e
(d) 97.6° F	4. d
(e) 98.6° F	5. e

12. Koch's postulates:

(a) Can always be satisfied	1. a and e
(b) Are important in establishing that a given organism causes a given disease	2. b, c, and d
	3. e
(c) Refer to Koch's procedure for isolating tubercle bacilli	4. b and e
	5. b and d
(d) Are the basis of experimental investigation of infectious disease	
(e) Are only of historical interest	

13. All the following diseases are commonly transmitted to man from animals except:
    (a) Rabies   1. a, b, and c
    (b) Bubonic plague   2. b, c, and e
    (c) Typhus fever   3. e
    (d) Psittacosis   4. d
    (e) Candidiasis   5. b
14. Hypersensitivity to horse serum protein may be induced by:
    (a) Injection of toxoid   1. d
    (b) Injection of immune serum human   2. d and e
    (c) Injection of bacterial vaccine   3. e
    (d) Injection of diphtheria antitoxin   4. c
    (e) Specific passive prophylaxis against   5. a and b
        tetanus
15. Interferon is described as:
    (a) An antibody to a virus   1. b
    (b) A substance in the complement system   2. b and c
    (c) A substance formed by host cells   3. c
    (d) A substance with action against viruses   4. c and d
    (e) A substance comparable to an antigen   5. a, b, c, d, and e
16. A tissue transplant from a person to his identical twin is designated:
    (a) Autograft   1. a
    (b) Homostatic graft   2. b
    (c) Isograft   3. c
    (d) Xenograft   4. d
    (e) Allograft   5. e
17. Complement:
    (a) Is found normally in circulation of man   1. a
    (b) Is an antibody   2. b
    (c) Is increased by immunization   3. a and b
    (d) Is made up of at least nine components   4. a, d, and e
    (e) Is inactivated in combination with anti-   5. a, c, and e
        gen and antibody
18. Phagocytic cells:
    (a) Are active producers of antibody   1. a and e
    (b) Are found only in the bloodstream   2. b and c
    (c) Can ingest bacteria in absence of anti-   3. a and d
        body   4. d
    (d) Are not related to immune mecha-   5. c
        nisms
    (e) Always digest ingested bacteria

Part II

1. Before each type of acquired immunity listed in column A, place the number of each method by which it may be conferred as listed in column B.

COLUMN A	COLUMN B
_____ Active	1. Injection of a vaccine
_____ Passive	2. Recovery from a given disease
	3. Injection of convalescent serum
	4. Injection of an antiserum
	5. Injection of an antitoxin
	6. Injection of a toxoid

2. Using the letter in front of the appropriate term from column A, designate the kind of immunity resulting from the situation indicated in column B.

COLUMN A
(a) Natural immunity
(b) Naturally acquired, passive immunity
(c) Naturally acquired, active immunity
(d) Artificially acquired, passive immunity
(e) Artificially acquired, active immunity

COLUMN B
_____ 1. Administration of tetanus antitoxin
_____ 2. Administration of Salk poliomyelitis vaccine
_____ 3. Administration of influenza vaccine
_____ 4. An attack of measles
_____ 5. Transfer of antibodies from the
mother to her newborn child
_____ 6. An attack of whooping cough
_____ 7. Administration of antivenin
_____ 8. Administration of gamma globulin
_____ 9. An attack of German measles
_____ 10. Administration of duck embryo
rabies vaccine
_____ 11. Administration of Sabin live poliomyelitis virus vaccine
_____ 12. Parenteral injection of convalescent serum
_____ 13. The Pasteur treatment
_____ 14. Vaccination with tetanus toxoid
_____ 15. Vaccination against smallpox
_____ 16. Injection of diphtheria toxoid
_____ 17. Resistance of Algerian sheep to
anthrax

3. On the line at the right side of the page, place the number of the term from the left side that does *not* bear a significant relationship to the other terms in the aspect mentioned.

(a) 1. Sporadic   4. Epidemic   (Occurrence) _____
    2. Endemic   5. Congenital
    3. Pandemic
(b) 1. Phagocytes   4. Lysins   (Action) _____
    2. Precipitins   5. Antitoxins
    3. Agglutinins

(c) 1. Folic acid   4. Nicotinic acid (Vitamins) _____
    2. Polysaccharide 5. Pantothenic acid
    3. Biotin
(d) 1. Scavenger cells 4. Lymphocytes   (Phagocytes) _____
    2. Polys   5. Microphages
    3. Macrophages
(e) 1. Antitoxin   4. Opsonin   (Do not prepare
    2. Agglutinin   5. Antigen   bacteria for
    3. Bacteriolysin   phagocytosis) _____

(f) 1. Hives   4. Serum sickness (Immediate-type
    2. Hay fever   5. Contact dermatitis allergic reactions) _____
    3. Asthma

4. Match the item in column B to the one in column A with which it is most closely associated.

COLUMN A

(a) Contact dermatitis
(b) Antiglobulin test
(c) Agglutination test
(d) Heterophil antibody test
(e) Skin test for fungous disease
(f) Skin test for bacterial allergy
(g) Skin test for chlamydial infection
(h) Skin test for susceptibility to scarlet fever
(i) Immunologic reaction requiring special microscope
(j) Precipitin reaction
(k) Serologic test for viruses
(l) Complement fixation test
(m) Skin test for susceptibility to diphtheria
(n) Serologic test for syphilis

COLUMN B

_____ 1. Widal test
_____ 2. Paul-Bunnell test
_____ 3. Fluorescent antibody test
_____ 4. Coombs' test
_____ 5. Weil-Felix reaction
_____ 6. Wassermann test
_____ 7. Histoplasmin test
_____ 8. Frei test
_____ 9. Dick test
_____ 10. Schick test
_____ 11. Coccidioidin test
_____ 12. Tuberculin test
_____ 13. Brucellergen
_____ 14. Gel diffusion
_____ 15. Patch test
_____ 16. Flocculation test
_____ 17. Neutralization test

5. Comparisons: Match the item in column B to the word or phrase in column A best describing it.

COLUMN A

Humoral immunity with cell-mediated immunity:

_____ 1. Graft rejection an example
_____ 2. Major role in resistance to viruses
_____ 3. Major role in resistance to bacteria
_____ 4. Presence of antigen not necessary
_____ 5. Passive immunization possible
_____ 6. Immune response within hours
_____ 7. Immune response within days
_____ 8. Contact with antigen the trigger mechanism
_____ 9. Antigen reaches lymphoid tissues
_____ 10. Related to cancer immunology
_____ 11. Mediating cells thymus-dependent lymphocytes
_____ 12. Mediating cells thymus-independent lymphoid cells
_____ 13. Part of body's defense against infection
_____ 14. Presence of demonstrable antibody
_____ 15. Runt disease

COLUMN B

(a) Humoral immunity
(b) Cell-mediated immunity
(c) Both
(d) Neither

Immediate-type allergic reaction with delayed-type allergic reaction:

_____ 16. Hay fever
_____ 17. Tuberculin hypersensitivity
_____ 18. Not associated with circulating antibody
_____ 19. Similar symptoms induced by histamine
_____ 20. Drug allergies
_____ 21. Passive transfer with cells or cell fractions of lymphoid series
_____ 22. Short duration
_____ 23. Prolonged duration
_____ 24. Contact dermatitis
_____ 25. Anaphylactic shock
_____ 26. Arthus phenomenon
_____ 27. Drug idiosyncrasy
_____ 28. Triggered by well-defined allergen usually
_____ 29. Emotional upsets
_____ 30. Asthma

(a) Immediate-type allergy
(b) Delayed-type allergy
(c) Both
(d) Neither

UNIT FOUR
MICROBES
PRECLUSION OF DISEASE

15 Physical agents in sterilization

Sterilization is the process of killing or removing all forms of life, especially microorganisms, associated with a given object or present in a given area. This includes bacteria and their spores, fungi (molds, yeasts), and viruses (which must be either destroyed or inactivated). An object on and within which all microbes are killed or removed is said to be *sterile*. The length of time an object remains sterile depends on how well it is protected from microorganisms after it is sterilized. For instance, the outside of a tube of culture medium soon becomes contaminated because it directly contacts microorganisms of the air, whereas inside the tube the culture medium, protected from microorganisms by the cotton plug, remains sterile indefinitely. Bacteria that have been killed cannot multiply, but their bodies are not necessarily completely destroyed. The dead bodies of certain bacteria retain their shape and staining qualities and even promote the production of antibodies when introduced into the bodies of man or lower animals.

Sterilization may be accomplished by mechanical means, by heat (moist or dry), or by chemicals.

MECHANICAL MEANS

Three key methods for removing microbes mechanically are (1) scrubbing, (2) filtration, and (3) sedimentation.

Scrubbing. Scrubbing is usually done with water to which some chemical agent such as soap, detergent, or sodium carbonate has been added. The process is both mechanical and chemical. Scrubbing, by itself, removes many microorganisms mechanically while the incorporated compound acts on them chemically. Scrubbing with soap (or detergent) and water is a process basic to any discussion of sterilization because the removal of dirt, debris, and extraneous matter from an area or object is crucial to the effective removal of microbes therefrom by any method. Hands and person, floors, walls, woodwork, furniture, utensils of all kinds, glassware, linens, clothing, instruments, thermometers—all must be clean!

Filtration. In bacterial filtration, liquid containing bacteria is passed through a material with pores so small that the bacteria are held back. The mechanical removal of the bacteria from the fluid sterilizes it. In the laboratory this process is used for sterilizing liquids and culture media that cannot be heated and for separating toxins, enzymes, and proteins from the bacteria that produced them. Certain pharmaceutical preparations are sterilized in this way. Materials most often used for bacterial filtration are unglazed porcelain, diatomaceous earth, compressed asbestos, sintered glass, and cellulose membranes. (The finest mesh filter paper of the best quality will not hold back bacteria, but the pore size of a cellulose membrane filter may be reduced so that even certain viruses are retained; viruses passing such filters are said to be *filterable*.) Bacterial filters are constructed so that the material to be filtered is made to pass through a disk or the wall of a hollow tube made of the filtering material.

Filtration is an important step in the purification of a city water supply (p. 658). Bacterial filtration by a plastic membrane technic is widely used as a laboratory procedure in sanitary microbiology (p. 656).

Sedimentation. The process by which suspended particles settle to the bottom of a liquid is sedimentation, an important event in the purification of water by natural or artificial means. In nature, large particles and suspended bacteria sink to the bottom of lakes, ponds, and flowing streams. In the water purification plant, sedimentation is a significant part of the artificial purification of the community's water supply (p. 658).

MOIST HEAT

The most widely applicable and effective sterilizing agent is heat. It is also the most economical and easily controlled. The temperature that kills a 24-hour liquid culture of a certain species of bacteria at a pH of 7 (neutral reaction) in 10 minutes is known as the *thermal death point* of that species. Since this represents the temperature at which all bacteria are killed, it is obvious that many are destroyed before this temperature is reached. In fact, the majority are destroyed within the first few minutes. For standardization, bacteria must be in a neutral medium when their thermal death point is determined because in either a highly acid or a highly alkaline medium they are more susceptible to heat. The *thermal death time* is the time required to kill all bacteria in a given suspension at a given temperature.

Aside from burning (really a chemical process), heat is applied as *moist* or *dry heat*. Moist heat may be applied as hot water or steam and is the method of choice in sterilization except for those things altered or damaged by it. See Table 15-1 for applications of heat sterilization.

Boiling. A commonly employed, although incompletely effective, method of sterilizing by moist heat is boiling. Boiling kills vegetative forms of pathogenic bacteria, fungi, and most viruses in a matter of minutes. Hepatitis viruses are probably destroyed at the end of 30 minutes; but for practical elimination of hepatitis viruses, boiling is *not* recommended. Spores are less readily destroyed. Although most of the spores of pathogenic bacteria can be destroyed in a boiling time of 30 minutes, boiling is not a reliable method when materials are likely to contain spores. Certain heat-loving saprophytes can survive at high temperatures, and their spores resist *prolonged* boiling for many hours.

Completely immerse objects in boiling water, and continue boiling long enough to ensure even distribution of heat through the object being sterilized. Remember that microbes cannot be eliminated from the interior of materials boiled until heat has penetrated there. Prolong boiling time 5 minutes for each 1000 feet above sea level. The addition of sodium carbonate to make a 2% solution in boiling water hastens destruction of spores and helps to prevent rusting of instruments. Surgical instruments, needles, and syringes that are boiled must be clean and free of organic material.

Sterilization by steam. Steam yields heat by condensing back into water. For instance, when a bundle containing fabrics is sterilized by steam, the steam contacts the outer layer, where a portion of it condenses into water and gives up heat. The steam then penetrates to a second layer, where another portion condenses into water and gives up heat. The steam thus moves centrally, layer after layer, until the whole package is sterilized.

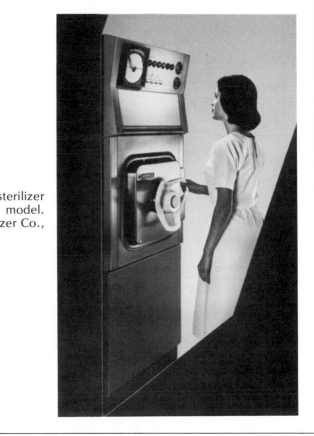

FIG. 15-1. Pressure steam sterilizer or autoclave, cabinet model. (Courtesy American Sterilizer Co., Erie, Pa.)

Steam may be applied as free-flowing steam or as steam under pressure. Free-flowing steam has about the same sterilizing action as boiling water. Steam under pressure is the most powerful method of sterilizing that we possess and is the preferred one unless the material being sterilized is injured by heat or moisture. The process is carried out in the *pressure steam sterilizer*, familiarly known as the *autoclave* (Figs. 15-1 and 15-2), a square sterilizing chamber surrounded by a steam jacket, the outside of which is insulated and covered. The chamber is loaded with supplies to be sterilized (the load) through a door that closes the front end of the sterilizing chamber. This is a safety steam-locked door made tight against a flexible heat-resistant gasket. The design is such that steam can be admitted to the closed chamber under pressure. The source of the steam varies; it may come from the central boiler supply of a large hospital or from an electrically heated boiler on the instrument itself. Valves on the autoclave control the flow and exhaust of steam. On some of the pressure steam sterilizers, these valves are operated by hand, but in the modern versions, the entire operation of the instrument is automatically designed.

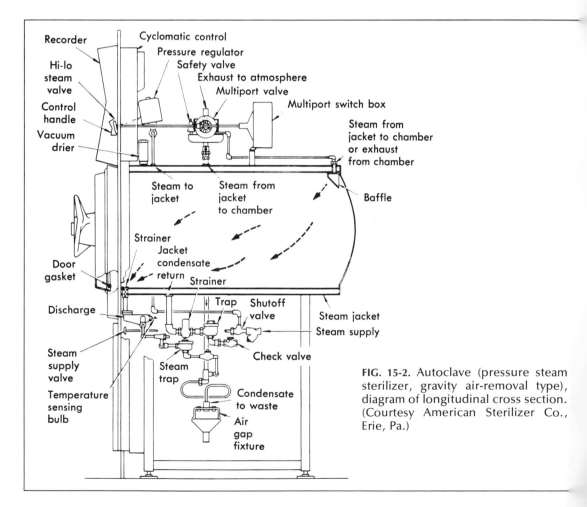

FIG. 15-2. Autoclave (pressure steam sterilizer, gravity air-removal type), diagram of longitudinal cross section. (Courtesy American Sterilizer Co., Erie, Pa.)

Steam under pressure is hotter than free-flowing steam, and the higher the pressure, the higher the temperature. The temperature of free-flowing steam (atmospheric pressure, sea level) is 100° C. At 15 pounds' pressure in the autoclave (atmospheric pressure, sea level), the temperature of steam is 121° C, and at 20 pounds' pressure it is 126° C. *Sterilization by steam under pressure is the result of the heat of the moist steam under pressure and not of the pressure itself.* Steam under a chamber pressure of 15 or 20 pounds will kill all organisms and spores in 15 to 45 minutes (depending on the materials involved). To maintain these temperatures at higher altitudes, the pressure must be increased 1 pound for each 2000 feet of increase in altitude.

OPERATION OF AN AUTOCLAVE. Steam sterilization in an autoclave (Fig. 15-2) has for a long time used a "downward displacement gravity system." (Steam is admitted to the sterilizing chamber in such a way as to drive air down and out, and steam being lighter than air displaces it downward.) In the operation of an autoclave, pressure is first generated in the steam jacket. The connection to the sterilizing chamber is kept closed until jacket pressure is constant at 15 to 17 pounds. (The pressure within the steam jacket is kept constant during the procedure to keep the walls of the chamber heated and dry.) The load is placed in the chamber and the door secured. Then steam is admitted to the sterilizing chamber and the load heated to the temperature of the steam in the chamber. At the time steam enters the chamber, the load and the chamber are both filled with air, which, if not evacuated, would reduce the moisture content of the autoclave and lessen its sterilizing capacity. Pressure steam sterilizers are vented so that the air can escape to the atmosphere as the temperature is raised. When all of the air has been evacuated, steam will contact the thermostatic valve, and it closes. The moisture that condenses on the door or back part of the sterilizing chamber drains downward from behind a steam deflector plate to the bottom part of the chamber and is then discharged to the waste line. After the end of the sterilizing cycle, the steam is exhausted from the chamber but not from the jacket. At this point a drying cycle is effected before the door is opened. This is done by creating a partial vacuum in the chamber with steam from the jacket through an ejector tube on the autoclave and at the same time admitting air through a pre-sterilized filter.

Since sterilization by steam under pressure is primarily a matter of temperature and the increase in pressure serves only to increase the temperature, the height to which the thermometer rises, rather than the reading on the pressure gauge, should be the guiding factor in the operation of the autoclave. This is particularly true because many pressure gauges are inaccurate, generally reading too high. (Autoclaves usually have two pressure gauges, one to indicate the pressure in the steam jacket and the other to indicate the pressure in the sterilizing chamber.) A thermometer or the sensing element of a thermometer at the bottom of the sterilizer is a better indicator of the efficiency of the sterilizing process than one placed at the top because, if any part of the autoclave fails to receive full benefit of the steam, it is the lower part. The thermometer in the discharge path of air and moisture coming from the sterilizing chamber can never indicate less than the lowest temperature in the system.

Modern autoclaves are equipped with a number of controls to increase the efficiency of sterilization and to remove insofar as possible the human factor. The *recording thermometer* is a clock-thermometer mechanism that indicates (1) the time at which the material being sterilized reaches the desired temperature, (2) whether the temperature remains stable, (3) how long the exposure lasts, and (4) how many times the autoclave is in operation during the day. The *indicating potentiometer* is an instrument for actually measuring the temperature of material in the autoclave. The *automatic time-temperature control* (1) operates the autoclave at the time and temperature for which it is set, (2) exhausts the steam from the chamber, (3) regulates drying, and (4) sounds an alarm indicating that the operation is complete.

Indicators are placed in with the load to be sterilized. An indicator changes in a predictable way its physiochemical properties or biologic nature when the prescribed temperature for sterilization has been reached. Indicators used are strips of paper impregnated with biologic material such as the dried spores of *Bacillus stearothermophilus*, the thermal death time of which is known. For steam sterilizers, *Bacillus stearothermophilus* is the monitor or challenge microorganism of choice; it grows readily and is more resistant to heat than the microbes usually found on the material being sterilized. For dry heat and ethylene oxide sterilizers (p. 286), *Bacillus subtilis* (var. *globigii* or *niger*) is preferred.

HIGH-PREVACUUM STERILIZER.* An improved pressure steam sterilizer, the high-prevacuum sterilizer, is in wide use (Fig. 15-3). With a vacuum system incorporated into the sterilizer unit, a precisely controlled vacuum is pulled at the beginning and end of the sterilizing cycle. Saturated steam at a temperature of 275° F (under a pressure of 28 to 30 pounds) enters the preevacuated chamber and instantly penetrates the load to be sterilized. Microbes present are killed within a few minutes. Under these high-temperature, high-pressure conditions of steam sterilization, there is a considerable shortening of the sterilizing (exposure) time (sometimes only 3 minutes). The vacuum pulled at the end of the sterilizing cycle dries the load. There is less damage to fabrics and to such items as rubber gloves because of reduced exposure.

Fractional sterilization (intermittent sterilization). When something that cannot withstand the temperature of an autoclave has to be sterilized, fractional or intermittent sterilization can be done. This procedure consists of exposing the material to free-flowing steam at atmospheric pressure for 30 minutes on 3 successive days; between times it is stored under conditions favorable for bacterial growth. With the first application of heat, all vegetative bacteria are killed, but the spores are not affected. Under conditions suitable for growth, the spores develop into vegetative bacteria, and the second application of heat kills them. The second incubation and third application of heat are added to ensure complete sterilization. Best results depend on the material being sterilized having such a nature as to promote the germination of spores. Therefore it is most useful in the sterilization of culture media. Sometimes referred to as *tyndallization*, it is infrequently used.

---

*Basic principles for sterilization are the same in both the gravity air-removal type of sterilizer (just described) and the high-prevacuum one.

FIG. 15-3. High-prevacuum sterilizer (Medium Rectangular Vacamatic). Panel of controls for manual operation behind white panel in upper right of instrument. (Courtesy American Sterilizer Co., Erie, Pa.)

The low-temperature method of sterilizing vaccines (p. 315) may be applied fractionally to biologic products that cannot withstand a temperature of 100° C. These may be sterilized by heating to a temperature of 55° to 60° C for 1 hour on 5 or 6 successive days.

Pasteurization. All nonsporebearing disease-producing bacteria and most nonsporebearing nonpathogenic bacteria are killed when exposed in a watery liquid to a temperature of 60° C for 30 minutes. This is the basis of *pasteurization* (p. 663), a special method of heating milk or other liquids for a short time to destroy undesirable microorganisms without changing the composition and food value of the material itself.

DRY HEAT

Dry heat (hot air) sterilization means baking the item to be sterilized in a suitable oven. Dry heat at a given temperature is not nearly so effective a sterilizing agent as is moist heat of the same temperature. Under controlled conditions of dry heat a temperature of 120° to 130° C kills all vegetative bacteria within 1½ hours, and one of 160° C kills all spores within 1 hour; but with moist heat a temperature of 120° C kills all vegetative bacteria and most spores within 15 to 20 minutes. Whereas moist heat sterilization is primarily a process of protein coagulation, dry heat sterilization is one of protein oxidation, and that oxidation goes on more slowly than coagulation. Moist heat also has greater penetrating power than dry heat.

An advantage of dry heat is that it does not dull cutting edges, but for most fabrics, even a moderate degree of dry heat is injurious. A temperature of more than 200° C causes cotton and cloth to turn brown. Therefore hot air is used mostly to sterilize glassware, metal objects, and articles injured by moisture or items such as petrolatum (Vaseline), oils, and fats that resist penetration by steam or water. In this form of sterilization the temperature should be slowly raised, and after sterilization is complete, the oven should be allowed to cool slowly to prevent breakage of glassware. Instruments to be sterilized must be clean and free of oil or grease films. For temperatures and times in practical dry heat sterilization, see Table 15-1.

There are two causes of ineffective dry heat sterilization: (1) the materials to be sterilized are too closely packed and (2) the temperature is not uniform within the sterilizing oven. An attempt to overcome the uneven distribution of heat has been made in sterilizers and sterilizing ovens constructed in such a manner that either gravity aids in the circulation of hot air through the sterilizer (gravity convection) or circulation is carried on by means of blowers (mechanical convection). Mechanical convection is more satisfactory than gravity convection.

Burning (incineration). Burning is a form of intense dry heat very effective in

TABLE 15-1. HEAT (PHYSICAL) STERILIZATION OF REUSABLE INSTRUMENTS AND SUPPLIES

Method	Administration		Application
	Temperature	Time*	
Autoclave	121°-123°C (250°-254°F) 15-17 lb pressure†	30 min	Gloves, drapes, towels, gauze pads, instruments, glassware, and metalware
Dry heat	170°C (340°F)	1 hr	Glassware, metalware, and dull instruments (any temperature listed)
	160°C	2 hr	Small quantities of powders, petrolatum (Vaseline), oils, and petrolatum gauze
	150°C	3 hr	Sharp instruments and metal-tip syringes
	121°C (250°F)	6 hr or longer	
Boiling	100°C (212°F)	30 min†	Method not recommended when dry heat and autoclave sterilization available

*With a high-prevacuum sterilizer, sterilizing (exposure) times are shorter; at a temperature of 132.8° to 135.5°C (271° to 276°F), sterilizing time is 4 minutes.
†Atmospheric pressure, sea level.

removing infectious materials of various kinds and generally applicable when materials and supplies are disposable or expendable. The platinum wire loop used to inoculate cultures is repeatedly and quickly sterilized by *flaming*—heating the wire in an open flame until it glows.

All contaminated objects that are of no value or cannot be used again are preferably burned!

STERILIZATION BY OTHER PHYSICAL MEANS

Natural methods. If a culture of bacteria is dried, the majority of the bacteria are quickly killed but some live for quite a while. Spores and encysted protozoa resist drying for a long time. Although drying is an important natural method of removing or destroying microbes, it is not applicable to "artificial" sterilization, except that sterile dressings and similar objects should be kept dry.

Sunlight has an inhibitory and destructive action on microbes. It will kill *Mycobacterium tuberculosis* within a few hours and many other bacteria in a shorter time. Sunlight is nature's great sterilizing agent but is so irregular in its presence that one cannot depend on its action. The antimicrobial action of both drying and sunlight is applied to advantage in the home drying of food.

Ultraviolet radiation. The purity of the air in the "wide-open spaces" has long been recognized, and it is well known that the sterilizing effect of sunlight there comes from the ultraviolet rays present. This fact has been applied to the construction of ultraviolet lamps in wide use to prevent the airborne spread of disease-producing agents, especially in public places, in hospitals (operating rooms, treatment rooms, nurseries), in microbiologic laboratories, and in quarters used to house animals.

Ultraviolet radiation is especially effective in killing organisms contained in the minute dried respiratory droplets that tend to disperse rapidly through the atmosphere of a building or hospital. It is not so active against dust-borne agents and microorganisms on surfaces. It inactivates certain viruses. Its bactericidal effect drops sharply when the humidity of the area exceeds 55% to 60%. In appropriate amounts it damages the skin or conjunctivae.

X rays and other ionizing radiations. X rays and other ionizing radiations are known to be lethal to microbes and to living cells as well, but there is no practical application for their use in routine sterilization. In industry, beta rays or electrons sterilize prepackaged materials such as sutures and plastic tubing. High-energy electrons have been proposed for the treatment of sewage and wastewater.

One interesting application is the combination of heat and gamma radiation for the sterilization of spacecraft. If heat alone is used, the spaceship is subjected to a temperature of 125° C (257° F) for 60 hours. The temperature has to be controlled carefully to prevent heat damage to and failure of components in such items as silver-zinc batteries and tantalum capacitors. If the spaceship with its equipment is sprayed with 150,000 rads of gamma radiation, the time for sterilization at the temperature of 125° C can be cut down to 2 hours.

Lasers. Recent investigation suggests the use of a laser to sterilize medical instruments, clear the air in operating rooms, and pick organisms off a wound surface.

The technical problem is that the laser beam must reach all parts of the item to be sterilized. If feasible, laser sterilization would be a split-second procedure.

Ultrasonics. Sound waves are mechanical vibrations. In the range of vibration (18,000 cycles/second or more) where they are no longer heard as sound (supersonic or ultrasonic), these waves have been demonstrated to coagulate protein solutions and to destroy bacteria. *Cold boiling* results from the passage of ultrasonic pressure waves through a cleaning solution. Very tiny empty spaces in the liquid form and collapse thousands of times a second. This type of scrubbing action can blast material from the surface of objects made of metal, glass, or plastic. The use of such vibrations (pitched too high to be heard) is not widely practical, but the principle has been incorporated into a commercially available dishwasher. Cleaning medical instruments is a common application. "Noiseless sound" is also used experimentally to treat sewage water with its disruptive effect on viruses, bacteria, and chemicals such as phosphates and nitrogen-containing compounds.

Action of fluorescent dyes. Certain dyes with the property of fluorescence, such as methylene blue, rose bengal, and eosin, are lethal to bacteria and viruses if in contact with these microbes in strong visible light.

QUESTIONS FOR REVIEW

1. Name three ways in which sterilization may be accomplished.
2. Define sterilization, bacterial filtration, sedimentation, thermal death point, pasteurization.
3. Name and describe briefly three mechanical means of removing or destroying microbes.
4. Why is moist heat more effective than dry heat as a sterilizing agent?
5. What is the effect of pressure on steam sterilization?
6. Explain intermittent or fractional sterilization.
7. What is an autoclave? Indicate basic principles of its operation.
8. Briefly describe the high-prevacuum sterilizer. State its advantage.
9. How is dry heat applied for sterilization? Cite examples of items that must be sterilized in this way.
10. Tabulate all physical agents used for sterilization.

REFERENCES. See at end of Chapter 17.

16 Chemical agents

EFFECTS OF CHEMICAL AGENTS ON MICROBES

Definitions. Certain definitions are necessary for discussions to follow. As we have learned, *sterilization* is an absolute term referring to the destruction or removal of all microorganisms present under given conditions. *Disinfection*, on the other hand, means death to disease-producing organisms and the destruction of their products, usually with chemical agents known as *disinfectants*. (A more precise definition might indicate that disinfection *halts* the spread of undesirable microorganisms by inducing structural or metabolic derangements in them.) Disinfection does not deal directly with saprophytic organisms present in a given setting that may or may not be killed. The terms *disinfection* and *disinfectant* are applied to procedures and chemical agents used to destroy microbes associated with inanimate objects. The term *antiseptic* is usually applied to an agent that acts on microorganisms associated with the living body to prevent their multiplication but not necessarily kill them. We should disinfect the excretions from a sick person but apply an antiseptic to his wounds. True, the terms are often interchanged.

Germicides are chemicals that kill microbes (not necessarily their spores). *Bactericides* kill bacteria, *viricides* destroy or inactivate viruses, *fungicides* destroy fungi, and *amebicides* destroy amebas, especially the protozoan *Entamoeba histolytica*. *Asepsis* means the absence of pathogenic microbes from a given object or area. To avoid infecting the patient, in aseptic surgery the field of operation, the instruments, and the dressings are rendered free of microorganisms by sterilization, and the operation is conducted in such a manner that it is kept as free of microbes as possible.

Bacteriostasis is that condition in which bacteria are prevented from multiplying (but in no other manner affected) by such agents as low temperature, weak antiseptics, and dyes. *Agents causing bacteriostasis are known as bacteriostatic agents*. Antiseptics and chemical bacteriostatic agents are synonymous terms. When used

275

to prevent the deterioration of foods, serums, and vaccines, such agents are *preservatives*.

Two terms with increasing applications are *degerm* and *sanitize*. To degerm is to remove bacteria from the skin by mechanical cleaning or application of antiseptics. To sanitize means to reduce the number of bacteria to a safe level as judged by public health requirements. It refers to the day-by-day control of the microbial population on utensils and equipment used in dairies and establishments where food and drink are served. In short, sanitization refers to a "good cleaning" and is basic to technics of sterilization. As used, the term *sterilization* implies a mechanically clean item. The word *decontamination* applies to the process of killing all microbes from an item known to be mechanically dirty and containing a heavy growth of microorganisms.

Fumigation is the liberation of fumes or gases to destroy insects or small animals. *Deodorants* are substances that destroy or mask offensive odors. They may have neither disinfectant nor antiseptic action and may generally tend to obscure infectious material rather than destroy it.

Qualities of a good disinfectant. Certain qualities specify the ideal disinfectant for general use. Unfortunately, at the present time no chemical agent possess all of them, and the selection of a disinfectant is often a workable compromise. The more of the following qualities that a disinfectant has, the more nearly it approaches the ideal.

1. Attack all types of microorganisms.
2. Be rapid in its action.
3. Not destroy body tissues or act as a poison if taken internally.
4. Not be retarded in its action by organic matter.
5. Penetrate material being disinfected.
6. Dissolve easily in or mix with water to form a stable solution or emulsion.
7. Not decompose when exposed to heat, light rays, or unfavorable weather conditions.
8. Not damage materials being disinfected such as instruments or fabrics.
9. Not have an unpleasant odor or discolor the material being disinfected.
10. Be easily obtained at a comparatively low cost and readily transported.

The most important feature of a disinfectant is its ability to form lethal combinations with microbial cells. *Remember that different species of microbes, especially bacteria, show much greater variation in their susceptibility to disinfectants than they do to sterilization by physical agents.*

Action of antiseptics and disinfectants. Antiseptics and disinfectants act by (1) oxidation of the microbial cell, (2) hydrolysis, (3) combination with microbial proteins to form salts, (4) coagulation of proteins, (5) modification of the permeability of the microbial plasma membrane, (6) inactivation of vital enzymes of microorganisms, and (7) disruption of the cell.

FACTORS INFLUENCING THE ACTION OF DISINFECTANTS. Factors influencing the action of disinfectants refer to (1) qualities of the disinfectant, (2) nature of the material to be disinfected, (3) concentration of the disinfectant, and (4) manner of application. A chemical in a solution of one strength may be a disinfectant, whereas in a weaker solution it may act as an antiseptic, and in certain very weak solutions it may actually stimulate microbial growth. The relative germicidal properties of the salts of a heavy metal are in proportion to their ionization.

The item for disinfection is evaluated as to (1) kind and number of microbes present, (2) presence of vegetative forms or spores (3) distribution of microbes in clumps or uniformly, and (4) presence of organic compounds or other chemicals that inactivate the disinfectant.

Most chemical disinfectants in common use are *germicidal but not absolutely sporicidal;* that is, they do not kill all spores present. As a rule, the process of disinfection is a gradual one, and a few microbes survive longer than the majority. To be effective, the disinfectant must be applied for a length of time sufficient to destroy all microorganisms. Many chemical disinfectants must be used for a long time to obtain the maximal effect; this often means 18 to 24 hours. An important factor relating to the disinfectant is the temperature at which it is applied. The higher the temperature, the more active it is. Remember that a disinfectant must penetrate all parts of the material being disinfected because it must contact microbes to destroy them. An article being disinfected must be *completely submerged* in the working solution. Also, a disinfectant should be properly chosen in accordance with the physicochemical nature of the material to be disinfected.

SURFACE TENSION IN DISINFECTION. The molecules lying below the surface of a liquid are pulled in all directions by the cohesive forces of neighboring molecules. Those at the surface are pulled downward and sideways only but not upward because there are no molecules above the surface to attract them. This phenomenon is surface tension, which is nontechnically defined as *that property due to molecular forces by which the surface film of all liquids tends to bring the contained volume into a shape having the least superficial area.* Since a sphere has the least area for a given volume, surface tension causes drops of liquid to become spherical.

If a drop of mercury is placed on a flat surface of metal or glass, it remains a distinct globule rolling about. If a drop of alcohol is placed on a surface, it does not form a globule but spreads out into a very thin layer over a large area. The surface tension of mercury is high; that of alcohol is low.

Surface tension is important in disinfection because liquids of low surface tension spread over a greater area and contact cells more intimately than liquids of high surface tension. Low surface tension liquids are often spoken of as *wetting agents*. A good wetting agent spreads over a surface rapidly and remains in a thin film. Chemicals that reduce the surface tension of water when dissolved in it are more concentrated on the surface of cells than they are throughout the solution. Thus some wetting agents are also effective antiseptics. Wetting agents thought of primarily as surface-cleaning agents are *detergents*. Many detergents act as both cleaning agents and inhibitors of bacterial growth. The classic example of a wetting agent or detergent is soap. However, soap does not have the antiseptic action of certain other detergents. Many synthetic organic detergents (liquids, granules, and such) on the market are sometimes spoken of as soapless soaps or nonsoap cleaners. Some detergents such as Tween 80 do not have an antiseptic action but promote bacterial growth.

Standardization. Various procedures have been designed to evaluate the antimicrobial activity of a given chemical agent as well as indicate its toxicity for tissues. Rideal and Walker in 1903 devised the original phenol coefficient test, which compared the disinfectant or antiseptic activity of a given compound with that of phenol under standard conditions. A phenol coefficient of greater than 1 indicated a stronger agent than phenol; a coefficient less than 1, a weaker one. Despite the number of methods existing today, inadequacies still remain and the tests fail to give all the information needed.

COMMON DISINFECTANTS AND ANTISEPTICS

Myriads of cleaning and disinfecting chemicals and combinations exist. With exceptions, there seems to be no general uniformity of opinion today as to which of these is best for application in any given situation, and the use of such chemical agents varies considerably, even in the same community. At the end of the listing of better known chemical agents, Table 16-1 provides a summary statement and compares the antimicrobial activity of some of the more important ones. As indicated, it does reflect material from a standard source.

Surface-active compounds

Soap (bar, liquid, granule, leaflet, or soap-impregnated tissue) is our most important cleaning agent. Although its utility as a disinfectant is limited, it is generally used before one is applied. The major action of soap is to aid the mechanical removal of microbes, primarily through scrubbing. In cleaning, soap and water separate particulate contamination of whatever kind from a given area, for example, the skin surface of the human body. This is an area constantly accumulating dead cells, oily secretions, dust, dried sweat, dirt, soot, and microorganisms. Soap breaks up the grease film into tiny droplets; water and soap acting together lift up the emulsified oily materials and dirt particles, floating them away as the lather is washed off.

Although the term *detergent* means any cleaning agent, even water, it is used to distinguish the synthetic compounds from soap, both of which lower the surface tension of water. Soap is made from fats and lye; detergents are made from fats and oils by complicated chemical processes and most contain a biodegradable linear sulfonate derivative of petroleum. Soaps depend for their cleaning action on their content of alkali, which suspends the grime from the surface of an object in water to be washed off. The detergent ionizes in water; its electrically charged ions attach themselves to the dirt. The washing action releases the ions, which carry the dirt away. Detergents dissolve readily in cold water and completely in even the hardest water. Soap combines with the calcium and magnesium salts in hard water to form an insoluble scum.

Enzyme detergent refers to laundry presoak products in which certain proteolytic enzymes from bacteria are incorporated (in amounts up to 1.0% active enzyme). Enzymes are obtained through a fermentation process from the widely distributed nonpathogenic soil organism, *Bacillus subtilis*. Enzyme detergent dissolves organic (protein) stains such as blood, feces, and meat juices without harming fabric or user.

TABLE 16-1. ANTIMICROBIAL ACTIVITY OF COMMONLY USED COLD "STERILANTS"*

Agent	Destructive action against				
	Bacteria	Tubercle bacilli	Spores	Fungi	Viruses
Alcohol—ethyl (70% to 90%)	+	+	0	+	±
Alcohol—isopropyl (70% to 90%)	++	+	0	+	±
Alcohol-iodine (2%)	++	+	±	+	+
Formalin (37%)	+	+	+	+	+
Glutaraldehyde (buffered, 2%) (Cidex)	++	+	++	+	+
Iodine (2% to 5% aqueous)	++	+	±	+	+
Iodophors (1%) (povidine-iodine complex)	+	+	±	±	+
Mercurials (Merthiolate)	±	0	0	+	±
Phenolic derivatives (O-syl, 1% to 3%)	+	+	0	+	±
Quats (benzalkonium chloride, 1:750 to 1:1000)	++	0	0	+	0

++, Very good; +, good; ±, fair (greater concentration or more time needed); 0, no activity.
*After DiPalma, J. R., editor: Drill's pharmacology in medicine, New York, 1971, McGraw-Hill Book Co.

Enzyme activity is quickly dissipated during the washing process and is inactivated by chlorine bleach.

Soap is mildly antiseptic because of its sodium and alkali content, but to remove most bacteria effectively, scrubbing must be followed by the application of a suitable disinfectant. Organisms susceptible to the germicidal action of soap are pneumococci, streptococci, gonococci, meningococci, *Treponema pallidum*, and influenza viruses.*

If soap is to be followed by some germicide, it should be thoroughly washed off with 70% alcohol before the germicide is applied, because soap and germicide might combine to form an inert compound. So-called germicidal soaps have little or no advantage over ordinary soaps. If soap is not properly handled and dispensed, it may become a source of infection within itself.

Benzalkonium chloride (Zephiran Chloride), a mixture of high molecular weight alkyl dimethylbenzylammonium chlorides, is one of the most important members of the surface-active chemical disinfectants known as quaternary ammonium disinfectants, or quats.† It is used in hospitals for the disinfection of hands and preparation

---

*Most pathogenic bacteria and viruses are removed or chemically killed by the soaps and detergents ordinarily used in the commercial self-service laundry machine. The temperature of the dryer usually is high enough to eliminate any remaining bacteria. If the clothes are then ironed, the heat of the hot iron destroys any microbes that possibly could have survived.
†A quaternary ammonium compound is built around a nitrogen atom of five valence bonds. Four of these bonds are attached to adjacent carbon atoms of organic radicals, and one is attached to an inorganic or organic radical. Soap reduces the germicidal action of the quats, as would hard water were it used to dilute a stock solution to make an aqueous preparation.

of the field of operation. A 1:1000 aqueous solution may be used for the disinfection of instruments, especially endoscopes and sharp-edged cutting instruments, the blades of which would be dulled by autoclaving. Such a solution kills vegetative bacteria (except tubercle bacilli) in 30 minutes. *It has no effect on tubercle bacilli and is not effective against spores.* The presence of serum or alkali retards its action. When Zephiran Chloride is used as a skin antiseptic, soap must be removed by thorough rinsing with 70% alcohol for 1 minute or more before Zephiran Chloride solution is applied. (Water usually does not remove all soap, since soap and water constitute a colloidal solution. Soap is completely soluble in 70% alcohol.) When metal instruments are to be stored in a solution of Zephiran Chloride, it is well to add an antirust agent or antirust tablets. One antirust tablet available commercially is a combination of sodium carbonate and sodium nitrite.

Diaparene Chloride is a quaternary ammonium compound particularly bacteriostatic toward *Brevibacterium ammoniagenes*, the intestinal saprophyte chiefly responsible for the production of ammonia in decomposed urine. It therefore helps prevent ammonium dermatitis in infants when used for disinfection of diapers.

Ceepryn Chloride closely resembles the quaternary ammonium compounds just described. A commercially available 1:1000 solution is recommended as a mouthwash or gargle, combining a foaming detergent cleaning action with antibacterial activity against certain pathogenic organisms of the mouth and throat.

Heavy metal compounds

Organic mercury compounds were developed to avoid the toxicity of inorganic mercurials and yet retain the disinfecting qualities of mercury. They are soluble in water and body fluids with a low toxicity for tissues, which is an advantage. They kill pathogenic bacteria that do not form spores other than tubercle bacilli. *Organic mercurials have no action against spores.* Among them are merbromin (Mercurochrome), nitromersol (Metaphen), thimerosal (Merthiolate), and Mercresin, a mixture instead of a definite compound.

Merbromin combines mercury with a derivative of fluorescein. A 1% solution is tolerated by the urinary bladder and kidney pelvis; a 2% solution is used to disinfect wounds, and stronger solutions may be used for skin antisepsis. *Nitromersol* is used for sterilization of instruments, for skin antisepsis, and for irrigation of the urethra. In strengths ranging from 1:10,000 to 1:1000, it is said to be comparatively nontoxic, nonirritating, and nondestructive to metallic instruments and rubber goods. *Thimerosal* is used for disinfecting instruments, skin, and mucous membranes and as a biologic preservative for vaccines, serums, and blood cells. Aqueous solutions are considered to be fungistatic also.

Mercresin combines the germicidal action of the mercurials with that of the phenolic derivatives, giving maximum disinfection with minimum tissue injury. Its action is not inhibited by the presence of serum, and it is widely used for local skin antisepsis.

Silver nitrate, a caustic, antiseptic, and astringent, is the inorganic silver salt most often used. Silver nitrate pencils cauterize wounds. Silver nitrate in a 1:10,000 solution inhibits growth of bacteria, and increasing the strength of solution, even to 10% in certain instances, enhances germicidal activity. It has a selective action for gonococci, and a 1% solution instilled in the eyes of newborn babies prevents ophthalmia neonatorum (p. 358). A 0.5% solution used in the continuously soaked dressings applied to burns reduces infection, but *Pseudomonas* can still grow beneath the dry black eschar that results. The action of silver nitrate is retarded by chlorides, iodides, bromides, sulfates, and organic matter. It is reduced on exposure to light and, because of its many incompatibilities, *should be used only with distilled water.*

Zinc salts are mild antiseptics and also astringents. *Medicinal Zinc Peroxide* is a mixture of zinc peroxide, zinc carbonate, and zinc hydroxide. The commercial powder should be sterilized at 140° C dry heat for 4 hours. In watery suspensions and ointments it is of special value in controlling infections caused by anaerobic and microaerophilic organisms in injuries such as gunshot wounds, bites, and deep puncture wounds. *Calamine Lotion* is mostly zinc oxide with some ferric oxide.

Copper sulfate is valuable chiefly for its destructive action on the green algae that often grow in reservoirs and render water obnoxious. A copper sulfate concentration of 1 part per million parts (ppm*) of water kills algae if the water does not contain an excess of organic matter. One part of copper sulfate added to 400,000 parts of water destroys typhoid bacilli in 24 hours. It is not harmful to drink water for a short time that contains this amount of copper sulfate. Copper sulfate is an important ingredient of sprays used to combat fungous diseases of plants.

Alcohols and aldehydes

Ethyl alcohol is one of the most widely used disinfectants and one of the best. For alcohol to coagulate proteins (its disinfectant action), water must be present. Because of this fact, 70% has long been considered the critical dilution of ethyl alcohol. However, there is good reason to think that against microorganisms in a moist environment alcohol acts over a range including 70% and up to 95%. Alcohol in a 70% dilution may be used to disinfect certain delicate surgical instruments, although it does tend to rust instruments and dissolves the cement from around the lights of endoscopes. It does not kill spores. Alcohol kills tubercle bacilli rapidly and is a tuberculocidal disinfectant of choice.

Isopropyl alcohol is slightly superior to ethyl alcohol as a disinfectant. It is also inexpensive, and its sale is not subject to legal regulations. Like ethyl alcohol, it acts against vegetative bacilli (not spores) and tubercle bacilli. Recent evidence indicates that this agent also is an effective disinfectant in dilutions stronger than the conventional 70% and that the most effective one may well be full strength (99%).

Formaldehyde, a gas, occurs in commerce in a watery 37% solution known as *formalin.* In addition to its disinfecting properties, formaldehyde serves as a deodorizer, as a preservative of tissues, and to convert toxins into toxoids. Its action depends on the presence of moisture, concentration of the gas, temperature, and

---

*1 ppm = 1 inch in 16 miles; 1 minute in 2 years; a 1-gram needle in a ton of hay; 1 penny in $10,000; 1 large mouthful of food when compared with the food a person will eat in a lifetime.

condition of the object to be sterilized. The presence of 1% of the gas in a room or chamber destroys all nonsporulating pathogenic bacteria. Cytoscopes and certain specialized instruments that would be damaged by heat are sometimes disinfected in airtight sterilizing cabinets equipped with electric terminals so that formaldehyde pastils can be vaporized in them. The instruments are left in the cabinet exposed to the formaldehyde fumes for 24 hours. Mixtures of formalin with alcohol and hexachlorophene make active germicidal solutions for sterilizing surgical instruments, speedily killing bacteria, spores, tubercle bacilli, and many viruses.

Glutaraldehyde (Cidex) is one of the latest additions to any list and one of the best. *Activated glutaraldehyde solution* is bactericidal, viricidal, and sporicidal in short exposure times. However, it corrodes rubber tubing and materials, etches plastic with repeated exposure, and irritates the skin. It is recommended for sterilization of anesthetic equipment, the intermittent positive-pressure breathing apparatus, and instruments containing optical lenses.

Phenol and derivatives

Phenol (carbolic acid), a corrosive poison, in a 1:500 solution inhibits the growth of bacteria. A 5% solution kills all vegetative bacteria and the less resistant spores in a short time. Contact with alcohol, ether, or soap decreases its action. The addition of 5% to 10% hydrochloric acid increases its efficiency. Since the action of phenol is not greatly retarded by the presence of organic matter, it is an excellent disinfectant for feces, blood, pus, sputum, and proteinaceous materials. It does not injure metals, fabrics, or painted surfaces. The crude acid may be used for woodwork because it is cheaper and more effective than the pure substance. The crystals or strong solutions should not touch the skin. If this happens, alcohol should be applied at once. Dilute solutions should not contact the skin or mucous membranes for more than 30 minutes or 1 hour because they injure tissues.

Cresol has a higher germicidal power than phenol and is less poisonous. *Saponated cresol solution (Lysol)*, an alkaline solution of cresol in soap, in a 2.5% solution makes a good disinfectant for feces and sputum. *Lysol*, a widely used and popular proprietary disinfectant, is most important in the disinfection of inanimate objects, including instruments, furniture, table surfaces, floors, walls, rubber goods, rectal thermometers, and contaminated objects of varied description, especially when these have been contaminated by *Mycobacterium tuberculosis*. Phenol derivatives such as Lysol can be used to disinfect excreta or contaminated secretions from patients with infectious diseases, but especially tuberculosis. They are of little value in antisepsis of the skin because in concentrations that would not injure the skin they possess little bactericidal activity. *They do not destroy all spores present.*

Amphyl, a modern and greatly improved phenolic disinfectant, has wide and varied applications as a germicide. (Chemically it is a mixture of orthophenylphenol and paratertiary amylphenol with potassium ricinoleate in propylene glycol and alcohol.) Nontoxic, noncorrosive to metals, and nonirritating to skin and mucous membranes, it can be mixed with soap and certain other antiseptics. Because it is an agent with low surface tension, it spreads and penetrates materials. Surfaces treated with it tend to retain an antimicrobial action for several days. It is effective in dilutions of 0.25% to 3% for a range of routine disinfecting procedures involving skin, mucous membranes, floors, walls, furniture, dishes, utensils, and surgical instruments. Fungi, bacteria (including the tubercle bacillus), and viruses, but *not* spores, are destroyed by its action; heating increases its germicidal properties. *Staphene*, a phenolic disinfectant related to Amphyl, is a mixture of four synthetic phenols* (paratertiary amylphenol, ortho-benzyl-para-chlorophenol, orthophenylphenol, and 2,2'-methylene-bis [3,4,6-trichlorophenol]).

O-syl, an antiseptic, germicide, and fungicide, is considered to be especially effective against the causative agent of tuberculosis. Chemically it is one synthetic phenol (orthophenylphenol).

Hexachlorophene (G-11), a diphenol, is one of the few antiseptics and disinfectants not affected by soap. Therefore in concentrations from 1% to 3% it has been incorporated into soaps and combined with detergents that have been used for the preoperative hand scrubs of the surgical team and for the preoperative and postoperative preparation of the skin of the patient. Hexachlorophene is long retained on the skin, from where it can be recovered more than 48 hours after use. Its benefit is attributed to the bactericidal film left on the skin after repeated application; the concentration on the skin is cumulative up to 2 or 4 days. (The skin cannot be completely sterilized.) Mechanical cleaning plus the use of germicidal agents removes the superficial growth of bacteria from the skin surface, which fortunately includes most of the pathogens. Hexachlorophene has been so widely used because of its bacteriostatic action against gram-positive microbes, notably the staphylococcus. Bathing of infants with a detergent containing hexachlorophene has been shown to reduce considerably the incidence of staphylococcal infection in a population of newborns.

Generally hexachlorophene is not irritating to the skin although it is sometimes associated with allergic manifestations. However, it is readily absorbed through the normal skin and more easily so through abraded or burned skin. Its use has been blamed for brain seizures developing in young burn victims who had been washed with it. Animal experiments relate the uptake through the skin over a period of days to subsequent brain damage and even paralysis. The United States Food and Drug Administration (FDA) warns against total body bathing of infants and adults with products containing 2% and 3% concentrations of hexachlorophene and bans over-the-counter sale of most products containing it.† Available only by prescription, it is still used in hospitals and similar institutions because of its effective antibacterial action.

---

*Some of the phenolic detergent germicides that have been widely used in cleaning solutions can cause depigmentation of the skin of the user.

†Hexachlorophene has been perhaps the most universal of the antibacterial agents, having been incorporated into a wide variety and number of consumer products—soaps, shampoos, toothpastes, deodorants, lotions, powders, ointments, cosmetics, and medicinal cleaners. According to the FDA, the ingredient replacing hexachlorophene in many antibacterial products—tribromsalan—should also be eliminated.

Currently the Committee on Fetus and Newborn of the American Academy of Pediatrics recommends that only with a serious outbreak of staphylococcal infection may one resort to total body bathing of newborns with hexachlorophene and, then, only if (1) a solution of not more than 3% is used, (2) it is applied to full-term infants only, (3) it is washed off thoroughly after each application, and (4) it is applied no more than two times to a given infant.

pHisoderm, a detergent cream, is not a single compound but a proprietary mixture of wool fat, cholesterol, lactic acid, and sulfonated petrolatum. It cleans faster and more effectively than does soap and is used for degerming the hands and other skin areas. It may be used alone, but it is most often combined with hexachlorophene. *pHisoHex* is the name given to the mixture of the detergent base pHisoderm and 3% hexachlorophene. It is still popular for surgical hand scrubs because regular use gives maximal bactericidal effect. It is nonirritating, only small amounts need to be used, and the time required for the surgical scrubbing is much shorter than with soap preparations. Its emulsifying action on oily material and its ability to form suds are desirable characteristics.

Halogen compounds

Iodine, one of the best known disinfectants, is a potent amebicide, bactericide in a wide spectrum, good tuberculocide, fungicide, and viricide. The ubiquitous *tincture of iodine* containing 2% iodine is one of the best disinfectants for small areas of skin, minor cuts, abrasions, and wounds. The strong tincture of iodine (7% in alcohol) is too toxic for most purposes. Iodine becomes freely soluble in water in the presence of soluble iodides such as those of sodium and potassium. *Iodine solution* (aqueous) containing 2% iodine is as effective as the alcoholic solutions (tinctures), but higher concentrations are sometimes used. Very strong and very old solutions of iodine burn the skin. *Iodine should not be applied under a bandage.* Although in some persons iodine compounds produce allergic skin rashes, iodine is less toxic than other routinely used germicides. Within the last few years iodine has been used for disinfection of water in swimming pools.

Iodophors are compounds in which iodine is carried by a surface-active solvent. The germ-killing action results from release of free iodine when the compound is diluted with water; only the free iodine has any appreciable effect. An iodophor enhances the bactericidal action of iodine and reduces odor. It does not stain the skin or items on which it is placed as does the tincture. The proprietary preparation *Wescodyne*, an iodine detergent germicide (also a good tuberculocide), is an example as are *Hi-Sine, Iosan,* and *Betadine* (povidone-iodine complex).

Chlorine, one of the most effective and widely used of all chemical disinfectants, is used in the disinfection of drinking water, in the purification of swimming pools, and in the treatment of sewage. It is applied by releasing the gas from cylinders or by the use of compounds that release free chlorine. The chlorine from chlorine-liberating compounds is spoken of as "available" chlorine. For effective disinfection the chlorine content of water must reach a concentration of 0.5 to 1 ppm. With no organic material present, 0.1 ppm destroys the poliovirus, but in a strength of 1 ppm, chlorine has no effect on the cercariae that represent one stage in the development of flukes (p. 598). (Nor are these minute worms removed by sand filtration or aluminum sulfate clarification of the water.) Cysts of *Entamoeba histolytica* (p. 581), the cause of amebic dysentery, can live as long as 1 month in water. The usual chlorination treatment for drinking water does not kill them; water containing an adequate amount of chlorine to destroy them would not be fit to drink. Cysts of *Giardia lamblia*, now a recognized pathogen, can survive cool or tepid water for as long as 3 months but quickly succumb to a temperature of 122° F (50° C). They can survive in 0.5% chlorinated water for 2 or 3 days. (However, they may be killed by iodine compounds in recommended doses for water purification.) The World Health Organization recommends that commercial airlines filter their water supply and overchlorinate it, 8 to 10 ppm of free available chlorine, in order to kill cysts of *Entamoeba histolytica* and viruses of infectious hepatitis. The water must then stand for at least 30 minutes. It is dechlorinated to eliminate the disagreeable taste and smell.

Sodium hypochlorite (NaOCl), made by the reaction of chlorine on sodium hydroxide, is a powerful oxidizing agent. It cannot be prepared as a powder but is manufactured in solutions of varying strength. The stronger solutions are used as bleaches by laundries and other establishments. The weaker solutions (example, Clorox) are used as household bleaches and for the bactericidal treatment of food-handling equipment. Hypochlorite solutions prepared fresh daily are active against viruses if they contain 5000 to 10,000 ppm of available free chlorine (0.5% to 1.0% solutions) and may be used in hemodialysis units, laboratories, and blood banks for disinfection of nonmetal equipment and surfaces likely to contain hepatitis viruses. *Dakin's solution* and modifications are weak neutral solutions of sodium hypochlorite, liberating from 0.5% to 5% available chlorine. The official *Sodium Hypochlorite Solution* (5%) is a valuable agent in root canal therapy.

Chloramines are organic chlorine compounds that slowly decompose and liberate chlorine. They are inferior when speed is required, for their value lies in their prolonged action. They may be used to sanitize glassware and eating utensils and to treat dairy and food-manufacturing equipment, but their practicality is mainly for emergency disinfection of drinking water, particularly of small amounts, and *Halazone* is the one applied.

Acids

Boric acid is a weak antiseptic most often used as an eyewash. However, when taken internally, boric acid is highly toxic. Because of the risk of accidental poisoning, its use in hospitals is condemned. It has no place in the pediatric division or in any nursery. If even a dilute solution of boric acid mistaken for distilled water should be used in an infant formula or in a parenteral fluid, the results could be disastrous. Dusting the powder repeatedly over the diaper area of a baby to remedy diaper rash can cause death. A so-called bland ointment applied to abraded areas of the skin can produce an unfavorable reaction.

Fuming nitric acid is the best agent for cauterizing the wounds inflicted by rabid animals.

Benzoic acid and *salicylic acid* are fungistatic agents. A mixture of benzoic acid, 6%, and salicylic acid, 3% (*Whitfield's ointment*), is used to treat fungous infections of the feet. Benzoic acid is an important food preservative (p. 673).

Undecylenic acid (Desenex) is used for athlete's foot and other fungous infections of the skin.

Oxidizing agents

Hydrogen peroxide, 3%, owes its disinfecting qualities to the free oxygen that it liberates. Its spectacular effervescence mechanically produces a weak cleaning effect in certain wounds. Not an overly reliable antiseptic, it deteriorates rapidly.

Potassium permanganate owes its antiseptic and antifungal effectiveness to strong oxidizing qualities. However, its action is weakened by the presence of organic matter, and it can irritate tissues if the concentration of the solution exceeds 1:5000.

Dyes

Dyes such as *gentian violet* and *crystal violet* in suitable dilutions inhibit the growth of gram-positive bacteria and fungi but have little effect on gram-negative bacteria, whereas dyes such as *acriflavine* and *proflavine* inhibit the action of gram-negative bacteria but have little effect on gram-positive bacteria. Dyes are used to obtain pure cultures in the laboratory. Gentian violet, 1%, is sometimes used to treat fungous infections. With the advent of the sulfonamide drugs, the need for dyes as therapeutic agents no longer existed.

Miscellaneous chemicals

Ethylene oxide is a gas with a broad range of antimicrobial activity that has been used in industry for years for such purposes as the disinfection of the furnishings of railway cars. It is an odorless, poisonous, and explosive gas that can be easily kept as a liquid. Since its vapor is highly inflammable in air, it is transported commercially in a noninflammable mixture with carbon dioxide. This gas is used especially to treat items that are damaged by heat, water, or chemical solutions and is a potent agent properly used. The article to be disinfected must be exposed to the gas in an appropriately designed sterilizing chamber in which the temperature can be raised and from which the air must be evacuated. Polyethylene and paper are suitable for packaging. A prolonged period of exposure is required, in one large model—at least 8 hours. Bacteria (including tubercle bacilli and staphylococci) and spores are killed; the hepatitis viruses are destroyed. Ethylene oxide properly used has an irreversible chemical reaction with all organisms including their spores, a reaction speeded by heat. Ethylene oxide sterilization is safe and effective provided adequate aeration time is allowed for the gas to elude from items so sterilized. This is important since the gas and its by-products are highly irritating to skin and mucous membranes (including those of the eye). This form of *cold* sterilization is widely used for delicate surgical instruments with optical lenses and for the tubing and heat-susceptible plastic parts of the heart-lung machine. Many prepackaged commercial items such as rubber goods and plastic tubes are so sterilized. Because of the penetration of the gas, this form of sterilization is effective for blankets, pillows, mattresses, and bulky objects.

Lime is one of the most common and, when properly used, one of the most effective germicidal agents. Limestone (calcium carbonate), which occurs plentifully in nature, is converted by heating into *quicklime* (calcium oxide), with release of carbon dioxide. When quicklime is treated with one-half its weight of water, *slaked lime* (calcium hydroxide) is formed. Slaked lime, a powerful disinfectant, is used as *milk of lime*, prepared by mixing 1 part of freshly slaked lime with 4 parts of water. Milk of lime is especially useful as a disinfectant for feces. A more dilute suspension of slaked lime is known as *whitewash*. Lime preparations must not be exposed to the air because they combine with the carbon dioxide of the air to form calcium carbonate without any antiseptic action.

Chlorinated lime (chloride of lime, bleaching powder, calcium hypochlorite), made by passing chlorine through freshly slaked lime, is one of the most important of the chlorine-liberating compounds used as disinfectants. An unstable compound, its antiseptic action results from its release of hypochlorous acid toxic to the bacterial cell. A 0.5% to 1% solution kills most bacteria in 1 to 5 minutes. A 1:100,000 solution destroys typhoid bacilli in 24 hours. Chlorinated lime bleaches and destroys fabrics and decomposes when exposed to the air. As a means of disinfecting excreta, chlorinated lime is probably without a rival.

Ferrous sulfate (copperas) is used in the form of the impure commercial salt. Besides being a good disinfectant, it is a good deodorant because it combines with both ammonia and hydrogen sulfide. It is an ideal disinfectant for use in damp musty places and around houses.

Chlorophyll in the form of its water-soluble components has been used in the local treatment of wounds. It cleans the wound, stops pus formation, destroys odors, and promotes healing.

THERAPEUTIC CHEMICALS FOR MICROBIAL DISEASES
Chemotherapeutic agents

Although it is possible to treat many local infections with chemicals, finding a chemical that, taken orally or parenterally, destroys microorganisms without injuring body cells is not easy. Examples of such chemicals are quinine, against the parasites of malaria, and emetine (the active principle of ipecac), against *Entamoeba histolytica*, the organism causing amebic dysentery. These are known as *chemotherapeutic agents*. The treatment of disease with chemotherapeutic agents is *chemotherapy*. The *chemotherapeutic index* compares the toxicity of a drug for the body with its toxicity for a disease-producing agent. It is obtained by dividing the maximal tolerated dose per kilogram of body weight by the minimal curative dose per kilogram of body weight. Within recent years there have been phenomenal developments in the field of chemotherapy, beginning with the discovery of the sulfonamide compounds. A brief discussion of these and certain specifically acting chemotherapeutic drugs follows.

Sulfonamide compounds contain the group $SO_2N<$. Many of them are derived from the compound known as sulfanilamide, which was among the first of the sulfonamides to be developed as a chemotherapeutic agent. Sulfanilamide was derived from a dye, known by the trade name of Prontosil, found to be effective in treating streptococcal infections. Prontosil was used for only a short time, and sulfanilamide itself was soon replaced by less toxic drugs.

The sulfonamide compounds are bacteriostatic in their mode of action; they interfere with certain enzymes in the affected bacterial cells. The sulfonamide compounds are not self-sterilizing, and their action is inhibited by pus. Sulfonamides are effective against gram-positive bacteria, some gram-negative diplococci and bacilli, chlamydiae, and *Actinomyces*. Generally sulfonamides have been replaced by antibiotics that are not so toxic and that are more certain and rapid in their action. Where an antibiotic and a sulfonamide are of equal value, it is customary to give the antibiotic. Sulfonamides have been used for treatment of epidemic meningitis because they pass into the cerebrospinal fluid more easily than do antibiotics. The highly soluble sulfonamides are important in the treatment of some infections of the urinary tract as well. Certain sulfonamides for intestinal use suppress the growth of the intestinal flora, an important measure in the preparation of the patient for surgery of the large bowel.

A patient receiving sulfonamide therapy should be continuously monitored for early toxic symptoms. If such appear, the drug must be discontinued.

Today there are many sulfonamide compounds with varying actions on bacteria and varying toxic effects on the body. The ones most readily available are sulfadiazine, sulfamerazine, sulfamethazine, sulfacetamide (Sulamyd), sulfadimethoxine (Madribon), sulfisoxazole (Gantrisin), sulfamethoxypyridazine (Kynex), sulfachloropyridazine (Sonilyn), sulfisomidine (Elkosin), sulfamethizole (Thiosulfil), sulfacytine (Renoquid), and sulfaethidole, all for systemic use. Other compounds include sulfaguanidine, succinylsulfathiazole (Sulfasuxidine), and phthalylsulfathiazole (Sulfathalidine) for intestinal use.

Co-trimoxazole (Septra), an important urinary tract antiseptic, is an oral synthetic antibacterial drug combining sulfamethoxazole with trimethoprim, a diaminopyrimidine. The combination results in a true synergistic action against common urinary tract pathogens with the exception of *Pseudomonas* by blocking two consecutive steps in the biosynthesis of nucleic acids and certain proteins essential to many bacteria.

Isoniazid (isonicotinylhydrazine) (INH) is a remarkably potent bacteriostatic agent against *Mycobacterium tuberculosis* but has little effect on most other microbes. Its mode of action is not known at the present time. Isoniazid is an effective drug not only in treating tuberculosis but also in preventing tuberculous infection from becoming active disease. The American Thoracic Society, American Lung Association, and the Center for Disease Control recommend ideally that all tuberculin-positive individuals receive a year of isoniazid prophylaxis (p. 444). Generally there have been few untoward reactions, but recently instances of drug toxicity have appeared. Some 10% of individuals receiving isoniazid prophylactically may develop evidence of liver dysfunction, rarely severe hepatitis. Many authorities feel this is no reason

288

to discontinue the regimen but rather an indication to screen the recipients initially and, once on the drug, to check them at monthly intervals for any sign of liver disorder.

Para-aminosalicylic acid (PAS) is another chemotherapeutic agent effective against the tubercle bacillus; it is often used along with isoniazid and streptomycin in the treatment of tuberculosis.

Ethambutol (2,2'-[ethylenediimino]-di-1-butanol dihydrochloride) is often used for the treatment of tuberculosis together with INH. It is valuable in the treatment of disease caused by tubercle bacilli resistant to other tuberculostatic drugs and works against certain atypical mycobacteria.

Nitrofurans are a group of synthetic antimicrobial drugs with activity against gram-positive and gram-negative bacteria, fungi, and protozoa but little toxicity for tissues. They interfere with basic enzyme systems within the microbial cell. An important one is *nitrofurazone* (Furacin), applied topically to prevent infections of wounds, burns, ulcers, and skin grafts. *Nitrofurantoin* (Furadantin) is used in bacterial infections of the urinary tract. *Furazolidone* (Furoxone) is effective against *Giardia lamblia* in intestinal infections; in a mixture with *nifuroxime* (Micofur) it is used in the treatment of vaginitis caused by bacteria, *Trichomonas vaginalis*, and *Candida albicans*.

Antibiotics

Experiments and observations from the last four decades have proved that numerous organisms produce substances with the power to inhibit the multiplication of other organisms or even to kill them. Such substances are known as *antibiotics*.* In 1941 the term *antibiotic* was defined by Waksman as follows: "An antibiotic is a chemical substance produced by microorganisms which has the capacity to inhibit the growth of bacteria and even destroy bacteria and other microorganisms in dilute solution."†

Antibiotics are biosynthesized by bacteria, fungi, and plants including certain flowering ones. Some organisms produce more than one, and certain antibiotics are produced by more than one organism. Table 16-2 gives such sources. Although thousands of species of microorganisms elaborate antibiotics and hundreds of them are known, only a few are practically useful in the treatment of disease. We consider them collectively, and yet their physical, chemical, and pharmacologic properties are divergent.

---

*This information is not so new as it might seem, because Pasteur in 1877 demonstrated that the growth of anthrax bacilli was retarded by the presence of common soil organisms, and several antibiotic-like substances (for example, pyocyanase) were prepared before 1900. The term *antibiosis* was introduced in 1889 by Vuillemin, who said, "No one considers the lion which leaps on its prey to be a parasite nor the snake which injects venom into the wounds of its victim before eating it. Here there is nothing equivocal; one creature destroys the life of another to preserve its own; the first is completely active, the second completely passive. The one is in absolute opposition to the life of the other. The conception is so simple that no one has ever thought of giving it a name. This condition, instead of being examined in isolation, can appear as a factor in more complex phenomena. In order to simplify words we will call it antibiosis . . . ." (Cited by Florey, H.: Antibiotics, Fifty-second Robert Boyle Lecture presented to the Oxford University Scientific Club, Springfield, Ill., 1951, Charles C Thomas, Publisher.)

†From Waksman, S. A.: Bull. N. Y. Acad. Med. **42:**623, 1966.

TABLE 16-2. SOURCES OF ANTIBIOTICS IN MICROBES

Microbe	Antibiotic
Bacteria	
Gram-positive rods	
Bacillus subtilis group *licheniformis*	Bacitracin
Bacillus polymyxa var. *colistinus*	Colistin
Bacillus polymyxa	Polymyxin
Bacillus brevis	Tyrothricin
Streptomycetes	
Streptomyces parvullus	Actinomycin D (Dactinomycin) (antineoplastic)
Streptomyces nodosus	Amphotericin B (Fungizone) (antifungal)
Streptomyces tenebrarius	Apramycin (Amyblan)
Streptomyces fragilis	Azaserine (antineoplastic)
Streptomyces verticillus	Bleomycin (antineoplastic)
Streptomyces griseus	Candicidin (antifungal)
Streptomyces capreolus	Capreomycin (Capastat)
Streptomyces venezuelae	Chloramphenicol (Chloromycetin)
Streptomyces aureofaciens	Chlortetracycline (Aureomycin)
Streptomyces bellus var. *cirolerosis,* var. *nova*	Cirolemycin (antineoplastic)
Streptomyces noursei	Cycloheximide (Actidione)
Streptomyces peucetius	Daunomycin (antineoplastic)
Streptomyces aureofaciens (mutant)	Demeclocycline (Declomycin)
Streptomyces peucetius var. *caesius*	Doxorubicin (Adriamycin) (antineoplastic)
Streptomyces erythraeus	Erythromycin (Ilotycin)
Streptomyces fradiae	Fosfomycin (antibiotic)
Streptomyces fradiae	Fradicin (antifungal)
Streptomyces tanashiensis	Kalafungin (antibiotic)
Streptomyces kanamyceticus	Kanamycin (Kantrex)
Streptomyces lincolnensis	Lincomycin (Lincocin)
Streptomyces melanogenes	Melanomycin
Streptomyces plicatus	Mithramycin (antineoplastic)
Streptomyces caespitosus	Mitomycin
Streptomyces carzinostaticus (variant)	Neocarzinostatin (antineoplastic)
Streptomyces fradiae	Neomycin
Streptomyces noursei	Nystatin (Mycostatin)
Streptomyces antibioticus	Oleandomycin
Streptomyces rimosus	Oxytetracycline (Terramycin)
Streptomyces rimosus	Paromomycin (Humatin)
Streptomyces lincolnensis	Ranimycin (antibiotic)
Streptomyces mediterranei	Rifampin (Rimactane)
Streptomyces spectabilis (variant)	Spectinomycin (Trobicin)
Streptomyces griseus	Streptomycin
Streptomyces lavendulae	Streptothricin
Streptomyces achromogenes	Streptozotocin
Streptomyces aureofaciens (mutant)	Tetracycline
Streptomyces tenebrarius	Tobramycin
Streptomyces orientalis	Vancomycin
Streptomyces antibioticus	Vidarabine (adenine arabinoside, ara-A) (antiviral)

TABLE 16-2. SOURCES OF ANTIBIOTICS IN MICROBES—cont'd

Microbe	Antibiotic
Bacteria—cont'd	
Streptomyces puniceus	Viomycin
Streptomyces bikiniensis (variant)	Zorbamycin
Actinomycetes	
Micromonospora purpurea	Gentamicin
Nocardia lurida	Ristocetin (Spontin)
Micromonospora inyoensis	Sisomicin
Fungi	
Arachniotus aureus	Aranotin (antiviral)
Cephalosporium acremonium	Cephaloridine (Loridine)
Cephalosporium	Cephalothin (Keflin)
Aspergillus fumigatus	Fumagillin
Penicillium griseofulvum dierckx and *Penicillium janczewski*	Griseofulvin (Fulvicin)
Aspergillus giganteus	Nifungin (antifungal)
Penicillium notatum and *chrysogenum*	Penicillin
Penicillium stoloniferum	Statolon (antiviral in animals)

Spectrum of activity. Some antibiotics affect mainly gram-positive bacteria, with little or no effect on gram-negative bacteria. Others seem to affect only certain species of bacteria, regardless of whether they are gram negative or gram positive. A few are active against rickettsias; a few attack fungi.

The susceptibility (or resistance) of bacteria to different antibiotics may be determined by suitable laboratory procedures (p. 119) designated susceptibility tests (Fig. 7-13), and there are also satisfactory ways to determine the amount of certain antibiotics in the blood.

The *spectrum* of a compound is its range of antimicrobial activity. A *broad-spectrum* antibiotic indicates a wide variety of microorganisms affected, including usually both gram-positive and gram-negative bacteria. A *narrow-spectrum* antibiotic limits only a few. Table 16-3 lists some antibiotics in either category.

Mode of action of antimicrobial drugs. Several mechanisms explain in part the harmful or lethal effects that antimicrobial drugs have for microbes. Most of the time the mechanisms of injury are subtle ones, usually reflecting an inhibitory effect on some aspect of microbial physiology. The first mechanism studied was that of the sulfonamides in competition for a place in the metabolic activities of the cell (*metabolic antagonism*). A drug utilized by mistake because of a close chemical resemblance to a vital cytoplasmic substance can halt growth and function.

The physiologic integrity of the microbial cell depends on key macromolecules and the processes in which they are involved. Certain ones vulnerable to antimicrobials include (1) bacterial cell wall synthesis, (2) nucleic acid synthesis, (3) protein synthesis, and (4) cell membrane function. (Table 16-4 lists antimicrobial compounds as to category of action.) The rigid cell wall jackets the bacterial cell cytoplasm

TABLE 16-3. ANTIBIOTICS—RANGE OF ACTIVITY

Narrow-spectrum antibiotics	Broad-spectrum antibiotics
Penicillin	Chloramphenicol
Streptomycin	Chlortetracycline
Dihydrostreptomycin	Demeclocycline
Erythromycin	Oxytetracycline
Lincomycin	Tetracycline and derivatives
Polymyxin B	Kanamycin
Colistin	Ampicillin
Vancomycin	Cephalothin
Nystatin	Rifampin
Spectinomycin	Gentamicin
	Tobramycin
	Paromomycin

TABLE 16-4. CELLULAR DISTURBANCES RELATED TO ANTIMICROBIAL AGENTS

		Inhibitory effect on			Metabolic antagonism
Cell wall synthesis	Nucleic acid synthesis	Protein synthesis		Cell membrane function	Competitive interference in cell metabolism
		Ribosome function	Other		
Bacitracin	Actinomycin D	Chloramphenicol	Chlortetracycline	Amphotericin B	Isoniazid
Cephalosporins	Griseofulvin	Dihydrostreptomycin	Erythromycin	Benzalkonium	Nitrofurans
Penicillins	Idoxuridine	Gentamicin	Lincomycin	chloride	Para-aminosalicylic
Vancomycin	Rifampin	Kanamycin	Methacycline	Colistin	acid
		Neomycin	Oxytetracycline	Nystatin	Sulfonamides
		Paromomycin	Tetracycline	Polymyxins	
		Streptomycin			

TABLE 16-5. SOME EXAMPLES OF PATHOLOGIC COMPLICATIONS TO ANTIBIOTIC THERAPY

Antibiotic	Therapeutic background	Physiologic derangement	Pathologic lesion
Neomycin	Prolonged administration	Changes in lining of small intestine (jejunum) with interference in absorption of glucose, iron, vitamin B_{12}	Malabsorption syndrome (deficiency states, weight loss, skin pigmentation, abdominal distention, and the like)
Tetracycline	Administration during pregnancy or to infants	Deposition of antibiotic in tooth enamel	Dental defects — pigmentation
Tetracycline	Large doses given intravenously	Toxic effect on liver — injury from hepatotoxin	Liver disease

with its characteristically high osmotic pressure. If the cell wall is faulty in construction and weakened because of the action of penicillin, the cytoplasmic membrane cannot withstand the internal pressure. In a medium of ordinary toxicity the bacterial cell explodes. Antibiotics can interfere with the synthesis of the nucleic acids within the cell. Some complex with DNA and block the formation of the messenger RNA. Other antibiotics inhibit the ribosomes (RNA) directly, thereby materially affecting the elaboration of proteins within the cell. As a result, "nonsense" proteins appear, indicating a chemical mixup. Since the biochemical activities relating to nucleic acid and protein synthesis are complicated, the possibilities for injury to the cell are many. One of the antifungal antibiotics combines with sterols in the cell membrane of the susceptible fungus, destroying membrane function. Cell contents leak out and the cell dies. Some antibiotics alter the permeability of the plasma membrane. Benzalkonium chloride disrupts a bacterial cell membrane by its detergent action.

An important factor in any consideration of the mode of action of antimicrobial compounds is that of *selective toxicity:* The agent, to be useful, must kill the microbial cells without damaging the cells of the animal host. Agents such as disinfectants that coagulate proteins in *all* cells are *nonselective.*

Side effects. There are four serious complications of antibiotic therapy: (1) hypersensitivity or allergy, (2) toxicity, (3) effects of the replacement flora, and (4) induction of bacterial resistance.

The turnover of drugs by the human body involves a series of complex, enzymatically mediated reactions. It is increasingly apparent that there can be vast individual differences. Signs of hypersensitivity to the antibiotics are variable in different persons but include skin rashes, joint pains, lymph node enlargement, lowered white blood cell count, and sometimes hemorrhages. Examples of toxicity* to antibiotic therapy include the aplastic anemia induced by chloramphenicol and the deafness sometimes produced by streptomycin. Table 16-5 gives more examples indicating the background and the lesion.

REPLACEMENT FLORA. The appearance of a replacement microbial flora is troublesome at times. Not being affected by the drug themselves, such organisms tend to increase during the period of therapy, being no longer held back by the organisms that succumb to the antibiotic. They can multiply in an uncontrolled way and, if potentially pathogenic, may start a disease process difficult to treat. The replacement flora may be quite resistant to standard antibiotics. Such secondary infections are often caused by *Proteus* or *Pseudomonas* species, and infections with *Candida albicans* easily follow prolonged and extensive antibiotic therapy, especially with broad-spectrum preparations. *Superinfection* as a complication is more likely with the broad-spectrum antibiotics than with the narrow-spectrum ones.

DRUG RESISTANCE. Bacteria vary in their capability to develop resistance to a given drug. Resistance means that the drug is no longer effective in suppressing

---

*Some antibiotics in doses ordinarily given are associated with psychiatric disturbances. Depression and hallucinations have been related to certain long-acting sulfonamides. Psychosis, especially in the elderly, can be precipitated with the antiviral amantadine. A buoyant state of well-being has been noticed in isoniazid-treated patients with tuberculosis.

their growth and multiplication. Remember that resistance is always a possibility whenever a bacterio*static* drug is given, since such a drug merely limits the activities of given microbes—body mechanisms dispose of them! Many antibiotics are only bacteriostatic within the human body. If only *bactericidal* (lethal) antibiotics were given, resistance would cause less concern.

The benefits to livestock of a few grams of antibiotic in a ton of feed grew out of the discovery of chlortetracycline (Aureomycin) in 1948. The fermentation process producing chlortetracycline yielded a needed food supplement, vitamin B_{12}, so that the crude residues given to farm animals for their content of vitamin B_{12} contained a small amount of chlortetracycline that was found to enhance noticeably the growth and well-being of the animals. By 1951 the practice of adding antibiotics to livestock feed began to spread rapidly over the world. Today some 16 antibiotics are used as additives. Farmers also rely on antibiotics to treat mastitis in cows giving milk for market. In industrial food processing, antibiotics help control spoilage. All these factors probably contribute significantly to the problem of bacterial resistance.

The synthesis of the adaptive enzyme *penicillinase*, inactivating penicillin G, by disease-producing staphylococci and coliform bacteria is the best known mechanism of drug resistance.

One of the most interesting mechanisms yet described concerns the *R*, or *resistance*, factors (also called resistance transfer factors, RTF). First detected in Japan in 1959, they were later identified in the United States as circular molecules of double-stranded DNA separate from the bacterial chromosome that could be passed by bacterial conjugation from one microorganism to another within a species or even from one member of a genus to another (Fig. 4-9). The particles replicate independently of the main genetic material, and the resistance is passed to successive generations. R factors are an example of bacterial plasmids, extrachromosomal genetic elements carrying information not normally necessary for the survival of the bacterial host. This genetic material, free-floating in the cytoplasm, endows its possessor with an all-inclusive type of resistance to diverse antibacterial agents—sulfonamides, streptomycin, tetracyclines, chloramphenicol, penicillin, neomycin, even experimental antibiotics—with obviously different modes of antibacterial activity. About two dozen gram-negative bacteria transfer the R factors. From Bergey's family Enterobacteriaceae they include members of the genera *Enterobacter, Escherichia, Klebsiella, Salmonella, Shigella, Serratia, Proteus, Citrobacter;* from Bergey's family Pseudomonadaceae, *Pseudomonas;* and from Bergey's family Vibrionaceae, *Vibrio.*

The roster. Short discussions of several well-known antibiotics follow.

Tyrothricin, one of the first antibiotics to be studied carefully, was isolated from an aerobic motile nonpathogenic bacillus *(Bacillus brevis)* growing in the soil. From tyrothricin were obtained two polypeptides—*gramicidin* (named in honor of the bacteriologist Gram) and *tyrocidine,* both highly antagonistic to gram-positive cocci and very toxic. Tyrothricin is without effect given orally and very dangerous given parenterally. It therefore must be used locally. It prevents bacterial growth and lyses bacteria.

Penicillin, an organic acid and a beta-lactam antibiotic discovered accidentally,

FIG. 16-1. *Penicillium chrysogenum,* giant colony. From this mutant form of the green mold almost all the world's supply of penicillin is obtained. (Courtesy Chas. Pfizer & Co., Inc., New York, N.Y.; from Arnett, R. H., Jr., and Bazinet, G. F.: Plant biology: a concise introduction, ed. 4, St. Louis, 1977, The C. V. Mosby Co.)

is isolated from certain molds, the most important of which are *Penicillium notatum* and *Penicillium chrysogenum* (Fig. 16-1). The first antibiotic to come into general use, it is still in many ways the best.* It may be administered intravenously, subcutaneously, into body cavities, orally, and locally, and in contrast to many antibiotics that are bacteriostatic, it is also bactericidal. It is one of the most important anti-infectives because it is bactericidal! Penicillin acts by interfering with synthesis of the bacterial cell wall and therefore affects actively growing bacteria. In the individual not allergic to penicillin its toxicity for human tissue is almost nonexistent. It is an effective agent, with some exceptions, against staphylococci, streptococci (group A), pneumococci, and meningococci, and its use revolutionized the treatment of syphilis. It does not have an effect on most gram-negative bacilli with the usual doses given. Most of the microorganisms originally sensitive to penicillin still are, with the notable exceptions coming from the penicillinase-producing staphylococci and gonococci. The enzyme penicillinase, accounting for penicillin resistance in

---

*Penicillin is an inexpensive drug. The chances are that a course of penicillin will cost only a fraction of that with another antibiotic.

both gram-positive and gram-negative microorganisms, inactivates penicillin by opening the beta-lactam ring. Its formation has been directed by the bacterial plasmid.

Penicillin has been isolated in six closely related forms called F, G, X, K, O, and V. Penicillin G is the most satisfactory type to manufacture and use. Therefore about 90% of commercial penicillin is of this type. The sodium, potassium, and procaine salts are the common ones. Penicillin G is the agent of choice in the nonallergic individual for the treatment of infections caused by group A hemolytic streptococci.

The strength of penicillin is designated in *units*. A unit (International) corresponds to the activity of 0.6 μg (0.000006 g) of pure sodium penicillin G.

The generic name *penicillin* includes all antibacterials isolated from the genus *Penicillium*. The natural penicillins are extracted from cultures.

Semisynthetic penicillins are the sequel to a basic observation on the chemical structure of penicillin made in Great Britain in 1959. When the chemist learns how the molecule can be modified, he can change it to give the new product more desirable properties than those of the natural substance. Consequently, more than 500 new semisynthetic penicillins have been made. Phenethicillin (Syncillin), an oral preparation, was the first synthetic one to be dispensed by prescription.

Penicillin resistance is most important from a pathogenic staphylococcus. In a general way the newer penicillins are indicated for pathogenic staphylococci resistant to penicillin G; otherwise the potency of these new drugs is less than that of the original penicillin. Compounds designed to counteract staphylococcal penicillinase are methicillin (Staphcillin), oxacillin (Prostaphlin), nafcillin (Unipen), cloxacillin (Tegopen), and dicloxacillin (Pathocil).

Ampicillin (Polycillin) is a broad-spectrum semisynthetic penicillin with activity against a number of gram-positive and gram-negative bacteria, including *Salmonella*, *Shigella*, *Escherichia coli*, and *Proteus*. If disease of the biliary tract is not present, ampicillin can be used to treat the typhoid carrier, and it has value in the treatment of intestinal amebiasis. It is ineffective in the presence of penicillinase. A close chemical relative, *amoxicillin* is also a broad-spectrum semisynthetic penicillin.

Carbenicillin (Pyopen) is a British semisynthetic penicillin with reported activity against *Pseudomonas*. *Disodium carbenicillin* (Geopen) is active against the gramnegative *Pseudomonas* and *Proteus*.

Cross allergenicity exists among the penicillins; that is, a person sensitive to penicillin G will be found sensitive to one of the semisynthetic penicillins. It is estimated that 5% to 6% of the population is allergic to penicillin.

Cephalothin, cephaloridine, cephaloglycin, and *cephalexin* are cephalosporins, a group of antibiotics of an entirely new class, though closely related chemically to penicillin (also beta-lactam antibiotics). They are semisynthetic derivatives from the fermentation products of the fungus *Cephalosporium* (Fig. 16-2), whose natural habitat is the seacoast of Sardinia in Italy. Cephalothin (Keflin) is a broad-spectrum antibiotic bactericidal against gram-positive and gram-negative bacteria. It must be given parenterally; it is unaffected by penicillinase and of low toxicity. Cephaloridine (Loridine) is similar to cephalothin. Cephaloglycin (Kafocin) and cephalexin (Keflex)

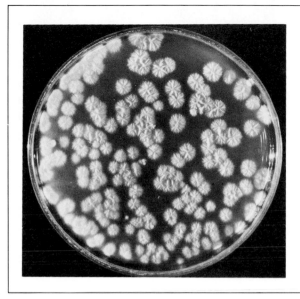

FIG. 16-2. *Cephalosporium,* colonies on agar plate. Important antibiotics are derived from this fungus. (From Profile of an antibiotic, Indianapolis, Ind., 1966, Eli Lilly & Co.)

can be given orally. An advantage to the cephalosporins is the general lack of cross allergenicity with penicillin.

Streptomycin, an organic base, was isolated from *Streptomyces griseus,* an organism from the soil, and is one of a large group of chemically related antibiotics, the aminoglycosides, whose discovery came as a result of a deliberate search. Its action in a general way is bactericidal, like that of penicillin. It is active against gram-negative and acid-fast bacilli. Although seldom given intravenously, it may be administered by many routes. There are three disadvantages to streptomycin: (1) toxicity—its use may induce injury to the auditory portion of the eighth cranial nerve, resulting in deafness; (2) drug resistance—bacteria sensitive to it can quickly become resistant during the course of therapy; and (3) poor oral absorption—if the drug is given orally, it is not significantly absorbed. A unit of streptomycin corresponds to the activity of 1 μg of pure crystalline streptomycin base.

Dihydrostreptomycin, a derivative of streptomycin, is much like it. Its use also may be followed by neurotoxic symptoms, but as a rule, it is less toxic than streptomycin.

Neomycin and *viomycin,* also aminoglycosides, are similar to streptomycin. Neomycin can damage the kidney. Viomycin is sometimes used in the treatment of tuberculosis, although it is also nephrotoxic.

Kanamycin (Kantrex), a broad-spectrum aminoglycoside antibiotic, must be given parenterally. Its activity is directed against many aerobic gram-positive and gram-negative bacteria, including most strains of staphylococci and many strains of *Proteus.* The major toxic manifestation of this drug is partial or complete deafness resulting from its action on the auditory portion of the eighth cranial nerve. It is also nephrotoxic. Similar to kanamycin is *amikacin,* a semisynthetic aminoglycoside produced by acylation of kanamycin A.

Gentamicin (Garamycin), an aminoglycoside, is a broad-spectrum antibiotic, chemically related to neomycin and kanamycin, which can be given parenterally. It is becoming a drug of choice for serious systemic infections caused by gram-negative bacteria. Like the other aminoglycoside antibiotics, it is nephrotoxic and ototoxic. Loss of hair including that of the eyebrows has been reported with its use.

Tobramycin (Nebcin) is a basic water-soluble aminoglycoside given parenterally. It has a wide range of antibacterial activity and is bactericidal against staphylococci and gram-negative bacteria, including *Pseudomonas*. It acts synergistically with cephalosporins and some penicillins and is useful in infections caused by organisms resistant to kanamycin and gentamicin. Ototoxicity, both vestibular and auditory, and nephrotoxicity account for complications.

Spectinomycin (Trobicin), another member of the streptomycin-kanamycin family, is designed for the one-dose intramuscular treatment of gonorrhea. Although an aminoglycoside antibiotic, it is not thought to be nephrotoxic or ototoxic.

Erythromycin (Ilotycin), a macrolide, is a valuable antibiotic effective when given orally against gram-positive bacteria and some gram-negative bacteria. It is especially recommended for the treatment of infections caused by organisms that have become resistant to penicillin. Its use is seldom followed by toxicity.

Lincomycin (Lincocin) is a medium-spectrum antibiotic that can be administered orally or by injection. It inhibits the growth of gram-positive cocci, including staphylococci that produce penicillinase. Although its action is similar to that of erythromycin, erythromycin is antagonistic to it (as are the cyclamate artificial sweeteners formerly used in various low-calorie preparations). Lincomycin is a parent drug to *clindamycin* (Cleocin), which may be the first antibiotic that is also an antimalarial drug. Clindamycin, an oral agent, is particularly useful in treatment of anaerobic infections, especially those of *Bacteroides*. A serious intestinal disorder, pseudomembranous colitis, may complicate lincomycin and clindamycin therapy.

Polymyxin is a general name for a number of polypeptide antibiotics derived from different strains of *Bacillus polymyxa*, known as polymyxin A, B, C, and D. Polymyxin is toxic, especially to the kidney* and brain. It is used locally and as an intestinal antiseptic. It is important in topical preparations, but its systemic use is hazardous. Polymyxin B is usually effective against infections with *Pseudomonas aeruginosa;* however, *Proteus* resists it.

Colistin (Coly-Mycin), a polypeptide antibiotic isolated in Japan, is chemically similar to polymyxin. It is effective against gram-negative bacteria in general and, like polymyxin B, against *Pseudomonas aeruginosa* in particular. Less toxic than polymyxin B and with fewer side reactions in the central nervous system and kidney, it is of special value in urinary tract infections. It is given intramuscularly.

Bacitracin, a polypeptide antibiotic, inhibits the growth of many gram-positive organisms, including penicillin-resistant staphylococci and that of the gram-negative gonococci and meningococci, but it can only be used locally.

Capreomycin, a polypeptide antibiotic, is related to viomycin. It is mentioned

---

*The polypeptide antibiotics—polymyxin, colistin, and bacitracin—are all nephrotoxic.

because of its activity against atypical mycobacteria and in combination with other drugs against tubercle bacilli resistant to standard antituberculosis drugs.

Vancomycin (Vancocin) must be given intravenously. It is a bactericidal drug promoted for the therapy of infections caused by antibiotic-resistant gram-positive bacteria, especially staphylococci.

Rifampin (Rifadin, Rimactane), one of the rifamycins, is a semisynthetic broad-spectrum antibiotic taken orally, with bactericidal activity notable against the tubercle bacillus. It is effective in the treatment of carriers of the meningococcus and has been used in the treatment of leprosy. It has an effect against the causative chlamydiae of trachoma and inhibits replication of the poxvirus by blocking synthesis of viral protein. Adverse reactions are infrequent. The urine, feces, saliva, sputum, tears, and sweat may take on an orange-red color because of therapy with the drug.

Tetracycline is one of a group of broad-spectrum antibiotics and represents the chemical moiety common to the group. It is produced chemically from oxytetracycline and has been isolated from a streptomycete found in Texas. As a group, tetracyclines sometimes irritate the alimentary tract. Tetracyclines sometimes pigment the teeth. If tetracycline gains access to the circulation of an unborn baby because the drug is given to the mother, it is deposited in areas of developing bones and teeth. The damage results in staining of tooth enamel (Fig. 16-3). Tetracyclines are given orally to treat intestinal amebiasis.

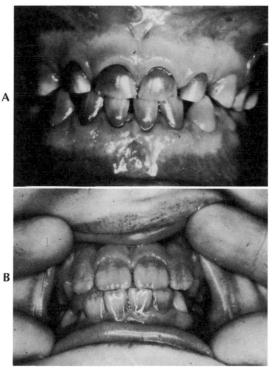

FIG. 16-3. Discoloration of teeth caused by tetracycline. **A,** Deciduous teeth (brownish gray). **B,** Permanent teeth (yellowish gray). (From Gorlin, R. J., and Goldman, H. M.: Thoma's oral pathology, ed. 6, St. Louis, 1970, The C. V. Mosby Co.)

Chlortetracycline (Aureomycin), a broad-spectrum antibiotic, is effective against a wider range of organisms than are such antibiotics as penicillin and streptomycin. It affects gram-positive bacteria, certain gram-negative bacteria, and rickettsias. It may be given orally or intravenously.

Oxytetracycline (Terramycin), like chlortetracycline, has a wide range of antibacterial activity. It is given orally and is comparatively nontoxic.

*Demeclocycline** (Declomycin), another member of the group of tetracyclines, was first prepared chemically but was later found in a mutant strain of *Streptomyces aureofaciens*. In general, its antibacterial activity is similar to that of the other tetracyclines, perhaps somewhat prolonged. Cases of photosensitization (sensitivity to light) have followed its use. Other derivatives of tetracycline and comparable to it pharmacologically are *methacycline* (Rondomycin), *minocycline* (Minocin), *rolitetracycline* (Syntetrin), and *doxycycline* (Vibramycin).

Chloramphenicol (Chloromycetin), the first of the broad-spectrum antibiotics, was isolated from *Streptomyces venezuelae* (found in the soil of Venezuela) but has now been crystallized. It is effective against certain bacteria and rickettsias and is the best antibiotic for the treatment of typhoid fever. It is usually given orally, but it may be given intramuscularly or intravenously. Two peculiar toxic disturbances occasionally complicate chloramphenicol therapy. One is hematologic; an aplastic anemia develops, with cessation of blood cell formation. The other is *gray syndrome*, cardiovascular collapse seen in babies 2 months of age and younger. Since chloramphenicol is toxic, it should not be used in those cases where tetracyclines are effective.

Nystatin (Mycostatin), an antifungous, polyene antibiotic, is used with favorable results in infections caused by *Candida albicans.* It is often given with a broad-spectrum antibiotic to suppress the growth of the resistant intestinal fungi (replacement flora) that might produce lesions in the bowel after prolonged antibiotic therapy.

Amphotericin B (Fungizone), a broad-spectrum antifungous drug, is a polyene macrolide that binds to sterols in eucaryotic cell membranes. Its avidity is for ergosterol, the primary sterol of the fungous cell membrane, rather than for cholesterol, that of the animal cell membrane. It is relatively specific for fungi but without effect on bacteria. It is the only polyene antibiotic sufficiently nontoxic to be useful in the treatment of systemic and deep-seated fungous infections, those involving internal organs and widely spread within the body. It should be given intravenously, and the initial infusion is regularly accompanied by a chill-fever reaction.

Fumagillin is produced during the growth of *Aspergillus fumigatus*. It is inactive against most bacteria, fungi, and viruses but is destructive to *Entamoeba histolytica* (the cause of amebic dysentery).

Griseofulvin (Fulvicin) is an oral antibiotic obtained from at least four species of *Penicillium.* A *fungistatic* drug, not a fungicidal one, it has a wide range of antifungous activity against practically all the dermatophytic fungi *(Microsporum, Epi-*

---

*Demethylchlortetracycline.

300

dermophyton, Trichophyton). Although systemically administered, it does not act on *Candida albicans,* on the deep systemic fungi, or on bacteria. The drug can be demonstrated to attack the advancing hyphal threads of invading fungi in skin, nails, and hair, causing these segments to shrivel and curl. Toxicity seems to be low; in anticipation of any effect on the blood-forming organs, periodic white blood cell counts are done.

Paromomycin (Humatin) is an amebicide and a broad-spectrum antibiotic. Because of its nephrotoxicity, it is given not by injection but orally, although it is poorly absorbed from the gastrointestinal tract. It is effective against *Entamoeba histolytica* and holds back the secondary infection of the bowel that complicates amebiasis. In addition to paromomycin, the antibiotics with amebicidal activity are neomycin, kanamycin, and the tetracyclines.

Antivirals

Although antiviral agents are still largely under investigation, there is a great deal of interest in them; many compounds are being studied, some released; and several major developments have occurred in the last few years. Currently three important categories have emerged: (1) thiosemicarbazone, (2) a group of pyrimidine nucleosides, and (3) amantadine.

Methisazone (N-methylisatin-β-thiosemicarbazone) has been shown to prevent smallpox in susceptible persons exposed in the epidemics studied in India.

Idoxuridine (5-iodo-2'-deoxyuridine, or IDU), an antimetabolite and pyrimidine analog that inhibits normal cellular DNA synthesis, has been found effective against eye infection caused by the herpes simplex virus. Idoxuridine represents an achievement—the results of treatment with it in herpetic keratitis have been spectacular.

Vidarabine, also known as adenine arabinoside or ara-A, is a purine nucleoside from *Streptomyces antibioticus.* It is a nonspecific inhibitor of viral replication, with a wide range of activity and almost no acute toxicity.

Cytarabine (Cytosar), also known as cytosine arabinoside or ara-C, is also a purine analog. It was studied as an antileukemic agent in the laboratory and found to be an antagonist of DNA synthesis, with potent antiviral properties in vitro. Both ara-C and idoxuridine are rapidly inactivated in vivo, and both can suppress bone marrow function.

Ribavirin (Virazole), a synthetic nucleoside, is active in tissue culture against a number of DNA and RNA viruses. Its action is to prevent replication, not to destroy.

Amantadine hydrochloride (Symmetrel) is a synthetic compound with a selective antiviral action. It has been used for the prevention of influenza due to A_2 strains of the virus only. It also has been used against rubella virus. Amantadine blocks the penetration of a virus into the host cell. It stimulates the central nervous system and can cause convulsions in large doses.

Interferon, a naturally occurring nonimmune protein, was the first substance to be associated with an antiviral action. An integral part of the host defense, it is a diffusible substance produced by a living cell infected with virus that blocks further viral infection of that cell (p. 503). It can also be induced by bacteria, bacterial products, rickettsias, and protozoa. Certain chemicals such as double-stranded RNA from a synthetic source can act as *interferon inducers*. Their use in enhancing resistance to viral infection is still experimental.

QUESTIONS FOR REVIEW

1. Define disinfection, germicide, bactericide, antiseptic, asepsis, fumigation, bacteriostasis, preservative, antibiotic, idophor, R transfer factors, penicillinase, interferon, detergent, cross allergenicity, enzyme detergent.
2. How do chemical disinfectants act?
3. Broadly classify antiseptics and disinfectants.
4. What is an antibiotic susceptibility test?
5. Name five sulfonamide compounds used in medicine.
6. What is meant by chemotherapy and chemotherapeutic agent?
7. Discuss ethylene oxide as a sterilizing agent.
8. Explain *cold* sterilization.
9. What are broad-spectrum antibiotics? Name five.
10. What agents are most effective in the disinfection of articles contaminated with *Mycobacterium tuberculosis?*
11. List chemotherapeutic drugs used in the treatment of pulmonary tuberculosis.
12. Briefly discuss harmful side effects of antibiotic therapy. What is meant by a replacement flora?
13. What is a narrow-spectrum antibiotic? Name three.
14. Briefly describe one antifungous agent and one amebicide.
15. Indicate the mode of action of antibiotics.
16. What is the current status of the antivirals?

REFERENCES. See at end of Chapter 17.

17 Practical technics

SURGICAL DISINFECTION AND STERILIZATION

Many different methods of sterilizing surgical instruments and supplies and of disinfecting wounds, the field of operation, and hands of persons taking part in surgical operations are used in different hospitals and related medical or health field settings. * An overview of the standard methods as evaluated in an experimental and testing laboratory is presented in Table 17-1. In the following paragraphs some of the ones commonly used are given.

The method of choice wherever applicable is sterilization by steam under pressure as usually carried out in a pressure steam sterilizer (preferably the modern high-prevacuum steam autoclave) (Figs. 15-1, 15-3, and 17-1). There are exceptional circumstances in which chemical disinfection, including sterilization with ethylene oxide gas, has to be substituted for autoclave sterilization. Certain items are damaged by the environment of the pressure steam sterilizer, such as surgical instruments with optical lenses, instruments or materials made of heat-susceptible plastics, articles made of rubber, certain delicately constructed surgical instruments, and nonboilable gut sutures.

Generally, the pressure steam sterilizer (gravity air removal type) for most routine purposes of sterilization is maintained at 121° to 123° C (250° to 254° F), 15 to 17 pounds' pressure, for 30 minutes (Table 15-1). For certain items of rubber subjected to steam sterilization the time is shorter, 15 to 20 minutes. (Note the shorter time with the higher temperatures in the high-prevacuum sterilizer.)

Gas sterilization. Ethylene oxide is widely applied today to the sterilization of

---

*An increasing number of commercially prepackaged, sterile, disposable items are in widespread use, the limits apparently set only by their cost. The advantages of disposables are obvious in most instances, and consequently, the use of disposables is remarkably changing patterns of sterilization in the health fields.

TABLE 17-1. OVERVIEW OF ANTIMICROBIAL METHODS*

Method	Temperature required	Minimum exposure time in minutes
Sterilization†		0' 2' 10' 12' 15' 20' 90' 120' 150' 180'
Saturated steam under pressure	285°F (140°C)	(Instantaneous)
	270°F (132°C)	
	250°F (121°C)	
Ethylene oxide gas (12% ETO, 88% Freon-12)	130°F (54°C)	
	105°F (40°C)	
Hot air	320°F (160°C)	or more
Chemical sporicide (solution)	Room temperature	or more
Disinfection		
Steam — free flowing or boiling water	212°F (100°C)	
Chemical germicide (solution)	Room temperature	
Sanitization		
Water, detergents, or chemicals aided by physical means for soil removal	Maximum 200°F (93°C)	

*Modified from chart copyrighted 1967, Research and Development Section, American Sterilizer Co., Erie, Pa.
†See pp. 265 and 275 for definitions.

TABLE 17-2. COMPARISON OF STEAM STERILIZATION WITH GAS STERILIZATION*

	Steam	Gas
Medical indications	Syringes, instruments, drapes, gowns, some plastics, rubber	Heat-sensitive plastics, rubber, bulky equipment (ventilators, telescopic instruments†)
Equipment	Automatic chamber (large and bulky)	Automatic chamber (large and bulky); also portable model
Penetration of material to be sterilized	Rapid	Very rapid
Time required	Minutes (15 minutes or less in some models)	Hours (2 to 12)
Efficiency	Excellent	100%
Technical snags	Removal of air	Prevacuum; correct humidification; temperature over 70°C (158°F)
Danger	None	Undiluted ethylene oxide explosive‡
Hazard	Sterilization inadequate if air not completely removed	Toxic residues if aeration inadequate
Contraindications	Many plastics; sharp instruments; rubber deteriorates	Previously irradiated plastics such as polyvinyl chloride

*Modified from Rendell-Baker, L., and Roberts, R. B.: Med. Surg. Rev. **5:**10 (fourth quarter), 1969.
†Modern-day cements used on optical lenses of such instruments have been sufficiently improved to permit sterilization with ethylene oxide gas.
‡The mixture of 12% ethylene oxide (ETO) and 88% Freon-12, which is generally used, is a safe one.

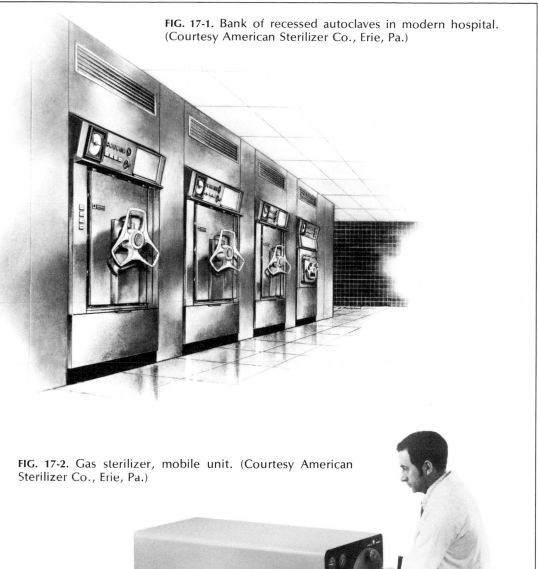

FIG. 17-1. Bank of recessed autoclaves in modern hospital. (Courtesy American Sterilizer Co., Erie, Pa.)

FIG. 17-2. Gas sterilizer, mobile unit. (Courtesy American Sterilizer Co., Erie, Pa.)

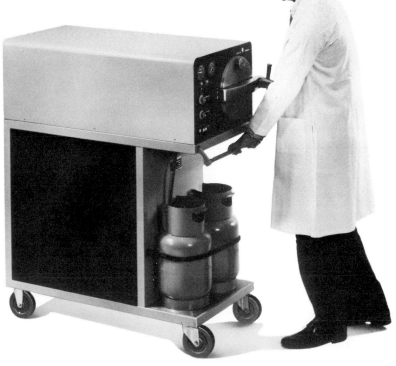

TABLE 17-3. TECHNICS OF STERILIZATION (OR DISINFECTION) OF COMMONLY USED ITEMS IN THE CLINICAL SETTING

Item to be sterilized	Autoclave (method of choice)	Chemical disinfection (choice indicated)*	Other technics applicable
Surgical instruments and supplies, mechanically clean			
Noncutting instruments	Yes	1. Amphyl, 2%, 20 min 2. Activated glutaraldehyde, 10 min to 10 hr 3. Ethylene oxide gas, 2 to 12 hr	Boil completely submerged 30 min
Sharp instruments	Yes, in certain instances	1. Amphyl, 2%, 20 min 2. Activated glutaraldehyde, 10 min to 10 hr 3. Ethylene oxide gas, 2 to 12 hr	1. Bake in hot-air sterilizer, 150°C, 3 hr 2. Boil 30 min (cutting part wrapped with cotton; instrument kept from tossing about; water boiling *before* instruments placed in it†)
Surgical needles	Yes, in packages	1. Amphyl, 2%, 20 min 2. Isopropyl alcohol, 70% to 90%, 20 min 3. Ethylene oxide gas, 2 to 12 hr Wash needles in sterile water	Hot air at 160°C, 2 hr
Hypodermic syringes and needles	Yes	Cold sterilization inadequate to remove viruses of hepatitis	1. Dry heat, 170°C, 2 hr 2. Boil 30 min
Hinged instruments	Yes, in certain instances	1. Isopropyl alcohol, 70% to 90%, 20 min 2. Activated glutaraldehyde (2% aqueous), 20 min to 10 hr 3. Ethylene oxide gas, 2 to 12 hr	—
Transfer forceps	Yes, for the container	Activated glutaraldehyde (change solution once a week)	—

*Sodium nitrite added to alcohols, formalin, formaldehyde-alcohol, quaternary ammonium, and iodophor solutions prevents rusting. Sodium bicarbonate added to phenolic solutions prevents corrosion. Antirust tables are available commercially.

†Boiling dulls knives and other sharp instruments because iron tends to enter solution in the ferrous state when heated. The site of corrosion—the part that is electropositive—is the sharp edge of the instrument. A "cathodic" sterilizer is so equipped electrically that such instruments are connected with the negative pole of a battery and corrosion prevented. This permits long-continued boiling without dulling.

‡Exposure to nos. 2 or 3 for 15 min eliminates tubercle bacilli and enteroviruses.

§To eliminate tubercle bacilli and enteroviruses, tubing must be filled.

TABLE 17-3. TECHNICS OF STERILIZATION (OR DISINFECTION) OF COMMONLY USED ITEMS IN THE CLINICAL SETTING—cont'd

Item to be sterilized	Autoclave (method of choice)	Chemical disinfection (choice indicated)	Other technics applicable
Surgical instruments and supplies, mechanically clean—cont'd			
Endoscopes and instruments with optical lenses (cystoscopes and others)	—	1. Ethylene oxide gas, 2 to 12 hr (method of choice) 2. Aqueous formalin, 20%, 12 hr 3. Activated glutaraldehyde (2% aqueous) 10 hr‡	—
Polyethylene tubing	—	Immerse tubing carefully filled with disinfectant: 1. Activated glutaraldehyde, 10 hr 2. Ethylene oxide gas, 2 to 12 hr 3. Aqueous formalin, 20%, 12 hr	—
Rubber tubing silk catheters, and bougies	—	1. Ethylene oxide gas, 2 to 12 hr (method of choice) 2. Iodophor, 500 ppm available iodine, 20 min§ 3. Amphyl, 2%, 20 min§ 4. Activated glutaraldehyde (2% aqueous), 15 min§	—
Inhalation equipment for anesthesia	—	1. Ethylene oxide gas, 2 to 12 hr 2. Activated glutaraldehyde (2% aqueous), 10 hr	—
Thermometers	—	1. Activated glutaraldehyde, 10 hr 2. Ethylene oxide gas, 2 to 12 hr (cold cycle only) 3. Isopropyl alcohol, 70% to 90%, plus iodine, 0.2%, 15 min (to eliminate tubercle bacilli and enteroviruses)	—
Surgical instruments and supplies, contaminated	Yes, wherever possible—as a rule, double exposure time, other conditions the same	Ethylene oxide gas, 2 to 12 hr	—

many items that cannot be steam sterilized. Gas sterilization is carried out in a specially devised chamber (Fig. 17-2). Table 17-2 compares this form with that of steam under pressure.

Surgical instruments and supplies

In Table 17-3 practical suggestions are given for the sterilization or chemical disinfection of specific instruments and supplies (see also Table 17-4). Although one chemical disinfectant may be indicated in Table 17-3 and for various purposes in this chapter, it does not follow that the one cited is necessarily the best. There are certainly many other good ones. Remember, to eliminate hepatitis viruses, chemical disinfection is used only when application of physical methods is impossible.

Syringes and needles contaminated by even minute traces of serum or plasma may be sterilized by one of the methods recommended in Table 17-4. Disposable needles, blood lancets, and syringes are available commercially, and their use is strongly recommended wherever feasible because one important method of spreading the hepatitis viruses is thereby blocked. *Disposable items are designed to be used once only and should never be sterilized and reused.*

The proper disposition of disposables must be stressed! Disposable syringes and needles must be kept away from children and drug addicts and must not wound the hand of the garbage collector. Once used, they can be incinerated, deformed in boiling water, or crushed in a machine.

One method for safe discard of disposable syringes and needles is as follows: The needle is placed inside the syringe, needle point upward toward the piston. The piston is inserted into the syringe and jammed down, locking the needle tightly between the barrel of the syringe and the plunger. The unit may then be thrown into the trash.

It is hard to render instruments with optical lenses completely free of microbes. To kill completely all microbes present might well loosen the cement of the lens system in the process or corrode the metal in some of the instruments. A recently developed, heat-resistant cement makes it possible to carry out a type of pasteurization wherein endoscopes are submerged in hot (not boiling) water at a temperature of 85° C for 1 hour. This method kills most of the ordinary bacterial contaminants.

TABLE 17-4. METHODS OF PHYSICAL STERILIZATION TO ELIMINATE HEPATITIS VIRUSES*

Method	Time	Temperature
Autoclave†	30 min	121.5° C (15 lb pressure)
Dry heat	2 hr	170° C
Boiling (active)	30 min	100° C

*Most human viruses are destroyed at 60° C for 30 minutes, with exception of the hepatitis viruses. At this temperature a 10-hour period is required for their inactivation.
†The autoclave is definitely the preferred method and strongly advised over the other two.

Telescopic instruments with improved cements in the lens systems can also be sterilized with ethylene oxide.

Formerly the preparation and care of rubber tubing used for intravenous medication was a difficult and time-consuming task. The use of plastic disposable units *should completely eliminate* the need for rubber tubing for intravenous medication and blood transfusion.

Dressings and linens. Towels, gowns, and other cloth articles may be sterilized by autoclaving at 15 pounds' pressure for 30 minutes. They may be arranged in packs 12 × 12 × 20 inches in size and wrapped in two layers of muslin or one layer of muslin and one or two layers of heavy paper.

Reusable gloves. When an operation is finished, the surgeon and his assistants should wash their gloves in cold water before removing them. After removal the gloves are washed outside and inside with tincture of green soap, thoroughly rinsed, and dried. They are then powdered, placed in glove envelopes, and wrapped. They are sterilized as follows: (1) autoclave at 15 pounds' pressure for 20 minutes, (2) follow with a vacuum for 5 to 10 minutes, (3) open the door of the sterilizer slightly and allow gloves to dry in the sterilizer for about 5 minutes, and (4) remove and separate the packages to allow rapid cooling (this prevents sticking). *Caution:* Gloves are often ruined by being left in the sterilizer too long.

Note: In some sterilizer models all these steps will not be necessary. (Note availability and practicality of prepackaged, sterile, or unsterile, disposable gloves.)

By putting an iodine solution in the water used to wash off glove powder and by taking advantage of the well-known color change, one can detect holes in gloves. If there is even a very tiny hole, a brownish discoloration of the hand is seen after the glove is removed.

Hand brushes. To sterilize hand brushes, (1) clean thoroughly, rinse free of soap under running water, and submerge in suitable disinfectant for 30 minutes (leave in the disinfectant until needed) or (2) autoclave (preferably) and store in a sterile dry container. (Note availability and practicality of prepackaged, sterile, disposable hand brushes.)

Materials of fat or wax, oils, and petrolatum gauze. To sterilize these heat-susceptible objects, place them in the sterilizing oven at 160° C for 2 to 2½ hours.

Water and saline solutions. The various intravenous solutions regularly used in the health field are available commercially in sterile, sealed bottles (bags or other receptacles) designed to be used with a sterile prepackaged unit of tubing in such a way that the likelihood of contamination is at a minimum. At times, watery solutions, saline solutions, and the like must be prepared specially. Such should be placed in suitable containers and autoclaved at 15 pounds' pressure for 30 minutes. If they are to be given intravenously, great care should be exercised to keep lint fibers and other particles out of the solution.

Surgical suite

The operative field is exposed to bacteria, both pathogenic and possibly pathogenic, floating in the air. These arise from the noses and throats of the surgical team and the occupants of the surgical suite. They may be destroyed by ultraviolet radiation.

At the end of the operation and when patient has been removed, the operating room (or theater of action) is disinfected. This is referred to as terminal disinfection. A phenolic derivative, iodophor, or quaternary ammonium preparation (germicidal detergent combination) compatible with tap water and used properly is satisfactory.

Body sites

Hands. There are many methods of disinfecting the hands. The following method gives good results:

1. Wash hands and arms well with soap and water.
2. With soap and a sterile brush, scrub nails and knuckles of both hands.
3. Scrub the hands, beginning with thumbs and in succession scrubbing inner and outer surfaces of thumbs and fingers, giving several strokes to each area.
4. Scrub palms and backs of the hands and then forearms to the arms 3 inches above elbow for 3 minutes; rinse well with running water.
5. With a sterile orange stick, clean under each nail thoroughly.
6. With a second sterile brush, scrub hands and forearms for 7 minutes in the same manner as with the first brush.
7. Thoroughly rinse off all soap with water and then rinse hands in 70% alcohol.
8. With one end of a sterile towel, dry one hand and, with a circular motion, dry forearm to arm above elbow. With the other end of the towel, dry the other hand and forearm in the same manner.

When one is scrubbing the hands for a surgical operation, plenty of soap should be used, and the scrubbing should be done with running water. When one is using cake soap, the soap is held between the brush and the palm until the scrubbing is complete. During the entire procedure the hands should be held higher than the elbows in order to prevent contamination.

Remember that soap dispensers should be emptied *regularly*, cleaned, and refilled with fresh solution. Cake soap, while not maintaining contamination directly (organisms inoculated on bar soap die quickly), may be left in pools of water supporting growth of microorganisms. If bar soap is desired, small bars that can be changed often and soap racks that allow for drainage are recommended. Because of such problems, many health care centers prefer an expendable form of dry soap such as the soap leaflets or soap-impregnated paper tissues (more expensive items, however).

At the present time the methods of disinfection of the hands are shortened by the habitual use of detergent soaps containing germicidal substances. The following method, known as the Betadine (p. 284) Surgical Scrub regimen, is widely used today. Because some members of the surgical team as well as some patients are allergic to iodine, pHisoHex (p. 284) can still be found in the health field setting.

1. Clean under nails. Keep fingernails short and clean.
2. Wet hands and forearms.
3. Apply to hands two portions, about 5 ml, of Betadine Surgical Scrub from a dispenser. (Depress Betadine dispenser foot pedal twice.)

4. Without adding water, wash hands and forearms thoroughly for 5 minutes. Use a brush if desired.
5. With a sterile orange stick, clean under nails. Add a little water for copious suds. Rinse thoroughly with running water.
6. Apply 5 ml from the dispenser to the hands. Repeat steps just given. Rinse with running water.

Actual sterilization of the hands is not possible. Microbes live not only on the surface of the skin but also in its deeper layers, in ducts of sweat glands, and around hair follicles. These resident microbes are chiefly nonpathogenic staphylococci and diphtheroid bacilli and are plentiful about the nails.

Site of operation. Although the deeper portions of the skin cannot be absolutely sterilized, enough bacteria can be removed to make infection unlikely. The selection of an antiseptic for preoperative skin preparation has had the attention of surgeons since the earliest times, and many different ones have been used.

The operative site may be prepared several hours ahead of time. It is scrubbed thoroughly with detergent soap containing hexachlorophene or green soap solution, and the area is shaved. However, many authorities object to the skin shave being done before the patient is taken to the surgical suite. They feel the best time is right before the surgical procedure, in the operating room, after the patient has been put to sleep and positioned. If lather from a dispenser can and a new razor blade are used, there is no danger of cross infection, and the lather traps the hair. When the skin shave is done many hours before surgery, there is a good chance bacteria will grow in the inflammatory reaction about small razor nicks. The incision the surgeon makes is then likely to go through minute abscesses that have formed in the interim.

The following outline utilizing the Betadine Surgical Scrub represents a satisfactory stepwise method of skin preparation. Although it uses a given iodophor, detergent soap with hexachlorophene or green soap solution may be used, and the method can be modified as indicated.

1. Shave field of operation. Wet with water.
2. Apply Betadine on a sponge to the umbilicus, if skin of the abdomen is being scrubbed, and discard sponge.
3. Scrub imaginary line of incision with Betadine on a sponge for 1 minute, using a rotating motion.
4. Carry the same sponge out from the line of incision and scrub outer portions of the surgical area for 1 minute; again use a circular motion and discard the sponge.
5. Repeat this entire maneuver for five separate instances. When skin area is prepared for operation, the imaginary line of incision is never painted with the same sponge more than once.
6. Use a little water to make suds. Use wet sterile gauze to rinse surgical area.
7. Pat dry with sterile towel.
8. Paint area with Betadine. Allow to dry. The surgical area is now ready for the drapes.

The operating room personnel and those who prepare patients for operations should not let the fact that many surgical infections may be controlled by antibiotics lull them into satisfaction with careless antiseptic or aseptic technic.

Other areas of skin. For rapid sterilization preceding routine hypodermic injections and venipunctures for intravenous medication, 70% to 90% ethyl or isopropyl alcohol or 2% iodine in 70% alcohol is effective when applied to a clean area of skin for 30 or 60 seconds. When lumbar puncture is to be done, the overlying skin should be cleaned with soap or detergent, rinsed, dried, and disinfected with a suitable preparation.

Mucous membranes. Absolute disinfection of mucous membranes is impossible. The mouth and throat may be cleaned with Dobell's solution, hydrogen peroxide, hexylresorcinol, or Ceepryn Chloride, 1:10,000 to 1:5000.

DISINFECTION OF EXCRETA AND CONTAMINATED MATERIALS FROM INFECTIOUS DISEASES

A patient with a communicable disease is not properly supervised until every avenue by which the infectious agent may be spread from the patient to others has been closed. These avenues are not closed until all excreta and objects that convey infection have been properly disinfected. A patient who is being cared for in such a manner that all avenues by which the infection may spread to others are closed is said to be *isolated*. A ward patient with a conscientious and capable nurse is more strictly isolated from the microbiologic point of view than is the patient in a room with plastered walls and airtight doors who has a careless or incompetent nurse. Isolation refers to avenues of infection, not to walls and doors. *Remember that pathogenic microbes are most virulent when first thrown off from the body.*

Disinfection in infectious diseases is of two types: concurrent and terminal. *Concurrent disinfection* means immediate disinfection of the patient's excreta or objects that have been contaminated by the patient or his excreta. *Terminal disinfection* means the final disinfection of the room, its contents, and environs after it has been vacated by the patient. The final chapter is written when it has been proved that patient and attendants are free of the agent that caused the disease; that is, they have not become carriers. In the following paragraphs are given methods of disinfecting excreta, materials, and objects by which infection is most often transmitted from the sick to the well.

Hands. Unless properly disinfected, the hands of nurses, doctors, or other persons in contact with the patient are almost certain to transfer the infectious agent to themselves or to others. Although disinfection of hands to prevent the spread of communicable diseases is not so time-consuming as for a surgical operation, it should be just as conscientiously done. The hands and forearms are submerged in a basin of suitable disinfectant, such as an iodophor- or hexachlorophene-containing preparation, and rubbed vigorously for 2 minutes. Ethyl (or isopropyl) alcohol, 70%, can be combined with a quaternary ammonium compound into an emollient mixture, effective and less irritating to the hands. Afterwards, the hands are washed with soap and running water (and of course with good mechanical friction). A brush is not necessary for scrubbing the hands and forearms, although in some institutions it is required.

The clean hands and forearms should be dried thoroughly. Hand lotion is advisable to keep the skin soft and in good condition. The iodophor-containing preparations of whatever kind and other disinfectants as well tend to dry the skin excessively. The nails should be kept clean at all times. The hands should be cleaned in this way every time they contact the patient for any purpose whatsoever. To allow one's hands to contact a person suffering from a communicable disease and then, without disinfection, to allow them to contact another person is little short of criminal neglect.

Soiled linens and clothing. All linens and clothing in contact with the patient should be considered contaminated and kept in the patient's room until ready for final disposal. In the hospital they are wrapped in sheets or placed in pillowcases and put into a special bag, care being taken that the outside of the bag does not become contaminated. Before the bag is full, it is closed tightly, marked to indicate that it contains infectious material, and sent to the laundry. On reaching the laundry, the bag itself (and contents) is put directly into the washer. The washer is then closed, and live steam is introduced for 15 to 20 minutes. After this the regular procedure of the laundry is carried out. In the home the linen or bedclothes are bundled and carried to the kitchen or home laundry to be placed in a boiler containing warm water and soapsuds and boiled from 10 to 15 minutes. After boiling, the process of laundering may be completed in the home or by a commercial laundry. On occasion, the boiler is carried to the patient's room to receive the soiled linen.

Mattresses, pillows, and blankets may be sterilized by heat in special sterilizing chambers, or they may be sterilized in a formaldehyde chamber. They also may be dried in the sunlight for a period of 24 hours. Mattresses and pillows are best covered with impervious plastic covers. Such plastic covers may be laundered readily or cleaned and disinfected with a germicidal detergent.

Shoes. Disinfection of shoes is best accomplished by placing a pledget of cotton saturated with formalin in each shoe and allowing the shoes to stay in a closed box for 24 hours. Shoes may also be disinfected by being sprayed with formalin from an atomizer for three successive evenings.

Feces and urine. Where connections exist to modern sewage disposal systems, feces and urine are disposed of in the commode without disinfection. Otherwise, feces and urine, especially from patients with typhoid fever, viral hepatitis, and similar disorders, should be mixed with either 5% phenol (carbolic acid) or preferably chlorinated lime. The volume of disinfectant is three times that of the material to be disinfected. Feces should be thoroughly broken up in order to ensure even penetration. After the urine and feces are thoroughly mixed with the disinfecting solution, the mixture is allowed to stand for at least 2 hours before disposal of it is made. Contaminated colon tubes and such may be disinfected in Amphyl, 2% solution, for 20 minutes.

Discharges from mouth and nose. Discharges from the mouth and nose are received on squares of tissue and placed in a paper bag pinned to the side of the bed. After the bag is wrapped in several layers of clean newspaper, it is burned in the incinerator.

Sputum. Sputum should not be allowed to dry so as to permit particles to float off in the air. It should be received in a paper cup held in a special metal holder. The paper cup is discarded and the metal holder sterilized at least twice daily (more often if necessary). To discard a paper cup, remove the cup from the holder with forceps and set it on several thicknesses of newspaper. The cup is then filled with sawdust to decrease the amount of moisture and make the cup burn more easily. The paper is wrapped around the cup and tied securely; next, the packet is placed in the incinerator or in a special receptacle to be carried to the incinerator later. After this has been done, the metal holder is autoclaved or boiled for 30 minutes. Ambulatory patients often use collapsible cups.

Sputum can be received into a waxed paper cup with a tight-fitting lid that can be secured before the cup is discarded.

Clinical thermometers. A clinical thermometer must be sterile each time it is used. After a thermometer is used and the temperature recorded, the thermometer must be washed with soap or detergent and water (with friction applied). It may be disinfected as indicated in Table 17-3.

After sterilization a thermometer is wiped with sterile cotton and placed in a sterile container for later use. Preferably it is stored in alcohol, in fresh daily solutions of tincture of iodine, 0.5%, or in 150 ppm of an aqueous iodophor. Alcoholic solutions of the quats must be used, since the aqueous solutions do *not* kill tubercle bacilli. Thermometers are rinsed in cold water before use.

A patient with hepatitis should have his own thermometer, which is destroyed when he leaves the hospital.

Eating utensils. A special covered container is placed in the sickroom, and the dishes are placed therein, preferably by the patient himself. Care to prevent contamination of the outside of the container must be exercised. After the dishes, knives, and forks are placed in the container, it is carried to the kitchen and boiled in soapy water for 5 minutes. Sometimes the eating utensils are carried to the kitchen before they are placed in a container. Leftover bits of food may be placed in a paper sack, wrapped well with newspaper, and burned in the incinerator. If this is not practical, the food remains are placed in chlorinated lime or phenol solution and left for 1 hour.

Terminal disinfection. When a patient is removed from a room, the floors, walls, woodwork, and furniture should be scrubbed with soapsuds and thoroughly aired. The phenol derivatives are valuable and widely used in terminal disinfection. Other good housekeeping disinfectants are 1% sodium hypochlorite and Wescodyne or other iodophor (500 ppm available iodine).

DISINFECTION OF ARTICLES FOR PUBLIC USE

Rest rooms and bathroom fixtures. The woodwork, floors, sinks, basins, and toilets of public rest rooms should be washed frequently with hot soapsuds or detergent and rinsed thoroughly. Phenolic disinfectants or one of the iodophors are suitable for disinfection.

Woodwork. The woodwork of schoolhouses, churches, theaters, or other places of public assembly should be washed frequently with a germicidal detergent solution.

Public conveyances. All parts of public conveyances touched by the hands of passengers should be frequently washed with a germicidal detergent solution.

FUMIGATION OF ROOMS AND DISINFECTION OF AIR

Fumigation. In former years it was an almost universal procedure to put a sickroom "into fumigation" as soon as the room was vacated. Fumigation at that time meant terminal gaseous disinfection with formaldehyde, the gas of choice. This gaseous method of disinfection has been abandoned because it does not destroy insects, small animals, or pathogenic microbes. The scrubbing of walls and floors to clean them thoroughly has replaced fumigation in the control of communicable disease; natural drying of surfaces effects the destruction of bacteria.

The term *fumigation* now denotes the destruction of disease-carrying animals, insects, or vermin. The fumigant of choice is hydrocyanic acid, a deadly poison that must be handled with extreme care. Sulfur dioxide is a splendid insecticide, but because of its deleterious effect on metals and household goods, it is seldom used in homes, public buildings, or hospitals.

Air disinfection. Two methods of destroying bacteria and other disease-producing agents as they are dispersed through the air are ultraviolet radiation and chemical disinfection by means of aerosols.

Aerosols are chemical vapors liberated into the air for disinfection. Two of the most potent ones are propylene glycol, effective in a dilution of 1 part in 100 million parts of air, and triethylene glycol, effective when diluted to 1 part in 400 million parts of air. The vapors may be introduced into rooms through ventilating systems, bombs, or special equipment. The importance of thorough cleaning and disinfection of air-conditioning units must be kept in mind.

STERILIZATION OF BIOLOGIC PRODUCTS

In the preparation of bacterial vaccines it is necessary to kill the bacteria at as low a temperature as possible because a temperature of more than 62° C coagulates the bacterial protein and renders the vaccine worthless. It has been found that an exposure for 1 hour to a water bath temperature of 60° C kills any of the ordinary bacteria used in making vaccines. Bacterial vaccines are sometimes sterilized by the addition of chemicals such as phenol or cresol. Mercury-vapor lamps as a source of ultraviolet radiation are used in the preparation of bacterial as well as viral vaccines, but ultraviolet radiation, ideal in theory, is practically ineffective at times. Ethylene oxide gas is also used for this purpose.

QUESTIONS FOR REVIEW

1. How may knives be sterilized?
2. What is the preferred method of sterilization for endoscopes? What is meant by pasteurization of endoscopes?
3. List three surgical items easily damaged by heat.
4. Give a method of preparing the skin for operation.
5. How may the air be kept free of pathogenic bacteria during the time of operation?
6. What is the difference between physical and microbiologic isolation?

7. How would you dispose of the feces and urine of a patient with typhoid fever?
8. Discuss the drugs used in wound disinfection.
9. Explain what is meant by concurrent disinfection. Terminal disinfection.
10. What precaution should be taken in handling sputum from a patient with tuberculosis? How is sputum sterilized?
11. How are clinical thermometers sterilized?
12. Briefly discuss the use of hydrocyanic acid as a fumigant.
13. How may hepatitis viruses be eliminated?
14. How are biologic vaccines sterilized?
15. Compare steam sterilization with gas sterilization.

REFERENCES FOR UNIT FOUR

AMA drug evaluations, ed. 2, Acton, Mass., 1973, Publishing Sciences Group, Inc.

American Academy of Pediatrics, Committee on Fetus and Newborn: Skin care of newborns, Pediatrics **54:**682, 1974.

Aronson, M. D., and others: Vidarabine therapy for severe herpesvirus infections, J.A.M.A. **235:**1339, 1976.

Ballinger, W. F., II, and others: Alexander's care of the patient in surgery, ed. 5, St. Louis, 1972, The C. V. Mosby Co.

Barlow, P. B., and others: Preventive therapy of tuberculous infection, Am. Rev. Respir. Dis. **110:**371, 1974.

Barlow, P. B., and others: Current trends: preventive therapy of tuberculous infection, Morbid. Mortal. Wk. Rep. **24:**71, 1975.

Bauer, D. J.: Antiviral chemotherapy: first decade, Br. Med. J. **3:**275, Aug. 4, 1973.

Bell, A. N.: Personal communication, 1976.

Bickel, L.: Rise up to life; a biography of Howard Walter Florey who gave penicillin to the world, New York, 1973, Charles Scribner's Sons.

Blumberg, P. M., and Strominger, J. L.: Interaction of penicillin with the bacterial cell: penicillin-binding proteins and penicillin-sensitive enzymes, Bacteriol. Rev. **38:**291, 1974.

Borick, P. M., editor: Chemical sterilization, Stroudsburg, Pa., 1973, Dowden, Hutchinson & Ross, Inc.

Bruun, J. N., and Solbert, C. O.: Hand carriage of gram-negative bacilli and *Staphylococcus aureus,* Br. Med. J. **2:**580, June 9, 1973.

Center for Disease Control (current trends): Isoniazid-associated hepatitis: summary of the report of the tuberculosis advisory committee and special consultants to the director, Morbid. Mortal. Wk. Rep. **23:**97, 1974.

Center for Disease Control: Isolation techniques for use in hospitals, ed. 2, DHEW Publication 75-8043, Washington, D.C., 1975, U.S. Government Printing Office.

Cooke, E. I., editor: Chemical synonyms and trade names, Cleveland, Ohio, 1971, CRC Press, Inc.

Copeland, E. M., III, and others: Prevention of microbial catheter contamination in patients receiving parenteral hyperalimentation, South. Med. J. **67:**303, 1974.

Culliton, B. J.: Penicillin G.: suddenly a shortage, Science **191:**1157, 1976.

DeHaen, P.: New drug introduction 1973-1974, J.A.M.A. **234:**728, 1975.

Deichmann, W. B., and Gerarde, H. W.: Toxicology of drugs and chemicals, New York, 1969, Academic Press, Inc.

DeSante, K. A., and others: Antibiotic batch certification and bioequivalence, J.A.M.A. **232:**1349, 1975.

Dineen, P.: Duration of shelf life—an evaluation, AORN J. **13:**63, Mar., 1971.

Dineen, P.: Personnel, discipline, and infection, Arch. Surg. **107:**603, 1973.

Disposables are here to stay, Lab. World **26:**10, Oct., 1975.

Duncalf, D.: Care of anesthetic equipment and other devices, Arch. Surg. **107:**600, 1973.

Easley, J. R.: Personal communication, 1976.

Finland, M., and Kass, E. H., editors: Trimethoprim-sulfamethoxazole: microbiological, pharmacological, and clinical considerations, Chicago, 1974, The University of Chicago Press.

Finland, M.: Changing patterns of susceptibility of common bacterial pathogens to antimicrobial agents, Ann. Intern. Med. **76:**1009, 1972.

Flower, M., and others: The role of the nurse epidemiologist in infection control and continuing education, Surg. Gynecol. Obstet. **141:**552, 1975.

Fox, M. K., and others: How good are hand washing practices? Am. J. Nurs. **74:**1676, 1974.

Gardner, P.: Reasons for "antibiotic failures," Hosp. Pract. **11:**41, Feb., 1976.

Garner, J. S., and Kaiser, A. B.: How often is isolation needed? Am. J. Nurs. **72:**733, 1972.

Gillespie, W. A., and others: Absorption of hexachlorophene from dusting powder on newborn infants' skin. J. Hyg. **73:**311, 1974.

Ginsberg, F.: Hair, long or short, must be covered in O.R., Mod. Hosp. **117:**106, July, 1971.

Goodman, L. S., and Gilman, A., editors: Pharmacological basis of therapeutics, New York, 1975, Macmillan, Inc.

Goth, A.: Medical pharmacology: principles and concepts, ed. 8, St. Louis, 1976, The C. V. Mosby Co.

Gowdy, J. M., and Ulsamer, A. G.: Hexachlorophene lesions in newborn infants, Am. J. Dis. Child. **130:**247, 1976.

Gröschel, D.: Fact and fiction in hospital disinfection, ASM News **38:**479, 1972.

Hartley, C. L., and Richmond, M. H.: Antibiotic resistance and survival of E coli in alimentary tract, Br. Med. J. **4:**71, Oct. 11, 1975.

Hill, R. M.: Will this drug harm the unborn infant? The doctor's dilemma, South. Med. J. **67:**1476, 1974.

Holloway, W. J.: Clindamycin-associated colitis, South. Med. J. **68:**1469, 1975.

Hussey, H. H.: Hexachlorophene bathing of neonates, J.A.M.A. **233:**172, 1975.

Hyams, P. J., and others: Staphylococcal bacteremia and hexachlorophene bathing, Am. J. Dis. Child. **129:**595, 1975.

Irey, N. S.: Adverse reactions to drugs and chemicals, a resume and progress report, J.A.M.A. **230:**596, 1974.

Jackson, G. G.: Influenza; the present status of chemotherapy, Hosp. Pract. **6:**75, Nov., 1971.

Jackson, G. G.: Perspective from a quarter century of antibiotic usage, J.A.M.A. **227:**634, 1974.

Johnston, R. F., and Wildrick, K. H.: State of the art review. The impact of chemotherapy on the care of patients with tuberculosis, Am. Rev. Respir. Dis. **109:**635, 1974.

Jukes, T. H.: Antibiotics in meat production, J.A.M.A. **232:**292, 1975.

Kitto, W.: Chloramphenicol today, J.A.M.A. **233:**325, 1975.

Kundsin, R. B., and Walter, C. W.: The surgical scrub—practical consideration, Arch. Surg. **107:**75, 1973.

LaDu, B. N., Jr.: The genetics of drug reactions, Mod. Med. **40:**10, Feb. 7, 1972.

Lal, S., and others: Effect of rifampicin and isoniazid on liver function, Br. Med. J. **1:**148, Jan. 15, 1972.

Laufman, H.: Cleanup techniques in the OR, Med. Surg. Rev. **7:**1, Oct-Nov., 1971.

Laufman, H.: Surgical hazard control. Effect of architecture and engineering, Arch. Surg. **107:**522, 1973.

Laufman, H., and others: Use of disposable products in surgical practice, Arch. Surg. **111:**20, 1976.

Lewis, J. R.: Standardization of drug names (editorial), J.A.M.A. **229:**559, 1974.

Lorian, V., and Atkinson, B.: Abnormal forms of bacteria produced by antibiotics, Am. J. Clin. Clin. Pathol. **64:**678, 1975.

Maddrey, W. C., and Boitnott, J. K.: Hepatitis induced by isoniazid and methyldopa, Hosp. Pract. **10:**119, Apr., 1975.

Maddrey, W. C., and Boitnott, J. K.: Isoniazid hepatitis, Ann. Intern. Med. **79:**1, 1973.

Mallison, G. F.: Housekeeping in operating suites, AORN J. **21:**213, 1975.

Maugh, T. H., II: Amantadine: an alternative for prevention of influenza, Science **192:**130, 1976.

Maugh, T. H., II: Chemotherapy: antiviral agents come of age, Science **192:**128, 1976.

McArthur, B. J., and others: Stopcock contamination in an ICU, Am. J. Nurs. **75:**96, 1975.

McHattie, J. C., and others: Comparison of hexachlorophene and Lactacyd on growth of skin flora in healthy term newborn infants, Can. Med. Assoc. J. **110:**1248, 1974.

McInnes, B.: Essentials of communicable disease, ed. 2, St. Louis, 1975, The C. V. Mosby Co.

McMahon, F. G.: The patient package insert (editorial), J.A.M.A. **233:**1089, 1975.

Medoff, G., and Kobayashi, G. S.: Amphotericin B, old drug, new therapy. J.A.M.A. **232:**619, 1975.

Meyer, C.: Don't wash labware: "incinerate" it, Lab. Management **13:**40, Jan., 1975.

Meyer, R. D., and others: Amikacin therapy for serious gram-negative bacillary infections, Ann. Intern. Med. **83**:790, 1975.

Mitchell, J. R., and others: N.I.H. Conference: Isoniazid liver injury: clinical spectrum, pathology, and probable pathogenesis, Ann. Intern. Med. **84**:181, 1976.

Modell, W., editor: Drugs of choice 1976-1977, St. Louis, 1976, The C. V. Mosby Co.

Morse, R. A.: Environmental control in beehive, Sci. Am. **226**:92, Apr., 1972.

Nelson, J. P.: OR clean rooms, AORN J. **15**:71, May, 1972.

Nichols, G. A., and Dobek, A. S.: Sterility of items sealed in plastic, Hosp. Top. **49**:84, Dec., 1971.

Notes on air hygiene: Summary of a conference on air disinfection, Arch. Environ. Health **23**:473, 1971.

Osol, A., and others, editors: United States dispensatory, Philadelphia, 1973, J. B. Lippincott Co.

Pavlech, H. M.: Reusable laboratory glassware or disposable glassware, ASM News **40**:275, 1974.

Peers, J. G.: Cleanup techniques in the operating room, Arch. Surg. **107**:596, 1973.

Perkins, J. J.: Principles and methods of sterilization in health sciences, Springfield, Ill., 1973, Charles C Thomas, Publisher.

Perlman, D.: Evolution of the antibiotics industry, 1940-1975, ASM News **40**:910, 1974.

Physicians' desk reference to pharmaceutical specialties and biologicals, Oradell, N.J., 1976, Medical Economics, Inc.

Prioleau, W. H.: Fashion hides its head in the OR, Hospitals **46**:107, May 1, 1972.

Richmond, A. S., and others: R factors in gentamicin-resistant organisms causing hospital infection, Lancet **2**:1176, 1975.

Roberts, R. B., editor: Infections and sterilization problems, Boston, 1972, Little, Brown & Co.

Roberts, R. B., and Rendell-Baker, L.: Ethylene-oxide sterilization, Hosp. Top. **50**:60, May, 1972.

Schiffman, D. O.: Evaluation of an anti-infective combination trimethorprim-sulfamethoxazole (Bactrim, Septra), J.A.M.A. **231**:635, 1975.

Selwyn, S., and Ellis, H.: Skin bacteria and skin disinfection reconsidered, Br. Med. J. **1**:136, 1972.

Shirkey, H. C.: The package insert dilemma, J. Pediatr. **79**:691, 1971.

Smylie, H. G., and others: From pHisoHex to Hibiscrub, Br. Med. J. **4**:586, 1973.

Snydam, D. R., and others: Prevention of nosocomial viral hepatitis, type B (hepatitis B), Ann. Intern. Med **83**:838, 1975.

Snyder, R. W., and Cheatle, E. L.: Alkaline glutaraldehyde, an effective disinfectant, Am. J. Hosp. Pharm. **22**:321, 1965.

Steere, A. C., and Mallison, G. F.: Handwashing practices for the prevention of nosocomial infections, Ann. Intern. Med. **83**:638, 1975.

Stiffler, P. W.: Mechanisms of bacterial susceptibility and resistance to antibiotics, Lab. Med. **7**:8, Apr., 1976.

Swenson, O., and Grana, L.: A new, improved sterilization set-up for the operating room, AORN J. **13**:80, Jan., 1971.

The cost of disposables (editorial), J.A.M.A. **217**:1859, 1971.

Walter, C. W., and Kundsin, R. B.: The airborne component of wound contamination and infection, Arch. Surg. **107**:588, 1973.

Weber, D. O., and others: Influence of operating room surface contamination on surgical wounds, Arch. Surg. **111**:484, 1976.

Wells, B. B.: Generic nomenclature, J.A.M.A. **229**:527, 1974.

Wettach, G. E.: Pseudomembranous colitis after the prophylactic use of clindamycin, Arch. Surg. **110**:1252, 1975.

Wilson, F. M., and others: Anicteric carbenicillin hepatitis, J.A.M.A. **232**:818, 1975.

Wilkins, T. D., and Appleman, M. D.: Review of methods for antibiotic susceptibility testing of anaerobic bacteria, Lab. Med. **7**:12, Apr., 1976.

Wise, R.: New penicillins—present and future, J.A.M.A. **232**:493, 1975.

LABORATORY SURVEY OF UNIT FOUR

PROJECT
Effects of physical agents on bacteria
Part A—Sterilization by filtration with demonstration by instructor and discussion

1. Method of filtration using either the Berkefeld or Seitz filter
2. Filtration of a test solution, which can be a broth culture of some nonpathogenic organism diluted 1:20 with sterile isotonic saline solution
3. Cultures taken before and after filtration of test material
4. Observations made on cultures after incubation period of 24 hours

Part B—Effects of heat on bacteria

1. Prepare five suspensions each of *Bacillus subtilis* and of *Escherichia coli* by placing in sterile cotton-plugged test tubes 1 or 2 ml amounts of a bacterial suspension furnished by the instructor.
2. Treat a properly labeled tube of each suspension as follows:
   a. Autoclave at 15 pounds' pressure for 15 minutes.
   b. Stand in a pan of boiling water for 30 minutes.
   c. Expose to free-flowing steam in an Arnold sterilizer for 1 hour.
   d. Stand in a pan of water at 60° C for 1 hour.
   e. Use the fifth pair as a control.
3. Transfer a loopful of the bacterial suspension from each tube to a culture tube of broth and incubate at 37° C for 24 hours.
4. Observe growth in cultures made. Record growth obtained.

Exposure to heat	Growth obtained	
	Bacillus subtilis	*Escherichia coli*
Autoclave, 15 pounds' of pressure, 15 min		
Boiling water, 30 min		
Free-flowing steam, 1 hour		
60° C, 1 hour		
Control		

5. Which tubes of the heated suspensions of *Bacillus subtilis* show growth? What does this indicate? What is the difference between the growth of *Bacillus subtilis* and that of *Escherichia coli*? What does this indicate? What part does spore formation play in this experiment?

Part C—Effects of boiling

1. Obtain test specimens of bacteria.
   a. Dip three sterile cotton swabs in a 5-day broth culture of *Bacillus subtilis*.

b. Dip three sterile cotton swabs into a 24-hour broth culture of *Escherichia coli*.

2. Expose test specimens to the effects of boiling water for 1 minute.
   a. Hold a swab from each suspension in a separate beaker of boiling water for 1 minute and then dip into a tube of nutrient broth.
   b. With the other swabs, repeat the procedure, using periods of 5 and 10 minutes.
   c. Incubate the six tubes of broth 24 to 48 hours.
   d. Indicate growth obtained in the following table:

Organism	Period of boiling		
	1 min	5 min	10 min
Bacillus subtilis			
Escherichia coli			

   e. What do the foregoing findings indicate?

Part D—Operation of autoclave with demonstration by instructor and discussion

1. Discussion of the autoclave or pressure steam sterilizer given on pp. 268-270.
2. Demonstration of the parts of the instrument and its operation
3. Direct supervision of students as they operate the autoclave
   Note: An autoclave or pressure steam sterilizer is filled with steam under pressure; therefore, if the pressure gets too high, there is always danger of an explosion.

PROJECT
Effects of chemical agents on bacteria
Part A—Effects of chemical disinfectants

1. Obtain from the instructor four sterile cotton-plugged tubes containing the following:
   a. 5 ml of 70% isopropyl alcohol
   b. 5 ml of 5% phenol or phenolic disinfectant
   c. 5 ml of some widely advertised antiseptic (one or several may be used)
   d. Broth suspension of staphylococci
2. Proceed as follows:
   a. To each tube of disinfectant, add 0.5 ml of the bacterial suspension. Do not let the broth run down side of tube. Mix by gently tapping the bottom of the tube.
   b. Place in water bath at 20° C.

c. At the end of 5, 10, and 15 minutes, transfer a loopful of material from each tube of disinfectant to a tube of sterile broth.

d. Label the tube, noting the organism and the time held in the water bath, and incubate 48 hours.

e. Record growth on the accompanying chart, using the plus sign to indicate growth and the minus sign to indicate absence of growth.

3. Repeat the preceding experiment using *Bacillus subtilis* or another spore-forming organism instead of staphylococci. Record on chart.

Disinfectant	*Staphylococcus*			Sporeformer		
	5 min	10 min	15 min	5 min	10 min	15 min
Isopropyl alcohol, 70%						
Phenol or phenolic derivative, 5%						
Commercial disinfectant						

Part B—Susceptibility of bacteria to penicillin

Note: Students may work in groups of three or four for this experiment.

1. Obtain from instructor the following:
   a. Petri dish of tryptose agar that has just been inoculated with an organism susceptible to the action of penicillin
   b. Petri dish of tryptose agar that has just been inoculated with an organism *not* susceptible to the action of penicillin
2. Use sterile forceps to drop filter paper disks* impregnated with different strengths of penicillin onto the inoculated surface of the agar plates. Place the disks about 2 cm apart.
3. Invert plates and incubate for 24 to 48 hours.
4. Note the presence or absence of bacterial growth around the filter paper disks.

Part C—Susceptibility of bacteria to other antibiotics

Note: Students may work in groups of three or four.

1. Obtain from the instructor the following:
   a. A Petri dish of tryptose agar that has just been inoculated with bacteria susceptible to some of the antibiotics to be used and *not* susceptible to other test antibiotics.
   b. Disks of filter paper* impregnated with different antibiotics to be used.
2. Use sterile forceps to place the antibiotic disks on the surface of the agar about 2 cm apart.

---

*Disks impregnated with graded amounts of penicillin or other antibiotics may be obtained from commercial sources.

3. Invert and incubate at 37° C for 24 to 48 hours.
4. Examine and record which antibiotics inhibited bacterial growth and which did not.
5. Perform a susceptibility test on an agar plate inoculated with *Pseudomonas aeruginosa*, using a number of different antibiotics.
6. Note the resistance exhibited by this organism.

PROJECT
Testing the efficiency of sterilization

Note: Students may work in groups of two or more to test the efficiency of some of the sterilizing processes outlined in this book.

Part A—Test of the effectiveness of methods for disinfecting the hands

1. Lightly touch the tips of the fingers of the recently disinfected hand to the surface of a Petri dish of nutrient agar.
2. Invert the dish and incubate. Examine the growth at the end of 24 hours.
3. Set up a control. Rub the palm of one hand before washing or disinfection with a moistened sterile cotton swab. Inoculate the swab onto sterile nutrient agar. Incubate for 24 hours and observe growth. Make comparisons.

Part B—Test of the sterility of recently sterilized solutions

1. Use a sterile pipette to remove a few drops of isotonic saline solution from a recently sterilized flask. (Sterile distilled water may be used. Commercially packaged solutions may also be used.)
2. Add the test solution to a tube of nutrient broth. Incubate for 24 hours and examine for growth.

Part C—Test of the sterility of a clinical thermometer

1. Dip the tip of a recently disinfected thermometer into a tube of nutrient broth.
2. Incubate the broth for 24 hours and examine for growth.

EVALUATION FOR UNIT FOUR

Part I

In the following exercise place in the blank space the letter from column B that represents the appropriate or preferred method of sterilizing or disinfecting each article named in column A.

COLUMN A—TASK

_____ 1. Transfer forceps
_____ 2. Dressings and linens
_____ 3. Silk catheters and bougies
_____ 4. Cystoscopes
_____ 5. Noncutting instruments
_____ 6. Hypodermic syringes and needles
_____ 7. Hand brushes
_____ 8. Sharp cutting instruments
_____ 9. Inhalation equipment
_____ 10. Water and saline solutions
_____ 11. Shoes
_____ 12. Sputum from patient with tuberculosis
_____ 13. Feces and urine
_____ 14. Air of nursery for newborn infants
_____ 15. Rectal thermometers
_____ 16. Hands of communicable disease nurse
_____ 17. Bacterial exotoxins
_____ 18. Glassware
_____ 19. Inoculating platinum wire loop
_____ 20. Blood-smeared tabletop
_____ 21. Petrolatum gauze
_____ 22. Furniture
_____ 23. Floors, walls, and woodwork
_____ 24. Endoscopes
_____ 25. Mucous membranes
_____ 26. Plastic parts of heart-lung machine
_____ 27. Polyethylene tubing
_____ 28. Suture needles
_____ 29. Eating utensils
_____ 30. Skin site of an operation
_____ 31. Powders
_____ 32. Hinged instruments
_____ 33. Contaminated surgical instruments
_____ 34. Plastic Petri dishes

COLUMN B—PROCEDURE

(a) Pressure steam sterilization (in an autoclave)
(b) Chemical disinfection
(c) Dry heat sterilization
(d) Incineration
(e) Seitz filtration
(f) Gaseous chamber sterilization (as with ethylene oxide)
(g) Boiling
(h) Ultraviolet radiation

Part II

Place the letter from column B in front of the corresponding definition in column A.

COLUMN A

_____ 1. Destroys or masks offensive odors (gas)
_____ 2. Prevents growth of bacteria, but does not destroy them
_____ 3. Prevents deterioration of food
_____ 4. Destroys disease-producing agents and their products
_____ 5. Destroys insects and small animals (gases or fumes)
_____ 6. Prevents multiplication of bacteria
_____ 7. Destroys bacteria (chemical)

COLUMN B

(a) Fumigant
(b) Disinfectant
(c) Preservative
(d) Deodorant
(e) Germicide
(f) Antiseptic
(g) Bacteriostatic agent

Part III

Circle the one item that best completes the introductory statement.

1. In using a germicidal agent it is best to know:
   - (a) The volatility of the agent
   - (b) The concentration needed and the time of contact
   - (c) The exact number of bacterial spores present
   - (d) The exact number of bacteria present

2. In the management of infection by an organism that has become highly resistant to tetracycline it would be best to:
   - (a) Withdraw all antibiotic treatment
   - (b) Test the causative microorganism for susceptibility to other antibiotics
   - (c) Increase the dose
   - (d) Switch to penicillin
   - (e) Do nothing

3. The development of resistance to an antibiotic by a microorganism comes from an:
   - (a) Alteration of cell wall permeability
   - (b) Alteration of plasma membrane
   - (c) Alteration of capsule
   - (d) Alteration of mitochondria
   - (e) None of these

4. The phenol coefficient of a germicide is determined by comparison of that germicide with phenol on the basis of:
   - (a) The concentration that kills the same number of bacteria
   - (b) Penetrating power of the agent
   - (c) Proportion of bacteria killed
   - (d) Zone of inhibition of bacteria
   - (e) None of these

5. If 0.1 ml of serum is added to 0.9 ml of saline solution and mixed and if one-half the mixture is transferred to a second tube containing 0.5 ml of saline solution, the dilution in the second tube (after mixing) would be:
   - (a) 1:5
   - (b) 1:10
   - (c) 1:20
   - (d) 1:40
   - (e) 1:100

6. Which of the following antimicrobial agents interferes with cell wall synthesis?
   - (a) Penicillin
   - (b) Colistin
   - (c) Isoniazid
   - (d) Tetracycline
   - (e) Actinomycin D

7. Which of the following antimicrobial agents does *not* inhibit protein synthesis?
   - (a) Chloramphenicol
   - (b) Tetracycline
   - (c) Kanamycin
   - (d) Polymyxin
   - (e) Streptomycin

8. The antimicrobial action of the polymyxins and colistin is related primarily to:
   - (a) Interference in cell membrane function
   - (b) Inhibition of cell wall synthesis
   - (c) Inhibition of nucleic acid synthesis
   - (d) An unknown mechanism
   - (e) Inhibition of protein synthesis

Part IV

When the paired statements are compared in a quantitative sense, one will be found to exert or imply a greater effect or value. Indicate which is the stronger statement in the following way:

"a" means that the first statement is the stronger one. "b" indicates that the second statement represents a greater effect.

_____ 1. Effectiveness of chemical disinfectant in saline solution (a)
           Effectiveness of chemical disinfectant in serum (b)

_____ 2. Lethal effect of temperatures 10° above maximum growth temperature (a)
           Lethal effect of temperatures 10° below minimum growth temperature (b)

_____ 3. Effect of chemical disinfectant at pH 7 (a)
Effect of chemical disinfectant at pH 4 (b)

_____ 4. Time required for sterilization in autoclave (a)
Time required for sterilization in hot air sterilizer (b)

_____ 5. Time required for sterilization in pressure steam sterilizer (a)
Time required for sterilization in high-prevacuum steam sterilizer (b)

_____ 6. Time required for sterilization in pressure steam sterilizer (a)
Time required for sterilization in ethylene gas sterilizer (b)

_____ 7. Sterilizing action of hot air at 170° C for 15 minutes (a)
Sterilizing action of free-flowing steam for 15 minutes (b)

_____ 8. Disinfectant action of ethyl alcohol against tubercle bacilli (a)
Disinfectant action of mercurials against tubercle bacilli (b)

_____ 9. Disinfectant action of iodophors against spores (a)
Disinfectant action of glutaraldehyde against spores (b)

_____ 10. Likelihood of drug resistance in tuberculosis if streptomycin is used alone (a)
Likelihood of drug resistance if drug combinations are used in tuberculosis (b)

_____ 11. Sensitivity of vegetative cells to chemical treatment (a)
Sensitivity of spores to chemical treatment (b)

_____ 12. Duration of active immunity (a)
Duration of passive immunity (b)

_____ 13. Temperature for sterilization by autoclaving (in pressure steam sterilizer) (a)
Temperature for sterilization by dry heat (in hot air sterilizer) (b)

_____ 14. Variability in gram-staining reaction of gram-positive organisms (a)
Variability in gram-staining reaction of gram-negative organisms (b)

_____ 15. Optimal growth temperature of a psychrophile (a)
Optimal growth temperature of mesophile (b)

_____ 16. Spectrum of microbes sensitive to penicillin (a)
Spectrum of microbes sensitive to tetracycline (b)

UNIT FIVE
MICROBES
PATHOGENS AND PARASITES

18 Pyogenic cocci

The pathogenic cocci are often called *pyogenic* cocci because of their ability to cause pus formation. The important gram-positive cocci are the staphylococci, the streptococci, and the pneumococci (genera *Staphylococcus* and *Streptococcus*).

The pneumococcus has long been considered as a streptococcus by many authorities who have even classified it in the same genus. Recognizing the biologic similarity of the two, the eighth edition of *Bergey's Manual* reclassifies the pneumococcus. Formerly *Diplococcus pneumoniae*, it is now *Streptococcus pneumoniae*, a name long used for it in some parts of the world. In this chapter, because of certain properties peculiar to it and because of its relation to an important disease, it is discussed separately.

STAPHYLOCOCCUS SPECIES (THE STAPHYLOCOCCI)

General characteristics. Staphylococci occur typically in grapelike clusters (Fig. 18-1). Under special conditions they may occur singly (Fig. 18-2), in pairs, or in short chains. They are gram positive, nonmotile, and nonsporeforming and grow luxuriantly on all culture media. Most grow best in the presence of oxygen, but they easily grow in its absence. A few are strictly anaerobic. They grow best between 25° and 35° C but may grow at a temperature as low as 8° C or as high as 48° C.

When staphylococci are grown on blood agar, there is characteristic pigment production, the colors ranging from a deep gold to lemon yellow to white. The deep golden color observed with growth was originally responsible for the species name *aureus*. We now know that there are white variants of the golden staphylococci.

Staphylococci are most resistant to the action of heat, drying, and chemicals. Although most vegetative bacteria are destroyed by a temperature of 60° C for 30 minutes, staphylococci frequently resist a temperature of 60° C for 1 hour, and some strains may resist a temperature of 80° C for 30 minutes. In dried pus they live for weeks or months. Another mechanism for survival of the staphylococcus is its tol-

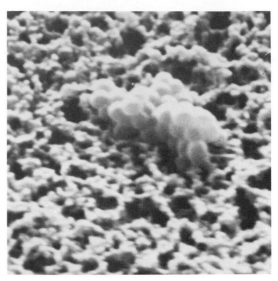

FIG. 18-1. Staphylococci, scanning electron micrograph to show typical grapelike cluster. (Courtesy Millipore Corp., Bedford, Mass.)

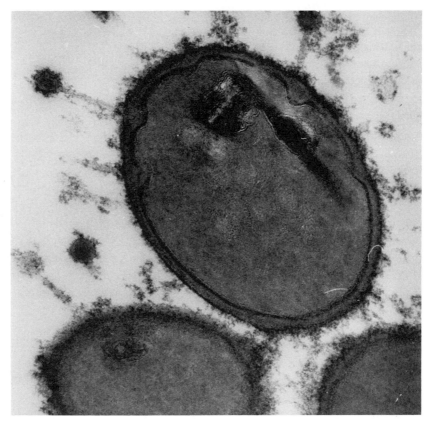

FIG. 18-2. *Staphylococcus,* electron micrograph. Note attached staphylophages. (Courtesy S. Tyrone, Dallas, Tex.)

erance to salt or a salty medium, as would be found in preserved foods. Staphylococci also tend to become resistant to the sulfonamides and antibiotics. They adapt quickly and easily to such agents. Many (about 80%) are penicillin resistant.

There are two species of note for the genus *Staphylococcus: Staphylococcus aureus (Staph. aureus)* and *Staphylococcus epidermidis (Staph. epidermidis)*, two organisms separated by a biochemical reaction and the presence of an enzyme. *Staph. aureus* ferments mannitol, which the other staphylococcus does not, and *Staph. aureus* is coagulase positive. *Staph. epidermidis* is coagulase negative.

Toxic products. Several important metabolic products are elaborated by staphylococci, some with toxic properties such as those that (1) destroy red blood cells (hemolysins, staphylolysins), (2) destroy leukocytes (leukocidin), (3) cause necrosis of tissue (necrotizing exotoxin), (4) produce death (lethal factor), and (5) cause gastroenteric symptoms (enterotoxin). Probably no single strain produces all of these poisons, and many produce none of them.

In addition, some staphylococci produce the enzyme *coagulase*, which causes the plasma of blood to clot. The *coagulase test* indicates the presence of coagulase, which is generally considered the best single bit of evidence that a given staphylococcus is a pathogen. Because of coagulase activity, a surface layer of fibrin accumulates on an individual staphylococcus and protects it from phagocytic attack. The production of

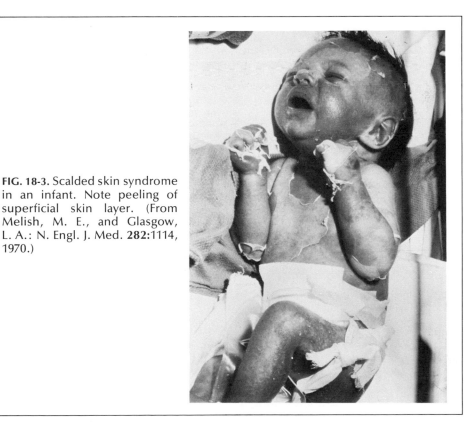

FIG. 18-3. Scalded skin syndrome in an infant. Note peeling of superficial skin layer. (From Melish, M. E., and Glasgow, L. A.: N. Engl. J. Med. **282:**1114, 1970.)

coagulase correlates with the production of other toxic products. Staphylococci also produce staphylokinase, which dissolves fibrin, and hyaluronidase.

Pathogenicity. Some staphylococci are nonpathogenic; others produce severe infections. Severe infections are caused by *Staph. aureus. Staph. epidermidis,* though sometimes responsible for very mild, limited infections, is generally nonpathogenic except under unusual medical circumstances.

The best known staphylococcal infections are those of the skin and superficial tissues of the body, the features of which are greatly influenced by the age of the patient. The scalded skin syndrome (Fig. 18-3) is an example of staphylococcal disease produced by certain strains with a predilection for the newborn and very young. Its manifestations are clearly depicted by the name. Staphylococci cause boils, pustules, pimples, furuncles, abscesses, carbuncles, paronychias, and infections of accidental or surgical wounds.

Staphylococci also produce systemic disease, and all organ systems in the body may be affected. They are one cause of pneumonia, empyema, endocarditis, meningitis, brain abscess, puerperal fever, parotitis, phlebitis, cystitis, and pyelonephritis. Staphylococci can infect a valve prosthesis in the heart. Staphylococcal pneumonia was a fatal complication of influenza in recent epidemics. Staphylococci are the commonest cause of osteomyelitis and impetigo contagiosa (Fig. 18-4). Systemic staphylococcal disease is often acquired in the hospital, especially in patients already ill with a serious disease, and staphylococcal pneumonia as a superinfection is a threat after large doses of antibiotics have been given.

Staphylococcal septicemia assumes two forms. The first is a fulminating, profound toxemia with death a few days later. The second and more frequent is of longer duration and marked by severe clinical disease, formation of metastatic abscesses in different parts of the body, and slightly reduced mortality. Staphylococcal septicemia may be a primary condition, but mostly it results from secondary invasion of the bloodstream by organisms from a localized site of infection in the skin (often a trivial one). It can come from an infected wound, a dental abscess, pneumonia, or an infected intravenous catheter site. An indwelling intravenous catheter should not remain in place longer than 3 to 4 days because of the likelihood of serious infection. Boils about the nose and lip are easily so complicated and for this reason should not be traumatized.

There is an area of the face, triangular in shape, lying with its base along the opening of the mouth and its apex in the region above the upper part of the nose. It is called the "dangerous triangle" (Fig. 18-5) because of the threat to a person's life if infection originating there spreads backward into the cranial vault. It is a peculiar area in that anatomic factors operate there to favor just such a disaster. (A comparable lesion in another part of the body would be inconsequential.) These factors include poor connective supports that provide no mechanical barriers, veins without valves connecting with veins that drain backward, and muscles of the face that are more or less constantly in motion.

If piercing of the ears (for cosmetic reasons) is done in an unsanitary setting, there is the danger of secondary infection with potentially deadly staphylococci.

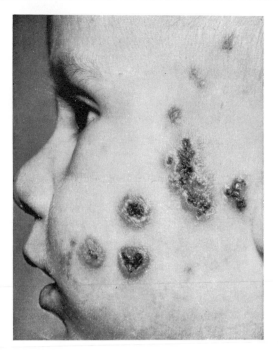

FIG. 18-4. Impetigo contagiosa, highly communicable skin disease. Note various sized lesions, some dark and encrusted. (From Top, F. H., Sr., and Wehrle, P. F., editors: Communicable and infectious diseases, ed. 8, St. Louis, 1976, The C. V. Mosby Co.)

FIG. 18-5. Triangular area of face where staphylococcal infection is very dangerous.

They can enter such a wound that is kept open for several days, and in rare instances the spread of infection to the bloodstream has been tragic.

On occasion, implantation of staphylococci in the intestinal tract after antibiotic therapy results in enteritis or enterocolitis causing dysentery. Staphylococcal food poisoning is the most frequent type of food poisoning from ingestion of a bacterial toxin.

Staphylococci are an important cause of suppurative conditions in cattle and horses. Mastitis of staphylococcal origin in cows can be transmitted to other cows by the hands of the milker. Staphylococcal bacteremia may follow tick bites in lambs.

Pathology. The hallmark of staphylococcal infection is the *abscess*, the type lesion. It reflects the excellent pus-forming ability of staphylococci and their limited capacity for spread. Since the microbes reside on the skin, this is the area most frequently involved. A boil is a skin abscess. The type lesion is modified by anatomic location and degree of involvement. Staphylococcal pneumonia means multiple abscesses in the lungs. Pyelonephritis means multiple abscesses extending down the tubular system of the kidney. Staphylococcal septicemia is the development of multiple abscesses over the body; the word *pyemia*, literally meaning "pus in the blood," is more appropriate. Staphylococci, as an important cause of wound infections, are responsible for pus formation therein.

Hospital-acquired (nosocomial) infection. Because of man's intimate contact with this microbe and because of its ubiquity, there are unique features to its infections at any time. For years, staphylococcal infections have been especially troublesome in hospitals all over the world. To begin with, there are several background factors. In one regard, the problem complicates modern antibiotic therapy. Important antimicrobial agents are freely given. Antibiotics generally are bacteriostatic, *not* bactericidal. Staphylococci are well endowed for survival, and consequently antibiotic-resistant strains develop. Moreover, with prepaid medical plans and medical advances favoring early detection of disease, more persons are being treated in hospitals. Modern surgery is expanded. Complicated surgical technics are done with a greater exposure of tissue at operation for a longer period of time than ever before. In certain instances drugs such as the corticosteroids that depress the patient's resistance to infection are indicated.

The hospital population is already large and complex, but it is growing with the increased demand for persons with specialized technical skills. Here, then, is a patient, sometimes seriously ill, confined within a large institution where the contacts with all kinds of persons are many and varied. Consider these factors in light of the fact that staphylococcus is everywhere. Is it any wonder that it is such a troublemaker!

Four major categories of disease caused by antibiotic-resistant staphylococci related to hospital-acquired infection are noted:

1. Skin abscesses (impetigo and pyoderma) in newborn infants (Many of these infants develop breast abscess. Fatal staphylococcal pneumonia or septicemia is prevalent. The nursing mother can pick up a virulent organism from her baby. Abscess formation in her breast may result.)
2. Wound infections, especially of surgical wounds
3. Secondary staphylococcal infections in hospitalized elderly and debilitated persons
4. Gastroenteritis, as a result of a change in the bacterial flora of the intestinal tract

Sources and modes of infection. Staphylococci are normal inhabitants of the skin, mouth, throat, and nose of man. They live in these areas without effect, but once past the barrier of the skin and mucous membrane, they can cause extensive disease. They pass through the unbroken skin via the hair follicles and ducts of the sweat glands under certain conditions. The natural invasive traits of staphylococci and the resistance of the body are so well balanced that infection probably never occurs unless a highly virulent organism is encountered or body resistance is lowered. As a rule, a localized process such as an abscess or a boil is first formed. Healing without dissemination of infection usually takes place, but in some cases organisms do escape from the localized process, invade the bloodstream, and affect distant parts of the body.

Crucial to the spread of staphylococcal infections are direct, person-to-person contacts. For example, in the hospital the hands of the doctor or medical attendant may carry the infection from one patient to another. The hospital staff generally has a higher carrier rate, and cross infection is significant. Staphylococci are abundant in hospitals. Nasal carriers are a source. Virulent organisms may also be passed to man from livestock and household pets.

Bacteriologic diagnosis. A bacteriologic diagnosis of a staphylococcal infection is made easily by smears and cultures. When blood cultures for staphylococci are made, special pains must be taken to exclude those inhabiting the skin. The coagulase test (slide and test tube methods) is performed to differentiate nonpathogenic from pathogenic staphylococci. A freshly isolated staphylococcus is most likely to be a virulent pathogen if it produces a yellow pigment, hemolyzes blood, ferments mannitol, elaborates deoxyribonuclease, and is coagulase positive.

The bacteriophage typing of staphylococci deserves special mention. Bacteriophages (p. 503) are viruses that attack bacteria and in certain instances dissolve the bacterial cell parasitized. The action of phages is specific; that is, only a certain phage or group of phages affects the given strain of bacteria. It has been found that specific bacteriophages (staphylophages, Fig. 18-2) react with about 60% of coagulase-positive staphylococci. Coagulase-negative strains of staphylococci are not so susceptible. Because of the specificity, phage typing of staphylococci can be done. For convenience, bacteriophages have been given identifying numbers, and the strains of staphylococci related to these particular phages are designated by the number of the bacteriophage or phages dissolving them. On this basis, *Staph. aureus* can be classified into lytic groups of staphylococcal typing phages* as follows:

Group I	29, 52, 52A, 79, and 80
Group II	3A, 3B, 3C, 55, and 71
Group III	6, 7, 42E, 47, 53, 54, 75, 77, and 83A
Group IV	42D
Not allotted	81 and 187

Immunity. Man possesses considerable natural immunity to staphylococci. Specific serum antibodies to staphylococci can be demonstrated in most human beings to suggest that almost everyone has had a staphylococcal infection at some time and that acquired immunity of a protective nature may exist. However, there is little reason to think that acquired immunity is practical against serious infection. Patients with debilitating diseases, especially diabetes, are especially vulnerable.

Prevention and control of staphylococcal infection. The key measures in the control of staphylococcal infections are the maintenance of good housekeeping standards† and adherence to strict aseptic technics. There is strong evidence that currently pathogenic strains of staphylococci are as susceptible to chemical germicides as are the ordinary nonpathogenic strains. The detection of nasal carriers, especially in the nurseries for newborn infants, is crucial, and phage typing of coagulase-

---

*Recommended for typing of *Staphylococcus aureus* of human origin by the International Subcommittee on Phage Typing of Staphylococcus.

†As Welton Taylor has said, "Perhaps the most important ally in the hospital's campaign against hospital-borne infection is the housekeeper, who merely has to employ the common-sense sanitation that any good housewife knows." (From Hosp. Tribune, **7:**3, Oct. 9, 1967.)

positive staphylococci has proved valuable in the epidemiologic study of hospital-acquired infections to track down the sources to carriers or other foci.

If one strain of coagulase-positive staphylococcus is biologically anchored to a given site in the human body, another coagulase-positive strain cannot implant there. The colonization of the second strain is blocked. There is *bacterial interference* between the two. In hospital nurseries, practical application of this phenomenon is sometimes made to protect newborn infants against current epidemic strains of staphylococci.

STREPTOCOCCUS SPECIES (THE STREPTOCOCCI)
General considerations

The term *streptococcus* is a morphologic one to include cocci that occur in pairs and chains. A biochemical feature distinguishing the genus *Streptococcus* is the fermentation of glucose by the hexose diphosphate pathway to yield mainly dextro-rotatory lactic acid. Within this genus of organisms are found significant variations in cultural characteristics and disease-producing properties. Some produce deadly disease, others do so only under special conditions, and still others are nonpathogenic. As a whole, streptococci are probably responsible for more illness and cause more different kinds of disease than any other group of organisms. They attack any part of the body and can cause primary as easily as secondary disease. They attack both man and animals. Some occur as saprophytes in milk and other dairy products.

A great deal of research is carried out on the biology of the streptococcus. The individual coccus is being studied in great detail as are the reactions induced on contact with a given host.

Characteristics. Streptococci of varying sizes are arranged in long or short chains. Long chains contain 50 or more cocci; short chains contain as few as four or six, the bacteria in pairs within the chains. Chains form when bacteria divide in one plane and still cling together. Streptococci are nonmotile, gram-positive organisms that do not form spores. Capsule formation is variable. Some species form distinct capsules; most do not.

The majority of streptococci grow best in the presence of oxygen but may grow in its absence. A few species are strict anaerobes. Streptococci grow well on all fairly rich media, and visible growth usually appears within 24 to 48 hours. Growth is especially luxuriant on hormone media or media containing unheated serum, whole blood, or serous fluid and occurs in milk. They grow best at body temperature but may grow through a temperature range of 15° to 45° C. Most are not soluble in bile and do not ferment inulin.

Streptococci may remain alive in sputum or other excreta for several weeks and in dried blood or pus for several months. They are destroyed at a temperature of 60° C within 30 to 60 minutes or in 15 minutes by 1:200 phenol solution, tincture of iodine, or 70% isopropyl alcohol. Penicillin is the most effective antibiotic against most types; it is the drug of choice against beta hemolytic streptococci but ineffective against the enterococci. Streptococci readily acquire resistance to other antimicrobial drugs.

Classification. Streptococci may be classified on the basis of their action on blood agar, biochemical properties, or serologic behavior (agglutination and precipitation).

To determine their action on blood agar, plates of blood agar are prepared. Nutrient agar is melted and cooled to 45° C; sterile defibrinated blood is added: 5 to 10 parts of blood to 100 parts of nutrient agar. The mixture is poured into Petri dishes and allowed to cool. The plates are streaked with the material containing the streptococci and incubated at 37° C for 24 hours or longer. Corresponding to their action on blood agar plates, streptococci are classified broadly as:

1. Alpha hemolytic or viridans—colony surrounded by green halo; hemolysis slight or incomplete
2. Beta hemolytic—colony surrounded by clear, wide, colorless zone of hemolysis (Fig. 18-6)
3. Gamma—colonies showing neither hemolysis nor color change (Fig. 18-7)

As a general rule, the beta hemolytic streptococcus is the most virulent and associated with acute fulminating infections in man; the viridans type is associated with low-grade chronic infections, including such nonlethal ones as tooth abscesses and sinus infections. However, infections with alpha hemolytic streptococci may be as serious as those with beta hemolytic streptococci—for example, subacute bacterial endocarditis. Many, but not all, strains of streptococci of the gamma type are nonpathogenic.

Another classification divides streptococci into the (1) pyogenic group, (2) viridans group, (3) lactic group, and (4) enterococcus group (Table 18-1). Enterococci, of which *Streptococcus faecalis* is representative, frequently infect the genito-

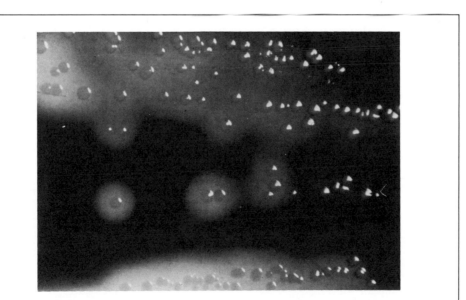

FIG. 18-6. Beta hemolytic streptococci on routine blood agar plate showing wide clear zones about small gray colonies.

TABLE 18-1. BIOLOGIC PROPERTIES OF STREPTOCOCCI

Division	Hemolytic streptococcus	Viridans streptococcus	Enterococcus	Lactic streptococcus
Serologic group	A, B, C, E, F, G, H, K, L, M, O	None	D	N
Hemolysis on blood agar	Usually beta	Usually alpha	Alpha, beta, or gamma	Alpha or gamma
Growth in 0.1% methylene blue milk	−	−	+	+
Growth in 6.5% salt broth	−	−	+	−
Growth on 40% bile blood sugar	−	±	+	+
Antibiotic susceptibility (bacitracin)	Usually sensitive	May be resistant	May be resistant	(Nonpathogenic)

TABLE 18-2. SEROLOGIC GROUPS OF STREPTOCOCCI

Group	Species	Significance
A	Streptococcus pyogenes	Important human diseases (infection by beta hemolytic streptococci initiates acute rheumatic fever); group sensitive to penicillin
B	Streptococcus agalactiae	Bovine mastitis
C	Streptococcus equi Streptococcus zooepidemicus Streptococcus equisimilis Streptococcus dysgalactiae	Animal diseases; mild respiratory infections in man
D	Streptococcus faecalis Streptococcus faecalis, subsp. liquefaciens Streptococcus faecalis, subsp. zymogenes Streptococcus faecium	Enterococci; genitourinary tract infections, endocarditis, wound infections in man; found in dairy products
E		Disease of swine; found in normal milk
F	Streptococcus minutus	Found in respiratory tract of man
G	Streptococcus anginosus	Mild respiratory infections in man; genital infections in dogs
H	Streptococcus sanguis	Found in respiratory tract of man
K	Streptococcus salivarius	Found in respiratory tract of man
L		Genital tract infections in dogs
M		Genital tract infections in dogs
N	Streptococcus lactis Streptococcus cremoris	Lactic group; found in dairy products
O		Viridans group; subacute bacterial endocarditis; found in upper respiratory tract in man
		Microaerophilic streptococci
		Anaerobic streptococci—13 species

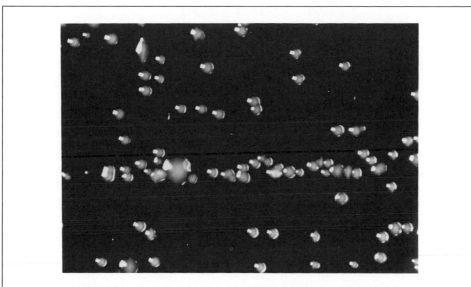

FIG. 18-7. Gamma hemolytic streptococci with no hemolysis about their colonies.

urinary tract and occasionally the respiratory tract; they are rare causes of subacute bacterial endocarditis. The lactic group, found in sour milk, includes *Streptococcus lactis* and *Streptococcus cremoris*. It is not pathogenic but is important in the dairy industry and in certain types of biologic assay. The representative of the viridans group is *Streptococcus mitis*, often referred to as *Streptococcus viridans*, and responsible for 90% of the cases of subacute bacterial endocarditis. The type organism of the pyogenic group is *Streptococcus pyogenes*, a serious pathogen.

By serologic (precipitin) methods, streptococci fall into 13 groups (the classification of Lancefield and others), which correspond in a general way to their pathologic action (Table 18-2). These groups may be further subdivided into types given Arabic numbers. For over 55 types in group A, the antigenic substance determining type specificity is the *M protein*. Superficially attached to the cell wall, it protects the virulent streptococci of this group from the phagocytes of the host. Most human infections are due to group A streptococci *(Strep. pyogenes)*. Streptococci of all other groups except those of N are indigenous to man and therefore potential pathogens. Certain members of groups B, C, D, H, K, and O and all members of group N are nonhemolytic. Although not listed in Table 18-2, groups P, Q, R, S, and T have been described.

Toxic products. Streptococci elaborate extracellular poisons, some of which can be called exotoxins. Among these are (1) hemolysins, (2) leukocidin, (3) streptokinase, (4) streptodornase, (5) hyaluronidase, and (6) erythrogenic toxin.

The hemolysins or *streptolysins* produced by streptococci are of two types, S and O. Streptolysin S is produced primarily by members of group A; streptolysin O is elaborated by most members of group A, by the "human" members of group C, and by certain members of group G. Most workers believe the leukocidin of *strep-*

tococci to be identical with streptolysin O. *Streptokinase* or *fibrinolysin* is notable in that it activates an enzyme that destroys fibrin (the framework of blood clots). Blood clots play an important part in wound healing and in blocking the spread of local infections.

Streptodornase (streptococcal deoxyribonuclease) acts to liquefy thick, tenacious exudates such as are seen in pneumonia. The enzymatic activity of the deoxyribonuclease is directed against the deoxyribonucleoprotein content of the exudate, the factor responsible for its viscosity. *Hyaluronidase* (the spreading factor, *invasin*) is a factor increasing the permeability of the tissues to bacteria and toxins by breaking down hyaluronic acid, one of the substances that cement tissue cells together. The *erythrogenic toxin* produces erythema or redness when injected into the superficial layers of the skin, and with a large enough dose, a generalized rash. This scarlet fever toxin of Dick occurs in two immunologic types, A and B.

Pathogenicity. Streptococci may be responsible for a localized inflammatory reaction, such as an abscess, or a generalized reaction, such as septicemia. The nature of the lesion depends on the virulence of the streptococci, the number introduced into the body, the mode of introduction, the tissue invaded, and the resistance of the host.

Pathology. Pathologically, the type lesion of hemolytic streptococci is the diffuse, ill-defined spreading lesion of *cellulitis*. The exudate contains few cells, consisting largely of fluid with little fibrin. The toxic products of the microbes greatly aid their extension through both natural tissue and inflammatory barriers, and they tend to infect lymphatic vessels at the site of invasion. Many well-known forms of streptococcal infection are an expression of cellulitis. Erysipelas is cellulitis with a specific anatomic pattern; septic sore throat is cellulitis of the throat.

Allergic manifestations may follow certain streptococcal infections (usually of the throat), for instance, acute rheumatic fever and one form of kidney disease (glomerulonephritis).

As a rule, the more virulent an infection, the more virulent are the streptococci isolated for animals of the same species. Usually their virulence is lowered when they are introduced into animals of another species. When transferred from one animal to another, streptococci tend to produce the same type of lesion in the new host as in the original one. Rabbits and white mice are more susceptible to streptococci of human origin than are other laboratory animals.

STREPTOCOCCI IN HUMAN DISEASES. In addition to being the cause of erysipelas and two epidemic diseases, scarlet fever and epidemic sore throat, streptococci belonging to group A *(Strep. pyogenes)* are the most common cause of acute endocarditis, septicemia, and puerperal sepsis. They may cause pneumonia, boils, abscesses, cellulitis, peritonitis, tonsillitis, lymphangitis, infection of surgical wounds, osteomyelitis, and empyema. From an infection of the middle ear (otitis media), streptococci may spread to the mastoid cells and cause mastoiditis. From either the middle ear or the mastoid cells, spread of infection to the meninges means streptococcal meningitis.

Strep. pyogenes is responsible for most bronchopneumonias complicating whooping cough, measles, and influenza. Such bronchopneumonias are often fatal and may reach epidemic proportions when outbreaks of whooping cough, measles, or influenza occur in communities containing numerous carriers of *Strep. pyogenes.* Streptococcal pneumonia is often the terminal phase of chronic diseases like tuberculosis and cancer. Streptococci as well as staphylococci (often acting together) are responsible for the highly communicable skin disease known as *impetigo contagiosa.*

STREPTOCOCCAL INFECTIONS IN LOWER ANIMALS. The majority of streptococcal infections in lower animals are caused by streptococci in Lancefield's groups B and C. Strangles, an acute communicable disease of the upper respiratory passages of horses, is caused by *Strep. equi.* Streptococcal mastitis, a serious disease of cows that renders milk unfit for use, is caused by *Strep. agalactiae* and most likely spread from cow to cow by the hands of the milker. *Strep. agalactiae* does not affect man.

Sources and modes of infection. Streptococci are normal inhabitants of the mouth, nose, throat, and respiratory tract. They may be conveyed from person to person by direct contact or by contaminated objects, hands, and surgical instruments. The hands are important conveyers of infection in puerperal sepsis and wound infections. Milk can be an important source. Streptococcal diseases, especially scarlet fever and septic throat, may be spread by milk that has been contaminated with the mouth and nose secretions of a carrier or by milk from a cow with mastitis caused by *Strep. pyogenes.*

Streptococci usually enter the body by the respiratory tract or through wounds of the skin. Only a minute abrasion is necessary, and streptococcal infections therein have many times in the past led to fatal septicemias in physicians and nurses. Streptococci leave the body by way of the mouth and nose and in the exudates from areas of infection. The nasal carrier is a source of infection. Enterococci are normal inhabitants of the intestinal canal and are excreted in the feces.

Laboratory diagnosis. Streptococci are detected by smears and cultures from the site of disease. In septicemias caused by beta hemolytic streptococci, the organisms can usually be detected by blood cultures. In subacute bacterial endocarditis, since the viridans organisms escape into the blood intermittently, repeated cultures may have to be made.

A presumptive test of value in the identification of streptococci is the *bacitracin disk test.* Even in low concentrations this antibiotic appears to be specifically active against members of Lancefield group A and without effect on other groups. A paper disk containing a known unit of antibiotic is placed on a blood agar plate previously streaked with the streptococcus in question. A zone of inhibition of growth found around the disk identifies it in group A.

When infection with streptococci that produce streptolysin O takes place, antibodies against the streptolysin O appear in the blood of the patient. The detection and quantitation of these antibodies in the *antistreptolysin O titer (ASTO)* form a useful procedure in diagnosis and management of streptococcal infections of a chronic and persistent nature such as acute rheumatic fever and acute hemorrhagic glomerulonephritis (allergic reactions to beta hemolytic streptococci). The course of these diseases relate to the titer, and the test helps to identify certain diseases similar to rheumatic fever where the streptococcus is not comparably implicated, since the titer in the latter is not significant.

Immunity. With the exception of scarlet fever, streptococcal infections are not followed by an immunity, and in scarlet fever the immunity is established only against scarlet fever toxin, not against the organisms.

Prevention. When caring for a patient with a streptococcal infection, nurses should remember that they are dealing with an infection that may be most virulent and easily spread. Physicians and nurses attending such infections, especially scarlet fever and erysipelas, should not attend an obstetric case or surgical operation until they are incapable of spreading the infection. Obstetric cases should be handled with strictest aseptic care because after delivery the uterus is extremely vulnerable.

The buccal and nasal secretions from a patient with streptococcal broncho-pneumonia should be handled in the same manner as those from a patient with diphtheria. A patient with streptococcal bronchopneumonia should be isolated from patients with measles or influenza. Conditions favoring contact infection should be avoided, and all wounds and abrasions on the body should be thoroughly disinfected.

The fact that more than 55 types exist among group A streptococci, all immunologically specific, explains the delay in the development of a clinically feasible vaccine, which centers around the highly antigenic M protein in the streptococcal cell membrane.

Scarlet fever

Scarlet fever (scarlatina) is an acute infection of childhood described by sore throat, severe constitutional manifestations, and a distinct skin eruption with massive exfoliation (Fig. 29-1). The nasopharynx and tonsils may be covered with a membrane, and the lymph nodes, especially those of the neck, are swollen and inflamed. The white blood count ranges from 15,000 to 30,000 cells/mm^3, of which 85% to 95% are neutrophils.* Scarlet fever is caused by streptococci that produce erythrogenic toxin. In almost all cases streptococci are of Lancefield's group A, and only rarely is scarlet fever produced by streptococci belonging in groups C and D.

The present opinion is that scarlet fever and streptococcal sore throat are different manifestations of the same basic disease. If the streptococci causing the sore throat produce erythrogenic toxin and the recipient of the infection is not immune to the toxin, scarlet fever results. If the streptococci do not produce erythrogenic toxin or if the recipient is immune to the toxin, then only the sore throat is present. There is a close relation between the streptococci causing scarlet fever, erysipelas, and puerperal sepsis.

In the past, scarlet fever was a disease of great severity and often fatal. Now it is relatively benign but may be complicated by suppurative otitis media and peritonsillar abscess or be followed by rheumatic fever and acute nephritis. It is much more common in European countries than in North America.

---

*Normal white cell count is 5000 to 9000 cells/mm^3 with 55% to 65% neutrophils.

Sources and modes of infection. The sources of scarlet fever are the nose and throat secretions of patients or carriers and pus from infected lymph nodes, ears, and other lesions. The organisms are present throughout the course of the illness and may persist in the nose and throat or in the exudates for weeks or months thereafter. As long as a person harbors the organisms, he is a dangerous source of infection. The desquamated scales are not infectious.

The organisms usually enter the body through the mouth and nose, less often through wounds, burns, and the parturient uterus. Infection may be transmitted by direct contact or by contaminated objects such as handkerchiefs, towels, pencils, toys, and dishes.

Erythrogenic toxin. The erythrogenic toxin of scarlet fever streptococci, released by the organisms at the primary site and absorbed into the body, brings about the rash and other constitutional effects. It may be found in the blood in a concentration as high as 300 units/ml and also in the urine. It is prepared artificially by growing scarlet fever streptococci for 5 days in broth, after which the bacteria are removed by filtration. The filtrate contains the toxin. When toxin prepared in this manner is injected in very small amounts into the skin of persons susceptible to scarlet fever, it gives rise to an inflammatory reaction, the basis of the *Dick test*. The unit of measurement is known as the skin test dose (STD), the smallest amount of scarlet fever toxin causing an inflammatory reaction in the skin of a susceptible person. Scarlet fever toxin can induce an active immunity with the formation of antitoxin when injected into a suitable animal.

Immunity. Immunity to scarlet fever relates to scarlet fever antitoxin in the blood. Infants inherit an immunity from their mothers that is lost within a year, and susceptibility increases until the sixth year. After that, susceptibility decreases until adult life, at which time most people are immune. An attack of scarlet fever is usually followed by permanent immunity. Remember that although an immune person will not be harmed by scarlet fever toxin, the streptococci themselves may invade his body and cause tonsillitis, abscesses, or otitis media. Immune persons may harbor the organisms for a long time and spread the infection widely.

DICK TEST. The Dick test is performed by injecting between the layers of the skin of the forearm 0.1 ml of scarlet fever toxin so diluted that one STD is given. In immune persons the antitoxin in the blood neutralizes the injected toxin, and no reaction occurs. In susceptible persons the toxin injures the cells around the injection site, producing within 24 hours a zone of inflammation and redness at least 1 cm in diameter. The test is positive at the beginning but becomes negative during the course of scarlet fever.

SCHULTZ-CHARLTON PHENOMENON. If a small amount of the serum of a person convalescent from scarlet fever or of an animal immunized against scarlet fever is injected intradermally into an area of scarlet fever rash, the skin blanches at the injection site because the toxin in the area is neutralized by antitoxin. This test differentiates scarlet fever from measles, German measles, and other skin diseases.

Prevention. The patient with scarlet fever should be isolated, and discharges from the mouth and nose as well as all contaminated articles should be disinfected.

The disinfecting procedures are the same as those for diphtheria (p. 431). Attendants on a patient should exercise every precaution to prevent the spread of infection to others, especially to obstetric or surgical cases. The patient should remain isolated until the discharges from the mouth and nose are free of scarlet fever streptococci and all complications have healed. Remember that during epidemics of scarlet fever persons with rhinitis and sinusitis may just as effectively spread the disease as those with a skin eruption. Pasteurization prevents milkborne epidemics.

Erysipelas

Erysipelas* (St. Anthony's fire) is an acute inflammation of the skin caused by hemolytic streptococci of Lancefield's group A or, infrequently, of group C or D. The streptococci grow at the periphery, not in the center, of the inflamed area at the site of infection, and they elaborate toxic substances effecting the constitutional state.

Mode of infection. The portal of entry is a wound, fissure, or abrasion. A hard, red thickening of the skin beginning at the infection site extends peripherally. The streptococci grow almost exclusively in the lymph channels of the inflamed area, and, as the disease progresses, they spread peripherally several centimeters beyond the line limiting the area of obvious inflammation. When streptococci enter the blood, the prognosis is bad. Erysipelas may be complicated by abscesses, pericarditis, arthritis, endocarditis, septicemia, and pneumonia. Patients with uncomplicated disease or without open wounds or superficial discharges will not transmit the infection to others. Erysipelas may be associated with other group A streptococcal infections.

Immunity. Instead of inducing immunity, an attack of erysipelas seems to render the patient more vulnerable to future attacks.

Streptococcal sore throat (septic sore throat)

Septic sore throat ("strep" throat) is an ulcerative inflammation of the throat with severe symptoms and a high mortality caused by hemolytic streptococci belonging to Lancefield's group A or, in a small proportion of cases, group C. It may be transferred by direct contact or droplet infection. Streptococcal sore throat may extend to the lungs to produce streptococcal pneumonia, and like scarlet fever, it may be followed by nephritis or rheumatic fever.

Rheumatic fever

Acute rheumatic fever, the forerunner of rheumatic heart disease, has a clinical course similar to an acute infection. In the acute phase there is fever, an increased pulse rate (tachycardia), carditis (inflammation of the heart), and a characteristic type of polyarthritis. Rheumatic fever is more common in the northern than in the southern climates and is infrequent in the tropics. It is most common between the sixth and twelfth years of life.

---

*Erysipelas should not be confused with erysipeloid, a localized infection of the skin caused by *Erysipelothrix rhusiopathiae* (gram-positive rod with a tendency to form long filaments) and occurring in those who handle fish or meats.

More than 50% of the attacks of acute rheumatic fever are preceded by tonsillitis or severe sore throat caused by hemolytic streptococci. An important diagnostic rise in the serum titer of antistreptolysin O occurs in more than 80% of the patients. Rheumatic fever tends to be a disease of long standing, with repeated attacks of the acute process. This favors the development of complications that add up to serious heart disease.

To prevent the harmful effects of rheumatic heart disease is to stop the recurrent attacks of acute rheumatic fever. Patients are protected as much as possible from streptococcal infections and even are given antibiotics prophylactically at times of the year when the incidence of such infections is high.

Puerperal sepsis

Puerperal sepsis (puerperal septicemia) is usually caused by a hemolytic streptococcus from the nose and throat of the patient herself or those in close association with her. It reaches the uterus via contaminated hands or instruments. Most of such streptococci belong to Lancefield's group A of organisms exogenous to the generative tract, but from 20% to 25% of the cases come from anaerobic streptococci, which are normal inhabitants of the vagina.

STREPTOCOCCUS (DIPLOCOCCUS) PNEUMONIAE (THE PNEUMOCOCCUS)

General characteristics. The organism *Streptococcus (Diplococcus) pneumoniae*, best known in relation to pneumonia, occurs in pairs of lancet-shaped diplococci with their broad ends apposed. Within the animal body or in excretions each pair is enclosed within a capsule. They are gram positive, nonmotile, and nonspore-forming.

Pneumococci grow equally well in the presence or absence of oxygen. Their optimum temperature is 37° C, and they grow best in a slightly alkaline medium. Growth is most abundant on such enriched media as hormone agar and blood agar (Fig. 18-8). On the latter medium the green zone of slight hemolysis surrounding the colony is a diagnostic feature. The power to form capsules is lost when pneumococci are cultivated for a long time on artificial media.

The pneumococcus is not very hardy and has no natural existence outside the animal body. In the finely divided spray thrown off from the nose and mouth, pneumococci live about 1½ hours in sunlight. In large masses of sputum they live for 1 month or more in the dark and about 2 weeks in sunlight. They are more susceptible to ordinary germicides than most other bacteria and are destroyed in 10 minutes by a temperature of 52° C.

Types. Although all pneumococci are much alike microscopically and culturally, they show distinct differences immunologically. This was discovered in 1910, when different cultures of pneumococci were used to immunize animals, and the serum of these animals was used to agglutinate pneumococci from various other sources. It was found that the majority of pneumococci fell into one of three rather distinct types (types I, II, and III) and that an antiserum prepared by immunizing an animal against

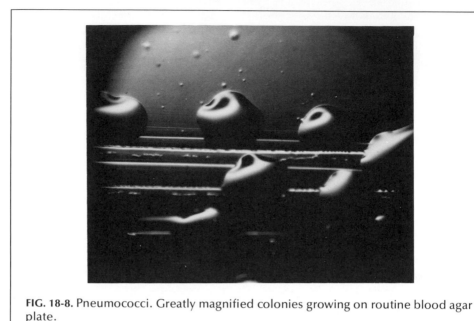

FIG. 18-8. Pneumococci. Greatly magnified colonies growing on routine blood agar plate.

pneumococci of one type agglutinated pneumococci of that type only. Pneumococci that did not fall into any of the three types were placed in group IV, which later was divided into 29 types. Other types were described in rapid succession until at the present time there are at least 82 types or subtypes. All can cause pneumonia, but 50% to 80% of adult cases are due to types I, II, and III. Type III pneumonias menace the aged. In the pneumonias of children the primary types of pneumococci are XIV, I, VI, V, VII, and XIX (in order of frequency).

Formerly, therapeutic serums made from rabbits were available for the treatment of pneumonia, and the determination of the type of pneumococcus causing the infection was necessary before serum therapy could be instituted. Today serum therapy has been replaced by antimicrobial therapy.

Pneumococcus type III differs somewhat from other pneumococci, with its wide capsule and slimy growth on culture media. It is not lancet shaped. There seems to be a direct relation between capsule development and virulence, thus explaining why type III infections have such a high mortality.

The soluble carbohydrates that give pneumococci their type characteristics are polysaccharides in the capsules, spoken of as *specific soluble substances* (SSS). They can be detected by the precipitin reaction in broth cultures of pneumococci and in the blood and urine of patients with pneumonia. In addition to these, a somatic antigen is common to pneumococci.

Toxic products. The clinical features of pneumococcal disease point to toxemia, but a toxin similar to that elaborated by the diphtheria bacillus has never been found. Pneumococci do, however, produce hemolysins, leukocidins, and necrotizing substances. Many strains produce hyaluronidase.

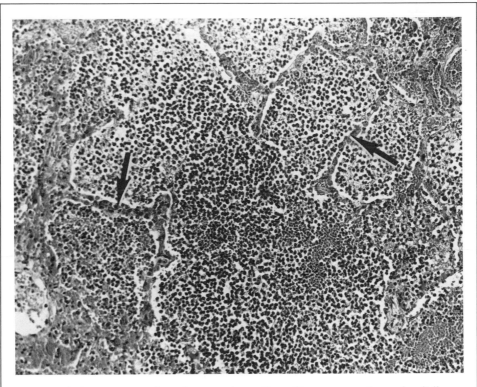

FIG. 18-9. Lobar pneumonia, microscopic section. Air sacs are plugged solidly as result of inflammatory process. The many small irregular, closely packed, dark nuclei are polymorphonuclear neutrophils. Arrows point to alveolar walls about consolidated air spaces. (×800.) (From Smith, A. L.: Microbiology and pathology, ed. 11, St. Louis, 1976, The C. V. Mosby Co.)

Pathology. The hallmark of pneumococcal infections is the presence of an abundance of fibrin in areas of inflammation. In lobar pneumonia there is much fibrin in the lungs; in pneumococcal meningitis, much fibrin is deposited in the subarachnoid space.

Pneumonia is an inflammatory condition of the air sacs (alveoli), bronchioles, and smaller bronchi of the lungs, in which these structures are filled with fibrinous exudate. *Consolidation* is the process whereby the air spaces of the lung are so plugged off (Fig. 18-9). The pneumococcus is the usual cause of the two kinds: lobar pneumonia and bronchopneumonia.

Lobar pneumonia is a severely toxic disease related to consolidation of one or more lobes of the lungs (by massive inflammatory exudation). Rapid shallow breathing, increased pulse rate, cyanosis, and nausea with vomiting are signs and symptoms, and the blood count usually shows a leukocytosis (30,000 to 40,000/mm^3) with 90% to 95% polymorphonuclear neutrophils. Pleurisy (inflammation of the pleura) is part of the disease. With recovery and resolution, the exudate in the lungs liquefies and is removed partly by absorption and partly by expectoration. Air reenters the affected lobe or lobes, and the lung completely returns to its former efficiency. Occasionally, delayed resolution leads to abscess formation or chronic organizing pneumonia. The disease has yielded so dramatically to the use of the antimicrobials that the classic pathologic stages are rarely seen today.

Lobar pneumonia is a primary disease, and 95% of cases are pneumococcal. It may also be caused by the Friedländer bacillus, the influenza bacillus, or other streptococci.

Bronchopneumonia is usually pneumococcal, but it may be caused by any one of a number of microbes, including other streptococci, staphylococci, and influenza bacilli, operating singly or in variable combinations.

Bronchopneumonia, more often secondary than primary, is a serious complication to measles, influenza, whooping cough, and chronic diseases of the heart, blood vessels, lungs, and kidneys. It peaks in the early and late years of life and frequently is the terminal event in debilitating diseases of the very young or extremely old. (It has long been called the "old man's friend.") Bronchopneumonia may follow the administration of an anesthetic or the aspiration of infectious material into the lungs during an operation. In newborn infants it is often caused by aspiration of infected amniotic fluid.

Unlike lobar pneumonia, bronchopneumonia consists of scattered small inflammatory foci, usually more numerous at the lung bases. The exudate consists of leukocytes, fluid, and bacteria, but less fibrin and few red blood cells. Pleurisy and empyema are complications. This kind of pneumonia does not resolve readily and chronic pneumonia often persists. *Hypostatic pneumonia* is bronchopneumonia complicating the hypostatic congestion of heart failure.

Pneumococci cause other diseases such as empyema, endocarditis, meningitis, arthritis, otitis media, peritonitis, and corneal ulcers. Some of these complicate pneumonia; others occur as primary conditions. Pneumococcal peritonitis is a primary disease in children. Pneumococcal otitis media tends to spread to the meninges, with resultant pneumococcal meningitis.

Sources and modes of infection. Lobar pneumonia is endemic in all centers of population. Epidemics seldom occur but may if conditions enhance exposure to infection with concomitant lowering of host resistance. The sources of infection are active cases and carriers.

Pneumococci enter and leave the body by the same route, the mouth and nose. Infection is usually transferred directly, most often by droplets from the mouth and nose, but indirect transmission by contaminated objects may also occur.

Practically every person becomes a carrier of pneumococci for a short time during the year. Those who have contacted a patient often carry them in their throats for a few days or weeks. Most carriers not in contact with pneumonia harbor comparatively avirulent pneumococci and are of little danger, although type III pneumococci may be found in such carriers. That carriers of type III pneumococci are common whereas type III infections are comparatively rare is difficult to explain.

Laboratory diagnosis. Pneumococci may be detected in sputum and other body fluids with some degree of certainty by direct microscopic examination of smears

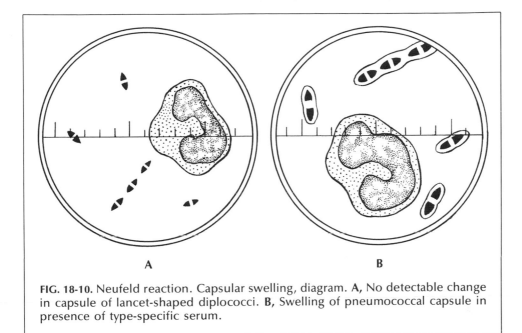

FIG. 18-10. Neufeld reaction. Capsular swelling, diagram. **A,** No detectable change in capsule of lancet-shaped diplococci. **B,** Swelling of pneumococcal capsule in presence of type-specific serum.

stained for capsules and by Gram's method. (Pneumococci and other streptococci bear a close microscopic resemblance to each other.) Confirmatory methods are cultures and the inoculation of white mice. (Rabbits and mice are very susceptible to pneumococci.) One milliliter of the emulsified sputum is injected into the peritoneal cavity of a white mouse, wherein the pneumococci outstrip all other organisms. The mouse becomes ill after 5 to 8 hours, at which time it is killed. The peritoneal cavity then contains many pneumococci to be identified by microscopic, cultural, and typing methods. Animal inoculation demonstrates pneumococci when direct smears fail to do so, and it always demonstrates them more quickly than do cultures.

Typing is done primarily to determine whether a highly virulent organism is present. There are several methods. All depend on the action of agglutinating and precipitating serums (typing serums) prepared by immunizing animals against the different types of pneumococci. In the method devised by Neufeld, flecks of sputum or other test material are mixed with the battery of type-specific serums. Where the type of pneumococcus matches that of the serum, the capsules of the pneumococci swell (Fig. 18-10). If the sputum contains too few pneumococci or if the typing is otherwise unsatisfactory, some of the specimen may be injected into a white mouse, as previously described, and in the course of a few hours, typing may be carried out on the yield from the peritoneal exudate.

DIFFERENTIATION OF PNEUMOCOCCI FROM OTHER STREPTOCOCCI. Practical differences existing between pneumococci and other streptococci are that: (1) on blood agar the colonies differ; (2) in animal tissues pneumococci have capsules whereas other streptococci seldom do; (3) when 1 part of bile is added to 3 parts of a liquid culture, pneumococci are dissolved whereas streptococci are not; (4) pneumococci ferment inulin, other streptococci do not; (5) pneumococci are inhibited by optochin (ethylhydrocupreine hydrochloride) whereas other streptococci are not; and (6) pneumococci are more pathogenic for mice than ordinary streptococci. Other differential aids are agglutination and precipitation tests with specific antiserums.

Immunity. Recovery from pneumococcal infection confers 6- to 12-month immunity to the type of pneumococcus causing the infection, but none to other types. (Instances have been reported in which a person has had pneumonia more than a dozen times.) The natural resistance of man against the pneumococcus is comparatively high, and a person probably never contracts pneumonia unless his resistance is lowered. Blacks are more susceptible than whites, and men more so than women.

Prevention. The number of persons contacting a patient with pneumonia should be restricted. The discharges from the mouth and nose of the patient should be burned or disinfected. The hands and all objects, such as spoons, cups, and other utensils, possibly contaminated by the patient should be disinfected. Measures should be taken to minimize droplet infection in the spray leaving the mouth of the patient when he talks or coughs.

Because of the antibiotic resistance encountered in pneumococci, vaccine therapy is coming back. The model is new. The design is for a polyvalent vaccine given in a single injection and incorporating the purified capsular polysaccharides from the 12 or 14 types of pneumococci most likely to cause disease. Vaccines under investigation comprise types I to IX, XII, XIV, XVIII, XIX, and XXIII.

QUESTIONS FOR REVIEW

1. Name important gram-positive cocci. Why are pyogenic cocci so called?
2. Characterize staphylococci.
3. List diseases caused by staphylococci.
4. Discuss the localization and invasion of the body by staphylococci.
5. How is pathogenicity of a staphylococcus established?
6. Comment on the background factors in hospital-acquired infections. How are they studied epidemiologically?
7. Outline the streptococci according to the following categories: morphology, general features, classification, pathogenicity.
8. Name 10 diseases of man caused by streptococci. Name two diseases of animals. State the Lancefield groups responsible.
9. How are streptococci and pneumococci transmitted from person to person?
10. What is the Dick test? Neufeld reaction? Bacitracin disk test? Coagulase test? Schultz-Charlton phenomenon? Optochin test?
11. How is the antistreptolysin O titer used clinically?
12. Briefly discuss septic sore throat, puerperal sepsis, and erysipelas.
13. What is the relation between streptococcal disease and rheumatic fever?
14. Compare lobar pneumonia with bronchopneumonia.
15. Contrast the type lesion of staphylococcal and streptococcal infections. State the notable feature of pneumococcal injury.
16. What is the current status of vaccines against pyogenic cocci?

REFERENCES. See at end of Chapter 27.

19 Neisseriae

Distinctive gram-negative cocci or plump coccobacilli, sometimes called neisseriae, are found in the family Neisseriaceae, which includes three parasitic genera— *Neisseria, Branhamella,* and *Moraxella.* The most important ones are the gonococcus (*Neisseria gonorrhoeae*) and the meningococcus (*Neisseria meningitidis*), two pathogenic neisseriae of genus *Neisseria,* whose members are aerobic or facultatively anaerobic, oxidase positive, and parasites of the mucous membranes of man. Members of the genera *Branhamella* and *Moraxella* and certain minor *Neisseria* species are important, not because of their pathogenicity but because of their habitat in the mouth and upper respiratory passages.

NEISSERIA GONORRHOEAE (THE GONOCOCCUS)

Neisseria gonorrhoeae is the cause of gonorrhea, one of the most prevalent diseases affecting man, the most common of the venereal diseases, and the number one communicable disease problem today in the United States. A disease known to the ancient Chinese and Hebrews, it was termed *gonorrhea* by Galen around AD 130. The gonococcus is sometimes called the diplococcus of Neisser after its discoverer, and its infections are referred to as neisserian infections.

General characteristics. The gonococcus is a gram-negative, nonmotile, nonsporeforming diplococcus (Fig. 19-1). In smears the opposing sides of the two cocci are flattened, and the cocci are placed like two coffee beans with their flat sides together.

Gonorrhea is a disease accompanied by a discharge from the genital tract that is at first thin and watery, then later purulent. The incubation period is 3 to 5 days. During the early stages gonococci are found free in the serous exudate or attached to epithelial cells, but when the exudate becomes purulent, phagocytosis takes place, and gonococci are found within the cytoplasm of the pus cells (polymorphonuclear neutrophilic leukocytes). A single white blood cell may contain from 20 to 100 micro-

351

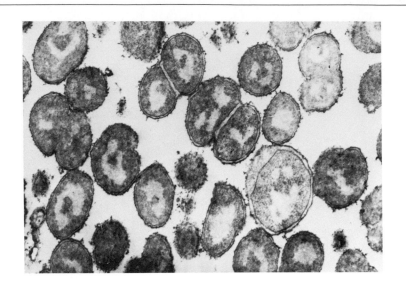

FIG. 19-1. *Neisseria gonorrhoeae,* cross section, electron micrograph. (×100,000.) (Courtesy Technical Information Services, State and Community Services Division, Center for Disease Control, Health Services and Mental Health Administration, Department of Health, Education and Welfare, Atlanta, Ga.)

organisms, gonococci that are not dead but still infectious. In later stages of the disease they may be found outside the white blood cells, and when the disease becomes chronic, they often cannot be found at all.

This fastidious microbe will not grow on ordinary culture media and is somewhat difficult to cultivate even on media prepared especially for it. Gonococci grow best at or slightly below body temperature (35° to 37° C) and in an atmosphere containing oxygen and carbon dioxide (3% to 10%). The appearance of colonies of gonococci is variable, and four types referable to the differences are described and related to virulence. Colony types 1 and 2 come from infective organisms; colony types 3 and 4 from gonococci, probably noninfective. The cocci of types 1 and 2 possess pili, which may help the bacteria to attach to epithelial cells and so resist phagocytosis. Gonococci produce the enzyme *oxidase,* as do other members of the genus *Neisseria.* The *oxidase reaction* is used to identify colonies of neisserian species in cultures.

Although gonococci are difficult to destroy within the body, they possess little resistance outside it. They are killed in a very short time by sunlight and drying. In pus or on clothing in moist dark surroundings they may live from 18 to 24 hours. They are very susceptible to disinfectants, especially silver salts, and are killed by a temperature of 60° C within 10 minutes. Although gonococci are susceptible to the modern antibiotics, drug resistance is an ever present problem, especially with the number of drug-resistant strains coming in from southeast Asia and the Philippine Islands. Gonococci are not exactly alike immunologically but cannot be typed. There is only one defined strain.

Pathogenicity. The gonococcus is parasitic specifically for man. Spontaneous infection does not occur. Nothing comparable to any of the clinical forms of gonorrhea had been produced artificially in lower animals until recently when gonococcal urethritis was produced experimentally in the chimpanzee and the male-to-female animal transmission demonstrated.

In typical cases of gonorrhea the sites of primary infection in the female are the urethra and cervix; in the male the site is the urethra. Gonococci injure columnar epithelium like that lining the cervix uteri and the rectum and the transitional (urothelial) epithelium lining the urinary tract. Vaginal infection is not seen because the epithelium lining the vagina of the adult woman is a cornified stratified squamous type resistant to infection with gonococci. Before the age of puberty, the vagina is lined with a softer, extremely susceptible epithelium. Gonorrheal vulvovaginitis in prepubertal girls may be epidemic and difficult to eradicate. The change in the epithelium with the onset of puberty usually eliminates this childhood infection completely.

A primary site seen more and more today is the conjunctiva of the eye, and the process (gonorrheal conjunctivitis and keratitis) is one that actively damages the tissues of the eye (Fig. 19-2). *Ophthalmia neonatorum*, gonorrheal conjunctivitis in

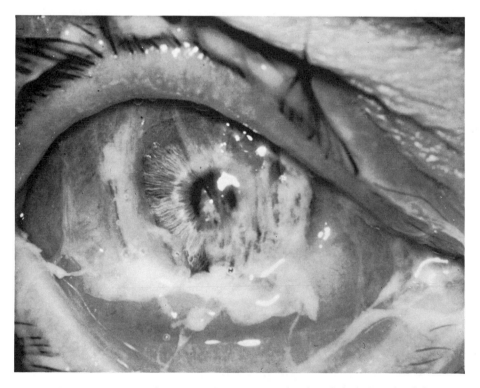

FIG. 19-2. Acute gonococcal conjunctivitis (gonorrheal ophthalmia of adults). Note inflammation, redness, irritation of conjunctivae, and copious pus. Smear of exudate showed pus cells with intracellular diplococci. Culture grew out gonococci. Patient also had gonococcal pelvic inflammatory disease. (From Donaldson, D. D.: Atlas of external diseases of the eye, vol. 1, St. Louis, 1966, The C. V. Mosby Co.)

newborn infants, results when the eyes are infected during the birth passage. A profuse purulent discharge in the eyes of a neonate can build up a considerable pressure behind the lids. If the lids are forced apart, pus spurts out. The physician and attendants of these babies must be careful to protect their own eyes. In babies or adults such infection easily results in blindness or serious sight impairment because of the inflammatory distortion of the structures in the eye. Ophthalmia neonatorum can be prevented by the Credé method, as is required by law for all babies.

From the urethra of the male, gonococcal infection may spread directly to the other parts of the male reproductive system. In the female it may likewise spread to other parts of the tract, especially to Bartholin's glands and the fallopian tubes. The lining of the uterus seems to resist the action of gonococci. Invasion of the fallopian tubes usually occurs with the first or second menstrual period after infection; however, in some cases it may not occur until later. Involvement of the fallopian tubes is associated with considerable distortion and scarring if the disease becomes chronic. Scarring of the urethra in the male may lead to stricture or closure of the urethral lumen at one or more focal points.

Gonococci sometimes pass from the genitourinary tract via the lymphatics or the bloodstream to set up distant sites of infection (examples, endocarditis, perihepatitis, meningitis). Gonococcemia is associated with varied skin lesions from which organisms may be identified. An important manifestation of extragenital gonococcal infection is a purulent, destructive arthritis. As the overall incidence of gonorrhea increases, extragenital lesions become more prominent.

Sources and modes of infection. Gonococci are never found outside the human body unless they are on objects very recently contaminated with gonorrhea discharges, and here they live only for a short time. Therefore gonorrheal infections are practically always spread by direct contact, and mostly the mode of contact is sexual intercourse. One cannot deny that gonorrhea is sometimes transmitted indirectly by contaminated objects, but this is rare.

Gonorrheal ophthalmia of adults is usually accidental. Infection from the genitourinary tract is inadvertently transferred to the eyes by the hands of the same or a different person. Vulvovaginitis in children is spread by the use of common bed linen, bathtubs, toilets, and such. It has been known to result from the use of contaminated rectal thermometers. It usually occurs where children live in closely crowded quarters.

Untreated gonococcal infections tend to become chronic. Females who are untreated or inadequately treated become infectious carriers for years after manifestations of disease have disappeared. An estimated 60% to 80% of females with the infection are asymptomatic. It is now recognized that infected males may also be asymptomatic, perhaps as many as two thirds of them are.

Laboratory diagnosis of gonorrhea. The microbiologist has at his command several procedures applicable to the diagnosis of gonorrhea. Smears, cultures, and the oxidase reaction are the *presumptive* tests. To confirm the results of these and to establish the diagnosis of gonorrhea he uses fluorescent antibody technics and carbohydrate fermentation reactions.

Direct smears of genital discharges may be stained with the Gram stain. There are rare exceptions to the rule that for all practical purposes the finding of gram-negative diplococci within the pus cells of an exudate from a genital infection strongly *suggests* that they are gonococci. This is especially true if the exudate is from the male urethra. In the male with typical acute purulent urethritis, the gram-stained smear of the exudate containing the distinctive intracellular diplococci ordinarily makes the diagnosis. In the female, early in the disease, typical diplococci may be seen in smears of material from Skene's and Bartholin's glands, but even a working diagnosis cannot be made on this basis alone. The reasons for this are several—gram-negative diplococci other than gonococci occur outside cells. Gonococci occur outside cells, singly or in pairs, and gram-positive organisms having the morphology of gonococci occur within cells. All that can be said about gram-negative diplococci found outside cells is that they *may be* gonococci. Very infrequently are gram-negative diplococci other than gonococci found within the pus cells of a genital ex-udate, but it is possible. The smear prepared from gonorrheal exudates should be quite thin because gonococci react to the Gram stain in an erratic way if the smear is thick and uneven. The microbes usually are not seen in the exudate of chronic gonorrhea.

Cultural methods are of special value in the diagnosis of chronic disease and in the determination of a cure. Cultures are incubated under increased carbon dioxide tension (Fig. 19-3). Enriched media such as chocolate agar, Hirschberg egg medium, and currently the Thayer-Martin medium* are used to cultivate the delicate gono-coccus. Thayer-Martin (T-M) medium contains hemoglobin, certain chemicals to enhance gonococcal growth, and antimicrobial agents to inhibit selectively fungi, gram-positive organisms, and many gram-negative ones. The plates of T-M medium

---

*Also called VCN medium for the three antibiotics it contains—vancomycin hydrochloride, colistimethate sodium, and nystatin.

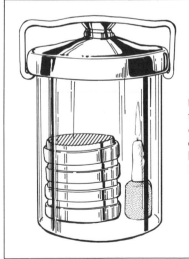

FIG. 19-3. Culture under increased carbon dioxide tension, simple method for partially anaerobic conditions. Cultures and lighted candle are placed in container and container made airtight. Candle burns until oxygen is almost completely exhausted, then goes out.

to be inoculated must be streaked in a specified way so as to spread the organisms out of the associated mucus, which tends to lyse them (Fig. 19-4). A modification of Thayer-Martin medium, Transgrow, has been developed so that suspect cultures can be sent into central laboratories. Transgrow comes in a screw-cap bottle containing a mixture of air and carbon dioxide. To inoculate Transgrow, one holds the bottle upright to prevent carbon dioxide from escaping and unscrews the cap (Fig. 19-5).

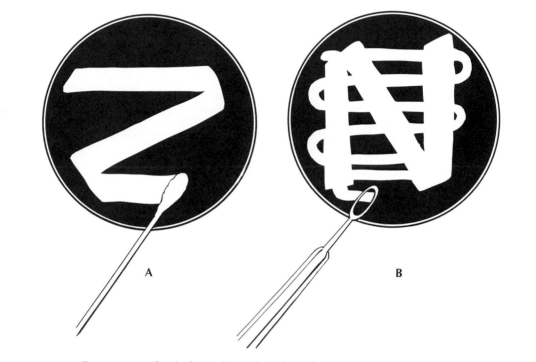

A   B

FIG. 19-4. Two-step method of streaking plate for culture of gonococci. **A,** Suspect secretions on swab rolled gently onto plate in Z pattern. **B,** Cross streaking of inoculum in A done with platinum loop.

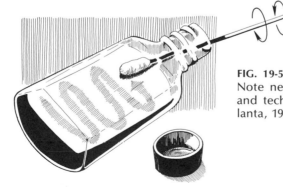

FIG. 19-5. Inoculation of Transgrow medium. Note neck of bottle elevated. (From Criteria and technics for diagnosis of gonorrhea, Atlanta, 1972, Center for Disease Control.)

The swab with test material on it is rolled over the medium. The cap is returned and the bottle is ready for shipping. The unopened bottle can be incubated directly at the receiving laboratory. Gonococci survive and grow 48 to 96 hours in this medium.

Public health authorities today recommend that suspect material for culture be obtained from the anorectal area and pharynx as well as from the urogenital tract (anterior urethra, endocervical canal). Rectal gonorrhea is easily overlooked. Carefully taken swabs are important in male homosexuals, and in females both cervical and rectal swabs are needed since half the women infected harbor the gonococcus in the rectum. Infection may persist there after it has been eliminated in the cervix.

Gonococci can be recovered in urine from the male if the first 10 ml of voided urine is centrifuged and the sediment cultured. In a simpler screening method, the first few drops of urine caught on a dry swab are immediately passed across Thayer-Martin medium. Cultures of urine in screening programs have helped to define a reservoir existing in the asymptomatic male.

The next steps in the bacteriologic study of the gonococcus are the determination of the biochemical reactions and identification of the organisms with fluorescein-labeled antiserums. The biochemical reactions of *Neisseria* species discussed in this chapter are given in Table 19-1. The fact that the gonococcus possesses a specific K-type antigen is the basis for the fluorescent antibody test to detect gonococci in direct smears of exudate or in smears made from cultures.

Several serologic tests under study are still in pilot stages, but results are discouraging. A serologic test to catch the large number of persons in the silent, communicable reservoir, although highly desirable, is associated with problems. One

TABLE 19-1. CHARACTERISTICS OF NEISSERIAE

Organism	Growth on Thayer-Martin medium	Growth on nutrient agar at 22° C	Oxidase	Fermentation of sugars with acid only					Pigment production	Habitat
				Glucose	Maltose	Sucrose	Fructose	Mannitol		
Neisseria gonorrhoeae	+	−	+	+	−	−	−	−	−	Genital infections of man
Neisseria meningitidis	+	−	+	+	+	−	−	−	−	Nasopharynx of man
Neisseria sicca	−*	+	+	+	+	+	+	−	Variable	Nasopharynx of man
Neisseria flavescens	−*	+	+	−	−	−	−	−	Golden yellow	Nasopharynx of man
Neisseria subflava	−*	+	+	+	+	−	+	−	Greenish yellow	Nasopharynx of man

*Growth with heavy inoculum.

of these is that the gonococcus seems to share its antigens with other, harmless neisseriae.

It is never within the province of the laboratory alone to say that a person is cured of gonorrhea, and in many cases the efforts of both laboratory and clinician do not determine whether the disease is eradicated. In dealing with a case of gonorrhea, physicians and nurses should consider its medicolegal potential.

Social importance of gonorrhea. In the United States today, gonorrhea is increasing in frequency to pandemic proportions. It is estimated that there are 3 million new cases annually, about 1, it is said, every 15 seconds. More than half the cases occur in teenagers and young adults under 25 years of age.

Generally, three males are treated and reported for each female, since in males the manifestations may be sufficiently disagreeable to motivate them to seek medical attention. Since females are often asymptomatic or relatively so, many of them are reported and diagnosed only because of information from male consorts. The silent reservoir of asymptomatic females constitutes a primary obstacle to control of disease, and if widespread screening of the female population is not done, it may remain so. Some health authorities recommend that a culture for *Neisseria gonorrhoeae* be considered an essential part of prenatal care and that a routine culture be taken more often at the time a pelvic examination is done in any woman.

The most destructive piece of misinformation that has been handed down from generation to generation is that gonorrhea is no worse than a cold. Such a fallacy both underestimates the danger of gonorrhea and creates the impression that colds are of no consequence. The medical, social, psychologic, and even medicolegal implications of gonorrhea are sizable.

Gonorrhea is said to be the most common cause of sterility in both sexes. In women, sterility results from occlusion of the fallopian tubes by scar tissue formed during the healing of gonorrheal salpingitis. In men, it results from occlusion of the vasa deferentia by a similar process of gonorrheal inflammation and healing, with scarring.

Immunity. An attack of gonorrhea confers little, if any, immunity to subsequent attacks.

Prevention. The general public should be warned of the dangers of gonorrhea and the difficulty of its cure. It is unfortunate that use of the wonder drugs has engendered an attitude that the disease is no problem. The dangers of quack doctors and folk remedies must be stressed. Patients must not allow their discharges to soil toilets or articles used by others, and they must be warned of the danger of transferring infectious material to the eyes by means of the hands.

Prophylactic treatment in newborn babies is as follows. Immediately after birth, the eyelids of the baby are cleaned with sterile water. A different piece of cotton is used for each eye, and the lids are stroked (or irrigated) from the nose outward. Next, the lids are opened, and one or two drops of 1% silver nitrate solution are instilled into each eye, care being taken that the conjunctival sac is completely covered with the solution. After 2 minutes the eyes are irrigated with isotonic saline solution. (A mild irritation of the lining membranes of the eye may be produced but is short lived.)

This is known as *Credé's* method, a procedure so important that it is required by law in most of the 50 states. It appears that penicillin and other antibiotics are as effective as silver nitrate in prevention of ophthalmia neonatorum, but where the law requires silver nitrate, such antibiotics are not used. A few states have passed laws allowing a suitable antibiotic to be used instead of silver nitrate. However, the National Society for the Prevention of Blindness (New York) is currently recommending that the silver nitrate method be continued as the standard and preferred procedure.

Vulvovaginitis in children may be prevented by proper care of bed linen, bathtubs, nightclothes, and wash water. All children should be examined for gonorrhea before admission to children's institutions or hospital wards with other children.

The U.S. Public Health Service, although presently concerned with the development of a vaccine for gonorrhea, is not optimistic. Because of the nature of the disease and the mode of transfer, every person who has gonorrhea should be serologically tested for syphilis.

The VD pandemic. Every 2 minutes somewhere in the United States a teenager contracts a venereal disease. Because of changing factors in our society, forms of disease contracted mainly through intimate sexual contact and some that may also be contracted this way are on the rise. This is a matter of great concern to public health officials and agencies. Table 19-2 gives a list of sexually transmitted diseases with causative agents. *

Venereal disease treatment information is available from a toll-free, nationwide "hot line," Operation Venus, financed by a volunteer organization, the United

---

*With the exception of gonorrhea, the ones due to microbes are discussed in subsequent chapters.

TABLE 19-2. SEXUALLY TRANSMITTED (VENEREAL) DISEASES

Disease		Agent
The "classic five"		
Gonorrhea	Bacterium	*Neisseria gonorrhoeae*
Syphilis	Bacterium—spirochete	*Treponema pallidum*
Chancroid (soft chancre)	Bacterium	*Haemophilus ducreyi*
Granuloma inguinale	Bacterium (Donovan body)	*Calymmatobacterium granulomatis*
Lymphopathia venereum	Bacterium—chlamydia	*Chlamydia trachomatis*
Newcomers and others		
Herpes genitalis	Virus	*Herpesvirus hominis* type II
Molluscum contagiosum	Virus (molluscum body)	Poxvirus
Mycoplasmosis	Bacterium—T strains mycoplasma	*Ureaplasma urealyticum*
Candidiasis	Fungus	*Candida albicans*
Trichomoniasis	Protozoan	*Trichomonas vaginalis*
Condyloma acuminatum (genital wart)	Virus	Papovavirus-like particle
Nongonococcal urethritis (NGU)	Bacterium—chlamydia	*Chlamydia trachomatis*
Pediculosis pubis	Insect—crab louse	*Phthirus pubis*
Scabies	Insect—itch mite	*Sarcoptes scabiei*

States Alliance for Eradication of Venereal Disease. The number is 800-462-4966 in Pennsylvania and 800-523-1885 for the rest of the nation.

NEISSERIA MENINGITIDIS (THE MENINGOCOCCUS)

The meningococcus *Neisseria meningitidis** is the cause of meningococcal septicemia with or without localization in the leptomeninges to produce epidemic cerebrospinal meningitis. Epidemic cerebrospinal meningitis is known also as cerebrospinal fever, spotted fever, and meningococcal meningitis and is one form of acute bacterial meningitis. During World War II more soldiers in the Army of the United States died as a result of meningococcal infection than of any other infectious disease.

General characteristics. Meningococci are gram-negative diplococci strikingly similar to gonococci but more irregular in size and shape. They are nonmotile and do not form spores. Capsules are not usually seen in ordinary smears but may be demonstrated by special methods. Meningococci produce endotoxins liberated when the cocci disintegrate; these toxins are partly responsible for the manifestations of meningitis. In the cerebrospinal fluid meningococci appear both within and without the polymorphonuclear neutrophilic leukocytes. (One must know the source of the specimen to determine whether gram-negative intracellular diplococci found in a smear are meningococci or gonococci.)

Meningococci grow best at body temperature in an atmosphere containing 10% carbon dioxide (Fig. 19-3). They do not grow at room temperature. Growth requires special media containing enriching substances such as whole blood, serum, or ascitic fluid. Agar containing laked rabbit blood and dextrose supports the growth of meningococci especially well, as does enriched hormone agar. Different strains of meningococci vary considerably in the ease with which they grow on artificial culture media. Salt has a toxic effect on meningococci and should be left out of media on which they are to be cultivated. Like gonococci, they are oxidase positive.

Meningococci are such frail organisms that they survive only a short time outside the body. Sunlight and drying kill them within 24 hours. Away from the body they are easily killed by ordinary disinfectants, but in the nasopharynx they are very resistant. They are quite susceptible to heat and cold. Meningococci quickly lyse in cerebrospinal fluid removed from the body. Therefore specimens of suspect fluid should be examined as quickly as possible.

Meningococci and gonococci compared. Meningococci and gonococci are alike in that (1) both are strict parasites and cause disease only in man, (2) they show little difference in resistance to injurious agents, (3) their distribution in the inflammatory exudate is the same, (4) they grow on artificial media with difficulty, and (5) their disease-producing properties are comparable. Skin lesions of gonococcemia are similar to those of meningococcemia, and gonococci have caused the Waterhouse-Friderichsen syndrome (see below). It may be said that they are morphologically, physiologically, and immunologically much alike.

---

*In 1887, Weichselbaum isolated and described this organism from the cerebrospinal fluid of a patient with meningitis.

Groups of meningococci. By serologic reactions, including a capsular swelling test similar to that used in typing pneumococci, meningococci are classified into four main groups—A, B, C, and D.* (The U.S. Public Health Service Center for Disease Control in Atlanta, Georgia, recognizes seven different types of meningococci with specific antigenic and epidemiologic differences—A, B, C, D, X, Y, and Z.)

Groups A and C are encapsulated and possess a specific capsular polysaccharide. There is a polysaccharide-polypeptide component in group B, but usually no capsule. In the past, group A meningococci have been responsible for 95% of cases of epidemic meningitis; group A was especially troublesome during World War II. Groups B and C have been the endemic organisms figuring in sporadic outbreaks between major epidemics. Until the mid sixties, group B was the prevalent one. Since then group C is causing an increasing number of epidemics at military bases. There are few meningococci of group D in this country.

Pathogenicity. The meningococcus is not very pathogenic for lower animals, and typical epidemic cerebrospinal meningitis occurs only in man. Invasion of the body by meningococci occurs in three steps: (1) implantation in the nasopharynx, (2) entrance into the bloodstream with septicemia, and (3) localization in the meninges (cerebrospinal meningitis). For most patients the invasion ends with implantation in the nasopharynx (the carrier state).

About one third of meningococcal infections are septicemias of such severity that without treatment the patient dies before meningeal infection can occur. Meningococci proliferate as massively in blood and tissues as though growing in laboratory broth culture. The events of meningococcal septicemia come together under the designation *Waterhouse-Friderichsen syndrome* (Fig. 19-6), an acute fulminating condition characterized by many small areas of hemorrhage in the skin and, within a period of hours, death in peripheral circulatory failure.† At autopsy large areas of hemorrhage are seen in the adrenal glands.

In only a few patients do meningococci in the bloodstream infect the meninges or other body tissues. When they do, purulent inflammation results. An exudate composed of leukocytes, fibrin, and meningococci forms in the subarachnoid space, the area between the two meningeal coverings, the arachnoid and the pia mater. Most prominent along the base of the brain, the exudate extends into the cerebral ventricles and down the spinal subarachnoid space. The cerebrospinal fluid may become so purulent that it scarcely flows through the lumen of the lumbar puncture needle.

Among the complications of epidemic cerebrospinal meningitis are arthritis, hydrocephalus, otitis media, retinitis, deafness from involvement of the eighth nerve, pericarditis, endocarditis, conjunctivitis, pneumonia, and blindness.

Sources and mode of infection. Since meningococci are such frail organisms and seldom found outside the body, the source of infection must be a patient or a

---

*Groups A, B, C, and D correspond respectively to groups I, II, II alpha, and IV in the formerly used classification.
†The rapid course of events possible with meningococcal disease—collapse and death within an hour— is terrifying to the social group. The disease upsets people to the point of mass hysteria.

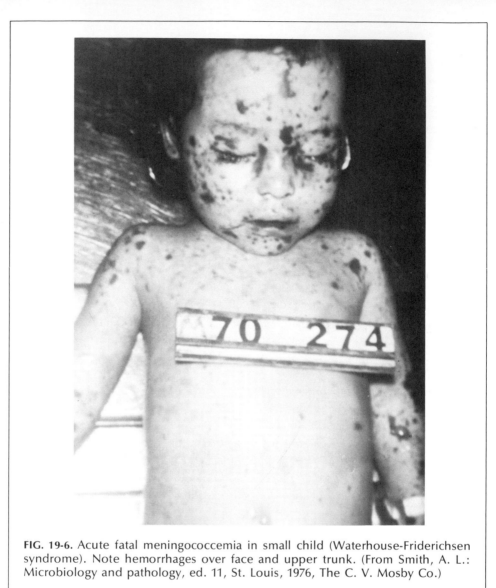

FIG. 19-6. Acute fatal meningococcemia in small child (Waterhouse-Friderichsen syndrome). Note hemorrhages over face and upper trunk. (From Smith, A. L.: Microbiology and pathology, ed. 11, St. Louis, 1976, The C. V. Mosby Co.)

carrier. The organisms reside in the nasopharynx of both patients and carriers and leave the body in the nasal and buccal secretions. The mode of transfer is by close contact, and the organisms enter the body by the nose and mouth. Venereal transmission is reported. When meningococci reach a new host, they localize in the nasopharynx and multiply. Infection traced to articles recently contaminated by infected nasal and buccal secretions seldom occurs.

Meningococcal infections are not highly communicable, and there seems to be a high degree of natural immunity to them. Many persons exposed become carriers but few develop disease. It is estimated that about 1 carrier in 1000 develops meningococcemia or meningitis. In the absence of immunity, the great number of carriers in the general population would maintain an epidemic at all times. Since meningococcal infection is endemic in all densely populated centers, it may become epidemic under overcrowded conditions, as in army camps. Overcrowding concentrates the carriers and promotes factors that lower individual resistance. When troops are mobilized, it is one serious disease to which the men seem to be exceedingly vulnerable.

Under ordinary conditions probably from 2% to 5% of the general population are carriers. When epidemics occur, the number of carriers, especially of group A meningococci, increases. In military establishments, 50% of the personnel may become carriers. About two thirds of those who contract meningitis harbor the organisms for a variable time after convalescence. The carrier state lasts about 6 months. However, healthy persons who have never had meningitis but nevertheless harbor meningococci in their nasopharynx play more of a part in the spread of meningitis than those who harbor the organisms after recovery from the disease.

At times the meningococcus has been recovered from the vagina and cervix in situations where its relation to disease was not clear. This finding emphasizes the importance of adequately identifying organisms from the female genital tract. Meningococci, although thought of as respiratory pathogens, on occasion may invade other parts of the body.

Laboratory diagnosis. To give the patient the advantage of early treatment, which means so much in meningococcal infections, septicemia should be diagnosed clinically. The diagnosis is confirmed by blood cultures. A valuable diagnostic adjunct is the demonstration by smears or cultures of meningococci in the hemorrhagic skin lesions associated with septicemia (Fig. 19-6). When cultures for meningococci are made, the specimen must not cool before inoculation onto media warmed to room temperature. The inoculated cultures are promptly incubated.

In epidemic meningitis the cerebrospinal fluid is first turbid and then purulent. In smears of cerebrospinal fluid the typical arrangement of gram-negative diplococci within pus cells is sufficient for practical purposes to make a working diagnosis of epidemic meningitis. However, cultures must confirm that. Because of the tendency of meningococci to undergo autolysis, the specimen must be examined as quickly as possible after it is withdrawn.

Fluorescent antibody technics detect the organisms, and immunoelectrophoresis can be used to test for the presence of meningococcal antigen in cerebrospinal fluid and serum. Group typing is done by immunoelectrophoresis.

Immunity. That epidemics of meningeal infection are not more common stems from the fact that whereas nasopharyngeal infection is prevalent (low immunity to localization), resistance to invasion of the bloodstream and body tissues is high. Some observers believe that a moderate degree of immunity results from an attack of meningitis. Others believe that no protection is acquired. Immunity seems to increase with age because the disease is more prevalent in children than in adults.

Before the advent of antimicrobial compounds the mainstay in the treatment of epidemic meningitis was antimeningococcal serum of two types, bactericidal and antitoxic. Although serum therapy reduced the mortality, its results did not equal those now obtained with antimicrobial drugs. The death rate of 40% to 50% was lowered to 5% to 10%.

Prevention. The patient with meningococcal infection should be isolated until cultures fail to show meningococci in his nasopharynx. All discharges from the mouth

---

TABLE 19-3. PATHOGENS RELATED TO MENINGITIS

A. Primary meningitis caused by
  1. Well-known microbes

Epidemic cerebrospinal fever	*Neisseria meningitidis*
Nonmeningococcal meningitis in young ages	*Haemophilus influenzae*
Endemic meningitis in adults (less so in children)	*Streptococcus pneumoniae*

  2. Unusual microbes

Meningitis of newborn infants	*Escherichia coli*
Purulent or nonpurulent meningitis (infection may be mixed)	*Mycobacterium tuberculosis*
	Neisseria species (other than meningococcus)
	Coliform bacteria
	Listeria monocytogenes
	Streptococcus pyogenes
	Klebsiella-Enterobacter-Serratia species
	Salmonella species
	Bacteroides species
	Pseudomonas species
	Proteus species
	Nocardia asteroides

B. Secondary meningitis caused by
  1. Well-known microbes

Meningitis complicating recent trauma, surgical procedure, lumbar puncture, congenital anomaly (spina bifida), and the like	*Staphylococcus aureus*
	Pseudomonas species
	Coliform bacteria
	Proteus species

  2. Unusual microbes

Complicating meningitis	Any invasive pathogen

C. Meningitis in generalized infection caused by
  1. Well-known microbes

Fulminating meningococcal septicemia	*Neisseria meningitidis*
Fulminating sepsis	*Streptococcus pneumoniae*
Miliary tuberculosis	*Mycobacterium tuberculosis*

  2. Unusual microbes

Generalized infection with bacteremia leading to meningitis	*Neisseria gonorrhoeae*
	Salmonella typhi
	Brucella species
	Staphylococcus aureus
	Bacillus anthracis
	Nocardia asteroides
	Cryptococcus neoformans

---

and nose and articles soiled therewith should be disinfected. The urine occasionally contains the organisms and therefore should be disinfected. The physician and nurse should use every precaution to avoid infection or the carrier state. Nurses should exercise care lest their hands convey infection. Dishes used by the patient should be properly sterilized. Persons who have been in close contact with a patient should not mingle with others until bacteriologic examination has proved them to be free of meningococci.

General preventive measures include the proper supervision of carriers. The wholesale isolation of carriers does not eliminate infection, and the trend presently is to isolate only those in immediate contact with a patient. Nose sprays are probably of little value. The antibiotic rifampin is used to eliminate the carrier state.

The control of epidemic meningitis is yet to be accomplished. For reasons that we do not understand, it suddenly becomes virulent in a community, attains epidemic form, persists for a time, and then disappears.

Three meningococcal vaccines—monovalent A, monovalent C, and bivalent A-C vaccines—are now licensed for *selective* use in the United States. The U.S. Public Health Service Advisory Committee does not recommend *routine* vaccination against group A and C disease because not enough is known about the benefits thereof. Antigens used for meningococcal vaccines are purified bacterial cell–wall polysaccharides (certain polymers of particular neuraminic acids). Vaccine is given as a single parenteral dose and adverse reactions have been insignificant. Duration of immunity is unknown. Group C vaccine has been given to American military recruits for the last several years with good results. Group B vaccine is still under investigation.

Microbial infection of meninges. Meningococci are responsible for the *epidemic* form of meningitis. *Nonepidemic meningitis* is caused by other organisms as outlined in Table 19-3. Bacteria reach the meninges (1) by penetrating wounds; (2) by passage from the nasopharyngeal mucosa through lymph spaces; (3) by extension of regional infections in the middle ear, mastoid process, bony sinuses, or cranial bones; and (4) by the bloodstream (bacteremia or septicemia). The diagnosis is made by the demonstration of the causal agents in smear and culture of cerebrospinal fluid.

OTHER GRAM-NEGATIVE COCCI (AND COCCOBACILLI)

Branhamella (Neisseria) catarrhalis. The organism *Branhamella (Neisseria) catarrhalis* is a normal inhabitant of the mucous membranes, especially those of the respiratory tract. It is important to the microbiologist because it may be confused with the meningococcus or the gonococcus. Like these organisms it is a gram-negative, biscuit-shaped diplococcus that on rare occasions may assume the intracellular position. Differentiation depends on agglutination tests and the ability of *Branhamella catarrhalis* to grow on ordinary culture media at room temperature.

Other neisseriae. Other gram-negative diplococci that may lead to errors in microbiologic diagnosis are *Neisseria subflava* and *Neisseria sicca*, both normal inhabitants of the pharynx and nasopharynx (Table 20-1). *Neisseria flavescens* has been recovered from the cerebrospinal fluid of patients with meningitis, and its colonies look like those of *Neisseria meningitidis*, but it does produce a golden yellow pigment when first isolated.

Moraxella lacunata (the Morax-Axenfeld bacillus). The oxidase-positive, gram-negative, strictly aerobic coccobacilli of the genus *Moraxella* are very similar to members of the genus *Branhamella* and, like *Branhamella*, are parasitic on mucous membranes of man and animals. *Moraxella lacunata* causes a subacute or chronic inflammation of the conjunctiva, eyelid, and cornea.

QUESTIONS FOR REVIEW

1. Name and describe the microbe causing gonorrhea.
2. What is the social importance of gonorrhea?
3. Briefly discuss the veneral disease pandemic.
4. List 10 diseases venereally transmitted. Give the causative organism for each.
5. Outline the laboratory diagnosis of gonorrhea.
6. What are the sources and modes of infection in gonorrhea?
7. How is gonorrheal ophthalmia contracted in adults? In infants?
8. What is Credé's method?
9. Compare gonococci with meningococci.
10. Name and describe the microbe causing epidemic meningitis.
11. What are the sources and mode of infection in epidemic cerebrospinal meningitis?
12. State the importance of the carrier in meningococcal infections.
13. Give the nursing precautions in epidemic meningitis.
14. How is the diagnosis of meningococcal infection made in the laboratory?
15. List 10 microorganisms causing meningitis.
16. Explain what is meant by the Waterhouse-Friderichsen syndrome.

REFERENCES. See at end of Chapter 27.

20 Enteric bacilli

The *enteric bacilli*, a large, heterogeneous group in the family Enterobacteriaceae, include several closely related genera of short, nonsporeforming, gram-negative rods, facultatively anaerobic, that inhabit or produce disease in the alimentary tract of man and warm-blooded animals. Here are the nonpathogenic bacteria that normally inhabit the intestinal canal and the highly pathogenic bacteria that invade and injure it. Heading any list of enteric pathogens are the consistent troublemakers, the *Salmonella* and *Shigella* species. Normal residents include many species in a number of genera, and if given the right opportunity, many of these supposedly benign bacteria can be awesome in their behavior. Not all members of Enterobacteriaceae are intestinal parasites; they have been placed here because of other similarities.

Today, members of this family of bacteria are notorious as causes of urinary tract infection and are recovered from a variety of clinical specimens taken from diseased foci other than in the gastrointestinal tract. The Enterobacteriaceae are probably responsible for more human misery than any other group.

Enteric microorganisms of this chapter are defined in the classification most widely used in microbiologic laboratories. The starting point is Bergey's Part 8 (Family I. Enterobacteriaceae). The breakdown as proposed by W. H. Ewing and approved by the subcommittee of the American Society for Microbiology is shown in Schema 5. *

Enteric bacilli may be studied as to disease production, physiologic behavior, and immunologic reactions. The practical identification of the enteric gram-negative bacilli, however, involves the use of an elaborate array of biochemical reactions. To demonstrate biologic activities of test organisms is usually a complex maneuver in the clinical laboratory. In a greatly simplified scheme, Table 20-1 presents key reactions for bacteriologic separation of enteric organisms to be discussed.

---

*Note departure from Bergey's classification in Schema 5.

SCHEMA 5
EWING'S CLASSIFICATION OF ENTERIC MICROORGANISMS

Genera

Tribe I: *Escherichieae* ---------------- { *Escherichia (Escherichia coli,* including
　　　　　　　　　　　　　　　　　　　　　　　　　　Alkalescens-Dispar)
　　　　　　　　　　　　　　　　　　　　{ *Shigella*

Tribe II: *Edwardsielleae* ------------- *Edwardsiella*

Tribe III: *Salmonelleae* ------------- { *Salmonella*
　　　　　　　　　　　　　　　　　　　{ *Arizona**
　　　　　　　　　　　　　　　　　　　{ *Citrobacter* (including Bethesda-Ballerup)

Tribe IV: *Klebsielleae* ---------------- { *Klebsiella*
　　　　　　　　　　　　　　　　　　　{ *Enterobacter* (formerly *Aerobacter,*
　　　　　　　　　　　　　　　　　　　　　　　　　　including *Hafnia)*
　　　　　　　　　　　　　　　　　　　{ *Pectobacterium†*
　　　　　　　　　　　　　　　　　　　{ *Serratia*

Tribe V: *Proteeae* ------------------ { *Proteus*
　　　　　　　　　　　　　　　　　　　{ *Providencia*

*The term *paracolon bacilli* was formerly used for enteric bacilli resembling *Escherichia coli* but fermenting lactose much more slowly. They were placed in the genus *Paracolobactrum* and separated into the Bethesda-Ballerup group, the Arizona group, the Providence group, and the Hafnia group. In Schema 5, paracolon bacilli are regrouped and the term *paracolon* discarded.

†A genus of plant pathogens (soft rot coliforms), it is not implicated in human infection. As might be expected, it is the only genus in the family liquefying sodium pectate.

SALMONELLA SPECIES*

General characteristics. The genus *Salmonella* † comprises the causative organisms of salmonellosis, some 250 species in four subgenera of motile enteric bacilli found everywhere that there are animals and man. All can produce disease in their natural hosts.

Although the members of the different species are similar in morphology, staining reactions, and cultural characteristics, they can be separated by fermentation reactions‡ (Table 20-2) and agglutination tests. Serologic typing is the ultimate

---

*The *Salmonella* genus was named for Daniel Elmer Salmon (1850-1914), American veterinary pathologist who first isolated the organisms in 1885.

†In sharp contrast to the complexity of the breakdown of the genus *Salmonella* found in the eighth edition of *Bergey's Manual* (Table 1-1, p. 6), Dr. Ewing recognizes only three species in genus *Salmonella* —*Salmonella cholerae-suis* as type species, *Salmonella typhi* as unique pathogen of man, and *Salmonella enteritidis* to comprise all other serotypes and serobiotypes. By this schema, *Salmonella enteritidis* would now be *Salmonella enteritidis* ser. *enteritidis,* and *Salmonella typhimurium* would be *Salmonella enteritidis* ser. *typhimurium.* Since this approach is not widely accepted, we shall continue with traditional species names.

‡In keeping with their role as intestinal pathogens, salmonellae do not ferment lactose.

TABLE 20-1. SIMPLIFIED SCHEME FOR THE SEPARATION OF SOME GRAM-NEGATIVE ENTERIC BACILLI

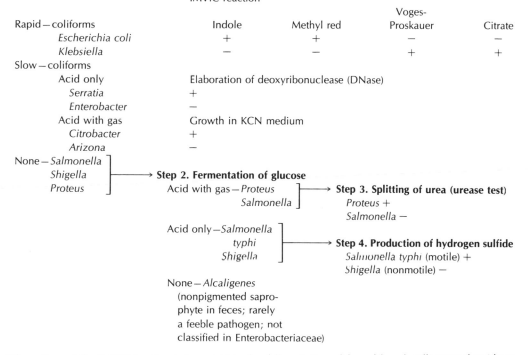

Step 1. Fermentation of lactose

	IMViC reaction*			
	Indole	Methyl red	Voges-Proskauer	Citrate
Rapid — coliforms				
Escherichia coli	+	+	−	−
Klebsiella	−	−	+	+
Slow — coliforms				
Acid only	Elaboration of deoxyribonuclease (DNase)			
Serratia	+			
Enterobacter	−			
Acid with gas	Growth in KCN medium			
Citrobacter	+			
Arizona	−			

None — *Salmonella*
    Shigella ⟶ Step 2. Fermentation of glucose
    Proteus   Acid with gas — *Proteus* ⟶ Step 3. Splitting of urea (urease test)
                                   Salmonella   *Proteus* +
                                                            Salmonella −

                         Acid only — *Salmonella*
                                     typhi ⟶ Step 4. Production of hydrogen sulfide
                                     Shigella   *Salmonella typhi* (motile) +
                                                           Shigella (nonmotile) −

                         None — *Alcaligenes*
                         (nonpigmented sapro-
                         phyte in feces; rarely
                         a feeble pathogen; not
                         classified in Enterobacteriaceae)

*The pattern of the IMViC reaction is important to the differentiation of the coliform bacilli (normal residents of the enteron). It refers to four biochemical color reactions. The first reaction, "I," is the production of indole from tryptophan (pink or red color). The next two, the methyl red reaction, "M," and the Voges-Proskauer reaction, "Vi," indicate a difference in the fermentation of glucose in a special test medium after 2 to 4 days' incubation. If the test organism ferments glucose with the accumulation of acid end products, the methyl red indicator gives a red color (pH below 4.5), a positive test. If glucose has been fermented with the accumulation of neutral end products, the Voges-Proskauer test is positive, a red-orange color showing the presence of acetylmethylcarbinol. The fourth reaction, "C," refers to the utilization of Simmons citrate agar and indicates that the test organism can use the citrate as its sole source of carbon.

step in classification, and more than 1600 serotypes have been identified. The immunologic classification of *Salmonella* is complex but important in tracing epidemiologic patterns of disease. Salmonellae are susceptible to the action of heat, disinfectants, and radiation. At refrigerator temperature they do not multiply but remain viable.

Sources and modes of infection. Since only a few species, such as *Salmonella typhi*, are indigenous to man, most infections come from species in lower animals. In nature, salmonellae occur in the intestinal tract of man and animals. Therefore they must be spread by feces either directly or indirectly from one host to another.

TABLE 20-2. BIOCHEMICAL PATTERNS FOR SALMONELLA SPECIES

Organism	Production of hydrogen sulfide	Reduction of nitrate	Production of indole	Liquefaction of gelatin
Salmonella typhi	+	+	−	−
Salmonella paratyphi-A	−	+	−	−
Salmonella schottmuelleri	+	+	−	−
Salmonella hirschfeldii	+	+	−	−
Salmonella typhimurium	+	+	−	−
Salmonella cholerae-suis	Variable	+	−	−
Salmonella enteritidis	+	+	−	−
Salmonella gallinarum	Variable	+	−	−

Excreta contain bacilli because the infected host has obvious disease or inapparent infection or is a carrier. The principal vehicle is water, milk, or food contaminated by feces. The most dangerous link in a chain of infection is the food handler who is a carrier.

If the food ingested contains only a small number of salmonellae, acid in the stomach of the host destroys them readily. If food is heavily contaminated, some of the bacilli escape the effects of the gastric acid to enter and injure the small bowel.

Pathogenicity. Infection with *Salmonella* is salmonellosis; its major site is the lining of the intestinal tract. Because of their toxic properties, every known strain of *Salmonella* can cause any one of three types of salmonellosis: (1) acute gastroenteritis of the food infection type, (2) septicemia or acute sepsis similar to pyogenic infections, and (3) enteric fever such as typhoid or paratyphoid fevers.

Most of the time, however, *Salmonella* infection is acute gastroenteritis, a condition marked by fever, nausea, vomiting, diarrhea, and abdominal cramps. Sometimes gastroenteritis is the forerunner of septicemia. Once the bacilli spread, they can cause a wide range of lesions in different parts of the body. Examples are abscesses of various organs, arthritis, endocarditis, meningitis, pneumonia, and pyelonephritis. Salmonelloses may be severe and death dealing, or they may be mild and even inapparent.

With regard to their natural host, salmonellae are divided into the following categories: (1) those that primarily affect man, (2) those that primarily affect the lower animals but may cause disease in man, and (3) those that affect lower animals only.

Salmonella typhi (the typhoid bacillus)

The organism *Salmonella typhi* (*S. typhi*) causes typhoid fever, which as the natural infection occurs only in man.

General characteristics. A short, motile, nonencapsulated bacillus, S. *typhi* grows luxuriantly on all ordinary media. It grows best under aerobic conditions but may grow anaerobically. The temperature range for growth is from 4° to 40° C, the optimum, 37° C. Typhoid bacilli can survive outside the body; they live about 1 week

| Fermentation of carbohydrates | | | | | |
Glucose	Lactose	Maltose	Sucrose	Mannitol	Dulcitol
Acid	—	Acid	—	Acid	—
Acid, gas	—	Acid, gas	—	Acid, gas	Variable
Acid, gas	—	Acid, gas	—	Acid, gas	Variable
Acid, gas	—	Acid, gas	—	Acid, gas	Acid, gas
Acid, gas	—	Acid, gas	—	Acid, gas	Variable
Acid, gas	—	Acid, gas	—	Acid, gas	Variable
Acid, gas	—	Acid, gas	—	Acid, gas	Variable
Acid	—	Acid	—	Acid	Acid

in sewage-contaminated water and not only live but multiply in milk. They may be viable in fecal matter for 1 to 2 months. Their pathogenic action relates to endotoxins.

Pathogenicity. Typhoid fever is an acute infectious disease with clinically continuous fever, skin eruptions, bowel disturbances, and profound toxemia. Except in the first few days, leukopenia is always present in uncomplicated cases, probably because typhoid bacilli depress the bone marrow, where normal production of white blood cells occurs. Leukocytosis in the course of the disease signals a complication. The incubation period is from 7 to 21 days.

Typhoid bacilli first penetrate the intestinal lining to attack the lymphoid tissue within the intestinal wall. Peyer's patches in the small bowel bear the brunt of the attack. From there they pass to the mesenteric lymph nodes draining the area, which they colonize, and on to the thoracic duct. They enter the bloodstream where many are lysed; the release of endotoxins brings about manifestations of the disease. The surviving bacilli tend to localize in the gallbladder, bone marrow, and spleen, disappearing from the bloodstream beyond the first week of illness. The rose spots seen on the skin of the abdomen early in the illness contain many bacilli. As the disease progresses, the bacilli confine their activities to the intestinal wall.

Pathology. Pathologically, typhoid fever is a granulomatous inflammation of the lymphoid tissue of the body. In the intestinal mucosa, hyperplastic lymphoid masses swell and plateau. When necrosis takes place with sloughing, an oval ulcer of varying depth is left. If ulceration extends deep enough, perforation of the intestinal wall results. In the spleen, its hyperplastic lymphoid tissue softens and enlarges it to two or three times normal size. If a blood vessel in the intestinal wall undergoes necrosis, hemorrhage may occur. The liver and kidneys show degenerative changes, and the gallbladder may be inflamed. Toxic injury to the heart muscle may precipitate acute heart failure and death.

COMPLICATIONS. Fully three fourths of the deaths in typhoid fever result from some complication. *Hemorrhage* occurs most often during the third week; the passing of a tarry stool or clotted blood may be the first sign. *Perforation* of the bowel is usually single, most often in the lower 18 inches of the small intestine, and the mortality is very high. *Cholecystitis* may follow the disease, and typhoid bacilli may be found in the gallbladder years later.

Spread. The fundamental basis of every typhoid infection is the same. *Typhoid bacilli from the feces or urine of a carrier or a person ill of typhoid fever have reached the mouth of the victim.* Any person contracting typhoid fever has swallowed typhoid bacilli. Bacilli entering the body by a route other than the alimentary tract do not infect.

Typhoid carriers are of two types: fecal and urinary. In the more common fecal carriers the bacilli multiply in the gallbladder and are excreted in the feces. (Infection of the gallbladder with stagnation of bile predisposes to the formation of gallstones.) In urinary carriers the organisms multiply in the kidney and are excreted in the urine. From 40% to 45% of patients become convalescent carriers for 3 to 10 weeks. About 5% carry the bacilli for 1 year, and about 2% become permanent carriers. The average carrier is female and more than 40 years of age.

The most dangerous factor in the spread of typhoid fever is the carrier food–handler who prepares foods that are served raw. Carriers contaminate their fingers with their discharges and then contaminate food with their fingers. It is hard to prevent the carrier state, and treatment for it can be difficult. Removal of the gallbladder cures selected cases.

With improvements in sanitation, widespread epidemics of typhoid fever from contamination of water supplies with sewage, once so common in large cities, are almost unknown. Strict laws regulate the cultivation of oysters and other shellfish eaten raw to eliminate contamination of these foods with sewage-laden water.

Laboratory diagnosis. The laboratory offers four procedures for the diagnosis of typhoid fever: (1) blood culture, (2) the Widal test, (3) stool culture, and (4) urine culture. Blood cultures are positive during the first week in 75% to 80% of the patients. The percentage falls 10% by the fourth week. Isolation of the organism from blood culture is diagnostic.

After 7 to 10 days of infection, agglutinins against typhoid bacilli appear in the patient's blood, increasing during the second and third weeks of disease. They may persist for weeks, months, or years after the patient recovers. Agglutinins also appear in the blood after typhoid vaccination and are usually found in the blood of carriers.

When an animal is immunized with typhoid bacilli and certain other actively motile organisms, two types of agglutinins are formed. One, acting on the flagella of the bacterium, is the *H (flagellar)* agglutinin. The other, acting on the body of the bacterium, is the *O (somatic)* agglutinin. The O agglutinins are thought to appear earlier in disease and to indicate actual infection with typhoid bacilli or *closely related organisms.* If they are present in the serum of persons vaccinated, they are in small quantities and only for a short time. In addition to H and O antigens typhoid bacilli responsible for active disease or the carrier state contain a third antigen, the *Vi (virulence)* antigen. Vi antibodies are not thought to occur significantly after vaccination. *Salmonella* species other than *S. typhi* possess O and H antigens; some possess Vi antigens. The species may be divided into groups on the basis of their O antigens and further subdivided on the basis of their H antigens.

The Widal test* is used to detect the presence of these agglutinins. It is said to be positive in 15% of typhoid patients during the first week of illness, with the incidence of positive tests rising to 90% or more during the third week. However, there are considerable difficulties in interpretation of the Widal test, and many authorities are thoroughly disillusioned with it.

Stool and urine cultures facilitate detection of carriers and indicate when a given patient ceases to be a source of infection. Typhoid bacilli may be cultivated from the stool in 50% of the cases by the third week. Positive urine cultures are obtained after the second week in 25% to 50% of the patients. Convalescent carriers excrete the bacilli in their feces or urine during convalescence only, but permanent carriers continue to excrete them.

Immunity. In 98% of cases an attack of typhoid fever gives permanent immunity. The production of an artificial immunity by the administration of typhoid vaccine has been a most important factor in the control of the disease.

Prevention. The prevention of typhoid fever is twofold—community and personal. Community prevention means those measures taken by the community as a whole to block spread of disease among its members; personal prevention means those measures taken to prevent the spread of infection from a person ill of the disease. The most important factors in the community program are (1) a supply of clean pasteurized milk, (2) pure water supply, (3) efficient disposal of sewage, (4) proper sanitary control of food and eating places, (5) detection and isolation of carriers, especially food handlers, (6) destruction of flies, and (7) vaccination. †

Personal prevention depends on isolation of the patient. Isolation does not mean merely putting the patient in a room and shutting the doors but must close all routes by which bacteria may be transmitted to others.

Preferably the patient is hospitalized and kept there until he is no longer infectious. Nurses attending a patient with typhoid fever should consider every secretion and excretion of the patient to be a living culture of typhoid bacilli and should use every means to protect themselves, members of the patient's family, and people in the community. They should have nothing to do with the preparation of food, and their hands should be disinfected after each contact with the patient or anything that either the patient or his secretions have touched. Feces and urine can be disinfected with chlorinated lime, 5% phenol, or 2% Amphyl (see also p. 282). Sputum should be received on disposable paper tissues and burned. Linen and bedclothes should be sterilized. Food remains should be burned and dishes boiled. The patient's bath water can be sterilized with chlorinated lime. The sickroom should be screened, and flies that accidentally gain access to it should be killed. All rugs, curtains, and similar fabric materials should be removed from the room. Pets should not be allowed to enter. Disinfection should continue through convalescence, and no patient should

---

*This agglutination test was devised in 1896 by Fernand Widal (1862–1929), French clinician and bacteriologist.

†Currently the U.S. Public Health Service is *not* recommending typhoid immunization within the United States. The Advisory Committee does not indicate typhoid vaccination for persons going to summer camp or even for those surviving a flood disaster. Typhoid immunization is discussed on p. 705.

be discharged as cured until repeated cultures of feces and urine fail to show typhoid bacilli.

That the procedures just outlined, together with vaccination, have materially reduced the incidence of typhoid fever is proved by the following facts. In 1900 the death rate in the United States from typhoid fever was 35.9 per 100,000 persons; today less than 400 cases are reported annually (0.18 cases per 100,000 persons reported in 1975). That universal vaccination alone will materially reduce the incidence of typhoid fever is proved by the experience of the United States Navy during the years 1911, 1912, and 1913, a period in which no revolutionary developments in sanitation took place. In 1911 there were 361 cases of typhoid fever per 100,000 men. In 1912 universal vaccination of naval personnel was instituted. In 1913 the rate fell to 34 per 100,000, a reduction of more than 90%.

Paratyphoid bacilli

Paratyphoid bacilli are so called because they have been isolated from paratyphoid fever in man, a disease resembling typhoid fever but milder in its manifestations and shorter in duration. Although associated with enteric fever, each of the three species causes other forms of salmonellosis.

The three are *Salmonella paratyphi*-A, *Salmonella schottmuelleri*, and *Salmonella hirschfeldii*, formerly known respectively as the paratyphoid bacilli A, B, and C. Infections with *Salmonella paratyphi*-A occur almost exclusively in man; *Salmonella schottmuelleri* occasionally infects lower animals. *Salmonella hirschfeldii* is rarely found in the United States; infections are frequent in Eastern Europe,

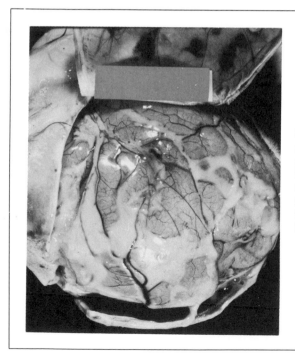

FIG. 20-1. *Escherichia coli* as cause of meningitis in 24-day-old premature infant. Note immature brain with coating of pus over outer surface and prominent vascular markings. Skull flaps are laid back; brain is seen in cranial cavity.

and mice are said to be its natural host. *Salmonella paratyphi*-A resembles *Salmonella typhi* more closely than does *Salmonella schottmuelleri*, but infections with the latter are more prominent.

The mode of infection, sources, laboratory diagnosis, nursing precautions, and prevention are the same as those in typhoid fever. The immunity produced by a paratyphoid infection or vaccination is uncertain (see also p. 705).

Other salmonellae

Among the many salmonellae that primarily affect lower animals but may infect man are *Salmonella typhimurium* (cause of a typhoidlike disease of mice; may cause gastroenteritis in man); *Salmonella cholerae-suis* (once thought to be the cause of hog cholera; can cause septicemia in man); and *Salmonella enteritidis*, known also as Gärtner's bacillus (found in hogs, horses, mice, rats, and fowls; infects man). An organism primarily affecting lower animals is *Salmonella gallinarum* (cause of fowl typhoid); it differs from other *Salmonella* species in that it is not motile.

Attacks of enteric infection are on occasion traced to salmonellae from unusual sources, such as the small pet turtles given to children who subsequently developed salmonellosis. Contaminated baby chicks have been a similar source for tiny tots.

SHIGELLA SPECIES (THE DYSENTERY BACILLI)

Dysentery (Greek *dys*, painful; *enteron*, intestine) is a painful diarrhea accompanied by the passage of blood and mucus and associated with abdominal pain and constitutional symptoms. It may be primary or secondary. When primary, it may be of protozoal, bacterial, or viral origin. Protozoal dysentery caused by an ameba is amebic dysentery (p. 581). Dysentery caused by members of the *Shigella* genus is known as bacillary dysentery. Bacillary dysentery may be epidemic or endemic, is usually acute but may be chronic, and is an important form of summer diarrhea of infants.

General characteristics. The *Shigella* are classified into four species, each of

---

TABLE 20-3. BIOCHEMICAL REACTIONS OF SHIGELLA SPECIES

Organism	Production of hydrogen sulfide	Liquefaction of gelatin	Reduction of nitrate	Production of indole	Fermentation of carbohydrates				
					Glucose	Lactose	Sucrose	Mannitol	Dulcitol
Shigella dysenteriae	−	−	+	−	Acid	−	−	−	−
Shigella flexneri	−	−	+	+	Acid	−	−	Acid	−
Shigella boydii	−	−	+	Variable	Acid	−	−	Acid	Variable
Shigella sonnei	−	−	+	−	Acid	−	Acid	Acid	−

which is divided into several types: *Shigella dysenteriae* (seven types), isolated by Shiga in the Japanese epidemic in 1898; *Shigella flexneri* (six types), isolated by Flexner in the Philippine Islands in 1900; *Shigella boydii* (eleven types); and *Shigella sonnei* (six types). *Shigella flexneri* and *Shigella sonnei* are worldwide. In recent years infection with *Shigella dysenteriae* has been limited to the Orient.

Dysentery caused by the Shiga bacillus *(Shigella dysenteriae)* is much more severe than that from the other organisms because this bacillus produces a powerful exotoxin-like substance in addition to an endotoxin. The exotoxin-like substance seems to be liberated by bacterial disintegration, and as a neurotoxin, it acts on the nervous system to paralyze the host. The endotoxin irritates the intestinal canal.

The dysentery bacilli are gram-negative, nonsporebearing rods that grow on all ordinary media at temperatures from 10° to 42° C but best at 37° C. They are aerobic and facultatively anaerobic. Unlike most other members of the enteric group, they are nonmotile. Table 20-3 gives biochemical reactions for *Shigella* species.

Pathogenicity. Dysentery is a human disease, and natural infections of lower animals do not exist. The incubation period is 1 to 7 days. Epidemic dysentery is chiefly an intestinal infection. Unlike typhoid bacilli, the organisms do not invade the bloodstream and are seldom if ever found in the internal organs or excreted in the urine. They are excreted in the feces.

Pathology. Pathologically, bacillary dysentery is recognized as diffuse inflammation with ulceration of the large intestine and sometimes the lower portion of the small intestine. Early in the disease shallow ulcers or extensive raw surfaces that may be covered by a pseudomembrane form in the mucous membrane. In mild cases healing is complete, but in severe cases there is extensive scarring.

Mode of infection. The mode of infection is practically the same as for typhoid fever; the bacilli enter the body of the victim by way of the mouth, having been transferred there from the feces of carriers or patients. Contaminated food, water, fingers, or other objects are the vehicles of spread. Contact with persons who have symptomless infections is especially significant in the spread of bacillary dysentery. Flies are a factor.

Laboratory diagnosis. The only practical method is the cultivation of the organisms from the stool. Unfortunately, this can be done only during the first 4 or 5 days of the disease.

Immunity. An attack of dysentery probably confers some degree of immunity, but the same person has been known to have two episodes in a single season.

Prevention. Bacillary dysentery may be checked by the sanitary measures that control typhoid fever, but it remains ready to rise in epidemic form when people are crowded together in unsanitary conditions. The feces and everything contaminated by patients should be handled exactly as for typhoid fever. Food or milk should not be carried from the premises, and the medical attendants should not handle food for others. The patient should not be dismissed as cured until repeated feces cultures have failed to show the causative organism. After the patient has recovered, the sickroom should be thoroughly cleaned.

Preventive vaccination is not yet available. An oral attenuated vaccine against

Shigella flexneri is being tested. One such is a "mutant hybrid," produced by "mating" *Escherichia coli* with a nonvirulent strain of *Shigella flexneri*. There is no serum therapy.

ESCHERICHIA COLI AND RELATED ORGANISMS (THE COLIFORM BACILLI)

General characteristics. Besides *Escherichia coli*, coliform bacteria comprise members of the genera in the following groups: (1) *Klebsiella-Enterobacter-Serratia*, (2) *Arizona-Edwardsiella-Citrobacter*, and (3) the "Providence" group.

As normal inhabitants of the intestinal tract,* these microbes share certain traits. They are gram-negative, short rods that do not form spores. They grow at temperatures from 20° to 40° C but best at 37° C. A few species are surrounded by capsules. Most are motile. They are not so susceptible to the bacteriostatic action of dyes as other bacteria. Therefore media for the isolation of coliform bacilli can contain inhibitory dyes. Table 20-4 presents bacteriologic features of coliforms.

In a sanitary water analysis the presence of coliform bacilli indicates fecal pollution of the water supply being tested.

Pathogenicity. Coliform bacilli usually do not penetrate the intestinal wall to produce disease unless (1) the intestinal wall becomes diseased, (2) resistance of the host is lowered, or (3) virulence of the organisms is increased greatly. Under one of these conditions, coliforms may pass to the abdominal cavity or enter the bloodstream. Once outside the intestinal canal and in the tissues of the body, their virulence is remarkably enhanced (Fig. 20-1). Among the diseases they may cause are pyelonephritis, cystitis, cholecystitis, abscesses, peritonitis, and meningitis. They play a part in the formation of gallstones; they are found in the cores of such stones. In peritonitis complicating intestinal perforation, the coliform group is joined by such organisms as streptococci and staphylococci. From any focus of inflammation coliform organisms may enter the bloodstream and produce a septicemia.

Escherichia coli

The signal member of the coliform group and long known as the colon bacillus (of man and other vertebrates) because of its natural habitat in the large bowel, *Escherichia coli (E. coli)* is unabashed as an opportunist when out of its natural setting. It is the commonest cause of pyelonephritis and urinary tract infections and an important cause of epidemic diarrhea in nurseries for newborn infants. The immunologic pattern for *Escherichia* is as complex as that for *Salmonella*. Immunologic subdivision of *E. coli* is made on the basis of O (somatic) antigens, K (capsular) antigens, and H (flagellar) antigens. More than 150 serologic types are known, of which 11 have been correlated with infantile diarrhea.

Infantile diarrhea. Infectious diarrhea is an extremely serious cause of infant death. In underdeveloped areas of the world over 5 million infants succumb to it each year. Infantile diarrhea including summer diarrhea is especially prevalent during

---

*In the adult, 30% of the dry weight of the feces is made up of bacteria.

TABLE 20-4. BIOCHEMICAL REACTIONS OF COLIFORMS AND PROTEUS BACILLI

Organism	Production of indole	Methyl red test	Voges-Proskauer reaction	Production of hydrogen sulfide	Liquefaction of gelatin (22° C)
Coliforms:					
Escherichia coli	+	+	−	−	−
Klebsiella species	−	−	+	−	−
Enterobacter species	−	Variable	Variable	−	Variable
Serratia marcescens	−	−	+	−	+
Arizona species	−	+	−	+	Slow
Edwardsiella tarda	+	+	−	+	−
Citrobacter species	−	+	−	+	−
Providencia species	+	+	−	−	−
Proteus bacilli:					
Proteus vulgaris	+	+	−	+	+
Proteus morganii	+	+	−	−	−

hot weather among bottle-fed babies who are reared in unhygienic surroundings, but it is also a problem in modern hospital nurseries. In institutional outbreaks (hospitals and orphanages) half the cases are caused by *Shigella* (especially *Shigella sonnei*), *Salmonella*, and enteropathogenic *E. coli. Proteus, Pseudomonas, Staphylococcus*, cholera-like vibrios, and enteroviruses may also be implicated with a significant number of cases related to enteroviruses.

When strains of enteropathogenic *E. coli* colonize the small intestine, they release an exotoxin, an "enterotoxin," which acts on the intact intestinal wall in such a way as to move large amounts of fluid and electrolytes across the mucous membrane. The losses are massive, and a severe watery diarrhea results.

In the hospital nursery, care in the preparation of infant formulas and the sterilization of bottles and nipples are essential to prevent this disorder, which is uncommon in the breast-fed baby. Fluorescent antibody technics detect the serotypes of *E. coli* in stools or rectal swabs from the babies and help to trace carriers among personnel.

Klebsiella-Enterobacter-Serratia

The biologic qualities of *Klebsiella* indicate a relationship to the coliforms; the disease-producing capacities indicate a kinship to respiratory tract pathogens. *Klebsiella* are nonmotile, gram-negative, aerobic organisms surrounded by a broad, well-developed polysaccharide capsule. They are differentiated from *E. coli* biochemically (Table 20-4). The type organism is *Klebsiella pneumoniae* (Friedländer's bacillus*). A hazard to the chronic alcoholic, pneumonia caused by Friedländer's bacillus may

---

*This microbe was discovered by Carl Friedländer (1847-1887) prior to the discovery of the pneumococcus and was thought by him to be the sole cause of pneumonia. Actually, it is responsible for less than 10% of all cases.

Use of Simmons citrate	Splitting of urea	Fermentation of carbohydrates				
		Glucose with gas	Lactose	Sucrose	Mannitol	Dulcitol
−	−	+	+	+	+	Variable
+	+	+	+	+	+	−
+	− (Variable)	+	Variable	+ (Variable)	+	−
+	Variable	+	Very slow	+	+	−
+	−	+	Slow	−	+	−
−	−	+	−	−	−	−
+	Variable	+	+	Variable	+	+
+	−	Variable	−	Variable	Variable	−
Variable	+	+	−	+	−	−
−	+	+	−	−	−	−

be either lobar or lobular, is very severe, and is often fatal. Middle ear infections and meningitis may complicate it. Friedländer's bacillus is also responsible for septicemia.

The genus *Enterobacter* includes the microbes formerly known as *Aerobacter aerogenes*. These bacilli are found in the soil, water, dairy products, and the intestines of animals, including man. As opportunists, they play a part in urinary tract infection and rarely cause more serious conditions— endocarditis, suppurative arthritis, and osteomyelitis. The type species is *Enterobacter cloacae*.

Serratia marcescens is a small, free-living, ubiquitous rod celebrated for the red pigment (prodigiosin) produced in cultures (Fig. 20-2). Since it has emerged as a sometimes formidable pathogen, we find that only a small number of the organisms are chromogenic. The pigment-producing strains do so at room temperature, seldom at incubator temperatures. As is true for other coliforms, infections with this organism (serratiosis) complicate surgical procedures and antibiotic therapy, especially in elderly, debilitated patients in the hospital. The results may be urinary tract infections, pneumonia, empyema, meningitis, or wound infections, to mention but a few.

Bacteriologically the organisms in the genera *Klebsiella*, *Enterobacter*, and *Serratia* are similar. However, when they are grown on a modified deoxyribonuclease (DNase) agar containing toluidine blue as indicator, colonies of *Serratia* may be distinguished from others. A bright pink zone around *Serratia* indicates that it elaborates deoxyribonuclease. The clear blue color of the other colonies indicates that these organisms do not.

Arizona-Edwardsiella-Citrobacter

Formerly known as paracolon bacilli, these organisms ferment lactose very slowly, if at all.

FIG. 20-2. *Serratia marcescens.* Colonies of pigmented strain on nutrient agar plate. Famous red suggested by dark appearance.

Arizonae are short rods similar to the salmonellae and in the eighth edition of *Bergey's Manual* are classified in genus *Salmonella*, subgenus III. From 7 to 10 days may be required for them to ferment lactose. The type species is *Arizona hinshawii* (or Bergey's *Salmonella arizonae*). They are found in human infections and cause disease in dogs, cats, and chickens.

Because of their pattern of biochemical reactivity, certain motile rods were separated into the genus *Edwardsiella*. The type species *Edwardsiella tarda* was given a name implying a kind of biochemical inactivity. *Edwardsiella tarda* has been isolated from man and animals, especially snakes. In fact, snake meat is postulated to be a source of infection. In man the infections are similar to the salmonelloses.

Citrobacter are motile rods fermenting lactose. The type species is *Citrobacter freundii*. They may be confused with *Salmonella* and *Arizona* organisms. Certain strains possess the Vi antigen found in *S. typhi*. *Citrobacter* is differentiated from *Arizona* and *Salmonella* by its ability to grow in the presence of potassium cyanide (positive KCN test).

"Providence" organisms

Also known as paracolons, these microbes are free living, lactose negative, and biochemically related to *Proteus*. The type species is *Providencia alcalifaciens*. (In the eighth edition of *Bergey's Manual*, they are classified within the genus *Pro-*

teus.) Ordinarily nonpathogenic, they have been recovered from cases of human diarrhea and urinary tract infections.

PROTEUS SPECIES (THE PROTEUS BACILLI)

Proteus bacilli comprise a genus of motile organisms that resemble and yet differ from other enteric bacilli in many ways (Table 20-4). They are gram negative, do not form spores, and are normal inhabitants of feces, water, soil, and sewage. They grow rapidly on ordinary media and, being very actively motile (Fig. 20-3), tend to "swarm" over the surface of cultures on solid media. As a consequence, colonies do not remain discrete. Growth from the colony edge in time extends over all available surface area as a thin, translucent, bluish film. The bacilli can be seen microscopically to break away, migrating over the agar surface. This property poses problems to the isolation of *Proteus* bacilli from mixed cultures, although generally with care it can be done.

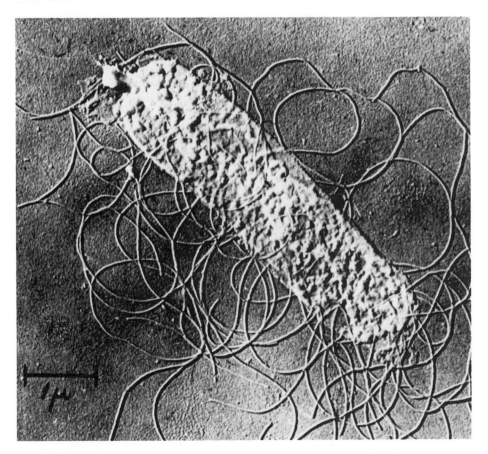

FIG. 20-3. *Proteus vulgaris.* Electron micrograph of actively motile bacterium. Note flagella all over bacterial cell. (Courtesy J. B. Roerig Division, Chas. Pfizer & Co., Inc., New York, N.Y.)

Proteus is pathogenic for rabbits and guinea pigs. Their primary pathogenicity is slight, but as secondary invaders they are vigorous. The type species is *Proteus vulgaris*, which is remarkably resistant to antimicrobial drugs. *Proteus bacilli* in the stool often increase in number after diarrhea caused by other organisms, especially when antibiotics have been given for the diarrhea.

Proteus species rank next to *E. coli* in importance as a cause of urinary tract infections and pyelonephritis. They are important in wound infections and a rare cause of peritonitis. *Proteus morganii* is thought to cause infectious diarrhea in infants.

It is a peculiar fact that although *Proteus* bacilli do not cause typhus fever nor act as secondary invaders, the serum of a patient with typhus fever agglutinates certain strains of *Proteus* bacilli. This is the basis of the Weil-Felix reaction for typhus fever.

ENDOTOXIN SHOCK

Endotoxin shock (bacteremic shock, gram-negative shock, gram-negative sepsis*) is a very serious complication of bacteremia with gram-negative enteric bacilli, notably *Pseudomonas aeruginosa*† (85% mortality), *Proteus*, *E. coli*, and coliforms of the genera *Klebsiella*, *Enterobacter*, and *Serratia*. The mortality is over 50%. The dramatic clinical picture is drawn by the precipitous development of chills, fever, nausea, vomiting, diarrhea, and prostration. There is a sharp drop in blood pressure, and the state of shock is profound.

The combination of events appears to result directly from the action of lipopolysaccharide endotoxin released into the circulation and there activated by certain components of the plasma including complement. Endotoxin effects constriction of small blood vessels and in so doing sets off a series of events leading to the shock state—increased peripheral vascular resistance, pooling of blood in the capillary circulation, hypotension, and shutdown of kidney function. The presence of the inciting gram-negative bacteria in the bloodstream also plays a part in pathogenesis.

This kind of bacteremia is especially prone to complicate infections where enteric bacilli are already entrenched such as those of the genitourinary tract, lungs, intestinal tract, and biliary tract. Instrumentation or an operative procedure on the infected area (2 to 24 hours before) favors release of organisms into the bloodstream. In diseases such as leukemia where host resistance is low, organisms flood the bloodstream from no apparent focus.

ENTERIC BACILLI IN HOSPITAL-ASSOCIATED (NOSOCOMIAL) INFECTIONS

Of all patients admitted to a modern hospital 3% to 5% acquire an infection there. Since around 30 million persons are admitted annually, this means that more than a million and a half patients develop an infection they did not themselves bring into the hospital.

---

*Bacterial shock rarely complicates bacteremia due to gram-positive cocci or systemic infections due to meningococci, *Clostridia*, viruses, rickettsias, and fungi.
†See p. 452.

The predominating organism in hospital-associated (acquired) infections has been *Staphylococcus aureus*. (It is still important, accounting for about one fifth of the infections.) Today most observers unanimously agree that the prime offender is gram-negative enteric bacilli as a group (*Escherichia coli*, other coliforms, *Proteus*, *Pseudomonas*). Such enteric bacilli easily account for two thirds of nosocomial infections. Other important agents are enterococci, pneumococci, other streptococci, and viruses.

Most hospital-associated infections with enteric bacilli seem to occur in the very young or very old patient. They tend to complicate chronic debilitating diseases, diseases that alter the patient's resistance to infection, and diseases treated with antibiotics to which the gram-negative bacilli have or easily acquire resistance. Most infections are endogenous (that is, caused by organisms available from a plentiful supply within the gastrointestinal tract, respiratory tract, or urinary tract of a given patient).

FIG. 20-4. Sources of infection in hospital setting. Note hazards in operating room. How many areas in this picture can you identify as potential foci of infection? What precautions would you advise?

Around 90% of the time there is a surgical wound infection (especially where the operation was done on the alimentary tract), a urinary tract infection, or respiratory tract infection such as bronchopneumonia.

It is to be reemphasized that varied pathogens are everywhere in the hospital environment (Fig. 20-4), in foci one would least suspect—faucets, flower vases, soaps, and lotions, for examples. The role of the more sophisticated items indigenous to a hospital is often overlooked—urethral catheters, intravenous catheters, respirators, reservoir nebulizers, and the like. Much current thought is directed to measures designed to reduce the hazards.

QUESTIONS FOR REVIEW

1. Classify the enteric bacilli.
2. Comment on the various ways to identify enteric bacilli.
3. Briefly characterize organisms of the genus *Salmonella*.
4. Name the three major types of salmonellosis.
5. Name and describe the organism causing typhoid fever.
6. Draw a diagram tracing the spread of typhoid infection from person to person.
7. Discuss the nursing precautions in typhoid fever and other infections of the alimentary tract.
8. What is the laboratory diagnosis of typhoid fever?
9. Name and describe the organisms causing bacillary dysentery.
10. Which of the *Shigella* causes the most severe disease and why?
11. Briefly comment on the pathogenicity of *Shigella* species.
12. Name the members of the coliform group of microorganisms.
13. Under what conditions can coliform bacilli invade the tissues of the body?
14. Discuss infantile diarrhea. Give causes.
15. What is the basis of the Weil-Felix reaction for typhus fever?
16. Compare *Proteus* bacilli with coliforms.
17. How does enteropathogenic *Escherichia coli* produce disease?
18. Briefly discuss nosocomial infections.
19. What is meant by gram-negative sepsis?
20. Make a list of opportunist pathogens you have studied.

REFERENCES. See at end of Chapter 27.

21 Small gram-negative rods

The microbes of this chapter are tiny gram-negative bacilli or coccobacilli, which considered together hardly measure 1 μm in greatest dimension. Three of the seven genera discussed are strict aerobes; the rest are facultative anaerobes. With one exception the genera presented are those of uncertain affiliation in Part 7 ("Gram-Negative Aerobic Rods and Cocci") and Part 8 ("Gram-Negative Facultatively Anaerobic Rods") of the eighth edition of *Bergey's Manual*. The exception from Part 8 is *Yersina* (Genus XI of Family I. Enterobacteriaceae).

BRUCELLA SPECIES (AGENTS OF BRUCELLOSIS)

The genus *Brucella* includes three intracellular parasites that attack lower animals primarily, from which they are transmitted to man to cause *brucellosis*. One, known as *Brucella melitensis*, produces Malta fever in goats and sheep, a disease with prolonged fever, arthritis, and a tendency to abort. Another, known as *Brucella abortus*, causes contagious abortion, or Bang's disease, in cattle. Contagious abortion of cattle is characterized by a tendency to abort, retention of placentas, sterility, and death of newborn offspring. The third, known as *Brucella suis*, causes contagious abortion in hogs. Male hogs are affected by brucellosis even more often than female hogs, and in females the tendency to abort is less than in cows. Brucellosis sometimes attacks animals other than the ones just mentioned. The disease known as poll evil in horses is caused by brucellae. A newly recognized strain, *Brucella canis*, infects dogs.

Brucella melitensis was discovered in Malta in 1887 by Sir David Bruce (1855-1931), a surgeon in the British army, and *Brucellis abortus* was discovered in Denmark by Bernhard L. F. Bang (1848-1932) in 1897. In 1918 and 1925, Alice C. Evans (1881-1975) proved that these organisms are closely related. *Brucellis suis*, discovered in infected hogs in the United States, was found to be closely related to the other two, and all three were placed in the genus *Brucella*, honoring Bruce, the discoverer

TABLE 21-1. DIFFERENTIAL PATTERNS OF BRUCELLA SPECIES

Characteristics	Species		
	Brucella abortus	*Brucella melitensis*	*Brucella suis*
Culture in medium containing dyes			
1. Thionine	Growth inhibited	No effect	No effect
2. Basic fuchsin	No effect	No effect	Growth inhibited
Exposure to atmosphere of 10% carbon dioxide	Required for best growth	Not required	Not required
Production of hydrogen sulfide	For 2 to 4 days	None	For 6 to 10 days
Hydrolysis of urea (urease test)	None (or very slowly)	None (or very slowly)	Rapid
Lysis by phage	Yes	None	None at routine test dilution (RTD); yes at 10,000 × RTD
Agglutination reaction	Differentiates this organism from *Brucella melitensis* but not from *Brucella suis*	Differentiates this one from other two	Differentiates this one from *Brucella melitensis* but not from *Brucella abortus*
Biotypes	Nine	Three	Four
Host reservoir	Cattle	Goats (also sheep)	Pigs (also hares, reindeer)

of *Brucella melitensis. Brucella melitensis* and *Brucella suis* probably represent variations of *Brucella abortus*, the result of its adaptation to the goat and hog, respectively.

General characteristics. Brucellae are small, nonmotile, ovoid, gram-negative, nonsporeforming coccobacilli. Some form capsules. Brucellae elaborate an endotoxin, and as is true of other gram-negative organisms, their cell wall is made up of a lipoprotein-carbohydrate complex. Being so closely related to each other, they cannot be differentiated by ordinary cultural methods. Special methods using to advantage the action of dyes such as basic fuchsin and thionine on the different species of *Brucella* effect separation. The production of hydrogen sulfide is another differential point. Brucellae grow best on enriched media, but growth is slow. *Brucella melitensis* and *Brucella suis* grow in atmospheric oxygen, but for the primary isolation of strains of *Brucella abortus* the presence of 5% to 10% carbon dioxide in the atmosphere is necessary. The agglutination test separates *Brucella melitensis* from *Brucella abortus* and *Brucella suis* but not *Brucella abortus* from *Brucella suis*. Table 21-1 summarizes important differential features in the genus *Brucella*.

Brucellae are destroyed within 10 minutes by a temperature of 60° C and fortunately are quickly destroyed by pasteurization. They can remain alive and virulent for as long as 4 months in dark, damp surroundings.

Pathogenicity. *Brucella melitensis* is more pathogenic for man than *Brucella suis*, and the latter is more pathogenic than *Brucella abortus*. Animals may be infected with any one of the three. Pregnant domestic animals are more likely to be

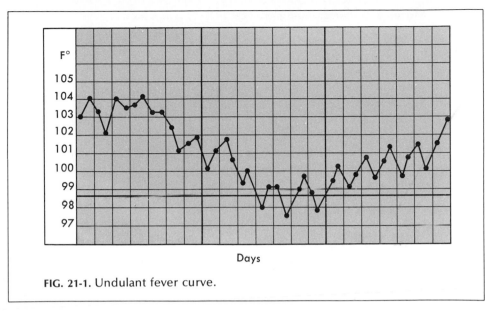

FIG. 21-1. Undulant fever curve.

infected than young animals or nonpregnant adults, and abortion is probable when the infection is acquired during the pregnancy.

Brucellae can attack every organ and tissue of the human body. For this reason a person with brucellosis may seek as his initial consultant almost any medical specialist—internist, surgeon, or other.

Infections in man follow five patterns. The commonest form of brucellosis is a long-continued fever or cycles of fever alternating with afebrile periods (Fig. 21-1), pronounced weakness, and profuse sweating. The other four types are variants. The incubation period varies from 5 to 30 days (sometimes longer). *Melitensis* infections are common in the Mediterranean basin from where imported goats brought the infection to the southwestern United States. *Brucella abortus* and *Brucella suis* infections are widespread in the United States.

Various names are applied to the disease in man—Malta fever (because of its prevalence on the island of Malta), undulant fever (because of its clinical fever curve —Fig. 21-1), and brucellosis. In southwest Texas it is goat fever and Rio Grande fever.

Sources and modes of infection. Infection is widespread among goats, cattle, and hogs. In cattle, brucellosis ranks with tuberculosis as a source of economic loss. Infection is transmitted from animal to animal by food contaminated with infectious urine, feces, or lochiae and by contact with infected placentas or fetuses. Suckling animals may be infected by the milk of infected mothers. Cows become carriers and excrete the organisms in their milk for as long as 7 years.

Human infections are common but often overlooked. In man, as in animals, infection enters via the gastrointestinal tract or through the skin. Most times the infectious material is derived from the excreta of living animals or the blood or tissues of dead animals, and the organism penetrates cuts or abrasions in the skin. It is thought that these virulent pathogens can pass the unbroken skin or mucous membrane.

Packinghouse workers, butchers, farmers, stockmen, and veterinarians can get the infection by handling infected cows or hogs and the meat of these animals. Man can become infected by eating dairy products from infected cows. Infection with *Brucella melitensis* is contracted by the consumption of unpasteurized milk from infected goats. Dust containing the organisms may be infectious. Transfer of the infection directly from person to person is very rare.

Laboratory diagnosis. Laboratory aids in the diagnosis of undulant fever are an agglutination test, blood cultures (about one-third positive), animal inoculation, feces cultures, urine cultures, cultures of bone marrow (taken by sternal puncture), and cultures of material aspirated from lymph nodes. The last two are important because, after the bacilli have spread through the body, they localize in the bone marrow, liver, spleen, and lymph nodes. Positive cultures confirm the diagnosis, but even with known disease *Brucella* often cannot be cultured. Therefore one must usually rely on the agglutination test done with a carefully standardized antigen. (It also may be negative, even with definite disease.) Persons who have received cholera vaccine may develop agglutinins against *Brucella,* a point to be noted in the investigation of suspected cases of undulant fever.

Patients recovering from one attack of brucellosis may continue to show agglutinins for months or years. If symptoms reappear, and the disease seems to be recurring, the determination of the nature of that person's immunoglobulins is significant. If immunoglobulins are 7S, even in low titers, active disease is a good bet. If they are 19S in type, it is not. Also used in the laboratory diagnosis of brucellosis are a modified Coombs' test (helpful in chronic cases), a complement fixation test, and fluorescent antibody technics. Veterinarians use the milk ring test, an agglutination test using the whole milk instead of serum as the source of antibodies.

Guinea pigs may be inoculated with blood or cream from a suspect cow. The skin test detecting hypersensitivity to *Brucella* is formed by injecting brucellergen, a crystalline polypeptide extract of killed organisms, between the layers of the skin. If positive, the skin test indicates infection at some time past or present. A negative skin test eliminates that possibility. The skin test is comparable to the tuberculin skin test and as informative. However, the antigen for intradermal testing is no longer commercially available.

Immunity. Children under 10 years of age seldom contract brucellosis. Men are more often attacked than women (4:1), probably because of greater exposure. One attack may confer permanent immunity, although the disease is not self-limited. After an acute attack, a person may be well, but there is reason to think that the organisms remain alive but quiescent in lymph nodes and the spleen.

Prevention. Vaccines may provide temporary protection for human beings but are still experimental. An avirulent live strain of *Brucella* can be used to vaccinate cattle.

All milk (including goat's milk) and milk products should be pasteurized. All dairy cattle should be tested for brucellosis and infected animals removed from the herd. Those who work with animals must know how the disease is spread and what precautions to take to avoid contact with the flesh or excreta of infected animals. The strict application of sanitary and hygienic measures helps to minimize the occupational hazards.

The excreta of a patient with undulant fever should be disposed of as though from a patient with typhoid fever.

HEMOPHILIC BACTERIA

The hemophilic (blood-loving) bacteria include diverse species (see Table 21-2 for comparisons) with unifying features such as (1) their growth depends on, or is aided by, hemoglobin, serum, ascitic fluid, or certain growth-accessory substances; (2) they are small; (3) they are gram negative and nonmotile; and (4) they are strict parasites. Two factors in blood aiding their growth are X and V. Factor X (hemin) withstands the temperature of the autoclave, but since factor V (phosphopyridine nucleotide) does not, that is, it is heat labile, it must be supplied by potatoes or yeasts in culture media. Some hemophiles depend on both factors for continued multiplication; others rely on only one; and still others require neither. One genus name for some of these organisms is *Bordetella*. Another genus in this category, *Haemophilus*, indicates by the name its members to be well known as "blood-loving" bacteria.

Bordetella species (the agents of pertussis)

The disease. *Whooping cough (pertussis)*, a communicable disease of 1 or 2 months' duration, affects children with a catarrhal inflammation of the respiratory tract productive of a paroxysmal cough that ends in a whoop. (In adults or in very young children, the whoop may be absent.) The incubation period is 5 to 21 days. A widespread and dangerous disease, it is one of the major causes of death in young children, and 90% of the deaths occur in children under 5 years of age. The danger lies in the frequency with which bronchopneumonia, malnutrition, or such chronic diseases as tuberculosis follow in its wake. In the 25% of children who have whooping cough before they are 1 year old (in 75% of cases the child is older), bronchopneumonia is the usual cause for the deaths that occur.

The agents. The causative organism was first observed in the sputum of patients

TABLE 21-2. BIOLOGIC FEATURES OF SOME HEMOPHILIC BACTERIA

Organism	Hemolysis	Capsule	Production of indole	Catalase	Factor required for growth	
					X	V
Bordetella parapertussis	+	+	−	+	−	−
Bordetella pertussis	+	+	−	+	−	−
Haemophilus aegyptius	−	−	−	?	+	+
Haemophilus ducreyi	Slight	−	−	?	+	−
Haemophilus influenzae	−	+	+ or −	+	+	+
Haemophilus parainfluenzae	−	+	+ or −	+	−	+
Haemophilus suis	−	+	−	+	+	+
Haemophilus vaginalis	+ or −	−	−	−	−	−

with whooping cough by Jules Jean Baptiste Bordet and Octave Gengou in 1900. Often spoken of as the Bordet-Gengou bacillus, it is properly known as *Bordetella pertussis* and is the agent in most cases of the disease.

In possibly 5% of the patients a disease clinically indistinguishable from whooping cough has been related to an organism other than *Bordetella pertussis*, known as *Bordetella parapertussis*. *Bordetella parapertussis* is similar to *Bordetella pertussis* in its growth patterns. It can be separated biochemically and by the production of a brown pigment, and it can be identified serologically.

The Bordet-Gengou bacillus. A small gram-negative bacillus, *Bordetella pertussis* (*Haemophilus pertussis*) shows polar staining, often forms capsules, is nonmotile, and does not form spores. It grows strictly in the presence of oxygen. When freshly isolated from the body, all pertussis bacilli are alike, but when cultivated, they resolve into four phases. Phase I represents the freshly isolated pathogen, phase IV is the completely nonpathogenic form, and phases II and III are intermediate.

Freshly isolated from the body, *Bordetella pertussis* grows only on special media containing blood. One especially suitable is glycerin-potato-blood agar, on which it grows slowly, with 2 or 3 days required for visible growth. If repeatedly cultured on media in which the amount of blood is decreased stepwise, *Bordetella pertussis* is able to grow on blood-free media. With prolonged cultivation, *Bordetella pertussis* loses its dependence on both X and V factors, and when this occurs, it loses its pathogenicity.

Pathogenicity. In the respiratory system of man the bacilli aggregate in large masses among the cells lining the wall of the trachea and bronchi. They do not invade the bloodstream. The action of *Bordetella pertussis* is, at least in part, the effect of a toxic substance it produces. This substance appears to be both endotoxin and exotoxin. One fraction is thermolabile, being destroyed by a temperature of 56° C for 30 minutes. The other is thermostable. *Bordetella pertussis* is also pathogenic for lower animals, especially rabbits and guinea pigs.

Modes of infection. Whooping cough is one of the most highly communicable of all diseases. It is usually transmitted by droplet infection and direct contact but may be spread by recently contaminated objects. The disease is communicable during any stage and often during convalescence. The child is most likely to spread the infection just before the whoop appears and for about 3 weeks thereafter. The bacilli enter the body by the mouth and nose and are thrown off in the buccal and nasal secretions and in the sticky sputum that is coughed up. With the exception of those closely associated with the disease or patients convalescent from the disease, carriers do not exist. Convalescent carriers become noninfectious soon after the disease subsides.

Laboratory diagnosis. Two culture methods for the bacteriologic diagnosis of whooping cough are the cough plate method and the nasal swab method. In the first, an open Petri dish containing glycerin-potato-blood agar is held in front of the child's mouth during a paroxysm of coughing. The organisms are sprayed on the medium in droplets from the mouth and nose. In the second, a nasal swab is passed through the nose until it touches the posterior pharyngeal wall. After the swab is withdrawn, it is passed several times through a drop of penicillin solution on the surface of a plate of Bordet-Gengou medium. (The penicillin solution destroys many species of contaminating bacteria.) Then the material is spread over the surface of the medium with a platinum loop. Cultures made by either method are incubated for about 3 days for visible growth. Identification of the bacilli may be made also in nasopharyngeal smears stained by fluorescent antibody technics.

A peculiar white blood count is found in pertussis (leukocytes 15,000 to 30,000/mm³, of which 80% are lymphocytes), and in many cases suggests the correct diagnosis.

Immunity. Man possesses no natural immunity to whooping cough. However, an attack is usually followed by permanent immunity. Second attacks, except in the aged, are usually mild. Even though immune herself, a mother does not transmit immune bodies to her offspring. Therefore the newborn child is susceptible. Some degree of immunity may be conferred on the baby if the mother is immunized during the latter months of pregnancy.

Prevention.* Whooping cough itself cannot be prevented completely by our present public health methods, but its death rate can be reduced to a remarkable degree. Mothers should be taught how deadly a disease whooping cough is among young children and the value of immunization. The unvaccinated child should be protected from exposure. The infected child should be isolated for several weeks after the whoop has disappeared. However, the child need not be kept in bed but may be allowed to play in the sunshine and fresh air.

Haemophilus influenzae (the influenza bacillus)

In 1892, Richard F. J. Pfeiffer (1858-1945) described a bacillus, now known as Pfeiffer's bacillus or *Haemophilus influenzae (H. influenzae)*, which he found in the sputum of patients with influenza. Until the 1918-1919 pandemic this bacillus was regarded as the sole cause of influenza. Now we know that the cause of influenza is a virus. *H. influenzae* maintains an important position as a secondary invader in influenza and many other respiratory diseases as well. It may be the primary cause of a purulent meningitis in young children and subacute bacterial endocarditis. It may be one of the primary pathogens in mixed infections of the upper respiratory tract and bronchopneumonia.

General characteristics. The smallest known pathogenic bacillus, *H. influenzae* grows best in the presence of oxygen but may grow without it. Nonmotile and nonsporeforming, it does not occur outside the body and is very susceptible to destructive influences. There are two forms, the encapsulated and the nonencapsulated. The encapsulated form has been divided into types a, b, c, d, e, and f. Most infections are caused by encapsulated type b organisms.

Influenza bacilli grow only on special media such as chocolate agar or hemoglobin oleate agar. Both factors V and X are required for growth, which is often more luxuriant about an organism such as *Staphylococcus aureus*. Some staphylococci and certain other bacteria produce factor V. The vigorous growth of one organism in

---

*For immunization, see pp. 682, 685, and 695-697.

proximity to colonies of another is the *satellite phenomenon. H. influenzae* forms a toxic substance that resembles an exotoxin though not a true one and is moderately pathogenic for lower animals, especially the rabbit.

By immunofluorescent microscopy, influenza bacilli may be identified in cerebrospinal fluid taken from children with meningitis. Diagnostic serologic tests include complement fixation and agglutination.

H. influenzae is found in the throats of about 30% of normal persons. Adult human beings show bactericidal substances for *H. influenzae* in their blood; such substances are not found in young children. Therefore infection is much more serious in children. *H. influenzae* meningitis, the most frequent form of nonepidemic meningitis, occurs in children between 2 months and 3 years of age in 85% of cases. Of these, 90% are due to type b bacilli, against which an anti–*Haemophilus influenzae* serum has been made in rabbits. Modern methods of treatment have considerably reduced the death rate in influenzal meningitis, which used to be uniformly fatal. *H. influenzae* is also responsible for an obstructive inflammation of the larynx, trachea, and bronchi that begins suddenly with great severity. Breathing is blocked, and if the obstruction is not relieved, death may occur in 24 hours.

A vaccine to *H. influenzae* type b is being field tested. The antigen is type b capsular polysaccharide.

Haemophilus ducreyi (Ducrey's bacillus)

Haemophilus ducreyi, or Ducrey's bacillus, causes *chancroid,* a local, highly contagious, venereal ulcer with no relation to the chancre, the initial lesion of syphilis. It begins as a pustule that ruptures, exposing an ulcer with undermined edges and a gray base. There are usually multiple ulcers that spread rapidly. Unlike chancres, they do not have indurated edges; hence they are spoken of as soft chancres. The infection also spreads to the inguinal lymph nodes to form abscesses known as buboes.

The chancroid must be differentiated from the chancre of syphilis, and the buboes from those of lymphopathia venereum. It is not uncommon for a chancroid and a syphilitic chancre to occupy the same site. The chancroid appears first and heals, after which the chancre presents, or the chancre may appear before the chancroid heals. About one half of venereal ulcers are syphilitic chancre mixed with chancroid. Chancroid is usually transmitted by sexual intercourse but may be transmitted by surgical instruments or dressings.

Haemophilus ducreyi may be detected in smears made directly from the edges of the lesions or cultivated from the lesions if some of the exudate is inoculated into sterile rabbit blood. However, both smear and culture sometimes fail to demonstrate the organisms.

A skin test for the diagnosis of chancroid has been devised and used extensively in European countries, consisting of the intradermal injection of a saline suspension of *Haemophilus ducreyi.* A positive result is an area of redness and induration at the injection site. The reaction reaches maximal intensity at the end of 48 hours.

Haemophilus vaginalis

Haemophilus vaginalis is a small, nonmotile, nonencapsulated, facultatively anaerobic, gram-negative rod, which although placed with other species in the genus *Haemophilus* is not comparable to them in its traits. Exacting in its growth requirements, the microbe grows slowly and forms small colonies. It is a low-grade pathogen related primarily to human vaginitis, but sometimes to cervicitis, and, in males, to mild prostatitis and nongonococcal urethritis.

The diagnosis is easily made from a gram-stained smear of the associated vaginal discharge (the leukorrhea). One notes that lactobacilli of normal vaginal flora have been replaced by the small rods and that squamous cells are covered with myriads of the tiny microorganisms (the "clue cells").

Haemophilus aegyptius

The Koch-Weeks bacillus, as *Haemophilus aegyptius* is also known, causes pink-eye, a highly contagious conjunctivitis prone to epidemics. It is well named because of the intense inflammation of the conjunctival linings, which imparts a brilliant pink color to the white of the eyes. There is intense itching, but rubbing the eyes aggravates the situation. A yellow discharge forms, dries, and crusts on the eyelids. The disease is transferred by hands, towels, handkerchiefs, and other objects that contact face and eyes. Other names for the bacillus are *Haemophilus Koch-Weeks* and *Haemophilus conjunctivitidis*.

Haemophilus suis

Haemophilus suis together with a virus causes swine influenza (p. 527).

Haemophilus parainfluenzae

Very similar to *Haemophilus influenzae* is *Haemophilus parainfluenzae*, which may cause subacute bacterial endocarditis, but usually it is a nonpathogenic microbe in the upper respiratory tract.

YERSINIA SPECIES

The genus *Yersinia* includes *Yersinia (Pasteurella) pestis*, the cause of plague, and two causes of yersiniosis, *Yersinia enterocolitica* and *Yersinia pseudotuberculosis*.

Yersinia pestis (the plague bacillus)

Plague is an infectious disease of rodents, especially rats, and is transferred from them to man. First described in Babylon, it has been a devastating pestilence for more than 3000 years. Pandemics in the past have swept over great areas of the world, with a terrifying mortality (50% or more). Hundreds of years ago the Chinese related the disease to rats by using the term *rat pestilence*. That this disease still exists over the world at all times must be remembered in these days of extensive travel and commerce. In our own country, plague has been encountered in Texas, Louisiana, and many states in the far West. The disease is endemic (enzootic) among wild rodents of the southwestern United States, especially ground squirrels, and this focus of

TABLE 21-3. BIOCHEMICAL IDENTITY OF SOME SMALL GRAM-NEGATIVE BACILLI

Organism	Optimal tempera-ture for growth	Fermentation of sugars				Production of indole	Production of hydrogen sulfide	Oxidase reaction
		Glucose	Maltose	Sucrose	Lactose			
Yersinia pestis	28° C	Acid	Acid	—	—	—	—	—
*Francisella tularensis**	37° C	Acid	Acid	—	—	—	+	—
Pasteurella multocida	37° C	Acid	—	Acid	—	+	+	+

*Note again hazard of handling cultures of this organism. In routine identification, biochemical differentiation is not necessary.

infection is an ever present and increasing threat. Among wild rodents it is known as sylvatic plague (Latin *silva*, forest).

General characteristics. *Yersinia (Pasteurella) pestis* (discovered in 1894) is small, aerobic or facultatively anaerobic, gram negative and does not form spores (Table 21-3). It grows on all ordinary media. Growth on agar containing from 3% to 5% salt is a specific feature in its identification.

Plague bacilli may live in the carcasses of dead rats, in the soil, and in sputum for some time. They retain their vitality for months in the presence of moisture and absence of light. Phenol, 5%, or Amphyl solution, 2%, destroys them in 20 minutes.

Plague bacilli are highly pathogenic for many animals—monkeys, rats, mice, guinea pigs, and rabbits. They owe their action to toxic substances, one of which is an endotoxin similar to other bacterial endotoxins in both structure and physiologic effects.

Clinical types. Plague occurs in three patterns: bubonic, septicemic, and pneumonic. The first is the most frequent. In bubonic plague the bacilli penetrate the skin and are carried by the lymphatics to the lymph nodes draining the site of infection, commonly in the inguinal region. Here they multiply and form abscesses of the nodes (buboes). The bubo is extremely painful; a severe cellulitis surrounds it. Secondary buboes are formed in the nodes draining the primary buboes, and the bacilli finally move into the bloodstream, causing septicemia. The hemorrhages in bubonic plague that produce black splotches in the skin gave the name "black death" to plague during the Middle Ages. (From AD 1347 to 1349 the black death killed 25 to 40 million persons.)

Pneumonic plague manifests as bronchopneumonia, and the bacilli are abundant in sputum. Only a very small percentage of cases in the average epidemic take the pneumonic form (or that of a meningitis), but epidemics strictly of pneumonic plague may occur. Septicemic plague is a highly virulent form in which the patient dies before buboes can develop.

Modes of infection. Although many rodent species may be infected by *Yersinia pestis*, the most important source of primary infection is the rat, and, as a rule, epidemics of human plague closely follow epizootics of rat plague. Plague is transmitted from rat to rat by the bite of the rat flea. Man contracts the bubonic form of plague by the bite of a flea that has previously fed on an infected rat or, rarely, on an infected person. The reason for epidemics of human plague occurring *after* epizootics of rat plague is that an epizootic destroys the rat flea's food source of first choice—the rat. The flea must then seek his food source of second choice—man. Unless infection is mechanically transferred by blood adherent to its mouthparts, the flea spreads the infection 4 to 18 days after biting an infected animal or person. Plague bacilli are found in the intestine of the flea, where they live for a long time and multiply rapidly, and the bacilli are introduced into the human body by material regurgitated from the flea while it is biting. The flea may die of the infection or become a carrier. The incubation period in man is 2 to 6 days.

Pneumonic plague may result from involvement of the lungs during the course of bubonic plague (primary pneumonic plague) or from inhalation of particles of sputum thrown off by a person with pneumonic plague (secondary pneumonic plague). Epidemic pneumonic plague is the secondary type.

Plague may occasionally be acquired by handling infected rodents. The disease is transmitted from locality to locality by infected rats. Dozens of wild and domestic rodents other than rats may contract plague naturally—ground squirrels, voles, prairie dogs, chipmunks, marmots, and guinea pigs. The black house rat is more dangerous to man than other types because this one lives in dwellings, and therefore its fleas are more likely to bite a human being.* The common gray sewer rat and the Egyptian rat can be infected. Bedbugs and the human flea may transmit plague from person to person. Human carriers (without symptoms) have been found to harbor *Yersinia pestis* in their throats.

Laboratory diagnosis. The important methods of laboratory diagnosis are the agglutination test on the patient's serum and the demonstration of the bacilli in lesions by smears, cultures, and animal inoculation. There is a fluorescent antibody test to identify the organisms in sputum.

Immunity. An attack of plague usually induces permanent immunity.

Prevention.† The prevention of plague depends on the eradication of rats and rat fleas. Patients with plague should be isolated in well-screened, vermin-free rooms. (The technic of quarantine was first used with plague.) Although seldom do patients with bubonic or septicemic plague transmit the infection to others, no chance should be taken. They should be nursed with the same precautions as patients with typhoid fever. The sputum from patients with pneumonic plague should receive special care. The sickroom should be dusted with a suitable insecticide to eliminate the vector flea. Attendants should be protected with plague vaccine. A live, attenuated plague vaccine is being tested in monkeys.

---

*Today in Southeast Asia conditions are favorable for outbreaks of plague. A large infected rat population with the right number of rat fleas exists alongside a dense nonimmunized human population in a humid, warm (not hot) climate. During the wet season, the monsoons cancel out the effect of insecticides and force rats into human shelters. The natives kill the rats with sticks and the fleas move onto their new, human hosts.

†For immunization procedures, see pp. 685, 706, and 707.

Measures designed to block transport of rats from infected to noninfected localities by trains, aircraft, and ships are imperative, and maritime rat-control measures are in force in most parts of the world today. Rat-infested ships can be fumigated with a potent rodenticide, or when rats can be trapped, fumigation may not be necessary. Fortunately, relatively few rats are found on ships today.

Other yersiniae

Yersinia pseudotuberculosis and *Yersinia enterocolitica* cause yersiniosis in man and animals, in which fever is prominent. Acute mesenteric lymphadenitis and enteritis with chronic arthritis are manifestations of infection with these two.

FRANCISELLA TULARENSIS (THE AGENT OF TULAREMIA)

Francisella (Pasteurella) tularensis causes tularemia (rabbit fever), an acute infectious disease of wild animals, especially rodents, transferred from animal to animal by the bite of an insect. It may be transmitted from animal to man by insect bite, but human infections come from contamination of the hands or conjunctivae by the tissues of body fluids of an infected animal or insect.

General characteristics. The organism *Francisella tularensis* derives its name from Tulare County, California, where the disease was first observed. It is a small, gram-negative, nonsporeforming, nonmotile organism varying considerably in size. It grows on cystine agar (not on ordinary agar) and is easily demonstrated in the lesions by animal inoculation. It behaves as an intracellular parasite, persisting for long periods in phagocytic and other body cells of the host. Enormous numbers fill these cells.

Clinical types. Tularemia is a febrile disease accompanied by severe constitutional manifestations, pain, and prostration. The incubation period is usually 3 to 10 days. The clinical types are (1) ulceroglandular (there is an ulcer at the infection site with regional lymph node involvement), (2) oculoglandular (pattern similar to ulceroglandular type with conjunctiva as primary site), (3) glandular (lymph nodes involved but ulceration absent), and (4) typhoid (neither ulceration nor lymph node involvement present). Death occurs in about 5% of the patients. Those who recover are incapacitated for weeks or months. (Clinically tularemia may be confused with cat-scratch disease.)

Modes of infection. Tularemia is most prevalent in cottontail rabbits, jackrabbits, snowshoe rabbits, and ground squirrels. Certain birds have tularemia, and the disease is not unknown in cats and sheep. The water of streams inhabited by infected animals (beavers and muskrats), once contaminated, may long remain a source of infection. Epidemics caused by pollution of streams have been reported. In Russia human infections are often contracted from a species of fur-bearing water rat. More than 45 species of wild animals are known to be infected. Tame rabbits are susceptible but ordinarily do not contract the disease, since they do not harbor the small parasites that transmit the infection among animals.

The insects that transmit tularemia in animals are wood ticks, rabbit ticks, lice, horseflies, and squirrel fleas. Wood ticks, dog ticks, horseflies, and squirrel fleas may

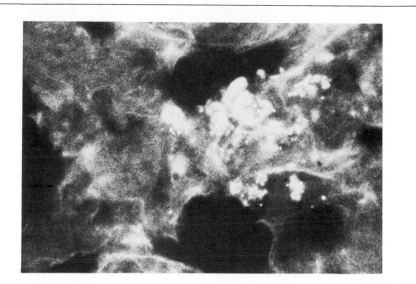

FIG. 21-2. *Francisella tularensis* identified by fluorescent antibody staining. Glow of bacterial immunofluorescence apparent even in black and white photomicrograph. (From White, J. D., and McGavran, M. H.: J.A.M.A. **194:**294, 1965.)

sometimes transmit the infection to man. Rabbit ticks, rabbit lice, and mouse lice, important agents in transferring the disease from rodent to rodent, do not bite man.

In our country, man usually contracts tularemia by handling infected rabbits; of these, about 70% are cottontails. Cold-storage rabbits remain infectious for 2 to 3 weeks, and if kept frozen, they may be a source of infection for 3½ years. The disease may be contracted if insufficiently cooked rabbit meat is eaten.

Inhalation of organisms can result in disease. The ocular type of tularemia usually results from a person rubbing his eye with contaminated fingers. Tularemia is not transmitted directly from man to man.

Francisella tularensis is the most easily communicable of all organisms, and many laboratory workers who have investigated tularemia have contracted the disease. Some have died.

Laboratory diagnosis. The laboratory diagnosis of tularemia depends on the agglutination test, immunofluorescence (Fig. 21-2), Foshay's antiserum test, the tularemia skin test, and the inoculation of guinea pigs or rabbits with material from lesions. Cultures are too dangerous to be handled routinely.

Visitors, secretaries, various employees, and the like should be carefully segregated and barred from the specific laboratory area where work is carried on with *Francisella tularensis*.

Immunity. An attack of tularemia is followed by permanent immunity.

Prevention. The nature of the infection and its modes of spread render the preventive measures obvious. Tularemia vaccine is a lyophilized, viable, attenuated strain of *Francisella tularensis*. Vaccination is indicated for those at risk.

PASTEURELLA SPECIES

The genus *Pasteurella* (named for Louis Pasteur) includes a group of pasteurellae causing disease in animals but seldom attacking man.

Pasteurellae of hemorrhagic septicemia

Certain members of genus *Pasteurella* cause hemorrhagic septicemia (pasteurellosis), a disease of cattle, horses, swine, bison, and poultry, characterized by septicemia, petechial hemorrhages of mucous membranes and internal organs, edema, and changes in the lungs. Mortality is high. The most important ones are *Pasteurella multocida* (cause of fowl cholera and shipping fever of cattle) and *Pasteurella haemolytica* (cause of pneumonia in sheep and cattle). *Pasteurella multocida* can infect man.

Microbiology of dog and cat bites. Over half a million persons, principally children younger than 12 years old, are bitten by animals each year in the United States. Most of the bites come from cats and dogs; dogs* head the list. Generally such wounds heal without complications, but they can be infected by microorganisms passed by the animal in biting.

Since it is normally found in the nose and throat of cats and dogs, *Pasteurella multocida* is an offender. Other organisms encountered, usually in mixed culture, include *Staphylococcus aureus* and alpha and beta hemolytic streptococci. The

---

*"No animal, including the human, is limitless in its love and good humor. A dog has only a few ways of expressing hostility, one of which is to bite and to claw. If sufficiently provoked, the dog cannot be blamed for obeying its instincts rather than its master or mistress. Let sleeping dogs lie and let strange dogs be!" Goldwyn, R. M.: Man's best friend (editorial), Arch. Surg. **111**:221, 1976.

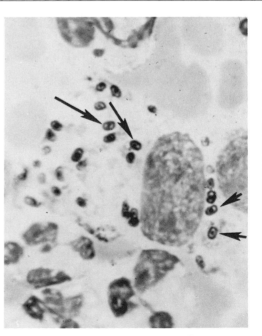

FIG. 21-3. Donovan bodies in cytoplasm of phagocytic cell. Note "safety pin" appearance of dark bipolar bodies. Clear surrounding areas correspond to capsules about microorganisms. (From Davis, C. M.: J.A.M.A. **211**:632, 1970.)

lesion produced may be a cellulitis, an abscess, or, about the extremities, a teno-synovitis. Osteomyelitis can complicate a deep-seated bite, and there is always the danger of bacteremic spread. Resolution of the tissue injury from the bite can be slow. Scarring results from damage to tissue, and the scar about the head and neck may disfigure.

CALYMMATOBACTERIUM GRANULOMATIS

Calymmatobacterium granulomatis, the only species of its genus, causes granu-loma inguinale, a venereal disease producing ulceration in the skin and subcutaneous tissues of the groin and genitalia. In the lesions, encapsulated, oval, rodlike bodies with a unique "safety pin" appearance are found within the cytoplasm of large mono-nuclear phagocytes (Fig. 21-3). These, the agents, are also known as Donovan bodies.

The laboratory diagnosis of the disease rests on microscopic demonstration of the Donovan bodies in Wright-stained smears made of material from the lesions.

QUESTIONS FOR REVIEW

1. Name the species of *Brucella*. How are they alike? How different?
2. Comment on the laboratory diagnosis of brucellosis, mentioning difficulties.
3. How is whooping cough transmitted? What are the dangers of this disease?
4. Give the laboratory diagnosis of pertussis.
5. List features shared by hemophilic bacteria.
6. What is the importance of factors V and X?
7. Briefly discuss pathogenicity of the influenza bacillus.
8. List five animals (other than man) contracting tularemia. Name five vectors associated.
9. Give clinical types of plague and tularemia.
10. What is the reservoir of plague? How is it conveyed to man?
11. What are Donovan bodies? Where are they found?
12. Briefly discuss microbiology of dog and cat bites.
13. Describe the condition pinkeye. What organism causes it?
14. What is chancroid? Its cause? How does it differ from chancre?
15. Briefly explain yersiniosis, clue cells, the satellite phenomenon, an undulant fever curve, milk ring test, bubo, "black death," hemorrhagic septicemia.

REFERENCES. See at end of Chapter 27.

22 Anaerobes

GENERAL DISCUSSION

Definition. Generally speaking, anaerobes are bacteria that must have a low or zero oxygen tension for their continued growth and are sensitive to the action of oxidizing agents such as peroxide. To them, oxygen is either inhibitory or toxic. Indeed, some are so fastidious that for their manipulation in the laboratory, oxygen must be meticulously eliminated every step of the way. However, within such a large group of complex microorganisms one would expect varying degrees of tolerance, and this is so.

The preceding definition suggests that anaerobes in their native habitats influence their surroundings. Studies show that most do synthesize regulatory substances like certain relatively strong organic acids in large amounts or particular enzymes, which act to keep the oxidation-reduction potential of the milieu at the required low level. For example, in many anaerobes one bacterial enzyme, superoxide dismutase, may be of importance.

The upsurge of interest in anaerobic bacteriology is another example of progress following hard on the heels of advancements in technology. Now that we can view anaerobes realistically, and with the greatly increased concern about them, the confusion that fogs their biologic attributes is clearing.

Pertinence. Important to a consideration of anaerobes in disease is the knowledge of their position in health. It is amazing to realize that this poorly understood category must make a sizable contribution to man's well-being, since in great abundance they are intimately bound to him. The vast majority are indigenous to man (in space age lingo—"Life on man is—anaerobic"). They are the predominant population group of his resident microflora, and their myriads occupy his intestinal tract, oral and nasal cavities, genital tract (especially in the female), respiratory tract, and skin. In most areas anaerobes outnumber aerobes 10 to 1; in the large intestine, 1000 to 1 (*Bacteroides* species, the anaerobe; *Escherichia coli*, the aerobe).

The more sophisticated technics of isolating, identifying, and studying anaerobes soon showed us that they had capabilities for severe disease and that under the right circumstances they expressed them. The restrained one of great potential quickly became the unleashed opportunist of great vigor.

Many observers state that anaerobes cause practically any type of disease known to man, and currently, of specimens from cases of infectious disease submitted to clinical laboratories properly equipped to deal with them, anaerobes are found in 40% or more.

Pathogenicity. The key to reversal of the biologic role of the normally saprophytic anaerobe is the level of tissue oxidation. As this level is crucial for normal growth, so it can, if lowered, further enhance growth. At the same time, the level of tissue oxidation is also important to the function of body defense mechanisms such as that of certain bactericidal systems. What expands microbial growth also diminishes body resistance.

The most common reason for lower tissue oxygen potential is the pathologic state of tissue anoxia regularly accompanying many pathologic processes that impair the circulation to a given area. Other factors operating are (1) the presence of dead tissue from whatever cause—trauma, injection of drugs, and the like; (2) the growth of aerobic microbes; and (3) the presence of ionized calcium salts. The last-named factor especially favors infections of wounds contaminated with well-tilled topsoil (containing the anaerobes).

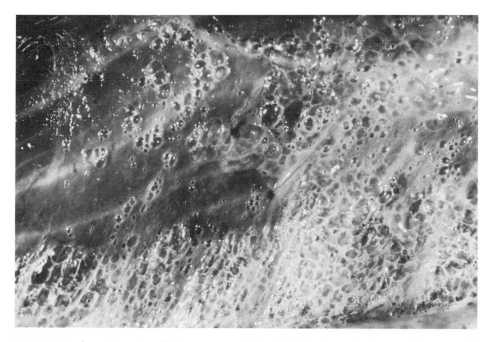

FIG. 22-1. Gas formation in soft tissues over belly of striated muscle, as seen in anaerobic infections. (From Smith, A. L.: Microbiology and pathology, ed. 11, St. Louis, 1976, The C. V. Mosby Co.)

Most anaerobes are destructive. The type lesion is properly tissue necrosis. Two other, sometimes dramatic, features of their infections are the foul odor of discharges and purulent drainages and manifestation of gas in soft tissues (crepitation) (Fig. 22-1).

The pattern for disease is obviously a diffuse one with a long list of possibilities. To emphasize the more important is to list septicemia with metastatic suppuration; intra-abdominal abscesses with peritonitis; infections of female genital tract like endometritis, tubo-ovarian abscess, pelvic cellulitis, and postsurgical wound infections; pleuropulmonary disease with empyema and abscess; and skin and soft tissue infections.

Laboratory diagnosis. Because of the ubiquity and profusion of anaerobic species, specimens for microbiologic study are taken directly only from a site normally sterile. Otherwise, special collection technics are used (p. 135). Important sources for test material include wounds, abscesses, blood, abdominal discharges, and such. Getting the specimen to the laboratory may necessitate transport in a special carbon dioxide–filled container; the specimen requires immediate attention on arrival at its destination.

For the isolation and identification of anaerobic species, three cultural methods are standard. (1) In the *roll-streak tube system*, PRAS (*p*rereduced, *a*naerobically *s*terilized) media are prepared in anaerobic culture tubes, stored, inoculated (rolled

FIG. 22-2. GasPak 100 anaerobic system (self-contained, evacuation-replacement system). Anaerobic conditions are induced when water is added to GasPak envelope to form hydrogen, which reacts with atmospheric oxygen on palladium catalyst to form water. Carbon dioxide is also generated to produce growth of fastidious anaerobes. (Courtesy BioQuest, Division of Becton, Dickinson & Co., Cockeysville, Md.)

to distribute inoculum), and manipulated under a stream of oxygen-free gas. (2) The *anaerobic glove box* or chamber houses cultures and cultural materials in an oxygen-free environment and yet allows the technician access to the contents of the chamber through openings (ports) sealed off with gloves. (3) In the *anaerobic jar system* (Fig. 22-2), plates are held during incubation in an atmosphere made appropriate by a specially devised elimination-replacement (of gases) scheme.

Colony growth is usually slow, several days required for visible signs. For species identification, organisms recovered are gram stained for morphology and subjected to selected biochemical reactions for metabolic patterns (all under oxygen-free gas, pH of reactions checked by pH meter). Gas-liquid chromatography is valuable in measuring acid and alcoholic products. Antibiotic susceptibility testing is done by a pour plate method.

The approach. A broad division of the total array of anaerobic microbes into two major categories on the basis of endospore formation is a convenient approach to the discussion of particular ones. The first such category would contain the spore-formers of a single genus, *Clostridium*, a fairly homogeneous group with several unifying features, a well-known and long-studied group, and one wherein the saprophytes exist in the external environment (although some are part of the intestinal flora of man and animals). A fund of knowledge has accumulated on these, their patterns for clinical disease are classic for the most part, and in some instances a great deal has been accomplished in related areas of preventive medicine.

The second category of nonsporulating microbes is not a straightforward one, gathering essentially all the rest. It is a more heterogeneous complex with few, if any, unifying features among the species other than their anaerobic specifications. There is no backlog of knowledge on these; in fact, up to recent past, information has been sketchy. Today technology makes the difference. (Technology, although designed for basic characteristics of both categories, has had a much greater impact for the nonsporeformers than sporeformers.) The microorganisms of the second category are from within, predominant in the internal environment of man. They operate from a more strategic position in the body than sporeformers. Their patterns for clinical disease are diffuse and diverse, ranging from minor superficial infections to septicemic dissemination.

Infections associated with anaerobes of whatever category are generally serious and carry a high mortality. Some of the more important genera and species now pass muster.

CLOSTRIDIUM SPECIES

Certain familiar sporulating anaerobes of genus *Clostridium* (Family Bacillaceae) are large bacilli peculiarly distorted by their heat-resistant spores and producing potent exotoxins and enzymes.

Clostridium tetani (the bacillus of tetanus)

The disease. Tetanus or lockjaw is a disease, worldwide in distribution, with manifestations in the central nervous system produced by the potent exotoxin of

TABLE 22-1. BACTERIOLOGIC PATTERNS OF CLOSTRIDIA

Organism	Growth in litmus milk	Growth in cooked meat	Gelatin liquefaction	Synthesis of lecithinase
Clostridium tetani	Soft cloth	Gas, slow blackening	+ (blackening)	No
Clostridium perfringens	Stormy fermentation	Gas	+ (blackening)	Yes
Clostridium novyi	Acid, no clot	Gas	+	Yes
Clostridium septicum	Slow clot, gas	Gas	+ (gas)	Yes
Clostridium botulinum	Acid	Gas, blackening, digestion	+	Yes

Clostridium tetani (Cl. tetani). There are classic muscular spasms in tetanus involving the face, neck, and other parts of the body that are provoked by the slightest sort of stimulation (noise, movement, touch). These are very painful. Severe ones may compromise respiration. The name *lockjaw* comes from the muscular contractions that rigidly close or lock the jaws together. The corners of the mouth are turned up and the eyebrows peaked, producing a unique facial expression referred to as *risus sardonicus* (sardonic grin).* Tetanus was known to Hippocrates and the ancients by the triad of wound, lockjaw, and death.

Tetanus bacilli do not invade the body but remain at the site of infection, where they elaborate their powerful poison. The disease comes from infection of a wound and is not transmitted from person to person.

General characteristics. The tetanus bacillus is an anaerobic sporeformer. Biologically a saprophyte, it probably never infects a wound that does not contain some dead tissue. *Cl. tetani* stains with ordinary stains and is gram positive. The vegetative forms are slightly motile; the sporulating forms are nonmotile. The spore is situated in one end, giving the bacillus an appearance of a roundheaded pin or drumstick. Growth is fairly luxuriant on all ordinary media if anaerobic conditions are maintained (Table 22-1). The optimum temperature for growth is 37° C. In both cultures and infected wounds the presence of certain aerobic bacteria accelerates the growth of tetanus bacilli. Young cultures contain numerous vegetative bacilli. Old cultures are composed chiefly of sporulating organisms. The bacilli are seldom demonstrated in cultures from infected wounds because so few are present.

The vegetative bacilli are no more resistant to destructive influences than other

---

*"They [the spasms] are characterized by a violent rigidity, usually sudden in onset but sometimes working up to a crescendo, with every single voluntary muscle in the body thrown into intense, painful tonic contraction. The eyes start, the jaw clenches, the tongue is bitten, the neck is retracted, the back arched, and opisthotonus is extreme. Often there is a muffled inspiratory cry, as the diaphragm contracts and draws air through the apposed vocal cords. Finally laryngeal spasm becomes complete, the chest fixed, and respiration ceases from muscle spasm. At the same time there is a gross outpouring of secretion, with profuse perspiration and foaming at the mouth." (From Ablett, J. J. L.: Tetanus and the anaesthetist; review of symptomatology and recent advances in treatment, Br. J. Anaesth. **28:**288, 1956.)

Nitrate reduction	Fermentation			Hemolytic	Spores	Motility
	Glucose	Lactose	Sucrose			
−	−	−	−	+	Round, terminal	+
+	+	+	+	+	Rare	−
−	+	−	−	+	Ovoid, eccentric	+
+	+	+	−	+	Ovoid, eccentric	+
−	+	−	−	±	Ovoid, eccentric	+

vegetative organisms. The spores, however, are very resistant. They withstand boiling for several minutes, pass through the intestinal canal unaffected, and when protected from the sunlight, remain infective for years.

Distribution. Tetanus bacilli are common residents of the superficial layers of soil. Since they are normal inhabitants of the intestines of horses, cattle, and other herbivores, they are always found where manure is freely used as fertilizer, and barn-yard soils are heavily contaminated. They are found in the intestinal canal of about 25% of human beings. Tetanus spores may be spread over a wide area by flies and high winds.

Extracellular toxin. If it were not for their toxin, infection with tetanus bacilli would be without effect. Tetanus toxin is one of the most powerful poisons known, second only to that of botulism in potency. This explains why a comparatively few bacilli at the infection site can induce such profound changes. Tetanospasmin, the toxin of tetanus, is a simple protein of a single antigenic type with a molecular weight of 68,000. It is a neurotoxin, that is, one with a specific affinity for the tissues of the central nervous system. From the infection site the toxin travels to the central nervous system along the axis cylinders of the motor nerves, where it is specifically and avidly bound to the gray matter. Its spasm-producing action is thought to result from the blocking of certain inhibitory synapses in the central nervous system.

Besides tetanospasmin, tetanus bacilli elaborate small amounts of a toxic sub-stance, tetanolysin, that destroys red blood cells and injures the heart. Taken orally, tetanus toxin is harmless.

Pathogenicity. Tetanus in man is death dealing. Horses, cattle, sheep, and hogs may become infected.

Tetanus occurs in two clinical forms: one form associated with a short incubation period (from 3 days to 3 weeks), an abrupt onset, severe manifestations, and a high mortality; the other is associated with a longer incubation period (from 4 to 5 weeks), less severe manifestations, and lower mortality. Rapidly fatal tetanus is more likely to follow wounds about the head and face, that is, near the brain.

Pathology. The toxin of *Cl. tetani* acts directly on the central nervous system to originate the characteristic muscle spasms and convulsive seizures. The impulses are transferred to the muscles by the motor nerves.

In this disease pathologic changes are functional, not structural. There is no typical lesion. Even after death, no organic lesions are seen, and the cerebrospinal fluid is normal.

Sources and modes of infection. Tetanus is practically always caused by spores introduced into a wound. Whether a given wound is complicated by tetanus depends on the type of wound, chance of contamination with tetanus spores, presence of dead tissue, and secondary infection. Deep puncture wounds are dangerous because they provide anaerobic conditions for growth of the bacilli. In lacerations, gunshot wounds, compound fractures, wounds containing foreign bodies, and infected wounds, the presence of dead tissue and certain other bacteria favors the growth of tetanus bacilli. If such wounds are contaminated by soil, especially heavily manured soil, the chances of tetanus infection are greatly increased. It is not surprising that war wounds are so easily complicated. Rusty nail wounds are a hazard, not because the nail is rusty but because rusty nails are usually dirty nails and tetanus spores are likely to be in the dirt.

In narcotic addicts injection-related tetanus is a public health problem. The rapidly progressive and highly fatal disease in the drug addict is another, even more terrifying expression of tetanus. It is seen in the large metropolitan centers of the United States, especially in black women. The narcotic, usually heroin, has been given by the subcutaneous route of injection. Quinine, the adulterant, favors the growth of the bacilli by promoting anaerobic conditions in the tissues, enhancing the disease.

Puerperal tetanus exists in tropical countries, where infection from bacilli inhabiting the intestinal canal sometimes complicates intestinal operations. Tetanus of the newborn, *tetanus neonatorum*, caused by infection through the navel is still seen in the Americas among black and Spanish-speaking groups with primitive medical care. At the extremes of life, mortality rates are high. Today in the United States most cases are seen in persons with an average age over 50. Elderly persons are exposed to tetanus-prone injuries in their gardens.

Although tetanus is more likely to develop under conditions such as those enumerated, it may also follow wounds inflicted by apparently clean objects.

Prevention of tetanus.* Any wound suggesting the least danger from tetanus should be treated surgically. Puncture wounds should be widely opened and thoroughly cleaned to remove tetanus bacilli and other organisms and to allow access of oxygen, antagonistic to the growth of tetanus bacilli. Mangled wounds should be thoroughly cleaned and all dead tissue removed. Proper wound care cannot be overstressed, since tetanus organisms do not multiply in a surgically clean, aerobic wound with a good blood supply. Antibiotics help to control associated pyogenic infections that damage tissues. *Cl. tetani* is sensitive to penicillin, but no amount of antibiotic has any effect at all on toxin released into the body from the wound site.

Tetanus is a nonimmunizing disease. No permanent or temporary active immunity develops in the patient who recovers. The amount of the very potent toxin

---

*See also pp. 678, 680, 681, 682, 683, 686, 695-697, and 706.

needed to produce disease is too small to trigger the immune mechanisms that would prevent a second attack.

Active immunization for every member of the community cannot be too strongly recommended. In special need of protection are those workers in industry or agriculture whose occupation predisposes them to wounds easily contaminated with tetanus bacilli, all allergy-prone individuals, and even recipients of a driver's license. All athletes must be fully immunized. (Football players are suffering more abrasions and minor injuries from sliding and falling on artificial turf than on sod fields, although artificial turf is a less likely source of tetanus spores.)

Active immunization is practiced on a large scale in the armed forces of both the United States and England. That this has been effective is proved by the fact that only 12 cases of tetanus appeared among 2,785,819 hospital admissions of military personnel for war wounds and injuries during World War II. Of the 12 patients, six had not been adequately immunized and two failed to get the booster dose of toxoid at time of injury.

Clostridium perfringens, novyi, and septicum (the clostridia of gas gangrene)

The organisms. Gas gangrene (clostridial myonecrosis, clostridial myositis) is a highly fatal disease caused by the contamination of wounds with one alone or any combination of certain anaerobic, toxin-producing, sporeforming, gram-positive bacteria (Table 22-1). (Less than half the time there is only one organism.) All exist as saprophytes in the soil and normal inhabitants of the intestinal canal of man and animals. Infection happens when soil contaminated by feces gets into a wound. Most important are *Clostridium perfringens* (Welch's bacillus*), *Clostridium novyi*, and *Clostridium septicum*. *Clostridium sporogenes*, considered a nonpathogenic organism, is often present also. The single most significant of the clostridia is *Clostridium perfringens* (*Cl. perfringens*), related to 75% of all cases of gas gangrene and responsible for practically all instances in civilian life. In one third of cases it is the sole pathogen.

Aerobic pus-producing microbes and certain proteolytic organisms without effect on clean wounds often produce considerable destruction of tissue in wounds infected with gas bacilli. Aerobes in the wound facilitate the germination of clostridial spores and their vegetative growth by reducing the oxidation-reduction potential. In most cases of gas gangrene two or more clostridial anaerobes are associated with one predisposing aerobe.

The disease. The organisms responsible for gas gangrene grow in the tissues of the wound, especially in muscle, releasing exotoxins and fermenting muscle sugars with such vigor that the pressure of accumulated gas tears the tissues apart. The air-filled (emphysematous) blebs of the wound give the name gas gangrene (Figs. 22-3 and 22-4). The exotoxins cause swelling and death of tissues locally, breakdown of

---

*William Henry Welch (1850-1934), first dean of the medical school of Johns Hopkins University, was well known for his original research in microbiology and pathology, especially for his studies (in 1892) of *Clostridium perfringens* and its relation to gas gangrene.

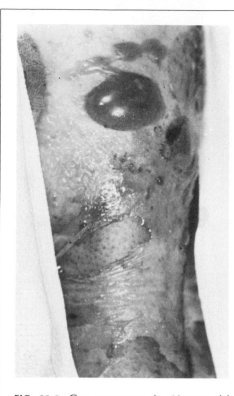

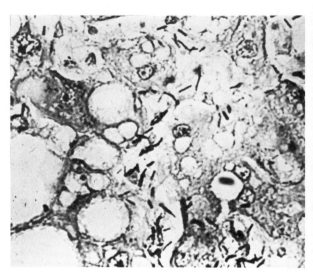

FIG. 22-3. Gas gangrene in 16-year-old boy, complicating compound fractures of leg sustained in motorcycle accident. Organism, *Clostridium perfringens*. Note blisters and discoloration. (From Altemeier, W. A., and Fullen, W. D.: J.A.M.A. **217**:806, 1972.)

FIG. 22-4. Gas gangrene in Gram-stained tissue section of liver. Large air-filled spaces disrupt structure. Large gram-positive rods are present. Patient had leukemia. (Courtesy Dr. R. C. Reynolds, Dallas, Tex.)

red blood cells in the bloodstream, and damage to various organs over the body. The bacteria enter the blood just before death. Clinically there is a profound toxemia. (The incubation period is 1 to 5 days.)

Cl. perfringens elaborates several hemolysins and an extracellular proteolytic enzyme, collagenase, that facilitates spread of gas gangrene organisms through tissue spaces, since it is active against the fibrous protein supports of the body.

Clostridium perfringens, novyi, and *septicum* elaborate lecithinase; species *histolyticum* and *sporogenes* do not. The potent lecithinase (the alpha-toxin) of *Cl. perfringens* is both hemolytic and necrotizing in its effects on tissues.

Lacerated wounds, compound fractures, wounds with extensive dead tissue, and war wounds are likely sites of gas gangrene. Injuries acquired in the vicinity of railroad tracks tend to be so complicated. Clostridial infection rarely complicates gangrenous appendicitis, strangulated hernia, and intestinal obstruction.

Skin samples taken from the thighs, groins, and buttocks of hospitalized patients sometimes yield a heavy growth of *Cl. perfringens*. To eliminate this kind of contamination, compresses of povidone-iodine or 70% alcohol may be applied. The presence of the organisms of gas gangrene in a wound does not invariably mean gas gangrene.

Clostridial infection and malignancy. *Clostridium septicum* is important in wartime gangrene but rarely complicates infection in otherwise healthy civilians. If the person's defenses are weakened, the situation may be reversed. This organism has acquired some notoriety by complicating cases of leukemia and intestinal cancer with clostridial septicemia.

Laboratory diagnosis. Smears and cultures are used to advantage. When *Cl. perfringens* is responsible for a gaseous putrid discharge, a gram-stained smear of material shows the presence of the large, gram-positive rods and the relative absence of neutrophils (Fig. 22-4). A differential medium such as lecithin-lactose agar containing neomycin and sodium azide to inhibit contaminating gram negatives allows for species identification based on whether lactose is fermented or lecithinase is produced. If, at the same time, a plain blood agar plate is inoculated with test material, a significant nonclostridial organism will be picked up.

Prevention. Prevention of gas gangrene depends on proper surgical care of wounds. With established disease there must be free incision to open the wound as widely as possible, all devitalized tissue must be excised, foreign bodies must be removed, and adequate drainage of the wound instituted. Gas gangrene antitoxin has been prepared against the main organisms causing gas gangrene, but its efficacy today is challenged. Toxoids for *Clostridium perfringens* and *Clostridium novyi* are experimental.

Clostridium botulinum (the bacillus of botulism)

Nature of botulism. Botulism is a specific intoxication caused by ingestion of foods in which *Clostridium botulinum* (*Cl. botulinum*) has grown and excreted its toxin. The toxin, not the bacilli, is responsible for the disease. Like tetanus, it is a poisoning, not an infection. The foods most often linked to botulism are sausage (the disease derives its name from the Latin word *botulus*, sausage), pork, and canned vegetables such as beans, peas, and asparagus. Cases have been traced to ripe olives and tuna. Most foods incriminated have shared one feature: they were processed (improperly!) in the home by canning or pickling weeks or months before.

In the last decade there have been 78 outbreaks of botulism in the United States, plus nearly 200 individual cases. The most serious ones of our times, those in 1963, resulted from the consumption of commercially processed foods. From eating such foods as liver paste, tuna fish, salmon eggs, and smoked whitefish products, 46 persons were poisoned, and 14 died of a disease ordinarily quite rare.

The disease develops as follows. Since the organism is widely distributed in nature, food is easily contaminated. If it is not canned or otherwise preserved properly, the spores are not destroyed. In the interval between preservation and use, the spores revert to the growing stage, and the bacilli multiply and excrete their toxin into the food. (Toxin remains potent in canned foods for 6 months or more.) If the contaminated food is not heated sufficiently for consumption, toxin has not been inactivated. When such food is ingested, the toxin is absorbed through the intestinal wall to exert its effects. Toxic signs usually appear within 24 to 48 hours and consist of generalized weakness, disturbance of vision (often double vision), thickness of speech, nausea and vomiting, and difficulty in swallowing. There is no fever. Death from asphyxia usually occurs between the third and seventh day. The mortality ranges from 50% to 100%.

Natural food poisoning of this type exists among certain wild animals (if toxin is swallowed with their food). Examples are limberneck of chickens, fodder disease of horses, and duck sickness.

The organism. The microbe *Cl. botulinum* is also an anaerobic, gram-positive, sporeforming bacillus (Table 22-1). A common inhabitant of the soil, the usual source from which foods are contaminated, *Cl. botulinum* is primarily a saprophyte. (Infections of laboratory animals have been caused experimentally.) Generally, the organism is unable to grow inside the body of a warm-blooded animal. *Cl. botulinum* and its toxin are destroyed in 10 minutes at the temperature of boiling water, but the "hard shell" spores must be held at a temperature of 120° C (249° F) for 15 minutes to be killed. Spores can withstand more than 2000 times the radiation lethal to a human being.

The toxin of *Cl. botulinum* is the most deadly of poisons, the mere tasting of food having been known to cause death. One ounce could exterminate all the people in the United States, and a mere half pound of botulinal toxin could wipe out the population of the world. (Botulinal toxin has been classified as an agent for chemical warfare.) It is 10,000 to 100,000 times more potent than diphtheria toxin or animal venoms. It differs from diphtheria and tetanus toxins in that it causes disease when swallowed and is more resistant to heat. A fast-acting neurotoxin, it affects the central nervous system, paralyzing the muscles of vision, swallowing, and respiration. It dilates blood vessels, and hemorrhages occur in different parts of the body. The toxin's effect is a specific one at the neuromuscular junction. Mental faculties are not impaired, and there are no sensory disturbances.

There are six toxigenic types of *Cl. botulinum:* A, B, C, D, E, and F. All these toxins resist the intestinal juices, and the action of type E toxin can be increased 50 times by trypsin. In man, botulism almost always results from ingestion of toxin from types A, B, E, and F. C and D toxins cause disease in animals. Most A and B toxins are found in home-preserved vegetables and fruit (food of plant origin). Recent outbreaks of botulism have pointed to the presence of *Cl. botulinum*, type E, in water and marine wildlife. Therefore type E toxin is found in fish and marine products. Type E spores can germinate at refrigerator temperatures and form toxin. Type F toxin has been reported in home-processed venison jerky.

There is good evidence that toxigenicity of clostridial species and even the type of toxin depend on the presence of specific bacteriophages or bacterial viruses (lysogeny). For instance, if *Cl. botulinum*, type C, is "cured" of its phage, it loses its toxin. A startling fact is that experimentally the right phage can then convert it to a different clostridial species, also toxigenic and producing that specific toxin.

Laboratory diagnosis. The laboratory diagnosis of botulism is made by (1) finding toxin in the patient's serum (toxemia may persist for prolonged periods), (2) isolating the microbes, and (3) identifying the toxin in food ingested. The mouse is a suitable test animal for inoculation. Being highly susceptible to the effects of the toxin, the mouse succumbs to even the small amounts of circulating toxin in the blood sample received from the patient. Serum, gastric contents, and feces from a patient may be examined for toxin by mouse toxin–neutralization tests. Identification of the toxin as to type in suspected foods may be made by mouse tests. For type A toxin there is also a radioimmunoassay.

Prevention.* Prevention of botulism depends primarily on (1) heating foods to a temperature of 120° C for at least 15 minutes (at sea level) in the canning process (or steam cooking under adequate pressure, especially for low-acid foods)† to destroy any spores present, (2) boiling canned foods for 15 minutes (at sea level) immediately before they are eaten to destroy the heat-labile toxins, and (3) proper refrigeration of foods after cooking. In the canning of high-acid foods such as tomatoes and fruits, botulinal spores can be killed by a temperature of 120° C for 15 minutes. For the processing of low-acid foods (beets, beans, corn, and meats) steam pressure methods are imperative. *Note the hazard for botulism at high altitudes— the boiling temperature is too low to destroy spores.*

Do not so much as taste canned food until it has been fully heated. Remember: the poison is odorless and tasteless! The telltale signs of food spoilage (swollen container lid, cloudy or slimy appearance to food, putrid odor, and the like) may not be present.

If fowls that have been eating discarded food develop limberneck and if the responsible food can be traced, persons known to have eaten the same food should receive botulinal antitoxin. If a case develops in a group of people who have eaten the same food, the members should be given the antitoxin.

Round-the-clock emergency help—available antiserums, consultation, laboratory assistance to establish the diagnosis—can be obtained from the U.S. Public Health Service Center for Disease Control in Atlanta, Ga. (Telephone during the day 404-633-3311, extension 3753-3756; at night, 404-633-2176.)

BACTEROIDES SPECIES

The Family Bacteroidaceae contains the obligate anaerobes, the natural inhabitants of the natural cavities of man, animals, and insects. (*Bacteroides fragilis* synthesizes vitamin K in the lower intestine.) The serious anaerobic infections in the hospital today come from the gram-negative members of this family.

Bacteria of the genus *Bacteroides* are mostly nonmotile, nonsporeforming, usually gram-negative, very pleomorphic rods, with fastidious growth requirements in the laboratory. They reside in the colon, oral cavity, genital tract, and upper respiratory tract in man. *Bacteroides* species make up 95% of the bacterial content of the stool and 20% by weight.

Outside their native haunts, these bacteria are astounding pathogens, feared for the hemolytic and destructive (necrotizing) lesions of bacteroidosis. The expected portal of entry is a break in the lining of the gastrointestinal tract, the female genital

---

*See p. 676 for passive immunization.
†If food is boiled, at least 3 hours are required.

tract, an area of decubitous ulceration (bedsore), or a focus of gangrene. A significant feature of bacteroidosis is bacteremia. *Bacteroides* species have a peculiar affinity for venous channels, wherein they induce clotting (thrombosis), which in turn leads to tissue damage over an expanding front from impairment of the circulation. If the blood clots formed fragment to release infected bits of blood clot (emboli) into the circulation, abscesses form at locations remote from the primary process. Typical lesions of bacteroidosis are associated with the blood clots, foul-smelling gas, and liquefaction of the dead tissue.

Bacteroides species are often found in mucosal ulcers and foul-smelling abscesses of the lungs and other organs together with other microbes, and they are especially prone to complicate abdominal surgery, alcoholic liver disease, diabetes mellitus, and malignancy.

Most human infections come from *Bacteroides fragilis*, the type species, and to a less extent from *Bacteroides melaninogenicus*. *Bacteroides melaninogenicus* elaborates a melanin-like pigment that makes its colonies brown or black. Its colonies also give off a red fluorescence when viewed under ultraviolet light. *Melaninogenicus* infections are less virulent and fulminant than those of *fragilis*.

FUSOBACTERIUM SPECIES

Members of the genus *Fusobacterium* (also in Family Bacteroidaceae) are typically long, slender, spear-shaped bacilli with tapered ends and are sometimes incriminated in suppurative and gangrenous lesions. The species of note include *Fusobacterium nucleatum*, the type species, and *Fusobacterium necrophorum* (*Sphaerophorus necrophorus*) (name means "necrosis producing"). *Fusobacterium fusiforme* has been reclassified to fit into the third genus *Leptotrichia* of the Family Bacteroidaceae as *Leptotrichia buccalis*.

OTHER ANAEROBES

The significance of a host of anaerobes is increasing in relation to disease. Mention is made of only a few being incriminated with greater regularity.

Gram-positive cocci of the genera *Peptococcus* and *Peptostreptococcus* are important in anaerobic infection of the female genitalia. Small gram-negative, capnophilic cocci of genus *Veillonella*, found ordinarily in the mouth, possess endotoxins. Species of gram-positive, nonsporeforming bacilli, notable for their output of organic acids, include *Propionibacterium acnes* and *Eubacterium limosum*. *Propionibacterium acnes*, based on the skin, is a troublesome contaminant of blood cultures. *Eubacterium limosum* resides in the intestinal tract and is known to synthesize vitamin B_{12}.

QUESTIONS FOR REVIEW

1. Define anaerobes.
2. Discuss anaerobes as organisms indigenous to man.
3. How do anaerobes produce disease? Give four factors relating to their pathogenicity.
4. List diseases produced by anaerobes.
5. Outline briefly the laboratory diagnosis of anaerobic infections.

6. What has been the effect of technologic developments in anaerobic microbiology?
7. Name and describe the causative agents for tetanus, gas gangrene, and botulism.
8. What is the derivation of the word botulism?
9. What is the effect of tetanus toxin? Of diphtheria toxin? Of botulinal toxin? Compare the three as to potency.
10. List types of wounds likely contaminated with tetanus bacilli.
11. What types of wounds are most likely infected with clostridia of gas gangrene?
12. What measures other than immunization help to prevent tetanus and gas gangrene?
13. Briefly describe tetanus.
14. What are the circumstances producing botulism? What are the main preventive measures?
15. Why is botulism a risk at high altitudes?
16. Briefly characterize *Bacteroides* species. What is their importance today in hospital infections?

REFERENCES. See at end of Chapter 27.

23 Spirochetes and spirals

Spirochetes are actively motile, flexible, spiral bacteria found in contaminated water, sewage, soil, decaying organic matter, and within the bodies of animals and man (Fig. 23-1). They may be free-living, commensal, or parasitic. Spiral microbes vary in length from only a few to 500 μm and move by rapidly rotating about their long axis, by bending, or by "snaking" along a corkscrew path. They are aerobic, facultatively anaerobic, or anaerobic, with no flagella and no endospores. Many of them are best visualized by phase-contrast and dark-field microscopy.

Within Bergey's Order I, Spirochaetales, Family I, Spirochaetaceae, contains slender spiral bacteria in five genera. They include nonpathogenic spirochetes often found on the mucous membranes of the mouth, about the teeth, and on the genitals, and three significant pathogens. The pathogenic spirochetes are found in Genus III, *Treponema*, containing the organisms responsible for syphilis and yaws; Genus IV, *Borrelia*, with the organisms responsible for relapsing fever; and Genus V, *Leptospira*, with the organism responsible for infectious jaundice (Weil's disease) and other forms of leptospirosis.

Other spiral-shaped bacteria with slightly different properties fall into another division (Part 6) of Bergey's classification. In Family 1, Spirillaceae, are the spirilla—spirally twisted rods. These are rigid, possess one flagellum or a tuft of flagella, and are actively motile, swimming in straight lines in corkscrew fashion. They may be saprophytes or pathogens. Genus I, *Spirillum*, contains the microbes of one form of rat-bite fever.

TREPONEMA PALLIDUM
(THE SPIROCHETE OF SYPHILIS)

Syphilis *(lues venerea)* is an infectious disease caused by *Treponema pallidum (T. pallidum)*. Clinically it may be either *acquired* or *congenital*, that is, incurred after or before birth; the former is more usual. The historic origin of syphilis is a de-

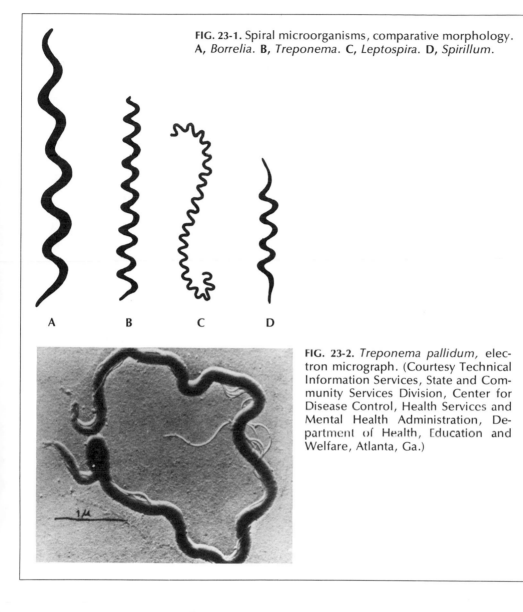

FIG. 23-1. Spiral microorganisms, comparative morphology. A, *Borrelia*. B, *Treponema*. C, *Leptospira*. D, *Spirillum*.

A   B   C   D

FIG. 23-2. *Treponema pallidum,* electron micrograph. (Courtesy Technical Information Services, State and Community Services Division, Center for Disease Control, Health Services and Mental Health Administration, Department of Health, Education and Welfare, Atlanta, Ga.)

bated question. Many observers believe that Columbus's sailors introduced it into Europe on their return from the New World. About that time the disease did spread over certain parts of Europe in virulent epidemic form.

T. pallidum (the pale spirochete) is an actively motile, slender, corkscrewlike organism that, when properly searched for, can be found in practically every syphilitic lesion (Fig. 23-2). It is especially abundant in chancres and mucous patches (lesions of the skin and mucous membranes found early in the disease). It has 6 to 14 spirals and rotates on its long axis. Usually rigid, it may bend on itself. In addition to its rotary motion, it has a slowly progressive one.

T. pallidum is very difficult to stain with ordinarily used bacteriologic dyes* and is best demonstrated by dark-field microscopy in the syphilitic chancres and mucous patches. Since the organisms are examined in the living state by this method, the diagnostic features of motility and shape are seen.

T. pallidum can live outside the body under suitable conditions for 10 to 12 hours but is killed within 1 hour by drying. It does not occur outside the body except very briefly on objects contaminated with syphilitic secretions. In whole blood or plasma at refrigerator temperature it remains viable for 24 hours but dies within 48 hours.

Under natural conditions *T. pallidum* infects man only, but anthropoid apes, monkeys, rabbits, and guinea pigs may be infected artificially. The disease in rabbits and monkeys resembles the human disease in many but not all respects.

Acquired syphilis

Modes of infection. Syphilis is caused by a microbe that is not borne by food, air, water, or insect. Man is its only reservoir. To contract the disease, a human being must be in close and intimate contact with an infectious person. Acquired syphilis therefore is contracted in most instances by sexual intercourse. In a few cases it is contracted by other types of direct contact of skin or mucous membranes, such as kissing, if the patient has lesions in his mouth. It is but rarely spread by contaminated objects such as drinking cups or towels. Physicians, dentists, and nurses may become infected in the examination of a syphilitic patient.

With modern blood banks and blood banking methods, the danger of transferring syphilis by blood transfusion is minimal. Blood in the modern blood bank is routinely tested by a standard serologic test for syphilis. If the result is positive, that blood is not released.

Evolution of a typical case. The course of syphilis is outlined in Table 23-1.

INCUBATION STAGE. *T. Pallidum* is a highly invasive organism. When introduced into the human body, it multiplies. Many of the treponemes migrate via the lymphatics to the regional lymph nodes. From the thoracic duct they enter the bloodstream and rather quickly spread over the body. Some continue to multiply at the original site. From 2 to 6 weeks later an inflammatory reaction at the site of inoculation develops, and the primary lesion, the *chancre*, forms.

PRIMARY STAGE. The chancre is the first clinical sign of infection. It usually appears on the genitals, where in the male it is readily observed. In the female the chancre, if on the cervix, is so located as to escape detection. Some 10% or more of chancres are extragenital, being found on the face, lips, tongue, tonsils, breasts, or fingers.

The chancre presents beneath the mucous membrane or skin as a small nodule having the feel of a shot. It breaks down, forming a shallow ulcer with indurated edges and a hard, clean base. There is little pain or discharge unless secondary infection occurs. Chancres are usually single but may be multiple. They vary in size

---

*Fritz Schaudinn (1871-1906), parasitologist, and Eric Hoffmann (1868-1959), clinician, described the causative agent of syphilis, *Treponema pallidum*, in 1905. Their discovery of an almost invisible parasite was the result of incomparable skill in technic and staining methods.

TABLE 23-1. EVOLUTION OF TYPICAL CASE OF SYPHILIS

Stage	Duration	Clinical disease	Activity of *Treponema pallidum*	Diagnosis	Tissue change
Incubation	2 to 6 weeks (most often 3 to 4 weeks)	None	Spirochetes actively proliferate at entry site, spread over body	Identification of *Treponema pallidum:* a. Dark-field microscopy b. Fluorescent antibody technic	Chancre appears at inoculation site
Primary	8 to 12 weeks	1. Chancre present at inoculation site 2. Regional lymphadenopathy	Chancre teeming with them	1. Dark-field microscopy of chancre 2. STS* become positive	Chancre present
Primary latent	4 to 8 weeks	None	Inconspicuous	STS positive	None demonstrable; chancre healed with little scarring
Secondary	Variable over period of 5 years (latent periods with recurrences)	1. Skin and mucosal lesions ("mucous patches") 2. Generalized lymphadenopathy	Skin and mucosal lesions rich in spirochetes (highly infectious)	1. Dark-field microscopy of lesions 2. STS positive	1. Infection active: a. Vascular changes b. Cuffs of inflammatory round cells about small blood vessels 2. Resolution spontaneous—little scarring
Latent	Few months to a lifetime (average 6 to 7 years)	None	Inconspicuous	STS positive (can be negative)	
Tertiary	Variable—rest of patient's life	Related to organ system diseased and the incapacity thereof	Paucity of spirochetes in classic lesions	1. STS positive or negative 2. Special silver stains of tissue lesions may show spirochetes	1. *Gumma* 2. Definite predilection to heal in lesions 3. Scarring 4. Tissue distortion and abnormal function

*Serologic tests for syphilis.

but are seldom more than one-half inch in diameter. After a few days the lymph nodes draining the site enlarge. Pain is absent, and the nodes do not suppurate.

After the chancre has persisted for 4 to 6 weeks, it heals. For the next 4 to 8 weeks (prior to the secondary stage) the patient shows no signs of disease. This is the *primary latent period*.

SECONDARY STAGE (STAGE OF SYSTEMIC INVOLVEMENT). The manifestations of secondary syphilis may recur over a period of 5 years. They are (1) skin lesions, (2) mucosal lesions, (3) generalized rubbery lymphadenopathy, and (4) an influenza-like syndrome. The variable skin eruption is usually symmetrically arranged, macular, copper colored, and seldom itches or burns. A patchy loss of hair, even that in the eyebrows, is associated. Known as mucous patches and most often found in the mouth, the mucosal lesions are painful superficial ulcers with a white raised surface and swarming with treponemes.

LATENT STAGE. After the secondary stage is a period during which the patient shows no signs, the disease recognized only by serologic tests. This stage may last a few months, but in 25% to 50% of the patients it lasts a lifetime. It may at anytime become active.

For some patients there is no latent period, the tertiary stage appearing right after systemic manifestations have disappeared. In those patients in whom the chancre and secondary manifestations are not present or not detected, the whole course of the disease may be of the latent type. Lesions of primary and secondary syphilis may be atypical and inconspicuous. Many patients are unaware or ignore the signs of the disease in these stages. Latent syphilis probably represents a biologic balance between the pathogenicity of *T. pallidum* and the defensive forces of the body.

TERTIARY STAGE. The hallmark of *tertiary* syphilis is the destructive *gumma*, a firm, yellowish white central focus surrounded by fibrous tissue. The tertiary lesions involve the deeper structures and organs of the body and interfere materially with their functions. Syphilis seems to have a predilection for the cardiovascular system and the central nervous system, but other organ systems are vulnerable. Treponemes are sparse in tertiary lesions, and the tissue reaction is usually attributed to some form of allergy.

Syphilis has been called the "great imitator." There is practically no organic disease whose manifestations it cannot copy, and no other disease assumes so many clinical expressions.

Cardiovascular syphilis is tertiary stage syphilitic involvement of the heart and blood vessels. It results in inflammation of the aorta (syphilitic aortitis), aneurysmal dilation of the thoracic aorta, and distortion of the aortic valve to produce aortic insufficiency (one form of valvular disease of the heart). Syphilis also predisposes to arteriosclerosis in the aorta.

Neurosyphilis is syphilitic involvement of the central nervous system. When syphilis becomes generalized during its early stages, the central nervous system seldom escapes. In many patients the treponemes die without causing any damage there, but in some 30% to 40% of patients they remain alive and initiate tertiary stage changes that come to light weeks, months, or years later. Depending on the anatomic site, neurosyphilis takes three forms: (1) syphilitic meningitis or meningovascular syphilis, (2) tabes dorsalis, and (3) general paresis. In addition, gummas occur as isolated lesions in various parts of the central nervous system.

A regular manifestation of neurosyphilis, *syphilitic meningitis* (more properly called *meningovascular syphilis*) is inflammation of the meninges with or without involvement of the brain itself. The meninges are variably thickened, and typical changes occur in the meningeal blood vessels. Although not all patients with syphilis develop meningovascular disease, a good percentage do. It usually develops within the first 5 or 6 years after infection but may appear as early as 2 months or as late as 40 years.

Tabes dorsalis or *locomotor ataxia* is a degeneration of the posterior columns of the spinal cord and the posterior nerve roots and ganglia. Since the sensory pathways of the spinal cord are involved, the disease is chiefly one of muscular incoordination and sensory disturbances.

General paresis (general paralysis of the insane; paralytic dementia) is a diffuse meningoencephalitis characterized by progressive mental deterioration, insanity, and generalized paralysis, terminating in death. About 3% of patients with syphilis are affected. Paresis prefers the highly civilized to the primitive races, better educated to less well-educated persons, and is 5 times more frequent in males than in females.

Neurosyphilis may simulate almost any disease of the central nervous system, but a careful history with a correct interpretation of laboratory tests on the blood and cerebrospinal fluid usually makes the differentiation.

Immunity. Immune mechanisms are poorly understood in syphilis. The victim does develop some kind of resistance for he is not susceptible to superimposed syphilitic infection. If he is completely cured, however, his susceptibility becomes as great as ever.

With syphilitic infection a heterogeneous group of antibodies result, but the exact nature of their contribution to immunity is unclear. It is postulated that invading treponemes interact with tissue cells of the host releasing fatty substances that may combine with protein from treponemes to form antigens. These, in turn, stimulate formation of antibodies to both lipids and the organisms. Two broad categories of syphilitic antibodies are defined: (1) nonspecific, *nontreponemal* antibodies directed against lipoidal antigens, and (2) specific *antitreponemal* ones against the spirochetes themselves.

It has been known for a long time that the serum of a patient with syphilis (cerebrospinal fluid in neurosyphilis) contains a nonspecific, nontreponemal antibody-like substance not found in normal blood or spinal fluid, which can combine with an antigen prepared as a lipid extract of normal tissue, for example, cardiolipin from the heart muscle of an ox. On this phenomenon are based the serologic tests for syphilis (STS), since this combination fixes complement and aggregates antigen from colloidal suspensions (complement fixation and flocculation tests). Because of the uncertain nature of the antibody-like material, it has been called "reagin."

There is no method of artificially inducing immunity against syphilis. Inadequate early treatment may be of greater harm than good because it fails to effect a cure and at the same time retards the establishment of any degree of immunity.

Laboratory diagnosis. The laboratory makes two important approaches to the diagnosis of syphilis: (1) the demonstration of the spirochetes in the lesions by dark-field microscopy or by immunofluorescence, and (2) serologic testing for syphilis, notably the use of complement fixation and flocculation tests. Serologic tests include nontreponemal ones to demonstrate reagin in serum and cerebrospinal fluid and treponemal tests to detect the antigens of *T. pallidum* in serum.

The dark-field microscope is most practical in the investigation of suspected chancres and mucous patches. Fluorescent antibody technics can also be used on exudates therefrom.

During the first few days of the chancre, the spirochetes are found in almost all patients. As it ages, *T. pallidum* gradually disappears. Serologic tests are seldom positive during the first few days of the chancre but usually become positive before the appearance of the secondary stage. In neurosyphilis, serologic tests on both blood and cerebrospinal fluid are positive in most patients.

The complement fixation (nontreponemal) test, regardless of the technic, is referred to as the Wassermann test because August von Wassermann (1866-1925) first applied in 1906 the principle of complement fixation to the diagnosis of syphilis. The test has been so much improved that of the original, only the principle and the name remain. Two modifications of the Wassermann test in routine use are that of Kolmer and that of Eagle.

Precipitation (or flocculation) tests are the Kline, Kahn, Eagle, Hinton, Mazzini, RPR (rapid plasma reagin), and VDRL (Venereal Disease Research Laboratory) tests. The VDRL is probably the most widely used and exclusively so for testing cerebrospinal fluid. The Kahn presumptive test and the Kline exclusion test are highly sensitive and can detect amounts of syphilitic reagin much less than the smallest amount that the so-called standard or diagnostic tests can. These sensitive tests have the disadvantage that positive results are often obtained in perfectly healthy persons or persons who at least do not have syphilis. Therefore they are more significant when negative than positive and are used for *screening*. A positive result is then followed by other serologic tests, and for the laboratory diagnosis of syphilis positive results are required from two or more tests less sensitive than the screening test.

Nontreponemal flocculation and complement fixation tests for syphilis may give negative results in the presence of syphilis and sometimes give positive results in its absence.* Such results are called false negative and false positive reactions. False negatives may be caused by technical errors or by undetectably small amounts of syphilitic reagin. False positives come from technical error, or they may be biologic false positives (BFP). Biologic false positives are occasionally or uniformly found in certain

---

*A serologic test for syphilis cannot be considered diagnostic of the disease. It only gives immunologic information, and none devised so far is absolutely specific. Two terms applied to serologic tests for syphilis, and other laboratory procedures as well, are *sensitivity* (adjective, *sensitive*) and *specificity* (adjective, *specific*). Sensitivity refers to the percentage of positive results in patients with the disease; specificity refers to the percentage of negative results among those who do *not* have the disease.

diseases such as yaws, infectious mononucleosis, malaria, leprosy, and rat-bite fever, and in drug addicts. Partly to eliminate the biologic false positive, serologic research has focused on tests utilizing either the organisms, dead or alive, or chemical extracts of them as antigens.

The *Treponema pallidum immobilization test* (TPI), a treponemal test in which *T. pallidum* itself is immobilized in the presence of guinea pig complement and syphilitic serum, is negative with serums giving BFP reactions. Its clinical usefulness depends on this fact. As specific as, and more sensitive than, the TPI test is the FTA-ABS (fluorescent treponemal antibody absorption) test, considered to be the best of the treponemal tests today. Some authorities believe it to be 99% accurate. As the name indicates, test serum is absorbed of nonspecific (confusing) antibodies. It is then brought into contact with *T. pallidum* and fluorescein-tagged antihuman globulin in a special way so that the combination may be viewed microscopically. Antibodies to the spirochetes, if present, attach to the organism, the antihuman globulin in turn uniting with them. When seen through the ultraviolet microscope, the result glows beautifully.

Other treponemal tests are the *Treponema pallidum* complement fixation test and the Reiter protein complement fixation test (RPCF).

Remember that the diagnosis of syphilis must be made only after careful evaluation of both clinical features and laboratory findings.

Prevention. The patient is most likely to convey syphilis during the primary and secondary stages because chancres and mucous patches are living cultures of syphilitic spirochetes. Attendants of patients with these lesions should be careful not to contact the patient's secretions. Certain manual examinations must always be made with gloves because a patient may show no evidence of syphilis but at the same time be capable of transmitting the infection.

People should still be educated as to the universal prevalence of syphilis and the danger of syphilis not only to the person who has it but also to his marriage partner and his children. Today a growing public health problem, the disease is significant among young adults and especially teenagers, in whom there has been better than a 200% increase in the incidence of syphilis over the last decade.

Considerable thought today goes into the development of a syphilitic vaccine. One under investigation is the gamma-irradiated Nichols strain of *T. pallidum*, nonvirulent but antigenic. Another is made from frozen spirochetes.

Congenital (prenatal) syphilis

Congenital syphilis refers to the disease acquired before birth. As a result of the upsurge in the incidence of venereal disease including syphilis, over 2000 babies will be born this year in the United States with congenital syphilis. For this form of syphilis to occur, the mother must be infected. The treponemes are blood borne to the maternal side of the placenta and deposited there. Syphilitic foci develop, and the organisms cross to the fetal circulation. A syphilitic father can transmit the infection to his child only indirectly, that is, by infecting the mother.

The shorter the time elapsing between infection of the mother and conception, the more likely the unborn baby will be infected. If adequate treatment of the mother is instituted before the fifth month of pregnancy, the child should be born free of syphilis. The requirements by law in many states for premarital examination for syphilis and for prenatal serologic tests for syphilis have been important public health measures in the prevention of this form of the disease.

Congenital syphilis usually appears at birth or within a few weeks thereafter, but it may appear years later. The child of a syphilitic mother may be (1) born dead, (2) born alive with syphilis, (3) born in apparently good health but show evidence of syphilis several weeks or months later, or (4) entirely free of the disease.

The placenta of a syphilitic child is large for the weight of the child. In children stillborn, the lungs fill the entire thoracic cavity, are grayish white, and are incompletely developed. This condition, known as white pneumonia or *pneumonia alba*, is pathognomonic for congenital syphilis. Congenital neurosyphilis simulates the acquired form and may appear in early life or be delayed to adolescence.

Infants born with active syphilis are undersized and have an appearance strikingly like that of an old man. A vesicular skin eruption and a persistent nasal discharge (*snuffles*) are often present. The child may have linear scars at the angles of the mouth (*syphilitic rhagades*). Among the late manifestations of congenital syphilis are poorly developed, small, peg-shaped permanent teeth. The upper central incisors are wedge shaped and show a central notch (*Hutchinson's teeth*). Other late manifestations are interstitial keratitis, anterior bowing of the tibia (saber shin), dactylitis, and neurosyphilis.

To detect congenital syphilis in a newborn baby with no outward signs, a modified FTA-ABS test is applied. Specific antihuman immunoglobulin tagged with fluorescein indicates the presence of 19S immunoglobulins (IgM antibodies). Unlike the smaller 7S immunoglobulins, which are formed by the mother, these macroglobulins cannot cross the intact placenta and must be formed by the baby. Their presence indicates the baby's reaction to *his* disease, not his mother's.

Syphilis and yaws

Yaws or frambesia, a tropical disease closely akin to syphilis, is caused by *Treponema pertenue*, a spirochete that cannot be differentiated serologically from *T. pallidum* and that, like *T. pallidum*, is susceptible to arsenic and penicillin. Yaws may be a special form of syphilis; however, most observers believe it is distinct.

Yaws is neither venereal nor congenital. Skin and bone lesions are prominent. When the yellowish crusts covering the large pustules are removed, the slightly bleeding surface looks exactly like a raspberry stuck on the skin—hence the name frambesia or "raspberry" disease.

BORRELIA SPECIES

Members of the genus *Borrelia* are helical cells with coarse, uneven coils. They are anaerobic parasites found on mucous membranes, and some cause disease in man and animals (for example avian and bovine spirochetosis in animals).

Disease in man. Relapsing fever is an acute infectious disease caused by several species of spirochetes in genus *Borrelia* (Fig. 23-1) and is described clinically by alternating periods of febrile illness with apparent recovery. The incubation period is 3 to 10 days. The spirochetes are found in the peripheral blood during the fever and are transferred from man to man by body lice *(Pediculus humanus* subsp. *humanus)* and from rodent to man by ticks *(Ornithodoros*)*. Thus there are two types of relapsing fever—that borne by the louse and that borne by the tick. The former occurs as epidemic relapsing fever in the general pattern of louse-borne disease, and the spirochete causing it is *Borrelia recurrentis*. Tick-borne relapsing fever does not occur in epidemics. It is endemic relapsing fever caused by other species of *Borrelia* and the only type found in North America. Infected ticks may transmit the infection to their offspring for generation after generation. Body lice do not thus transmit it. Rodents such as the ground squirrel and prairie dog are the main reservoir of the tick-borne disease. A wide variety of animals may be infected with borreliae, including armadillos, bats, dogs, foxes, horses, rabbits, and porcupines.

Laboratory diagnosis. Actively motile spiral organisms, *Borrelia* species differ from *T. pallidum* in that they take the usual laboratory stains. They may be detected in the Wright-stained smears of peripheral blood or by dark-field illumination. Blood from a patient may be inoculated intraperitoneally into a white mouse or young guinea pig. Wright-stained films made from the tail blood 1 to 4 days later will show the borreliae. *Borrelia* species can be grown in the chick embryo.

LEPTOSPIRA SPECIES

Leptospires are tightly and finely coiled spiral bacteria with one or both ends bent typically to form a hook. Some are free-living; others are parasitic or pathogenic in vertebrates. Only one species of the genus *Leptospira* is recognized in Bergey's recent classification (1974)—*Leptospira interrogans (Leptospira icterohaemorrhagiae)*. There are serotypes as yet incompletely defined. Taxonomic data in this genus are incomplete at present.

The disease. Leptospirosis is an acute febrile disease caused by spirochetes in the genus *Leptospira*, the best known of which is *Leptospira interrogans*. There are various names for this condition, including swamp fever, swineherd's disease, infectious jaundice, spirochetal jaundice, and Weil's disease. Weil's disease represents the most severe form.

Clinically, leptospirosis is marked by high fever, muscular pains, redness of the conjunctivae, jaundice (not invariable), and aseptic meningitis. An attack is followed by a lasting immunity. Inapparent infections also occur.

Leptospires (Fig. 23-1) are found in wild and domestic animals all over the world. Disease is mainly in animals; man is only accidentally infected. In animals the spirochetes localize in the kidney and are excreted profusely in the urine. They can survive if urine is discharged to neutral or slightly alkaline water, sewage, or mud. Man probably acquires the infection through the skin from soil contaminated with the urine of

---

*Note that *Ornithodoros*, which transmits borreliae, usually bites painlessly, takes a short blood meal, and leaves the host after 30 minutes to an hour. The host may be unaware of the tick bite.

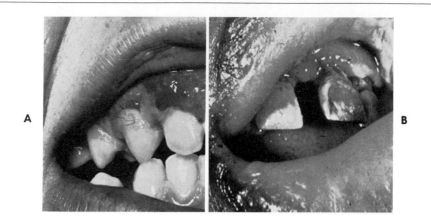

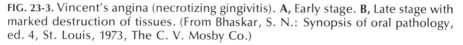

FIG. 23-3. Vincent's angina (necrotizing gingivitis). **A,** Early stage. **B,** Late stage with marked destruction of tissues. (From Bhaskar, S. N.: Synopsis of oral pathology, ed. 4, St. Louis, 1973, The C. V. Mosby Co.)

infected rats or through the mouth by food or water that has been contaminated in the same way. The organisms enter the animal or human body through the abraded skin and the mucous membranes of the eye, nose, and mouth. Leptospirosis is an occupational disease. Especially vulnerable to it are workers in rat-infested mines, rice fields, and sewage disposal plants.

Laboratory diagnosis. The spirochetes are found in the blood early in the disease and in the urine after the seventh day. Leptospires of typical shape and motility are best seen by dark-field examination of blood and urine. They may be recovered from blood cultures and from the peritoneal cavity of a guinea pig inoculated with blood or urine. A microagglutination test is diagnostic.

OTHER SPIROCHETES AND ASSOCIATED ORGANISMS

Treponema buccale and *Treponema denticola* are nonpathogenic saprophytes found in the mouth. Their presence may cause confusion in the examination of material from the mouth suspect for the spirochete of syphilis. *Treponema refringens* is part of the normal flora of the male and female genitalia and may be associated with *Treponema pallidum* in various syphilitic lesions.

Vincent's angina (fusospirochetal disease). In Vincent's infection (fusospirochetal disease), a grayish white pseudomembrane forms in the throat or mouth, beneath which ulceration occurs. When the gums and mouth are primarily involved, the disease is known as *trench mouth*. When the throat and tonsils are ulcerated, it is called *Vincent's angina* (Fig. 23-3). The disease is more properly called acute necrotizing ulcerative gingivitis. The disease is important within itself and because the membrane may be mistaken for a diphtheritic membrane. The extensive ulceration may cause the disease to be confused with syphilis.

Associated together in the lesions are two microbes: (1) a large, gram-negative, cigar-shaped, anaerobic "fusiform" bacillus and (2) a gram-negative spirochete,

Treponema (Borrelia) vincentii. Vincent's infection is thought by some to be caused by *Bacteroides melaninogenicus,* found in the mouth and pathogenic usually if associated with other kinds of organisms. Early the cigar-shaped bacilli are more numerous, whereas later on in the disease, spirochetes are. These two do not cause this condition, and most workers believe that they are only secondary invaders. They grow symbiotically here as opportunists, being already in the mouth. Their mere presence is not enough to cause disease. This must be triggered by some unusual circumstance, such as injury to the mouth or decreased oral resistance. Vincent's angina (fusospirochetal disease) accompanies malnutrition, debilitating states, viral infections, and poor oral hygiene.

Fusospirochetal organisms are identified directly on a crystal violet–stained smear of membranous exudate.

SPIRILLUM MINOR

There are two clinical entities known as rat-bite fever that are conveyed by the bite of a rat or other rodent. One, known in Japan as sodoku, is caused by *Spirillum minor.* This small and rigid spiral microbe (Fig. 23-1) is found primarily in wild rats and is spread from rat to rat and from rat to man by the bite of the rat (the healthy carrier). The features of the disease are ulceration of the bite, fever, and a skin eruption. The spirilla may be seen in material from the ulcer examined in stained smears or by dark-field illumination, and a guinea pig may be inoculated with blood or tissue from lymph nodes to isolate the microbes. Rat-bite fever of this type is uncommon but worldwide.

The other entity is streptobacillary rat-bite fever (Haverhill fever), caused by *Streptobacillus moniliformis* (formerly *Actinomyces muris ratti*), an actinomycete-like, necklace-shaped bacterium found in the mouth and nasopharynx of normal rats, among which it causes widespread epidemics. Man contracts the infection from the rat bite. The disease is a febrile one with manifestations similar to those of rat-bite fever caused by *Spirillum minor.* Both resemble tularemia clinically. *Streptobacillus moniliformis* is identified by fluorescent antibody technics.

QUESTIONS FOR REVIEW

1. Name and describe the causative agent of syphilis.
2. Outline the evolution of a typical case of syphilis.
3. Characterize congenital syphilis. What are its hazards?
4. How is syphilis transmitted?
5. Outline the laboratory diagnosis of syphilis. Briefly discuss the serologic tests.
6. Classify spiral bacteria. Present salient features.
7. Comment on the pathogenicity of *Treponema pallidum.* Why is it called the "great imitator"? What is a gumma?
8. How does the causative agent of yaws compare with that of syphilis?
9. Name two vectors of relapsing fever and possible rodent reservoirs.
10. Characterize leptospirosis.
11. Give the names of the two microbes causing rat-bite fever.
12. How is Vincent's angina diagnosed bacteriologically?

REFERENCES. See at end of Chapter 27.

24 Actinomycetes
(also corynebacteria)

CORYNEBACTERIA

The coryneform group of bacteria takes the lead position in the discussion of microbes gathered into Part 17 of *Bergey's Manual*, eighth edition, which is entitled "Actinomycetes and related organisms." The term *coryneform* is a working concept to sidestep for the time being unresolved problems in classification. Genus I, *Corynebacterium* (club bacteria), includes one dreaded pathogen—*Corynebacterium diphtheriae*, the type species.

Corynebacterium diphtheriae (the bacillus of diphtheria)

Six or seven decades ago diphtheria, or "membranous croup" as it was called, was a major cause of death. Since the causative organism, *Corynebacterium diphtheriae (C. diphtheriae)*, was first discovered by Edwin Klebs (1834-1913) in 1883, the epidemiology has become known, methods of producing a permanent immunity have been devised, and the mortality has been reduced from nearly 50% to such a level that death seldom occurs in patients who are adequately treated during the early days of the disease.* Few diseases have been so well studied and are today so well understood as diphtheria.

The causative organism belongs to the genus *Corynebacterium* of gram-positive, unevenly staining bacteria with clubbed or pointed ends. *(Coryne* signifies clubbed.) *C. diphtheriae* is often called Klebs-Löffler bacillus because it was identified by Klebs and first grown in 1884 in pure culture by F. A. J. Löffler (1852-1915)..The word *diphtheria* is derived from a Greek word meaning leather. The disease was so named because of the leathery consistency of the diphtheritic membrane.

General characteristics. *C. diphtheriae* is distinct for its variation in size, shape,

---

*An especially severe type of diphtheria, which is difficult to treat and often seen in adults, has appeared in different parts of the world within the last 3 decades.

and appearance (pleomorphism). The bacteria may be straight or curved and swollen in the middle or clubbed at one or both ends. When division occurs, the bacteria remain attached at the point of separation in a V-shaped pattern (snapping). Stained organisms have a granular, solid, or barred appearance, and deeply staining granules (metachromatic granules) are characteristic. Granules at the ends of the bacilli are known as *polar bodies*.

C. *diphtheriae* is a gram-positive, nonmotile, strictly aerobic microbe that does not form spores. It grows best at 35° C on almost all ordinary media, but growth is luxuriant on Löffler blood serum and media containing potassium tellurite. Potassium tellurite inhibits the growth of many organisms found in the throat, and colonies of diphtheria bacilli assume a typical appearance on media containing it. The morphologic features of the organisms are best preserved on Löffler blood serum.

Based on cultural characteristics, fermentation tests, and immunologic reactions, C. *diphtheriae* has been divided into three types: *gravis* (meaning severe), *intermedius* (meaning of intermediate severity), and *mitis* (meaning mild). Epidemics are most often the *gravis* type, and its manifestations are the most severe.

For man and some laboratory animals C. *diphtheriae* is highly pathogenic. In nature the disease is restricted to man. The incubation period of diphtheria is 2 to 6 days.

Diphtheria bacilli are fairly resistant to drying but easily destroyed by heat and chemical disinfectants. Boiling destroys them in 1 minute. They may remain alive in bits of diphtheritic membrane for several weeks.

Extracellular toxin. Diphtheria bacilli owe their pathogenicity to their extracellular toxin. They do not invade tissues but grow superficially in restricted areas, usually on a mucous membrane. The exotoxin formed is released into the bloodstream and circulated. As a result of its primary action in blocking the cellular synthesis of protein, toxin produces cellular injury, leading to systemic disturbances and degenerative changes in organs of the body. It affects certain nerves, the heart muscle, the kidneys, and the cortex of the adrenal gland. The features of the disease relate directly to the exotoxin; that is, it is a toxemia and a molecular disease.

In 1951 the remarkable discovery was made that exotoxin can be elaborated only by lysogenic strains of C. *diphtheriae;* that is, strains infected with certain bacteriophages (bacterial viruses) carrying the *tox* gene. If the specific phage is lost to the bacterium, the quality of toxigenicity goes also, and conversely a nontoxigenic strain may be converted to a lysogenic and toxigenic strain if treated (infected) with the proper phage. The tox gene programs the structure of exotoxin, but biosynthesis is enacted by the bacterial cell. The amount of inorganic iron in the external and internal milieu is a critical factor regulating toxin production, which is depressed until that level of iron is critically reduced.

Soluble exotoxin released into media is concentrated to yield the diphtheria toxin of commerce. Such preparations contain 200 to 1000 MLD (minimum lethal dose) (p. 675) per milliliter. The toxin can be separated from the medium and purified. It deteriorates with age and is destroyed by a temperature of 60° C.

Not all diphtheria bacilli produce toxin. Toxigenic organisms cannot be distinguished from nontoxigenic ones by microscopic appearance or cultural characteristics. Differentiation is by animal inoculation. A small amount of a liquid culture of the bacilli is injected either beneath or between the superficial layers of the skin of two guinea pigs, *only one* of which has been protected by a dose of diphtheria antitoxin. If the bacilli are toxigenic, a zone of inflammation appears at the inoculation site in the *un*protected pig, but there is no reaction at the site in the protected pig. If the bacilli do not produce toxin, neither pig shows a reaction. This is the *guinea pig virulence test.* Virulence tests may be carried out on rabbits. The technic differs somewhat from that in guinea pigs, but the underlying principle is the same. Eight or ten tests may be carried out at one time on the same rabbit. A cultural method, the in vitro gel diffusion test, has also been devised to determine the virulence of diphtheria bacilli.

Pathogenicity. Diphtheria is an acute infectious disease induced by the extracellular toxin of *C. diphtheriae* with a characteristic type of inflammatory change at the site of infection and systemic disturbances to ccompany it.

Pathologically the type lesion of diphtheria is the *pseudomembrane,* a superficial lesion occurring on mucous membranes (Fig. 24-1). The first stage in its formation is degeneration of the epithelial cells of the affected area. This is followed by an abundant fibrinous exudation onto the surface. As the fibrin precipitates, it entraps leukocytes, red blood cells, bacteria, and dead epithelial cells to form a thick, tough membranelike structure anchored to the underlying tissues. If the pseudomembrane is pulled off, a raw bleeding surface is left, but a new pseudomembrane soon forms. In the absence of antitoxin it persists for 7 to 10 days and then disappears. It may obstruct breathing, and in some patients tracheotomy or intubation is required to prevent suffocation.

The most frequent sites for pseudomembrane formation are the tonsils, pharynx, larynx, and nasal passages. The diphtheritic membrane usually begins on one or both tonsils and spreads to the uvula and soft palate. Less often, diphtheria attacks the

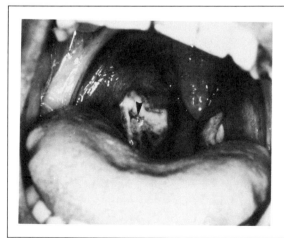

FIG. 24-1. Diphtheria in 43-year-old man; fourth day of disease. Arrow is over membrane in tonsillar area of throat. (Tongue in foreground.) (From McCloskey, R. V., and others: Ann. Intern. Med. **75:**495, 1971.)

vulva, conjunctiva, middle ear, and skin and infects wounds. Skin infections were prevalent among soldiers of World War II. Organisms other than *C. diphtheriae* may produce pseudomembranes, and in a few patients with diphtheria pseudomembranes do not form.

Diphtheria occurs in three clinical patterns: (1) *faucial*, in which the membrane appears on the tonsils and spreads to other parts of the pharynx; (2) *laryngeal*, in which the membrane may easily cause suffocation; and (3) *nasal*, in which the membrane is rarely associated with severe disease because toxin is poorly absorbed by the lining of the nose. Bronchopneumonia is an important complication.

The really serious effects of diphtheria stem from the action of the toxin. The damage done to the heart may precipitate heart failure, even after other manifestations of the disease have subsided and sometimes in comparatively mild cases. Sudden death has been reported several weeks after apparent recovery. Degeneration of peripheral nerves induced by toxin leads to late paralysis, particularly of the soft palate.

Sources and modes of infection. The sources of infection in diphtheria are persons with typical cases, persons with mild undetected cases, and carriers. Ordinarily the bacteria enter and leave the body by the same route, the mouth and nose. They may be transferred directly from person to person by droplets expelled from the mouth and nose or indirectly by cups, toys, pencils, dishes, and eating utensils contaminated by buccal or nasal secretions. The direct method of transfer is the more important.

Nasal diphtheria is an important source of infection because it is often overlooked. Diphtheria bacilli can be spread from skin infections appearing to be less serious ailments. In epidemics such skin sources figure significantly. In fact, cutaneous diphtheria in indigent adults is an important reservoir. A few milkborne epidemics are reported, usually caused by contamination of the milk by a person working in the dairy who either is a carrier or may have a mild case of disease. He transfers his buccal or nasal discharges by his hands. The idea that cats convey diphtheria is not true.

Over the last decade or so in this country diphtheria has continued to occur in the economically depressed areas of the South where living conditions are crowded, medical care lacking, hygiene poor, and immunization inadequate.

DIPHTHERIA CARRIERS. In about one half of patients with diphtheria, the bacilli leave the body within 3 days after the membrane disappears, and in four fifths of patients they have disappeared within 1 week, but sometimes they persist and the patient becomes a carrier. Also, certain contacts of a diphtheria patient or a carrier become carriers themselves without contracting the disease. It has been estimated that from 0.1% to 0.5% of the population carry virulent diphtheria bacilli. The percentage increases during epidemics and in crowded communities during cold weather.

Most carriers harbor the bacilli a short time only (from a few days to a few weeks), but a few harbor them permanently even with intensive treatment. Because a high percentage of organisms with the morphology and cultural characteristics of diphtheria bacilli in normal throats are nontoxigenic and therefore not dangerous, viru-

TABLE 24-1. BACTERIOLOGIC DIAGNOSIS OF DIPHTHERIA

Organism	Morphology	Growth on blood tellurite	Hydrolysis of urea
Corynebacterium diphtheriae	Pleomorphic	Gray to black colonies	−
Corynebacterium pseudo-diphtheriticum (example of a diphtheroid)	More uniform	Opaque grayish	+

lence tests should always be done on diphtheria bacilli from suspected carriers. If a suitable antibiotic such as erythromycin or penicillin is given in conjunction with antitoxin during the acute stage of the disease and continued during convalescence, the carrier rate is reduced.

Bacteriologic diagnosis. The laboratory diagnosis of diphtheria is made by culture of the organism (Table 24-1). Some of the pseudomembrane is removed with a sterile swab; a slant of Löffler serum or tellurite medium is inoculated and incubated from 12 to 14 hours. Smears made from the culture are stained with Löffler methylene blue or one of the special stains for diphtheria bacilli. The bacilli may sometimes be found in smears made directly from the pseudomembrane, but failure to find them in no manner indicates that the patient does not have the disease. Antiseptics, gargles, or mouthwashes must not be used before cultures are taken because they may prevent proper growth, and antibiotics given 5 to 7 days previously may inhibit bacterial growth. If possible, cultures should be taken before antibiotics are given. Care should be taken to bring the swab in contact only with the site of disease, and cultures should be made from both throat and nose.

In a patient with ulcerative or membranous inflammation of the throat both cultures for *C. diphtheriae* and smears for the organisms of Vincent's angina should be made. The two conditions may be easily confused. If only a culture is made, an infection with the organisms of Vincent's angina could be missed because these organisms do not grow in cultures, and diphtheria could be missed if smears alone are examined. Moreover, the two diseases can coexist.

The finding of diphtheria bacilli in the throat does not necessarily mean the patient has diphtheria; he may be a carrier. Remember that membranous infections of the throat may be caused by organisms other than *C. diphtheriae* and that nonmembranous infections with *C. diphtheriae* occasionally occur. If streptococcal infection is suspected, a blood agar plate is inoculated.

DIPHTHEROID BACILLI. The heterogeneous diphtheroid bacilli bear a close microscopic resemblance to *C. diphtheriae* and are sometimes confused with it on throat smears. They are quite numerous and have been isolated from varied sources such as the skin, nose, throat, urethra, bladder, vagina, and prostate gland. They reside in soil and water as saprophytes, do not produce toxins, and are nonpathogenic for man except under exceptional circumstances. They have been reported as rare causes of wound infections, meningitis, osteomyelitis, and hepatitis.

Reduction of nitrate	Fermentation of carbohydrates				Toxigenicity
	Trehalose	Glucose	Maltose	Sucrose	
+	−	+	+	−	+
+	−	−	−	−	−

Immunity. Immunity to diphtheria results from the presence of diphtheria antitoxin in the blood. Newborn babies of immune mothers receive a passive immunity because of the transfer of antitoxin from the maternal to the fetal circulation via the placenta. In breast-fed infants this immunity is augmented by antibodies in the milk of the mother. This immunity is usually lost by the end of the first year, and from this time to the sixth year most children are susceptible. Minor infections with *C. diphtheriae* reestablish an immunity during late childhood.

Possibly 50% of adults are immune, a figure once considerably higher. The reduction in the incidence of diphtheria, the decreased likelihood of minor infections, and the institution of vaccination during childhood (which does not give the permanent immunity that repeated minor infections do), has meant a decrease in adult immunity. The presence of from $1/500$ to $1/250$ unit of diphtheria antitoxin per milliliter of blood renders a person immune.

The *Schick* test purports to determine whether a person has sufficient diphtheria antitoxin in his blood for immunity. One-fiftieth MLD of diphtheria toxin (0.1 ml of diluted toxin) is injected into the skin of one arm of a subject and a control dose of toxoid into the other. The skin sites on both arms are read 4 days later. In principle, if the subject's blood contains sufficient antitoxin for protection against diphtheria, no reaction occurs (negative test). An insufficient amount of antitoxin (positive test) is indicated by the appearance within 24 to 36 hours of a firm red area, 1 to 2 cm in diameter, persisting 4 or 5 days. In the past the Schick test has been used to detect persons lacking immunity and to determine the efficacy of active immunization once given. Since the highly purified, adult-type toxoids have become available, the need for routine Schick testing is largely eliminated. Active immunization is desirable, and difficulties encountered with the Schick test can be bypassed.

An attack of diphtheria is usually followed by a fair degree of immunity, which may be temporary or permanent, but in a few patients immunity does not seem to develop. Most carriers of virulent diphtheria bacilli are immune to the toxin.

Prevention and control of diphtheria.* All persons ill of diphtheria should be isolated, and neither they nor their close contacts should be released until it is proved that they harbor no virulent diphtheria bacilli in their noses and throats. The mouth secretions and all objects so contaminated should be disinfected. The patient's eating

---

*For immunization in diphtheria, see pp. 677, 680, 686, 695-698, and 706. For disinfection, see p. 312.

utensils should be boiled. Nurses must be careful that they do not contaminate their hands with the mouth and nose secretions of patients so as to infect themselves. Persons known or presumed susceptible who contact a diphtheria patient should receive diphtheria toxoid (for active immunization) and antibiotics. If they cannot be seen daily by the physician, they may be temporarily immunized with 10,000 units of diphtheria antitoxin intramuscularly. Remember that the administration of antitoxin can set the stage for an anaphylactic reaction should another agent containing horse serum be given that person.

The general measures to reduce the incidence of diphtheria are (1) detection and treatment of carriers and (2) production of an active immunity in all susceptibles — children under 6 years of age and all older children and adults, especially physicians and nurses not previously immunized.

ACTINOMYCETES

Actinomycetes are now classified as bacteria. For a long time these microbes, because of similarities to both the true bacteria and the true fungi, were thought of as intermediates between the two, were sometimes called "higher bacteria," but were usually included in a discussion of medical mycology.

Within Part 17 of the eighth edition of *Bergey's Manual*, Order I, Actinomycetales, contains gram-positive, sometimes acid-fast, mostly aerobic bacteria forming branching filaments that in some of the families develop into a mycelium. Some members are pathogens of man, animals, and plants.

In this classification of funguslike bacteria, there are four families of note. The family Mycobacteriaceae comprises the genus *Mycobacterium* of acid-alcohol–fast microbes,* of which *Mycobacterium tuberculosis* is the type species (Chapter 25). The family Streptomycetaceae is medically very significant because the actinomycetes of genus *Streptomyces* provide us with many important pathogenic antibiotics. Two families of actinomycetes, each with an important genus, are Actinomycetaceae of nonacid-fast, diphtheroid-shaped bacteria with no mycelium and Nocardiaceae of variably acid-fast organisms with variable mycelial development. The genera are *Actinomyces* and *Nocardia*, respectively, and the diseases, actinomycosis and nocardiosis. Species of genus *Actinomyces* are anaerobic or microaerophilic, whereas species of genus *Nocardia* are aerobic. The acid-fast species of genus *Nocardia* may be confused with *Mycobacterium tuberculosis*.

The actinomycetes (including streptomycetes) resemble the fungi in that they may have a mycelium of masses of branched filaments, but their "hyphae" are much slenderer than those of true fungi. The filaments of actinomycetes fragment into spherical or rod-shaped segments that function as spores and in turn develop into new hyphae. Spore formation comparable to that in bacteria is seen in the hyphae of the genus *Streptomyces*, wherein development of a mycelium is complete. Unlike fungi, actinomycetes lack a nuclear membrane.

Actinomycetes are widely distributed in nature and play a vital part in changes

---

*Mycobacteria are not referred to as actinomycetes as are members of the other three families.

in organic material of the soil. This activity is more crucial to man than any disease-producing capacity.

Actinomycosis

Actinomycosis is an infectious disease of lower animals (especially cattle) and man caused by several species of bacteria belonging to the genus *Actinomyces*. Of these, *Actinomyces bovis* (bovine actinomycosis) and *Actinomyces israelii* (human actinomycosis) are the most important. Actinomycosis is typified clinically by formation of nodular swellings that soften and form abscesses, discharging a thin pus through multiple sinuses. The disease in man is in three patterns: (1) cervicofacial, (2) thoracic, and (3) abdominal. The cervicofacial type is described by swelling and suppuration of the soft tissues of the face, jaw, and neck. It is the usual type (about half the cases) and the least dangerous. It appears to have a special association with dental defects. The thoracic type is characterized by multiple small cavities and abscesses in the lungs. The abdominal type usually begins about the appendix or cecum. In advanced thoracic and abdominal disease, sinus tracts extend to the surface. The disease in cattle is known as lumpy jaw.

The etiologic agent is found in the pus or in the walls of abscesses as small, yellow granules about the size of a pinhead, the sulfur granules. When a sulfur granule is placed on a microslide and a cover glass pressed down on it, a distinct microscopic picture is seen—a central threadlike mass from which radiate many clublike structures (Fig. 24-2). For this reason *Actinomyces* have been called ray fungi. The clubbed appearance is not very pronounced in cultures. The laboratory diagnosis of actinomycosis is made by finding objects in the discharges that are both grossly and micro-

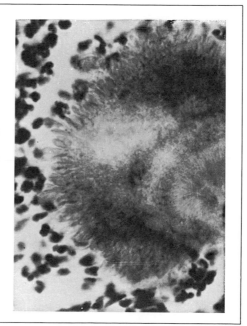

FIG. 24-2. *Actinomyces* sulfur granule (colony) in a microscopic field of pus. Club-shaped processes are at periphery. (From Bauer, J. D., and others: Clinical laboratory methods, ed. 8, St. Louis, 1974, The C. V. Mosby Co.)

scopically sulfur granules or by demonstrating the organisms in sections of tissue taken from the lesions. Cultural methods help, but serologic and skin tests contribute little.

Pathogenic actinomycetes are normal inhabitants of the mouth of cattle and man, probably existing in an attenuated state. There is no evidence that the organisms are saprophytic outside the animal body. Infection occurs in the event of injury to the mouth, tooth decay, or some other abnormal state. Such conditions favor invasion of the tissues. There is no evidence of direct transmission from animal to animal or from animal to man. People who pursue nonagricultural occupations are as likely to contract actinomycosis as are farmers and stockmen.

Nocardiosis

Nocardia asteroides, an aerobic actinomycete found free in nature, is responsible for infection in man, designated nocardiosis, which is usually a primary disease of the lungs (pulmonary nocardiosis) simulating tuberculosis. Bloodstream dissemination of infection to the rest of the body leads to abscesses in subcutaneous tissues and other internal organs. Brain abscess commonly complicates systemic infection. Nocardiosis (like cryptococcosis), with an affinity for the lungs, accompanies malignant diseases of the reticuloendothelial system—Hodgkin's disease, lymphosarcoma, and leukemia. Steroid hormones enhance infection.

Nocardia asteroides is one cause of mycetoma or "Madura foot" (Fig. 24-3). *Mycetoma* is a generic name for a localized but destructive infection involving skin,

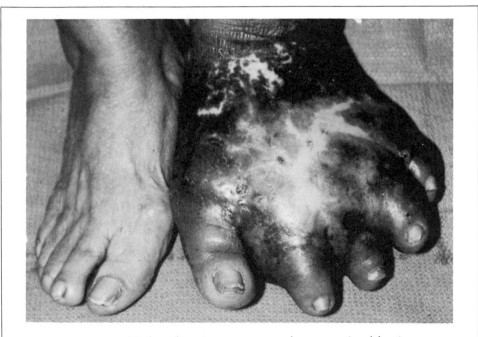

FIG. 24-3. Madura foot (mycetoma, maduromycosis of foot).

subcutaneous tissue, bone and fascia, usually of the foot and especially prevalent in the tropics. It may be actinomycotic if caused by actinomycetes or maduromycotic if certain fungi are operative. It is related to the custom of walking barefoot and is an occupational hazard to farmers and field-workers. After initial injury, the etiologic agents invade tissues to produce in time a network of interlocking abscesses and granulomas in the substance of the foot. End stage disease may necessitate amputation. The organisms are identified in the granules found in the purulent drainages from the sinus tracts.

QUESTIONS FOR REVIEW

1. What are actinomycetes? Their importance?
2. Briefly indicate what is meant by actinomycosis, toxigenicity, sulfur granule, Madura foot, mycetoma, nocardiosis, mycelium, lumpy jaw, ray fungus, diphtheroid bacilli, tox gene.
3. What does the word diphtheria mean?
4. Name and describe the bacterium causing diphtheria.
5. What are the clinical patterns in this disease? Characterize the pseudomembrane.
6. Explain the pathogenicity of the diphtheria bacillus.
7. Briefly discuss virulence tests in diphtheria.
8. Outline the laboratory diagnosis of diphtheria.

REFERENCES. See at end of Chapter 27.

25 Acid-fast mycobacteria

Nonacid-fast bacteria are easily stained with basic dyes, but the color is quickly removed when the microorganisms are treated with acid-alcohol. Acid-fast bacteria, on the other hand, are so resistant to the penetration of basic dyes that in order to color them the stain has to be either gently heated or applied for a length of time. Once stained, however, they are not easily decolorized with acid-alcohol.

The acid-fast bacteria are classified in the genus *Mycobacterium*, the members of which are straight or curved acid-fast rods that may show branching or irregular forms. Those of greatest importance and pathogenic to man are *Mycobacterium tuberculosis,* the cause of tuberculosis in man*; *Mycobacterium bovis,* the cause of tuberculosis in cattle, also in man; *Mycobacterium leprae,* the cause of leprosy; and certain species of atypical, formerly "unclassified" mycobacteria, the cause of mycobacteriosis and various other infections. In addition to these acid-fast organisms, 40 or more species exist. Most are saprophytes and nonpathogenic, but they may gain access to milk, butter, or other dairy products and be mistaken for *Mycobacterium tuberculosis.* Some are pathogenic for lower animals, for example, one, *Mycobacterium paratuberculosis* (Johne's bacillus), causes a granulomatous enteritis in cattle, known as Johne's disease.

TUBERCULOSIS

Importance. In 1900, tuberculosis (a preventable disease) was the leading cause of death in the United States as it was throughout the civilized world. (In the tropics it ranked second to malaria.) Since then it has dropped from first place in this country but still remains a major cause of chronic disability and ill health produced by a com-

---

*Tuberculosis has claimed the lives of many great writers, painters, and musicians—Keats, Chopin, Goethe, Poe, Gauguin, Paganini, Molière, and many others. In fact, no other disease has had such a significant relation to literature and the arts.

436

municable disease. In the United States each year some 30,000 or more new cases and around 4000 deaths are reported.

Tuberculosis, a lifelong disease, is a health hazard of lower socioeconomic groups, densely populated areas, persons over 50 and under 5 years of age, and the chronically ill. Over the world it is still a principal cause of death, ranking with malaria and malnutrition; there are at least 15 million persons with active tuberculosis, and 3 million die annually of the disease. Throughout the world as well as in this country, it is the most frequent *infectious* cause of death. The mortality is highest in the Orient, Asia, Africa, and Latin America.

The infectious nature of tuberculosis was suspected 5 centuries before the tubercle bacillus was discovered by Koch in 1882, and it had been produced by artificial inoculation 40 years before that time.

Etiologic agents. Three species of genus *Mycobacterium* are important causes of tuberculosis. They are *Mycobacterium tuberculosis* (the human bacillus; primary host, man), *Mycobacterium bovis* (the bovine bacillus; primary host, cattle), and *Mycobacterium avium* (the avian bacillus; primary host, birds)—the human, bovine, and avian variants. The three resemble each other rather closely but may be differentiated by animal inoculation and cultural procedures.*

Mycobacterium tuberculosis (the human tubercle bacillus)

Mycobacterium tuberculosis (M. tuberculosis)—the tubercle bacillus of common parlance—attacks all races of man, other primates, and some domestic animals, such as swine† and dogs. Wild animals living in their natural surroundings do not have tuberculosis but may contract it when placed in captivity.

General characteristics. The organism *M. tuberculosis* is a slender, rod-shaped, nonmotile, nonsporeforming bacillus, often beaded or granular in appearance on acid-fast smears. Tubercle bacilli are more resistant than other nonsporeforming organisms to the deleterious effects of drying and germicides. They remain alive in dried sputum or dust in a dark place for weeks or months and in moist sputum for 6 weeks or more. Direct sunlight kills them in 1 or 2 hours. Sufficiently susceptible to heat, however, they are destroyed by the temperature of pasteurization. Phenol, 5%, kills them in sputum in 5 to 6 hours. Tubercle bacilli are not affected by routinely used antibiotics but are susceptible to streptomycin, dihydrostreptomycin, para-aminosalicylic acid (PAS), rifampin, and isoniazid (INH). *M. tuberculosis* tends to develop a resistance to these agents, and toxic effects are sometimes observed after their use. However, their value in treatment and in modifying the course of the disease has been considerable.

M. tuberculosis grows only on special media. Even then, growth is slow; 2 to 4 weeks often elapse before any growth is visible (ordinary bacteria display colonies on

---

*In 1898 Theobald Smith (1850-1934) of Albany, New York differentiated the human and bovine forms of *Mycobacterium tuberculosis*. America's foremost bacteriologist, he made several important contributions to microbiology.
†A small percentage of hogs slaughtered for food shows evidence of tuberculosis. Hogs are susceptible to the bovine, avian, and human bacilli.

routinely used media within 24 to 48 hours). Body temperature is best (37° C), but the tubercle bacilli may grow at a temperature as low as 29° or as high as 42° C. Although tubercle bacilli are aerobic, 5% to 10% CO_2 enhances growth, and the presence of glycerin in the medium accelerates it for the human bacillus. A bacteriostatic dye such as malachite green added to the culture medium suppresses the more rapid growth of contaminants. *Note:* the human tubercle bacillus is more readily cultivated than the bovine bacillus.

Toxic products. Tubercle bacilli do not produce exotoxins, hemolysins, or comparable substances, but poisonous products partly responsible for the clinical features of tuberculosis are liberated when the bacilli disintegrate.

When *M. tuberculosis* is grown artificially, the culture medium contains a product known as *tuberculin,* without effect on a nontuberculous animal (no history of contact with the tubercle bacillus) but with powerful effects in the body of a tuberculous animal. These effects come from surprisingly small doses, and if the dose is large enough, the results are disastrous.

There are more than 50 methods of preparing tuberculin, and the nature of each tuberculin depends to some extent on the method. The best known and, in the past, the most extensively used tuberculin is Koch's original (or old) tuberculin, often spoken of simply as OT. It is prepared from a culture of tubercle bacilli in 5% glycerin broth that is concentrated and filtered. The bacteria-free filtrate of tuberculin contains bacterial disintegration products, substances formed by the action of the bacilli on the culture medium, and concentrated culture medium.

The tuberculin used today in the tuberculin test is known as PPD (purified protein derivative). It consists chiefly of the active principle of tuberculin without extraneous matter. It is more stable than OT and gives less variable results.

Sources and modes of infection. The sources of infection in tuberculosis are the sputum of patients with pulmonary disease and discharges from other tuberculous foci. The patient discharging tubercle bacilli in the sputum is by far the most important reservoir of infection. The bacilli have no natural existence outside the body and are transmitted from source to destination by some form of direct or indirect contact. Droplet infection is the most common mode of spread. Droplets from the mouth of the tuberculous patient or dust containing the partly dried but still living bacilli are inhaled. The size of the infectious droplet in tuberculosis has been determined at 5 to 10 μm; such a particle can remain suspended indefinitely. It is thought that just one bacillus can cause disease.

The next most common mode is the transfer of the bacilli to the mouth by contaminated hands, handkerchiefs, or objects. The infant crawling on the floor may contract tuberculosis by contaminating his hands with tuberculous material and then placing them in his mouth. Unless properly washed and sterilized, the eating utensils used by a tuberculous patient may be a source of infection. The milk of a tuberculous mother may rarely convey the infection to her nursing child.

The avenues of exit for the tubercle bacilli depend on the part of the body infected. In pulmonary tuberculosis they are cast off in the sputum, although tubercle bacilli may sometimes be found in the feces because of swallowed sputum. In intestinal tuberculosis they are discharged in the feces, and in tuberculosis of the genitourinary system they appear in the urine. Tubercle bacilli may be found in exudates from abscesses and in lesions of the lymph nodes, bones, and skin.

Pathogenicity. Tuberculosis is the chronic granulomatous infection caused by *M. tuberculosis.* Although almost any tissue or organ of the body may be affected, the parts most often involved are the lungs, intestine, and kidneys in adults and the lungs, lymph nodes, bones, joints, and meninges in children.

Pathology. The reticuloendothelial system of the body provides the main line of cellular defense, and the term *tuberculosis* is derived from the small nodules of reticuloendothelial cells (tubercles), the unit lesions produced by *M. tuberculosis.*

PRIMARY TUBERCULOSIS (CHILDHOOD TYPE). When tubercle bacilli first enter the body, the ensuing infection lasts several weeks, during which time the individual develops considerable immunity and a state of specific hypersensitivity to the tubercle bacillus. As a result, his tuberculin test becomes positive and remains so for his lifetime. The sequence of events in the first infection is termed the *primary complex,* and it may occur in the lungs (most commonly), in the intestinal tract, in the posterior pharynx, or in the skin (rarely). In the average person the primary complex is benign on the whole. The lesions in the lung heal or become latent without treatment. Most show residual calcification.

A feature of the primary complex is the extension of tubercle bacilli to the regional lymph nodes. With a primary complex in the lung the chest nodes are infected as are the mesenteric nodes from the mucosal focus of the small bowel and the cervical lymph nodes from the mucosal focus in the tonsils, throat, or nasopharynx.

THE LESIONS. As a consequence of the hypersensitive (allergic) state developed, lesions in sensitized tissues are seen that are typical of tuberculosis and not exactly duplicated in any other disease.

The reticuloendothelial system responds to the presence of tubercle bacilli by sending out macrophages to engulf them and to form restricting barriers around them. In their attack on the bacilli the macrophages form firm, round or oval, white, gray,

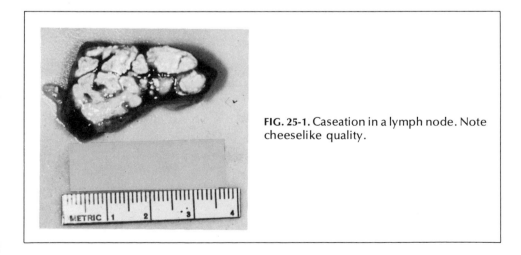

FIG. 25-1. Caseation in a lymph node. Note cheeselike quality.

or yellow nodules from 1 to 3 mm in diameter, which are the *tubercles*. Microscopically a tubercle contains tubercle bacilli, epithelioid cells, and giant cells surrounded by a narrow band of lymphoid cells and encapsulated by fibrous tissue. Epitheloid cells are macrophages altered on contact with the fatty substances contained in the tubercle bacillus. The merging together of many macrophages results in a giant cell.

A tubercle may enlarge singly, or a number of small tubercles may coalesce to form a *conglomerate* tubercle. Because of the cellular reaction at the periphery of the tubercle, blood vessels are compressed and the circulation impaired. This, plus the lethal action of the tubercle bacilli themselves, leads to a peculiar type of change in the center of the tubercle known as *caseation* (Fig. 25-1), from the dry, granular, cheesy quality of the dead tissue.

Such a caseous focus may remain unchanged for a long time, or it may be altered pathologically. It may calcify. Calcification, the deposit of lime salts in dead tissue, is a reparative process. It may be organized, in which event the dead tissue is partially or completely replaced by scar. On the other hand, the caseous focus may soften, enlarge, expand, and even dissect through the adjacent tissues. In the lung, progression of the focus in time erodes the wall of a bronchus. The soft caseous material sloughs into its lumen, leaving an ovoid, scooped-out area termed a *cavity*. Near a body surface, such as the mucosal surface of the intestinal tract, a similar progression of a caseous focus with slough of its contents onto that surface results in the formation of an ulcer.

CHRONIC TUBERCULOSIS. Chronic tuberculosis (adult type) is not thought to come from reinfection from the outside (exogenous) but from progression of the primary infection (primary complex). In some instances, after a period of latency following primary infection, there is reactivation of dormant foci.

Pulmonary tuberculosis (Fig. 25-2) is the usual form in the hypersensitive adult, involving the upper and posterior portion of the upper lobe of the lung, especially on the right side. This is *apical tuberculosis*. A cavity may form, enlarge, and become secondarily infected. Secondary infection, probably caused by streptococci and staphylococci, contributes to the hectic fever occurring in tuberculosis. Hemorrhage comes from erosion of a blood vessel in the wall of the cavity.

The body's attempts to check and heal a progressively destructive tuberculous focus in the lung (and elsewhere) are reflected in a buildup of fibrous and scar tissue. Scar tissue not only replaces and eliminates functioning tissue, but as it contracts, it further distorts and damages, oftentimes even relatively uninvolved tissue.

Tuberculous pneumonia results from the sudden spilling of tuberculous exudate into the sensitized air sacs of a large lung area. It is an acute dramatic manifestation of tuberculous allergy, clinically giving the picture of "galloping consumption."

Tuberculous pleurisy or pleuritis is secondary to tuberculosis of the lungs. It is an inflammation of the pleural membrane of the lung accompanied by a collection of fluid in the pleural cavity (pleural effusion). *All unexplained pleural efffusions should be considered tuberculous until proved otherwise.*

SPREAD. Tubercle bacilli may be spread in the lymph or bloodstream to different parts of the body. This is the usual way. Tuberculous infection may also permeate

FIG. 25-2. Pulmonary tuberculosis, section of lung, with large cavity at apex and numerous grayish white tuberculous foci throughout the rest of the lung. (From Anderson, W. A. D., and Scotti, T. M.: Synopsis of pathology, ed. 9, St. Louis, 1976, The C. V. Mosby Co.)

adjacent tissues, move along natural passages (from kidney to bladder via ureter), and expand over a surface. Occasionally material heavily laden with tubercle bacilli (example, the liquefied center of a tubercle) is discharged into a blood vessel and infection seeded widely over the body. Many small tubercles resembling millet seeds form in the lungs, spleen, liver, and various organs. This is *miliary tuberculosis*.

Tuberculosis of the intestine and regional mesenteric lymph nodes in adults is secondary to pulmonary tuberculosis, developing because human tubercle bacilli are in sputum that is swallowed. (In children intestinal tuberculosis is usually a primary infection from consumption of bovine bacilli in unpasteurized milk.) Tubercles and caseous foci involve the lymphoid tissue of the lower end of the ileum and cecum, leading to the formation of ulcers on the mucosal surface that tend to encircle the bowel wall. Perforation is uncommon, but adhesions form on the peritoneal surface of the bowel at the ulcer sites.

Tuberculous meningitis is a well-marked, acute allergic inflammation of the meninges associated with the formation of tubercles, especially in those covering the base of the brain. It is usually a disease of childhood but may occur in adults. It may appear after generalized miliary tuberculosis or result from bacilli brought to the meninges by the bloodstream from distant foci. Tubercle bacilli may extend directly from adjacent tuberculous foci in the brain or the bones of the skull and spinal column.

Laboratory diagnosis. Direct microscopic examination of suspect material after it has been stained with an acid-fast stain is the first method to be used in the laboratory diagnosis of tuberculosis. *M. tuberculosis* is the only acid-fast organism consistently found in sputum. The failure to find tubercle bacilli in sputum does not rule out pulmonary disease, since the bacilli may not appear there until the disease is advanced. Moreover, they may occur plentifully in one specimen and be scanty or absent in the next. Tubercle bacilli are especially hard to find in smears of urine, cerebrospinal fluid, pleural fluid, joint fluid, and pus because of the relatively small number of organisms present, sometimes even with advanced disease. Various methods of concentration are applied to the test material, especially sputum, if direct smears fail to reveal the bacilli.

The second step in the laboratory diagnosis of tuberculosis is the culture of suspicious material on special media devised for the growth of the tubercle bacillus, regardless of whether acid-fast organisms were demonstrated in the stained smear. The best media are (1) the egg-containing ones, which are opaque, and (2) those made of oleic acid agar, which are translucent. Earlier detection of colonies is possible on translucent media. Cultural methods require a number of days or weeks for the dry, crumbly, colorless colonies to form, sometimes even 10 to 12 weeks. When bacilli are not demonstrated directly in suitably prepared smears, they may frequently be found in cultures or by animal inoculation, the third step in laboratory diagnosis.

Some of the test material is injected subcutaneously into the groin of the extremely susceptible guinea pig.* If tubercle bacilli are present, the guinea pig be-

---

*Guinea pigs are very susceptible to both the human and bovine tubercle bacilli but are unaffected by the avian bacilli.

comes infected; the tissues about the site of inoculation thicken and may ulcerate; the inguinal lymph nodes enlarge; a generalized tuberculosis develops; and the animal dies about 6 weeks later. As a rule, the guinea pig is not allowed to die but is killed at a stated time and an autopsy performed when the disease is known to be far advanced.

Infection of a guinea pig with a given acid-fast organism is used to indicate the virulence of that organism; this is the *guinea pig virulence test.* If it produces disease in the animal, it is, for practical purposes, the pathogenic *M. tuberculosis.* If it fails, it is a nonpathogenic acid-fast organism, with certain important exceptions (p. 449).

Note: the bovine bacillus *(Mycobacterium bovis)* is more pathogenic for ordinary laboratory animals than is the human bacillus. When inoculated into a rabbit, the bovine bacillus kills the animal in 2 to 5 weeks, whereas the human bacillus kills it in about 6 months; in some cases death does not occur at all.

There is presently a fluorescent antibody test in human tuberculosis that is applied to the serum (indirect fluorescent antibody).

Tuberculin tests. The *tuberculin* test depends on the fact that persons infected with tubercle bacilli are allergic to the tubercle bacillus and its products (tuberculin). If a scratch is made on the arm of a person at some time infected and tuberculin rubbed into the scratch (the *von Pirquet test*) or if some of the diluted tuberculin is injected *between the layers* of his skin (the *Mantoux test*), an area of redness and swelling appears at the site. If the person has never been infected, no reaction occurs. In the patch test (the *Vollmer test*) a drop of ointment containing tuberculin or a small square of filter paper saturated with tuberculin is placed on the properly cleaned skin and held in place with adhesive tape for 48 hours. A positive test is indicated by redness and papule formation at the site of application.

The multiple-puncture technic (tuberculin *tine test*) applies tuberculin *transcutaneously.* Tuberculin four times the standard strength of old tuberculin is dried onto the tines of a small, specially constructed metal disk, backed by a plastic holder. It is packaged commercially as a sterile disposable unit. After the skin has been

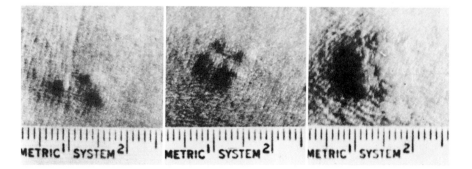

FIG. 25-3. Tine test—typical reactions with old tuberculin (OT). In good light and with forearm slightly flexed, make readings at 48 to 72 hours. Inspect and gently palpate site to determine induration. Measure diameter in millimeters of largest single area. Results: 5 mm or more of induration—positive reaction; 2 to 4 mm—doubtful reaction; and 2 mm or no induration—negative. (Courtesy Lederle Laboratories, Pearl River, N.Y.)

cleaned, the disk is applied briefly to the test area with firm downward pressure, allowing the prongs to pierce the skin. Reactions to this technic are seen in Fig. 25-3.

The Mantoux test, performed with serial dilutions of tuberculin beginning with a high dilution (less tuberculin) and descending to lower dilutions (more tuberculin), is considered the most reliable of the tuberculin tests. The dose used in the Mantoux test may be given by jet gun. The intradermal wheal produced should be 6 to 10 mm in diameter. In case-finding, multiple-puncture tests are practical and satisfactory for screening large groups, but a positive tine test, unless strongly reactive, should be confirmed with a Mantoux test.

The appearance of a positive tuberculin reaction indicates that tuberculous infection has occurred. Since many reach adulthood today in this country with a negative reaction, the circumstances under which there is conversion to a positive reaction may be known and the time of infection approximated. Timing is important because in the early stages tubercle bacilli proliferate very actively. Since infection in the infant or young child would be a recent event, a positive tuberculin reaction is a serious finding. For all persons of any age with a positive tuberculin reaction, the best thinking is that a course of antituberculous therapy is advisable for at least 1 year. After a preliminary chest film, the individual is given isoniazid as a single drug (see also p. 288). In an adult where the presence of calcified lymph nodes in the chest indicates a process of long standing, such therapy is recommended because of the 5% hazard that this person will have active tuberculosis during his lifetime.

Positive reactions in the tests just described are known as *local* reactions. If tuberculin is given *beneath the skin,* two other reactions may occur, the *focal* and the *constitutional.* By focal reaction one means acute inflammation around a tuberculous focus in the body. This may have serious consequences; for instance, if the tuberculous focus is in the lung, a hemorrhage may result. By constitutional reaction one means systemic reaction with a sharp rise in temperature and a feeling of malaise lasting for several hours.

Immunity. Whites possess considerable inherent resistance to tuberculosis but never a complete immunity. The incidence in blacks is high, averaging in proportion to population eight cases to one in whites. Not only is this true, but when the black contracts tuberculosis, the average time that he will live, if untreated, is about one-sixth that of the white who contracts it. The American Indian and Mexican are also very susceptible. The incidence of tuberculosis tends to be increased among doctors, nurses, and persons working directly with the disease.

Persons who live in isolated communities seem to be more vulnerable when exposed for the first time during adult life than those who have been reared in closely crowded and highly infected communities.

Not very long ago it was thought that tuberculosis was hereditary and ran in families, but tuberculosis runs in families because closely associated members of the family pass it to each other. Children of tuberculous parents may inherit certain predisposing factors, but a child born of tuberculous parents and at once removed to infection-free surroundings has a better chance of escaping the disease than one born of healthy parents but reared in contact with tubercle bacilli.

The defensive factors that operate in a given infection depend on whether infection has occurred before. Most observers think that a degree of protection against subsequent infections is afforded the individual by the first infection. Whether a given infection means active disease depends on (1) the number of bacilli, (2) their virulence, (3) the state of allergy, and (4) resistance of the subject. Conditions that lower body resistance are malnutrition, crowded housing, and predisposing diseases such as measles, whooping cough, and diabetes mellitus.

Prevention.* Since tuberculosis sputum is the chief source of infection with human bacilli, it should be disposed of carefully. It should not be allowed to dry, for then it may be blown from place to place spreading the germs over a wide area. Sputum should be received in suitable, covered containers and burned. Promiscuous spitting should be taboo. The patient must cover his mouth when he coughs. A tuberculous mother should not nurse her child. The woodwork of a room occupied by a tuberculous patient can be washed with soap and water, and a suitable disinfectant, usually one of the phenol derivatives, applied.

Eradication of tuberculous cows and the universal use of pasteurized milk control infection with the bovine bacillus.

In communities with the best type of health supervision there has been a decided decrease in the incidence of tuberculosis because of several factors, among which are better living conditions and medical prophylaxis. Mass surveys of the population by chest x-ray examination have been of great value in detecting pulmonary tuberculosis and other lung lesions as well. Radiographs of the chest at regular intervals are recommended for individuals whose work brings them into contact with active cases of tuberculosis.

Mycobacterium bovis (the bovine tubercle bacillus)

Mycobacterium bovis (M. tuberculosis var. *bovis)* is a bit shorter and plumper than *M. tuberculosis.* It grows more slowly in culture media than the human bacillus, forming slightly smaller colonies and is niacin test negative. However, its infection cannot be distinguished from that with the human bacillus by the tuberculin test.

Mycobacterium bovis is highly virulent for man. Human infections caused by bovine bacilli were once very common in children and young people and affected the cervical lymph nodes, intestines, mesenteric lymph nodes, and bones. *Scrofula,* tuberculosis of the lymph nodes of the neck, is typically part of the primary complex produced by the bovine bacillus in this area. Bovine bacilli cause pulmonary tuberculosis in cattle and can cause it in man. However, practically all pulmonary infections in children and adults result from the human organism.

Infection is acquired from the milk of tuberculous cows. As a rule, the bacilli get into milk by fecal contamination. In the cow tuberculosis usually attacks the lungs, but since the cow swallows her sputum, the bacilli are excreted in the feces. The udder and flanks of the cow become contaminated, and the bacilli gain access to the milk, which, if consumed unpasteurized, is the source of infection. Sometimes the

*For BCG vaccination, see pp. 684 and 702.

bacilli are excreted directly into the milk as it comes from an infected udder. This may occur without demonstrable tuberculous lesions in the udder.

The tuberculin reaction is of inestimable value in detecting tuberculous cows. Routine tuberculin testing of cattle by the U.S. Department of Agriculture and elimination of infected cattle from the herd, plus the pasteurization of milk have practically eliminated bovine infection in the United States. In countries where no such safeguards exist, the bovine bacillus is responsible for 15% to 30% of the cases of tuberculosis in young ages.

LEPROSY (HANSENOSIS)

Leprosy is a chronic communicable disease caused by *Mycobacterium leprae* (Hansen's bacillus)* that involves skin, mucous membranes, and nerves. Leprosy has affected man from the beginning of history, and descriptions of it occupy a prominent place in the Old Testament and other ancient writings. It is estimated that presently there are about 15 million leprosy patients† in the world with approximately 2000 in the United States.

Mycobacterium leprae (the leprosy bacillus)

Mycobacterium leprae (M. leprae) is an acid-fast microbe closely resembling *M. tuberculosis*. It occurs abundantly in the lesions of leprosy. The bacillus of leprosy has been cultivated in the ears and rear footpads of white mice and hamsters, sites chosen as the largest and coolest parts of the experimental animal. After many months mild changes detectable with the microscope occur at the sites of inoculation, but there is no gross advanced disease. The bacillus has also been grown in cell culture.

Mode of infection. The imagination of man has clothed leprosy with many attributes it does not possess. One is that it is a highly communicable disease. This is not true. Man contracts leprosy only after prolonged and intimate contact, and even then he often escapes infection. The bacilli leave the body in great numbers from degenerating lesions of the skin and mucous membranes, and spread is directly from person to person. Nose and mouth discharges are especially dangerous because lesions are very common in these locations. The portals of entry are probably skin and mucous membranes. Small breaks in the skin may admit organisms discharged from a patient with the disease. Although the bacteria prefer certain cool sites, they are spread widely over the body. They may be found in feces and urine. The disease is not hereditary, but since children are unusually susceptible, the percentage of infection in children associated with leprous parents is very high (30% to 40%). The highly variable incubation period is estimated at from 5 to 15 years.

Pathogenicity. Leprosy belongs with the infectious granulomas (diseases with a defensive multiplication of reticuloendothelial cells at the sites of infection). Al-

*Discovered in Norway by Gerhard A. Hansen (1841-1912) in 1874, *Mycobacterium leprae* was actually the first bacterium identified as a cause of human disease. For nearly a century it remained the only one known to infect man that could *not* be cultured in the laboratory and that did not produce progressive disease if inoculated into a test animal.

†The International Leprosy Association has unanimously resolved that the term *leper* be abandoned and that the person suffering with the disease be designated the "leprosy patient."

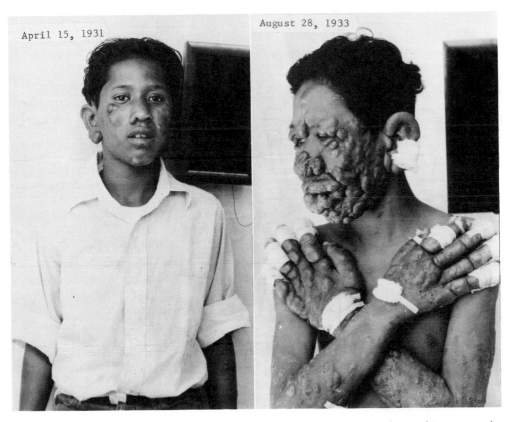

April 15, 1931

August 28, 1933

FIG. 25-4. Leprosy. Progression of disease in a teenaged Filipino. Note dates; this occurred before sulfone drugs were available. (Courtesy Dr. C. Binford, Washington, D.C.)

though leprosy is generalized with a variety of changes, there are two conspicuous patterns: (1) the *lepromatous* or *nodular* form, marked by tumorlike overgrowths of the skin and mucous membranes, and (2) *tuberculoid* or *anesthetic* leprosy, manifest by involvement of peripheral nerves with localized areas of skin anesthesia. The end stages of the disease are associated with extensive deformity and destruction of tissue (Fig. 25-4).

The earliest lesion in leprosy may be diagnostic in that it contains acid-fast bacilli. Since it may not indicate the subsequent course, it is referred to as an *indeterminate* lesion. *Dimorphous leprosy* refers to the coexistence of the two forms—*lepromatous* and *tuberculoid.*

Laboratory diagnosis. The organism *M. leprae* occurs abundantly in the lesions of leprosy. A diagnosis of leprosy is most often made with scrapings from the nasal septum or biopsies of the skin of the ear that have been stained for acid-fast bacilli. Fluorochrome staining (Fig. 25-5) can be done on leprae bacilli. The bacilli in large numbers are found within phagocytic cells packed together like packets of cigars. Bacilli may also be demonstrated in acid-fast stained material from lesions elsewhere in the skin and lymph nodes. Leprosy and tubercle bacilli are separated by guinea pig inoculation, since leprosy bacilli have no effect on the animal.

447

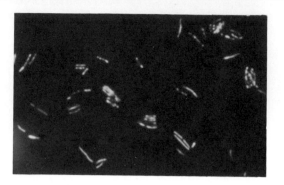

FIG. 25-5. *Mycobacterium leprae,* fluorescent staining. Note morphology. (×1000.) (Courtesy Dr. R. E. Mansfield, Canton, Ill.)

FIG. 25-6. Armadillo (nine-banded), *Dasypus novemcinctus* Linn. (Courtesy Dr. Eleanor E. Storrs, New Iberia, La.)

No animal shares a susceptibility to leprosy comparable to man except the armadillo (Fig. 25-6). The discovery has been made only recently that this creature develops leprosy (analogous to lepromatous leprosy in man) some 3 years after initial infection, and it plays host to an enormous profusion of mycobacteria. The number of lepra bacilli in its infections is 1000 to 10,000 times greater than that in advanced human disease. The experimental animals also succumb to a more severe disease, showing involvement of the central nervous system and lungs, areas spared in human beings.

Lepromin is a suspension of killed leprosy bacilli used in a skin test (the Mitsuda test) to detect susceptibility to this disease. The armadillo is a good source of lepromin.

Prevention. Leprosy is best prevented by good living conditions. Prolonged and intimate contact with leprosy patients is hazardous. Strict isolation with segregation has not proved completely successful as a method of control, but most persons in the United States discharging *M. leprae* are referred to the national leprosarium, the United States Public Health Service Hospital at Carville, Louisiana.

A patient who has improved and has failed to discharge the bacilli for a period of 6 months may be paroled. Paroled patients should be examined twice yearly. Back home the patient should be semisegregated with his own room, linens, and dishes. His discharges, cooking utensils, and linens should be carefully disinfected. No children or young people should live in the house. Children of leprosy patients should be separated from their parents at birth because the chances of infection are much greater in infants and young children. All persons who contact a patient should be examined every 6 months.

Results of clinical trials in eastern Uganda in Africa indicate that BCG vaccine protects children against leprosy. The possibility of such a surrogate vaccine strengthens the case for a strong cross immunity among mycobacteria.

MYCOBACTERIOSIS

Mycobacteriosis, or atypical tuberculosis, refers to any tuberculosis-like disease caused by mycobacteria other than *M. tuberculosis.*

Other mycobacteria (atypical, anonymous, unclassified mycobacteria)*

The organisms. In tuberculosis hospitals unusual acid-fast bacilli are sometimes found in the sputum of patients with a typical setting for tuberculosis and cavitary disease by x-ray examination. Although they cause pulmonary disease mimicking tuberculosis *(mycobacteriosis)*, these acid-fast organisms differ sharply from the tubercle bacillus. They also cause infection in lymph nodes and skin. Infections are usually chronic, variably destructive granulomas.

Classification. These mycobacteria are differentiated from *M. tuberculosis* by their colonial characteristics, biochemical reactions,† susceptibility to antituberculous drugs, and animal pathogenicity. Although species identification is important here, these organisms are commonly considered in four categories (Table 25-1). In group I the photochromogens produce yellow-orange colonies in the light. *Mycobacterium kansasii* ("yellow bacillus"), the representative of this group, is a respiratory pathogen. Group I members are most closely associated with human pulmonary disease.

The scotochromogens of group II produce a yellow to orange pigment in the dark as well as in the light. *Mycobacterium scrofulaceum* ("tapwater bacillus") is the example. It causes cervical lymphadenitis in children.

Not affected by light, the nonphotochromogens of group III form buff-colored colonies, soft in consistency by comparison with the rough ones of the tubercle bacillus. These include the Battey bacillus *(Mycobacterium intracellulare)* and other respiratory pathogens.

The rapid growers of group IV form colonies on simple media within a short

---

*"MOTT" bacilli (*m*ycobacteria *o*ther *t*han *t*ubercle bacilli). They have been referred to as anonymous or unclassified because most of them up until recently were not adequately identified to permit assignment to a definite species.
†In the clinical setting, production of niacin (p. 125) by an acid-fast bacillus identifies it as the human tubercle bacillus.

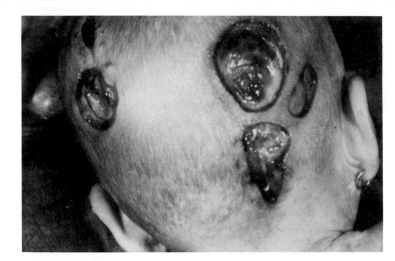

FIG. 26-2. Skin ulcers on head of young child with systemic *Pseudomonas aeruginosa* infection.

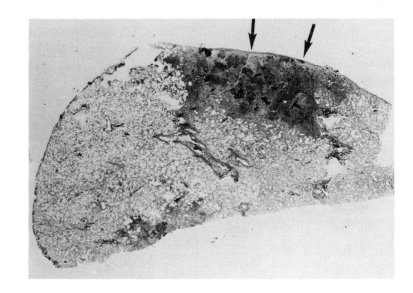

FIG. 26-3. Bronchopneumonia caused by *Pseudomonas aeruginosa*. A well-defined, solidified, darkened portion is sharply distinct in thin slice of air-containing lung. Focus of disease is hemorrhagic and partly necrotic.

are susceptible. Since *Pseudomonas aeruginosa* is resistant to the action of most antibiotics, it tends to become the dominant organism in a diseased area after extended antimicrobial therapy has eliminated the primary ones. Then it takes on a vigorous disease-producing capacity and is responsible for significant destruction of tissue, especially with lowered host resistance. Microscopically, there is massive overgrowth of the organisms in the areas of dead tissue.

The sources of *Pseudomonas aeruginosa* in hospital-acquired infections are many, but especially the pieces of equipment that are hard to clean and sterilize, such as face masks, certain kinds of humid rubber tubing, water containers, nebulizers, and parts of the intermittent positive-pressure breathing machines. A source easily overlooked is the hospital supply of distilled water. This is a serious hazard because of the widespread need for distilled water in preparation of various hospital solutions—detergents, disinfectants, and even parenteral medications. Pseudomonads have been demonstrated not only to survive but also to multiply rapidly in *distilled* water and seem to have a greater resistance to antimicrobial agents than the ones recovered from laboratory media.

Patients with 40% or more burned surface area are vulnerable to *Pseudomonas* sepsis. Administration of a heptavalent lipopolysaccharide *Pseudomonas* vaccine to these seriously burned persons is a great advance in the management of their injury.

Pseudomonas pseudomallei

The pseudomonad, *Pseudomonas pseudomallei*, is a gram-negative, motile, nonpigmented bacillus that causes *melioidosis*, an uncommon tropical disease of man and animals. Its pulmonary manifestations are easily confused with those of tuberculosis. There may be a septicemia, with widespread lesions over the body. Nicknamed the "Vietnamese time bomb," the disease can be dormant for a number of years only to appear suddenly, become active, and produce death within days or weeks. The disease is endemic in Southeast Asia, where the organism is found in surface waters, soil (notably the rice fields), and organic matter. The epidemiology and transmission of the disease are not known. The organisms are thought to enter the body by way of the mouth and nose or through open wounds in the skin.

Growth of the bacilli is typically crinkly in cultures with a characteristic odor. A hemagglutination test is available for diagnostic study.

Pseudomonas mallei

Glanders (farcy) is chiefly a disease of horses, mules, and donkeys, which may be transmitted to man. Cattle are immune. It is characterized by the formation of ulcerating, tubercle-like nodules in the lungs, superficial lymph nodes, and mucous membranes. If lymph nodes are affected, the disease is known as *glanders;* if mucous membranes are affected, it is *farcy.* With the replacement of the horse by the automobile, this once prevalent disease is now rare.

The cause, *Pseudomonas (Actinobacillus) mallei*, is a narrow, sometimes slightly curved, small bacillus that is gram negative, nonmotile, nonencapsulated, and nonsporeforming, with little resistance to physical and chemical agents. Man is easily infected if he contacts tissues or excreta of diseased animals, since these contain virulent bacilli. The microbes enter the body by a wound, scratch, or abrasion of the skin.

Laboratory diagnosis. Glanders is diagnosed in man and animals by the isolation of the bacilli from the lesions or in blood cultures. There is a complement fixation test. The skin test utilizes *mallein,* a product of *Pseudomonas mallei,* injected subcutaneously in animals. Mallein tablets are placed in the conjunctival sac. If infectious material is injected intraperitoneally into a male guinea pig, the testicular swelling and generalized reaction within 3 or 4 days give the *Straus reaction*.

Prevention. No immunity exists to glanders. Control depends on destruction of animals with clinical or occult disease and on disinfection of stables, blankets, harnesses, and drinking troughs used by sick animals.

VIBRIO SPECIES

In the family Vibrionaceae are rigid, gram-negative, straight or curved rods, facultatively anaerobic, found in fresh water and seawater. In its member genus *Vibrio* are the agents of cholera and vibriosis.

Vibrio cholerae (the comma bacillus)

The organism. Cholera is caused by *Vibrio cholerae (V. cholerae)*, the comma bacillus, a small, comma-shaped, motile, gram-negative rod. It multiplies rapidly in the lumen of the small bowel and produces a powerful exo-enterotoxin that acts on the lining of the bowel to induce copious loss of water and essential salts. This explains the fecal discharges being described as "rice water"—clear, not malodorous, with flecks of mucus. Organisms abound in the stools. There are three immunologic types, the Inaba, Ogawa, and Hikojima.

The cholera vibrio grows aerobically on routine laboratory media. If a few drops of sulfuric acid are added to a growth of cholera vibrios in nitrate-peptone broth, a red color develops, the *cholera red reaction.*

The disease. Asiatic cholera is a specific infectious disease that affects the lower portion of the intestine and is described by violent purging, vomiting, burning thirst, muscular cramps, suppression of urine, and rapid collapse. Untreated, it can be a terrifying disease with a 70% or more mortality. Only plague causes as much panic. With massive diarrhea the patient's fluid losses are enormous—10 to 20 quarts a day. With severe rapid dehydration death comes within hours. Four great pandemics spread over the world during the eighteenth century, and on two occasions in the nineteenth century the disease invaded the United States: in 1832 in New York City and in 1848 in New Orleans, whence it spread up the Mississippi Valley. The disease is endemic in India and China. The "scourge of antiquity," it originated in India in the vicinity of Calcutta and on the delta of the Ganges River, where it was known at the time of Alexander the Great.

Mode of infection. The disease is contracted by the ingestion of water or food contaminated by the excreta of persons harboring the bacilli.* Man is the only host. The bacilli leave the body in the feces, urine, and secretions of the mouth. As a rule, the feces become free of bacilli during the last days of the disease, but some patients become convalescent carriers. Permanent carriers do not occur.

---

*John Snow (1813-1858), English epidemiologist, was the first to recognize the transmission of cholera by contaminated water, and by the middle of the nineteenth century he had formulated his views. He correctly observed in 1854 that the Broad Street Pump in London was a source of infection in an epidemic of cholera killing some 11,000 persons.

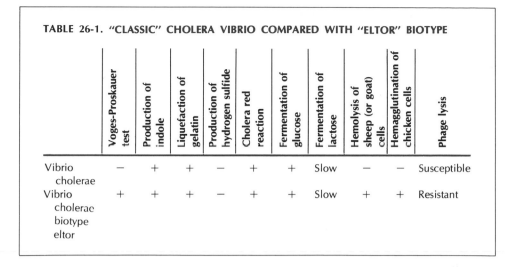

TABLE 26-1. "CLASSIC" CHOLERA VIBRIO COMPARED WITH "ELTOR" BIOTYPE

	Voges-Proskauer test	Production of indole	Liquefaction of gelatin	Production of hydrogen sulfide	Cholera red reaction	Fermentation of glucose	Fermentation of lactose	Hemolysis of sheep (or goat) cells	Hemagglutination of chicken cells	Phage lysis
Vibrio cholerae	−	+	+	−	+	+	Slow	−	−	Susceptible
Vibrio cholerae biotype eltor	+	+	+	−	+	+	Slow	+	+	Resistant

Cholera today is rampant in the areas of the world occupied by better than half the world's population where, because of low standards of sanitation, human excrement easily contaminates the waterways and surface wells that are sources of drinking water. Vibrios may live in water for as long as 2 weeks.

As a disease cholera should be only a historical note. Modern sanitation can eliminate waste-contaminated water supplies, the primary source of infection, and modern medical treatment can effectively deal with it.

Immunity. The immunity occurring after an attack of cholera is short-lived. The risk of reinfection is only slightly less than that of the initial infection. Although vaccination is done against cholera,* to date, a truly effective vaccine is still not widely available. There is an experimental toxoid in field trial.

Vibrio cholerae biotype eltor

A biotype to *V. cholerae,* more resistant to physical and chemical agents and in itself quite virulent, is *Vibrio cholerae* biotype *eltor (Vibrio El Tor).* Unlike the classic one, the El Tor vibrio produces a hemolysin that lyses the red cells of sheep and goats. (See Table 26-1 for comparison of the two.) It causes cholera and in recent times has been more significant than the classic vibrio as the cause of pandemics in South and Southeast Asia. The current one is the seventh resulting from this biotype.

Vibrio parahaemolyticus

A vibrio like the classic one, *Vibrio parahaemolyticus* is a major cause of a gastroenteritis with manifestations similar to the classic disease (a vibriosis). In countries such as Japan it is responsible for 50% to 60% of cases of "summer diarrhea." Widely distributed as a marine microorganism, it is an important cause of "food poisoning" where seafood has been consumed, and it can infect wounds incurred along the sea-

*For cholera immunization, see pp. 684, 706, and 707.

shore. It is a small halophilic vibrio giving a negative cholera red reaction. Present vaccines are of no effect against it.

BACILLUS ANTHRACIS (THE ANTHRAX BACILLUS)

Known since antiquity, anthrax (charbon) is an acute infectious disease caused by *Bacillus anthracis (B. anthracis)*. It is primarily a disease of lower animals, especially cattle and sheep, but it is easily communicated to man. Horses and hogs may become infected. Anthrax is worldwide but, with the exception of certain restricted areas, is unusual in the United States.

In man anthrax presents in two forms: *external* (malignant pustule or carbuncle and anthrax edema) and *internal* (pulmonary anthrax or woolsorters' disease and intestinal anthrax). External anthrax is the more common. Because of the fever and enlargement of the spleen that accompany the disease, anthrax is often called splenic fever. The incubation period for external anthrax is 1 to 5 days.

In the body of an animal dead of acute disease, there is little change other than the dark blood and a swollen spleen.

General characteristics. The organism *B. anthracis* is quite large, in fact one of the largest bacterial pathogens. It forms spores and is gram positive. The swollen and concave ends of the bacilli give the chains of bacilli the appearance of bamboo rods. The organism grows in the presence of oxygen and also in its absence. It is the only sporeforming aerobic pathogenic bacterium and, unlike most sporeforming aerobic bacteria, is nonmotile. Since spores are not formed in the absence of oxygen, they are not formed in the animal body. In the blood and tissues of infected animals, the anthrax bacillus is surrounded by a capsule. Growth is luxuriant on all ordinary culture media. Spores retain their vitality for years and are extremely resistant to heat and chemical disinfectants in the concentrations ordinarily used. They are fairly susceptible to sunlight.

B. anthracis is of historic interest because it was the first pathogenic organism to be seen under the microscope, the first one *proved* to be the cause of a specific disease, the first one grown in pure culture, and the organism that Pasteur used in his classic experiments on artificial immunization.

Modes of infection. Animals usually become infected via the intestinal route while grazing in infected pastures. Buzzards carry the infection long distances and contaminate soil and water with organisms on their feet and beaks. Dogs discharge spores in their feces after eating the carcasses of infected animals. This disease is sometimes transmitted from animal to animal and from animal to man by greenhead flies and houseflies.

In man, anthrax is primarily an occupational disease confined to those who handle animals and animal products, such as hair and hides. Imported animal products are of special danger. The commonest route of infection is through wounds or abrasions of the skin (cutaneous route of infection). In numerous cases of anthrax, infection of the face by the bristles of cheap shaving brushes has been reported. The pulmonary form of anthrax is caused by inhalation of dust containing the spores (respiratory route of infection) and endangers those who handle dry hides, wool, or hair. The intestinal form of anthrax is acquired by ingestion of infected milk or insufficiently cooked food (intestinal route of infection). The bacilli leave the body in the exudate of the local lesion (malignant pustule) and in the sputum, feces, and urine.

Prevention. Patients with anthrax should be isolated. The dressings of external lesions should be burned, and the person doing the dressings should wear gloves. The feces, urine, sputum, and other excreta should be disinfected at once to prevent the formation of spores. The disinfectant should be strong and applied for a long time. Phenol, 5%, is probably the best one. The local lesions of anthrax should not be traumatized as by squeezing because this may cause the bacilli to invade the bloodstream.

The possibility of confusing a malignant pustule with an ordinary boil is so important that a description of the former is not out of place. A malignant pustule begins as a small, hard, red area; a small vesicle soon develops in the center. The area increases in size, vesicles develop at the periphery, and the surrounding tissues become swollen. The center of the lesion softens, and a dark eschar forms. Pain is not present. The draining lymph nodes are swollen.

Infected animals should be separated from the herd. The bodies of animals dead of anthrax should be completely burned to ashes. Cremation prevents coyotes, bobcats, dogs, and other scavengers from catching the disease. Their blood should not be allowed to escape, and their bodies should not be opened for autopsy except by an experienced veterinarian because the bacilli form spores when exposed to the air. If the body is buried, it should be packed with lime and buried at least 3 feet in the ground. When infection occurs in a herd, the pasture must be changed, and the herd quarantined. A sharp watch should be kept for new cases, and the quarantine should not be lifted until 3 weeks after the last case. Cattle that have been vaccinated and survive an epidemic must not be taken to slaughter for 42 days. Milk from infected herds should not be used. Hides, hair, and shaving brushes should be sterilized. If the soil becomes contaminated, it remains infectious for years. The best soil disinfectant is lye. Anthrax spores from buried animals have been brought to the surface by earthworms.

Prophylactic vaccination of animals against anthrax must be carried out with vigor. Cattle and sheep may be actively immunized with a vaccine that contains living but attenuated bacteria. Dead bacteria are without effect. A therapeutic serum has been prepared by immunizing horses against anthrax bacilli.

An anthrax vaccine prepared from a culture filtrate of an avirulent, nonencapsulated strain is available for laboratory workers at risk and for workers in occupations exposed to the disease.

LACTOBACILLUS SPECIES (THE LACTOBACILLI)

The lactobacilli are gram-positive, nonsporulating, microaerophilic rods that produce lactic acid from simple carbohydrates and preferentially grow in a more highly acid environment (pH 5) than most other bacteria. They are widely distributed in nature and out of character as disease producers. Many are normal inhabitants of the alimentary tract; most are nonmotile. Since they require many essential

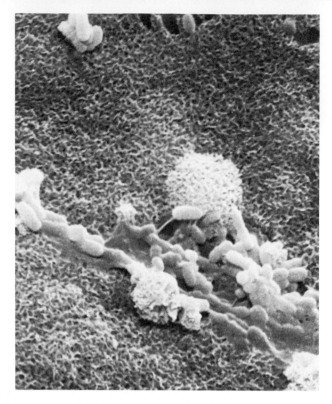

FIG. 26-4. *Streptococcus mutans* in experimental cariogenesis, scanning electron micrograph. Note cocci attached to toothlike surface, "mesh" of which corresponds to salivary pellicle, the thin surface film of adsorbed glycoproteins from saliva. After bacteria caught to pellicle were incubated in 1% glucose, they synthesized glucan, the extracellular glue of the micrograph. (Courtesy Dr. W. B. Clark, Boston, Mass.)

Plaque (microbial growth, debris, desquamated cells)

Dentin (ivory of tooth)

Thin layer of cementum covering dentin

FIG. 26-5. Early dental decay (plaque), microscopic section. Microbial growth is associated with debris, mucus, and desquamated cells. Action of acid-producing bacteria, including lactobacilli, on sugars in the food results in an accumulation of lactic acid, believed to trigger process of decay by decalcifying enamel of tooth. (From Bhaskar, S. N.: Synopsis of oral pathology, ed. 3, St. Louis, 1969, The C. V. Mosby Co.)

growth factors for their metabolism, they are used commercially in the bioassay of the B complex vitamins and certain amino acids. They are a more sensitive index than current chemical methods. They are also used extensively in the fermentation and dairy industries.

Some noteworthy members* of the genus are *Lactobacillus delbrueckii, Lactobacillus acidophilus, Lactobacillus bulgaricus,* and *Lactobacillus casei. Lactobacillus delbrueckii* is the type species. *Lactobacillus acidophilus,* a normal inhabitant of the intestinal tract, is increased by a diet rich in milk or carbohydrates. The *Boas-Oppler bacillus,* found in the stomach in gastric cancer, is most likely the same organism. *Lactobacillus bulgaricus* was isolated from Bulgarian fermented milk and is the organism that Metchnikoff thought would prevent intestinal putrefaction and thereby increase length of life. *Lactobacillus casei* is most often used in the bioassay of vitamins and amino acids. *Lactobacillus* species make up the normal flora of the vagina during a woman's active reproductive years, where they are referred to collectively as *Döderlein's bacilli.*

Dental caries. Dental caries (tooth decay) is one of the most widespread disorders; it is estimated that every American has at least three. The factors implicated in the pathogenesis of caries and determining their severity are (1) individual susceptibility, (2) presence of bacteria capable of producing organic acids, and (3) presence of carbohydrates to support the cariogenic activities of acid-producing microorganisms. Some lactobacilli and certain acid-producing streptococci, notably *Streptococcus mutans,* are found in the mouth associated with tooth decay, and most observers believe that they play a role, especially *Streptococcus mutans,* whose cariogenic property is nicely shown in experimental animals (Fig. 26-4). The first attack on tooth structure is postulated to come from the acid-producing bacteria. *Streptococcus mutans* produces an enzyme, dextran-sucrase, that converts sucrose of food eaten to dextran. Dextran combines with salivary proteins to create on tooth surfaces a sticky, colorless film called a *plaque* (Fig. 26-5). Plaque piles up continuously on teeth, re-forming after removal at any given time. The plaque provides a haven for bacteria from which they can undermine and demineralize the enamel of a tooth. Subsequently with breakdown of tooth enamel, the typical cavity appears. Fragments of debris found therein support continued bacterial growth and thus tend to perpetuate the process of decay.

LISTERIA MONOCYTOGENES

Listeria monocytogenes is a small (less than 2 μm long), gram-positive, aerobic to microaerophilic, nonsporeforming, motile coccobacillus that affects man and a great variety of animals. It causes disease *(listeriosis)* noted for an increase in large mononuclear leukocytes (monocytes) in the bloodstream. It produces an encephalitis or encephalomyelitis, especially in ruminant animals. In man the most prominent manifestation is a purulent meningitis. Infection can also be widely disseminated.

---

*Lactobacillus bifidus, found in the intestine of breast-fed infants, is now classified as *Bifidobacterium bifidum,* the type species of that genus.

Listeriosis is seen as encephalitis, conjunctivitis, endocarditis, urethritis, or septicemia with multiple abscesses. Untreated, it is fatal in 90% of cases. Involving the fetus and newborn infant, *granulomatosis infantiseptica* is an intrauterine listerial infection associated with widespread necrosis of the internal organs and hemorrhagic areas in the skin. The newborn infant usually dies within 2 or 3 days. Although human listeriosis is rare, the number of cases in man and in poultry and livestock is increasing. It is a hazard to the compromised host, especially the baby and the old person. Its infection complicates renal transplantation.

From specimens of blood, spinal fluid, and pus the organism can be identified by cultural methods in the laboratory. It produces beta hemolysis and is novel in that it can grow at the temperature of the refrigerator. A highly specific test for identification utilizes the rabbit; the everted eyelid of the animal is swabbed with the culture. Typically, the reaction to *Listeria* is the development of purulent keratoconjunctivitis. *Listeria monocytogenes* is widely distributed in nature and can be recovered from such diverse sources as man and animal feces, ferrets, insects, sewage, silage, and decaying vegetation. The transmission of listeriosis in man and animals is unknown.

MYCOPLASMA SPECIES

Mycoplasmas are out of the ordinary. They do not have a cell wall, and because they do not, they demonstrate a set of remarkable and unusual features. In the recent edition of *Bergey's Manual*, they are segregated into Part 19. The classification begins with Class Mollicutes to encompass the delicate coccoid to filamentous procaryotic microbes with a "pliable cell boundary." Gram negative, nonmotile, tiny, some are so small as to be ultramicroscopic (about 200 nm). In fact, they are considered to be the smallest free-living units, that is, forms of life capable of independent existence. Some of the spherical ones with a diameter of about 0.35 μm are of a size that would allow 8000 of them to fit inside a human red blood cell. Order I, Mycoplasmatales, and Family I, Mycoplasmataceae, gather the ones needing sterol for continued growth. Mycoplasmas are everywhere. They are saprophytes, parasites, and pathogens for a wide range of hosts.

General characteristics. Mycoplasmas (pleuropneumonia-like organisms, PPLO's) are smaller than ordinary bacteria and about the size of the larger viruses; they are the smallest microorganisms that can be cultivated on cell-free media (Fig. 26-6). They possess soft, fragile cell bodies and lack a rigid cell wall.* Hence they are highly pleomorphic. They contain both DNA and RNA and some metabolic enzymes. Like viruses, they are filtrable and, in the animal body, intracellular parasites. Also they are sensitive to ether and insensitive to many antibiotics. Their ability to induce cytopathogenicity in cell cultures is a nuisance. The type species is *Mycoplasma mycoides*, the cause of pleuropneumonia in cattle.

Pathogenicity. In addition to bovine pleuropneumonia, mycoplasmas cause contagious mastitis in sheep and goats and respiratory disease in poultry. Dogs, rats, and

---

*Since they have no cell wall, they would have to be gram negative.

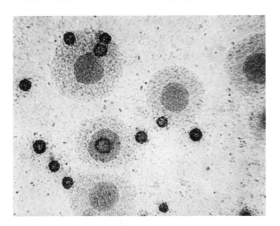

FIG. 26-6. Mycoplasmas. Mixture of classic *(Mycoplasma hominis)* colonies and smaller, very dark colonies of T (T-strain) mycoplasmas. Direct test for urease (that of Shepard and Howard) has been applied to this preparation. Colonies of T-strain mycoplasmas are urease positive, showing dark bronze color of reaction product. Note fried-egg appearance of classic colonies. (×300.) (From Shepard, M. C., and Howard, D. R.: Ann. N.Y. Acad. Sci. **174:**809, 1970.)

mice may become infected with them. In man they are linked to infections of the genitourinary tract and the upper and lower respiratory passages, and rarely to other systemic disorders.

PRIMARY ATYPICAL PNEUMONIA. *Mycoplasma pneumoniae* (the Eaton agent*) is the smallest known pathogen that can live outside of cells. It causes primary atypical pneumonia, an acute febrile, self-limited disease of man that begins as an upper respiratory infection and spreads to the lungs. It was among the most common respiratory infections of World War II, and the Eaton agent can still be recovered from a large number of military recruits. The manifestations of the disease including headache and malaise are fairly severe. The cough is paroxysmal, but little sputum is raised. The lungs may be extensively involved, although in most patients only one lobe is affected. The attack lasts 2 to 3 weeks. Recovery is gradual. The disease, although prevalent, does not occur in widespread epidemics. Epidemics are confined to persons living in crowded conditions and have been reported among newborn infants. Milder respiratory disease or even inapparent infection ("walking pneumonia") may be associated with *Mycoplasma pneumoniae.*

Primary atypical pneumonia is not very contagious, although apparently it is spread directly from one person to another by oral or nasal secretions, and the portal of entry is the upper respiratory tract. The disease is not always followed by immunity.

---

*The Eaton agent was discovered by Monroe Davis Eaton of Harvard University in 1944. Originally grown on chick embryo lung culture, it was thought to be a virus until it was cultured on an artificial agar medium. Dr. Eaton suspected its relation to primary atypical pneumonia, but it was only after fluorescent antibody technics were applied years later that the Eaton agent was proved a cause of this disease.

The classic case of primary atypical pneumonia is associated wtih an increased number of cold agglutinins * in the patient's serum. In the serums of some convalescent patients, agglutinins develop to *Streptococcus MG* (from the name McGuinness of the patient from whom it was originally recovered), an alpha hemolytic streptococcus now classified as *Streptococcus anginosus*. There is also a complement fixation test.

An alum vaccine prepared from formalin-killed organisms has been tested experimentally in animals and man. An experimental intranasal vaccine against *Mycoplasma pneumoniae* has been prepared from temperature-sensitive mutants of a virulent wild strain; it is being field tested.

T-strain mycoplasmas. "T" strains of mycoplasmas, "tiny forms," or "T-form" PPLO colonies represent human mycoplasmas with a distinct growth pattern. T strains produce very small colonies on their agar media (only a few micrometers in diameter). Colonies of the "classic" organisms, first studied, are much larger, many micrometers in diameter, even up to 0.75 mm (Fig. 26-6). (Both "tiny form" and "large form" colonies must be viewed with a microscope.) Moreover, T strains are unique among mycoplasmas in being sensitive to therapeutic agents not affecting the classic ones and in possessing an active urease system. Since their ability to hydrolyze urea sets them apart, a reclassification of T strains has been proposed. A new genus *Ureaplasma* in Family I, Mycoplasmataceae, would recognize this property, and the designation *Ureaplasma urealyticum* would presently contain a single human species with at least eight serotypes.

Mycoplasmas are common parasites of the genital tract. Colonization is related therein to sexual activity. In fact, they are considered part of the normal flora of sexually active males and females. However, there is reason to think that genital mycoplasmas—both T strains and *Mycoplasma hominis*—are not always benign. They are postulated to be agents of venereal infection, producing urethritis and prostatitis in the male and causing cervicitis, cystitis, endometritis, and contributing to infertility and premature births in the female. Classic mycoplasmas are also implicated in so-called reproductive failure.

Laboratory diagnosis. The laboratory diagnosis for mycoplasmas is made by cultural methods adapted to their size and peculiar growth requirements. Mycoplasmas tend to grow down into a solid medium and in a distinctive fashion produce a colony with a "fried-egg" appearance. They proliferate also in cell culture and in the developing chick embryo. Serologic identification is practical because mycoplasmas are vulnerable to neutralizing antibodies and induce hemagglutination and hemadsorption reactions. Since many normal persons demonstrate circulating antibodies to them, it is the *rising* titer that is of diagnostic significance.

---

*These are agglutinins that clump human type O, Rh-negative erythrocytes at low temperatures (5° to 20° C) but not at body temperature. They are found in several unrelated diseases.

QUESTIONS FOR REVIEW

1. What is the relation of acid-producing bacteria such as lactobacilli to dental caries?
2. What is the significance of *Pseudomonas aeruginosa* in clinical medicine? Why is it so hard to deal with?

3. What is melioidosis? Name and describe the causative microbe.
4. Briefly characterize mycoplasmas. How is the laboratory diagnosis made for mycoplasmas?
5. List diseases caused by *Pseudomonas*.
6. State the practical importance of lactobacilli. List notable members of the genus *Lactobacillus*.
7. Define listeriosis. What forms are significant in humans?
8. Compare mycoplasmas with viruses.
9. Give the postulated role of T-strain mycoplasmas in human disease.
10. Outline salient features of primary atypical pneumonia.
11. What are Döderlein bacilli?
12. Give the method of spread for glanders. How may this disease be controlled?
13. Compare the classic cholera vibrio with its biotype eltor.
14. How is Asiatic cholera transmitted? How can it be prevented?
15. Explain the pathogenic mechanism for the diarrhea in cholera. State the serious consequences of this.
16. Describe the organism causing anthrax.
17. Discuss the proper disposal of bodies of animals dead of anthrax.
18. Why is the anthrax bacillus of historic interest?

REFERENCES. See at end of Chapter 27.

27 Rickettsias
(also chlamydiae*)

GENERAL DISCUSSION

Rickettsias are known for a distinct, selective type of parasitism of cells in disease and for a special relation to an arthropod, be it vector or host. Long thought of as occupying an intermediate position between bacteria and viruses, these procaryotic microbes are now recognized as bacteria.

In Bergey's Part 18, one finds Order I, Rickettsiales, with two such families. The members of Family I, Rickettsiaceae, parasitize tissue cells (excepting the red blood cells) of a vertebrate host; within Tribe I, Rickettsiaeae, of microbes adapted to existence in arthropods, there are three genera of medical importance—*Rickettsia*, *Rochalimaea*, and *Coxiella*. The members of Family II, Bartonellaceae, affect both red blood cells and tissue cells; one genus, *Bartonella*, causes disease in man.

Special properties. Rickettsias of Family I are small, pleomorphic, bacillary or coccobacillary forms named in honor of Dr. Howard T. Ricketts † of Chicago. They occur singly or in pairs, chains, or irregular clusters. The electron microscope indicates that they have an internal structure much like that of bacteria and that they divide by binary fission. Most rickettsias are held back by bacteria-retaining filters. They are close to the size of some of the larger viruses, being $0.3 \, \mu$m in their smallest dimension. They are nonmotile, gram negative, and stain with difficulty. They are obligate intracellular parasites. With rare exception, they do not multiply in the absence of living cells; the pathogenic forms grow only in the cells of infected animals. Some grow only in the cytoplasm, but others grow in both cytoplasm and nucleus of

---

*Within Part 18, the Rickettsias, of the recent Bergey classification, there are two orders. The first classifies microbes traditionally thought of as rickettsias; the second, those better known as chlamydiae or bedsoniae.

†Dr. Ricketts (1871-1910) first observed rickettsia bodies in a case of Rocky Mountain spotted fever in 1909 and demonstrated that it was transmitted by the wood tick. Later on he showed that tabardillo (Mexican typhus) was transmitted by the body louse, but in the investigation he contracted the disease and died, a martyr to the disease that he was studying.

the infected cell. The fragments of enzyme systems that rickettsias possess allow them a range of metabolic activity. Their resistance to deleterious influences such as heat, drying, and chemicals is about the same as that of most bacteria.

The rickettsial diseases are transmitted to man by insects, the natural and primary hosts of the rickettsias.

Pathogenicity. When rickettsias invade man, they attack the reticuloendothelial system, colonizing the endothelial lining cells of the walls of small blood vessels (small arteries, arterioles, and capillaries). Within these cells they induce a vasculitis (inflammation of the blood vessel) that is distributed all over the body, just as small blood vessels are. The consequence is disease in many anatomic areas in the body. A classic skin rash and a whole host of pathologic changes ensue. Rickettsial diseases can be quite serious and many times life-threatening in spite of the best available therapy.

An attack of rickettsial disease is usually followed by a lasting immunity.

Laboratory diagnosis. Rickettsias can be cultivated in the yolk sac of the chick embryo (provided the hen did not receive antibiotics) or in cell cultures. Recovery of the microorganisms may be made when blood (rickettsemia present) and suitable specimens from a patient are inoculated into laboratory animals such as guinea pigs, mice, and rabbits.

The most important rickettsial diseases are associated with a positive *Weil-Felix reaction.* This is an agglutination test similar to the Widal test for typhoid fever except that the test serum is mixed with different types of *Proteus* bacilli. The Weil-Felix reaction is a heterophil antibody reaction, since *Proteus* bacilli *do not* cause any of the rickettsial diseases or even act as secondary invaders. Complement fixation tests are of value in differentiating the rickettsial diseases.

Classification. Rickettsial diseases may be divided into the following groups:
1. Typhus fevers
2. Spotted fevers
3. Scrub typhus
4. Trench fever
5. Q fever

Table 27-1 indicates epidemiologic features of these groups. Rocky Mountain spotted fever, rickettsialpox, and Q fever are the three prevalent in North America.

RICKETTSIAL DISEASES
Typhus fever group

General considerations. The typhus fever group includes epidemic typhus, murine typhus, and Brill-Zinsser disease. Clinically the different types of typhus fever closely resemble each other but vary in severity. They usually begin with severe headache, chills, and fever. A rash develops about the fourth day and persists throughout the course of the disease. Mentally the patient is dull and stuporous. This feature gave origin to the name "typhus," which is derived from the Greek word *typhos,* meaning vapor or smoke. Disease in this group extends from 3 to 5 weeks. The mortality varies from 5% to 70%.

467

TABLE 27-1. RICKETTSIAL INFECTIONS IN MAN*

Group	Agent	Weil-Felix reaction			Complement fixation*
		OX-19	OX-2	OX-K	
Typhus fevers					+
Epidemic typhus	*Rickettsia prowazekii*	++			(specific)
Murine typhus	*Rickettsia typhi (mooseri)*	++			+ (specific)
Brill-Zinsser disease	*Rickettsia prowazekii*				
Spotted fevers					
Rocky Mountain spotted fever	*Rickettsia rickettsii*	+	+	−	+
Rickettsialpox	*Rickettsia akari*	−	−	−	+
Fièvre boutonneuse	*Rickettsia conorii*	+	+	+	
Scrub typhus	*Rickettsia tsutsuga-mushi*	−	−	++	±
Trench fever	*Rochalimaea quintana*				
Q fever	*Coxiella burnetii*	−	−	−	+

*Cross reactions between Rocky Mountain spotted fever and rickettsialpox.

EPIDEMIC (CLASSIC, EUROPEAN, OLD WORLD, LOUSE-BORNE) TYPHUS. Epidemic typhus is caused by *Rickettsia prowazekii** and is spread from person to person by the body louse *(Pediculus humanus corporis)* (Fig. 27-1). Head lice *(Pediculus humanus capitis)* can transmit the disease but seldom do so. No animal reservoir has been found. Epidemic typhus is an acute and severe disease with a high mortality (10% to 40%) that has been known to spread over the world in devastating epidemics. It is a disease of overcrowding, famine, filth, and war. (It is also known as jail fever, war fever, and famine fever.)

When a louse bites a person with epidemic typhus, the rickettsias are taken into the stomach of the louse and invade the cells lining the intestinal tract. They multiply to such an extent that the cells become greatly swollen, burst, and liberate the rickettsias into the feces of the louse. When a louse bites, it defecates at the same time. The site of the bite itches, and that person introduces the infectious material into the skin by scratching. Infected lice die within 8 to 10 days after infection from intestinal obstruction caused by parasitism of the lining cells.

MURINE (ENDEMIC, NEW WORLD, FLEA-BORNE) TYPHUS. Murine typhus fever, caused by *Rickettsia typhi*, is relatively mild. A natural infection of rats, less often of mice, it is transmitted from rat to man by the rat flea (Fig. 27-2) and from rat to rat

*The species name *Rickettsia prowazekii* is derived from Stanislas von Prowazek (1876-1915), an early investigator who lost his life in the study of typhus.

Reservoir in nature	Vector	Geography
Man	Body lice	Worldwide
Rats	Rat fleas	Worldwide
Man		Worldwide
Wild rodents	Ticks	Western hemisphere
House mice	Mites	United States, Russia, Korea
Small wild mammals	Ticks	Mediterranean coast, Middle East
Wild rodents	Mites	Asia, Australia, Pacific Islands
Man	Body lice	Europe, Mexico, Africa
Ruminants, small mammals, domestic livestock	Ticks — animals Airborne animal products — man	Worldwide

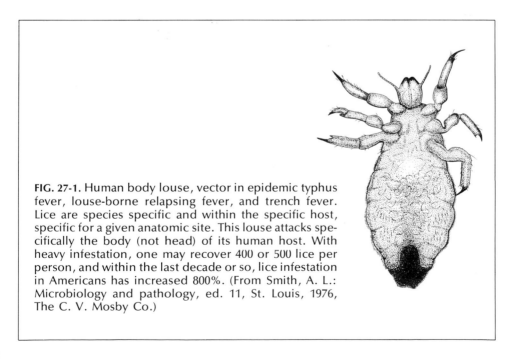

FIG. 27-1. Human body louse, vector in epidemic typhus fever, louse-borne relapsing fever, and trench fever. Lice are species specific and within the specific host, specific for a given anatomic site. This louse attacks specifically the body (not head) of its human host. With heavy infestation, one may recover 400 or 500 lice per person, and within the last decade or so, lice infestation in Americans has increased 800%. (From Smith, A. L.: Microbiology and pathology, ed. 11, St. Louis, 1976, The C. V. Mosby Co.)

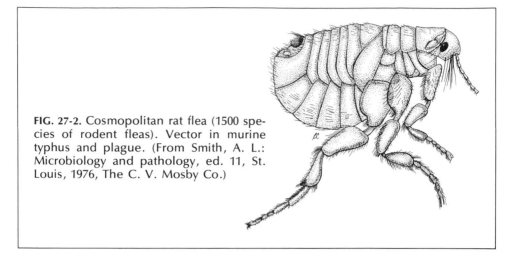

FIG. 27-2. Cosmopolitan rat flea (1500 species of rodent fleas). Vector in murine typhus and plague. (From Smith, A. L.: Microbiology and pathology, ed. 11, St. Louis, 1976, The C. V. Mosby Co.)

by the rat louse and the rat flea. Rats do not die of the disease, nor do the fleas or lice that infest them. Murine typhus may be transmitted from man to man by the body louse and become epidemic in louse-infested communities. In man the mortality is about 2%. The endemic typhus fever of Mexico is known as *tabardillo* (from the Spanish word *tabardo,* meaning a colored cloak, to designate the mantlelike spotted rash of the disease).

BRILL-ZINSSER DISEASE (RECRUDESCENT TYPHUS). Brill-Zinsser disease is a mild form of typhus existing along the Atlantic Coast and occurring in persons who have had classic typhus fever many years before.*

Laboratory diagnosis. The typhus fevers all are identified by a positive Weil-Felix reaction (agglutination in serum of the bacteria known as *Proteus OX-19*). Although the Weil-Felix reaction may be detected as early as the fourth day, it is strongly positive about the eighth day. The Weil-Felix reaction does not differentiate the typhus fevers from Rocky Mountain spotted fever, but it does separate them from certain other rickettsial infections. Fluorescent antibody technics can also be used in the diagnosis of the typhus fevers. The epidemic and endemic forms may be distinguished and the rickettsias demonstrated if blood from a patient is inoculated into a male guinea pig.

Immunity. If a person recovers from either epidemic or endemic typhus, he is immune to both, but a vaccine prepared against one type protects against only that one. Vaccines seem to be rather effective in preventing the typhus fevers, and when disease does occur in spite of vaccination, the severity is lessened and mortality reduced.† In only a few cases is recovery from epidemic or endemic typhus not followed by permanent immunity.

---

*The disease was first observed by Nathan E. Brill (1860-1925) in 1898 in immigrants to the United States from various countries of Eastern Europe. The exact nature of the disease was not known until 1934, when Hans Zinsser (1878-1940) advanced the hypothesis that the disease was a mild second attack of classic typhus, an idea now amply confirmed.

†For immunization, see p. 685.

Prevention. Prevention of epidemic typhus fever depends on (1) isolation of the patient in a vermin-free room, (2) use of insecticides on clothing and bedding, with destruction of insect eggs (nits) attached to hair, (3) quarantine of all susceptible persons if many lice are in the vicinity of the patient, (4) systematic delousing and vaccination of susceptible persons, and (5) general improvement of living and sanitary conditions. The control of the endemic form depends on ratproofing buildings and destroying rodents with their louse and flea populations.

Spotted fever group

Rocky Mountain spotted fever. Recognized as one of the most severe of all infectious diseases, Rocky Mountain spotted fever is like typhus fever and is caused by a similar organism, *Rickettsia ricketsii*. In areas where it is well known it is often referred to as tick fever. (This should not cause it to be confused with relapsing fever, which also is known as tick fever.) It occurs in the Rocky Mountain states, many cases being found in Idaho and Montana, but it is increasingly prevalent on the Atlantic seaboard, chiefly east of the Appalachian Mountains. For this reason two types of the disease are designated: the western type (the original Rocky Mountain spotted fever) and the eastern type. The difference rests mainly on geographic distribution and mode of spread. The disease does show, however, great variation in severity. In Idaho the mortality varies from 3% to 10%, whereas in the Bitter Root Valley of Montana it may reach 90%. In the eastern states the disease is comparatively mild, and the mortality is about 6%.

The western variety is transmitted by the Rocky Mountain wood tick *(Dermacentor andersoni)*, and the eastern variety is transmitted by the American dog tick *(Dermacentor variabilis)*. In the southwestern United States the disease is transmitted by the Lone Star tick *(Amblyomma americanum)*. Several species of hard-shell ticks (Ixodidae) (Fig. 27-3) are infected. The male tick can infect the female, who is able to transmit the infection to her offspring. Fluorescent antibody technics show that infected female ticks may pass rickettsias to all their progeny, an example of *transovarian* infection. The infection may be retained in ticks through many generations. The tick is therefore able to maintain the disease in nature without either rodent or human help. A reservoir of infection appears to exist among a variety of rodents, particularly jackrabbits and cottontails. The disease is transmitted from rabbit to rabbit by the rabbit tick, a tick that does not bite man. It is also maintained in birds and in large wild and domestic animals.

The rickettsias of Rocky Mountain spotted fever differ from those of typhus fever in that they invade both the cytoplasm and nucleus of the infected cell. An attack renders a person immune for life. The serum from persons with Rocky Mountain spotted fever agglutinates *Proteus OX-19* (positive Weil-Felix reaction), but the agglutination is not so strong as in typhus fever. *Proteus OX-2* also may be agglutinated. A complement fixation test is highly specific, and there is an indirect hemagglutination test. The patient's blood may be inoculated into a male guinea pig and the rickettsias recovered from the animal.

Suspect ticks may be examined microscopically for the presence of rickettsias.

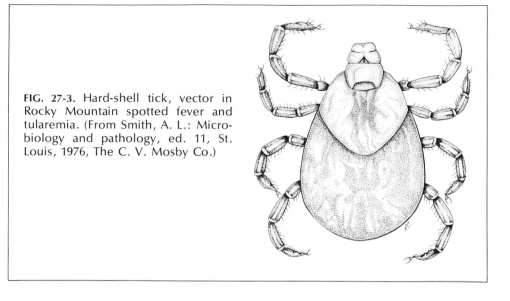

FIG. 27-3. Hard-shell tick, vector in Rocky Mountain spotted fever and tularemia. (From Smith, A. L.: Microbiology and pathology, ed. 11, St. Louis, 1976, The C. V. Mosby Co.)

A tick engorged from feeding is dissected from its shell, fixed in a suitable solution, and stained with Giemsa stain. If infectious, the tick will display rickettsias within the nuclei of cells of the salivary glands.

Health officials advise persons going into tick-infested areas to wear protective clothing such as high boots and as soon as possible afterward to examine clothing and their bodies carefully for ticks. A tick may wander over a person's body for an hour or two before feeding, and it is said that the time required for rickettsias to be transmitted from the attached tick to the human being on which it is feeding is 2 hours or more. Dogs and other pets should be inspected daily. Infective ticks are likely to be found in underbrush, tall grasses, weeds, and wet woods, especially in spring and early summer. It is advisable to clear out such from an area where children may be playing or else to keep the children away from that area. Parents should inspect their children for ticks, particularly their hair, after they have played in grassy or wooded areas or with pets.

Rickettsialpox. Rickettsialpox is an acute mild febrile disease described by the appearance of a red papule at the site of inoculation followed later by a skin rash. The disease is caused by *Rickettsia akari.* The house mouse is the reservoir of infection, and the bite of the house mouse mite conveys the infection from mouse to mouse and from mouse to man. The serum of those ill with or convalescing from the disease does *not* give a positive Weil-Felix reaction. Clinically, rickettsialpox may be mistaken for chickenpox. The primary lesion at the site of inoculation often resembles the vesicle of a smallpox vaccination. The destruction of mice and their mites eliminates the disease.

Other fevers. Included with the spotted fevers are tick-borne typhus fevers— fièvre boutonneuse, Queensland tick typhus, South African tick-bite fever, and tick-bite rickettsioses of India.

These illnesses are nonfatal and only moderately severe. They are caused by rickettsias related to the agent of spotted fever, and each can be traced to the bite of an ixodid tick. Fièvre boutonneuse is the most widespread and the prototype; it is caused by *Rickettsia conorii*. In this disease there is an initial lesion, the *tache noire* (an ulcer covered over by a black crust is the "black spot"), which is followed by a generalized rash. The Weil-Felix reaction is positive but not strongly so.

Scrub typhus

Scrub typhus, known also as tsutsugamushi (Japanese, dangerous bug) disease and mite typhus, is so called because the mites that transmit the disease to man are found in scrubland, land covered by a stunted growth of vegetation. The disease attacked thousands of soldiers in the Southwest Pacific area during World War II. It is an acute febrile disease with a rash, generalized lymphadenopathy, and lymphocytosis. The disease may simulate infectious mononucleosis. The mortality is from 0.5% to 60%. The disease resembles Rocky Mountain spotted fever clinically, except that in scrub typhus there is a primary sore at the site of the mite bite. The causative rickettsia is named *Rickettsia tsutsugamushi.*

Scrub typhus is a disease primarily of mites and wild rodents, transmitted from rodent to rodent and from rodent to man by mites. Two types of mites are known to transmit the disease to man. The rodents and mites that propagate the disease probably differ in various parts of the world. The mites grow on short blades of grass and in top layers of the soil. Persons who contact the soil are therefore likely to contract the infection. There is no evidence that the disease spreads from man to man. The serum of persons who have the disease agglutinates *Proteus* OX-K, but not OX-19. There is an indirect fluorescent antibody test, and inoculation of a mouse may be done.

Prevention depends on the destruction of the mites in the soil by insecticides and the application of insecticides to clothing. There is as yet no effective preventive vaccine.

Trench fever

Trench fever is a remittent or relapsing fever that affects soldiers on trench duty, but it may occur in civilian life when people live under conditions comparable to those of the trenches. It was prevalent during World War I but has since almost passed out of existence. It was infrequent during World War II. *Rochalimaea quintana*, the causative agent, is transmitted by the body louse. Unlike other rickettsias, man is its primary host, and the rickettsias of the genus *Rochalimaea* can be cultured in host cell–free media.

Although trench fever is never fatal and recovery is usually complete, it may recur, and in some cases it is followed by a state of chronic ill health with pain in the limbs, mental depression, and circulatory disorders. An attack does not confer immunity.

Q fever

Q fever is a febrile disease of short duration caused by *Coxiella burnetii*. Rickettsias of the genus *Coxiella* are set apart from other rickettsias by their amazing resistance outside cells to heat, drying, and sunlight. The usually mild respiratory symptoms of the disease make it similar to primary atypical pneumonia or influenza. It differs from most other rickettsial diseases in that it is not transmitted to man by an insect vector, there is no rash, and the serum of an infected person does not agglutinate *Proteus* bacilli. The disease was first observed in Queensland, Australia, but the place of discovery did not give origin to the term *Q fever*—Q indicates "query."

Q fever is worldwide in distribution and prevalent in the United States, especially in certain areas along the west coast. It is a disease of animals, a zoonosis. Man becomes infected by residence or occupation that brings him in contact with infected livestock—cattle, sheep, and goats. Infection is especially common among packing-house employees and laboratory workers. Most human infection is acquired by breathing contaminated air such as the dust from dairy barns and lambing sheds. Mass aerial infection of large groups has occurred. Such an instance happened to more than 1600 soldiers who were returning from Italy to the United States and who apparently received their infection from winds blowing at an airport. Infected cattle and sheep may shed infected placentas and fluids at time of parturition. About one half of the cows harboring this rickettsia discharge it in their milk, and ingestion of the milk of infected cows or goats is a source of infection. Although the rickettsias of Q fever are fairly resistant to heat, they are effectively destroyed in milk by the high-temperature, short-time method of pasteurization.

A capillary tube agglutination test detects antibodies in serum from human beings and animals. Blood from a patient may be inoculated into a male guinea pig.

Rickettsial diseases in animals

A list of rickettsial diseases in animals follows.
Heartwater disease of cattle
Rickettsiosis of cattle
Febrile disease of dogs and sheep
Salmon disease of dogs and foxes

BARTONELLA BACILLIFORMIS

The intracellular parasites of the genus *Bartonella* (in Family II) are small, gram-negative, flagellated, and very pleomorphic coccobacilli. They can be cultivated on cell-free media, unlike other rickettsias. All infect red blood cells as well as endothelial cells of liver, spleen, and lymph nodes. They are transmitted by the night bite of the sandfly *(Phlebotomus)*. The one species of the genus, *Bartonella bacilliformis*, is the cause of the South American disease known as Carrión's disease (Oroya fever, verruga peruana), which is characterized by an acute anemia with fever followed by a verrucous, or wartlike, skin eruption. *Salmonella* species are prone to complicate. Infection with *Bartonella*, a South American parasite, is endemoepidemic in the Andean valleys of the Peruvian sierra.

Bartonella bacilliformis is easily recognized in Wright-stained films of peripheral blood and in material from skin lesions because of its unique microscopic appearance.

CHLAMYDIAE (BEDSONIAE)

In Bergey's Order II of Part 18, Chlamydiales, are placed gram-negative, intracellular pathogens of vertebrates with an intracellular pattern for reproduction peculiarly their own.

Description. The name *chlamydiae* (bedsoniae)* is applied to a large group of microorganisms (psittacosis group) with comparable design and makeup and causing several important diseases. Their life cycle is a complex one. Chlamydiae multiply within the cytoplasm of the cell attacked and form typical inclusion bodies there. The infectious particle is known as an elementary body, and the mature inclusion body contains a host of them.

For a long time chlamydiae were thought of as large viruses and were classified and referred to as basophilic viruses and mantle viruses. Like viruses they are obligate intracellular parasites, but there are important differences (see also Tables 28-1, 28-2, and 28-3). Unlike viruses they possess enzymes and both DNA and RNA. They have no capsid symmetry and are susceptible to some antimicrobial drugs (very sensitive to tetracyclines). The patterns of their diseases differ in many ways from those of viruses.

Recent studies indicate that these parasites are better separated from true viruses and considered as small bacteria (Fig. 27-4). They are comparable to rickettsias but still distinct. The name *Bedsonia* was given in honor of Sir Samuel Bedson (1886-1969), who studied them extensively. However, many microbiologists believe the genus name should be *Chlamydia*, and therefore these microbes are most often called chlamydiae. They are nonmotile, reproduce by binary fission, and have an outer cell wall chemically similar to that of gram-negative bacteria, from which they were probably derived.

Laboratory diagnosis. Many procedures used in the laboratory diagnosis of viral infection apply to these organisms (p. 500). They grow in the fertile hen's egg and express in cell culture their distinct cytopathogenicity. They are easily stained with basophilic dyes (hence the term *basophilic viruses*), and because of their unique staining qualities, they can be visualized readily in suitable specimens by light or fluorescent microscopy. There are diagnostic serologic tests, and the Frei test for lymphopathia venereum is a well-known skin test.

Importance. The genus *Chlamydia* of Family I, Chlamydiaceae, comprises only two known species, but their diseases sort out ecologically into three categories: (1) man—oculogenital and respiratory; (2) birds—respiratory and generalized; and (3) nonprimate mammals—respiratory, placental, arthritic, enteric, and so on. *Chlamydia trachomatis* is responsible for trachoma, lymphopathia venereum, inclusion conjunctivitis, urethritis,† and proctitis in man. *Chlamydia psittaci* causes ornithosis and psittacosis in birds and various other diseases in animals.

---

*From the Greek *chlamys*, cloak. Each organism is surrounded by a dense cell wall in part of its life cycle.

†*Chlamydia trachomatis* is thought to be the major cause of nongonococcal urethritis (NGU) in men and certain chlamydiae are commonly found in the genital tracts of sexually active women. Chlamydial infections may be more important in venereal disease than previously noted.

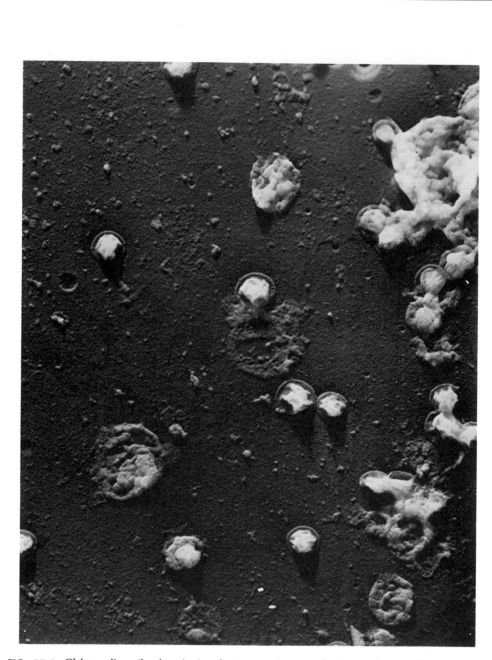

FIG. 27-4. Chlamydiae (bedsoniae), electron micrograph. Extensive studies on this microbe (meningopneumonitis agent) were important to the conclusion that these basophilic organisms were not viruses. (×28,000.)

The as yet undefined agent of cat-scratch fever is postulated to belong with the chlamydiae.

CHLAMYDIAL INFECTIONS
Trachoma

Although rare in the United States, trachoma, one of the oldest diseases known to man, afflicts more than 500 million persons in the world. The causative organism, one of the chlamydiae, specifically attacks the lining cells of the cornea and conjunctival membrane of the eye. Growth in these cells produces characteristic inclusion bodies. As is often the case with viral diseases, bacterial infection is superimposed to intensify the injury already done. The inflammatory changes are pronounced, and the scar tissue formed over the cornea, ordinarily transparent, results in impaired vision. The name "trachoma" is derived from a Greek word meaning rough to set forth the pebble-like appearance of the infected conjunctival membrane (Fig. 27-5). This disease is the world's leading cause of blindness (in some 20 millions or more with the disease).

Spread. Trachoma is endemic in the large underprivileged areas of the tropics and subtropics; in some countries 75% of the population suffer. Transmission requires close personal contact, and for this reason the disease is often passed from mother to

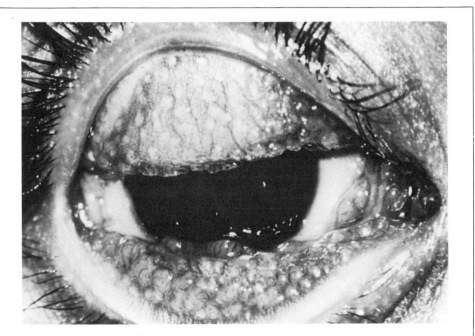

FIG. 27-5. Trachoma. Inflammatory nodules are spread over thickened conjunctiva of eye of Pima Indian (from tribe of southern Arizona and northern Mexico). Trachoma is prevalent among Indians of Southwest. (Courtesy Dr. Phillips Thygeson, Los Altos, Calif.)

child. Flies carry the infection mechanically. Chlamydiae may also be spread by contaminated fingers and articles of clothing.

Laboratory diagnosis. The diagnosis is made by identification of the typical inclusion bodies in scrapings from the conjunctival membranes and by a complement fixation test.

Control. The agent has been grown in fertile eggs and an effective vaccine prepared. The prognosis in this disease has been greatly improved with the use of antibiotics alone. The incidence of the disease has been shown to drop simply with improvements in living conditions.

Lymphopathia venereum

Lymphopathia venereum (venereal lymphogranuloma, lymphogranuloma venereum, lymphogranuloma inguinale), formerly known as climatic bubo, is a venereal disease of chlamydial etiology. Formerly considered a disease of the lower echelon of society, it is appearing today in the affluent and seems to be increasing not only in this country but also over the world.

Pathology. In the male it starts with a primary ulceration on the external genitalia. The infection extends to the inguinal lymph nodes where buboes are formed. Inguinal lymphadenopathy is a salient feature of the disease. In the female, infection extends from the primary lesion along the lymphatic drainage to lymph nodes within the pelvis, where the chronic inflammation set up may in time lead to stricture of the rectum. In either sex the lymphatics are mainly involved, and the severe scarring associated with long-standing lymphangitis results in a dramatic deformity of the external genitalia in both.

Lymphopathia venereum should not be confused with *granuloma inguinale,* also a venereal disease, but one caused by a different agent (p. 399).

Laboratory diagnosis. Lymphopathia venereum (past or present infection) is detected by the Frei test, which consists of the intradermal injection of material prepared by growing the agent in the yolk sac of a developing chick embryo. The development of a bright red papule at the inoculation site in 48 hours constitutes a positive test. There is a diagnostic complement fixation test. Inclusion bodies may be seen in suitably stained smears of pus from the buboes.

Psittacosis (parrot fever) and ornithosis

Psittacosis (parrot fever) is a chlamydial infection so named because it affects psittacine birds, most often parrots and parakeets. Ornithosis is the corresponding disease in domestic fowl and birds other than those of the parrot family—chickens, turkeys, pigeons, and sea birds. Psittacosis is important because of its worldwide distribution. The number of human beings attacked, however, is small.

Transmission. Psittacosis is widespread among the birds just named, the sources of most human infections. One contracts the disease from birds by (1) handling of sick birds or their feathers, (2) transmission of the agent through the air, (3) contact with material contaminated by infected birds, and (4) bites or wounds inflicted by sick birds. Sick birds show the chlamydia in their nasal discharges and feces and may

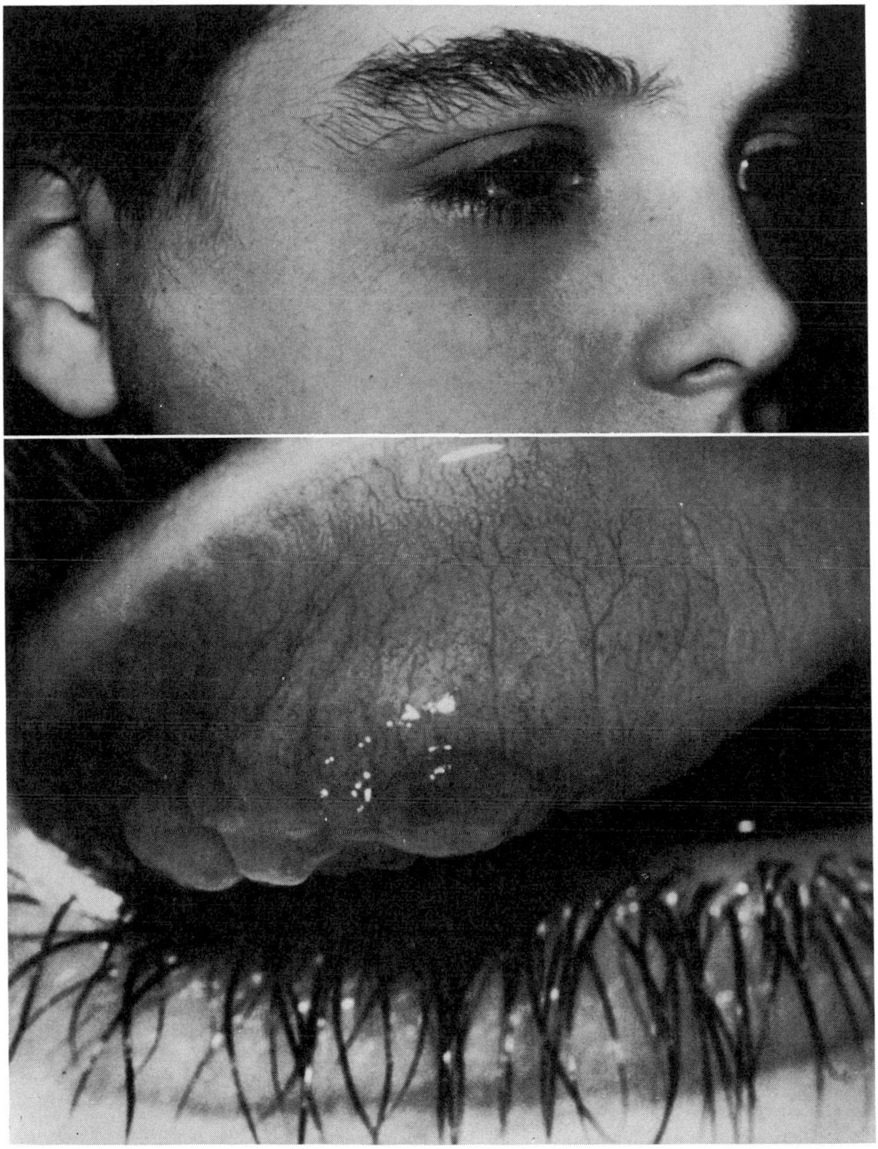

FIG. 27-6. Cat-scratch disease. A, Lymph node regional to inflamed right eye (preauricular node) is swollen and tender. B, Upper eyelid everted to show inflammatory swelling and congestion of blood vessels. Purulent exudate is present. Not only did this patient have a cat, but cat-scratch antigen test was positive. (From Donaldson, D. D.: Atlas of external diseases of the eye, vol. 1, St. Louis, 1966, The C. V. Mosby Co.)

become carriers. In birds the gastrointestinal tract is primarily affected, but in man the infection enters via the respiratory tract and localizes there. In a few cases infection is transmitted directly from man to man. The onset of disease in man may be sudden and the mortality high. Explosive outbreaks sometimes occur in poultry-processing plants.

Laboratory diagnosis. Diagnosis may be made by observing the development of the disease in a mouse following injection of a sample of the patient's sputum. A complement fixation test is also available.

Prevention. The patient must be isolated and all discharges disinfected. A vaccine has been prepared against psittacosis, but its effectiveness is undetermined.

Cat-scratch disease

Cat-scratch disease (benign lymphoreticulosis) is a febrile illness that may come after scratching, biting, or licking by a cat. A purulent lesion appears at the site of infection, and 2 or 3 weeks later the lymph nodes draining the area enlarge (Fig. 27-6). Incubation time, from injury to illness, is usually 3 to 10 days. As a rule, the disease is mild. Cats are mechanical vectors and do not show related illness.

It is possible that animals other than cats carry the infection, including birds and insects, but the monkey is the only animal artificially infected.

Laboratory diagnosis. Antigen for the skin test is prepared from the purulent material aspirated from the lesions or from a lymph node extract. It is evaluated in 48 hours. A positive test indicates previous or present infection. The agent has not been isolated, but elementary bodies similar to those of psittacosis may be seen in stained sections of affected lymph nodes.

Prevention. The only preventive measure known is the avoidance of cats.

Chlamydial infections in animals

A list of chlamydial (bedsonial) diseases in animals follows.

Feline pneumonitis

Conjunctivitis and keratitis of sheep, cattle, goats, and chickens

Pneumonitis in mice

Stiff-lamb disease (polyarthritis) — one form

Sporadic bovine encephalomyelitis

QUESTIONS FOR REVIEW

1. What are rickettsias? How did they get their name? Give characteristics.
2. What is the Weil-Felix reaction? Its importance?
3. Into what five groups may rickettsial diseases be classified?
4. Give the laboratory diagnosis of rickettsial infections.
5. Briefly discuss Q fever.
6. Explain the pathologic changes in rickettsial diseases.
7. State special properties of chlamydiae. Compare them with other microbes.
8. Give examples of chlamydial diseases with diagnostic inclusion bodies in parasitized cells.
9. What disease is responsible for most of the world's blindness? Give its salient features.
10. What is the cause of Carrión's disease?
11. Briefly explain Frei test, tabardillo, Ixodidae, transovarian infection, benign lymphoreticulosis, "Q" of Q fever, the word trachoma, the word tsutsugamushi.

480

REFERENCES FOR CHAPTERS 18 TO 27

Adams, M. M. E.: Cholera: new aids in treatment and prevention, Science **179:**552, 1973.

Allison, F., Jr.: Anaerobic bacterial infection in man, South. Med. J. **68:**1088, 1975.

Allison, F., Jr., and Sanders, C. V.: The extragenital manifestations of gonococcal infection, South. Med. Bull. **59:**38, Apr., 1971.

Alston, J. M., and Broom, J. C.: Leptospirosis in man and animals, Edinburgh, 1958, E. & S. Livingstone, Ltd.

Altemeier, W. A.: The significance of infection in trauma, AORN J. **15:**92, Mar., 1972.

Anigstein, L., and Anigstein, D.: Review of evidence in retrospect for rickettsial etiology in Bullis fever, Tex. Rep. Biol. Med. **33:**201, (no. 1) 1975.

Armstrong, D., and Kaplan, M. H.: Sepsis: a pictorial guide to some important clinical signs and initial laboratory studies, Hosp. Med. **8:**22, July, 1972.

Artenstein, M. S.: Prophylaxis for meningococcal disease, J.A.M.A. **231:**1035, 1975.

Astor, G.: Plague, American-style, Today's Health **50:**54, Aug., 1972.

Bailey, W. R., and Scott, E. G.: Diagnostic microbiology, ed. 4, St. Louis, 1974, The C. V. Mosby Co.

Balentine, J. D., and others: Infection of armadillos with *Mycobacterium leprae*, Arch. Pathol. Lab. Med. **100:**175, Apr., 1976.

Balows, A., and others, editors: Anaerobic bacteria: role in disease, Springfield, Ill., 1975, Charles C Thomas, Publisher.

Barrett-Connor, E.: Gonorrhea and the pediatrician, Am. J. Dis. Child. **125:**233, 1973.

Bartlett, J. G., and Finegold, S. M.: Anaerobic pleuropulmonary infections, Medicine **51:**413, 1972.

Bartlett, R. C.: Medical microbiology, New York, 1974, John Wiley & Sons, Inc.

Barton, F. W.: Venereal disease (editorial), J.A.M.A. **216:**1472, 1971.

Barua, D., and Burrows, W., editors: Cholera, Philadelphia, 1974, W. B. Saunders Co.

Beck, A. M., and others: The ecology of dog bite injury in St. Louis, Missouri, Public Health Rep. **90:**262, 1975.

Benenson, A. S., editor: Control of communicable diseases in man, ed. 2. Washington, D.C. 1975, American Public Health Association, Inc.

Bessman, A. N., and Wagner, W.: Nonclostridial gas gangrene, J.A.M.A. **233:**958, 1975.

Bisno, A. L., and Ofek, I.: Serologic diagnosis of streptococcal infection, Am. J. Dis. Child. **127:**676, 1974.

Bloch, M.: The royal touch: sacred monarchy and scrofula in England and France, London, 1973, McGill-Queens University Press.

Blount, J. H.: A new approach for gonorrhea epidemiology, Am. J. Public Health **62:**710, 1972.

Bode, F. R., and others: Pulmonary diseases in the compromised host, Medicine **53:**255, 1974.

Bowen, W. H.: Prospects for the prevention of dental caries, Hosp. Pract. **9:**163, May, 1974.

Branson, D.: Timely topics in microbiology, mycobacteria, 1968-1971, Am. J. Med. Technol. **38:**13, 1972.

Burgoon, C. F., Jr.: Acne, Mod. Med **40:**57, Apr. 17, 1972.

Burnett, G. W., and Schuster, G. S.: Review of pathogenic microbiology, St. Louis, 1974, The C. V. Mosby Co.

Butler, T., and others: Bubonic plague: detection of endotoxemia with the limulus test, Ann. Intern. Med. **79:**642, 1973.

Caldwell, J. G.: Congenital syphilis: a nonvenereal disease, Am. J. Nurs. **71:**1768, 1971.

Chapman, J. S.: Ecology of atypical mycobacteria, Arch. Environ. Health **22:**41, 1971.

Chow, A. W., and Guze, L. B.: Bacteroidaceae bacteremia: clinical experience with 112 patients. Medicine **53:**93, 1974.

Cherington, M.: Botulism, Arch. Neurol. **30:**432, 1974.

Cohen, J. O., editor: The staphylococci, New York, 1972, John Wiley & Sons, Inc.

Convit, J., and Pinardi, M. E.: Leprosy: confirmation in the armadillo, Science **184:**1191, 1974.

Cooke, E. M.: *Escherichia coli* and man, New York, 1974, Longman, Inc.

Corman, L. C., and others: The high frequency of pharyngeal gonococcal infection in a prenatal clinic population, J.A.M.A. **230:**568, 1974.

Costerton, J. W., and others: Structure and function of the cell envelope of gram-negative bacteria, Bacteriol. Rev. **38:**87, Mar., 1974.

Crowder, J. G., and White, A.: Teichoic acid antibodies in staphylococcal and nonstaphylococcal endocarditis, Ann. Intern. Med. **77:**87, 1972.

Darrow, W. W., and Wiesner, P. J.: Personal prophylaxis for venereal diseases, J.A.M.A. **233:** 444, 1975.

Davis, B. D., and others: Microbiology including immunology and molecular genetics, New York, 1973, Harper & Row, Publishers.

DeShazo, R. D., and others: Early diagnosis of Rocky Mountain spotted fever, J.A.M.A. **235:**1353, 1976.

Dillon, H. C., Jr., and Derrick, C. W., Jr.: Streptococcal complications: the outlook for prevention, Hosp. Pract. **7:**93, Sept., 1972.

Dryden, G. E.: Sources of infection in the surgical patient, AORN J. **13:**64, May, 1971.

Duma, R. J.: *Pseudomonas* and water, with a new twist (editorial), Ann. Intern. Med. **76:**506, 1972.

Edwards, P. R., and Ewing, W. H.: Identification of Enterobacteriaceae, rev. ed. 3, Minneapolis, 1972, Burgess Publishing Co.

Edsall, G.: The inexcusable disease (editorial), J.A.M.A. **235:**62, 1976.

Ehrenkranz, N. J., and Kicklighter, J. L.: Tuberculosis outbreak in a general hospital: evidence of airborne spread of infection, Ann. Intern. Med. **77:**377, 1972.

Eisenberg, M. S., and Bender, T. R.: Botulism in Alaska, 1947 through 1974, J.A.M.A. **235:**35, 1976.

Eklund, M. W., and others: Interspecies conversion of *Clostridium botulinum* type C to *Clostridium novyi* type A by bacteriophage, Science **186:**456, 1974.

Erdtmann, F. J., and others: Skin testing for tuberculosis, J.A.M.A. **228:**479, 1974.

Faden, H. S. and others: Nursery outbreak of scalded-skin syndrome, Am. J. Dis. Child. **130:**265, 1976.

Farmer, J. J.: Pseudomonas in the hospital, Hosp. Pract. **11:**63, Feb., 1976.

Faur, Y. C., and others: Isolation of *Neisseria meningitidis* from the genito-urinary tract and anal canal, J. Clin. Microbiol. **2:**178, 1975.

Favero, M. S., and others: *Pseudomonas aeruginosa:* growth in distilled water from hospitals, Science **173:**836, 1971.

Feeney, R.: Preventing rheumatic fever in school children, Am. J. Nurs. **73:**265, 1973.

Feldman, H. A.: Some recollections of the meningococcal diseases, J.A.M.A. **220:**1107, 1972.

Feldman, S., and Pearson, T. A.: *Limulus* test and gram-negative bacillary sepsis, Am. J. Dis. Child. **128:**172, 1974.

Felsenfeld, O.: Borrelia, St. Louis, 1971, Warren H. Green, Inc.

Finegold, S. M., and Rosenblatt, J. E.: Practical aspects of anaerobic sepsis, Medicine **52:**311, 1973.

Finegold, S. M., and others: Scope monograph on anaerobic infections, Kalamazoo, Mich., 1974, The Upjohn Co.

Fisher, M. W.: Development of immunotherapy for infections due to *Pseudomonas aeruginosa*, J. Infect. Dis. **130** (suppl.):149, 1974.

Fleming, W. L., and others: "Clinical gonorrhea" in the female: laboratory findings, South. Med. J. **65:**890, 1972.

Fossieck, B., Jr., and others: Counterimmunoelectrophoresis for rapid diagnosis of meningitis due to *Diplococcus pneumoniae*. J. Infect. Dis. **127:**106, 1973.

Francis, D. P., and others: *Pasteurella multocida,* infections after domestic animal bites and scratches, J.A.M.A. **233:**42, 1975.

Freedman, S. O.: Tuberculin testing and screening: a critical evaluation, Hosp. Pract. **7:**63, May, 1972.

Gilardi, G. L.: Practical schema for the identification of nonfermentative gram-negative bacteria encountered in medical bacteriology, Am. J. Med. Technol. **38:**65, 1972.

Gleckman, R.: The EWYM* syndrome—asymptomatic gonorrhea (*embarrassed, worried young man), J.A.M.A. **234:**321, 1975.

Goldmann, D. A., and others: Guidelines for infection control in intravenous therapy, Ann. Intern. Med. **79:**848, 1973.

Goldwyn, R. M.: Man's best friend (editorial), Arch. Surg. **111:**221, 1976.

Goodgame, R. W., and Greenough, W. G., III: Cholera in Africa: a message for the West, Ann. Intern. Med. **82:**101, 1975.

Gorbach, S. L.: The toxigenic diarrheas, Hosp. Pract. **8:**103, May, 1973.

Gorbach, S. L., and Bartlett, J. G.: Anaerobic infections, N. Engl. J. Med. **290:**1177, 1974.

Gorbach, S. L., and Bartlett, J. G.: Anaerobic infections: old myths and new realities (editorial), J. Infect. Dis. **130:**307, 1974.

Greenberg, J. H., and Madorsky, D. D.: Young physicians' knowledge of venereal disease (editorial), J.A.M.A. **220:**1736, 1972.

Greenwood, B. M., and others: Counter-current immunoelectrophoresis in the diagnosis of meningococcal infections, Lancet **2:**519, 1971.

Handsfield, H. H., and others: Neonatal gonococcal infection. I. Orogastric contamination with *Neisseria gonorrhoeae*, J.A.M.A. **225:**697, 1973.

Hassell, T. A., and Stuart, K. L.: Rheumatic fever prophylaxis: three-year study, Br. Med. J. **2:**39, Apr. 6, 1974.

Hattwick, M. A., and others: Surveillance of Rocky Mountain spotted fever, J.A.M.A. **225:**1338, 1973.

Holdeman, L. V., and Moore, W. E. C., editors: Anaerobe laboratory manual, Blacksburg, Va., 1972, Virginia Polytechnic Institute Education Foundation–Anaerobe Laboratory Fund.

Hume, J. C.: New developments in our knowledge of gonorrhea; their implications for controlling disease, Tex. Med. **72:**45, May, 1976.

Isenberg, H. D.: The ecology of nosocomial disease, ASM News **38:**375, 1972.

Jacobs, N. F., and Kraus, S. J.: Gonococcal and nongonococcal urethritis in men, Ann. Intern. Med. **82:**7, 1975.

Jawetz, E., and others: Review of medical microbiology, Los Altos, Calif., 1976, Lange Medical Publications.

Jelinek, G.: Gonorrhoeae, Nurs. Mirror **137:**30, July 20, 1973.

Johanson, W. G., Jr., and others: Nosocomial respiratory infections with gram-negative bacilli, the significance of colonization of the respiratory tract, Ann. Intern. Med. **77:**701, 1972.

Judson, F. N.: Update in sexually transmitted diseases, J. Am. Med. Wom. Assoc. **31:**11, 1976.

Kaiser, A. B., and Schaffner, W.: Prospectus: the prevention of bacteremic pneumococcal pneumonia, a conservative appraisal of vaccine intervention, J.A.M.A. **230:**404, 1974.

Kampmeier, R. H.: Final report on the "Tuskegee syphilis study," South. Med. J. **67:**1349, 1974.

Karandanis, D., and Shulman, J. A.: Recent survey of infectious meningitis in adults: review of laboratory findings in bacterial, tuberculous, and aseptic meningitis, South. Med. J. **69:**449, 1976.

Kauffmann, F.: Serological diagnosis of *Salmonella* species, Kauffmann-White Schema, Baltimore, 1972, The Williams & Wilkins Co.

Kaufman, R. E., and Wiesner, P. N. J.: Nonspecific urethritis, N. Engl. J. Med. **291:**1175, 1974.

Knittle, M. A., and others: Role of hand contamination of personnel in the epidemiology of gram-negative nosocomial infections, J. Pediatr. **86:**433, 1975.

Krolls, S. O., and others: Oral manifestations of syphilis, Hosp. Med. **8:**14, July, 1972.

Krugman, S., and Ward, R.: Infectious diseases of children and adults, ed. 5, St. Louis, 1973, The C. V. Mosby Co.

Lasagna, L.: The VD epidemic: how it started, where it's going, and what to do about it, Philadelphia, 1975, Temple University Press.

Leino, R., and Kalliomaki, J. L.: Yersiniosis as an internal disease, Ann. Intern. Med. **81:**458, 1974.

Lennette, E. H., and others, editors: Manual of clinical microbiology, Washington, D.C., 1974, American Society for Microbiology.

Lenz, P. E.: Women, the unwitting carriers of gonorrhea, Am. J. Nurs. **71:**716, 1971.

Lewis, R. J., and others: *Diplococcus pneumoniae* cellulitis in drug addicts, J.A.M.A. **232:**54, 1975.

Lincoln, E. M., and Gilbert, L. A.: Diseases in children due to mycobacteria other than *Mycobacterium tuberculosis*. Am. Rev. Respir. Dis. **105:**683, 1972.

Liu, P. V.: Biology of *Pseudomonas aeruginosa*, Hosp. Pract. **11:**138, Jan., 1976.

Maki, D. G.: Preventing infection in intravenous therapy, Hosp. Pract. **11:**95, Apr., 1976.

Maki, D. G., and others: Infection control in intravenous therapy, Ann. Intern. Med. **79:**867, 1973.

Marlow, F. W., Jr.: Syphilis then and now, J.A.M.A. **230:**1320, 1974.

Matsen, J. M.: The sources of hospital infection, Medicine **52:**271, 1973.

McCormick, J. B., and Bennett, J. V.: Public health considerations in the management of meningococcal disease, Ann. Intern. Med. **83:**883, 1975.

McCracken, G. H., Jr.: Neonatal septicemia and meningitis, Hosp. Pract. **11:**89 Jan., 1976.

McGowan, J. E., and others: Meningitis and bacteremia due to *Haemophilus influenzae:* occurrence and mortality at Boston City Hospital in twelve selected years, 1935-1972, J. Infect. Dis. **130:**119, 1974.

McLelland, B. A., and Anderson, P. C.: Lymphogranuloma venereum, outbreak in a university community, J.A.M.A. **235**:56, 1976.

Medical News: Rocky Mountain spotted fever case numbers rise, J.A.M.A. **230**:1503, 1974.

Medoff, G., and others: Listeriosis in humans: an evaluation, J. Infect. Dis. **123**:247, 1971.

Merson, M. H., and others: Current trends in botulism in the United States, J.A.M.A. **229**:1305, 1974.

Moffet, H. L.: Clinical microbiology, Philadelphia, 1975, J. B. Lippincott Co.

Morehead, C. D., and Houck, P. W.: Epidemiology of *Pseudomonas* infections in pediatric intensive care unit, Am. J. Dis. Child. **124**:564, 1972.

Moser, R. H.: Ruminations: holy man and his disease, J.A.M.A. **228**:78, 1974.

Moulder, J. W.: Intracellular parasitism: life in an extreme environment, J. Infect. Dis. **130**:300, 1974.

Munford, R. S., and others: Diphtheria deaths in the United States, 1959-1970, J.A.M.A. **229**:1890, 1974.

Murray, H. W., and others: The protean manifestations of *Mycoplasma pneumoniae* infection in adults, Am. J. Med. **58**:229, 1975.

Musher, D. M., and Schell, R. F.: The immunology of syphilis, Hosp. Pract. **10**:45, Dec., 1975.

Nahmias, A. J., and others: Newer microbial agents in diarrhea, Hosp. Pract. **11**:75, Mar., 1976.

Norins, L. C.: The case for gonococcal serology (editorial), J. Infect. Dis. **130**:677, 1974.

Owen, R. L., and Hill, J. L.: Rectal and pharyngeal gonorrhea in homosexual men, J.A.M.A. **220**:1315, 1972.

Owens, D. W.: General medical aspects of atypical mycobacteria, South. Med. J. **67**:39, 1974.

Pappenheimer, A. M., Jr., and Gill, D. M.: Diphtheria, Science **182**:353, 1973.

Pennington, J. E., and others: *Bacillus* species infection in patients with hematologic neoplasia, J.A.M.A. **235**:1473, 1976.

Peterson, J. C.: Congenital syphilis: a review of its present status and significance in pediatrics, South. Med. J. **66**:257, 1973.

Poindexter, H. A., and Washington, D.: Bacteroidosis, South. Med. J. **68**:995, 1975.

Powell, S., and McDougall, A. C.: Clinical recognition of leprosy: some factors leading to delays in diagnosis, Br. Med. J. **1**:612, Mar. 30, 1974.

Price, D. L., and others: Tetanus toxin: direct evidence for retrograde intraaxonal transport, Science **188**:945, 1975.

Qadri, S. M. H., and others: Incidence and etiology of septic meningitis in a metropolitan county hospital, Am. J. Clin. Pathol. **65**:550, 1976.

Quinn, R. W.: Epidemiology of gonorrhea, South. Med. Bull. **59**:7, Apr., 1971.

Randolph, M. F., and others: Streptococcal pharyngitis, Am. J. Dis. Child. **130**:171, 1976.

Recommendation of the P.H.S. Advisory Committee on Immunization Practices: Meningococcal polysaccharide vaccines, Ann. Intern. Med. **84**:179, 1976.

Recommendation of the P.H.S. Advisory Committee on Immunization Practices: Meningococcal polysaccharide vaccines, Morbid. Mortal. Wk. Rep. **24**:381, 1975.

Read, S. E., and Reed, R. W.: Electron microscopy of the replicative events of A_{25} bacteriophages in group A streptococci, Can. J. Microbiol. **18**:93, 1972.

Reichman, L. B.: Tuberculosis care: when and where? Ann. Intern. Med. **80**:402, 1974.

Rief, J. S., and Marshak, R. R.: Leptospirosis: a contemporary zoonosis (editorial), Ann. Intern. Med. **79**:893, 1973.

Reimann, H. A.: The pneumonias, St. Louis, 1971, Warren H. Green, Inc.

Rensberger, B., and Roueché, B.: When Americans are a swallow away from death, Today's Health **49**:40, Sept., 1971.

Rhamy, R. K.: Gonococcal complications in genitourinary system of the male, South. Med. Bull. **59**:29, Apr., 1971.

Riley, H. D., Jr.: Meningococcal disease and its control, South. Med. J. **66**:107, 1973.

Rocky Mountain spotted fever (editorial), J.A.M.A. **225**:1372, 1973.

Rodman, M. J.: Drugs for treating tetanus, RN **34**:43, Dec., 1971.

Rose, H. D., and Babcock, J. B.: Colonization of intensive care unit patients with gram-negative bacilli, Am. J. Epidemiol. **101**:495, 1975.

Rosebury, T.: Microbes and morals: the strange story of venereal disease, New York, 1976, Ballantine Books, Inc.

Rosenblatt, J. E., and Stewart, P. R.: Anaerobic bag culture method, J. Clin. Microbiol. **1:**527, 1975.

Rothenberg, R. B., and others: Efficacy of selected diagnostic tests for sexually transmitted diseases, J.A.M.A. **235:**49, 1976.

Rowley, D.: Endotoxins and bacterial virulence, J. Infect. Dis. **123:**317, 1971.

Rubio, T., and Riley, H. D., Jr.: Serious systemic infection associated with the use of indwelling intravenous catheters, South. Med. J. **66:**633, 1973.

Rudolph, A. H.: Control of gonorrhea, guidelines for antibiotic treatment, J.A.M.A. **220:**1587, 1972.

Runyon, E. H.: Identification of mycobacterial pathogens utilizing colony characteristics, Am. J. Clin. Pathol. **54:**578, 1970.

Runyon, E. H.: Whence mycobacteria and mycobacterioses? (editorial), Ann. Intern. Med. **75:**467, 1971.

Russell, P., and Altshuler, G.: Placental abnormalities of congenital syphilis, Am. J. Dis. Child. **128:**160, 1974.

Rytel, M. W.: Counterimmunoelectrophoresis in diagnosis of infectious disease, Hosp. Pract. **10:**75, Oct., 1975.

Sanders, D. Y.: Complications of ear piercing, Wom. Physician **26:**459, 1971.

Sanders, W. E.: Diphtheria: "From miasmas to molecules" revisited (editorial), Ann. Intern. Med. **75:**639, 1971.

Sayeed, Z. A., and others: Gonococcal meningitis, J.A.M.A. **219:**1730, 1972.

Schachter, J., and others: Are chlamydial infections the most prevalent venereal disease? J.A.M.A. **231:**1252, 1975.

Sexton, D. J., and Burgdorfer, W.: Clinical and epidemiologic features of Rocky Mountain spotted fever in Mississippi, 1933-1973, South. Med. J. **68:**1529, 1975.

Shirai, A., and others: Indirect hemagglutination test for human antibody to typhus and spotted fever group rickettsiae, J. Clin. Microbiol. **2:**430, 1975.

Shubin, H., and Weil, M. H.: Bacterial shock, Emergency Med. **8:**51, Apr., 1976.

Silberg, S. S., and others: Epidemiologic aspects of nosocomial infections, South. Med. J. **69:**312, 1976.

Skinsnes, O. K.: Histochemical analysis of leprosy tissue leads to cultivation of leprosy bacilli, Arch. Pathol. Lab. Med. **100:**173, 1976.

Smilack, J. D.: Group Y meningococcal disease, Ann. Intern. Med. **81:**740, 1974.

Smith, L.: The pathogenic anaerobic bacteria, Springfield, Ill., 1975, Charles C Thomas, Publisher.

Southern, P. M., and Sanford, J. P.: Relapsing fever, Medicine **48:**129, 1969.

Spitz, B.: In the hospital, a bug is no joke, Hosp. Top. **50:**42, Jan., 1972.

Status of immunization in tuberculosis in 1971: report of a conference on progress to date, future trends and research needs, by the John E. Fogarty International Center for Advanced Study in the Health Sciences, (Bethesda, Oct., 1971), Washington, D.C., 1972 U.S. Government Printing Office.

Sterne, M.: Pathogenic *Clostridia*, Reading, Mass., 1975, Butterworth (Publishers), Inc.

Storrs, E. E., and others: Leprosy in the armadillo: new model for biomedical research, Science **183:**851, 1974.

Subak-Sharpe, G.: The venereal disease of the new morality, Today's Health **53:**42, Mar., 1975.

Sullivan, R. J., Jr., and others: Adult pneumonia in general hospital, Arch. Intern. Med. **129:**935, 1972.

Sutter, V. L., and others: Anaerobic bacteriology, Los Angeles, 1972, UCLA Extension Division.

Syphilis: the same old disease (editorial), South. Med. J. **65:**250, 1972.

Taylor, A., Jr.: Botulism and its control, Am. J. Nurs. **73:**1380, 1973.

Thomas, B. J.: Leprosy, an ancient scourge—a continuing problem, RN **38:**47, Mar., 1975.

Thompson, T. R., and others: Gonococcal ophthalmia neonatorum, J.A.M.A. **228:**186, 1974.

Thorsteinsson, S. B., and others: Clinical manifestations of halophilic noncholera vibrio infections, Lancet **2:**1283, 1974.

Tindall, J. P., and Harrison, C. M.: *Pasteurella multocida* infections following animal injuries, especially cat bites, Arch. Dermatol. **105:**412, 1972.

Todd, J. K., and Bruhn, F. W.: Severe *Haemophilus influenzae* infections, Am. J. Dis. Child. **129:**607, 1975.

Top, F. H., Sr., and Wehrle, P. F., editors: Communicable and infectious diseases, ed. 8, St. Louis, 1976, The C. V. Mosby Co.

Turk, D. C., and Porter, I. A.: A short textbook of medical microbiology, Chicago, 1975, Year Book Medical Publishers, Inc.

Tyeryar, F. J., Jr., and others: DNA base composition of rickettsiae, Science **180:**415, 1973.

Van Beek, A., and others: Nonclostridial gas-forming infections, Arch. Surg. **108:**552, 1974.

Vandermeer, D. C.: Meet the VD epidemiologist, Am. J. Nurs. **71:**722, 1971.

Von der Muehll, E., and others: A new test for pathogenicity of staphylococci, Lab. Med. **3:**26, Jan., 1972.

Wannamaker, L. W., and Matsen, J. M., editors: Streptococci and streptococcal diseases; recognition, understanding, and management, New York, 1972, Academic Press, Inc.

Weg, J. G.: Tuberculosis and the generation gap, Am. J. Nurs. **71:**495, 1971.

Weinstein, L.: Tetanus, N. Engl. J. Med. **289:**1293, 1973.

White, P. C., Jr., and others: Brucellosis in Virginia meat-packing plant, Arch. Environ. Health **28:**263, 1974.

Woodward, T. E.: A historical account of the rickettsial diseases with a discussion of unsolved problems, J. Infect. Dis. **127:**583, 1973.

Yow, M.: Group B streptococci: a serious threat to the neonate (editorial), J.A.M.A. **230:**1177, 1974.

28 Viruses

GENERAL CONSIDERATIONS

Definition. Viruses are agents that cause infectious disease and that can only be propagated in the presence of living tissues (that is, they are obligatory intracellular parasites). If we say that the least requirement for life is the ability of a living thing to duplicate or reproduce itself, viruses are the smallest known living bodies. Most are so small that they cannot be seen with an ordinary light microscope, and some are so small that they approximate the size of the large protein molecules. Viruses seem to lie in a partially explored twilight zone between the cells the biologist studies, on the one hand, and the molecules with which the chemist deals, on the other, being smaller than the smallest known bacterial cells and just larger than the largest macromolecules.* (It would take 2500 poliovirus particles to span the point of a pin.) Viruses can pass through filters that retain all ordinary bacteria and for many years were called filtrable viruses.

Our knowledge of the nature and structure of viruses and their disease-producing potential has increased greatly in the last several decades, especially since the invention of the electron microscope, the introduction of the ultracentrifuge, the development of modern cell (tissue) culture technics, and the applications of immunofluorescence and cytochemistry. As a result, many viruses previously unknown have been isolated and identified—in the last 15 to 20 years, over 100. Also, more has been learned about the natural history or *life cycle* of a virus.

Structure. Viruses are particulate and vary considerably in size (Table 28-1) and shape (Fig. 28-1). Generally, the viruses of man and animals are spherical, those of plants are rod shaped or many sided, and those of bacteria (bacterial viruses or *bac-*

---

*Dr. Wendell Meredith Stanley (1904-1971), biochemist and virologist, defined viruses as follows: "The virus is one of the great riddles of biology. We do not know whether it is alive or dead, because it seems to occupy a place midway between the inert chemical molecule and the living organism." (From Alvarez, W.: Mod. Med. **35**:78, Jan. 30, 1967.)

TABLE 28-1. RELATIVE SIZES OF VIRUSES

Biologic unit	Approximate diameter (or diameter × length) in nm (mμ)*
Red blood cell†	7500
Chlamydiae†‡	300-800
Poxvirus	230×300
Rhabdovirus	60×225
Coronavirus	80-160
Paramyxovirus	100-300
Myxovirus	80-120
Herpesvirus	110-200
Bacterial virus	25-100
Adenovirus	70-80
Leukovirus	100
Reovirus	70-75
Rubella virus	50-60
Papovavirus	40-55
Arbovirus (togavirus)	40
Picornavirus	18-30
Parvovirus (picodnavirus)	18-24
Certain plant viruses	17-30
Serum albumin molecule†	5

*1nm (mμ) = 1/1000 μm (μ) = 1/1,000,000 mm.
†These are not viruses; they are shown here for comparison.
‡See p. 475.

TABLE 28-2. COMPARISON OF VIRUSES WITH OTHER INFECTIOUS MICROBES*

Agent	Presence of nucleic acids	Reproduction	Possession of metabolic enzymes	Obligate intracellular parasite	Rigid cell wall	Antibiotic susceptibility
Viruses	Either DNA or RNA (not both)	Replication (synthesis, then assembly of subunits)	None	Yes	None	None
Chlamydiae (bedsoniae)	DNA and RNA	Binary fission	Yes (parts of enzyme systems present)	Yes	May be present	Yes
Rickettsias	DNA and RNA	Binary fission	Yes	Yes	Yes	Yes
Mycoplasmas	DNA and RNA	Binary fission	Yes	No	None	Yes
Other bacteria†	DNA and RNA	Binary fission	Yes	No	Yes	Yes

*After Moulder, J. W.: Hosp. Pract. 3:35, June, 1968.
†Note that chlamydiae, rickettsias, and mycoplasmas are now classed as bacteria.

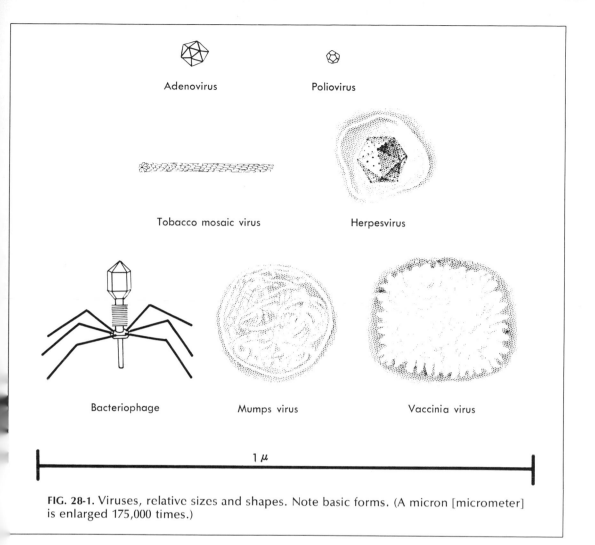

FIG. 28-1. Viruses, relative sizes and shapes. Note basic forms. (A micron [micrometer] is enlarged 175,000 times.)

teriophages) are tadpole shaped. Once thought to be fairly simple, viruses are known to be highly complex structures in which viral components are fitted together into rigid geometric patterns with mathematical precision. In fact, a highly purified preparation of virus is referred to as a virus crystal (Fig. 28-2). There are three basic forms in viral structure—the icosahedral, helical, and complex enveloped. The *icosahedron*, a crystal, is a solid, many-sided geometric form with 20 triangular faces and 12 apexes. The *helical* form indicates a spiral tubular structure bound up to make a compact, long rod. The *complex-enveloped* form is enclosed by a loose covering envelope, and because the envelope is nonrigid, the virus is highly variable in size and shape.

In the simplest viruses there is a long coil of nucleic acid sufficient for several hundred genes, the *chromosome* of the virus, tightly folded and packed within a protein coat called the *capsid*. The capsid is made up of subunits or *capsomeres* (aggregates of oligopeptides) arranged in precise fashion around the nucleic acid core. In the larger viruses the more complicated chemical envelope is seen, wherein fatty

489

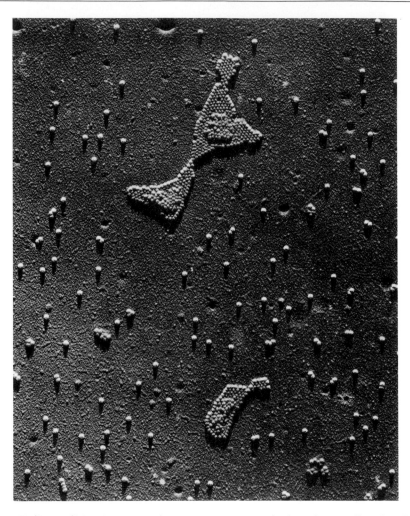

FIG. 28-2. Poliomyelitis virus crystal (type 1 strain purified and crystallized), electron micrograph. Individual viral particles are 28 nm in diameter. In the absence of any immunity one crystal of poliovirus contains enough particles to infect population of the world. (×64,000.) (Courtesy Parke, Davis & Co., Detroit, Mich.)

substances and complex sugars (but no enzymes) are linked to the protein coat. The viral particle as a unit is a *virion*.*

Cells of bacteria, plants, and animals contain both deoxyribonucleic acid (DNA) and ribonucleic acid (RNA). True viruses, however, contain only one. DNA serves as the genetic material in complex viruses, such as vaccinia and bacteriophages, as it does for all other living things. RNA is the genetic material not only in complex viruses

---

*A *viroid* is a new kind of agent recently described. Said to be like a virus, it is a self-replicating, infectious entity 1/80 the size of the smallest known virus and is thought to be single-stranded, free RNA of low molecular weight. It may be the responsible agent in some forms of cancer, but presently its importance is as cause for certain plant diseases (examples: potato spindle tuber disease and citrus exocortis disease).

such as the influenza viruses but also in some of the simplest and smallest such as the polioviruses. The DNA or RNA may be either single stranded or double stranded.

Table 28-2 highlights some of the differences between viruses and other microbes.

Life cycle. The life cycle of a virus may be briefly sketched as follows: A virus contacts and parasitizes a susceptible cell in a given host. In some way the protein coat of the virion seems to be able to find just the right cell. Besides protecting the viral particle during its extracellular existence, the protein coat can attach to the cell under attack. Next the protein coat is stripped off, and viral nucleic acid only (the chromosome of the virus) penetrates the parasitized cell, probably aided by enzymes from that cell. This kind of phagocytosis is called *viropexis*. The dissembled virus appears to vanish into the cytoplasmic maze of the cell.

Invasion of the cell by the virus is an act of piracy. Because a virus contains such a small number of proteins, it cannot do much, if any, metabolic work. It is not capable of independent reproductive activity. It must seek out an environment in which chemical building blocks and energy for work are provided, conditions existing only inside a cell. A virus must take nourishment from the parasitized cell, and it can multiply only within the confines of that cell. Since its genetic material is carried into the cell, the information needed to make new viruses inside that cell is available.* In fact, viral RNA can initiate the life cycle of the virus from which it was isolated even when chemically separated from the protein coat. The entry of viral nucleic acid into the host cell results in an infected cell that is then immune; it cannot be reinfected by the same or related viruses.

A variable period of time (the eclipse) elapses after the resources of the cell have been commandeered before new viral particles are released. At the end of this phase, swarms of full-grown viruses appear and escape from the cell. (In the case of poliovirus [Fig. 28-2] in only a few hours a single parasitized cell has produced 100,000 poliovirus particles.) When released, the new generation of viruses is able to survive outside the cell until it can reach the susceptible cells of another host, but it must find new living quarters.

Pathogenicity. In many instances, little or no harm is occasioned to the host in the life cycle of a virus, and many viral infections are silent, inapparent ones. For example, unrecognized infection with polioviruses is hundreds of times more common than the clinical disease. Some infections are latent ones, undetectable until activated by the proper stimulus, for example, infections with herpes simplex. However, if the cells are damaged by the viral attack, disease exists, and the signs of viral infection naturally reflect the anatomic location of the cells.

The pathology of viral diseases relates primarily to the visible effects of intracellular parasitism. When cells that have been specifically attacked by a virus show changes that directly reflect cell injury, the sum total of these is designated the *cytopathogenic* or *cytopathic effect* (CPE), an especially useful concept in cell (tissue) cultures.

---

*Viruses act as independently existing genes.

TABLE 28-3. INCLUSION BODIES IN INFECTIONS WITH INTRACELLULAR PARASITES

Disease	Etiologic agent	Location of inclusion*	Name of inclusion body or comment
Rabies	Virus	Cytoplasm	Negri bodies
Yellow fever	Virus	Cytoplasm	Councilman bodies — areas of necrosis in cell
Cytomegalic inclusion disease	Virus	Nucleus, also cytoplasm	Cytomegalic inclusions; prominent in enlarged cells
Molluscum contagiosum	Virus	Cytoplasm	Molluscum bodies; elementary bodies are Lipschütz bodies
Varicella-zoster	Virus	Nucleus	Prominent inclusions
Herpes simplex	Virus	Nucleus	Prominent inclusions
Vaccinia-variola	Virus	Cytoplasm	Guarnieri bodies; elementary bodies are Paschen bodies
Adenovirus pneumonia	Virus	Nucleus	Inclusions of rosette type
Measles giant cell pneumonia	Virus	Nucleus, also cytoplasm	Prominent within syncytial giant cells
Trachoma	Chlamydia	Cytoplasm	Prominent
Psittacosis	Chlamydia	Cytoplasm	Psittacosis bodies
Lymphopathia venereum	Chlamydia	Cytoplasm	Gamna-Favre bodies
Granuloma inguinale	Bacterium	Cytoplasm	Donovan bodies

*Note that what is called an inclusion body is a body visible with the *compound light microscope*. The electron microscope visualizes viral particles throughout the cell even in instances where there is a characteristic inclusion body in only one area.

In some cells small, round, or oval bodies known as *inclusion bodies* will be formed as a manifestation of the cytopathic effect. Inclusion bodies are typical in size, shape, and intracellular position for a given virus. They are emphasized by the pathologist, since they are usually easily visualized in the compound light microscope. In stained preparations inclusion bodies are commonly about the size of a red blood cell and rounded. Some viruses produce inclusion bodies in the cytoplasm of the parasitized cell and some in the nucleus; others produce them in both cytoplasm and nucleus. At times within the inclusion bodies still smaller units known as *elementary bodies* are seen. The elementary bodies are the virus particles, and an inclusion body with its elementary bodies may be likened to a colony of bacteria. Table 28-3 lists some infections where an intracellular parasite is related to an inclusion body.

Another reflection of cytopathogenic effect of virus is the formation of a giant cell or syncytium. This is prominent in measles and cytomegalic inclusion disease (Fig. 28-3).

Mechanisms of viral injury to the infected cell are not well understood. There are several possibilities. Viral particles may interfere with cell function by altering protein synthesis, or they may cause chromosomal changes, even to incorporating viral nucleic acid into the host cell genome. The presence of virus may mean a direct injurious effect. Certain viruses such as poliovirus can activate and release enzymes contained in the lysosomes of the cell. In whatever way sustained, progressive injury to a parasitized cell may mean that the cell is either partly or completely destroyed.

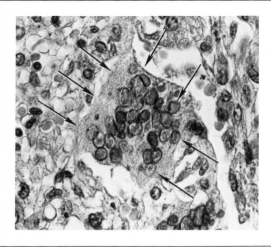

FIG. 28-3. Inclusion bodies of measles virus within nuclei of multinucleated giant cell (arrows), as seen in cell cultures and lymphoid tissues of patients with measles. (Microscopic section of lung from patient with rare measles giant cell pneumonia.) Note many nuclei piled on one another in murky cytoplasm. In each, chromatin is displaced to nuclear membrane by inclusion body. Viral particles in cytoplasm are not shown in routine hematoxylin-eosin stain. (×800.)

There is an opposite effect that viruses at times produce on parasitized cells, an effect not seen in relation to any other known living agent. Viruses can stimulate the parasitized cells to increase their numbers in a way the cells would not do if they did not contain virus. Such an increase in number of cells is referred to as *hyperplasia* (virus induced). It is this tissue response of cells parasitized by certain viruses that implicates viruses as a possible cause of cancer in human beings. Viruses may not only destroy cells or stimulate them to proliferate but may also induce a reaction in parasitized tissue cells that combines features of both destruction and hyperplasia.

Two factors influencing the response of cells to viral injury are (1) the rapidity with which a virus produces its effects—the more rapid its action, the more likely for cells to be damaged severely, and (2) the power of the cells injured to multiply. Nerve cells regulating sensory and motor function in the body cannot duplicate themselves *(regenerate);* if parasitized, they tend to be killed in the process. Cells lining the skin possess great powers of regeneration; therefore viral lesions in the lining cells of the skin may result in a piling up of increased numbers of epithelial cells at the site of injury. If not followed by any kind of cellular deterioration, virus-induced hyperplasia is not accompanied by any detectable inflammation.

There is often little in the way of measurable inflammatory response in viral infection. Where there is cell destruction, the tissues may show a minimal to moderate infiltration of inflammatory cells (lymphocytes and mononuclear cells, a few neutrophils) and vascular changes (edema and hyperemia) of limited extent.

Viruses are responsible for more than 80 diseases of plants and many diseases of the lower animals. Practically all plants of commercial importance are attacked by them.* Viruses also attack insects. Of more than 550 viruses known, better than 200 produce over 50 viral diseases in man, some of which are the most highly communicable and dangerous diseases known. Today the prevalent infections in man are viral.

---

*Viruses have been known to attack the fungus from which penicillin is derived, thereby interfering with its manufacture.

Classification. At present, most virologists seem to favor a classification system based on major biologic properties such as nucleic acid present, size, structure (geometric pattern of capsid and number of capsomeres), sensitivity to physical and chemical agents (ether and chloroform), immunologic aspects, epidemiologic features, and pathologic changes. Consistent with this, most animal viruses fall into 13 groups, eight with a DNA core (deoxyriboviruses) and five with an RNA one (riboviruses). The important groups (corresponding to genera) are outlined as follows.

1. Poxviruses
2. Herpesviruses
3. Adenoviruses
4. Papovaviruses
5. Parvoviruses (picodnaviruses)
6. Myxoviruses
7. Paramyxoviruses
8. Rhabdoviruses
9. Arboviruses (togaviruses)
10. Reoviruses
11. Picornaviruses
12. Leukoviruses
13. Coronaviruses
14. Unclassified viruses

POXVIRUSES. The largest and most complex of the animal viruses, these brick-shaped, enveloped, double-stranded DNA viruses form characteristic inclusions in the cytoplasm of parasitized cells. Poxviruses may be categorized into three groups: (1) poxviruses of mammals, (2) poxviruses of birds, and (3) oncogenic poxviruses (myxoma and fibroma). There are 22 members whose action may either destroy cells or stimulate them to proliferate. The skin is a prime target, and viruses included are those of smallpox (variola), vaccinia, molluscum contagiosum, and cowpox.

HERPESVIRUSES. These medium-sized ether-sensitive, enveloped, double-stranded DNA viruses, like poxviruses, often parasitize lining cells of skin and can pass from cell to cell without killing. The characteristic viral inclusions are found in the nucleus. The protein shells of herpesviruses show cubic symmetry, with 162 capsomeres. Latent infections with these agents may endure a lifetime. There are 20 members including varicella-zoster virus, herpes simplex virus types I and II, Epstein-Barr (EB) virus, and cytomegaloviruses. (Some oncogens are found here.)

ADENOVIRUSES. These medium-sized, ether-resistant, double-stranded DNA viruses may persist for years (latent) in human lymphoid tissue. Adenoviruses have cubic symmetry, with 252 capsomeres. There are 31 human, 12 simian, and 2 bovine types. Adenoviruses produce characteristic cytopathogenic changes in cell culture, and their inclusion bodies are within the nucleus. These agents produce a range of diseases; some cause tumors in animals.

PAPOVAVIRUSES. These small ether-resistant, nonenveloped, double-stranded DNA (circular) viruses are noted for producing neoplasms in animals (oncogenic). Their growth cycles are relatively slow, and they replicate in the nucleus of the parasitized cell. Capsid symmetry is cubic. Papovaviruses include the *pa*pilloma viruses of man, rabbits, cows, and dogs; the *po*lyoma virus of mice; and the *va*cuolating virus of monkeys (SV/40), all known oncogenic (tumor-producing) agents. There are at least 11 of them.

PARVOVIRUSES (PICODNAVIRUSES). These are small, ether-resistant DNA viruses. Capsid symmetry is cubic. Certain adeno-associated or adenosatellite viruses (defective viruses not able to replicate without adenovirus) and certain viruses of hamsters, rats, and mice are found here. They can produce latent infections in animals.

MYXOVIRUSES. These medium-sized, spherical, ether-sensitive, enveloped, single-stranded RNA viruses have helical symmetry and replicate in the nucleus of the parasitized cell. Viral particles are pleomorphic, sometimes filamentous. Members of the myxovirus group agglutinate red blood cells of many mammals and birds. This hemagglutination is associated with the viral particle itself and inhibited by antibody acting against it. Some contain an enzyme, neuraminidase, capable of splitting neuraminic acid from mucoproteins. Myxoviruses (or orthomyxoviruses) include the influenza viruses A, B, and C, the virus of swine influenza, and that of fowl plague.

PARAMYXOVIRUSES. These medium-sized, ether-sensitive, enveloped, single-stranded RNA viruses have helical symmetry. They are similar in appearance to but somewhat larger than myxoviruses. Paramyxoviruses appear to be antigenically stable, and certain ones hemagglutinate red blood cells, as do myxoviruses, with or without hemolysis. Replication in cell cultures occurs within cytoplasm. Some cause multinucleated giant cells to form in tissue cultures and sometimes in human tissues (for example, measles virus). Distinctive inclusion bodies are seen with certain ones. Within this category are the parainfluenza viruses (four types), respiratory syncytial virus, the viruses of measles and mumps in man, and Newcastle disease and distemper in animals.

RHABDOVIRUSES. These ether-sensitive, single-stranded RNA viruses have helical symmetry. Members of this group have an unusual appearance. Mature virions are shaped like bells or bullets. Intracytoplasmic inclusions, the Negri bodies, are seen with the rabies virus. Rhabdoviruses include the viruses of rabies and vesicular stomatitis of cattle, some insect viruses, and three important plant viruses.

ARBOVIRUSES (TOGAVIRUSES*). These small, ether-sensitive, enveloped, RNA viruses (*ar*thropod-*bo*rne) have a complex life cycle involving biting (hemophagous) insects, especially mosquitoes and ticks. Arboviruses (togaviruses) can multiply in many species—man, horses, birds, bats, snakes, and insects. With the exception of dengue fever and urban yellow fever, man is only an accidental host. They are most prevalent in the tropics, notably in the rain forests. Three disease patterns are prominent; one is fairly mild, and the other two are severe and often fatal. They include (1) fever (denguelike), with or without skin rash; (2) encephalitis; and (3) hemorrhagic fever, as its name suggests, with skin hemorrhages and visceral bleeding.

Presently, on the basis of antigenicity, togaviruses are laid out in groups A and B. Other arthropod-borne viruses resembling togaviruses such as the Bunyamwera supergroup and the arenaviruses are tentatively aligned with them. Groups A and B togaviruses are best known. Included in group A are the viruses of western equine encephalitis, eastern equine encephalitis, and Venezuelan equine encephalitis, and in group B are Japanese B encephalitis, St. Louis encephalitis, Murray Valley encephalitis, West Nile fever, dengue fever, and yellow fever. Bunyamwera supergroup comprises 100 viruses or more, including that of California encephalitis. *Arenaviruses* (enveloped, single-stranded RNA viruses) were so designated because of their unique

---

*Note the new taxonomic designation for arboviruses.

electron-microscopic appearance. They include the Tacaribe viruses of South American hemorrhagic fevers and the viruses of Lassa fever and lymphocytic choriomeningitis.

The ecologic hodgepodge of this total group with over 220 members is reflected in the names of many exotic diseases such as Bwamba fever, Singapore splenic fever, Kyasanur Forest disease of India, and O'nyong-nyong infection in Uganda.

REOVIRUSES. These are medium-sized, ether-resistant RNA (uniquely double-stranded as a group) viruses. Capsid symmetry is cubic, with 92 capsomeres. Reoviruses replicate within the cytoplasm. The three members in this group were originally so named because of their presence in the respiratory tract and in the enteric canal and because of their orphan status (*respiratory* *enteric* *orphans*). (Orphans are not known to produce disease.) The term *diplornavirus* is suggested to take in the reoviruses of mammals, wound tumor virus of plants, Colorado tick fever virus, and certain arboviruses with double-stranded RNA. A number of reovirus strains have been recovered from patients in Africa with Burkitt's lymphoma. They are excellent inducers of interferon.

PICORNAVIRUSES. These include the smallest known, simplest, and most readily crystallizable RNA viruses. The term *picorna* was coined to designate enteric and related viruses. *Pico* is for very small viruses, and *rna* indicates their nucleic acid. In addition, *p* is for polioviruses, the first known members of the group; *i* is for insensitivity to ether, a distinguishing feature of the group; *c* is for coxsackieviruses; *o* is for orphan or echoviruses; and *r* is for rhinoviruses, further indicating their membership. Capsid symmetry is cubic, with 32 capsomeres.

The nearly 200 members are subdivided into (1) enteroviruses, including polioviruses (three types), coxsackieviruses (29 types), and echoviruses (31 types); and (2) rhinoviruses (100 types), the major causes of the common cold. A rhinovirus in cattle causes foot-and-mouth disease. Picornaviruses produce a wide range of diseases in many areas of the human body, the best known of which is poliomyelitis.

LEUKOVIRUSES. Also called C-type particles, these small, ether-sensitive, enveloped RNA viruses of known structure are member viruses of the avian leukosis complex and the viruses of murine and feline leukemias. No such viruses have yet been identified in man.

CORONAVIRUSES. Ether-sensitive, enveloped RNA viruses, the members of this group are similar to myxoviruses. Symmetry is helical, and the surface projections are petal shaped. Replication is cytoplasmic. In this group are included the viral agents of avian infectious bronchitis, mouse hepatitis, and certain human respiratory viruses.

UNCLASSIFIED VIRUSES. These include viruses on which pertinent information is lacking. This category includes the agents of well-known diseases such as rubella and viral hepatitis, types A and B. Rubella virus, a small, ether-sensitive, enveloped RNA virus, does not seem to fit readily into existing schemes. Only one antigenic type has been detected.

Categories as to source. When viruses are emphasized as to their source, the following three categories are named: (1) enteroviruses, or those isolated from the alimentary tract, (2) respiratory viruses, or those isolated from the respiratory tract, and (3) arboviruses (arthropod-borne), or those isolated from insects. (The term *arbovirus* emphasizes a biologic feature, that of insect transmission.)

Sometimes, in the classification of viruses, emphasis is given to the anatomic area in the body where the virus produces its dramatic effects. Viruses so distinguished include (1) those whose typical lesions appear on the skin and mucous membranes— *dermotropic viruses* of smallpox, measles, chickenpox, and herpes simplex; (2) those related to acute infection of the respiratory tract—*pneumotropic viruses* of the common cold, influenza, and viral pneumonia; (3) those that primarily affect the central nervous system—*neurotropic viruses* of rabies, poliomyelitis, and encephalitis; and (4) those in a miscellaneous group with no common organ system (each virus having its own special affinity for a given organ—*viscerotropic viruses* of viral hepatitis (liver), mumps (salivary glands), and other diseases. This restricted scheme of classification, although of some interest to the pathologist, is far from rigid, since certain viruses or groups of viruses induce various disease processes. In some instances, a virus may invade the body without necessarily attacking the part that it ordinarily

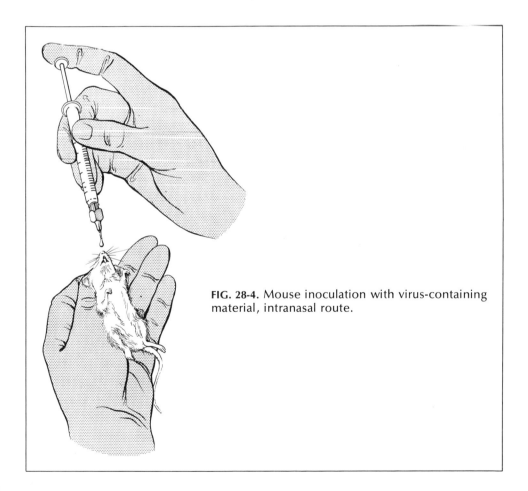

FIG. 28-4. Mouse inoculation with virus-containing material, intranasal route.

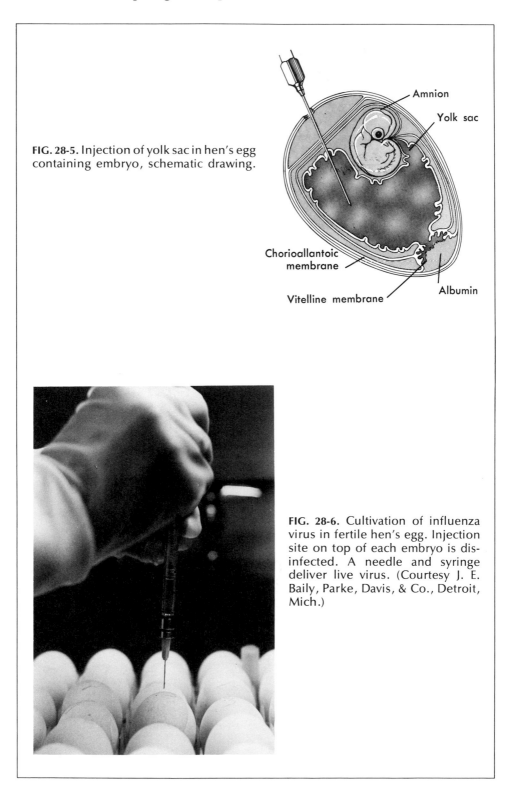

FIG. 28-5. Injection of yolk sac in hen's egg containing embryo, schematic drawing.

Amnion

Yolk sac

Chorioallantoic membrane

Vitelline membrane

Albumin

FIG. 28-6. Cultivation of influenza virus in fertile hen's egg. Injection site on top of each embryo is disinfected. A needle and syringe deliver live virus. (Courtesy J. E. Baily, Parke, Davis, & Co., Detroit, Mich.)

does. Further knowledge of viruses is revealing the fact that certain ones thought limited in their preference for certain cells can infect a number of different cells in man and animals. The viruses of poliomyelitis, thought to be highly neurotropic for many years, are now known to be more flexible, with all three types growing well in tissue cultures of many different cells.

Viral diseases may be either generalized or localized. In *generalized* infection the virus is disseminated by the bloodstream throughout the body but without significant localization, even though a skin rash may be present. Examples of generalized infections include smallpox, vaccinia, chickenpox, measles, yellow fever, rubella, dengue fever, and Colorado tick fever. In some viral diseases there is restriction of viral effect to a particular organ to which the virus travels by the bloodstream, peripheral nerves, or other body route. Examples of *localized* viral infections include poliomyelitis, the encephalitides, rabies, herpes simplex, warts, influenza, common cold, mumps, and hepatitis.

Cultivation. To repeat, viruses multiply only inside living cells. Viruses in virus-containing material may be artificially brought into contact with living cells by (1) inoculation of a susceptible animal such as a suckling mouse (Fig. 28-4), (2) inoculation of cell (tissue) culture systems, and (3) inoculation of the membranes or cavities of the developing chick embryo (Fig. 28-5).

The chick embryo method is carried out as follows: A fertile egg is incubated for 7 to 15 days. With a syringe and needle (Fig. 28-6), the virus-containing material is injected into the membranes of the embryo or into one of the cavities connected with its development (yolk sac, amniotic cavity, or allantoic sac). This may be done through a window made in the shell of the egg over the embryo or through a hole drilled in the shell. Of course, *aseptic technic is mandatory*. After inoculation, the egg is incubated 48 to 72 hours, time for the virus to multiply. The material in which the virus has multiplied is then removed, used for the study of the virus, or sometimes purified, processed, and used as a vaccine.

One of the most widely used laboratory methods of growing viruses today is the cell (tissue) culture method, which has come into its own within the last 4 decades. It has been made possible because of discoveries in basic technics such as the preparation of improved culture media for living cells. The use of antibiotics has eliminated bacterial contamination, and the recognition of the cytopathogenic effect has furnished a measure of viral injury in growing cells. Cytopathic changes require virus living and growing in the culture.

Formation of *plaques* (precise areas of cell degeneration) in the cell culture under a layer of agar may also reflect viral growth as does the phenomenon of *hemadsorption*. Hemadsorption is displayed under specified conditions when either guinea pig or chicken erythrocytes incubated in the culture system adhere in clumps to the infected host cells.

There are three standard cell culture systems, using (1) human fibroblasts from embryo lung, (2) a neoplastic line of cells, or (3) primary monkey kidney.

Both chick embryo and cell cultures enable the laboratory workers to obtain large quantities of virus for study, for vaccine production, or whatever purpose.

Laboratory diagnosis of viral infections. Many procedures are available for the laboratory diagnosis of viral infections, although some are most efficiently carried out in specially equipped laboratories. These include (1) technics for isolation and identification of the virus, (2) serologic tests done serially during the course of infection, (3) fluorescent antibody technics to detect viral antigen directly in lesions, (4) microscopic examination of the lesions, and (5) skin tests.

Test material may be inoculated into the embryonated hen's egg, cell culture, or a susceptible laboratory animal (mouse, guinea pig, cotton rat, rabbit, monkey), from which the virus is recovered and identified (Fig. 28-7). The appearance of typical lesions or inclusion bodies may be diagnostic.

Important in the identification of viruses are several serologic procedures, of which the virus neutralization, hemagglutination-inhibition, complement fixation, and precipitin reaction tests are especially valuable. For the serologic evaluation of a patient's serum it is important that *paired samples* of blood be taken. The first is collected as early as possible in the acute stages of disease, and the second is drawn 2 to 4 weeks later during convalescence. The significant finding is at least a fourfold increase in the content of antibodies between the specimens.

Viruses contain good antigens in their makeup (that is, in the capsid) and, as a result, stimulate antibody formation. One of the most important antibodies formed in man and animals with viral infection is the virus-neutralizing antibody, an antibody that neutralizes or obliterates the destructive capacity of the virus. This humoral antibody is an important constituent of immune serum given for viral infection. To measure its activity, suitable mixtures of serum and virus preparation are inoculated into white mice susceptible to the pathogenic effect of the virus, and the protective quality of the test serum is titrated. To gauge the protective effects of the combination, either cell cultures or chick embryos may also be used. This is the virus neutralization test.

The phenomenon of hemagglutination, important in the identification of a given virus, is valid because most disease-producing viruses exert a direct or indirect action

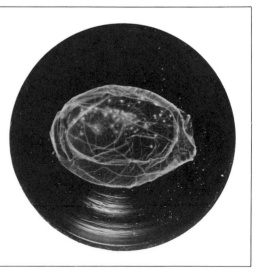

FIG. 28-7. Chorioallantoic membrane, showing whitish plaques (pocks) produced by smallpox virus. (From Hahon, N., and Ratner, M.: J. Bacteriol. **74:**696, 1957.)

TABLE 28-4. LABORATORY DIAGNOSIS OF SELECTED VIRAL DISEASES IN MAN

Virus	Test specimen	Practical diagnostic tests							
		Animal Inoculation	Chick embryo culture	Cell culture – cytopathic effect	Complement fixation	Hemagglutination-inhibition	Virus neutralization	Fluorescent antibody	Inclusion bodies
Viruses in respiratory disease									
Influenza viruses	Nose and throat secretions		x		x	x		x	
Adenoviruses	Throat secretions, fecal material, eye fluid, cerebrospinal fluid			x	x	x	x		
Viruses in central nervous system disease									
Polioviruses	Throat secretions, rectal swabs, blood, urine, cerebrospinal fluid	x		x	x		x	x	
Encephalitis viruses	Blood, throat swabs, cerebrospinal fluid, urine, brain, other tissues if illness fatal	x		x	x	x	x		
Rabies virus	Saliva, throat swabs, eye fluid, cerebrospinal fluid	x			x		x	x	x
Viruses in skin disease									
Variola virus	Material from vesicle, pustule, or scab		x	x	x	x	x	x	x
Rubeola virus	Blood, urine, throat swabs, eye fluid			x	x	x	x		
Rubella virus	Blood, urine, throat swabs	x	x		x		x	x	
Other viruses									
Colorado tick fever virus	Blood, throat swabs, fecal material	x	x	x	x	x	x		

on red blood cells of man and animals to clump them. Viral hemagglutination may be blocked by the action of specific antibodies found in immune or convalescent serum. This is the basis of the hemagglutination-inhibition (HI) test.

Many viruses induce the formation of complement-fixing antibodies. Precipitin reactions associated with several viruses are analogous to those observed with soluble products and toxins of bacteria. Precipitation is measured by immunodiffusion (p. 225).

The fluorescent antibody technic can be used to demonstrate viral antigens in specimens from patients or diseased animals. A diagnosis may sometimes be made within a few hours. In rabies, for example, the fluorescent antibody is applied to thin sections or smears of brain tissue or salivary gland, and fluorescence is observed within hours if viral antigen is present.

Direct microscopic examination may be made of suitable preparations such as smears, imprints, scrapings, or tissue sections. In certain viral diseases inclusion bodies (viral aggregates) and cellular reactions are easily recognized in thin sections of diseased tissue with the compound light microscope (Table 28-3). The electron microscope reveals viral particles both in body fluids and secretions and in *ultra*thin sections of tissue. Negative staining of viruses with an electron-dense material such as phosphotungstic acid is a technic used in the preparation of electron micrographs (Fig. 5-8).

Skin tests are available to aid in the diagnosis of some viral diseases. Vaccinia virus vaccine administered in a skin test indicates immunity to smallpox. Chick embryo antigens prepared for skin testing in mumps and herpes simplex infections indicate past or present infection.

Table 28-4 indicates how the laboratory is used in the diagnosis of viral disease.

Spread. The most significant of the viral diseases are caused by organisms that have as their natural host man himself. Spread from man to man is either by direct or indirect contact, especially by means of nose and throat secretions, fecal material, and articles so contaminated. Droplet infection is very common. Some viral diseases are spread to man by insect vectors (flies, cockroaches, mosquitoes, or ticks) from a reservoir of infection in either a lower animal or in the insect. One viral disease (rabies) is transmitted directly by the bite of the infected animal. Viral diseases are transmitted by water (hepatitis, enterovirus infections), and instances of milk- or food-borne epidemics are occasional (hepatitis, poliomyelitis, and enterovirus infections). Human carriers, except in enteric infections, are not important in the spread of viral diseases.

Immunity. In many viral diseases (mumps, smallpox, measles) one attack confers lifelong immunity. In others (common cold, influenza) no immunity of any appreciable duration results. Where immunity is short-lived, the incubation period has been short, viruses have not circulated in the bloodstream, and antibody-forming tissues have failed to receive adequate stimulation. In most persons in whom immunity lasts, antibodies in the serum of that person may be demonstrated for many years, and it has been postulated that virus, inactivated and nondisease-producing, has remained within his body.

The mechanisms of natural resistance are poorly understood. Newborn mice are extremely susceptible to coxsackievirus and easily infected, whereas adult mice are quite resistant under ordinary conditions. The virus of chickenpox is pathogenic only for man; other animals are completely resistant.

An *interference* phenomenon unlike immunologic mechanisms for other microbes is observed with viruses. A plant or animal cell exposed to a given virus subsequently develops a resistance to infection by a closely related strain of the same virus or another similar virus. There is this kind of interference between the viruses of yellow fever and dengue fever in the body of their insect vector, the mosquito *Aedes aegypti*. Such a mosquito cannot spread more than one of these diseases at the same time.

Interferon or the interferon system is thought of as a broad-spectrum antiviral agent, a soluble, nontoxic, nonantigenic protein (or family of proteins), smaller in size than antibodies. It is elaborated in small amounts by a normal body cell under attack from an invading virus. Interferon is cell-specific (including species specificity), *not* virus-specific. Present within the specific cell, it is able to block the effect of the virus by stopping the synthesis of nucleic acid for the virus and thereby breaking into its life cycle. The action of interferon is a manifestation of viral interference and a most important part of the body's defense against viral infection—part of nature's first line of defense. A patient with agammaglobulinemia who is plagued with repeated infections caused by bacteria seems to recover uneventfully from those caused by viruses, perhaps because of interferon. Practically every class of animal virus has been associated with its formation. Only a few viruses are resistant to it.

There is a great deal of interest in interferon inducers, substances stimulating the endogenous production of interferon and thereby active against viral infection. A well-known one in the investigational field is polyinosinic: polycytidylic acid (Poly I:C), a synthetic, double-stranded RNA.

Prevention of viral disease. Immunization procedures prevent viral diseases (Chapters 36 and 37) as do proper technics of sterilization (Chapter 17). Effective disinfectants to inactivate or destroy viruses are alkaline glutaraldehyde, formalin, dilute hydrochloric acid, organic iodine, and phenol, 1%. Roentgen rays and ultraviolet light destroy viruses, but the effective dose varies with different viruses. Most viruses, except those of hepatitis, are destroyed in pasteurized milk. Influenza viruses are readily destroyed by soap and water.

Except to treat bacterial complications that are prone to follow viral diseases, antimicrobial drugs are generally of no value in the management of diseases caused by true viruses.

Bacteriophages (bacterial viruses)

General characteristics. In 1917 d'Herelle* discovered that a bacteria-free filtrate obtained from the stools of patients with bacillary dysentery contained an

---

*Two investigators independently discovered bacteriophage—F. W. Twort (1877-1950) in 1915 and F. H. d'Herelle (1873-1949) in 1917.

agent that when added to a liquid culture of dysentery bacilli dissolved the bacteria. If but a minute portion of the dissolved culture were added to another culture of dysentery bacilli, the bacteria in the second culture were likewise dissolved. This transfer from culture to culture could be kept up until the bacteria in hundreds of cultures were lysed, which proved that the agent was not consumed in the process but apparently increased in amount. This agent d'Herelle called *bacteriophage* meaning bacteria eater (Fig. 18-2). We know bacteriophages today as viruses that attack bacteria—or *bacterial viruses.*

Viruses infecting many strains of bacteria have been isolated, and it has been shown that a given phage acts only on its own particular species or group of species. In fact, their highly specific nature makes them useful to the epidemiologist in classifying bacteria; for example, phage typing of pathologic staphylococci is crucial to the epidemiologic study of hospital-acquired staphylococcal infections.

Tending to occur in nature with their specific hosts, bacteriophages are found most plentifully in the intestinal discharges of man and the higher animals or in water and other materials contaminated with these discharges. They are also found in pus and even in the soil. When phages are named, reference is made to the specific hosts, as with coliphages, staphylophages, cholera phages, and typhoid phages. A very few bacteria such as the pneumococci do not possess phages. Of all the viruses known, bacterial viruses are the most easily studied in the research laboratory, and the ones most thoroughly investigated have been those related to the enteric group of microorganisms. As obligate intracellular parasites, they closely resemble the other viruses in their biologic properties.

Life cycle. With the aid of the electron microscope, phages are seen to be tiny tadpole units possessing a head, which is either rounded or many-sided, and a tail, which is a specialized structure for attachment. Like other viruses, a phage particle is composed of nuclear material (DNA) encased in a protein coat (Fig. 28-8).

Although appearing to be harmless, a bacteriophage in its virulent form can literally blow its bacterium to bits in a matter of minutes. The attack on the specific bacterium occurs in a fantastic series of steps. Seeking out the susceptible bacterial cell, the virus fixes itself tail first to the cell (Fig. 28-9). By chemical action it drills out a tiny hole in the cell wall, and the tail penetrates the plasma membrane to the interior of the cell. The head changes shape, and soon the DNA of the virus flows through the tail into the cell. The bacteriophage seems to work like the world's smallest syringe and needle when it injects its nuclear DNA into the bacterial cell. Once the fatal injection is made, drastic changes occur. The DNA takes command of the vital forces of the microbe and in a matter of minutes imposes the synthesis of hundreds of bacterial viruses exactly like the one originally invading the injured bacterial cell. The deranged cell swells and shatters, setting free a multitude of new viruses able to seek out a new host and repeat the cycle. This rapid destruction of a bacterium is *lysis.*

Lysis is not the invariable result of bacteriophage action. When certain phages infect, they do not destroy but seem to be able to establish a relatively stable symbiotic relation with the host they have parasitized. The host bacterium continues to grow and multiply, carrying the virus in its interior in a noninfective condition more or less

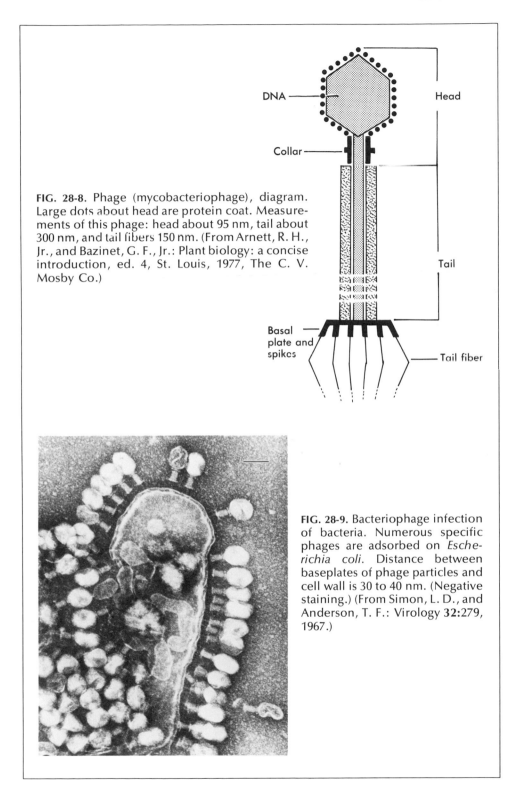

FIG. 28-8. Phage (mycobacteriophage), diagram. Large dots about head are protein coat. Measurements of this phage: head about 95 nm, tail about 300 nm, and tail fibers 150 nm. (From Arnett, R. H., Jr., and Bazinet, G. F., Jr.: Plant biology: a concise introduction, ed. 4, St. Louis, 1977, The C. V. Mosby Co.)

DNA

Collar

Head

Tail

Basal plate and spikes

Tail fiber

FIG. 28-9. Bacteriophage infection of bacteria. Numerous specific phages are adsorbed on *Escherichia coli*. Distance between baseplates of phage particles and cell wall is 30 to 40 nm. (Negative staining.) (From Simon, L. D., and Anderson, T. F.: Virology **32**:279, 1967.)

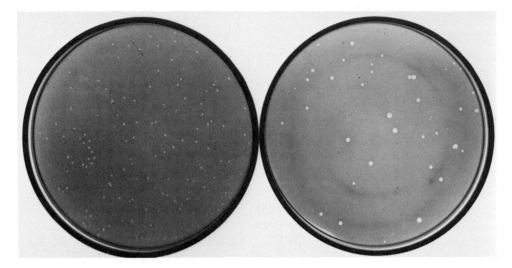

FIG. 28-10. Phage plaques. Small plaques (zones of clearing) in assay plate on left are *Staph. aureus* phages; those on right are *Staph. epidermidis* phages. In preparation of an assay plate, a mixture of test phages and microorganisms is poured over nutrient base. After plate has been incubated, plaques are evaluated and counted. (Courtesy Drs. E. D. Rosenblum and B. Minshew, Dallas, Tex.)

indefinitely, and the virus meantime multiplies at the same rate. This kind of phage is referred to as a *prophage*, and this condition of mutual tolerance is termed *lysogeny* or *lysogenesis*. Lysogeny is widespread in nature. That the process is not without effect is seen in changes in physiologic characteristics in the infected bacteria. Phages may be responsible for the conversion of an avirulent strain to a virulent one.

Laboratory diagnosis. Bacteriophages may be isolated and studied in the laboratory. If they are taken from a source in nature, they must be separated from bacteria by filtration. When they are added to a growing culture of specific bacteria, their action to block bacterial growth may be nicely shown. In a liquid culture medium, clearing indicates bacterial lysis. On a solid culture medium, lysis by virulent phages is seen in zones, usually circular, where bacterial growth has disappeared (clearing of bacterial growth). These zones are called *plaques* (Fig. 28-10). Each plaque contains many particles, which in turn can form plaques. Pure preparations of phage may be obtained by picking material from well-isolated plaques.

Importance. Bacteriophages may be of practical importance. In certain of the fermentation industries in which the commercial product is dependent on bacterial action (streptomycin, acetone, and butyl alcohol, for examples), an "epidemic" of viral infection in the large vats used to grow the microorganisms can be of grave economic concern.

Viruses and teratogenesis

Currently the role of viruses in teratogenesis (the production of physical defects in the offspring in utero) is being carefully studied. When the pregnant woman contracts a viral infection accompanied by a viremia, the infection often crosses the placenta to the susceptible embryo or fetus. Notable examples are measles, smallpox, vaccinia, western equine encephalitis, chickenpox, poliomyelitis, hepatitis, and coxsackievirus infection. Viral infections involving the offspring within the uterus may be more common than suspected, especially in the lower socioeconomic brackets. Although the possibilities for fetal infection are many, only three viruses fully qualify as teratogens. For rubella virus, cytomegalovirus, and herpesvirus, the evidence is clear-cut.

Birth defects with rubella virus (see also p. 515) have been more thoroughly examined than those with other viruses, and most of what is known about viral teratogenesis comes from such investigations. In rubella, the teratogenic mechanism seems to stem from a direct interaction between virus and parasitized cell that continues throughout the length of gestation. That plus viral damage to blood vessels disrupts normal organ development.

The cytomegalovirus (salivary gland virus) produces a mild infection in the mother but can cause extensive and widespread damage in the neonate. (See also p. 522.) In a few instances herpesvirus has crossed the placenta to produce generalized infection with documented malformations in the central nervous system and in the eye. (See also p. 520.)

Viruses and cancer

Experimental evidence definitely indicates that filtrable viruses do cause several kinds of cancerous growths in lower animals, that is, in rabbits (Fig. 28-11), mice, chickens, hamsters, rats, dogs, frogs, monkeys, horses, squirrels, and deer (Table 28-5). There are more than 60 known viral oncogens. Cancer is induced in all major groups of animals (including subhuman primates). Leukemia is an important form of

---

TABLE 28-5. EXAMPLES OF VIRAL NEOPLASMS IN ANIMALS

Animal species	Date reported	Cancerous growth (neoplasm)
Chicken	1908	Chicken leukemias
Chicken	1910	Rous sarcoma
Rabbit	1933	Shope papilloma-carcinoma
Mouse	1936	Breast carcinoma*
Frog	1938	Kidney carcinoma
Mouse	1951	Spontaneous leukemia
Mouse	1957	Salivary gland tumors of polyoma virus
Hamster, rat, and rabbit	1958	Great variety of solid tumors produced by polyoma virus
Hamster	1962	Chest and liver tumors (adenovirus responsible isolated from human cancer)

*This most common form of cancer in the most commonly used laboratory animal was shown by John Bittner to be related to a virus, long referred to as Bittner's milk factor. Bittner's discovery stimulated greatly the study of the role of viruses in the induction of cancerous growths.

TABLE 28-6. PATTERNS OF TUMOR VIRUSES

Category	Nucleic acid	Approximate size (nm)	Member viruses
Papovaviruses	DNA	45	Polyoma
			SV 40
			Papilloma Human Rabbit Bovine Dog
Poxviruses	DNA	250 × 200	Yaba Fibroma
			Molluscum
Adenoviruses	DNA	80	Types 3, 7, 12, 18, 31
Herpesviruses	DNA	100	Lucké
			Marek's disease
			Herpesvirus saimiri
			Herpesvirus of rabbit EB (Epstein-Barr)
			Herpesvirus hominis, type II
Myxovirus-like particles	RNA	70-110	Rous sarcoma
			Avian leukosis complex* Murine leukemia complex: 14 strains, including Gross, Friend, Graffi, Rauscher, Moloney Milk factor or Bittner virus (mouse mammary tumor)

*Leukoviruses.

Natural host	Tumors produced	Experimental hosts
Mouse	Solid tumors in many sites	Mouse, hamster, guinea pig, rat, rabbit, ferret
Rhesus monkey	Sarcomas (malignant tumors of connective tissue)	Hamster
Man	Common warts (papillomas)	Man
Rabbit	Papillomas	Rabbit
Cow	Papillomas	Cow, hamster, mouse
Dog	Papillomas	Dog
Monkey	Benign histiocytomas	Monkey
Rabbit, squirrel, deer	Fibromas, myxomas	Rabbit, squirrel, deer
Man	Molluscum contagiosum	Man
Man	Sarcomas, malignant lymphomas	Hamster, mouse, rat
Leopard frog	Renal adenocarcinoma	Leopard frog
Chicken	Neurolymphoma (Marek's disease)	Chicken
Squirrel monkey	Melendez' lymphoma of owl monkey	Owl monkey, marmoset
Rabbit	Hinze's lymphoma	Cottontail rabbit
Man	Burkitt's lymphoma(?), nasopharyngeal carcinoma (?)	
Man	Carcinoma of cervix of uterus (?)	
Chicken	Sarcomas, leukemias, adenocarcinoma of kidney	Chicken, turkey, rat, monkey, hamster, guinea pig
Chicken	Leukemias, sarcomas	Chicken
Mouse	Leukemias, malignant lymphoma	Mouse, rat, hamster
Mouse	Mammary carcinoma	Mouse

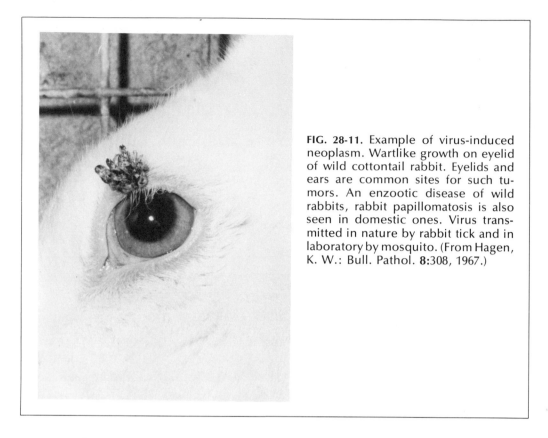

FIG. 28-11. Example of virus-induced neoplasm. Wartlike growth on eyelid of wild cottontail rabbit. Eyelids and ears are common sites for such tumors. An enzootic disease of wild rabbits, rabbit papillomatosis is also seen in domestic ones. Virus transmitted in nature by rabbit tick and in laboratory by mosquito. (From Hagen, K. W.: Bull. Pathol. **8**:308, 1967.)

virus-induced neoplasm in animals. Many different solid cancers in the skin, breast, lungs, gastrointestinal tract, and salivary glands of several species of laboratory animals are caused by the polyoma virus.

Oncogenic viruses are known to lie dormant for a long time in animals before they are activated. It is postulated that either external factors such as x-rays and other forms of radiant energy or internal factors of metabolic or hormonal nature trigger the mechanism of neoplasia (new growth). John Bittner demonstrated that breast cancer in mice is caused by an interplay of the virus (milk factor), hormones, and hereditary background. Table 28-6 emphasizes certain features of viruses known to cause tumors.

The role of viruses in human neoplasms is unknown, but all experimental reports strongly suggest that viruses are indeed involved. In man, the common wart, verruca vulgaris, a benign or harmless tumor, has been clearly demonstrated to be induced by a virus. Viruslike particles have been isolated from human cancers, most recently from human breast cancer and from the milk of breast cancer patients. Table 28-7 presents viruses indicted as oncogens in man.

Burkitt's lymphoma. Burkitt's African lymphoma* is a malignant disease of the jaw and abdomen affecting children between the ages of 2 and 14 years. The striking

---

*Lymphoma is a malignant neoplasm (cancer) of the lymphoid tissue.

TABLE 28-7. VIRUSES INDICTED FOR ONCOGENICITY IN MAN

Virus or viruslike particle	Tumor
Type C and related particle*	Leukemia
	Lymphoma
	Sarcomas
	Bowen's disease (precancerous skin change)
Type B particle (RNA particle) (mouse mammary tumor virus)	Breast cancer
EB virus (Epstein-Barr virus, herpes type virus)	Burkitt's lymphoma
	Hodgkin's disease
	Leukemia
	Lymphoma
Herpesvirus hominis type 2 (genital strain of herpes)	Carcinoma of cervix uteri
Reovirus type 3	Burkitt's lymphoma

*Leukoviruses with RNA core and double membrane, a complex of oncogenic viruses, the first ones isolated from animal neoplasms. Referred to as oncogenic RNA viruses, or oncornaviruses, they have been divided into three classes, A, B, and C (the most important). Crucial to their oncogenic potential is their relatively large genome as compared with other viruses and their possession of an RNA-directed (-dependent) DNA polymerase (reverse transcriptase). This enzyme mediates the synthesis of DNA from an RNA template and indicates a biochemical mechanism for perpetuation of viral genome when the host cell divides.

---

feature of the disease is its sharp geographic distribution. It is found in Central Africa limited to a malarious belt where the conditions of rainfall, temperature, altitude, vegetation, and humidity are the same. It also appears to be a geographic equivalent of lymphomas of children in other parts of the world. But, unlike lymphomas elsewhere, it has a pattern for a specific infectious disease, strongly suggesting a viral etiology. Several viruses have been found in tumor tissue and in cell culture made of tumor. The Epstein-Barr virus was first isolated from such a cell culture.

At first it was postulated that the viral agent causing the disease was spread by a vector mosquito. The lymphoma is prevalent in areas where malaria is endemic and rare outside tropical Africa. Now it is thought that the mosquito is not more implicated than in the transmission of malaria, for it is believed that the damage to the lymphoid system from chronic malaria is the factor that determines in some way the oncogenic potential of a virus. The EB virus implicated here is also found in nonneoplastic disease elsewhere in the world.

QUESTIONS FOR REVIEW

1. State briefly the salient features of viruses. Compare them with other microbes.
2. Sketch the life cycle of viruses (including bacterial viruses).
3. What are inclusion bodies? Their importance? Their specificity? Cite examples.
4. What is meant by cytopathic effect of viruses? How is this used in virology?
5. Give the two major pathologic effects viruses produce on cells they parasitize.
6. How may viruses be cultivated in the laboratory?
7. Discuss briefly the spread of viral diseases.

8. Outline the laboratory diagnosis of viral infections.
9. Comment on the nature and importance of interferon and viral interference.
10. Classify viruses. Indicate the most widely used system.
11. Discuss the role of viruses in teratogenesis.
12. What is the importance of the nucleic acids in viruses?
13. State the case for viral oncogenesis.
14. Define or briefly explain bacteriophage, arbovirus, enterovirus, virion, viroid, viropexis, capsid, picorna, virology, lysogeny, papovaviruses, viral hemagglutination, plaque, icosahedron, hyperplasia, dermatropic, prophage, capsomere, life cycle, helix, regeneration, virus neutralization, elementary body.

REFERENCES. See at end of Chapter 29.

29 Viral diseases

SKIN DISEASES
Measles (rubeola)*

Measles is an acute communicable disease associated with a catarrhal inflammation of the respiratory passages, fever, constitutional symptoms, a skin rash (Fig. 29-1), and a distinct predilection for grave complications. Among these are streptococcal or pneumococcal bronchopneumonia, encephalitis, otitis media, and mastoiditis. The incubation period is 10 to 12 days. One of the most common diseases, measles is said to be responsible for about 1% of the deaths occurring in the temperate zones.

Only one immunologic type of measles virus (a paramyxovirus) is known. The virus of measles is thrown off in the lacrimal, nasal, and buccal secretions and enters the body by the mouth and nose. It is found in the blood, urine, secretions of the eyes, and discharges of the respiratory tract. Infection is usually transmitted directly from person to person. Healthy carriers are unknown. The time that an object contaminated with the secretions of a patient with measles remains infectious is short. Measles is not transferred by the scales from the skin. Measles virus may be spread a considerable distance through the air. Measles is most highly communicable during the 3 or 4 days preceding the skin eruption. It is not transmitted after the fever has subsided. Epidemics tend to recur every 2 or 3 years and are likely to break out when young adults from rural communities come together in large groups, such as when armies are mobilized. The disease is especially virulent in populations native to tropics or in primitive races anywhere with no ethnic history of previous exposure. Children of mothers who have had measles are immune to the disease until they are about 6 months old. An attack of measles almost invariably produces a permanent immunity.

---

*For a discussion of measles vaccines and immunization procedures, see pp. 680, 688, 696, 704, and 705.

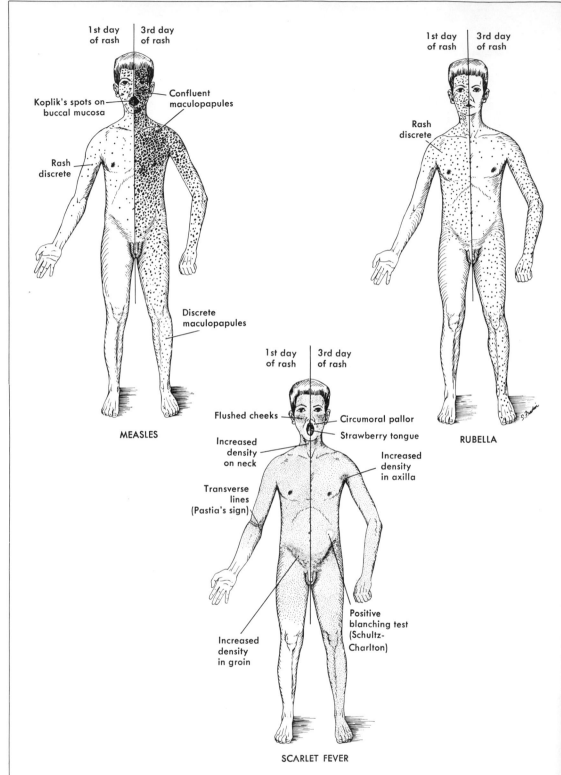

FIG. 29-1. Skin rashes in measles, rubella, and scarlet fever, sketch. (From Krugman, S., and others: Infectious diseases of children, ed. 6, St. Louis, 1977, The C. V. Mosby Co.)

514

Measles depresses certain allergic conditions and immune processes. It renders the tuberculin test and agglutination tests less positive or even negative. The Dick and Schick tests may become more strongly positive, and eczema and asthma often disappear during or after the attack.

The measles patient should be isolated and protected against streptococcal infections, staphylococcal infections, and common colds. Discharges from the nose, mouth, and eyes should be disinfected. When measles appears in army camps, daily inspections of personnel should be made, and persons having conjunctivitis, colds, or fever should be isolated. Bear in mind that patients with measles are vulnerable to pneumonia and other serious complications.

Rubella (German measles)*

German measles (3-day measles) is a mild but highly prevalent disease of viral origin described by fever, lymphadenopathy, mild catarrhal inflammation of the respiratory tract, and a skin rash similar to that of measles (rubeola) or scarlet fever (Fig. 29-1). The incubation period is 14 to 21 days. Transmission is airborne from person to person. A common source is someone with an inapparent infection. The cause is a medium-sized, spherical unit, an extremely pleomorphic virus that, isolated from throat washings and the blood of patients, has been grown in cell cultures. This peculiar virus, as yet unclassified, with an RNA core is about the size of a myxovirus, but it behaves somewhat like a togavirus. There is no insect vector. Only one antigenic type is known.

Rubella is important because approximately one out of four children born to mothers contracting German measles during the first 4 months of pregnancy has congenital defects or is malformed. If rubella is contracted during the first 4 weeks of pregnancy, approximately half the children born will be deformed.

From the mother's blood, the rubella virus crosses the placenta into the developing tissues of the new individual, where it can persist throughout gestation and into the neonatal period. The unborn child is infected at the same time as his mother, and since the viral injury leads to irregularities in the development of one or more organs, malformations result. The defects are common in the eyes, ears, heart, and brain. Examples are microcephaly (extremely small head), deaf-mutism, cardiac defects, and cataracts. †

Viral injury in rubella is unique. In keeping with the mild nature of the infection, the virus neither destroys nor invigorates. It merely slows things down. The cells it parasitizes continue to grow and multiply but at a decreased rate. The embryo and fetus of the first 3 to 4 months of pregnancy are in a period of development during which the anatomic units define their shapes, take their places, and lay out their interconnections. Timing here is as critical as it is with a trapeze artist.

When infection with this virus persists after birth, the baby is born with *congenital rubella,* the manifestations of which may be mild or severe. There may be a

---

*For immunization, see pp. 680, 690, 696, and 700.
†As a result of the epidemic of 1964 in the United States, there were 20,000 stillbirths and 30,000 babies born with congenital anomalies to mothers whose pregnancies were complicated early by rubella.

single serious defect or multiple ones in a small undersized baby. The *rubella syndrome* indicates active infection in the newborn, which is easily demonstrated by recovery of the rubella virus from nasopharyngeal washings, conjunctivae, urine, and cerebrospinal fluid.

Laboratory diagnosis. As a routine, virus isolation is impractical. The following serologic procedures are available for the laboratory diagnosis of rubella infection: virus neutralization, complement fixation, hemagglutination inhibition, and immunofluorescence.

The hemagglutination-inhibition test is the most sensitive and most widely done. With a single specimen of serum this test indicates immunity in the presence of HI antibody. With paired serums spaced a week or two apart and a fourfold increase in the HI titer, it can make the diagnosis. If a woman is exposed to a possible case of rubella during her pregnancy, the HI test may be used to determine both in the pregnant woman and in the contact whether the exposure is to rubella and what the woman's immune status is to rubella. For the diagnosis of congenital rubella there is the determination of rubella-specific immunoglobulin M in the baby. (This is an antibody that does not cross the placenta and that therefore could not be derived from the baby's mother.)

Immunity. An attack of German measles is followed by permanent immunity. Girls should have German measles if possible before childbearing years, and a pregnant mother should avoid exposure to the disease. This is true even in women supposed to have had an attack because an erroneous diagnosis is often made. Although the newborn baby possesses virus-neutralizing antibodies in high titer, he may continue to harbor virus and to shed it, even for several years, thus constituting an important reservoir and source of infection to susceptible individuals in his environment.

Smallpox (variola)*

Smallpox, one of the most highly communicable diseases, is characterized by severe constitutional symptoms and a rash (Figs. 29-2 and 29-3) that goes through a typical evolution to become hemorrhagic in the severest cases. The incubation period is 12 days. Smallpox is the most infectious of diseases, and the death rate in unprotected individuals is high. Before the days of vaccination, smallpox spread over the world in devastating epidemics. It has been estimated that in some of these, 95% of the population were attacked and 25% died. Today, thanks to the heroic efforts of the World Health Organization (WHO), smallpox as a disease has virtually been eradicated worldwide, except possibly in some of the small villages in Ethiopia.

There are two types of smallpox virus (a poxvirus). One causes a severe form, *variola major. Variola minor*, or *alastrim*, the other type, is mild.

Transmission. Smallpox is transmitted directly from person to person by droplet infection. Man alone carries the infection. It may sometimes be conveyed by objects such as handkerchiefs and pencils that have been contaminated with the nasal and

---

*For a discussion of immunization, see pp. 690, 698-700, and 706.

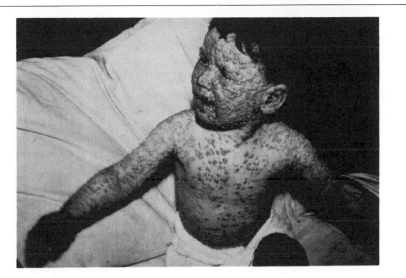

FIG. 29-2. Smallpox in unvaccinated 2½-year-old boy on eighth day of illness. Attack was severe, but boy recovered. (Courtesy Dr. Derrick Baxby, University of Liverpool, England.)

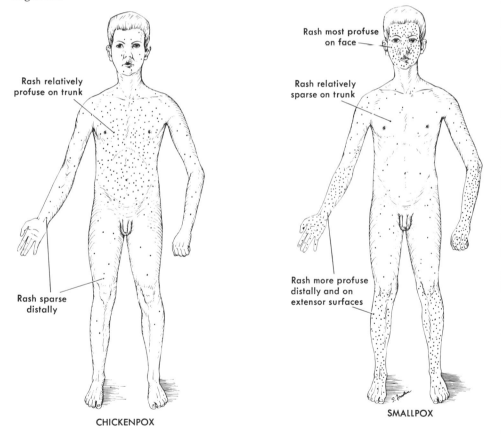

Rash most profuse on face

Rash relatively profuse on trunk

Rash relatively sparse on trunk

Rash sparse distally

Rash more profuse distally and on extensor surfaces

CHICKENPOX

SMALLPOX

FIG. 29-3. Skin rashes in chickenpox and smallpox. (From Krugman, S., and others: Infectious diseases of children, ed. 6, St. Louis, 1977, The C. V. Mosby Co.)

517

buccal secretions of a patient. It may also be transmitted from the pustules by the hands. Persons immune by virtue of vaccination or an attack of the disease may become contact carriers and disseminate virus for a short period. The infectious agent is generally thought to enter the body by the respiratory tract and leave by the buccal and nasal secretions. The virus may be found in the blood, skin lesions, and secretions of the mouth and nose. It may remain active for a long time in the dried crusts of the skin lesions. Even the dead body may be a source of infection.* A mother with smallpox may infect her child in utero causing the child to be born with a typical smallpox skin eruption.

An attack of smallpox usually renders the patient immune for the remainder of his life.

Laboratory diagnosis. The elementary bodies of the smallpox virus may be seen microscopically within cells of a stained preparation of scrapings taken from a lesion. In material suitably prepared the virus may also be identified in an electron micrograph, usually within a few hours. A smallpox gel precipitin test has been developed.

Prevention. Patients with smallpox should be isolated. The nurse should be isolated and of course vaccinated or revaccinated at once. All objects in contact with the patient should be sterilized, preferably by heat (burning, high-pressure steam, or boiling). If this cannot be done, they should be soaked in a 2.5% saponated cresol solution. The feces and urine should be disinfected with chloride of lime. Sputum and discharges from the mouth and nose should be received on tissues and burned. The patient should not be released until desquamation is complete.

Vaccinia

Smallpox and vaccinia or cowpox are caused by viruses with similar biologic properties. Both can be grown on the chorioallantoic membrane of the chick embryo. The virus of vaccinia produces a mild disease either in man or in cattle (its natural host), and its importance is that it can be used to produce immunity against the more severe disease, smallpox.

Molluscum contagiosum

Molluscum contagiosum is a skin disease associated with small, pink, wartlike lesions on the face, extremities, and buttocks. It is spread from person to person by direct and indirect contacts. The cause is a large poxvirus that produces very dramatic intracytoplasmic inclusions in the squamous cells lining the affected skin site.

Chickenpox and shingles (varicella—herpes zoster)

The two diseases, chickenpox (varicella) and shingles (herpes zoster), represent two phases of activity of a single herpesvirus, referred to as the varicella-zoster or

---

*In the 1960s an epidemic of smallpox was initiated in the United Kingdom by a Pakistani girl with an unsuspected case entering the country. In this epidemic the body of the victim was thoroughly sprayed with Lysol, 5%, placed in a plastic bag, wrapped in a rubber sheet, placed on a Lysol-soaked bed of sawdust, and sealed in an airtight and watertight casket. Cremation was urged. The health officers caring for the body wore full protective clothing.

VZ virus. The first invasion of the body by the virus produces chickenpox, the generalized infection. Shingles or zoster is the localized infection in a partially immune host. It is most often a recurrence of latent infection activated by exogenous factors such as trauma, intercurrent disease, or drugs, or with exposure to chickenpox.

Varicella, usually a mild disease of childhood, presents a typical (teardrop) vesicular (blisterlike) rash, which, although generalized in the skin and mucous membranes, is concentrated on the trunk (Fig. 29-3). The incubation period of chickenpox is 14 to 16 days. Herpes zoster, on the other hand, is a disease of adults defined by the appearance of a vesicular eruption similar to that of varicella but with a quite different distribution. In shingles, vesicles occur on one side of the chest, following the course of the peripheral nerves supplying that part of the chest. They may occur in other parts of the body, but since the primary involvement is in the ganglion of the posterior nerve root, they always follow the course of the nerve or nerves supplying the affected part. Shingles may cause prolonged suffering because the skin lesions are associated with intense burning pain.

In the vesicles of the skin in either disease there are typical reddish inclusion bodies in the nuclei of injured epithelial cells, and virus is present in the fluid (Fig. 29-4). The moist crusts of chickenpox are infectious, whereas the dry ones are not. Chickenpox is transmitted by direct contact with a patient. Frequently it has been observed that herpes zoster in adults has served as a source of varicella in children.

No immunity for chickenpox is conferred by the mother on her newborn infant, and convalescent serum is of little value in preventing or modifying the disease. A

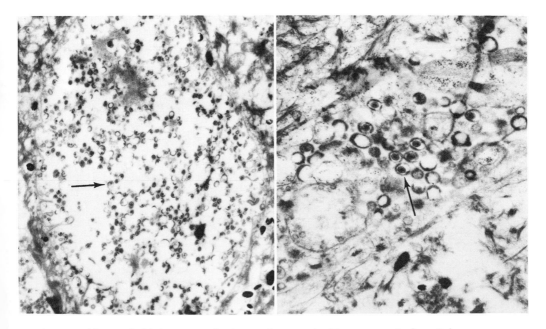

B

FIG. 29-4. Virus of chickenpox, electron micrograph. Numerous viral particles are seen in segment of squamous cell from skin. (A, ×15,000; B, ×30,000.)

vaccine for chickenpox is in the experimental stage, but any licensed vaccine goes far into the future.

Herpes simplex

Infection with herpes simplex virus (HSV; *Herpesvirus hominis*) is related to either of two recognized serotypes, each with unique biologic features. The more common HSV type I is responsible for the familiar fever blisters (herpes simplex) and cold sores found about the mouth and lips *(herpes labialis)*. Type I lesions include recurrent labialis, gingivostomatitis, corneal lesions in the eye, and eczema herpeticum. This type is probably spread via the respiratory route and occurs in older children and adults.

Type II is found about the genital organs. Spread as a venereal disease, it produces skin and mucosal lesions below the waist *(herpes progenitalis)*. (Type I tends to produce lesions above it.) Because patients with cancer of the mouth of the womb (cervix uteri) show statistically significant evidence for having had type II infection previously, this herpesvirus is increasingly implicated in the causation of this cancer.

In most instances the first infection with herpesvirus is an inapparent one; clinical signs of disease occur in only about 10% to 15% of persons infected. Saliva and genital secretions are probable sources of infection. The incubation period is 2 to 12 days. It is believed that thereafter the virus exists in the body as a latent infection and that when body resistance is weakened from any cause, the virus is reactivated. Recurrent herpes simplex often accompanies febrile illness, exposure to cold or sunlight, fatigue, mental strain, or menstruation (Schema 6). Unlike other viral infections, the host's humoral antibodies apparently give no protection against subsequent lesions.

One very serious complication of herpes progenitalis is infection of the newborn. By contacting a herpetic lesion in his mother's birth canal, the baby acquires the infection. In 1 to 3 weeks he becomes gravely ill with generalized herpesvirus infec-

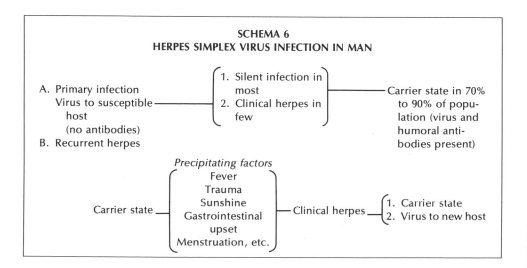

SCHEMA 6
HERPES SIMPLEX VIRUS INFECTION IN MAN

tion and usually succumbs. Herpetic lesions can be found in all the organs including the brain.

Herpes simplex virus forms characteristic intranuclear inclusion bodies in multinucleated giant cells present within the lesions. The intranuclear inclusions may be identified by the light or electron microscope in suitable preparations (smears, imprints, Pap smears of cervix uteri, and the like) of fluid from the superficial vesicles or parasitized tissue cells (Figs. 29-5 and 29-6). The virus can be cultivated in cell cultures and recovered from newborn mice. Fluorescent antibody technics as well as neutralization and complement fixation tests are used in the diagnosis of this infection.

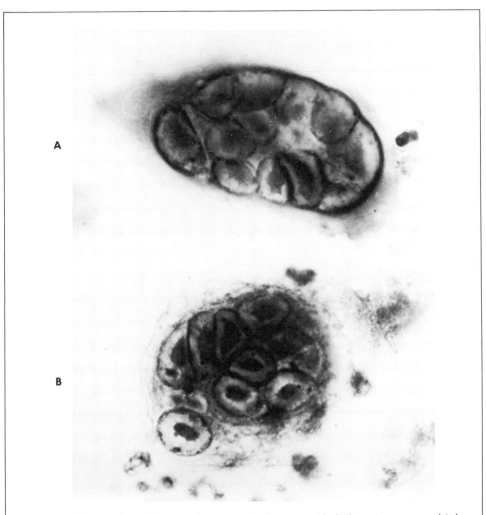

FIG. 29-5. Herpesvirus *(Herpesvirus hominis)* seen with light microscope, high-power photomicrograph. Inclusion bodies within nuclei of multinucleated squamous cell from skin. **A,** Early stage in formation of inclusion bodies. **B,** Inclusion bodies well defined.

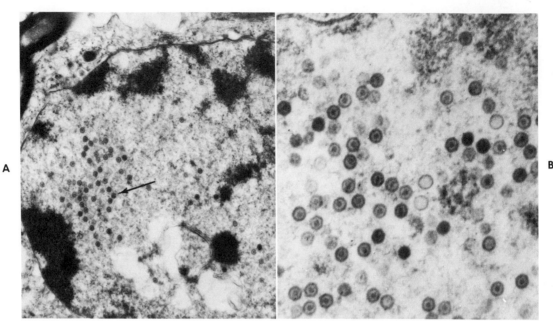

FIG. 29-6. Herpesvirus *(Herpesvirus hominis),* electron micrograph. Numerous viral particles in one nucleus of squamous cell from skin. (**A,** ×20,000; **B,** ×56,000.)

Cytomegalic inclusion disease

Cytomegalic inclusion disease (salivary gland disease) is infection caused by the ubiquitous cytomegaloviruses (salivary gland viruses). Cytomegaloviruses are widely distributed among mankind. It is said that 80% of the population in the United States have antibodies indicative of infection by the age of 35 or 40 years.

The pathologic lesions are striking in their nature. Large, well-defined viral inclusions are seen in the nucleus, and smaller ones are found in the cytoplasm of the injured cells, which are enlarged *(cytomegaly* means cell enlargement) (Fig. 29-7). In fatal cases the cell changes are seen in the gastrointestinal tract, lung, liver, spleen, and other organs. In nonfatal cases, inclusions may be found in epithelial cells from the kidney shed into the urinary sediment.

In its overt form, cytomegalic inclusion disease is seen in the newborn as a congenital infection acquired from a mother who was probably asymptomatic (latent infection) and in an older individual as a complication of a preexisting disease state. In the infected newborn cytomegalovirus is a major cause of birth defects, especially in the central nervous system (for example, microcephaly, hydrocephaly, blindness, mental retardation, deafness). Malformed babies exhibit the viral inclusions. The older patient on immunosuppressive therapy after organ transplant and the leukemic patient receiving cancer chemotherapy are vulnerable for cytomegalovirus pneumonia. Cytomegalic inclusion disease is associated with the postperfusion syndrome. This is an infectious mononucleosis–like syndrome following perfusion of fresh blood in patients undergoing open-heart surgery with cardiopulmonary bypass.

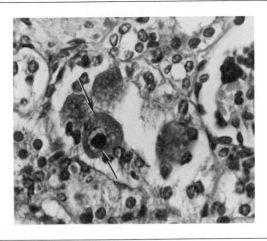

FIG. 29-7. Cytomegalic inclusion disease of kidney, photomicrograph. Arrows spot enlarged renal tubular cell with large intranuclear inclusion body of cytomegalovirus. (×800.)

The viral inclusions are readily visualized and diagnostic. Serologic technics used are those of neutralization, complement fixation, and immunofluorescence. If paired acute and convalescent sera show a fourfold increase in titer of antibodies to virus, a current or very recent infection is presumed. There are two distinct antigenic types of cytomegalovirus (a herpesvirus) from man. Virus has been propagated in cell culture. There is an experimental vaccine.

RESPIRATORY DISEASES
Influenza

Influenza* is a highly communicable disease occurring in epidemics that are described by explosive onset, rapid spread, involvement of a high percentage of the population, and frequency of serious secondary bronchopneumonia. An influenza pandemic occurs about every 10 to 14 years. The name *influenza* comes from Italian astrologers of long ago who believed that the periodic appearance of the disease was in some way related to the *influence* of the heavenly bodies.

One of the world's greatest catastrophes was the pandemic of 1918-1919. Although other epidemics have had higher death rates, on the basis of total number of deaths this was the worst pestilence civilization had ever experienced. It is estimated that there were 200 million cases with 20 million deaths over the world. In the United States alone there were 850,000 deaths, a figure greater than that for the combined battlefield losses of World War I, World War II, and the Korean War. And this occurred within the twentieth century! †

The agent. Influenza is caused by a myxovirus ‡ (so tiny that 29 to 30 million of

---

*For immunization, see pp. 687, 702, and 706.

†A guess is that the lives of 1 billion or more persons were affected or half the population of the world at the time. Only two places in the world allegedly escaped the pandemic—St. Helena, an island in the South Atlantic, and Mauritius, one in the Indian Ocean.

‡It was originally believed that *Haemophilus influenzae* was the cause of influenza. This organism is an important *secondary* invader in this and other respiratory diseases.

523

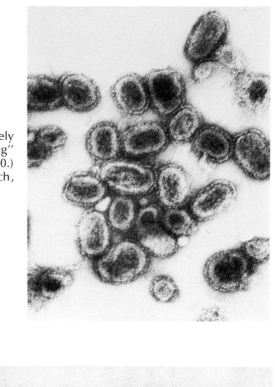

FIG. 29-8. Influenza virus negatively stained (A$_2$/Aichi/68 or "Hong Kong" virus), electron micrograph. (×303,100.) (Courtesy C. A. Baechler, Virus Research, Parke, Davis & Co., Detroit, Mich.)

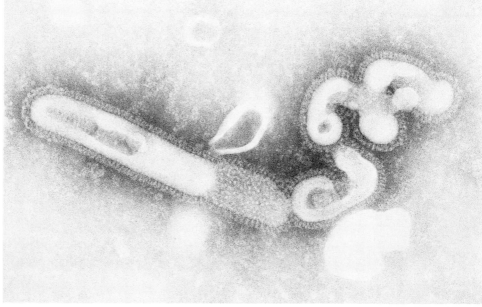

FIG. 29-9. Influenza virus, negatively stained (A/Hong Kong/1/68), electron micrograph. Note filamentous and pleomorphic particles. Virus isolated from throat of newborn baby. (×200,000.) (Courtesy Bauer, C. R., and others: J.A.M.A. 223:1233, 1973.)

them could rest comfortably on the head of a pin), and, like many viral diseases, it seems to potentiate serious bacterial infection, mostly bronchopneumonia. The mortality of the 1918 pandemic was largely the result of the severe, complicating bronchopneumonia produced by virulent streptococci.

Three types of influenza virus are known: A, B, and C. Of these, A and B have been best studied. They are alike in many ways but differ serologically. Each includes numerous distinct strains. Influenza A strains (Figs. 29-8 and 29-9) have figured much more frequently in epidemics than those of type B, the strains of which are considered to be less virulent than those of type A. B strains have recently been reported in scattered outbreaks.

The epidemic of 1947 was caused by a strain of virus A designated as A'; Asian strains of virus A were responsible for the pandemic of 1957, and a variant also in the A group—A/Hong Kong/68—caused the Hong Kong pandemic of 1968. An unexpected event was the emergence of an A variant—the A/England/42/72 strain— as the prevalent epidemic virus and the cause of "London flu" in 1972-1973. It was superseded by the Port Chalmers (A/Port Chalmers/1/73) strain, named after the town in New Zealand from where it spread slowly since being first isolated in 1973. Toward the end of the summer of 1975, the A/Victoria strain (thought to be an offshoot of the Hong Kong strain) was first isolated in Victoria, Australia. A/Victoria flu reached the United States in early 1976 at which time A/Victoria-like strains were reported from all over the world as the predominant flu strains.

The killer flu virus of the unprecedented 1918-1919 pandemic is postulated to have been an A/swine-type flu virus. (Flu in the United States at the time was sometimes referred to as hog flu.) There is renewed interest in such a virus because an A/swine-type virus (A/New Jersey/76), possibly similar to that killer virus, has recently been isolated in the United States, an event suggesting its reappearance after over half a century. It is a peculiar virus, difficult to detect, growing poorly in hens' eggs and not at all in standard kidney cell cultures. Like the agent, the disease of 1918-1919 had unusual features. For one thing its high mortality was among persons 15 to 50 years of age, whereas in the flu epidemics of today, the highly susceptible persons are the very old or the very young. Serologic studies shed some light on the nature of this swine-type flu virus. Antibodies to it are widespread in persons tested who were alive in 1918 and 1919. In fact the pattern of antibody distribution in the older age groups suggests the presence of antigenically similar viruses diffuse in the human population through 1930.

Epidemiology. The clinical manifestations of influenza remain fairly constant throughout the years although caused by an agent capable of remarkable changes. It has been said that it is an "unvarying disease caused by a varying virus." The infectious agent of influenza enters the body by the mouth and nose and leaves by the same route. Recently influenza virus has been recovered from anal swabbings. Influenza has been produced artificially in chimpanzees, ferrets, and mice by injection of filtrates of nose and throat washings from known human cases.

The disease is spread by direct and indirect contact, including droplet infection. An epidemic is of short duration and quickly subsides, to be followed after several

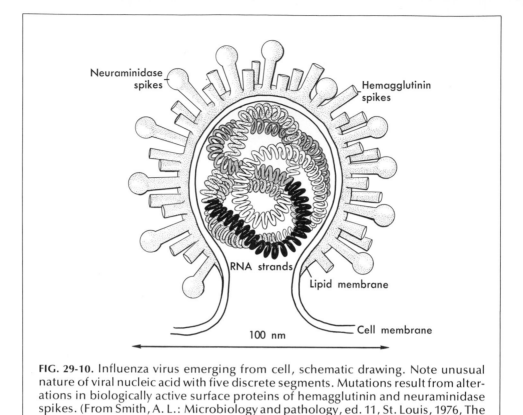

FIG. 29-10. Influenza virus emerging from cell, schematic drawing. Note unusual nature of viral nucleic acid with five discrete segments. Mutations result from alterations in biologically active surface proteins of hemagglutinin and neuraminidase spikes. (From Smith, A. L.: Microbiology and pathology, ed. 11, St. Louis, 1976, The C. V. Mosby Co.)

weeks by a secondary wave, a free period, and a tertiary wave. Never does the infection spread faster than people travel. What happens to the virus between epidemics still remains a mystery, although the relationships of influenza viruses to animals are becoming more apparent. There is good reason to think that in the interval the virus resides in animals, multiplies, and changes genetically, possibly by genetic recombination of human strains with those from lower animals, thus becoming infective for man again. An epidemic seems to depend on the emergence of the new strain; since this organism possesses a continuing and dramatic tendency to change its chemical and genetic nature, that is, to mutate, new variants of the virus types arise constantly from the major antigenic shifts. Flu virus mutates to a magnitude not found for any other infectious agent. If the majority of persons do not possess the immune mechanisms to meet the new variation, then far-reaching epidemics and pandemics threaten (Fig. 29-10).

The explosive outbreak of an epidemic may be explained by the high communicability of the disease, the great number of susceptible people, and the fact that during the early days of the attack the patient is not confined to bed but mingles freely with other people. The average duration of an epidemic in a community is from 6 to 8 weeks. In secondary and tertiary outbreaks the number of people attacked is less than in a primary outbreak, but the disease tends to be severe, complications are more common, and the mortality is higher. Interepidemic cases are usually mild.

Recent epidemics. In early 1957 influenza appeared out of north China and within the next several months spread all over the world. There were millions of cases over a wide area of the Far East. Referred to as Asian flu, it quickly spanned the extent of the United States.

The virus was found to comprise new strains of influenza virus, type A, designated the Asian strains, and cultures were established in embryonated eggs. On the whole, the mortality in this pandemic was low; in areas where notable, it was associated with severe staphylococcal pneumonia.

In the summer of 1968 another pandemic of influenza came out of the Orient, starting in Hong Kong with close to half a million cases. Hong Kong influenza was caused by a virus not completely different from the Asian flu virus of 1957, and the clinical disease was similar to that of the prior decade. During the last 3 months of 1968 it was estimated that 30 million persons were stricken. There were more deaths with Hong Kong influenza than with Asian influenza, especially among chronic invalids. *Staphylococcus aureus* and *Pseudomonas aeruginosa* were the complicating bacterial infections.

Since the 1968 pandemic, lesser epidemics have tended to occur annually. Asian strains could dominate the influenza picture for a period of time, until the level of immunity in the general population has risen significantly.

Immunity. Part of the population apparently possesses natural immunity to influenza because during an epidemic some persons do escape infection. There is reason to think that natural immunity acquired from an attack of the disease protects against a repeat attack by the same or very closely related strain. However, there is no protection against the new strains that are appearing. Little success has been obtained in producing passive immunity by means of convalescent or immune serum.

Laboratory diagnosis. The diagnosis of influenza can be confirmed only by isolation of the virus or serologic studies. The virus can be grown in cell cultures or in the embryonated egg. The serologic tests include hemagglutination-inhibition, neuraminidase-inhibition, hemadsorption-inhibition, neutralization, and complement fixation.

Prevention. When an epidemic of influenza strikes, all the methods known to preventive medicine fail to check it. Wholesale isolation seems to be of little value and almost impossible to achieve. Nurses attending patients with influenza should disinfect the mouth and nasal secretions of the patient and should avoid exposing themselves to droplet infection.

Swine influenza

Swine influenza is a severe respiratory infection of hogs caused by the combined action of a myxovirus and a bacterium, *Haemophilus suis*. The bacterium is closely related to *Haemophilus influenzae,* and the virus is closely related to that of human influenza. The diseases differ in that the human virus can produce the complete disease, whereas the virus of swine influenza alone produces only a mild form. Swine may become infected with the strains of type A influenza virus that happen to be prevalent in the human population at a given time. The Asian strains of the 1957 epidemic were recovered from hogs in Japan.

Acute respiratory syndromes

As reflected in a national health survey, acute viral infection of the respiratory tract is one of the most common causes of illness in man. It is responsible for approximately one third of all days lost from work and two thirds of days missed from school. Acute respiratory infection of viral type is not a single disease but a spectrum of such entities as rhinitis, pharyngitis, tonsillitis, laryngitis, and bronchitis (Table 29-1). These inflammatory processes can occur singly, but combinations are frequent.

Clinical features. Viral infection of the lining of the nose results in a reddened

---

TABLE 29-1. RESPIRATORY ILLNESSES RELATED TO KNOWN RESPIRATORY VIRUSES

Illness	Virus
Influenza	Influenza viruses A, B, C
Common cold of adults	Rhinovirus (Salisbury) — 100 or more types Coronavirus — 3 types
Acute respiratory syndromes* 　Common cold syndrome in children, some adults 　Rhinitis 　Pharyngitis and nasopharyngitis 　Tonsillitis 　Laryngitis 　Tracheitis 　Bronchitis 　Bronchiolitis 　Bronchopneumonia 　Pneumonitis 　Croup (acute laryngotracheobronchitis)	Influenza viruses A, B, C Parainfluenza viruses 1, 2, 3, 4 Respiratory syncytial virus Adenovirus — 10 types Echovirus — 6 types Coxsackievirus A — 23 types Coxsackievirus B — 6 types Poliovirus, types 1, 2, 3 Reovirus, types 1, 2, 3
Viral pneumonia (primary atypical pneumonia)	Influenza viruses A, B, C Parainfluenza viruses 1, 2, 3, 4 Adenovirus — 2 types Respiratory syncytial virus (in children) Measles virus Varicella virus Vaccinia virus Rubella virus
Pharyngitis (part of poliomyelitis)	Poliovirus, types 1, 2, 3
Pharyngitis (part of infectious mononucleosis)	Epstein-Barr virus

*No specific constant relation between given disorder and viral agent. See text.

mucous membrane and a hypersecretion of mucus—hence the well-known runny nose. A sensation of stuffiness comes from the nasal congestion and blockage. When the throat is involved, it is reddened, tonsils and other lymphoid masses are enlarged, and there is soreness. Hoarseness may result from inflammatory swelling of the larynx. Other clinical features associated in whole or in part with acute respiratory disease include low-grade fever, cough, pain and discomfort in the chest, headache, sensations of chilliness, and sometimes enlargement of cervical lymph nodes. Droplet infection accounts for spread; respiratory diseases are prevalent in cool months, and the epidemic pattern is a familiar one.

The disease influenza was originally considered to be part of the general complex of respiratory disorders. With a typical pattern for recurrent epidemics, influenza could be easily separated when the causative virus was discovered. Primary atypical pneumonia emerged in part from the respiratory assortment in World War II because of the characteristic findings in the lungs of the soldiers. A mycoplasma, *Mycoplasma pneumoniae*, not a virus, is responsible for a significant number of these cases (p. 463). In Great Britain certain viruses have been recovered from adults with the common cold (acute coryza), and the infection transferred to the chimpanzee, the only animal susceptible to the particular agents. The experimental studies suggest that the common cold may well have unique features.

Relation of viruses. By the early 1950s, tissue (cell) culture technology was developed to a high degree of efficiency. Viruses were recognized and recovered in such rapid-fire succession as to incur a kind of viral explosion. So many were identified that proper assimilation and definition of medical importance are yet incomplete. Regarding respiratory infections, it soon became clear that an array of viral agents was associated with a variety of clinical infections with no fixed or specific relationship between the ailment and the virus. Viruses were recovered singly or together from the respiratory tract, both in health and in disease. Many observers prefer the term *syndrome*, the sum total of the clinical features of an illness, to emphasize the lack of specific correlation between any precise agent and clinical manifestations. In the light of present knowledge of viruses associated with respiratory disease, the best we can do is to present certain viruses as being found in respiratory secretions and tissues and to point out the variety of associated syndromes. Table 29-1 lists the viral agents recovered from acute respiratory syndromes.

Respiratory viruses. From the known acute respiratory infections, viral agents have been recovered in most instances. Viruses that cause respiratory disease are spoken of collectively as *respiratory viruses*, and as an entity they encompass several major groups of viruses, including the myxoviruses (influenza viruses), the paramyxoviruses (parainfluenza viruses and the respiratory syncytial virus), the picornaviruses (enteroviruses and rhinoviruses), the adenoviruses, and the reoviruses.

Parainfluenza viruses, classified as paramyxoviruses, were first recognized in 1957. The four antigenic types are more stable than the influenza viruses; they do not mutate so often. Parainfluenza viruses are present in the community most of the year. The typically mild first contact infection usually comes early in life. These viruses enter by way of the respiratory tract, and except in infants and young children, inflammation is usually confined to the upper part. They cause a number of illnesses, primarily in infants and young children, ranging in severity from mild upper respiratory infection to croup and pneumonia. The clinical features of their infections are not distinct; the majority are clinically inapparent.

Respiratory syncytial (RS) virus, also a paramyxovirus, infects a child before the age of 4 years. It is probably the most important cause of acute respiratory disease in infants and very young children. A formalin-killed vaccine containing the respiratory syncytial and types 1, 2, and 3 parainfluenza viruses from kidney cell cultures is being field-tested.

Among the *picornaviruses*, certain of the *enteroviruses* are respiratory pathogens. All of the coxsackieviruses and some of the echoviruses (p. 541) are believed to cause respiratory disease. Coxsackieviruses have been isolated from nasal and pharyngeal secretions, and coxsackievirus A, type 20, has been studied in its relation to an acute upper respiratory coldlike illness. The respiratory agents among the echoviruses probably infect through the respiratory tract. They are known to multiply in the pharynx, producing a pharyngitis, and most likely are transmitted in respiratory secretions. Upper tract infections of echovirus are generally mild; the syndrome is many times that of the common cold. Echovirus types 11 and 20 have been specifically observed in such infections.

Adenoviruses, * also known as adenoidal-pharyngeal-conjunctival (APC) viruses, were first discovered in adenoids removed surgically from persons without any detectable clinical disease. To date, 31 types have been isolated from human beings. Although these viruses may be recovered from persons without disease, some do cause acute infection of the respiratory and conjunctival mucous membranes. Adenoviruses are related to a striking variety of clinical conditions such as acute respiratory disease, febrile pharyngitis or pharyngoconjunctival fever (especially in children), acute follicular conjunctivitis, epidemic keratoconjunctivitis (shipyard eye), tracheobronchitis, bronchiolitis, and pneumonitis. Types 3, 4, and 7 are related to epidemics of respiratory disease. Type 8 virus is the major cause of epidemic keratoconjunctivitis. Type 3, sometimes found in swimming pools, is the cause of pharyngoconjunctival fever.

In civilian groups, only a small percentage of acute viral respiratory diseases is caused by adenoviruses, but this group, particularly types 3, 4, 7, 14, and 21, is of special importance to the Armed Forces. It incapacitates many of the recruits in the fall and winter about 8 weeks later. Immunization of military inductees therefore is highly desirable.

Man is the only known reservoir of adenoviruses, and the virus is transmitted in respiratory and ocular secretions. The fact that the conjunctival inflammation often precedes the respiratory infection indicates the eye to be a significant portal of entry. Epidemics of pharyngoconjunctival fever and conjunctivitis have been traced to dissemination of the virus in swimming pools.

The respiratory and ocular involvement, often combined in adenovirus infection,

---

*For immunization, see p. 687.

TABLE 29-2. COMPARISON OF VIRAL INFECTION OF THE RESPIRATORY TRACT IN CHILDREN AND ADULTS

	Children	Adults
Type of infection	Primary contact often	Inactivation of latent virus often; rarely, primary contact
Degree of involvement	Severe; infection may be fatal; mild cases also seen	Mild (except in aged and debilitated)
Clinical picture	Spectrum of respiratory disease	Common cold and related syndromes prominent
Causative viruses	Parainfluenza viruses, adenoviruses, and respiratory syncytial virus (viruses of childhood)	Cold viruses (coronaviruses and rhinoviruses) important; parainfluenza viruses, adenoviruses, and enteroviruses implicated

is associated with enlargement of the lymphoid tissue in the respiratory passages and of the lymph nodes in the neck.

Adenoviruses cannot be studied by their effects in animals, since they do not produce disease in the routinely used laboratory animals, nor do they grow on the membranes of the fertile egg. However, they can be demonstrated nicely in cell cultures in which certain human and monkey cells are used. In virus-infected cells examined with the electron microscope, viral particles can be seen. This group shares a common antigen demonstrable in the serum of an infected person by means of a complement fixation test. The breakdown of the group into the different types is done by means of the virus neutralization test.

Reoviruses, viruses of undetermined pathogenicity and ubiquitous in nature, have been recovered from the respiratory and alimentary tract in man in health and disease. Respiratory disease from all three types is usually low grade.

In the early stages of certain viral diseases—for example, poliomyelitis and infectious mononucleosis—involvement of the upper respiratory tract figures prominently.

Influence of age. The patterns of respiratory disease in adults and children are basically similar in many respects. Influenza, for example, is much the same at any age. Yet there are points of contrast. These are presented in Table 29-2.

Common cold (acute coryza)

The common cold is said to be the most commonly occurring ailment of mankind and to temporarily disable more people than any other infectious disease. The typical nasal discharge of only a few days' duration is spread by direct contact, and a cold is most communicable in the early stages. The incubation period is 2 to 3 days. The agent enters the body by way of the upper respiratory tract. In sneezing and coughing, the affected person spreads not only his disease but also the bacteria he may be carrying in his throat. As with other viral diseases, the cold is often complicated by more serious bacterial diseases such as pneumonia.

Some persons appear comparatively resistant to colds, whereas others are relatively susceptible. Factors said to predispose to colds are exposure to chilling and dampness, sudden changes in temperature, dusty atmospheres, drafts, loss of sleep, overwork, and lowered bodily resistance. Most colds occur between October and May, and preschool children have the greatest number. Immunity after an attack is brief. An individual may average two to four colds a year. The development of a vaccine is still experimental.

Cold viruses. In patients with the syndrome of the common cold, adenoviruses, certain enteroviruses, respiratory syncytial virus, influenza viruses, and parainfluenza viruses have at times been incriminated specifically. A true cold may be caused by one or more than one virus. The designation "cold viruses," however, usually applies to rhinoviruses and coronaviruses.

Echovirus 28, reclassified as rhinovirus type 1 (p. 529), was the first to be implicated as a "cold virus." Later, pathogenic strains referred to as *rhinoviruses* or *Salisbury viruses* (100 or more rhinoviruses recognized, 80 characterized) were found in persons with colds and were grown in cell cultures containing cells from lungs of human embryos. Rhinoviruses are responsible for more colds than any other known agent. Another group of cold viruses, the *coronaviruses,* is made up of at least 10 serotypes. It is so named because of a segmented ring about the virion. The cold caused by coronaviruses has a longer incubation period and a shorter sequence of illness than that seen with rhinoviruses.

Viral pneumonia*

Viral pneumonia is a clinical syndrome—*not* a single, specific disease—acute, infectious, and self-limited. It is similar in its manifestations to primary atypical pneumonia, also a syndrome. Viruses implicated include the influenza viruses, the parainfluenza viruses, the adenoviruses, and in infants the respiratory syncytial virus.

DISEASES OF THE CENTRAL NERVOUS SYSTEM
Rabies (hydrophobia)†

The disease. Rabies is an acute, paralytic, ordinarily fatal, infectious disease of warm-blooded animals, including man. Rabies has been recognized for 2000 years and has changed little in that period of time. It is primarily a disease of the lower animals, and dogs are chiefly responsible for its propagation in civilized communities. Other domestic animals that contract rabies are cats, horses, cows, sheep, goats, and hogs. When the disease occurs in wildlife, it is referred to as sylvatic rabies or wildlife rabies. Wild animals that often contract it are wolves, foxes, skunks, coyotes, raccoons, and hyenas. The rabid skunk is most dangerous to man or other animals, since the saliva of the skunk carries 100 to 1000 times the amount of virus carried by the dog. Skunks

---

*See the discussion on primary atypical pneumonia, p. 463.
†For immunization, see pp. 679, 680, 689-690, and 700-702.

are more susceptible than are dogs and spread the disease among themselves. The skunk is said to be the most common rabies carrier in the United States today. In certain parts of the world (South America, West Indies, Central America, and Mexico) the vampire bat transmits the disease to horses and other livestock. This is a serious veterinary problem because the vampire bat differs from other animals in that it does not succumb to the disease but becomes a symptomless carrier, remaining infective a long time. Bats are dangerous carriers of rabies because of their habits and close contact with man's dwellings. They are more likely to be found in barns and in deteriorating houses than out in the wilds. They migrate constantly and thus evade man. The disease has been recognized in certain insect-eating bats in parts of the United States (Florida, Texas, Pennsylvania, California, and Montana), in the Caribbean area, and in Europe. No instance of the transmission of the disease from man to man has been recorded. Rabies in dogs is declining in incidence, but the disease in wild animals is definitely on the increase. It is rare in rodents (rats, mice, squirrels) and almost unheard of in laboratory animals (hamsters, gerbils, guinea pigs).

Rabies in animals occurs in two forms, the *furious* and the *dumb*. In the former, a stage of increasing excitability is followed by a stage of paralysis ending in death. The animal fearlessly* attacks anything that it encounters; it drools saliva because its throat is paralyzed. In dumb rabies, paralysis and death supervene without a preceding stage of excitement. The animal neither bites nor attacks. Except that its lower jaw droops, it may show little sign of illness. It often hides and may be found dead. In either form of rabies there is no fear of water, and the animals do attempt to drink. The human patient, however, avoids fluids because of painful spasms in the throat muscles induced with the act of swallowing. This explains the name *hydrophobia* for the disease. It has been said that the agony of the spasms of the victim of rabies possibly exceeds all other forms of human suffering.

The virus. The bullet-shaped rabies virus (rhabdovirus) (Fig. 29-11) is found in the nervous system and saliva of infected animals. It is transmitted in the saliva introduced into the body through a wound that is usually the bite of a rabid animal. If a wound such as a cut or abrasion becomes accidentally contaminated with the saliva of a rabid animal, infection is as likely to occur as if the animal had inflicted the wound. Rabid animals can transmit the infection to others for several days before they show signs of disease. When a person or animal is infected, the infectious agent passes from the inoculation site along nerve trunks to the central nervous system. When the virus reaches the brain, the manifestations of rabies appear. About 30% to 40% of the persons and 50% of the dogs bitten by rabid animals develop the disease. *But once rabies develops, it is almost invariably fatal.*†

Rabies virus in the brains of animals naturally infected is known as street virus. If some of the emulsified brain containing street virus is inoculated into the brain of a rabbit, that rabbit will develop rabies within 15 to 30 days. If some of the emulsified

---

*Absence of fear in a wild animal is abnormal.
†In 1970, in Lima, Ohio, the first case of rabies to survive the ordeal was given extensive coverage by the news media.

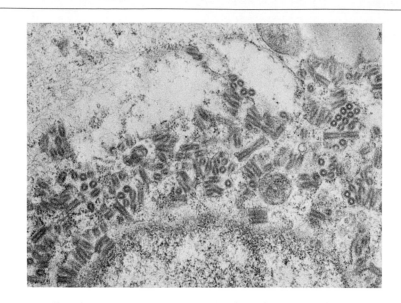

FIG. 29-11. Bullet-shaped rabies virus particles from hamster kidney cells shown in clusters, electron micrograph. (×32,000.) (Courtesy Dr. Klaus Hummeler, Philadelphia, Pa.)

brain of this rabbit is injected into a second rabbit, the second rabbit will contract rabies in a slightly shorter time. When the virus has been passed through a series of some 90 rabbits, it becomes highly virulent for the rabbit, producing rabies in 6 days. At the same time it has to some extent lost its virulence for all other species of animals. It is now *fixed* virus because of this loss of virulence. Further passage through rabbits does not reduce the time required for rabies to occur. This virus, treated with phenol, is the Semple vaccine used in the Semple method of antirabies vaccination. It has been replaced by improved preparations.

Incubation period. The period of incubation of rabies is remarkable for its length. In man it can vary from 10 days to 1 year, with an average of 2 to 6 weeks. In dogs it varies from 8 days to 1 year, with an average of 2 to 8 weeks. The nearer the site of inoculation to the brain, the shorter the period of incubation. It is also shorter in children than in adults.

Laboratory diagnosis. Adelchi Negri (1876-1912) discovered certain bodies in the brain cells of animals with rabies. The *Negri bodies* store ribonucleoprotein and virus antigen and are found in almost 100% of the patients with fully developed disease. By finding them the laboratory diagnosis of rabies is most quickly made. The fluorescent antibody examination of brain tissue is the preferred diagnostic method. Fluorescent antibody examination may also be made on smears or imprints of the epithelial cells of the cornea of the eye.

Infant mice are susceptible to rabies virus and may be infected by intracerebral inoculation of suspect material; the virus is then identified by immunologic methods or by demonstration of Negri bodies in brain tissue.

534

Prevention. *If a person is bitten by an animal suspected of having rabies, the animal should not be killed but should be placed in the hands of a competent veterinarian for observation.* If the animal has rabies, the disease will be sufficiently developed within a few days for a definite diagnosis to be made. If the dog is well 10 days after biting a person, the person is in no danger from the bite. If, on the other hand, the animal is destroyed at the time of the bite, examination of the brain may fail to show Negri bodies because they are often absent in the early stages of the disease.

If examination of the brain of the animal fails to show Negri bodies, six or eight mice should be inoculated intracerebrally with an emulsion of the brain. If the animal had rabies, the mice may show Negri bodies in their brain tissue as early as 6 days and will exhibit signs of the disease on the seventh or eighth day. If the mice survive the twenty-first day, the animal may be considered not to have had rabies.

Local treatment of wounds inflicted by rabid animals is very important in prevention of the disease. Adequate first aid treatment is crucial. Immediate and thorough cleansing of superficial wounds with a tincture of green soap or benzalkonium chloride solution may inactivate the virus. (Even tap water has merit.) An antiseptic may be applied and the wound dressed. If a wound is deeply placed and washing by soap and water is not feasible, it is then considered advisable to cauterize the wound with fuming (concentrated) nitric acid. Antiserum may be injected into the base of the wound. Dogs, cats, and other pets bitten by rabid animals should be destroyed at once or, otherwise, vaccinated and kept in strict isolation for 6 months.

Persons who work with this disease are often confronted with the question of what course to take when a person drinks the milk or eats the flesh of a rabid animal. Eating or handling infected flesh or drinking contaminated milk can produce rabies if there is an open lesion on the skin or in the alimentary tract.

The nursing precautions in rabies are rather simple. All that is necessary is to sterilize the secretions from the mouth and nose of the patient and articles so contaminated. Rabies can be eradicated from civilized communities by measures designed for intensive control of stray dogs, immunization of resident dogs, and elimination of any reservoir of infection in the wild animals of the area.

Viral encephalitis and encephalomyelitis

The term *encephalitis* (*pl.*, encephalitides) means inflammation of the brain, but by common usage it is applied to inflammatory conditions of the brain accompanied by degenerative changes instead of suppuration (pus formation). When the brain is involved, the spinal cord usually is also, hence the term *encephalomyelitis*.

Togaviral (arboviral) encephalitides

Animal viruses carried in the body of an insect vector are *arboviruses*, a simplification of "*ar*thropod-*bo*rne viruses." Arboviruses, now togaviruses, are the most numerous of the viruses infecting man. Important forms of encephalitis are caused by neurotropic arboviruses (togaviruses), the *arboviral or togaviral encephalitides*. A noteworthy form of encephalitis is *postinfection encephalomyelitis*, which is neither due to a neurotropic virus nor transmitted by an insect, but it resembles viral encephalitis clinically (not pathologically); viral infection is usually implicated in its development.

Togaviral infections occur principally in mammals or birds, and for most, man is only an accidental host. They are the largest group of zoonoses. The mosquito is an important insect vector, although togavirus is found in ticks and mites. The togaviral encephalitides are similar in many respects, but the causative agents are immunologically distinct, and the geographic distribution of disease is defined.

Equine encephalomyelitis, a disease of horses and mules, secondarily of man, occurs in three types, each with its own virus: the eastern type (EEE) is seen in the southern and eastern United States, the western type (WEE) in the western United States and Canada, and the Venezuelan type (VEE) in South America and Panama. *St. Louis encephalitis* (SLE), so named because it was first recognized in an epidemic in the vicinity of St. Louis in 1933, is a disease widespread in the United States. It occurs only in man. *Japanese B encephalitis* is found in the Far East, and *Murray Valley encephalitis* is found in Australia.

The eastern type of equine encephalitis is a severe form with a mortality up to 70%. The death rate is much lower in the western form and in St. Louis encephalitis. Young persons, particularly infants, have an increased susceptibility to western equine encephalomyelitis, and St. Louis encephalitis has its greatest incidence in persons of middle age or older. The incubation period for the different encephalitides ranges from 5 to 15 days.

The epidemiologic pattern for the different forms of togaviral encephalitis is reasonably consistent, with but slight variations. Togavirus resides and multiplies in wild birds, its natural and primary hosts, and occasionally in domestic fowl. Venezuelan equine encephalomyelitis multiplies best in a reservoir of mammals; the eastern equine encephalomyelitis virus multiplies in both mammals and birds. From its natural reservoir togavirus is transmitted from fowl to fowl (or mammal to mammal) and from bird to horses and man (terminal hosts) mainly by female mosquitoes of the genus *Culex*. There is no known instance of direct person-to-person transmission of encephalitis.

Encephalitis is a disease of warm weather and the summertime. In nontropical areas snakes and other cold-blooded animals harbor the virus for the cold months of the year. When snakes come out of hibernation, they seek out areas that also shelter large numbers of wild birds. Virus circulates in the bloodstream of snakes, as it does in birds, so that in the early spring mosquitoes can pick up the virus from such an overwintering host and carry it straight to the wild bird reservoir. Once infected, a mosquito remains so for life.

For mosquito-borne encephalitis to reach epidemic proportions, a combination of factors is required, including a large wild bird population, favorable breeding sites for mosquitoes in stagnant pools and puddles, a high temperature, and susceptible terminal hosts.

The laboratory diagnosis of encephalitis is made on recovery of the virus from suitable specimens and its proper identification. Clotted whole blood or serum, throat washings, cerebrospinal fluid, urine, and, in fatal cases, brain tissue are specimens for virus isolation. Young or newborn mice are inoculated and observed. Serologic tests for identification of virus or its presence are the neutralization test, complement fixation test, and a hemagglutination-inhibition test.

Effective vaccines made by growing the equine viruses in a chick embryo confer immunity in horses lasting from 6 months to 1 year or longer. A vaccine prepared against eastern equine encephalitis is suitable for use in human beings, but the vaccine for the western form of the disease is given only in special circumstances to individuals at high risk because of contact with the exotic and virulent form of the disease. The practical approach to the control of the disease is eradication of the mosquito vector.

Postinfection (demyelinating) encephalomyelitis

Postinfection encephalomyelitis (allergic encephalomyelitis) is an acute disease of the central nervous system that occasionally arises during convalescence from infectious diseases or occurs after vaccination against such. The related diseases are most often viral, notable among which are measles, German measles, smallpox, and influenza. Vaccination for smallpox and rabies can be so complicated.

Encephalomyelitis following smallpox vaccination (postvaccinal encephalomyelitis) threatens children and young adults who have never been vaccinated. Infants seem to be resistant. The incidence of postvaccinal encephalomyelitis is less than 1 in 33,000 vaccinations, but with the virtual elimination of smallpox from the earth, even this low incidence becomes a hazard.

The following theories are offered to explain postinfection encephalomyelitis: (1) it is caused by the virus of the primary disease; (2) vaccination activated some latent virus; and (3) the disease reflects an allergic reaction either to the virus of the preceding infection or to the patient's nervous tissue now altered by it. Currently, this last, the autoimmune, approach is favored.

Slow virus infections

In some chronic diseases of animals and man, there is experimental evidence that the end-stage changes in the tissues, particularly in the central nervous system, are the result of a slowly progressive and damaging proliferation of a virus. In such diseases there is a long incubation period (years), slow start, protracted course, fatal outcome, no demonstrable formation of antibodies, no fever, and no inflammatory changes. *Scrapie*, a slowly developing, fatal neurologic disease of sheep, is an example of a "slow virus" infection. An example in man is *kuru*, a neurologic disease found in certain cannibals in New Guinea, transmitted by the ingestion of undercooked brain tissue containing the agent. In the chronic disorder in man, subacute sclerosing panencephalitis, an agent resembling measles virus, has recently been cultured from affected brain tissue. Slow viruses may also be implicated in arthritic and rheumatic diseases and in the autoimmune diseases. Slow viruses are so designated because of their very long developmental cycles during which they are masked or inapparent. There is a definite possibility that some so designated may not be viruses at all. To date they have not been isolated. The current interest in slow virus infections stems

537

from the suggestions they give as to the cause of many poorly understood central nervous system diseases in man.

Poliomyelitis (infantile paralysis)*

Poliovirus infection. Poliomyelitis is an acute infectious disease that in its severest form affects the brain, spinal cord, and certain nerves (Fig. 29-12).

Poliomyelitis occurs in four forms:

1. Silent or asymptomatic infection—any symptoms present are so mild as to be overlooked. Such a person is a healthy carrier. Poliovirus infection of the human alimentary tract is exceedingly common all over the world. From this area the virus at times enters the blood or lymph stream. Silent infection exists among the members of the patient's family.

2. Abortive infection—findings referable to the nervous system are absent, although there may be a brief febrile illness, such as a mild respiratory infection or a simple gastrointestinal upset. Most cases are abortive.

3. Nonparalytic infection—findings indicate disease of the nervous system but without residual paralysis.

4. Paralytic infection—paralysis persists.

Polioviruses. There are three serologic types of poliovirus, designated type I (Brunhilde), type II (Lansing), and type III (Leon). These viruses, among the smallest in size, may be grown in monkeys and chimpanzees. Cell cultures of monkey kidney sustain a generous growth, as do certain human cell cultures. Polioviruses are pathogenic for man, monkeys, chimpanzees, and apes. As far as we know, man is the only animal subject to the disease; no reservoir of infection has been found in animals.

Polioviruses can be inactivated by ultraviolet radiation, by drying, and, if in a *watery* suspension, by a temperature of 50° to 55° C for 30 minutes. Inactivation is slow with disinfectant alcohol and unsatisfactory with many bacterial germicides in wide use.

---

*For immunization, see pp. 680, 688-689, 696, 703-705, and 706.

FIG. 29-12. Poliomyelitis, cross section of spinal cord from patient with paralytic poliomyelitis. (Paralysis results when motor neurons of anterior horns of spinal cord are destroyed by poliovirus.) Arrow points to anterior horn on left, a softened, depressed area of dead tissue, focally hemorrhagic. Anterior horn on right is similarly involved. (Courtesy Dr. B. D. Fallis, Dallas, Tex.)

Transmission. The incubation period ranges from 3 to 35 days. The infected person is most likely to pass the virus to a noninfected person during the latter part of the incubation period and the first week of the clinical illness, the time at which virus is present in his throat. Before the onset of symptoms in infected persons, poliovirus is found in the secretions from the mouth and throat and in the feces. Poliovirus is believed to enter the body at the upper part of the alimentary track and to leave through either the upper or the lower end.

Poliovirus has been recovered in large amounts from sewage, and milkborne epidemics have been recorded. But the epidemiologic pattern for poliomyelitis is not that for the enteric infections. The favored mode of spread seems to be the direct one from person to person. It has been said that poliomyelitis "travels with a crowd." Houseflies, filth flies, and cockroaches can be contaminated with virus, especially during an epidemic, but the role of these agents in transmission is undefined.

Epidemiology. Poliomyelitis occurs sporadically but tends to be epidemic. Infantile paralysis is not a good name for the disease because it occurs in adults and paralysis may be absent. In the early epidemics the disease chiefly attacked children less than 5 years of age, but then, in time, the children affected were older, and the number of adult cases increased.

The disease is worldwide. All races and classes of people are affected. In temperate climates, poliomyelitis appears in early summer. The number of patients and the severity of the disease increase, a peak is reached in late summer and early fall, and the disease subsides after the first frost. Occurrence in more than one member of a family is frequent. Pregnant women are more susceptible to the disease, and there have been cases of congenital poliomyelitis in which the mother had the disease late in pregnancy.

Radical changes have taken place in the epidemiology of poliomyelitis with the use of the polio vaccines. In 1957, for the first time, all states and territories of the United States were free of epidemics. For the year 1957, just less than 6000 cases of poliomyelitis were reported to the United States Public Health Service, of which about 2500 were paralytic. A decade later in 1967, there were 44 cases (29 were paralytic), mostly in unimmunized or inadequately immunized children. Compare these figures with over 57,000 cases (21,000 paralytic) reported for the prevaccine year of 1952!

Immunity. Infection with one type of virus does not confer immunity for the other types. Poliomyelitis confers lasting immunity only for the viral type responsible. A high percentage of adults have virus-neutralizing substances in their blood. Infants can inherit an immunity from their mothers by placental transfer.

Prevention. Preferably, the patient with poliomyelitis is isolated, although with widespread distribution of the virus, real insulation from contacts is impossible. Cross infections between patients probably are inconsequential in hospitals where no attempt to isolate is made. However, the disease is sometimes spread from patient to attendant, nurse, or physician. On the whole, contacts, especially between children, should be minimized during epidemics, and in an epidemic, such public health measures as the closing of public swimming pools are advisable. Tonsillectomies should not be done during epidemics or in the season of the year when the incidence of the disease is highest. There is a greater risk for the severe form of the disease to develop in persons who have recently had their tonsils removed. Healthy children may carry the virus in their throats, and virus has been demonstrated in surgically removed tonsils.

OTHER ENTEROVIRUS DISEASES
Infections with coxsackieviruses

The coxsackieviruses are worldwide in distribution and have been frequently associated with epidemics of poliomyelitis. They were named after the town in New York where the first virus was identified as research work was done on poliomyelitis. They resemble the viruses of poliomyelitis in many respects, including their epidemiology. They are found in the nasopharynx and feces and at times in other parts of the body. They have never been found in the cerebrospinal fluid.

They are divided into groups A and B. Within each group is a number of types, of which at least 30 are known. These viruses possess an unusual pathogenicity for infant mice and hamsters but none for the adult animals. Group B is the more important in man. Coxsackieviruses produce such conditions as aseptic meningitis, epidemic pleurodynia (group B), herpangina (group A), vesicular pharyngitis, encephalitis, hepatitis, orchitis, an influenza-like disease, "three-day fever," peri-

FIG. 29-13. *Aedes aegypti* mosquito, larval forms in Petri dish. (From Med. World News **6:**168, Nov. 12, 1965; courtesy Eli Lilly & Co., Indianapolis, Ind.)

carditis, and a severe form of myocarditis in infants. Group A viruses usually are associated with infections of the mouth. Most illnesses produced by coxsackieviruses present in children. Neutralizing antibodies to the virus are found in the convalescent serum of patients recovering from the diseases listed.

Modern classification of viruses assigns the coxsackieviruses to the category of picornaviruses, along with polioviruses and echoviruses. Most strains of coxsackieviruses and echoviruses can cause a disease closely resembling either paralytic or nonparalytic poliomyelitis.

Infections with echoviruses

Echoviruses were found by accident during epidemiologic studies of poliomyelitis. They were recovered from human fecal material in cell cultures. Their destructive effect on tissue culture cells and the fact that any disease to which they were related was unknown at first led to the designation of *echo*—enteric, cytopathogenic, *human orphan* viruses. Some 31 members of this group have been classified so far. Many of these viruses may be harmless parasites, but some cause epidemics of aseptic meningitis, summer diarrhea in infants and young children, and febrile illnesses, with or without rash. To date, the echoviruses have been found as etiologic agents only in clinical syndromes (not specific diseases) that may be produced also by a number of other viruses (and bacteria as well). For example, aseptic meningitis, their chief central nervous system manifestation, is a syndrome, not a distinct disease.

OTHER ARTHROPOD-BORNE VIRAL DISEASES
Yellow fever

Yellow fever,* a mosquito-borne hemorrhagic fever, is an acute infectious disease defined by an abrupt onset, a rapid course, and a high mortality. The most striking pathologic feature is the rapid and extensive destruction of the liver. Prominent clinical features are jaundice (yellow color to skin and mucous membranes), albumin in the urine, hemorrhage, and vomiting.

The virus of yellow fever is transmitted by the mosquito *Aedes aegypti* (Fig. 29-13), which also transmits dengue fever. The mosquito bites a person during the first few days of illness and becomes infected. The virus multiplies in the body of the mosquito and reaches the salivary glands at the end of about 12 days; the mosquito remains infective the remainder of *her* life. (Only the female transmits the disease.) When a nonimmune person is bitten by an infected mosquito, manifestations of yellow

---

*For immunization, see pp. 690-691, 706, and 707. Whenever yellow fever (once known as yellow jack) is discussed, eight names are brought to mind: Carlos Juan Finlay (1833-1915), who first accused the mosquito of spreading yellow fever; Walter Reed (1851-1902); James Carroll, Jesse W. Lazear, and Aristides Agramonte, who, in 1900, formed the Commission of the United States Army to study yellow fever in Cuba; Privates John R. Kissinger and John J. Moran, who permitted themselves to be inoculated with yellow fever; and William C. Gorgas, who applied the knowledge obtained to make the tropics more habitable for man. The members of the commission went to Cuba, lived in the tents of those who had had yellow fever, wore their clothes, and were bitten by infected mosquitoes. Carroll and Lazear contracted the disease; Lazear died. To these men, all who live in tropical and temperate zones owe a debt of gratitude.

fever develop in 3 to 5 days. The virus is found in the bloodstream during the first 3 days of the illness.

Yellow fever still remains endemic in many parts of the world. We must always guard against it because an epidemic requires but three things: a person ill of the disease, mosquitoes to be infected, and susceptible recipients. The transport of infected mosquitoes by airplanes and ships must be prevented.

A source of infection in tropical Africa and Latin America is jungle yellow fever found in monkeys. It is transmitted from animal to animal and from animal to man by forest mosquitoes (genus *Haemagogus*) living high in treetops. Epidemics appear in monkeys. The causative virus seems to differ in no respect from that producing ordinary yellow fever. If man becomes infected, the infection is transmitted from him by *Aedes aegypti*.

An active immunity to yellow fever can be established. The value of convalescent serum is undetermined.

Dengue fever

Dengue or "breakbone" fever is an acute disease lasting 10 days and depicted by sudden onset of paroxysmal fever, intense joint pain, skin rash, and mental depression. It is caused by a virus that is found in the bloodstream during the early days of the attack and transmitted by the *Aedes aegypti* mosquito, which, once infected, remains so the rest of *her* life. When dengue fever is introduced into a community, a high percentage of the population contracts the disease. An attack is followed by an immunity that may persist for 1 to 2 years or the remainder of the patient's life.

There are four distinct immunologic types of the virus, of which types 1 and 2 are most often responsible for the disease in the Western Hemisphere. The dengue viruses can be grown in cell cultures. Type 1, live virus, weakened by repeated passages through mouse brain is an effective vaccine in the clinical trials to date. Neutralizing and complement-fixing antibodies may be demonstrated in the serum of a patient after infection.

Colorado tick fever

A togaviral infection transmitted by an insect vector other than the mosquito is Colorado tick fever. The single, small virus is transmitted by the wood tick, *Dermacentor andersoni*. A mild febrile, diphasic illness, it is found largely in the Western states, the natural habitat of the tick vector. Patients usually have been in a tick-infected area 4 to 6 days before becoming ill, and ticks are even found still attached to their bodies. The onset of disease is sudden, with chills, fever, and bodily aches that continue for 2 days or so. Then follows a period wherein the patient is essentially free of clinical manifestations. The fever returns to last for several days more before the disease has run its course. There are no complications. The disease has not been fatal, and the pathology is not known.

Laboratory diagnosis. Colorado tick fever virus may be recovered from the patient in specimens of blood, throat washings, and stool. It grows in tissue culture and in the fertile hen's egg. When inoculated into young mice, it produces paralysis. There are complement fixation, virus neutralization, and hemagglutination-inhibition tests for serologic diagnosis.

Immunity and prevention. Infection is thought to produce lasting immunity. Prevention depends on the avoidance of tick-infested areas or the wearing of adequate protective clothing. The body should be regularly inspected for ticks, and any present should be detached. Effective vaccines have been prepared by growing the virus in chick embryos.

DISEASES OF THE LIVER (VIRAL HEPATITIS)*

There are two viral diseases of the liver of special interest. They closely resemble each other in their clinical manifestations, and the pathologic changes in the injured liver are the same. Because of certain supposed differences in epidemiology and mode of onset, they have been defined as distinct entities. One, viral hepatitis type A, has been known as infectious hepatitis, catarrhal jaundice, epidemic jaundice, and short-incubation hepatitis; the other, viral hepatitis type B, as serum hepatitis, homologous serum jaundice, transfusion jaundice, posttransfusion jaundice, and long-incubation hepatitis.† The agent postulated to cause the first has been referred to as virus A and the agent for the second as virus B.

As a frame of reference, a traditional consideration of these two is presented in Table 29-3. With the impact of new discoveries related to viral hepatitis and its causation, our concepts are drastically changing. Conventionally both are acute or subacute infections of the liver found naturally only in man.‡ The true overall incidence of viral hepatitis is unknown, since subclinical infection occurs without jaundice. It has been estimated that 80% to 90% of the cases go unrecognized. Immunity from an attack of one form of disease has been thought to result in immunity to that form only.

There is now good reason for believing that the agents, presumably viruses, causing both type A and type B hepatitis have at last been identified. The development of sophisticated methods in immunology and their application to the study of viral hepatitis has contributed much to the delineation of the agents causing both types and to a clear-cut serologic differentiation between the two. The viruses cannot be grown in cell culture.

Viral hepatitis type A

Viral hepatitis type A is usually spread directly from person to person. The principal pathway is the fecal-oral route. Poor sanitation favors its spread. Water,

---

*Viruses that can cause hepatitis include togavirus, myxovirus, adenovirus, herpesvirus, coxsackievirus, reovirus, cytomegalovirus, and the Epstein-Barr virus.

†There may also be a viral hepatitis type C with an incubation period intermediate to those of type A and type B. The incidence of posttransfusion hepatitis is still high with multiple blood transfusions, even where careful screening of would-be donors is carried out to eliminate type B infection and where infection cannot be tied in serologically to A or B. Whether forms of viral hepatitis are caused by two distinct viruses (A and B), or perhaps a whole family of viruses, is not known.

‡Hepatitis A and B can be transmitted to subhuman primates and certain other experimental animals.

543

TABLE 29-3. COMPARISON OF THE TRADITIONAL TWO TYPES OF VIRAL HEPATITIS

	Viral hepatitis type A	Viral hepatitis type B
Agent		
Virus	A	B (see discussion)
Epidemiology		
Transmission	Fecal-oral route; also parenteral	From blood and blood products; close personal contact; parenteral; also oral; insect (?)
Incubation period	Around 25-30 days (15-50 days)	30-180 days
Age preference	Young adults, 15-24 yr; also children	15-24 yr (but all ages)
Duration of infectivity	Virus in feces and blood 1-2 weeks before disease; remains 3-4 weeks longer	Virus in blood 3 months before disease; asymptomatic carrier for as long as 5 years
Virus present	Feces, blood	Blood (thought to be in feces)
Clinical features*		
Onset	Acute	Slow, usually insidious
Fever	Common before jaundice	Less common; low-grade
Jaundice	Rare in children, more frequent in adults	Rare in children, more frequent in adults
Severity of disease	Less severe	More severe (fatality rate three times higher)
Prognosis	Good	Less favorable
Laboratory evaluation		
Thymol turbidity†	Increased	Normal
Abnormal SGOT‡	Transient, 1-3 weeks	Prolonged; 1-8 or more months
IgM levels	Increased	Normal
Prevention and control		
Prophylactic effect of gamma globulin	Good	Present but poor (questionable)
Nursing precautions and control	1. Isolation of patient 2. Sterilization of contaminated items	1. No isolation of patient needed 2. Sterilization of permanent medical apparatus in contact with blood 3. Use of disposables (needles, syringes, tubing) for all patients

*Many clinical features are the same.
†Test of liver function.
‡The enzyme serum glutamic oxaloacetic transaminase is elevated with liver disease.

milk, and food can be sources of infection. Recent outbreaks have occurred in persons who have eaten raw shellfish contaminated by sewage. The cockroach and fly may be vectors. The disease occurs sporadically or in epidemics, which tend to recur cyclically every 7 years. It is thought to be a commonplace infection, with clinical illness only occasional.

The virus is present in the feces and blood of infected persons. With technics of immune electron microscopy, the antigen HA and the antibody anti-HA have been identified. A serologically distinct, exceedingly small picornavirus-like agent, one millionth of an inch in diameter, has been recovered from a case of hepatitis A. It is thought to be an RNA virus and the etiologic agent, virus A.

The mortality of type A hepatitis is not high. Widespread immunity does exist, probably gained from inapparent childhood infection. Gamma globulin given as late as 6 days before onset of illness may protect a person for as long as 6 to 8 weeks.

The patient with recognized disease should be isolated. Diligent handwashing, wearing of protective gowns, and autoclaving of articles contaminated by the patient are necessary.

Viral hepatitis type B

Viral hepatitis type B is carried by human serum (or plasma) and may complicate blood transfusion or the administration of convalescent serum, vaccines, and other biologic products containing human serum. Needles, syringes, and tubing sets for transfusions and stylets for finger puncture are important conveyors of the infection when soiled by blood or blood products.* *As little as 0.000025 ml of blood contaminated with B virus (0.01 ml with A virus) has been shown to cause disease.* For this reason the disposable needles, disposable stylets for finger puncture, and disposable syringes, all now commercially available, are strongly recommended. Disposable units of plastic tubing suitable for blood transfusion are now in wide use.

Hepatitis B represents an occupational hazard among medical personnel working with blood or serum, including laboratory technologists, blood bank workers, physicians, dentists, and nurses. Surgeons are at high risk because of the possibility of accidental self-inoculation. Persons so exposed should observe rigid asepsis. The attack rate is high in the renal dialysis unit, and paradoxically, the hepatitis is much more severe among the nurses and other personnel than it is in the patients. The disease is also prevalent among drug addicts who share their unsterilized and contaminated hypodermic needles. Heroin addicts are especially notorious for "passing the needle." Epidemiologic changes in viral hepatitis within the last 5 years reflect the increase of illicit drug use and its patterns. There have been shifts in seasonal and age incidences, a trend from rural to urban cases, and more of hepatitis B than of hepatitis A. Vaccines are experimental.

Australia antigen. A population geneticist, working with the blood of an Australian aborigine in 1964, found a new antigen, which he called the *Australia antigen (Au)*. As it turned out, his was a most significant discovery.†

Shortly after its discovery, Australia antigen was visualized in infectious serum with the electron microscope as very small, spherical and filamentous viruslike

---

*For sterilization technics to prevent the spread of the hepatitis viruses, see p. 308.
†The research from which this discovery "fell out" was a basic type, not goal-oriented at all. The impact of the finding and the way in which it was made reemphasizes the continuing need for the "basic" approach to scientific matters.

particles about 20 nm in diameter with knoblike units on their surface. Its identity as a virus was challenged in part because of its rather simple outlines. In 1970 the Dane particle was discovered, a structurally more complex particle with a 7 nm outer coat and a 42 nm internal core, containing circular, double-stranded DNA. The core has antigenic specificity of its own designated $Hb_c Ag$ and is coated by the Australia antigen, which, as the surface antigen of the Dane particle, is referred to as $HB_s Ag$. (There are subtypes of $HB_s Ag$.) The evidence is strong that the Dane particle plus $HB_s Ag$ represents the complete virus of hepatitis B (HBV).

$HB_s Ag$ (Australia antigen) is tightly linked to acute and chronic hepatitis (although it is also found in diseases such as Down's syndrome and leukemia, wherein immune mechanisms have gone awry). More than half the patients with hepatitis B have the antigen in their blood. It can be recovered from liver cell nuclei, urine, feces, and various body fluids. It is found clinically before the onset of jaundice but is usually undetected during convalescence. It tends to persist into the carrier state more with the patient who does not have jaundice. The antibody to the surface antigen, anti-$HB_s Ag$, usually is not present until late in convalescence, but it may persist for years in the serum after the acute episode.

Implications for blood banking. The immediate application of the knowledge gained about $HB_s Ag$ and the hepatitis B virus is in the blood bank to screen donors. The incidence of viral hepatitis among patients receiving blood containing $HB_s Ag$ is five times greater than it is among those patients given $HB_s Ag$-negative blood. Already a number of tests have been designed to detect the surface antigen in blood for transfusion. In rapid succession many are still coming onto center stage. Technics include those of complement fixation, immunoelectrophoresis, hemagglutination-inhibition, reverse passive hemagglutination, and the widely acclaimed radioimmunoassay. The very serious risk involved has been minimized to the extent that presently only about one fifth of cases of posttransfusion hepatitis are attributable to the agent of hepatitis B.

MISCELLANEOUS VIRAL INFECTIONS
Mumps (epidemic parotitis)*

Mumps, or epidemic parotitis, is an acute contagious disease prevalent in winter and early spring and accompanied by a painful inflammatory swelling of one or both parotid glands. It is easily recognized because of the typical appearance of the patient. (A child may look like a chipmunk with nuts in its cheeks.) Caused by a paramyxovirus, it occurs most often between the fifth and fifteenth years. Adult epidemics may erupt in military organizations and are extremely difficult to control.

The period of contagion begins before the glandular swelling and persists until it subsides. The incubation period is usually 14 to 21 days. Mumps is mostly transferred directly from person to person by droplets of saliva, but indirect transfer by contaminated hands or inanimate objects occurs. An estimated 30% to 40% of persons infected with the mumps virus have a silent infection followed by permanent im-

---

*For immunization, see pp. 688, 696, and 700.

munity. During the course of the inapparent infection, these individuals may pass the virus to others.

·The virus is thought to enter the body by way of the mouth and throat and probably reaches the salivary glands via the bloodstream. Viremia is responsible for such complications as orchitis, oophoritis, encephalitis, and pancreatitis. Inflammation of the testicle (orchitis), usually one sided, occurs in 20% of adult males with this disease. If, as is rarely the case, both testicles are involved, sterility may result. The mumps virus is an important cause of the aseptic meningitis syndrome. Inflammation in the central nervous system can be present with and without parotitis.

Laboratory diagnosis. Mumps virus may be recovered from the urine, saliva, or blood; in cases of central nervous system disease, it is found in the cerebrospinal fluid. The chick embryo and cell cultures are used for its identification. The virus agglutinates red blood cells of the chicken and of the human blood group O. A skin test, not widely used, is available. The antigen for the test is killed virus from fertile eggs. Inflammation at the site of injection 18 to 36 hours later indicates immunity in the absence of sensitivity to egg protein.

Immunity. A passive immunity is given by an immune mother to her baby and persists for about 6 months. Convalescent serum given from 7 to 10 days after exposure protects a high proportion of children from infection. An attack of mumps is usually followed by permanent immunity, and, contrary to lay belief, the same immunity follows either unilateral or bilateral involvement.

Infectious mononucleosis

Infectious mononucleosis, or glandular fever, is an acute infectious disease with enlargement of the lymph nodes and spleen, sore throat, and mild fever. The total number of leukocytes is increased, and characteristic white blood cells (lymphocytes of unusual or atypical appearance) appear in the peripheral circulation. Rarely fatal, it may occur in epidemic form or sporadically. It is transmitted by mouth and throat secretions. Babies may pass it via the oral-fecal route. It usually attacks children and young adults (to 30 years of age). The incubation period is unknown but is thought to be 4 to 14 days or longer; some say 5 weeks. The disease may last 1 to 6 weeks, sometimes longer.

Infectious mononucleosis is important because it may be mistaken for diphtheria (because of sore throat) and such a serious disease as lymphatic leukemia. Nonsyphilitic patients may give positive serologic tests for syphilis during the disease and for several weeks after recovery. The heterophil antibody or Paul-Bunnell test is important in the diagnosis.

Epstein-Barr virus (EB virus). Infectious mononucleosis has long been considered a viral disease, the course of the disease suggesting that it should be. But a specific virus was not pinpointed for a long time.

In the mid-1960s a brand new member of the herpesvirus group was found under circumstances linking it first to Burkitt's lymphoma (p. 510) and then even more significantly to infectious mononucleosis. The virus was named Epstein-Barr virus (EBV) after the cell culture cell line from which it was first isolated. (It is freely called the EB virus.) Specific antibodies to the EB virus are regularly associated with both Burkitt's lymphoma and infectious mononucleosis.

There are good reasons for thinking that the EB virus causes infectious mononucleosis with only a few reservations. For one thing, serologic studies comparing antibodies to EB virus with the diagnostic heterophil antibodies in the course of the disease correlate well. One can demonstrate that antibodies to EB virus and heterophil antibodies come in and peak at about the same time in a given case. Three months later heterophil antibodies drop out; EB virus antibodies remain. One important reservation is that Koch's postulates have not been satisfied. Infectious mononucleosis has not as yet been transmitted to healthy human volunteers.

Infection with EB virus must be widespread. The virus is everywhere. Most infections are postulated to be either silent inapparent ones or at most mild upper respiratory ailments. The virus is thought to replicate in the oropharynx, the tonsillar and adenoidal lymphoid tissue being the milieu for the virus.

There are immunofluorescent and complement fixation tests for identification of the EB virus, but these are not offered in routine clinical laboratories. A vaccine is investigational.

Condyloma acuminatum

Condyloma acuminatum is a cauliflower-like mass of coalescent warty escrescences formed on the external genitalia and about the anal region in either sex (Fig. 29-14). The growths are often multiple and vary considerably in size. The lesion is called a venereal wart and is transmitted as a venereal disease. The incubation period is 1 to 6 months. Large and bulky warts cause considerable discomfort and are associated with a disagreeable odor. The cause of this condition is a virus, either the same or an allied strain of the one causing the human wart so familiar in its extragenital

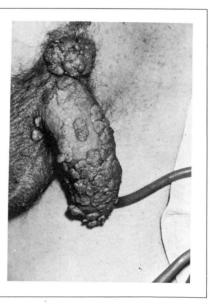

FIG. 29-14. Condylomata acuminata of skin of penis. Note confluence of wartlike growths. A giant condyloma acuminatum is present at penile-abdominal junction. (From Winter, C. C.: Practical urology, St. Louis, 1969, The C. V. Mosby Co.)

setting. Encouraging results in treatment are reported with the use of autogenous vaccines.

Foot-and-mouth disease (rhinovirus of cattle)

Foot-and-mouth disease, worldwide in distribution, is an acute infectious process in animals described by the formation of vesicles in the mouth, on the udder, and on the skin about the hoofs. One of the most contagious of all diseases, it primarily affects cloven-footed animals (cattle, sheep, swine, and goats). Horses are immune, and man seldom contracts the disease. Within the last few years it has been present in Mexico.

The disease is readily transmitted by direct contact or indirectly by contamination of fodder with infectious discharges. The disease has been transferred to man by contact with infected animals, by milk from infected animals, and by contaminated material.

Control of the disease depends on the slaughter of all exposed animals, with burial or cremation of their bodies, disinfection of pens, and proper quarantine. There is no specific treatment. A diagnosis may be established by injecting suspicious material into the footpads of a guinea pig. Typical lesions will be produced if the rhinovirus is present.

Distemper of dogs and cats

Canine distemper and feline enteritis (cat distemper) are important diseases of young dogs and cats. They are caused by two entirely different viruses. The distemper of dogs affects primarily the respiratory and nervous systems. In cat distemper the intestinal tract is primarily involved. Canine distemper is spread by food and water contaminated with the virus and to a lesser degree via the respiratory tract. Feline distemper is spread by direct contact and probably by fleas. Canine distemper affects dogs and related species. Feline distemper affects cats and racoons. Neither affects man. Highly successful vaccines have been prepared against both diseases.

Additional diseases caused by viruses

In the following lists are viral diseases that affect man, animals, plants, and insects, respectively.

Man

Hemorrhagic fevers (Omsk, Crimean, Bolivian, Indian, and Korean types, hemorrhagic nephrosonephritis, Kyasanur Forest disease)
Warts (verrucae)
Lymphocytic choriomeningitis
Phlebotomus (sandfly) fever

Bwamba fever
Lassa fever
Rift Valley fever
West Nile fever
Semliki Forest disease
Louping ill
Russian spring-summer encephalitis

Animals

Pox of horses, sheep, goats, swine, buffalo, and camels
Rinderpest
Ectromelia

Louping ill of sheep
Fowl plague
Fowlpox
Newcastle disease of chickens

Animals—cont'd

Vesicular stomatitis	Mengo fever
Orf	Visna in sheep
Infectious canine hepatitis	Infectious pancreatic necrosis of trout and
Aleutian disease of mink	Atlantic salmon
Hog cholera	Duck virus enteritis
Rift Valley fever of sheep	

Plants

Tobacco mosaic	Barley stripe mosaic
Tomato bushy stunt	Turnip yellow mosaic
Tobacco necrosis	Tomato spotted wilt
Peach yellows	Lettuce necrotic yellows
Curly top of sugar beet	Potato X
Swollen shoot	Striate mosaic of wheat
Wound tumor	Alfalfa mosaic
Tobacco rattle	

Insects (mostly the larval stage actively diseased)

Chronic bee paralysis	Polyhedrosis of spruce sawfly
Silkworm jaundice	Other polyhedroses
Tipula iridescence	Granuloses

QUESTIONS FOR REVIEW

1. Make a chart showing the causative agent, clinical features, laboratory diagnosis, transmission, and prevention of smallpox, measles, German measles, mumps, influenza, poliomyelitis, rabies, yellow fever, dengue fever, Colorado tick fever, infectious mononucleosis.
2. Discuss briefly the relation of chickenpox to shingles.
3. Outline the logical procedure to follow when a person is bitten by an animal suspected of having rabies.
4. Briefly compare hepatitis A with hepatitis B.
5. Explain the occurrence of fever blisters.
6. Briefly characterize adenoviruses, coxsackieviruses, echoviruses, togaviruses, rhinoviruses, reoviruses, enteroviruses, coronaviruses, cytomegaloviruses, herpesviruses, cold viruses.
7. Give the significance of the Australia antigen and the Dane particle.
8. What is meant by a syndrome? Briefly discuss acute respiratory syndromes.
9. Define or briefly explain alastrim, variola major, variola minor, microcephaly, pustule, coryza, rhinitis, encephalomyelitis, yellow jack, silent infection, SGOT, vesicle.
10. What is the current position of the Epstein-Barr virus?
11. What serious effect may rubella have in a pregnant woman? Explain congenital rubella.
12. It is superstition among laymen that if the eruption of shingles encircles the body death occurs. Explain why the eruption does not do this.
13. Discuss sylvatic or wildlife rabies and its implications for the spread of the disease.
14. What is meant by slow virus infection? What is the implication?

REFERENCES FOR CHAPTERS 28 AND 29

A shedding of light (editorial), J.A.M.A. **212:**1057, 1970.
Adams, J. M.: Persistence of measles virus and demyelinating disease, Hosp. Pract. **5:**87, May, 1970.
Anderson, F. D., and others: Recurrent herpes genitalis, Obstet. Gynecol. **43:**797, 1974.
Andrewes, Sir C., and Pereira, H. G.: Viruses of vertebrates, ed. 3, Baltimore, 1972, The Williams & Wilkins Co.
Austin, D. F., and others: Excess leukemia in cohorts of children born following influenza epidemics, Am. J. Epidemiol. **101:**77, 1975.
Baltimore, D.: Viruses, polymerases, and cancer, Science **192:**632, 1976.

Behbehani, A. M.: Laboratory diagnosis of viral, bedsonial and rickettsial disease, Springfield, Ill., 1972, Charles C Thomas, Publisher.

Behbehani, A. M.: Human viral, bedsonial, and rickettsial diseases, a diagnostic handbook for physicians, Springfield, Ill., 1972, Charles C Thomas, Publisher.

Behbehani, A. M., and Marymont, J. H., Jr.: The role of the hospital laboratory in the diagnosis of viral diseases. 1. Basic concepts, Am. J. Clin. Pathol. 53:43, 1970.

Bishop, R. F., and others: Virus particles in epithelial cells of duodenal mucosa from children with acute nonbacterial gastroenteritis, Lancet 2:1281, 1973.

Blumberg, B. S., and others: Australia antigen and hepatitis, Cleveland, 1972, CRC Press, Inc.

Boffey, P. M.: Anatomy of a decision: how the nation declared war on swine flu, Science 192:636, 1976.

Boffey, P. M.: Swine flu campaign; should we vaccinate the pigs? Science 192:870, 1976.

Burkitt, D. P., and Wright, D. H.: Burkitt's lymphoma, Baltimore, 1970, The Williams & Wilkins Co.

Carter, W. A., and De Clercq, E.: Viral infection and host defense, Science 186:1172, 1974.

Chang, T-W, and others: Genital herpes, some clinical and laboratory observations, J.A.M.A. 229:544, 1974.

Cohen, E. P.: What you should do when the flu bug bites, Today's Health 53:16, Feb., 1975.

Collier, R.: The plague of the Spanish lady, New York, 1974, Atheneum Publishers.

Conrad, M. E., and Knodell, R. G.: Viral hepatitis—1975, J.A.M.A. 233:1277, 1975.

Culliton, B. J.: Cancer virus theories: focus of research debate, Science 177:44, 1972.

Cunningham, C. H.: A laboratory guide in virology, Minneapolis, 1973, Burgess Publishing Co.

Dietzman, D. E., and others: The occurence of epidemic hepatitis in chronic carriers of Australia antigen, J. Pediatr. 80:577, 1972.

Dmochowski, L.: Viruses and breast cancer, Hosp. Pract. 7:73, Jan., 1972.

Docherty, J. J., and Chopan, M.: The latent herpes simplex virus, Bacteriol. Rev. 38:337, 1974.

Dowdle, W. R., and others: Simple double immunodiffusion test for typing influenza viruses, Bull. WHO 51(3):213, 1974.

Downham, M. A. P. S., and others: Role of respiratory viruses in childhood mortality, Br. Med. J. 1:235, Feb. 1, 1975.

Dueñas, A., and others: Herpesvirus type 2 in a prostitute population, Am. J. Epidemiol. 95:483, 1972.

Faulkner, R. S., and Gough, D. A.: Rubella 1974 and its aftermath, congenital rubella syndrome, Can. Med. Assoc. J. 114:115, Jan. 24, 1976.

Feldman, R. E., and Schiff, E. R.: Hepatitis in dental professionals, J.A.M.A. 232:1228, 1975.

Fenner, F.: Classification and nomenclature of viruses: the current position, ASM News 42:170, 1976.

Fenner, F., and others: The biology of animal viruses, ed. 2, New York, 1973, Academic Press, Inc.

Finter, N. B., editor: Interferons and interferon inducers, New York, 1973, American Elsevier Publishing Co.

Fisher, M. M., and Steiner, J. W., editors: Proceedings of the Canadian Hepatic Foundation 1971 International Symposium on viral hepatitis, Can. Med. Assoc. J. 106(suppl.):417, 1972.

Fox, J. P., and Kilbourne, E. D.: Epidemiology of influenza—summary of influenza workshop. IV, J. Infect. Dis. 128:361, 1973.

Fraenkel-Conrat, H., and Wagner, R. R.: Comprehensive virology, New York, 1974, Plenum Publishing Corp.

Garibaldi, R. A., and others: Hospital-acquired serum hepatitis, J.A.M.A. 219:1577, 1972.

Gocke, D. J.: A prospective study of posttransfusion hepatitis, the role of Australia antigen, J.A.M.A. 219:1165, 1972.

Green, M.: Viral cell transformation in human oncogenesis, Hosp. Pract. 10:91, Sept., 1975.

Gregg, M. B.: The current status of influenza in the United States (editorial), South. Med. J. 66:1085, 1973.

Gunther, M.: Don't feed the rats, Today's Health 53:48, Jan., 1975.

Henry, J. B., and Widmann, F. K.: The Australia antigen: where do we stand? Part 1, Postgrad. Med. 50:167, Dec., 1971; Part 2, 51:257, Jan., 1972.

Hewetson, J. F.: The role of Epstein-Barr virus (EBV) in various human disorders, South. Med. J. 68:1271, 1975.

Hollinshead, A. C., and others: Antibodies to Herpesvirus novirion antigens in squamous carcinomas, Science 182:713, 1973.

Horne, R. W.: Virus structure, New York, 1974, Academic Press, Inc.

Horoshak, I.: As flu season approaches: a call to battle, RN **38:**17, Oct., 1975.

Horstmann, D. M.: Rubella: the challenge of its control, J. Infect. Dis. **123:**640, 1971.

Horstmann, D. M.: Medicine and the two faces of virology, ASM News **42:**120, 1976.

Horta-Barbosa, L., and others: Chronic viral infections of the central nervous system, J.A.M.A. **218:**1185, 1971.

Interferon: 1973 (editorial), South. Med. J. **67:**1, 1974.

Jacobs, J. W., and others: Respiratory syncytial and other viruses associated with respiratory disease in infants, Lancet **1:**871, 1971.

Kaplan, A. S., editor: The herpesviruses, New York, 1973, Academic Press, Inc.

Katz, S. L., and Griffith, J. F.: Slow virus infections, Hosp. Pract. **6:**64, Mar., 1971.

Kaufman, R. H., and Rawls, W. E.: Herpes genitalis and its relationship to cervical cancer, CA **24:**258, 1974.

Kilbourne, E. D.: The molecular epidemiology of influenza, J. Infect. Dis. **127:**478, 1973.

Kilbourne, E. D., editor: The influenza viruses and influenza, New York, 1975, Academic Press, Inc.

Knight, V., and others: Immunofluorescent diagnosis of acute viral infection, South. Med. J. **68:**764, 1975.

Koch, F. J., and others: Diagnosis of human rabies by the cornea test, Am. J. Clin. Pathol. **63:**509, 1975.

Kohn, A., and Klingberg, M. A., editors: Immunity in viral and rickettsial diseases, New York, 1972, Plenum Publishing Corp.

Krech, U., and others: Cytomegalovirus infections of man, White Plains, N.J., 1971, Albert J. Phiebig.

Krugman, S.: Viral hepatitis and Australia antigen, J. Pediatr. **78:**887, 1971.

Krugman, S.: Hepatitis: current status of etiology and prevention, Hosp. Pract. **10:**39, Nov., 1975.

Krugman, S., and others: Viral hepatitis type B. DNA polymerase activity and antibody to hepatitis B core antigen, N. Engl. J. Med. **290:**1331, 1974.

Langmuir, A. D.: Influenza: its epidemiology, Hosp. Pract. **6:**103, Sept., 1971.

Lindsay, M. I., Jr., and Morrow, G. W., Jr.: Primary influenzal pneumonia, Postgrad. Med. **49:**173, May, 1971.

Lindsay, M. I., Jr., and others: Hong Kong influenza, J.A.M.A. **214:**1825, 1970.

Liu, C.: Rapid diagnosis of viral infections, South. Med. J. **68:**679, 1975.

Marx, J. L.: Cytomegalovirus: a major cause of birth defects, Science **190:**1184, 1975.

Marymont, J. H., Jr., and Behbehani, A. M.: The role of the hospital laboratory in the diagnosis of viral diseases. II. Technics, Am. J. Clin. Pathol. **53:**51, 1970.

Maugh, T. H., II.: Hepatitis: a new understanding emerges, Science **176:**1225, 1972.

Medical News: New studies tighten link between herpesvirus, cervical cancer, J.A.M.A. **234:**1101, 1975.

Medical News: Influenza virus mutation evokes memories of past epidemics, J.A.M.A. **235:**1193, 1976.

Medical Staff Conference, University of California, San Francisco: Arthritis caused by viruses, Calif. Med. **119:**38, Sept., 1973.

Melnick, J. L., editor: Progress in medical virology, White Plains, N.Y., 1974, Albert J. Phiebig, vol. 17.

Melnick, J. L., and others: Recent advances in viral hepatitis. Part I, South. Med. J. **69:**468, 1976.

Metz, D. H.: The mechanism of action of interferon, Cell **6:**429, 1975.

Miller, G.: The oncogenicity of Epstein-Barr virus, J. Infect. Dis. **130:**187, 1974.

Miller, L. W., and others: Poliomyelitis in a high-risk population, Pediatrics **49:**532, 1972.

Monif, G. R. G., and others: The correlation of maternal cytomegalovirus infection during varying stages in gestation with neonatal involvement, J. Pediatr. **80:**17, Jan., 1972.

Nahmias, A. J.: Herpes simplex virus infection—present status of diagnosis and management (editorial), South. Med. J. **68:**1191, 1975.

Oswald, N. C., and others: Review of routine tests for respiratory viruses in hospital inpatients, Thorax **30:**361, 1975.

Pattison, C. P., and others: Epidemiology of hepatitis B in hospital personnel, Am. J. Epidemiol. **101:**59, 1975.

Paul, J. R.: A history of poliomyelitis, New Haven, Conn., 1971, Yale University Press.

Perspectives on the control of viral hepatitis, type B, Morbid. Mortal. Rep. 25(suppl.):3, May 7, 1976.

Phillips, D. F.: Hepatitis. 2. The scientific advances, Hospitals 45:48, Jan. 1, 1971.

Probert, M., and Epstein, M. A.: Morphological transformation in vitro of human fibroblasts by Epstein-Barr virus: preliminary observations, Science 175:202, 1972.

Redeker, A. G.: Hepatitis B, risk of infection from antigen-positive medical personnel and patients, J.A.M.A. 233:1061, 1975.

Rhodes, A. J., and Van Rooyen, C. E.: Textbook of virology, Baltimore, 1968, The Williams & Wilkins Co.

Rogers, R. S., III, and Tindall, J. P.: Herpes zoster in the elderly, Postgrad. Med. 50:153, Dec., 1971.

Rose, H. M.: Influenza; the agent, Hosp. Pract. 6:49, Aug., 1971.

Rosen, P., and Hajdu, S.: Cytomegalovirus inclusion disease at autopsy of patients with cancer, Am. J. Clin. Pathol. 55:749, 1971.

Rosenburg, J. L., and others: Viral hepatitis: an occupational hazard to surgeons, J.A.M.A. 223:395, 1973.

Rovozzo, G. C., and Burke, C. N.: A manual of basic virological techniques, Englewood Cliffs, N.J., 1973, Prentice-Hall, Inc.

Sauer, G. C.: Skin diseases due to viruses, Hosp. Med. 7:82, Aug., 1971.

Sever, J. L.: Viral teratogens: a status report, Hosp. Pract. 5:75, Apr., 1970.

Sinclair, J. C., and others: Hepatitis B surface antigen and antibody in asymptomatic blood donors, J.A.M.A. 235:1014, 1976.

Snydman, D. R., and others: Prevention of nosocomial viral hepatitis, type B (hepatitis B), Ann. Intern. Med. 83:838, 1975.

Subak-Sharpe, G.: The venereal disease of the new morality, Today's Health 53:42, 1975.

Sulkin, S. E., and Allen, R., editors: Virus infections in bats, White Plains, N.Y., 1974, Albert J. Phiebig.

Szmuness, W., and others: On the role of sexual behavior in the spread of hepatitis B infection, Ann. Intern. Med. 83:489, 1975.

Timbury, M. C.: Notes on medical virology, ed. 5, New York, 1974, Longman, Inc.

Vaisrub, S.: Expect the unexpected in hepatitis (editorial), J.A.M.A. 230:1020, 1974.

Vaisrub, S.: Identifying hepatitis type A (editorial), J.A.M.A. 233:987, 1975.

Vianna, N. J., and Hinman, A. R.: Rocky Mountain spotted fever on Long Island: epidemiologic and clinical aspects, Am. J. Med. 51:725, 1971.

Villarejos, V. M., and others: Role of saliva, urine, and feces in the transmission of type B hepatitis, N. Engl. J. Med. 291:1375, 1974.

Weller, T. H.: Cytomegaloviruses: the difficult years, J. Infect. Dis. 122:532, 1970.

Wenzel, R. P., and others: Acute respiratory disease; clinical and epidemiologic observations of military trainees, Milit. Med. 136:873, 1971.

Werner, B., and London, W. T.: Host responses to hepatitis B infection: hepatitis B surface antigen and host proteins, Ann. Intern. Med. 83:113, 1975.

Williams, R. C., and Fisher, H. W.: An electron micrographic atlas of viruses, Springfield, Ill., 1974, Charles C Thomas, Publisher.

Williams, S. V., and others: Dental infection with hepatitis B, J.A.M.A. 232:1231, 1975.

Zeman, W., and Lennette, E. H., editors: Slow virus diseases, Baltimore, 1974, The Williams & Wilkins Co.

30 Fungi

FUNGI IN PROFILE

Molds, yeasts, and certain related forms constitute the organisms in the plant kingdom known as *fungi*. From this group come those whose presence is a common sight on stale bread, rotten fruit, or damp leather. Be it fuzzy or sooty, green, black, or white, the growth on moldy food and clothing is familiar to everyone. Fungi (100,000 species or more*) are among the most plentiful forms of life—they powder the earth and dust the atmosphere. Their science is *mycology*.

Fungi do not contain chlorophyll and are probably degenerate descendants of chlorophyll-bearing ancestors, most likely the algae. Being unable to make their own food by photosynthesis as higher plants do, they must either exist on other living organisms as parasites or avail themselves of the dead remains as saprophytes. Within the protoplasm of saprophytic fungi are elaborated chemical substances and enzymes that diffuse into the environment, changing what complex substances are there (wood, leather, clothing, bread, and dead organic plant or animal matter) to simpler substances that can be used for their food. The chemical processes of digestion are completed outside the organism, and the end products are then absorbed by the fungus.

Structure. Fungi vary in size. Some, the mushrooms and toadstools, are large and easily visible to the naked eye, but most of the medically important ones are microscopic or at least so small that the microscope is necessary for their complete investigation. A given fungus may be a single cell, or it may be composed of many cells laid out in a definite pattern. This distinction is not always sharp, for the two forms may represent different phases of fungous growth. Certain important pathogenic fungi exhibit this quality of *dimorphism* (having two forms). They produce disease in the body tissues as single cells but, when cultured in the laboratory, present a complex

---

*Less than 100 can invade man or animals, and less than a dozen can infect and kill.

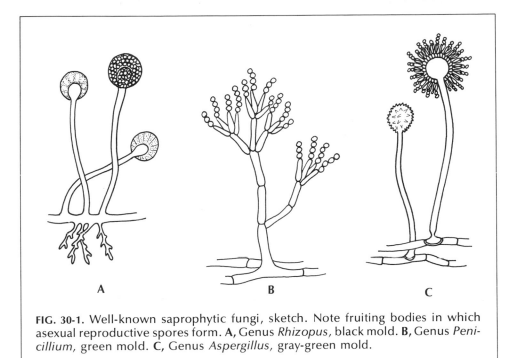

FIG. 30-1. Well-known saprophytic fungi, sketch. Note fruiting bodies in which asexual reproductive spores form. **A,** Genus *Rhizopus,* black mold. **B,** Genus *Penicillium,* green mold. **C,** Genus *Aspergillus,* gray-green mold.

multicellular arrangement. In this chapter, however, to describe fungi, we shall separate them loosely on this basis, designating the unicellular forms as *yeasts* and *yeastlike* organisms and the multicellular ones as *molds* and *moldlike* forms.

The unicellular fungi or yeasts are nonmotile, round or oval organisms most of which reproduce by a characteristic process of budding. They vary considerably in size, depending on age and species, but all are microscopic. Nuclei may be demonstrated by a suitable stain.

The multicellular fungi, the molds and related forms, present typical structures associated with nutrition and reproduction. Most molds are made up of a *mycelium* (Fig. 30-1), a network or matlike growth of branched threads bearing fruiting bodies. A rudimentary plant known as a *thallus* (see also p. 5) (no root, stem, or leaf) is formed. The individual threads of the mycelium are known as *hyphae*. In some fungi nonseptate hyphae consist of single threads containing many nuclei spaced along the thread. In most, septate hyphae are divided by cross walls, or *septa*, into distinct cells, each containing a nucleus. The hyphae have thin walls to allow for ready absorption of food and water. This fact helps to account for the rapid growth of fungi. The portion of the mycelium concerned with nutrition is referred to as the *vegetative* mycelium. The part that usually projects into the air is the *aerial* or *reproductive* mycelium.

Reproduction. Multicellular fungi reproduce by the conversion of a spore into a vegetative fungus. Spores are formed in a great variety of ways from the reproductive mycelium, depending on the species. In some molds the spores are simply attached to the hyphae; in some the hyphae bear little pods or sacs in which the spores rest

(Fig. 30-1, *A*); in others the hyphae are branched to form brushlike processes, each of which bears a spore (Fig. 30-1, *B*); in still others the hyphae produce heads from which radiate fine chains of spores (Fig. 30-1, *C*). In most the spores separate from the hyphae before reproduction.

If a spore is formed after the nuclei of two hyphae have contacted and fused or after an association of a specialized structure on the mycelium with the nucleus of another specialized structure developed close by, that spore is designated a *sexual spore*. When there is no fusion of nuclei and the spore is simply formed as a swollen body at the end of the hypha, it is called an *asexual* spore. Asexual spores are formed in great numbers, sexual spores only occasionally. The fungi that produce only asexual spores are referred to as *Fungi Imperfecti*, or imperfect fungi. (Fungi producing sexual spores are perfect fungi.) The Fungi Imperfecti are important because the pathogenic fungi are concentrated in this division.

Spores* are resistant to drying, cold, and moderate heat and may maintain their vitality for a long time. They are constantly present in the air. When they contact food or other material supplying the necessary elements for growth, they develop into new fungi.

Unicellular fungi thought of as yeasts may reproduce by spore formation. The cell enlarges slightly, and the nucleus becomes converted into a definite number of spores. However, *budding*, considered by some as a simple type of spore formation, is the usual way. In the process of budding the nucleus moves toward the edge of the cell and divides into two daughter nuclei. A knoblike protrusion of the cytoplasm forms at this point, and one of the nuclei passes into it. The protrusion increases in size and becomes constricted at its base until there is only a narrow connection between the protrusion and the parent cell. Finally, the two separate, and the protrusion or bud, now a small yeast cell, continues to grow until it reaches full size; then the process is repeated.

Multiplication by simple fission, characteristic of some yeasts, resembles the process in bacteria. The yeast develops to its full size, and a membrane forms across the middle of the cell. A dividing wall forms, and the two parts separate.

Conditions affecting growth. Fungi grow best under much the same conditions as do bacteria, that is, in warm, moist surroundings. They grow in the presence of much acid and large amounts of sugar, which bacteria cannot do. Most do not grow in the absence of free oxygen, and large amounts of carbon dioxide are harmful to them. Only a few are anaerobic. Many grow best at temperatures somewhat lower than body temperature. At low temperatures metabolic activities may be slowed, but the organisms do not die. In fact, they are rather resistant to cold; many of the commonly encountered ones survive freezing temperatures for long periods of time (months or years). Some can even grow at temperatures below freezing. To prevent growth of mold, meats and certain food products must be refrigerated at temperatures less than 20° F ($-6.67°$ C). On the other hand, fungi are quite susceptible to heat, being

---

*The spores of fungi should not be confused with those of bacteria, the resistant bodies formed by bacteria for survival, not for reproduction. Only one spore forms from one bacterial cell.

FIG. 30-2. Black bread mold, *Rhizopus nigricans,* on nutrient agar plate. Each tiny dot is a sporangium (fruiting body) containing hundreds of spores. (From Noland, G. B., and Beaver, W. C.: General biology, ed. 9, St. Louis, 1975, The C. V. Mosby Co.)

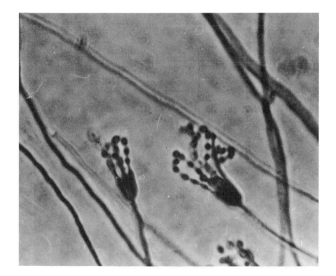

FIG. 30-3. Blue-green mold *Penicillium* (wet mount examined microscopically). Note conidia (specialized hyphae) and conidiospores (asexual spores). (From Noland, G. B., and Beaver, W. C.: General biology, ed. 9, St. Louis, 1975, The C. V. Mosby Co.)

easily killed at high temperatures. Most species are little affected by light, being able to grow either in the light or in the dark.

Classification. The *Eumycetes* (p. 5) are the true fungi, organisms without chlorophyll ordinarily thought of as yeasts, molds, and related forms. On the basis of morphology—the appearance of the colony, spores, and mycelium—the true fungi are placed in three classes and one form-class (the imperfect fungi) as follows:

1. *Phycomycetes* or *water molds*. These grow on water plants and fish. Included are the ubiquitous bread molds of genera *Rhizopus* (Fig. 30-2) and *Mucor* and other widespread contaminants. (The hyphae of water molds are nonseptate; in the other three classes the hyphae of the mycelium, where present, are septate.)

2. *Ascomycetes* or *sac fungi*. These constitute the largest order. Sexual spores are formed within a specially developed sac or *ascus*. Included here are single-celled yeasts of the genus *Saccharomyces*, crucial to the brewing, baking, and wine industries, as well as the multicellular molds *Aspergillus* and *Penicillium* (Fig. 30-3). The common contaminant *Aspergillus* is a sometime pathogen, and *Penicillium* species are the source of the antibiotic penicillin. Certain pathogenic ascomycetes can cause maduromycosis. (See Fig. 24-3, p. 434.)

3. *Basidiomycetes* or *club fungi*. These encompass the large fleshy toadstools, puffballs, and mushrooms as well as small plant smuts and rusts. Certain members produce poisons toxic to man. None is infectious.

4. *Fungi Imperfecti* or *Deuteromycetes*. These lack sexual spores. Asexual spores are formed on or from filaments in a special way. The important pathogens here will be discussed in their role as disease producers.

Laboratory study. Fungi may be studied and identified in a number of ways.

DIRECT VISUALIZATION. Stained or unstained material may be observed directly with the microscope. Usually a wet mount is prepared. The specimen, a bit of fungous growth, sputum, pus, skin scrapings, or infected hairs, is placed in a drop of mounting fluid on a glass slide, covered with a cover glass, and examined under the microscope. It is important that the illumination of the specimen be reduced by lowering the intensity of light from the source, or the phase-contrast microscope may be used.

Yeasts may simply be suspended in water. Material containing molds is commonly placed in a 10% to 20% solution of potassium hydroxide. The specimen is cleared when left in contact with the alkali for 10 to 15 minutes or longer; that is, it becomes more transparent and more easily defined microscopically.

Microscopic identification of a given fungus rests on the study of its structure, especially the kind of spores and their relation to the hyphae.

The phenomenon of fluorescence is important in the study of superficial fungous infections. For example, infected hairs often fluoresce under a filtered source of ultraviolet light.

CULTURE. Both pathogenic and saprophytic fungi are highly resistant to acid environments, and both prefer large amounts of sugar in their food supply. Consistent with these requirements, the French mycologist Sabouraud, around the turn of the century, devised a culture medium of maltose, peptone, and agar still widely used today. A fungous culture on Sabouraud agar is incubated at room temperature (20° C). Blood agar may also be inoculated but is incubated at body or incubator temperature (37° C). Littman oxall agar is frequently used. Fungi do particularly well when portions of raw or cooked vegetables are added to nutrient media. Potato and carrot combinations with cornmeal agar are valuable. Antibiotics added to fungous culture media suppress bacterial growth. Fungi grow slowly and cultures must be kept a week or two.

If set up as a slide culture, fungous growth may be observed daily under the microscope. If the very center of the agar medium in a Petri dish is inoculated, a giant colony grows out to the edge, covering the surface of the dish. The gross appearance of the colony is usually typical for a specific organism. In a liquid culture medium bacterial growth is dispersed; few bacteria form the characteristic sheet or pellicle on the surface of the liquid that is seen with the growth of fungi.

As with bacteria, fermentation reactions are important in identification of fungi, and animal inoculations are performed.

IMMUNOLOGIC REACTIONS. Serologic studies in the laboratory evaluation of fungous diseases include agglutination, precipitation, and complement fixation tests, and immunofluorescent and immunoelectrophoretic technics. These are generally positive with active disease. Skin testing is of great value.

Pathology of fungous disease. Fungi are important causes of disease in man and animals and one of the chief causes in plants. In man, *mycoses* (fungal infections) are of two types: superficial and systemic. Superficial fungi (in the skin, hair, and nails) causing the *dermatomycoses* spread from animal to man, or man to man, even cause epidemics, but do not invade. Systemic fungi contact man from his environment— from the soil, the vegetation, bird droppings, and so on. Ordinarily they are very insidious in their approach, gaining a foothold in the body but progressing rather leisurely. (Regression seems slow also in mycoses.)

The body's response to the intrusion is granulomatous inflammation, that is, a reaction in which the macrophages of the reticuloendothelial system are seen microscopically to be numerous and conspicuous in characteristic arrangements and appear to be the main participating cells. Tissue damage comes after the state of allergy has been set up in the host to the proteins of the fungus. In many respects the pathology of the mycoses is similar to that of tuberculosis. Usually the etiologic agent can be demonstrated in sections of infected tissues or in fluids therefrom; *its presence makes the diagnosis.*

Systemic mycoses fall into three categories: (1) *primary* infections, usually with a geographic pattern; (2) *secondary* infections or "superinfections"; and (3) *complicating* infections. Most of the fungous diseases discussed in this chapter are systemic, including the important primary mycoses. The four major ones are blastomycosis, coccidioidomycosis, cryptococcosis, and histoplasmosis.

Secondary mycoses develop during the course of a bacterial or viral disease for which antibiotics are being given. (Bacteria help to control fungi in nature.) The single most important example is candidiasis. Superinfection of this kind, often hospital acquired, is produced by both fungi and a number of bacteria—many gram-negative bacilli, including the enterics, and staphylococci.

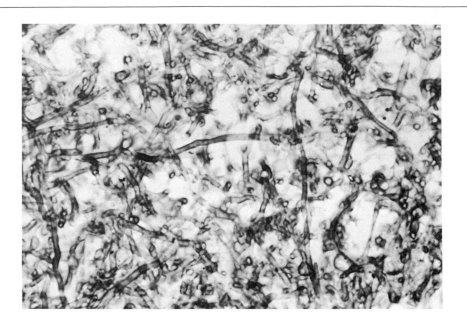

FIG. 30-4. *Aspergillus* (branching hyphae), demonstrated in microsection of lung with special stain. Abundant growth of this opportunist occurred in diabetic patient. (From Zugibe, F. T.: Diagnostic histochemistry, St. Louis, 1970, The C. V. Mosby Co.)

Complicating infections appear after special therapeutic procedures such as peritoneal dialysis or the prolonged use of a catheter indwelling a blood vessel. They are a distinct hazard in the management of the "compromised host," that patient with severe, chronic, debilitating disease for which antibiotics, steroid hormones, or immunosuppressive agents are required. They also follow close on disturbances of the immune mechanisms such as are seen in cancer of the lymphoid system and bone marrow failure.

Certain fungi (example, *Cryptococcus*) are readily pathogenic either as primary or secondary invaders. Others (examples, *Candida*, *Aspergillus* [Fig. 30-4], *Mucor*, and *Rhizopus*), benign in nature, seize the "opportunity" with inordinate vigor.

Importance. Fungi are important in the processes of nature, agriculture, manufacturing, and medicine. Commonly encountered molds can injure woodwork and fabrics and spoil food. They destroy food during its growth, in its manufacture, and after it has reached the consumer. Foods most vulnerable are bread, vegetables, fruits, and preserves. Molds are an important factor in the decay of dead animal and vegetable matter, complex organic compounds broken down into simple ones are returned to the soil to be used as food by green plants. Some fungi (example, certain mushrooms) are food for man. Molds are used commercially in the manufacture of beverages and to give flavor to cheeses (Roquefort, Camembert). Penicillin, an effective antibacterial substance, is derived from a common mold, *Penicillium notatum*.

Yeasts are economically important because they ferment sugars. The fact that they convert sugars by enzymatic action into alcohol and carbon dioxide is practically applied in the manufacture of alcoholic beverages and in baking. In the manufacture of alcoholic beverages carbon dioxide is a by-product, whereas in baking it is the essential factor (p. 648). In the preparation of commercial yeast the cells are grown in a suitable liquid medium, separated out from the liquid portion by centrifugation, mixed with starch or vegetable oil, and then molded and cut into cakes. Yeast is a source of vitamin B and of ergosterol, from which vitamin D is obtained.

DISEASES CAUSED BY FUNGI

Medical mycology treats of the fungi that bring about disease.

Superficial mycosis
Dermatomycoses

Superficial fungous infections of the skin, hair, and nails, generally called *ringworm* or *tinea*, are *dermatomycoses* or *dermatophytoses*. Fungi causing dermatomycoses and showing no tendency to invade the deeper structures of the body are called *dermatophytes*. There are three important genera: (1) *Microsporum*, (2) *Trichophyton*, and (3) *Epidermophyton*. Dermatophytes are closely related botanically.

The genus *Microsporum* is the most frequent cause of ringworm of the scalp and may give rise to ringworm in other parts of the body. Hairs removed from the affected regions are surrounded by a coat of spores, and scales of skin show many branched mycelial threads. *Trichophyton* causes ringworm of the scalp, beard, skin, or nails. The organisms are found as chains of spores, inside or on the surface of affected hairs, or as hyphae and characteristic spores in skin scrapings. *Trichophyton schoenleini* is the cause of almost all cases of favus.* The spores and mycelial threads are found in the favus crusts. Hairs in the affected areas are filled with vesicles and channels from which the mycelia have disappeared. *Epidermophyton* is largely responsible for ringworm of the body, hands, and feet. Epidermophyta appear as interlacing threads in the skin and do not invade the hairs.

Ringworm of the scalp *(tinea capitis)* seen most often in children is a common and highly communicable disease. It may be spread directly from person to person or by articles of wearing apparel. It occurs in domestic animals, from which it may be transmitted to man. Ten to thirty percent of ringworm infections occurring in cities and 80% of those in rural areas are thus transmitted, either by direct or indirect contact with the animal. Pets (dogs and cats) readily pass the ringworm fungi onto their human masters.

Favus usually affects the scalp with the formation of yellowish cup-shaped crusts or *scutula* about the mouths of the hair follicles. These crusts consist of masses of spores and mycelial threads mixed with leukocytes and epithelial cells. Favus may be transmitted directly or indirectly from person to person and tends to run in families.

Ringworm of the beard *(tinea barbae)* is known as *barber's itch*. Ringworm of the

---

*In 1839, *Johann Lukas Schönlein* (1793-1864) isolated the fungus causing favus. A leading figure in German clinical medicine of the early nineteenth century, Schönlein is better known for his clinical teaching and his general influence on German medicine than for his individual contributions to medical progress.

groin is known as *tinea cruris* or *dhobie itch*. Ringworm of the feet is known as *tinea pedis* or *athlete's foot*. It was thought for years that athlete's foot is contracted from footwear, lockers, and floors, but experiments indicate that exposure to the pathogenic dermatophytes in public swimming pools or shower stalls plays a minor role. These fungi are everywhere, even on the feet of noninfected individuals. The lesions of athlete's foot most probably appear because of a factor of decreased resistance in the skin to contact with the causative fungi. Uncomplicated tinea pedis is a dry scaly lesion of the skin with no symptoms. However, feet sweat, and in such a moist environment bacteria normally resident on the feet invade the fungous lesion producing the white, soggy, malodorous, itching changes between the toes usually called athlete's foot. When these are present, the infection is more properly considered both fungal and bacterial.

Laboratory diagnosis of the dermatomycoses relies on demonstration of fungi in hair and skin scrapings by direct microscopy.

Control of the dermatomycoses is very difficult. It consists of the proper sterilization of clothing, bathing suits, and objects subjected to frequent handling. Hygiene of the feet is extremely important.

Systemic mycosis
Aspergillosis

Aspergillosis is an infection most often produced by *Aspergillus fumigatus*, a gray-green mold growing in the soil. This fungus may cause various types of infection in chickens, ducks, pigeons, cattle, sheep, and horses. Animals usually contract the disease from moldy feed.

In man the disease generally is an infection of the external ear (otomycosis). The infection may be superficial and mild, or it may cause ulceration of the membrane lining the ear and perforate into the middle ear. Infection most likely comes from fungi living saprophytically on the earwax. Other types of infection in man are pulmonary infections, sinus infections, and infections of the subcutaneous tissues. Aspergillosis as a superinfection complicating antibiotic therapy affects the lungs (Fig. 30-4). (Aspergilli can no longer be passed off simply as weeds in the laboratory likely to contaminate any culture.)

Aspergillosis may be caused by aspergilli other than *Aspergillus fumigatus*, including *Aspergillus nidulans*, *Aspergillus niger*, and *Aspergillus flavus*. Certain strains of *Aspergillus flavus* (and also *Penicillium puberulum*) elaborate *aflatoxins*, toxic substances that can cause severe damage or even cancer in the liver of animals ingesting them. Aflatoxins are exceedingly potent in this respect, cancer formation requiring an amount no more than 0.05 ppm, and they are also natural mutagens. Animals contact these mycotoxins (fungal toxins) in the mold produced by *Aspergillus* on peanuts and other harvest feeds.

The aflatoxin content of such foods as peanut butter is greatly reduced by careful removal of infected peanuts before processing. Currently the Food and Drug Administration has established the safe level in peanuts as 20 parts per billion (ppb), but it may lower this value to 15 ppb aflatoxin. An FDA survey has demonstrated that 93%

of peanut butter samples are below 20 ppb. There is no direct evidence that aflatoxins are cancer-producing in man.

Serologic aids to the diagnosis of aspergillosis include a complement fixation test, a double diffusion in agar gel technic, and the indirect fluorescent antibody determination.

Blastomycosis

There are two kinds of blastomycosis—the *North American* and the *South American*. North American blastomycosis, known as *Gilchrist's disease* after its discoverer, is a granulomatous and suppurative inflammation. Multiple abscesses form in the skin and subcutaneous tissues (blastomycetic dermatitis) or in the internal organs of the body (systemic blastomycosis). The cutaneous form is more often seen than the systemic (Fig. 30-5). The lesions of blastomycetic dermatitis may be mistaken for cancer or tuberculosis. With systemic disease pulmonary blastomycosis is most likely present, closely resembling pulmonary tuberculosis.

North American blastomycosis, confined almost exclusively to the United States and Canada, is caused by *Blastomyces dermatitidis*. The organisms are demonstrated by direct microscopic examination of pus from the lesions or by cultural methods (Fig. 30-6). In pus they are round or oval granular yeast forms, varying from 8 to 15 μm in diameter. They are surrounded by a thick refractile wall, which makes them doubly contoured, and single budding forms are present (Fig. 30-7). The organisms grow typically in the mycelial phase on all media, but isolation is often complicated by overgrowth of bacterial contaminants from the lesion.

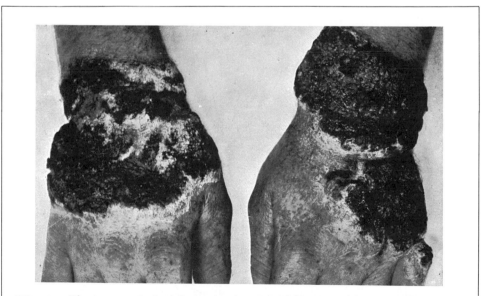

FIG. 30-5. Blastomycosis in Minnesota farmer. Note verrucous or warty nature and sharp margins. (Courtesy Dr. John Butler; from Sutton, R. L., Jr.: Diseases of the skin, ed. 11, St. Louis, 1956, The C. V. Mosby Co.)

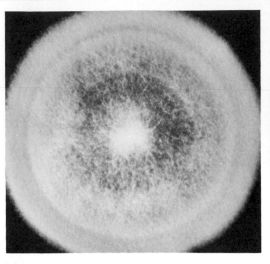

FIG. 30-6. *Blastomyces dermatitidis,* giant colony, showing abundant fluffy mycelial growth. (From Musgnug, R. H.: Med. Tribune **3**:16, May 28, 1962.)

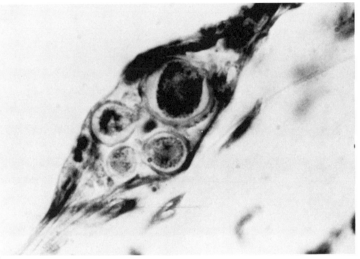

FIG. 30-7. *Blastomyces dermatitidis,* microscopic appearance of budding yeast forms in stained smear of sputum.

Transmission of human infection is unsolved. Accumulated evidence indicates that cutaneous blastomycosis results from infection through wounds. The finding of a primary focus in the lungs in systemic blastomycosis indicates that systemic infections are probably acquired via the respiratory tract. In some cases of systemic blastomycosis, infection may come from the skin.

Although fungi are weak antigens, their infections are associated with a degree of allergy useful in laboratory diagnosis. For North American blastomycosis there is both a complement fixation test and a skin test (the blastomycin test). The first test is specific, but cross reactions may occur between the skin test for blastomycosis and that for histoplasmosis. (An individual with histoplasmosis may appear to give a positive test for blastomycosis.) Skin testing may be done with *Blastomyces dermatitidis* vaccine, considered a more effective antigen than blastomycin.

South American blastomycosis is similar to the North American type. Caused by *Blastomyces (Paracoccidioides) brasiliensis*, it is known also as *paracoccidioidal granuloma*. The fungus enters the body by way of the mouth, where it localizes and causes ulcers and granulomas. From these lesions the fungi spread to the lungs and other parts of the body. The disease may terminate fatally. Most cases have been reported in Brazil.

Candidiasis (candidiosis, moniliasis)

Candida (Monilia) albicans is a budding yeastlike organism, worldwide in distribution. Its relationship to disease is often hard to determine. It is found on the mucous membranes of the mouth, intestinal tract, and vagina and on the skin of normal persons with no disease.* It is found also in association with known pathogens in persons with known illness but in whom there is no reason to suspect its pathogenicity. In making the diagnosis of candidiasis (candidiosis), *Candida albicans*, the chief pathogen of the genus, must be repeatedly isolated from the lesions to the exclusion of better defined agents.

Two infections of *Candida albicans* have been with us for a long time. One, on the mucous membranes of the mouth, is known as *thrush*. The other, involving the mucous membranes of the female genitalia, is *vulvovaginitis* or *vaginal thrush*.

Thrush is seen as many small milklike flecks that may coalesce and cover the entire lining of the mouth. Beneath these patches on the inside of the lips, on the hard palate, and on the tips and edges of the tongue are areas of catarrhal inflammation. Thrush is especially troublesome in newborn infants in hospital nurseries. The baby acquires the organism from the vagina of the mother during the birth process, and the organisms are spread from person to person by contaminated fingers, utensils, and nipples. Thrush tends to be rare after the newborn period in healthy subjects of any age but is comparatively common in poorly nourished children and in chronically ill and aged adults.

Vulvovaginitis is a thrushlike infection associated with a typical vaginal discharge. The sugar content of the urine in pregnancy and uncontrolled diabetes may be a contributing factor, and vaginal thrush is associated with oral contraceptives. Today it has taken on the proportions of venereal disease.

Candidal infections are found in individuals, such as fruit canners, who by occupation must keep their hands constantly soaked in water. *Candida albicans* is an important cause of chest disease. The manifestations of bronchopulmonary or pulmonary candidiasis vary from those of mild inflammation to those of a severe infection like tuberculosis.

Systemic candidiasis is seen more often today. It is prone to follow persistent skin or mucosal lesions in a person with lowered resistance or altered immunologic

---

*Only a few fungi such as *Candida albicans* are normal inhabitants of the human body; most belong to the environment.

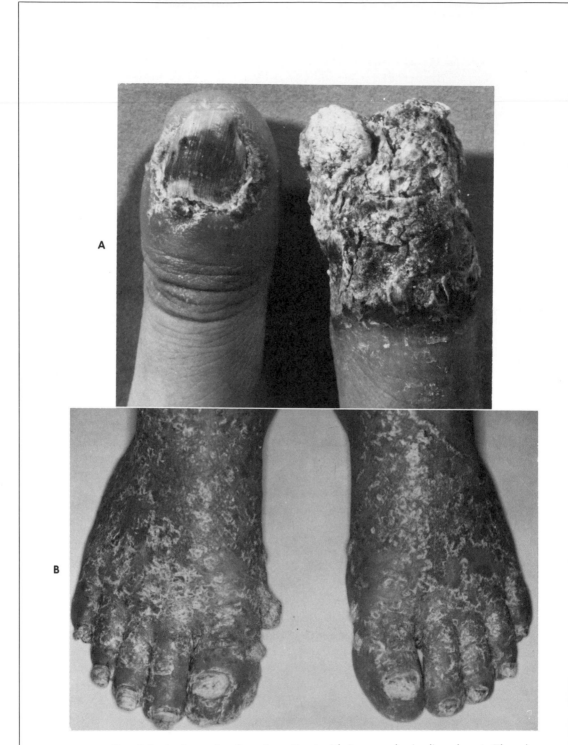

FIG. 30-8. Candidiasis, long standing, in patient with immunologic disorder. A, Thumbs —destruction of nails with accumulation of horny material. B, Feet—changes in skin and nails. (From Kirkpatrick, C. H., and others: Ann. Intern. Med. **74**:955, 1971.)

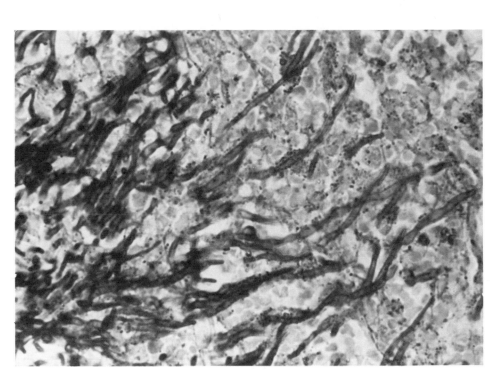

FIG. 30-9. Candidiasis of spleen in disseminated disease, microsection. Note abundant mycelial filaments of *Candida albicans*. (×800.)

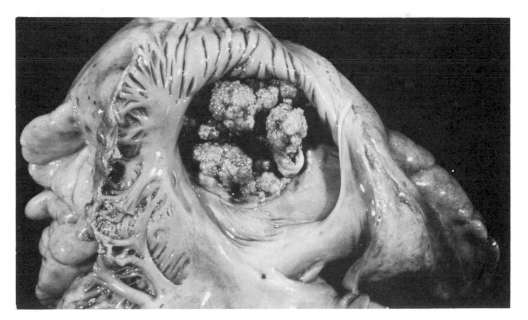

FIG. 30-10. Candidiasis of heart valve (candidal endocarditis). Atrium of the heart has been opened to expose valvular opening. Note large cauliflower-like excrescences (vegetations) on leaflets of tricuspid valve. This was a complication in a drug addict.

mechanisms (Fig. 30-8). *Candida* easily gains the ascendancy when dosage with a broad-spectrum antibiotic is prolonged, since the normal bacterial flora is thereby depressed. It is considered the major hospital-acquired, pathogenic fungus.

Overgrowth of *Candida* for any reason is an ominous event. If the organisms circulate in the bloodstream (a fungemia), they set up a serious toxic reaction and are widely disseminated in the body (Fig. 30-9). The patient at greatest risk is the one with leukemia, some kind of bone marrow failure, or an organ transplant. Other pathologic conditions over which the threat of candidiasis hangs are diabetes, chronic alcoholism, endocrine disorders, malnutrition, and certain kinds of cancers. Under the right circumstances *Candida* strikes as a formidable pathogen (Fig. 30-10).

The laboratory demonstration of *Candida (Monilia) albicans* is easily made either by direct microscopic examination of unstained or stained material or by culture. In the exudate from a lesion or in sputum the organisms appear as oval, budding yeasts with scattered hyphal segments. Serologically there are the hemagglutination test, the precipitin reaction, an indirect immunofluorescent technic and counterelectrophoresis. Inoculation of the chorioallantoic membrane of the chick embryo, with visible lesions appearing 72 to 96 hours later, is a good test for the pathogenicity of *Candida*. With oidiomycin, an extract of the organism, a skin test may be done to indicate past or present infection.

A *Candida* vaccine is still experimental.

Coccidioidomycosis (coccidioidal granuloma)

Coccidioidomycosis, one of the most infectious of the fungous diseases, exists in two forms—the primary (usually self-limited) and the progressive. In the primary form the lesions are confined to the lungs, giving pulmonary symptoms of varying severity, sometimes with cavitation. As a rule, the infection ends in recovery, but in a small percentage of cases the process spreads from the lungs to produce the progressive form. This happens more in blacks and the darker pigmented races than in whites.

In the progressive form the disease spreads to the skin, subcutaneous tissues, bones, meninges, and internal organs. The lesions in the skin resemble those of blastomycosis. In other parts of the body they resemble those of tuberculosis. The mortality is high. This is the form of the disease designated *coccidioidal granuloma*.

Coccidioidomycosis is endemic in the desert valleys of California and the dry dusty areas of Southwestern United States, especially in Arizona, New Mexico, Texas, and parts of Mexico. It is most likely that man and animals are infected by the inhalation of spore-bearing dust. The infection is found in cattle, sheep, dogs, and certain wild rodents. There may be a reservoir of infection in small wild rodents, these animals passing the spores in their feces to contaminate the soil, after which the spores would spread by the wind. The rather frequent dust storms of the Southwest might carry these organisms long distances.

The cause is *Coccidioides immitis*. Diagnosis is made by finding it in the disease. The appearance of the fungus in lesions differs from its appearance on culture media. In body tissues and exudates one sees yeastlike forms, large thick-walled, nonbudding

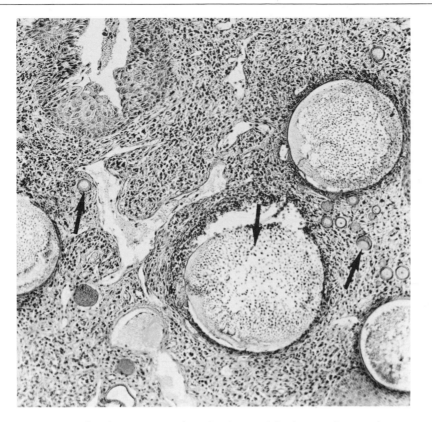

FIG. 30-11. *Coccidioides immitis* spherules (arrows) in tissue microsection; reproduction by endosporulation. Large balloonlike structures (central arrow) contain myriad endospores that they release into tissue spaces. Doubly refractile capsule seen about varying-sized spherules.

spherules, 20 to 60 μm in diameter, filled with endospores, 2 to 5 μm in diameter (reproduction by endosporulation) (Fig. 30-11). As many as 1000 endospores may be released from a single spherule. In laboratory cultures growth is that of a mold, and one sees a fluffy, cottony white colony. The *coccidioidin test*, a test of sensitivity to an extract of the organism, is of value. (Cross reactions may occur with skin tests for histoplasmosis and blastomycosis, however.) *Spherulin*, an extract of the spherule phase of growth, is also used in sensitivity testing and may prove to be a superior skin test agent to coccidioidin. Immunofluorescence can be used to identify the infection as can precipitation, latex agglutination, and complement fixation tests, and quantitative immunodiffusion. These serologic tests are used to follow the course of the infection in a given patient. Currently a coccidioidal vaccine made from formalin-inactivated spherules has been successful in trial runs on human beings.

Cryptococcosis (torulosis)

Cryptococcus neoformans (Torula histolytica) is a yeastlike organism that usually infects the lungs and central nervous system but may attack other parts of the body. It

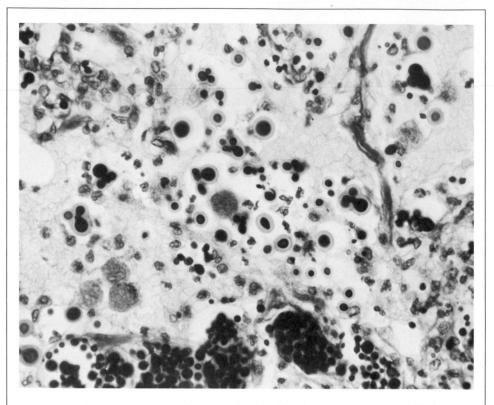

FIG. 30-12. *Cryptococcus neoformans (Torula histolytica):* among partially hemolyzed red cells in air sacs of lung. Wide gray capsules encompass budding yeast forms in this specially stained microsection. (From Kent, T. H., and Layton, J. M.: Am. J. Clin. Pathol. **38:**596, 1962.)

is the only encapsulated yeast to invade the central nervous system. With infection, multiple small nodules form, with the gross and microscopic appearance of tubercles. In the central nervous system, the meninges are thickened and matted together, and the brain is invaded. To the patient with preexisting malignancy of the reticuloendothelial system, this agent is an important opportunist.

Man becomes infected through the skin, mouth, nose, and throat. Transmission from man to man or from animal to man has not been recorded. This fungus, saprophytic in nature, has been found in cattle, horses, dogs, and cats. Birds are not its hosts, but it is a saprophyte in pigeon droppings, and cases of cryptococcal meningitis have been traced to the vast pigeon populations found in many large cities. Pigeons are mechanical vectors, carrying the organisms on their feet and beaks. They are not affected probably because of their high body temperature. *Cryptococcus neoformans* has a definite predilection for pigeon droppings, which are rich in creatinine. Creatinine is assimilated by this organism only and not by other species of cryptococci or other fungi.

Since *Cryptococcus neoformans* is known also as *Torula histolytica*, infections with it may be referred to as either *cryptococcosis* or *torulosis*. Torulosis or crypto-

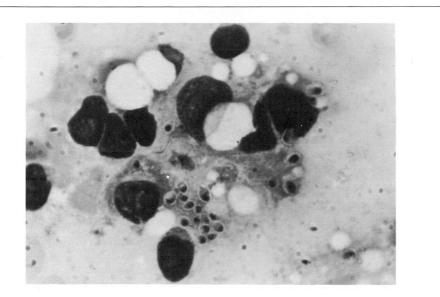

FIG. 30-13. Histoplasmosis of bone marrow, Wright-stained smear examined microscopically. (×950.) (From Anderson, W. A. D.: Pathology, ed. 6, St. Louis, 1971, The C. V. Mosby Co.)

coccosis can be diagnosed with certainty only by finding the budding organisms in the affected tissues, pus, sputum, or cerebrospinal fluid (Fig. 30-12). In wet mounts (prepared with nigrosin) cryptococci are ovoid to spherical, budding yeast forms 5 to 15 μm in diameter. With fluorescent antibody technics, the diagnosis may be made within hours. Agglutination tests are sensitive and specific. There is a *cryptococcin* skin test.

Histoplasmosis

Histoplasmosis, sometimes called *Darling's disease,* is caused by *Histoplasma capsulatum,* a diphasic organism—a single budding yeast at body temperature and a mold at room temperature and in nature. The fungus attacks primarily the reticuloendothelial system, parasitizing the component cells. Like coccidioidomycosis, the disease exists in the primary and progressive forms. The primary form involves the lungs but usually heals, leaving many small calcified foci in the lungs and lymph nodes of the chest. In the progressive disseminated form, ulcerating lesions are found in the nose and mouth, and there is enlargement of the spleen, liver, and lymph nodes. The progressive form is generally fatal.

Man probably contracts the disease by inhalation of spores from fungi growing in the soil. Infection of the soil comes from the excreta of a variety of birds and bats in which the microbes have been found. No intermediate host is identified. Spores may be carried by prevailing winds and even by tornadoes. Outbreaks of the disease have been traced to inhalation of dust from caves. Histoplasmosis is also referred to as cave sickness or *speleonosis.* Victims of histoplasmosis and blastomycosis as well tend to be outdoor types—construction workers, farmers, spelunkers, and the like.

571

The disease is encountered in the Central Mississippi Valley and the Ohio Valley and in widespread areas of the world.

To identify the fungi, stained smears and imprints (as well as cultures) are made of peripheral blood, bone marrow, aspirated material from lymph nodes, and sputum (Fig. 30-13). To aid in the diagnosis and follow-up of histoplasmosis, the laboratory offers a skin test (the histoplasmin test or the histoplasmin tine test) and five serologic tests—precipitation, agglutination, complement fixation, fluorescent antibody detection, and immunodiffusion.

Phycomycosis

Phycomycosis or *mucormycosis* can be an overwhelmingly acute and fatal infection. It is caused by species of *Mucor* and *Rhizopus* and other normally harmless phycomycetes of the soil and decaying organic matter.

Diabetes, untreated and out of control, is the most important forerunner; the ketoacidosis rather than the hyperglycemia is thought to trigger the process. Along with disease in the lungs and central nervous system, an intraorbital cellulitis is a prominent feature. In the tissues, the hyphae abound, very broad and branching, especially within walls and lumens of blood vessels. No spores are seen, and there is little if any inflammation. In tissue section, hyphae are easily identified as belonging to the Phycomycetes because they are nonseptate and coenocytic, that is, contain many nuclei within a continuous mass of cytoplasm. The term *mucormycosis* is often used simply to indicate infections in which such hyphae are seen.

Sporotrichosis

Sporotrichosis, caused by *Sporotrichum schenckii*, may affect man, lower animals, or plants. *Sporotrichum* is widely distributed in nature as a saprophyte on vegetation, and man usually acquires the infection from plants, especially barberry shrubs and certain mosses that seem to harbor the fungi. The fungi are introduced into wounds (inoculation infection) by infected plants or vegetable matter. The agent is thought to be a normal inhabitant of the alimentary and respiratory tracts in man, and in a few instances the disease has been transferred from man to man. Transmission of the infection from lower animals to man by bites or indirect routes has been noted. Recently sporotrichosis was reported as a complication of catfish stings. Animals most often affected are horses, mules, dogs, rats, and mice. The majority of cases occur in the United States, especially in the Missouri and Mississippi valleys.

The disease presents as a chronic infection usually limited to the skin and underlying tissues and is accompanied by the formation of nodular masses that slowly undergo softening and ulceration. In typical cases the first evidence of disease is seen at the site of some trivial injury, usually on the fingers. The wound does not heal, and an ulcer appears, to be followed by nodular corklike swellings in chains up the forearm. The disease seldom extends farther than to the regional lymph nodes, but secondary foci sometimes crop up in other parts of the body such as the lungs, spleen, liver, and other organs.

The oval or cigar-shaped fungus of sporotrichosis, resident within mononuclear

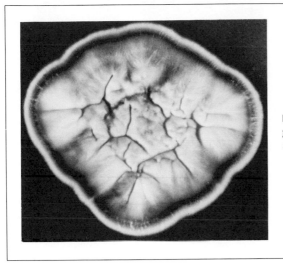

FIG. 30-14. *Sporotrichum schenckii,* giant colony. (Courtesy Dr. R. H. Musgnug, Haddonfield, N.J.)

cells, is very rarely found in smears from the pus of a skin lesion or in sections of tissue taken from the lesions. As a rule, it is demonstrated only in culture (Fig. 30-14). Fluorescent antibody technics detect the microorganisms in exudates from the lesions.

Other fungous diseases of man

Other fungous infections, infrequent in man, are the following:
1. Rhinosporidiosis (caused by *Rhinosporidium seeberi*)
2. Geotrichosis (caused by one or more species of *Geotrichum*)
3. Chromoblastomycosis (caused by three different fungi)
4. Penicilliosis (caused by certain species of *Penicillium*)
5. Piedra (caused by two different species of fungi)
6. Otomycosis (caused by species of *Penicillium* and other fungi)

Fungous diseases of lower animals

Fungous infections are very important to the veterinary microbiologist. The following are some common ones:
1. Ringworm of the horse (caused by *Trichophyton equinum*); may occur in man
2. Ringworm of horses, cattle, dogs, and possibly sheep and hogs (caused by *Trichophyton mentagrophytes*); infection possible in man
3. Ringworm of cats and dogs (caused by *Microsporum canis*)
4. Epizootic lymphangitis of horses (caused by *Blastomyces farciminosus*)
5. Favus of chickens (caused by *Trichophyton gallinae*)
6. Coccidioidomycosis of cattle, dogs, horse, and sheep (caused by *Coccidioides immitis*); same organism causes the disease in man
7. Aspergillosis of wild and domestic fowls (caused by *Aspergillus fumigatus*)
8. Candidiasis of poultry (caused by *Candida albicans*)
9. Histoplasmosis of dogs, cats, cattle, sheep, swine, poultry, and horses (caused by *Histoplasma capsulatum*)
10. Cryptococcosis (mastitis) of cattle (caused by *Cryptococcus neoformans*)

Fungous diseases of plants

Molds and yeasts are economically important for they cause so many diseases of plants. Plants and vegetables imported from other countries are rigidly inspected on arrival in our seaports and points of entry. If a fungous infection is found, the plant or plant product is not allowed entry. Among important plant diseases caused by molds are:

1. Brown rot of peaches and plums
2. Chestnut blight
3. Mildew of grapes
4. White pine blister rust
5. Rust of oats, wheat, and barley
6. Smuts of various grains
7. Potato rot
8. Corn leaf blight

Ergotism

Ergot, a drug whose derivatives are widely used to check hemorrhage after childbirth, is composed of several alkaloidal poisons (mycotoxins) produced by the growth of a mold *Claviceps purpurea* in the grains of rye, wheat, and barley. The fungus (also referred to as ergot) grows as a purple-black, slightly curved mass that replaces the infected grain and converts it to a black sclerotium, from which the drug is extracted. An enzyme secreted by the hyphae is contained in a thick honeydew that attracts insects and helps to spread the fungus. When bread made from infected grain is eaten, the condition known as *ergotism* develops. Ergotism is characterized by gangrene of the extremities, abortion, and convulsions. It was at one time most prevalent in Central Europe.

A specific component of the pharmacologically potent alkaloids produced by the fungus ergot is lysergic acid. A well-known derivative is the hallucinogen lysergic acid diethylamide, or LSD.

QUESTIONS FOR REVIEW

1. Give the general characteristics of fungi.
2. Name the science that treats of fungi.
3. What is a fungous infection called? Cite five examples.
4. Make the distinction between molds and yeasts. Is it a clear one?
5. How do fungi perpetuate themselves?
6. What are dermatomycoses? Name three major dermatophytes.
7. Outline the laboratory diagnosis of fungous disease.
8. Name and briefly describe the chief systemic diseases caused by fungi.
9. What is ergotism? What is the pharmacologic nature of LSD?
10. Classify fungi.
11. Make pertinent comments regarding the pathology of fungous disease.
12. What threat does fungous infection pose to the "compromised host"? Cite the common offenders.
13. Briefly define speleonosis, torulosis, mycotoxin, maduromycosis, tinea, superinfection, complicating infection (in fungous disease), mycelium, aflatoxins, favus, otomycosis, thallus, cave sickness.

REFERENCES FOR CHAPTER 30

Artman, S. J.: Permanent slides from fungus preps, Lab. Med. **3:**36, Sept., 1972.

Baker, R. D., editor, and others: Human infection with fungi, actinomycetes, and algae, New York, 1971, Springer-Verlag New York, Inc.

Baker, R. D.: The primary pulmonary lymph node complex of cryptococcosis, Am. J. Clin. Pathol. **65:**83, 1976.

Basler, R. S. W., and Friedman, J. L.: Mucocutaneous histoplasmosis, J.A.M.A. **230:**1434, 1974.

Buechner, H. A., and others: The current status of serologic, immunologic and skin tests in the diagnosis of pulmonary mycoses, Chest **63:**259, 1973.

Caporael, L. R.: Ergotism: the Satan loosed in Salem? Science **192:**21, 1976.

Conant, N. F., and others: Manual of clinical mycology, ed. 3, Philadelphia, 1971, W. B. Saunders Co.

Deppisch, L. M., and Donowho, E. M.: Pulmonary coccidioidomycosis, Am. J. Clin. Pathol. **58:**489, 1972.

Dolan, C. T.: Evaluation of various media for growth of selected pathogenic fungi and *Nocardia asteroides*, Am. J. Clin. Pathol. **58:**339, 1972.

Dolan, C. T., and Stried, R. P.: Serologic diagnosis of yeast infections, Am. J. Clin. Pathol. **59:**49, 1973.

Edds, G. T.: Acute aflatoxicosis: review, J. Am. Vet. Med. Assoc. **162:**304, 1973.

Emmons, C. W., and others: Medical mycology, Philadelphia, 1977, Lea & Febiger.

Gervasi, J. P., and Miller, N. G.: Recovery of *Cryptococcus neoformans* from sputum using new technics for the isolation of fungi from sputum, Am. J. Clin. Pathol. **63:**916, 1975.

Goldblatt, L. A.: Aflatoxin, scientific background, control, and implications, New York, 1969, Academic Press, Inc.

Goodwin, R. A., Jr., and Des Prez, R. M.: Pathogenesis and clinical spectrum of histoplasmosis, South. Med. J. **66:**13, 1973.

Hatcher, C. R., Jr., and others: Primary pulmonary cryptococcosis, J. Thorac. Cardiovasc. Surg. **61:**39, Jan., 1971.

Hoffman, H-P., and Avers, C. J.: Mitochondrion of yeast: ultrastructural evidence for one giant, branched organelle per cell, Science **181:**749, 1973.

Hussey, H. H.: Athlete's foot (editorial), J.A.M.A. **233:**539, 1975.

Johnston, W. W.: The cytopathology of mycotic infections, Lab. Med. **2:**34, Sept., 1971.

Jones, H. E., and others: Apparent cross-reactivity of airborne molds and dermatophytic fungi, J. Allergy Clin. Immunol. **52:**346, 1973.

Katzenstein, A., and others: Bronchocentric granulomatosis, mucoid impaction and hypersensitivity reactions to fungi, Am. Rev. Respir. Dis. **111:**497, 1975.

Kepron, M. W., and others: North American blastomycosis in central Canada: review of 36 cases, Can. Med. Assoc. J. **106:**243, 1972.

Koneman, E. W., and Fann, S. E.: Practical laboratory mycology, New York, 1971, Medcom Books, Inc.

Kozinn, P. J., and others: The precipitin test in systemic candidiasis, J.A.M.A. **235:**628, 1976.

Lenoir, E., and Carson, P.: A rapid screening test for the identification of *Candida albicans*, Lab. Med. **4:**28, Dec., 1973.

Miller, D. L., and others: Preparation of permanent microslides of fungi for reference and teaching, Am. J. Clin. Pathol. **59:**601, 1973.

Miller, G. G., and others: Rapid identification of *Candida albicans* septicemia in man by gas-liquid chromatography, J. Clin. Invest. **54:**1235, 1974.

Moore-Landecker, E.: Fundamentals of the fungi, Englewood Cliffs, N.J., 1972, Prentice-Hall, Inc.

Moss, E. S., and McQuown, A. L.: Atlas of medical mycology, Baltimore, 1969, The Williams & Wilkins Co.

Myrvik, Q. N., and others: Fundamentals of medical bacteriology and mycology for students of medicine and related sciences, Philadelphia, 1974, Lea & Febiger.

Orr, E. R., and Riley, H. D., Jr.: Sporotrichosis in childhood: report of 10 cases, J. Pediatr. **78:**951, 1971.

Richards, R. N., and Talpash, O. S.: Sporotrichosis, Can. Med. Assoc. J. **106:**1097, 1972.

Richter, M. W., and Amsterdam, D.: Plate-slide for the culture and morphological observation of fungi, Appl. Microbiol. **24:**667, 1972.

Rippon, J. W.: Medical mycology. The pathogenic fungi and the pathogenic actinomycetes, Philadelphia, 1974, W. B. Saunders Co.

Roberts, G. D.: Detection of fungi in clinical specimens by phase-contrast microscopy, J. Clin. Microbiol. **2:**261, 1975.

Serstock, D. S., and Zinneman, H. H.: Pulmonary and articular sporotrichosis, J.A.M.A. **233:** 1291, 1975.

Smith, J. W.: Coccidioidomycosis, a review, Tex. Med. **67:**117, Nov., 1971.

Smith, J. W., and Utz, J. P.: Progressive disseminated histoplasmosis, Ann. Intern. Med, **76:**557, 1972.

Snell, W. H., and Dick, E. A.: A glossary of mycology, ed. 2, Cambridge, Mass., 1971, Harvard University Press.

Spickard, A.: Diagnosis and treatment of cryptococcal disease, South. Med. J. **66:**26, 1973.

Stevens, D. A., and others: Spherulin in clinical coccidioidomycosis, Chest **68:**697, 1975.

Sutaria, M. K., and others: Focalized pulmonary histoplasmosis (coin lesion), Chest **61:**361, 1972.

Taschdjian, C. L., and others: Serodiagnosis of candidal infections, Am. J. Clin. Pathol. **57:**195, 1972.

Vaisrub, S.: Beware, the Sporothrix (editorial), J.A.M.A. **215:**1976, 1971.

Warintarawej, A., and others: Maduromycosis (Madura foot) in Kentucky, South. Med. J. **68:**1570, 1975.

Wolf, P. L., and others, editors: Practical clinical microbiology and mycology, New York, 1975, John Wiley & Sons, Inc.

Young, R. C., and others: Species identification of invasive aspergillosis in man, Am. J. Clin. Pathol. **58:**554, 1972.

Young, R. C., and others: Fungemia with compromised host resistance. A study of 70 cases, Ann. Intern. Med. **80:**605, 1974.

31 Protozoa

Parasites are generally defined as organisms that require living matter for their nourishment; that is, they must live within or on the bodies of other living organisms. According to this definition, a parasite may be a bacterium, virus, rickettsia, protozoon, plant (example, mistletoe) or animal. However, by common usage, *medical parasitology* refers to certain animal parasites of medical interest and their diseases. Although parasitic infections are commonly treated as exotic diseases of another world, and while the parasitic burden of North Americans is not great, it is nonetheless true that practically every parasitic disease known to man has been recognized in recent times in the United States.

The animal within or on which a parasite lives is the *host*. All stages of the parasite's development may take place in the same animal host. On the other hand, a parasite may have one or more hosts. It undergoes its larval stage in the *intermediate* host and its adult stage in the *definitive* host. A parasite that lives within the body of the host is known as an *endoparasite;* one that lives on the outside of the body is an *ectoparasite*. A tapeworm is an example of an endoparasite; a louse is an ectoparasite.

GENERAL CHARACTERISTICS

The animal kingdom is divided into two great divisions: the *Protozoa*, unicellular organisms and the lowest form of animal life, and the *Metazoa*, multicellular organisms (Chapter 32). Protozoa are more complex in their functional activities than bacteria or the average cell of a multicellular organism. Each is a complete, self-contained unit, with special structures known as *organelles* to carry out such functions as nutrition, locomotion, respiration, excretion, and attachment to objects. The vast majority are of microscopic size. As a rule, pathogenic ones are smaller than nonpathogenic ones. They may be spherical, spindle, spiral, or cup shaped. In medical parasitology, identification of a given animal parasite is of paramount importance. Practically, this is done by the recognition of specific structural (morphologic)

577

details in the makeup of the given parasite. There are many species, but only about 30 affect man.

Structure. Protozoa are units of protoplasm differentiated into cytoplasm circumscribed by the cell or plasma membrane and a nucleus encased by the nuclear membrane. Some have more than one nucleus. The cytoplasm is separated into a homogeneous *ectoplasm* and a granular *endoplasm.* The ectoplasm helps form the various organs of locomotion, contraction, and prehension, such as pseudopods, flagella, cilia, and suctorial tubes. In certain species of protozoa the ectoplasm contains a definite opening or portal for intake of food. The endoplasm digests food materials and surrounds the nucleus.

Many protozoa, especially pathogenic ones, absorb fluid directly through the plasma membrane. The majority take in solid particles, such as small animal or vegetable organisms, and digest them enzymatically. Because their food consists chiefly of bacteria, protozoa may be important in limiting the bacterial population of the universe. Waste material is excreted through the cell membrane or, in some cases, through an ejection pore.

Locomotion. All protozoa possess some type of motility. It may be by pseudopod formation (Fig. 31-1) or by the action of flagella or cilia. For locomotion by *pseudopod* (false foot) formation a sharp or blunt ectoplasmic process flows forward, pulling the rest of the organism after it. *Flagella* are whiplike prolongations of protoplasm that propel the organism by their lashing motions. Some protozoa have only one flagellum; others have several. Some of the flagellate protozoa also have an *undulating membrane* to help in locomotion. This is a fluted membranous process attached to one

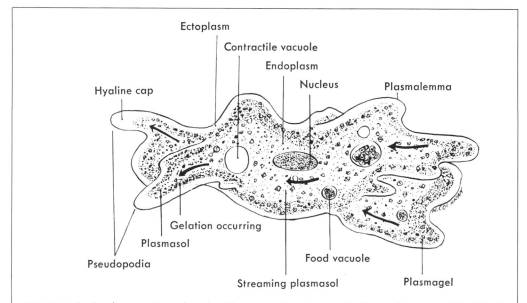

FIG. 31-1. Active locomotion sketched in an ameba. Arrows indicate direction. (Modified from Hickman, C. P., and others: Integrated principles of zoology, ed. 5, St. Louis, 1974, The C. V. Mosby Co.)

side of the organism. *Cilia* are similar to flagella except that they are shorter, more delicate, more plentiful, and are attached to the entire outer surface of the microbe. Individually they are less powerful than flagella, but the synchronous action of the many cilia accomplishes the most rapid motion of which unicellular organisms are capable.

Cyst formation. When protozoa are subjected to adverse conditions, they become inactive, assume a more or less rounded form, and surround themselves with a resistant membrane (cell wall) within which they may live for a long time and resist various destructive agents in their environment. This is *cyst* formation. When conditions suitable for growth are reestablished, the cyst imbibes water, and the protozoan returns to the vegetative state. Sometimes cyst formation precedes reproduction. Since vegetative protozoa are very susceptible to deleterious influences and cysts are very resistant, it is the cysts that are usually responsible for the spread of protozoan infections.

Reproduction. In protozoa, reproduction may be either sexual or asexual. In some (example, *Plasmodium* of malaria) the sexual cycle occurs in one species of animal and the asexual cycle in another. The sexual cycle occurs in the *definitive* host, the asexual cycle in the *intermediate* host. Protozoan cells capable of sexual reproduction are known as *gametes*. The cell formed by the union of two gametes is a *zygote*. Asexual reproduction occurs in amebas and flagellates. Lengthwise or crosswise division of the protozoon yields two new members of the species.

Classification.* In the phylum Protozoa there are six classes of organisms of medical interest in man. The method of locomotion varies in each.

---

*Classification taken from Faust, E. C., and others: Craig and Faust's clinical parasitology, ed. 8, Philadelphia, 1970, Lea & Febiger.

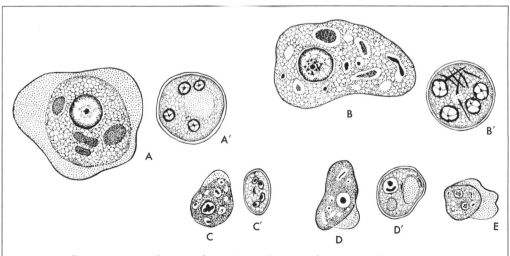

FIG. 31-2. Protozoa—amebas, pathogenic and nonpathogenic, sketch. Note rounded cyst to right of trophozoite in **A'**, **B'**, **C'**, and **D'**. **A-A'**, *Entamoeba histolytica* (the pathogen). **B-B'**, *Entamoeba coli*. **C-C'**, *Endolimax nana*. **D-D'**, *Iodamoeba bütschlii (williamsi)*. **E**, *Dientamoeba fragilis*.

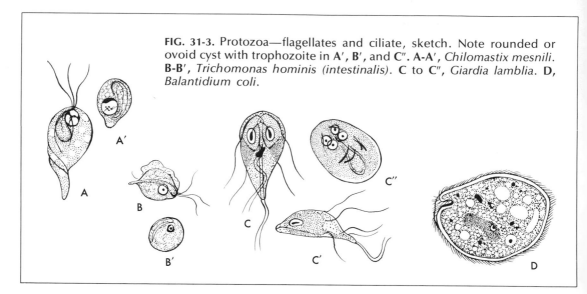

FIG. 31-3. Protozoa—flagellates and ciliate, sketch. Note rounded or ovoid cyst with trophozoite in **A'**, **B'**, and **C''**. **A-A'**, *Chilomastix mesnili*. **B-B'**, *Trichomonas hominis (intestinalis)*. **C** to **C''**, *Giardia lamblia*. **D**, *Balantidium coli*.

RHIZOPODEA. Locomotion is characterized by pseudopod formation. The cytoplasm is divided into ectoplasm and endoplasm. This class includes the pathogenic and nonpathogenic amebas (Fig. 31-2).

ZOOMASTIGOPHOREA (commonly called flagellates). Movement is by means of flagella and an undulating membrane. Flagellates have two nuclei, and the cytoplasm is not differentiated into endoplasm and ectoplasm. Cell bodies are often pear shaped and fixed in outline. The most important flagellates medically are in the genera *Trypanosoma*, *Leishmania*, *Trichomonas*, *Giardia*, and *Chilomastix* (Fig. 31-3, *A* to *C*).

TELOSPOREA. There are no external organs of locomotion. These organisms live within the cells, tissues, cavities, and fluids of the body and are represented by the *Plasmodium* of malaria.

CILIATEA. Cilia are present for locomotion. The only pathogenic member of this group is *Balantidium coli* (Fig. 31-3, *D*).

TOXOPLASMEA. There are no external organs of locomotion. The protozoa move by bending and gliding movements of their bodies. The representative pathogen here is *Toxoplasma gondii*.

HAPLOSPOREA. There are no flagella but pseudopodia may form. The parasite here is *Pneumocystis carinii*. (With some authorities the taxonomic position of *Pneumocystis* is still uncertain. It has been considered both a protozoan and a fungus.)

Laboratory diagnosis. The structure of many protozoan parasites makes it easy for them to be identified microscopically in suitably prepared clinical specimens such as blood or stool. At times, however, a morphologic diagnosis is not possible in parasitic infections.

Fortunately, parasites possess a variety of antigens within their makeup and therefore lend themselves nicely to serologic testing. The immunodiagnostic tests that have been standardized for the identification of the protozoa of this chapter and

TABLE 31-1. IMMUNODIAGNOSTIC TESTS OF VALUE IN PROTOZOAN DISEASES*

	Amebiasis	Chagas' disease	African trypano-somiasis	Leish-maniasis	Malaria	Toxo-plasmosis	Pneumo-cystosis
Serologic test							
Complement fixation	×	×		×		×	×
Indirect hemagglutination	×	×		×	×	×	
Indirect fluorescent antibody	×	×	×		×	×	×
Precipitin			×				
Immunoelectrophoresis	×						
Double diffusion	×		×		×		
Intradermal test				×		×	

*From Kagan, I. G.: Hosp. Pract. **9:**157, Sept., 1974.
× indicates an accepted test for routine use.

the metazoa of the next include complement fixation, indirect hemagglutination (most sensitive), latex agglutination, precipitation, indirect fluorescent antibody detection, countercurrent electrophoresis, bentonite flocculation, double diffusion, and intradermal ones. For most situations there is not necessarily one best test; it may be that two or more may have to be used for clinical evaluation. Tables 31-1 and 32-2 show the applications of various ones to diseases for which they are useful. Within the last few years a number (at least 30) of reagents and tests have been packaged commercially for the serodiagnosis of major parasitic diseases. Best known, perhaps, are the kits and reagents for the protozoan diseases, amebiasis and toxoplasmosis, and for the metazoan diseases, echinococcosis and trichinosis.

PROTOZOAN DISEASES
Amebiasis

The term *amebiasis* indicates an infection with *Entamoeba histolytica*. The disease occurs in two forms: acute amebiasis (amebic dysentery), characterized by an intense dysentery with bloody, mucus-filled stools, and chronic or latent amebiasis, described by vague intestinal disturbances, muscular aching, loss of weight, even constipation. In some chronic cases, manifestations are absent. The chronic form is more common than the acute. An estimated 5% to 10% of persons in the United States are affected.

The organism. The organism *Entamoeba histolytica* exists as a vegetative ameba or *trophozoite* and as a *cyst*. Vegetative trophozoites possess an active type of ameboid motion on a warm microscopic stage. Microscopically one sees the pseudopods, a distinctive nucleus, and red blood cells within the cytoplasm of the trophozoite. Vegetative amebas are very susceptible to injurious agents. In an unfavorable environment they quickly succumb; therefore they do little to transmit the disease. The cysts are smaller than vegetative amebas, nonmotile, and surrounded by a resistant wall.

Life history. The life cycle of *Entamoeba histolytica* begins with the cysts by which the disease is transmitted from person to person. After the cysts are passed in the feces, they remain infectious for several days if not destroyed by heat and drying. When the cysts are swallowed by a new host, they pass through the stomach unchanged. The shells are dissolved by juices of the small intestine, and the vegetative forms are liberated. The trophozoites pass to the large intestine to attack the mucous membrane and produce ulceration. The vegetative amebas multiply in the ulcers; some escape into the lumen of the intestine. If diarrhea is present, they are swept out of the intestinal tract. If diarrhea is not present, they multiply one or more times and then encyst. Encystment does not occur outside the body. Cysts are excreted in the feces.

Sources and modes of infection. The life cycle of *Entamoeba histolytica* reveals three facts: (1) infection can be acquired only by swallowing cysts, (2) infection comes from the feces of a person excreting cysts, and (3) acute cases are of little danger. The feces of patients with acute amebiasis contain largely vegetative parasites that die quickly; they could not survive the acid gastric juice should they accidentally be ingested. Infection is usually acquired by eating uncooked food contaminated with feces containing cysts. The most important single source of infection is the food handler with chronic amebiasis, especially the one preparing uncooked foods. Other sources of infection are vegetables fertilized with human excreta and drinking water contaminated with sewage. Apparently the latter was the cause of the Chicago epidemic of 1933, in which there were 1409 cases with 98 deaths. The water in two hotels had been contaminated by sewage. Flies and other insects may spread the cysts mechanically.

Lesions. In the majority of cases there seems to be a state of balance between the amebas and the host. The patient experiences mild disturbances or none at all and is able to repair the ulcers almost as fast as they are formed. This is chronic or latent amebiasis.

If the resistance of the host is lowered or massive infection occurs, the host is unable to repair the ulcers as fast as they are formed, and the increasing ulceration causes a violent dysentery in which the stools consist entirely of blood and mucus. This is acute amebiasis or amebic dysentery. Occasionally, intestinal perforation occurs.

Sometimes amebas penetrate deeper into the intestinal wall and enter tributaries of the portal vein to be carried to the liver, where they produce amebic hepatitis or liver abscess. Amebic abscesses can occur in the lungs or brain (Figs. 31-4 and 31-5).

Laboratory diagnosis. The laboratory diagnosis of amebiasis necessitates the examination of *fresh warm* stools for vegetative amebas and the examination of ordinary specimens for cysts. The examination of iron hematoxylin–stained smears of specimens is helpful. Although morphologic recognition of *Entamoeba histolytica* under

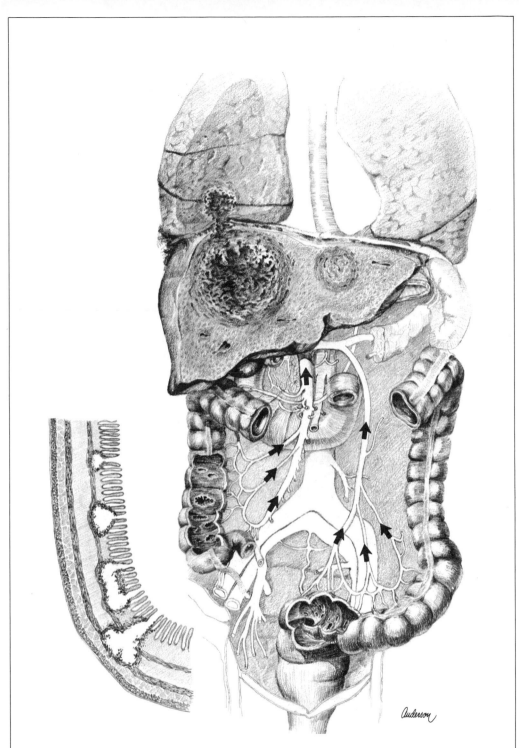

FIG. 31-4. Major pathology of amebiasis. Invasion of intestinal mucosa occurs most commonly in the cecum and next most commonly in rectosigmoid area. Passage of trophozoites via portal circulation may result in liver abscess formation. Metastasis through diaphragm may result in secondary abscess formation in lungs. Trophozoites carried in bloodstream may cause foci of infection anywhere in body. (From Beck, J. W., and Barrett-Connor, E.: Medical parasitology, St. Louis, 1971, The C. V. Mosby Co.)

583

FIG. 31-5. Amebic abscess of liver, cross section of the organ.

the microscope is the prime concern of the laboratory, serologic tests are used to advantage to identify this parasite. These include a complement fixation test, hemagglutination tests, an indirect fluorescent antibody test, and an agar gel double diffusion technic.

Prevention. The prevention of amebiasis depends on the proper control of carriers, proper sanitary supervision of foods, and general cleanliness.

In addition to *Entamoeba histolytica*, several other amebas may be found in the intestinal canal, but *Entamoeba histolytica* is the only one that causes disease. *Entamoeba coli* is notable because it must be distinguished from *Entamoeba histolytica*.

Trypanosomiasis

Trypanosomes (hemoflagellates) (Fig. 31-6), of which there are many species, are spindle-shaped protozoa that enter the bloodstream of many different species of animals. They are found in the plasma, not within the blood cells. Infection with trypanosomes is known as *trypanosomiasis*. The types important to man are African trypanosomiasis or African sleeping sickness and South American trypanosomiasis or Chagas' disease.

Abastrin, a substance elaborated by *Trypanosoma*, has some antimicrobial activity.

African trypanosomiasis. African trypanosomiasis is seen in two forms: Gam-

584

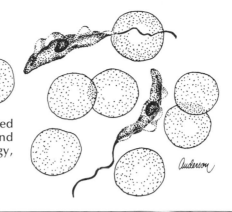

FIG. 31-6. *Trypanosoma gambiense* sketched in blood smear. (From Beck, J. W., and Barrett-Connor, E.: Medical parasitology, St. Louis, 1971, The C. V. Mosby Co.)

bian trypanosomiasis (agent, *Trypanosoma gambiense*) and Rhodesian trypanosomiasis (agent, *Trypanosoma rhodesiense*). Each is transmitted by a species of the tsetse fly. The fly becomes infected by ingesting the blood of a person with the disease, and the parasite undergoes a cycle of development in its body. When the parasites develop to a certain point, they invade the salivary glands of the fly, whence they are transferred to persons bitten. Cattle, swine, and wild game animals, especially antelope, may harbor the parasites and be a source of human infection. Rhodesian trypanosomiasis is more virulent than the Gambian form. Early in the course of either form of trypanosomiasis there are acute episodes of fever and inflammation of lymph nodes as the trypanosomes multiply in the bloodstream. The Rhodesian form is usually fatal within a matter of months and rarely progresses to the chronic stages of the Gambian form. In the end stages of the disease, invasion of the brain and its coverings produces the celebrated and uncontrollable sleepiness.

South American trypanosomiasis. South American trypanosomiasis, caused by *Trypanosoma cruzi*, is transferred to man by small, bloodsucking, cricketlike insects from a reservoir in man and in domestic and wild animals such as dogs, cats, rats, armadillos, and opossums. Infection results from the contamination of the skin with insect feces and is not transferred by the actual insect bite. South American trypanosomiasis differs from African trypanosomiasis in that the parasites multiply in the tissues rather than in the blood. They reappear in the blood to be picked up by the vector. If the patient survives the acute stage, the disease becomes chronic, with the trypanosomes localized in various organs.

An experimental vaccine has been prepared by killing the microorganisms of a culture by physical means—subjecting the trypanosomes to high-frequency sound waves, to pressure, or to mechanical forces evoked when the culture is shaken with glass beads. It has been used only in mice.

Laboratory diagnosis. During the fever, trypanosomes of the African disease may be demonstrated in Giemsa-stained films of peripheral blood. Concentration technics for peripheral blood facilitate the search for parasites. Smears and imprints of lymph nodes may contain them. *Trypanosoma cruzi* is identified in aspirated material from spleen, liver, lymph nodes, and bone marrow.

Leishmaniasis

Leishmaniasis is a protozoan disease caused by what is probably man's most ancient parasite. It exists in two forms: the visceral and the cutaneous.

Visceral leishmaniasis (kala-azar, dumdum fever). The visceral form of leishmaniasis is characterized by fever, enlargment of the spleen and liver, progressive emaciation, weakness, and, in untreated patients, death. The agent, *Leishmania donovani*, is transmitted from man to man by the bite of sandflies of the genus *Phlebotomus*. The disease is endemic among dogs, which may be a source of infection. It occurs in the countries bordering the Mediterranean Sea, in India, in the Middle East, in China, and in parts of Africa.

Cutaneous leishmaniasis. Cutaneous leishmaniasis is described by the presence of nodular and ulcerating lesions in the skin. There are two types. One, known as Oriental sore, Aleppo button, or Delhi boil, is caused by *Leishmania tropica*. The other, known as American leishmaniasis or espundia, is caused by *Leishmania braziliensis*. Cutaneous leishmaniasis is transmitted as is the visceral disease by sandflies. The individual lesions on the skin represent the bites of insects or the mechanical transfer of infection by scratching or some form of abrasion. This disease is seen in the same parts of the world as visceral leishmaniasis, but the two forms of leishmaniasis are said not to occur in exactly the same localities. Cutaneous leishmaniasis is occasionally seen in the United States in persons coming from endemic areas.

Laboratory diagnosis. In all forms of leishmaniasis the diagnosis is made by demonstrating the organisms in smears from lesions or in biopsies of involved tissues.

Trichomoniasis

Trichomoniasis occurs as a widespread infection of the genitourinary tract caused by *Trichomonas vaginalis*. In women it is an intractable vaginitis with a profuse, cream-colored, foul-smelling discharge in which the trichomonads abound. In men they are found in the prepuce and prostatic urethra, but symptoms seldom occur. The infection is transmitted by sexual intercourse and is a venereal disease of generally unrecognized significance. The trichomonads are readily identified in vaginal discharges from the female and in urine or prostatic discharges of the male.

Infection with intestinal flagellates

The most important intestinal flagellates found as cysts and trophozoites in stools are *Giardia lamblia*, *Trichomonas hominis*, and *Chilomastix mesnili* (Fig. 31-3). The latter two are not considered pathogenic by most protozoologists.

Giardiasis. Although *Giardia lamblia* (Fig. 31-7) rarely invades the intestinal mucosa, its presence in the upper small intestine in man is regularly associated with bowel disturbances, and increasing evidence points to this flagellate as a significant pathogen. Giardiasis, its infection, means persisting diarrhea, malabsorption, and inflammatory changes in the small bowel lining. Endemic in the United States, it is an important explanation for traveler's diarrhea, the most likely source of infection being contaminated tap water or ice and iced beverages consumed on the trip.

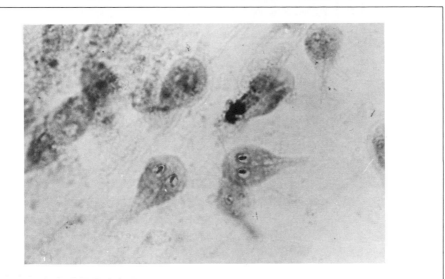

FIG. 31-7. *Giardia lamblia* in gastric aspirate processed for cytologic examination, photomicrograph. Many are present.

Diagnosis is made by examination of feces either in direct smear or in preparations obtained by concentration technics. *Giardia* are also found in material aspirated from the duodenum. No serologic test is yet available.

Malaria

Malaria is an acute febrile disease caused by the malarial parasite, a protozoon of the genus *Plasmodium*. The word *malaria* is derived from Italian for "bad air," and the disease got its name in the eighteenth century because of its association with the ill-smelling vapors from the marshes around Rome.

Malaria, one of the most widely prevalent diseases in the world, remains the number one public health problem globally. Worldwide it results in a greater morbidity and mortality than any other infectious disease. More people have malaria than any other disease. It is a constant threat to more than 1 billion human beings. Prior to America's civil and military involvement in Southeast Asia, malaria was infrequently seen in the United States (60 cases reported in 1961). In 1970 there were over 3500 cases, mostly in individuals returning from Southeast Asia. Malaria has been a scourge of war and of mankind throughout the ages. The Army Medical Corps states that this disease can put more men out of action than battle casualties. In World War II there were over 490,000 cases of malaria with 8 million man-days lost.

Types. Malaria exists in three distinct types* (each caused by its own species of *Plasmodium*) as follows:

    1. *Tertian*—a paroxysm of chill and fever every 48 hours; cause, *Plasmodium vivax* (the most common type)

---

*A fourth parasite is *Plasmodium ovale*, whose appearance and life cycle are much like those of *Plasmodium vivax*. Infection with this parasite, tertian malaria, is usually mild and not widely distributed over the world.

2. *Quartan*—a paroxysm of chill and fever every 72 hours; cause, *Plasmodium malariae* (the least common type)
3. *Estivoautumnal* or *malignant*—irregular paroxysms; cause, *Plasmodium falciparum**

The first two types are known as *regular intermittent types.* After the paroxysm of chill and fever, temperature returns to normal, and the patient is fairly comfortable until the next paroxysm. In the third or *remittent* type the fever varies in intensity, but the patient does not become completely afebrile. Estivoautumnal (also called pernicious) malaria is the most severe in its consequences and is the treacherous form. Its clinical picture is diverse, sometimes obscure, oftentimes dramatic, and it can be rapidly fatal. Death, it is said, may come within a matter of hours. If there is mixed infection, falciparum is the dominant type.

Modes of infection. The different species of malarial parasites are closely related and transmitted in the same way—by the bite of a female mosquito of the genus *Anopheles.* Of this genus, nearly 100 species may transmit the infection naturally. Man is the main reservoir of infection. In malaria-infected countries many of the inhabitants become asymptomatic carriers. They harbor the parasites in their blood (parasitemia) and tissues without manifesting the disease. Repeated attacks seem to give the patient some immunity. Unrecognized infections and insufficient treatment lead to the carrier state, so important in the spread of malaria.

The disease is occasionally transferred by the use of contaminated hypodermic syringes, as is common among heroin addicts, or by blood transfusion. There is an increased awareness of this hazard in blood banking partly because of the likelihood of the source of infection being a blood donor who is a drug addict.

A reservoir of malaria exists in monkeys, from which the infection is transmitted to man and other primates by certain forest species of *Anopheles.*

Life story of the parasite. There are two major events in the complicated life story of the malarial parasite.

ASEXUAL DEVELOPMENT IN MAN. When young malarial parasites *(sporozoites)* are introduced into the bloodstream by the mosquito bite, they localize in the cells of the liver where they multiply. This is the *preerythrocytic phase.* (Some 500,000 parasites must be injected by the mosquito for a human being to be infected.) After 6 to 9 days young parasites *(merozoites)* are released into the bloodstream. The *erythrocytic phase* begins when each one bores into a red blood cell on which it feeds and therein develops. The parasite does not fully utilize the hemoglobin of the red cell, and so granules of pigment (an iron porphyrin hematin) accumulate within its cytoplasm. This residual product is the malarial pigment and not a normal breakdown product of hemoglobin. When the parasite reaches maturity, it is known as a *schizont.* Within the red cell the mature schizont arranges itself into a number of segments.

---

*Ninety percent of malaria in Southeast Asia is caused by *Plasmodium falciparum.* Although this is true, 85% of malaria in returnees is caused by *Plasmodium vivax.* In the Korean conflict malaria was almost entirely vivax. This is less severe clinically but more likely to lie dormant only to recur many months after the initial episodes. There is little tendency for relapses with falciparum malaria.

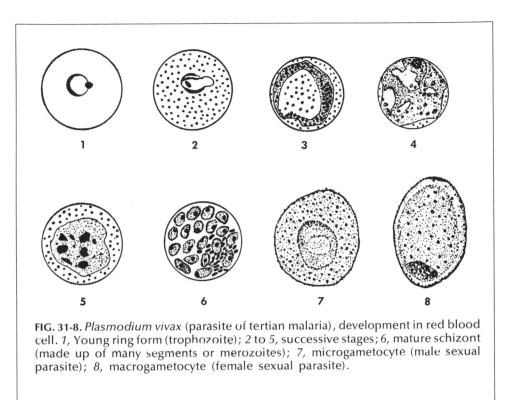

FIG. 31-8. *Plasmodium vivax* (parasite of tertian malaria), development in red blood cell. *1,* Young ring form (trophozoite); *2* to *5,* successive stages; *6,* mature schizont (made up of many segments or merozoites); *7,* microgametocyte (male sexual parasite); *8,* macrogametocyte (female sexual parasite).

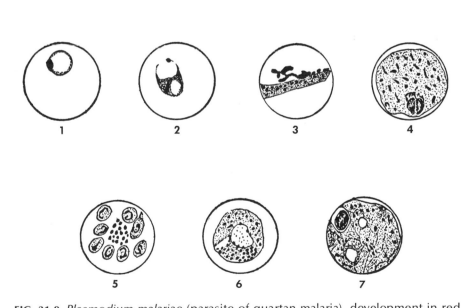

FIG. 31-9. *Plasmodium malariae* (parasite of quartan malaria), development in red blood cell. *1,* Young ring form (trophozoite); *2* to *4,* successive stages; *5,* mature schizont (made up of merozoites); *6,* microgametocyte; *7,* macrogametocyte.

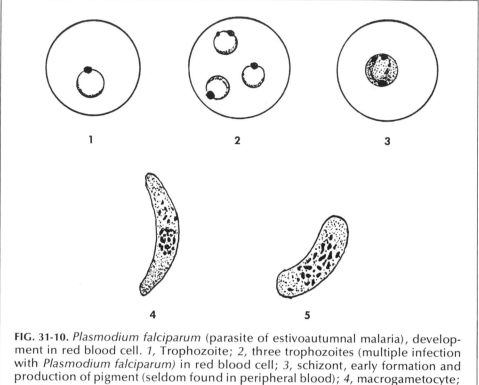

FIG. 31-10. *Plasmodium falciparum* (parasite of estivoautumnal malaria), development in red blood cell. *1*, Trophozoite; *2*, three trophozoites (multiple infection with *Plasmodium falciparum)* in red blood cell; *3*, schizont, early formation and production of pigment (seldom found in peripheral blood); *4*, macrogametocyte; *5*, microgametocyte.

Suddenly the segments separate, release another generation of merozoites, and destroy the red cell. This is *segmentation*. Some of the merozoites are destroyed by the white blood cells, but the majority bore into red blood cells to repeat the process of asexual growth and segmentation. In estivoautumnal malaria two or even three or four parasites invade a single red cell (Figs. 32-10 to 32-12).

About 2 weeks (sometimes longer) after the infecting mosquito bite, enough parasites are present for red blood cell destruction to cause trouble. The *incubation period* covers the initial preerythrocytic phase and the first 2 weeks or so of the erythrocytic phase. The time elapsing between the entrance of a parasite into a red blood cell and its segmentation is the *periodicity* of the parasite. For *Plasmodium vivax* it is 48 hours and for *Plasmodium malariae*, 72 hours. For *Plasmodium falciparum* it is usually 48 hours, but not regularly so. (See Figs. 31-8 to 31-10.)

The paroxysms of chills and fever in malaria stem from the liberation of metabolic by-products from the parasite and toxic breakdown products from the disrupted blood cell. Sharp paroxysms occur in tertian and quartan malaria because all parasitized cells rupture at about the same time. In estivoautumnal infections some parasitized cells rupture ahead of time, and some rupture behind time, so that several hours are required for the whole brood of parasites to be released. Thus chills are usually absent and fever may be continuous.

590

Some parasites do not repeat the asexual phase of development but produce male and female sexual forms or gametocytes, the sexual development of which is completed within the stomach of an *Anopheles* mosquito. Sexual forms do not appear in the blood until infection is of 2 or 3 weeks' duration.

SEXUAL DEVELOPMENT IN THE MOSQUITO. When a female *Anopheles* mosquito ingests male and female parasites from the blood of an infected person, a rather complicated sexual cycle begins in its stomach. First, the female parasite or *macrogamete* is fertilized by flagellar structures, *microgametes*, that break away from the male parasite or *microgametocyte* by a process of exflagellation. These structures correspond to spermatozoa in higher forms of life. The fertilized parasite or *ookinete* bores into the stomach wall, encysts (the *oocyst*), and divides into many small spindle-shaped parasites or *sporozoites*. The cyst ruptures, and the sporozoites are carried by the lymphatic system to the salivary gland of the mosquito, so constructed that the parasites are ejected in the saliva when the mosquito bites.

Pathology. The chief pathologic changes in malaria relate to destruction of red blood cells (hemolysis). Hemolysis leads to different degrees of anemia and jaundice. The parasitic infection seems to make the blood more viscous, and the sticky parasitized cells plug and obstruct small blood vessels. This is prone to occur with falciparum malaria and accounts for its worst complication, cerebral malaria. The liver and spleen enlarge (hepatosplenomegaly), partly because the presence of the malarial pigment stimulates the reticuloendothelial system to activity. Reticuloendothelial cells ingest the pigment and deposit it in liver, spleen, and bone marrow. In acute malaria the spleen is moderately enlarged, soft, and friable. In chronic malaria it is enlarged and markedly fibrotic. Such a spleen easily ruptures as the result of a blow or fall or even spontaneously.

A dreaded complication, blackwater fever, develops in certain cases with acute and massive hemolysis. Large amounts of hemoglobin pass into the plasma (hemoglobinemia), spilling over into the urine (hemoglobinuria). There is a severe kind of acute renal failure associated with the passage of reddish black urine.

Laboratory diagnosis. Malarial parasites are easily seen in the red cells of properly prepared (and stained) thin smears of peripheral blood taken just before or at the peak of the paroxysm. Wright's and Giemsa's stains are most often used. The young parasites, seen as blue rings with a red chromatin dot attached, have a signet-ring appearance. The quartan ring is thicker than the tertian, and the estivoautumnal ring is thin and hairlike. The latter often has two chromatin dots. Full-grown malarial parasites almost completely fill the red cell and contain numerous red granules. Just before red cell rupture, parasites assume a segmented or rosette pattern. So-called malarial crescents are sausage-shaped gametocytes with a chromatin mass near their center and are the sexual parasites of estivoautumnal malaria. They are frequently seen in the blood, whereas the fully developed parasite or schizont is seldom seen in this form of malaria.

At times, when organisms are too few to be seen in thin smear, a sample of blood may be smeared thickly on a glass slide, treated to remove hemoglobin from the erythrocytes, and then stained. The disadvantage of the thick smear is that the shape and appearance of the parasites are altered in preparation. Its advantage is that a much greater volume of blood may be examined within a given length of time.*

Fluorescent antibody technics are also used to stain the malarial parasite specifically. A soluble antigen fluorescent antibody (SAFA) diagnostic test fills a need in the blood bank in the screening of donors.

A small dose of quinine or other antimalarial drug drives parasites out of the peripheral blood; therefore it is practically useless to examine blood for malaria right after such drugs are given.

The laboratory diagnosis of malaria *does not consist only in finding the parasites but includes determination of species.*

Mosquitoes transmitting malaria. Malaria is transmitted by various species of *Anopheles* mosquitoes and only by the female because the male lives on fruits and vegetables (not on blood). Since the common house mosquito *(Culex)* is not a vector, one must distinguish it from *Anopheles*. The *Culex* mosquito bites during the daytime; the *Anopheles* at night or about dusk. The wings of *Anopheles* are spotted, whereas those of *Culex* are not. When *Culex* is resting on a wall, its body is almost parallel to the wall; the body of *Anopheles* stands at an acute angle (Fig. 10-2, p. 172).

Prevention. Prevention of malaria depends on blocking the transfer of infection from person to person by mosquitoes. Recommended for this purpose are (1) screening of houses, (2) draining and oiling of ponds of water to prevent mosquito breeding and using minnows to destroy larvae, (3) proper treatment of patients with antimalarial drugs, and (4) detection and cure of carriers. The presence of an animal reservoir greatly complicates the problem of malarial control in those countries where jungles swarm with monkeys and other primates.

Development of a vaccine for malaria is still experimental. Partially purified material from the malarial plasmodium is used.

Balantidiasis

Balantidium coli is the most important intestinal ciliate and the largest protozoon to invade man (Fig. 31-3, *D*). It is seen in two life stages: the cyst and the motile trophozoite. In some cases it seems to be a harmless inhabitant of the large intestine, but usually its presence is associated with diarrhea. It may invade the intestinal wall and produce ulcerations or abscesses with intense dysentery that may even cause death (Fig. 31-11). *Balantidium coli* is a normal inhabitant of the large intestine of the domestic hog. Man is probably infected by ingesting cysts passed by the hog. The laboratory identification of cysts and trophozoites in the stools or in the exudate from intestinal ulcers makes the diagnosis.

Toxoplasmosis

Toxoplasma gondii, the cause of toxoplasmosis, is a delicate, boat-shaped, obligate intracellular parasite, somewhat similar to *Leishmania* and easily killed by

---

*In falciparum malaria thick smears may not show parasites for several days after onset.

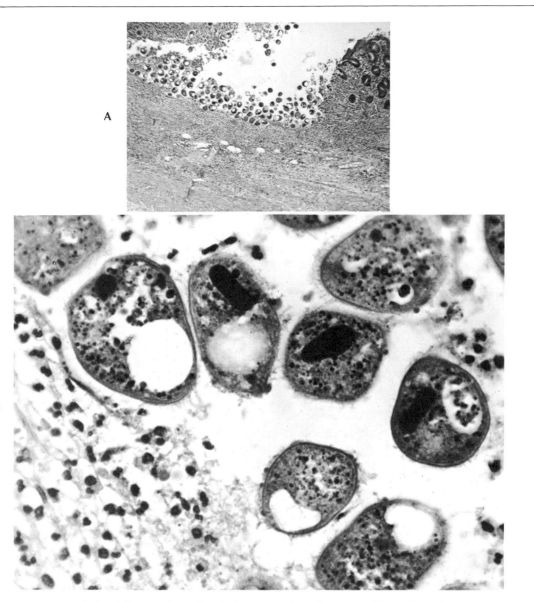

FIG. 31-11. *Balantidium coli,* stained tissue section (hematoxylin and eosin [H & E stain]). **A,** Balantidiasis of appendix. In crater of an ulcer of appendiceal wall are numerous ciliates cut in cross section (×35). **B,** Cross-sectional areas of *Balantidium coli* as seen with higher microscopic magnification (×430). (From Anderson, W. A. D., and Kissane, J. M.: Pathology, ed. 7, St. Louis, 1977, The C. V. Mosby Co., vol. 1.)

physical agents. It is a cosmopolitan sporozoan, being found in animals and birds all over the world.

The disease. The organism can invade practically any tissue cell and is especially prone to affect cells of the reticuloendothelial organs, including the lining cells of blood vessels. Within a cell the organisms rapidly proliferate in typical fashion to

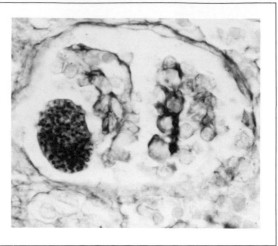

FIG. 31-12. Toxoplasma cyst (left), in microsection of kidney from infant dead of disease, photomicrograph. (Courtesy Dr. J. S. Remington, Palo Alto, Calif.; from Feldman, H. A.: Hosp. Pract. **4:**64, Mar., 1969.)

form a rosette that may become a cyst containing some 3000 parasites (Fig. 31-12). Inflammation may be present in the lungs, lymph nodes, eyes, and brain. Characteristic are the focal deposits of calcium in nervous tissue (especially in the fetus). In man, the disease occurs in two forms: acquired and congenital. The acquired or adult form often passes unnoticed. One third to one half of adults in the United States have been infected at some time. The congenital form is acquired in utero from a mother most probably with no history of previous infection. The newborn baby becomes very ill, develops a skin rash, turns yellow, and may convulse. If the baby survives the damage done in the brain (associated with the calcium deposits), he may be born with microcephaly, hydrocephalus, and mental retardation. *Toxoplasma gondii* produces changes in the eye designated as chorioretinitis, and the lesion produces blindness in these infants. (No effective form of treatment is known.)

Toxoplasmosis is associated with the formation of tumors in birds; in man it is seen with neoplasms of the central nervous system.

Sources and mode of infection. This sporozoan has a life cycle in cats similar to that of the malarial parasite in mosquitoes. Cats pick up the parasites when they consume the intermediate hosts—infected birds and mice. In the intestine of the cat, which provides a peculiarly suitable habitat for the parasite, the organisms go through asexual (schizogony) and sexual (gametogony) stages of development in the epithelium, and the oocysts are passed in the feces. After being passed (into soil, sand, or litterbox), oocysts become infectious after 3 to 4 days in warm, moist surroundings. They can be recovered after many months from water or wet soil and are generally resistant to many chemical agents including ordinary disinfectants. However, drying and heat kill them. The common house cat is implicated as the primary host and a human reservoir.

The parasite is universal. Many animals harbor it, and it may persist in the raw flesh of the slaughtered animal until killed by heat, drying, or freezing. Eating raw or undercooked meat is a principal source of human infection. The National Livestock and Meat Board recommends that all meat be heated to at least 140° F through-

594

out to kill the toxoplasmas. (This is the "rare" reading on meat thermometers designed for home use.) Pregnant women should be very careful in handling cats, cat feces, or articles contaminated with cat feces, and should avoid altogether any contact with a strange cat or one newly brought into the household.

Toxoplasmas are taken into the human body by way of the mouth either because the individual has been in contact with an infected cat or because he has consumed infected meat. They are released from the oocysts and migrate into body tissues and fluids. For the congenital form to develop, the organisms must get into the bloodstream of the mother sometime after the first trimester of pregnancy. They establish a focus of infection in the placenta that enables them to penetrate the fetal circulation and so infect the fetus.

Laboratory diagnosis. These microorganisms, possessing a delicate crescent-shaped body with tapered ends, are easily stained and identified in smears made from body fluids, exudates, and diseased tissues. Toxoplasmas can be cultured in cells or in a fertile hen's egg. Serologic examinations include an indirect hemagglutination test, a complement fixation test, a neutralization test, and an immunofluorescence test. The antibodies detected by the Sabin-Feldman dye test are those preventing parasites of a laboratory culture from taking up methylene blue dye. A

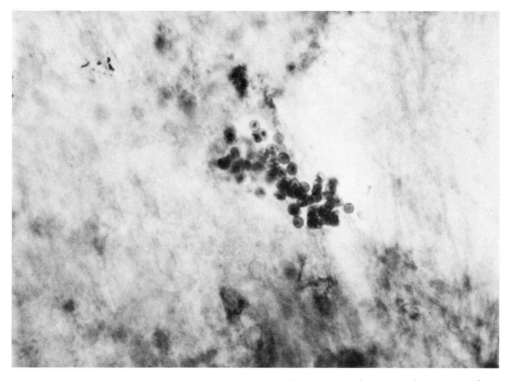

FIG. 31-13. *Pneumocystis carinii* in direct smear of pulmonary lavage sediment (methenamine silver stain). (×500.) (From Drew, W. L., and others: J.A.M.A. **230:**713, 1974.)

toxoplasmin skin test is available. A mouse can be inoculated with suspicious material so that the organisms can be recovered and identified from that animal.

Toxoplasmas may be detected in the feces of a cat. Direct smears are not adequate; the fecal flotation method must be used.

Pneumocystosis

Pneumocystis carinii causes pneumocystosis (diffuse interstitial pneumonitis, interstitial plasma cell pneumonia), an inflammatory process unique in the lungs. Although thought to be a widespread saprophyte, little is known about the organism, which was not recognized as pathogenic until the advent of immunosuppression. Nothing is known as to mode of infection. In pneumocystosis the air sacs of the lungs are filled with a foamy, semiliquid, lightly staining material representing clustered masses of oval, minute organisms about 1 μm in diameter, surrounded by a thin homogeneous capsule. The walls of the air sacs are permeated by inflammatory cells.

Pneumocystosis is another, not too rare, disorder of the "compromised" host, particularly the immunosuppressed one. It was first encountered in debilitated infants and is said to be one of the worst, fastest moving, and cruelest killers of such children. The vulnerability of much older patients stems from prolonged steroid hormone therapy, extended immunosuppression, or the presence of leukemia or other cancer of the lymphoid system. Cortisone, a steroid hormone, is believed to enhance the growth of these organisms.

The organisms are identified and studied in smears of aspirates and imprints from diseased lungs stained by the Papanicolaou or other suitable technic and also in tissue sections of lung obtained either from antemortem biopsy or autopsy. Special silver stains demonstrate them effectively (Fig. 31-13).

QUESTIONS FOR REVIEW

1. Describe protozoa. What are their salient features?
2. Give the six classes of protozoa of medical interest. Delineate each.
3. Define intermediate host, definitive host, organelle, pseudopod, cyst, flagella, cilia, trophozoite, parasite, endoparasite, ectoparasite.
4. How is amebiasis spread? How may it be prevented?
5. Outline the development of the malarial parasite in (a) man and (b) the mosquito.
6. Discuss the prevention of malaria.
7. Compare *Anopheles* mosquito with *Culex*.
8. Give the laboratory diagnosis for:
   a. Malaria
   b. Amebiasis
   c. Trypanosomiasis
   d. Leishmaniasis
   e. Toxoplasmosis
   f. Pneumocystosis
9. Characterize briefly Chagas' disease, African trypanosomiasis, trichomoniasis, giardiasis, balantidiasis, toxoplasmosis, pneumocystosis.
10. Compare the life cycle of *Plasmodium* with that of *Toxoplasma*.

REFERENCES. See at end of Chapter 32.

32 Metazoa

PERSPECTIVES

Among multicellular animal parasites, or Metazoa, three phyla are medically note-worthy in man: (1) *Platyhelminthes* or flatworms, which include the two classes of *Trematoda* (flukes) and *Cestoidea* (tapeworms), (2) *Nematoda* (roundworms), and (3) *Arthropoda*. Worms are elongated, invertebrate animals without appendages or bilateral symmetry, and the study of the pathogenic ones is *medical helminthology*. Arthropoda includes mites, spiders, ticks, flies, lice, and fleas, important as vectors in the transmission of disease* (Table 32-1).

The laboratory diagnosis of metazoan, as with protozoan, infections depends most of the time on morphologic identification of the parasite or its ova (Fig. 32-1). Table 32-2 presents immunodiagnostic tests in relation to the diseases for which they are applicable. (See also p. 581)

REGISTRY OF PATHOGENS
Trematodes (flukes)

Trematodes are flat, leaflike, nonsegmented parasites provided with suckers for attachment to the host. All species, except those that inhabit the bloodstream, are hermaphrodites and have operculate eggs (Fig. 32-1, *A, J, L-N*). Some have the most complicated life histories in the animal kingdom.

Life history. The cycle of development of flukes is briefly as follows. The egg is passed from the body of the host, and the contained embryo develops into a ciliated organism, the *miracidium* (*pl.*, miracidia). If water is present, the *miracidium* escapes from the egg and swims (about 5 or 6 hours) until it reaches an intermediate host.

---

*If the parasite is carried unchanged, the vector is a *mechanical* one. The parasite undergoes a series of developmental changes in the body of the *biologic* vector.

TABLE 32-1. OVERVIEW OF ARTHROPODA IN SPREAD OF DISEASE

| Examples | Vectors | |
	Common names	Genus names
Metazoan diseases		
Filariasis	Mosquito	*Culex*
River blindness	Black fly	*Simulium*
Ascariasis	Housefly	*Musca*
Protozoan diseases		
Malaria	Mosquito	*Anopheles*
Sleeping sickness	Tsetse fly	*Glossina*
Leishmaniasis	Sandfly	*Phlebotomus*
Amebiasis	Housefly	*Musca*
Chagas' disease	Triatomid bug	*Triatoma; Rhodnius*
Bacterial diseases		
Tularemia	Tick	*Dermacentor; Ambylomma; Rhipicephalus*
Relapsing fever	Louse	*Pediculus*
	Tick	*Ornithodoros*
Plague	Flea	*Xenopsylla; Pulex*
Salmonellosis	Housefly	*Musca*
Bacillary dysentery	Housefly	*Musca*
Cholera	Housefly	*Musca*
Rickettsial diseases*		
Spotted fevers	Tick	*Dermacentor; Ambylomma; Rhipicephalus; Ornithodorus*
Scrub typhus	Mite	*Trombicula*
Rickettsialpox	Mite	*Liponyssoides*
Typhus fever	Louse	*Pediculus*
Murine typhus	Flea	*Xenopsylla*
Bartonellosis	Sandfly	*Phlebotomus*
Trachoma†	Housefly	*Musca*
Viral diseases		
Yellow fever	Mosquito	*Aedes*
Encephalitis	Mosquito	*Culex*
Poliomyelitis	Housefly	*Musca*

*Rickettsias are now classified as bacteria.
†Chlamydiae, the cause of trachoma, are now classified with rickettsias.

Certain species of freshwater snails, prevalent in rice paddies and irrigation ditches, are important in the life cycle (Fig. 32-2). Hatching miracidia are phototrophic; that is, they swim toward light. (This phenomenon can sometimes be demonstrated in fecal or urinary specimens.) The miracidium penetrates the host snail and forms a cyst in its lungs, in which many organisms develop. These wander to other parts of the snail and develop into minute worms called *cercariae* (Fig. 32-3).

One infested snail alone can release 169 million cercariae in 1 month's time. Being phototrophic, the cercariae emerge during the hours of sunlight, and the

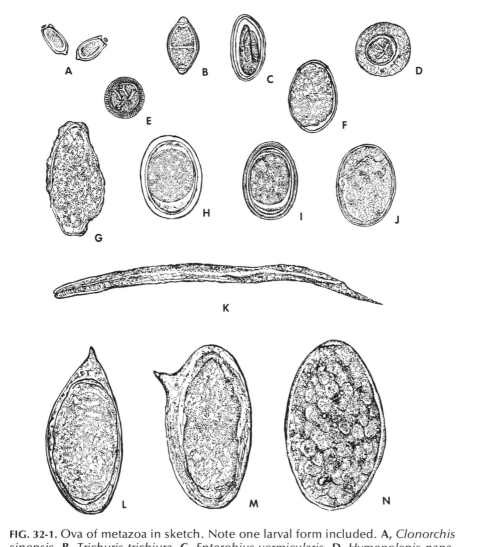

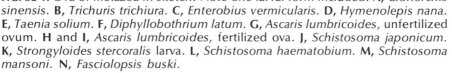

FIG. 32-1. Ova of metazoa in sketch. Note one larval form included. **A,** *Clonorchis sinensis.* **B,** *Trichuris trichiura.* **C,** *Enterobius vermicularis.* **D,** *Hymenolepis nana.* **E,** *Taenia solium.* **F,** *Diphyllobothrium latum.* **G,** *Ascaris lumbricoides,* unfertilized ovum. **H** and **I,** *Ascaris lumbricoides,* fertilized ova. **J,** *Schistosoma japonicum.* **K,** *Strongyloides stercoralis* larva. **L,** *Schistosoma haematobium.* **M,** *Schistosoma mansoni.* **N,** *Fasciolopsis buski.*

greatest number (hence the best chance of getting the disease) is at high noon. (They seldom live more than 1 day.) They swim around until they attach themselves to blades of grass, where they encyst. Sometimes they enter other aquatic animals such as certain fishes and crabs to encyst. When the encysted organisms are swallowed by man, the definitive host, they develop into adult flukes in his tissues. The cercariae of the blood flukes gain access to the body of man through the skin, especially between the toes. They make their way via the bloodstream to the portal venous system, where they mature into adult organisms.

TABLE 32-2.   IMMUNODIAGNOSTIC TESTS OF VALUE IN METAZOAN DISEASES*

	Schisto-somiasis	Cysticercosis	Echino-coccosis	Ancylo-stomiasis	Ascariasis	Trichinosis	Filariasis
Serologic test							
Complement fixation	×	×	×			×	×
Bentonite flocculation			×		×	×	×
Indirect hemagglutination	×	×	×	×	×	×	×
Latex agglutination			×			×	
Indirect fluorescent antibody	×		×			×	
Precipitin			×			×	
Intradermal test	×		×			×	

*Modified from Kagan, I. G.: Hosp. Pract. **9:**157, Sept., 1974.
× indicates an accepted test for routine use. Commercial reagents are available.

FIG. 32-2. *Australorbis glabratus,* snail vector for schistosomiasis in Western Hemisphere. (From Med. World News **6:**35, Dec. 3, 1965; photograph by Pete Peters.)

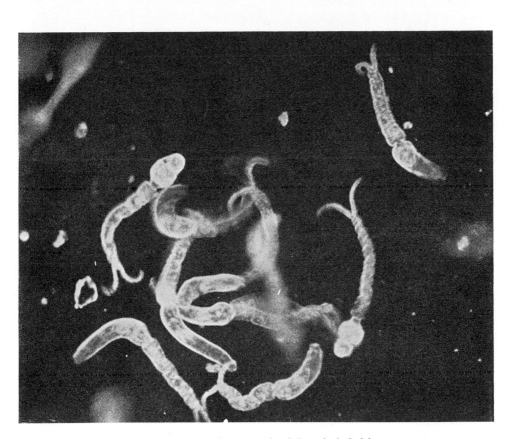

FIG. 32-3. Schistosome cercariae photographed by dark-field illumination. (Photograph by D. M. Blair; published by permission of F. Goodliffe, Southern Rhodesian Public Relations Department; from Gradwohl, R. B. H., and others: Clinical tropical medicine, St. Louis, 1951, The C. V. Mosby Co.)

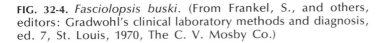

FIG. 32-4. *Fasciolopsis buski.* (From Frankel, S., and others, editors: Gradwohl's clinical laboratory methods and diagnosis, ed. 7, St. Louis, 1970, The C. V. Mosby Co.)

Classification. Flukes are classified according to the area of the body in which their development into adults is completed and their eggs deposited. From the standpoint of habitat there are flukes that live in the intestine, liver, lungs, and the portal venous system with its tributaries. The last are called blood flukes; of special note are three: *Schistosoma japonicum*, *Schistosoma mansoni*, and *Schistosoma haematobium*. The large intestinal fluke is *Fasciolopsis buski* (Fig. 32-4). *Clonorchis sinensis* and *Fasciola hepatica* are liver flukes, and *Paragonimus westermani* is the lung fluke.

Pathogenicity. The pathologic changes in the human body center about the eggs trapped in the tissues. Female blood flukes may migrate to the terminal vessels of the bladder and rectum to lay their eggs. (A single worm can deposit eggs in a given area for up to 20 years.) There the eggs set up an inflammatory reaction in the mucous membrane. This results in papillomatous thickenings. The eggs may escape into the lumen of the bladder or rectal canal to be passed in the urine or feces. The ova of flukes that migrate to the lungs may be found either in sputum or with swallowed sputum in feces. Because of the anatomic relation of the liver to the intestine, we would expect to find the ova of flukes of this organ in the feces. Ova from all types of flukes may be found in stool specimens, regardless of the habitat of a given fluke.

Trematode infections are prevalent in the Orient and in the tropics where contamination of fresh water by human feces is widespread. It is estimated that blood flukes or schistosomes infest 250 million persons. Infection with blood flukes is *schistosomiasis* or *bilharziasis*. It is one of man's oldest diseases; calcified ova have been found in Egyptian mummies. Also called snail fever, schistosomiasis is one of the world's most important medical problems. As a global disease, it is second only to malaria in the geographic extent of the incapacity and morbidity produced.

As an aside, schistosomes recovered from human beings have been found to be parasitized, in turn, by salmonellae, an instance of bacterial parasitism of a parasite in man. The association and interaction may explain why chronic salmonellosis often accompanies schistosomiasis.

Prevention. Measures of control have been largely directed toward the elimination of the intermediate host, the snail.

Experimentally it has been found that if a small innocuous dose of *Klebsiella pneumoniae* is given to an animal infected with *Schistosoma mansoni*, the flukes die out and the host gets well. This, if feasible worldwide, would open a new era of biologic control of parasites.

Cestodes (tapeworms)

Tapeworms *(Taenia)* are typically intestinal parasites producing digestive disturbances of variable degree.

Anatomy. Adult tapeworms have a small head, which buries itself in the intestinal mucosa and anchors the worm, and a nonsegmented neck (head and neck, collectively spoken of as the *scolex*) to which are attached in line a variable number of segments *(proglottids)*. New ones are formed by a process of segmentation from the scolex—the youngest segment is the one joining the scolex; the last segment is the oldest. Up to a certain point the farther the segment from the scolex, the larger it is.

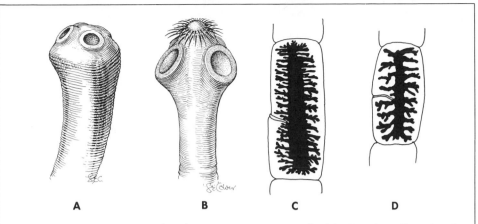

FIG. 32-5. *Taenia saginata* (beef tapeworm) compared with *Taenia solium* (pork tapeworm). **A,** Scolex of *Taenia saginata* with suckers only. **B,** Scolex of *Taenia solium* with apical hooks and suckers. **C,** Uterus of *Taenia saginata* with more than 14 primary lateral branches. **D,** Uterus of *Taenia solium* with less than 12 primary lateral branches.

There is no alimentary canal. Each segment obtains its nourishment from the host's intestinal juices by osmosis.

The head is extremely small by comparison with the remainder of the body, often the size of a pinhead (Fig. 32-5). It is provided with hooklets or suckers, or both, for attachment to the intestinal wall. The hooklets are arranged in one or more rows around a small prominence *(rostellum)* situated on the head. In at least one species, attachment is accomplished by suctorial grooves on the sides of the head. Treatment that fails to recover the head, regardless of the number of segments removed, is valueless because the head immediately replaces the lost segments. The peculiar shape, arrangement, and deeply embedded position of the hooklets often make removal of the head from the intestine extremely difficult.

Each fully developed segment is a sexually complete hermaphrodite. From the scolex to the other end of the worm, there are the following:

1. Undeveloped segments: *immature proglottids*
2. Segments with both male and female elements: well-developed, *mature proglottids*
3. Segments filled by the egg-laden uterus: *gravid proglottids*
4. Degenerating gravid proglottids

In a few species the ova are extruded from the segment through a birth pore. In most species, however, no birth pore is present, and the ova escape from the proglottid through a longitudinal slit. The gravid proglottids toward the end of the tapeworm separate and may be passed in the feces. A person can harbor a tapeworm without ova appearing in the stool because the segments may be expelled before the ova are liberated.

Life cycle. Tapeworms have a larval and an adult cycle of existence (Figs. 32-6 and 32-7). As a rule, the cycles take place in different species of animals. The adult cycle occurs within the intestinal canal, the larval cycle within the tissues of the host.

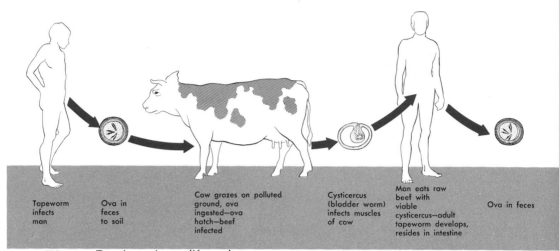

FIG. 32-6. *Taenia saginata,* life cycle.

| Tapeworm infects man | Ova in feces to soil | Cow grazes on polluted ground, ova ingested—ova hatch—beef infected | Cysticercus (bladder worm) infects muscles of cow | Man eats raw beef with viable cysticercus—adult tapeworm develops, resides in intestine | Ova in feces |

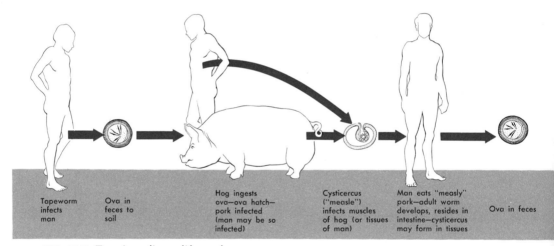

FIG. 32-7. *Taenia solium,* life cycle.

| Tapeworm infects man | Ova in feces to soil | Hog ingests ova—ova hatch— pork infected (man may be so infected) | Cysticercus ("measle") infects muscles of hog (or tissues of man) | Man eats "measly" pork—adult worm develops, resides in intestine—cysticercus may form in tissues | Ova in feces |

The egg develops into an adult tapeworm as follows. Through a series of changes, an embryo is formed within it. After the eggs are swallowed by a susceptible intermediate host, larvae are set free. By means of their hooklets the larvae penetrate the intestinal wall and pass into the tissues where the hooklets are lost. They then reach the bloodstream to be carried to different parts of the body to lodge and each to develop into a scolex. The irritation caused by formation of the scolex sets up a tissue reaction, and a cyst wall is formed around it. The scolex encased in the cyst is termed a *cysticercus.* When the raw or insufficiently cooked flesh of the animal containing the cyst is eaten by a susceptible host, the cyst wall is digested, the scolex attaches itself to the intestinal wall of the new host, and an adult tapeworm develops.

In worms (example, *Diphyllobothrium*) whose eggs (Fig. 32-1, *F*) escape by a birth pore, ciliated embryos escape from the egg after it is passed. They take up an aquatic existence and swim until they gain access to certain species of freshwater fish. They parasitize the fish with the help of *Cyclops* species, or water fleas, which act as transferring hosts. Consumption of the raw or poorly cooked fish transfers the parasite to man.

Taenia saginata (beef tapeworm, unarmed tapeworm)

Infection by *Taenia saginata* is quite common. The cow and giraffe are the intermediate hosts. Typically only one worm is present in man. The segments have independent motility and may escape through the anal canal.

Anatomy. The organism *Taenia saginata* ranges in length from 4 to 10 meters. The head is small (1.5 mm in diameter), pear shaped, and somewhat quadrangular. It has four suckers but no hooklets. The absence of hooklets gives it the name "unarmed" tapeworm. The neck is rather long and slender. The mature segments measure 5 to 7 mm by 18 to 20 mm. The uterus extends along the midline and gives off 20 to 30 delicate branches on each side. The eggs are spherical or ovoid in shape and yellow or brown in color, measuring 20 to 30 μm by 30 to 40 μm.

Taenia solium (pork tapeworm)

Infection with *Taenia solium*, rare in America, is acquired by eating measly pork. As a rule only one worm is present, but occasionally two or more are found.

Biologic features. Of the two worms, *Taenia solium* is shorter than *Taenia saginata*, measuring 2 to 8 meters in length. The head is very dark in color and globular or quadrangular in shape. It is provided with four suckers and two rows of hooklets projecting from a rostellum. The neck is threadlike. The mature segments measure 5 to 6 mm by 10 to 12 mm.

The ova (Fig. 32-1, *E*) resemble those of *Taenia saginata*. Practically, it is impossible to distinguish the two. The differentiation of the worms therefore depends on the characteristics of the uterus in the terminal proglottids. There are more branches coming from the sides of the uterus in the beef tapeworm than in the pork tapeworm (Fig. 32-5). Man may also become infected by swallowing the eggs of *Taenia solium* because it is possible for both cycles of development to occur in the human being.

Hymenolepis nana (dwarf tapeworm)

The dwarf tapeworm, *Hymenolepis nana*, the smallest and the one most frequently found in man, measures from 1 to 4 cm in length. The head is round and provided with four suckers and a single row of 24 to 30 hooklets. Ova are distinctive (Fig. 32-1, *D*). As a rule, many worms and ova are present, but the worms are so degenerated as to be unrecognizable.

Infection is spread directly from one person to another with no intermediate host. Eggs containing a fully developed embryo are released from a disintegrating end segment and passed in the feces. Within the new host, the eggs hatch in the stomach or small intestine, and the resultant larval forms move in the lumen to attach to the intestinal wall at a lower site, where they develop in the mucous membrane. In about 2 weeks adult worms appear.

Dipylidium caninum (dog tapeworm)

The dog tapeworm, *Dipylidium caninum,* is common in cats and dogs. It is 15 to 30 cm in length with a small head displaying 4 suckers and about 60 hooklets arranged in 4 rows. Hooked ova remain within the ripe proglottid until it has broken loose and migrated from the anus.

Diphyllobothrium latum (fish tapeworm or broad Russian tapeworm)

Diphyllobothrium latum usually measures 3 to 6 meters in length, but occasionally a length of 12 meters is reached. The head, flattened and almond shaped, is provided with two lateral grooves for attachment to the intestinal mucosa of the host. The segments show a characteristic brown or black rosette pattern (uterus filled with ova).

The presence of this worm gives rise to irregular fever, digestive disturbances, and a blood picture identical to that of pernicious anemia. Manifestations promptly subside after its removal.

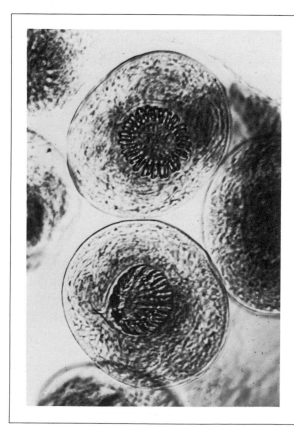

FIG. 32-8. Hydatid sand *(Echinococcus granulosus).* (From Frankel, S., and others, editors: Gradwohl's clinical laboratory methods and diagnosis, ed. 7, St. Louis, 1970, The C. V. Mosby Co.)

Echinococcus granulosus

Echinococcus granulosus is a tiny worm 3 to 6 mm long with a pear-shaped head. It has four suckers, 30 to 36 hooklets, but only one each of the immature, mature, and gravid proglottids. The adult host is the dog and other canines. By ingesting the eggs man can become the larval or intermediate host, and the unique development of the scolex within the cystic cavity can occur in his organs. The liver is a favored site. The result is termed the *hyatid cyst.* Free-floating scolices and brood capsules, structures within which scolices are formed, in cyst fluid are known as *hydatid sand* (Fig. 32-8). The gravity of the infection in man is determined by the location and the progression of the cyst with time.

Nematodes (roundworms)

Nematodes are nonsegmented worms with a flattened cylindric body tapering toward both ends. The mouth is frequently surrounded by thick lips or papillae, and there is a complete digestive tract. The sexes are distinct. The male is shorter and more slender than the female. The adults inhabit the intestinal tract of man, and as a rule, there is no intermediate host. The nematode life cycle passes through a series of stages from the larval forms to the adult worm.

The severity of the digestive disorders related to nematode infection depends on the load of worms carried.

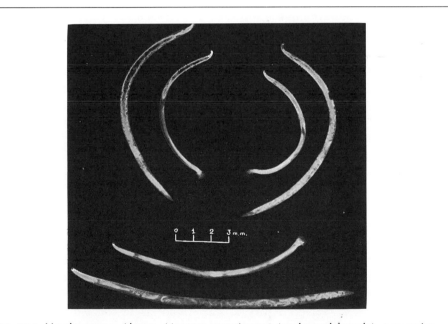

FIG. 32-9. Hookworms. Above, *Necator americanus* (male and female), two pairs; below, *Ancylostoma duodenale* (male and female), one pair. Note millimeter scale. (Original figure of P. Kourí.) (From Frankel, S., and others, editors: Gradwohl's clinical laboratory methods and diagnosis, ed. 7, St. Louis, 1970, The C. V. Mosby Co.)

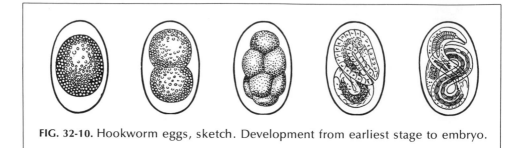

FIG. 32-10. Hookworm eggs, sketch. Development from earliest stage to embryo.

Ancyclostoma duodenale and Necator americanus

Ancylostoma duodenale and *Necator americanus* are, respectively, the Old World and New World hookworms. They resemble each other closely but differ in several important details (Fig. 32-9).

Anatomy. Both species are pale red and pointed at both ends. As a rule, the adult *Ancylostoma duodenale* is larger than *Necator americanus*. Its mouth has a pair of ventral hooks on each side of the midline and a pair of dorsal hooks. The mouth of *Necator americanus* is provided with plates instead of ventral hooks and has a distinct dorsal conical toothlike structure.

The ova of the species are practically identical, except that those of *Necator americanus* are larger. They are oval or oblong in shape but in certain positions appear spherical. They have three distinct parts—the shell, the yolk, and a clear space between the yolk and shell (Fig. 32-10). The thin, smooth shell appears as a distinct line. Eggs that have been passed for 24 hours or more show well-developed embryos. The ova adhere noticeably to glass or other surfaces. Advantage is taken of this trait in certain diagnostic procedures, but it makes thorough washing of laboratory glassware imperative.

Habitat. The adult worms live in the small intestine attached to the mucous membrane, where their presence produces a characteristic train of events. Great numbers are usually present. They tear the tissues to get to small blood vessels, from which blood is pumped into their intestines. However, they extravasate wastefully much of the blood into the lumen of the intestinal tract. A profound anemia is secondary to their bloodsucking activities. Hookworm disease is sometimes referred to as *uncinariasis*.

Life history. The life history of the hookworm is as follows (Fig. 32-10). After the ova are passed, development begins with the proper temperature and moisture. The larvae hatch and undergo certain developmental changes whereby they can infect a new host. The first-stage larvae emerging are the free-living ones of distinct shape, the so-called *rhabditoid* larvae. Rhabditoid larvae can metamorphose into long, delicate, threadlike forms, the *filariform* larvae of the infective stage. When the filariform larvae contact the skin of man, they penetrate it, producing a dermatitis (ground itch). They travel by the lymph and bloodstream to the lungs. Here they gain access to the bronchi and are carried by the bronchial secretions to the pharynx, where they are swallowed. After they reach the small intestine, they develop into

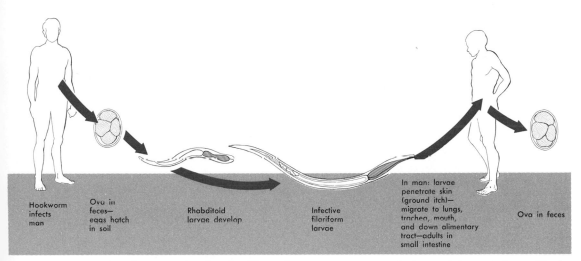

Hookworm infects man — Ova in feces—eggs hatch in soil — Rhabditoid larvae develop — Infective filariform larvae — In man: larvae penetrate skin (ground itch)—migrate to lungs, trachea, mouth, and down alimentary tract—adults in small intestine — Ova in feces

FIG. 32-11. *Ancylostoma duodenale,* life cycle.

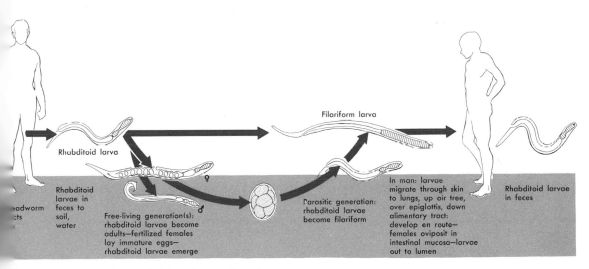

Rhabditoid larva — Filariform larva

...adworm ...cts — Rhabditoid larvae in feces to soil, water — Free-living generation(s): rhabditoid larvae become adults—fertilized females lay immature eggs—rhabditoid larvae emerge — Parasitic generation: rhabditoid larvae become filariform — In man: larvae migrate through skin to lungs, up air tree, over epiglottis, down alimentary tract: develop en route—females oviposit in intestinal mucosa—larvae out to lumen — Rhabditoid larvae in feces

FIG. 32-12. *Strongyloides stercoralis,* life cycle.

adult worms (Fig. 32-11). The adult worm is seldom found in the feces without antihelminthic treatment. Hookworms parasitize an estimated 456 million persons.

Strongyloides stercoralis

The adult *Strongyloides stercoralis* is only about 2 mm long. It has a four-lipped mouth and an esophagus that extends through the anterior one-fourth of its body. Male worms have not been found in man. The adult females live deep in the intestinal mucosa, where the ova are deposited. The larvae (Fig. 32-1, *K*) hatch in the intestines and are passed in the stool (Fig. 32-12). Neither adult worms nor ova appear in the feces unless active purgation is present. The larvae are 250 to 500 μm in length.

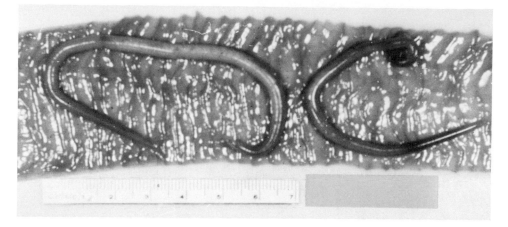

FIG. 32-13. *Ascaris lumbricoides* in lumen of intestine.

They are actively motile, and, in fresh specimens, are constantly wiggling and bending but have little progressive motion. The disturbance produced in the laboratory preparation is often noticed under the microscope before the larvae are seen. Infection is acquired when the larvae penetrate the skin or are accidentally swallowed.

Ascaris lumbricoides (eelworm or roundworm)

Ascaris lumbricoides, the largest intestinal nematode, is harbored by approximately 644 million persons. It is fusiform in shape and yellow or reddish in color. The male measures 15 to 20 cm in length, the female 20 to 40 cm. The head is relatively small, and the oral cavity has three serrated lips. This worm looks like the ordinary earthworm but is not so red (Fig. 32-13).

The habitat of *Ascaris lumbricoides* is the upper end of the small intestine, but it may be found in any part of the intestinal tract, free in the peritoneal cavity, or in the trachea and bronchi. Several are usually present (as many as 100 can be), clinging together to form palpable masses or even cause intestinal obstruction. Infection is mostly in children under 10 years of age.

The fertilized eggs (Fig. 32-1, *H* and *I*) are oval in shape and average 48 μm in diameter and 62 μm in length. If only female worms are present, unfertilized eggs (Fig. 32-1, *G*) will be found. They are elongated, irregular in shape, and bear little resemblance to the fertilized egg. They frequently escape detection.

Adult worms appear in the feces only with active purgation. After the eggs are passed, segmentation takes place, and an embryo develops. If these mature eggs are swallowed, the embryos escape from the eggs and travel to the lungs by the bloodstream. They reach the intestines in the same manner as do hookworm larvae (Fig. 32-14).

Enterobius vermicularis (Oxyuris vermicularis, threadworm, pinworm, or seatworm)

Infection with *Enterobius vermicularis* (enterobiasis) is the most prevalent worm infection of children and adults in the United States. Over the world some 200 million

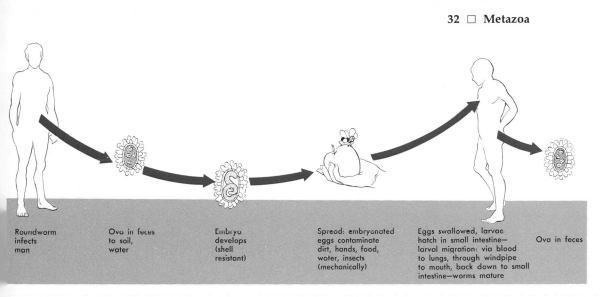

FIG. 32-14. *Ascaris lumbricoides,* life cycle.

| Roundworm infects man | Ova in feces to soil, water | Embryo develops (shell resistant) | Spread: embryonated eggs contaminate dirt, hands, food, water, insects (mechanically) | Eggs swallowed, larvae hatch in small intestine— larval migration: via blood to lungs, through windpipe to mouth, back down to small intestine—worms mature | Ova in feces |

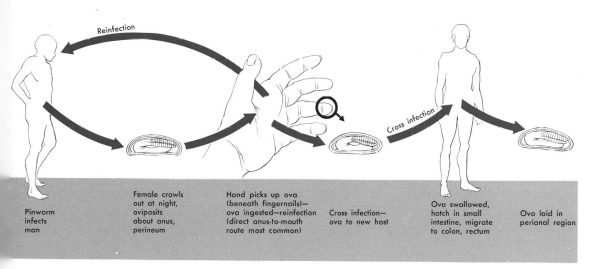

FIG. 32-15. *Enterobius vermicularis,* life cycle.

| Pinworm infects man | Female crawls out at night, oviposits about anus, perineum | Hand picks up ova (beneath fingernails)— ova ingested—reinfection (direct anus-to-mouth route most common) | Cross infection— ova to new host | Ova swallowed, hatch in small intestine, migrate to colon, rectum | Ova laid in perianal region |

persons are affected. The males of this species measure 3 to 5 mm in length, the females about 10 mm. The adult female worms migrate through the anus and deposit their eggs on the perianal region, most frequently at night. Because of the peculiar laying habits of the female, the eggs (Fig. 33-1, *C*) seldom occur in the feces but are present around the anal region. They are best found by scraping this region and examining the scrapings. Eggs may be picked up from the skin of the perianal region by means of swabs made of cellulose adhesive tape, the sticky side applied to the skin. The eggs are removed from the tape by toluene and identified under the microscope. An enema may be given and the adult worms identified in the stool that is passed.

Deposition of eggs on the perianal skin results in intense itching. In children, pinworms should be suspected from this finding alone. The small child may maintain the infection from ova collected under his fingernails when he scratches himself (Fig. 32-15).

Infection comes from swallowing the eggs, after which male and female parasites hatch out at the lower end of the small intestine. After fertilizing the females, the males die, and the females migrate to the colon and rectum. The patient may continually reinfect himself, and parents may acquire the infection from their children. Eggs may be widely disseminated in a household or in an institution, in the dust, in the clothing, bedding, on furniture, doorknobs, and the like. To eliminate them is an exasperating and almost hopeless task.

The United States Public Health Service recommends that families with pinworms pay careful attention not only to general household cleanliness but also to personal hygiene among the members. It stresses the following health measures in the control of the parasites: (1) regular daily morning showers, (2) frequent handwashing, especially before food is eaten or prepared, (3) keeping of fingernails short and clean, (4) vacuuming of household surfaces and floors thoroughly and often, and (5) washing of bed linens two or three times a week in a machine in which the temperature exposure is at least 150° F for several minutes or more.

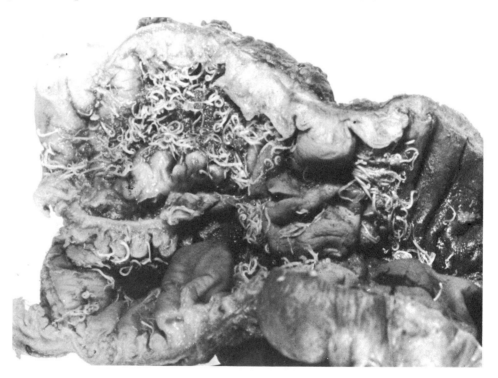

FIG. 32-16. *Trichuris trichiura,* massive infestation in young child producing severe hemorrhagic diarrhea and death. Segment of bowel is opened to show the many whipworms on mucosal surface. (From Anderson, W. A. D., and Kissane, J. M.: Pathology, ed. 7, St. Louis, 1977, The C. V. Mosby Co.)

Trichocephalus trichiurus (Trichuris trichiura, Trichocephalus dispar, whipworm)

Trichocephalus trichiura is characterized by a long threadlike neck that makes up about one half the length of the body. The male is 30 to 45 mm in length; the female is somewhat longer, 45 to 50 mm. The worms live in the cecum and large intestine, with the slender end of the worm embedded in the mucosa. The worms themselves are rare in the feces, and the eggs (Fig. 32-1, *B*) are not abundant. Generally, symptoms are related to the number of worms in the bowel (Fig. 32-16).

Trichinella spiralis

Trichinella spiralis is the cause of trichinosis. It is the smallest worm, with the exception of *Strongyloides stercoralis*, found in the intestinal canal and barely visible with the unaided eye. The males are about 1.5 mm in length, the females 3 to 4 mm. The posterior end of the male is bifid and has two tonguelike appendages.

The infection is primarily one of rats, propagated because rats eat their dead. Hogs acquire the infection from rats, and man becomes infected by eating insufficiently cooked pork. Pork is not the only source. Man has acquired the infection from eating bear meat. Polar bears are said to be heavily infected. The fact that it is found in the arctic region indicates that the parasite possesses unique tolerance for cold.

Life history. The cycle of development of *Trichinella spiralis* (Fig. 32-17) is practically the same in man as in other animals. Pork containing the encysted larvae is eaten and the cyst capsule digested away. The larvae pass to the small intestine, where they mature. After copulation the males die, and the females embed in the mucous membrane, where they give birth to as many as 1000 to 1500 larvae. These larvae migrate by the lymph and bloodstream to the skeletal muscles to encyst, become encapsulated, and subsequently calcify (Fig. 32-18).

The free larvae measure 90 to 100 μm in length and 6 μm in diameter. They may be found in the blood and spinal fluid during the period of migration (6 to 22 days after infection).

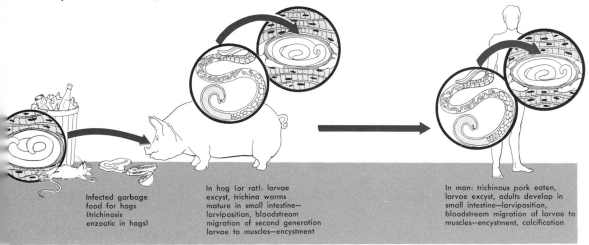

Infected garbage food for hogs (trichinosis enzootic in hogs)

In hog (or rat): larvae excyst, trichina worms mature in small intestine—larviposition, bloodstream migration of second generation larvae to muscles—encystment

In man: trichinous pork eaten, larvae excyst, adults develop in small intestine—larviposition, bloodstream migration of larvae to muscles—encystment, calcification

FIG. 32-17. *Trichinella spiralis,* life cycle.

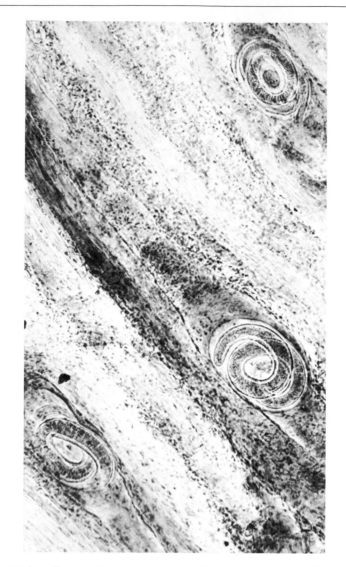

FIG. 32-18. *Trichinella spiralis* encysted in muscle, microscopic section. Larvae may live 10 to 20 years in these cysts. (From Hickman, C. P., and others: Integrated principles of zoology, ed. 5, St. Louis, 1974, The C. V. Mosby Co.)

After encystment occurs, the coiled embryos may be found with the low power of the microscope in a teased portion of muscle and the diagnosis made. The cysts are most frequently found at the tendinous insertions of the muscle, and the muscles most frequently infected are the pectoralis major, the outer head of the gastrocnemius, the deltoid, and the lower portion of the biceps. The cysts appear as white specks, measuring 250 by 400 μm. The long axis of the cyst extends in the same general direction as the fibers of the muscle. Muscle biopsy is the surest method in diagnosis.

Trichinosis. When the parasites are developing in the intestines, gastrointestinal disturbances are prominent. These appear 2 to 3 days after ingestion of the contaminated pork. During this time the adult worms may be found in the feces. When the larvae migrate, fever, delirium, rheumatic pains, and labored respiration are present. This period begins at the end of 1 week after infection and lasts 1 or 2 weeks. When encystment begins, edema and skin eruptions appear. This period lasts about 1 week. After the disease becomes chronic, muscular pains of a rheumatic character may be present for months.

Laboratory diagnosis. Intense eosinophilia, commonly over 500 eosinophilic leukocytes per cubic millimeter of blood, is a feature of all stages of the disease. A skin test is available. A flocculation test, the Sussenguth-Kline test, becomes positive 2 to 3 weeks after infection and remains so for 10 months or longer. There are other serologic tests including complement fixation, latex agglutination, hemagglutination, and fluorescent antibody technics.

Identification of the worms in feces or larvae in other clinical specimens is usually impractical because of the course of the disease, and biopsy of a tender muscle poses difficulties. Generally serologic tests are helpful, since they indicate recent or current infection. The skin test is consistent with more remote infection.

Prevention. There is no simple inspection method at a slaughterhouse for the detection of trichinas in a carcass of meat. The elimination of the disease rests practically with *adequate cooking of pork*. For example, one should cook a pork roast at an oven temperature of at least 350° F, allowing 35 to 50 minutes per pound. Smoking, pickling, heavy seasoning, or spicing, does not make uncooked pork products safe. Freezing meat at −15° C for 30 days or at −28.9° C for 6 to 12 days eliminates the larvae.

Wuchereria bancrofti (Bancroft's filaria worm)

Filariasis is infection with the filarial worms, which incorporate certain unique features in their life cycle. The best known is *Wuchereria bancrofti.* Man is its reservoir of infection and the mosquito its vector. Of historical note is the fact that the first demonstration of the mosquito as vector of disease was in connection with Bancroft's filaria worm, and the work suggested to Sir Ronald Ross the possibility that malaria might be similarly transmitted.*

Life cycle. The life cycle of *Wuchereria bancrofti* is sketched as follows (Fig. 32-19). When man, the definitive host, is infected, slender white adult male and female worms reside in the lymphatic system. The females are 80 to 100 mm in length and 0.24 to 0.3 mm in diameter; the males are 40 mm long and 0.1 mm in diameter. Within the uterus of the adult female, embryos develop as tightly coiled threads within eggshells. About the time an egg is laid, the embryo uncoils into a tiny, delicate, eel-like form. The eggshell remains applied about the elongated embryo and it is said to be sheathed. In some species of filarial worms, a naked or unsheathed embryo

---

*In 1878, Sir Patrick Manson, in Amoy, China, showed *Culex* mosquitoes to be the natural transmitters of filarial worms. In Great Britain, Manson is known as the father of tropical medicine.

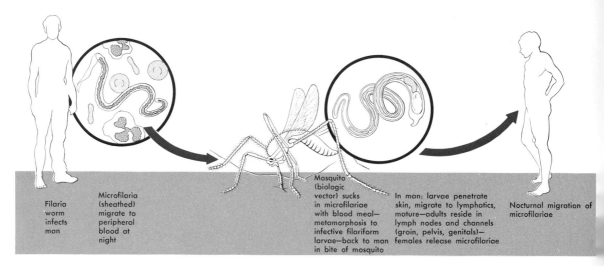

FIG. 32-19. *Wuchereria bancrofti,* life cycle.

is discharged. The embryos in the lymph and bloodstream become *microfilariae* (Fig. 32-20) with remarkable habits. One is their migration into the peripheral blood at night at a time that coincides with the feeding time of the vector mosquito. During the day, they are hidden away in an undetermined site.

Taken into the body of the mosquito, the microfilariae lose their sheath and develop in a larval series. In time, the infective larvae escape from the mosquito as it takes a blood meal, returning to the definitive host in whose lymphatic channels they continue their growth. Adolescent worms gather within sinuses of lymph nodes of the groin and pelvis, where they mature and mate to repeat the cycle.

Disease. In localizing within the human lymphatic system the adult worms start an inflammatory process that progresses from acute to chronic stages. It is associated with tissue changes causing obstruction of lymphatic vessels, stagnation of lymph flow, and proliferation of connective tissue. The skin of the lower extremities and external genitalia becomes thickened, coarse, and redundant. The tremendous enlargement of the affected part becomes a great burden to the victim. The results of long years of infection, the end stage lesion, is referred to as *elephantiasis.* This disorder was well known to the ancient Hindu physicians around 600 BC.

Filariasis is a widely distributed disease in tropical areas of the world and extends into some of the subtropical areas. Man is the only known definitive host. The appropriate mosquitoes breeding close by the dwelling places of human beings with microfilariae circulating in their bloodstream are two factors that favor the continuous spread of this disease. Therefore methods of prevention and control must be designed for treatment of the human carrier and elimination of the insect vector.

Laboratory diagnosis. The diagnosis of filariasis lies in the identification of the microfilariae found in blood films. In a wet mount of blood, they are seen to move gracefully, pushing the red blood cells gently aside. They may be seen also in a dried,

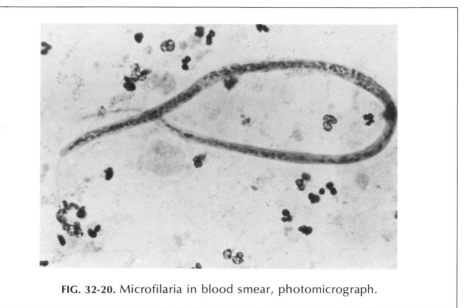

FIG. 32-20. Microfilaria in blood smear, photomicrograph.

fixed blood smear stained with Wright's stain. For best results with *Wuchereria bancrofti*, peripheral blood films are made between the hours of 10 PM and 2 AM.

Onchocerca volvulus

Onchocerciasis or river blindness is a disease afflicting an estimated 300 million persons in Africa and Latin America. It is produced by the filarial worm, *Onchocerca volvulus*. A hardy black gnat of the genus *Simulium*, which breeds in fast-flowing, aerated mountain streams, transmits the infection as intermediate host. In the bite of the fly, larvae of the threadlike worm are deposited beneath the skin, the site of residence of the adult worms. The female worm gives birth to unsheathed microfilariae, which pass into the lymphatic system. They are rarely found in peripheral blood. The microfilariae migrate throughout the body via the lymphatics. They die on reaching the eyes, but not until they have induced serious changes in the tissues of the eye that can in time lead to complete blindness. They can also produce subcutaneous nodules on the head or trunk, wherein a group of adult worms is walled off by fibrous connective tissue.

MACROSCOPIC EXAMINATION OF FECES FOR PARASITES

Sometimes the nurse or medical attendant must examine feces macroscopically for parasites. In such a case, the procedure is as follows.

If need be, improvise a sieve by removing both ends of a can and tying one or two layers of gauze over one opening. Allow water to run slowly over the feces and break up the masses with a glass rod or wooden applicator. Note the presence of worms. If necessary, remove worms and examine with a hand lens or the eyepiece of the microscope. Small worms may be mounted in a drop of water for macroscopic examination. If a tapeworm is present, be sure to search carefully for the head. After

617

tapeworm treatment, the entire quantity of feces passed should be saved in order that the search for the head may be thorough. If it is suspected that the patient is suffering from typhoid fever or any other condition in which pathogenic bacteria appear in the feces, the washings should be received in a vessel containing disinfectant. Needless to state, the sieve, glassware, and utensils used should be properly sterilized or carefully discarded.

Objects likely to be mistaken for intestinal worms. Segmented strands of mucus may be mistaken for tapeworms. The same is true of banana fibers because of their segmented structure and oval cells. Fibers of celery and green vegetables may be mistaken for roundworms and orange fibers for pinworms.

QUESTIONS FOR REVIEW

1. Define microfilariae, hydatid sand, mechanical vector, biologic vector, ova, scolex, proglottid, ground itch, definitive host, cysticercus, hermaphrodite, operculate, phototrophic.
2. List five arthropod vectors in disease transmission. Give the diseases so spread.
3. Briefly describe the three phyla of metazoa concerned in disease production in man.
4. Outline the life cycle of the following:
   a. Blood flukes
   b. Beef tapeworm
   c. Pork tapeworm
   d. Fish tapeworm
   e. Hookworm
   f. Pinworm
   g. Roundworm
   h. Filarial worms
5. Discuss the pathogenicity of flukes.
6. What is elephantiasis? River blindness?
7. Give the laboratory diagnosis for
   a. Schistosomiasis
   b. Trichinosis
   c. Filariasis
   d. Onchocerciasis
   e. Enterobiasis
8. List objects mistaken for worms in a *macro*scopic stool examination.
9. Comment on the control and prevention of metazoan diseases.

REFERENCES FOR CHAPTERS 31 AND 32

Anderson, S. E., and Remington, J. S.: The diagnosis of toxoplasmosis, South. Med. J. **68**:1433, 1975.

Barrett-Connor, E.: Amebiasis, today, in the United States, Calif. Med. **114**:1, Mar., 1971.

Barrett-Connor, E.: Human fluke infections, South. Med. J. **65**:86, 1972.

Beck, J. W., and Barrett-Connor, E.: Medical parasitology, St. Louis, 1971, The C. V. Mosby Co.

Brown, H. W.: Basic clinical parasitology, New York, 1975, Appleton-Century-Crofts.

Carter, J. P.: Nutrition and parasitism, South. Med. Bull. **59**:31, Oct., 1971.

Chandler, F. W., and others: Pulmonary pneumocystosis in nonhuman primates, Arch. Pathol. Lab Med. **100**:163, 1976.

Chitwood, B. G., and Chitwood, M. B.: Introduction to nematology, Baltimore, 1975, University Park Press.

Danilevicius, Z.: A call to recognize PCP (editorial), J.A.M.A. **231**:1168, 1975.

DeFord, J. W.: Amebiasis: newer methods of diagnosis and treatment, South. Med. J. **66**:1149, 1973.

Faust, E. C., and others: Craig and Faust's clinical parasitology, Philadelphia, 1970, Lea & Febiger.

Faust, E. C., and others: Animal agents and vectors of human diseases, ed. 4, Philadelphia, 1975, Lea & Febiger.

Fletcher, J. R., and others: Acute *Plasmodium falciparum* malaria, Arch. Intern. Med. **129**:617, 1972.

Garcia, L. S., and Ash, L. R.: Diagnostic parasitology: clinical laboratory manual, St. Louis, 1975, The C. V. Mosby Company.

Gould, S. E.: Trichinosis in man and animals, Springfield, Ill., 1970, Charles C Thomas, Publisher.

Gould, S. E.: The story of trichinosis, Am. J. Clin. Pathol. **55**:2, 1971.

Grell, K. G.: Protozoology, New York, 1973, Springer-Verlag New York, Inc.

Hammond, D. M., and Long, P. L., editors: The Coccidia. *Eimeria, Isospora, Toxoplasma,* and related genera, Baltimore, 1973, University Park Press.

Healy, G. R.: Laboratory diagnosis of amebiasis, Bull. N.Y. Acad. Med. **47**:478, 1971.

Heineman, H. S.: The clinical syndrome of malaria in the United States: a current review of diagnosis and treatment for American physicians, Arch. Intern. Med. **129**:607, 1972.

Hoare, C. A.: The trypanosomes of mammals, Philadelphia, 1972, F. A. Davis Co.

James, M. T., and Harwood, R. F.: Herms' medical entomology, New York, 1969, The Macmillan Co.

Jokipii, L., and Jokipii, A. M. M.: Giardiasis in travelers: a prospective study, J. Infect. Dis. **130**:295, 1974.

Juniper, K., Jr., and others: Serologic diagnosis of amebiasis, Am. J. Trop. Med. **21**:157, 1972.

Kagan, I. G.: Current status of serologic testing for parasitic diseases, Hosp. Pract. **9**:157, Sept., 1974.

Kagen, L. J., and others: Serologic evidence of toxoplasmosis among patients with polymyositis, Am. J. Med. **56**:186, 1974.

Krogstad, D. J., and others: Toxoplasmosis with comments on risk of infection from cats, Ann. Intern. Med. **77**:773, 1972.

Lau, T., and Pierson, K. K.: Dog heartworm infection in man, Lab. Med. **3**:41, Aug., 1972.

Marcial-Rojas, R. A.: Pathology of protozoal and helminthic disease, Baltimore, 1971, The Williams & Wilkins Co.

Meyer, M. C., and Olsen, O. W.: Essentials of parasitology, Dubuque, Iowa, 1975, William C. Brown Co., Publishers.

Miller, J. H., and Abadie, S. H.: Common intestinal parasites of the United States, South. Med. Bull. **59**:11, Oct., 1971.

Moser, R. H.: Trichinosis: from Bismarck to polar bears (editorial), J.A.M.A. **228**:735, 1974.

Ormerod, W. E.: Ecological effect of control of African trypanosomiasis, Science **191**:815, 1976.

Peterson, D. R., and others: Human toxoplasmosis prevalence and exposure to cats, Am. J. Epidemiol. **96**:215, 1972.

Quinn, R. W.: The epidemiology of intestinal parasites of importance in the United States, South. Med. Bull. **59**:20, Oct., 1971.

Rein, M. F., and Chapel, T. A.: Trichomoniasis, candidiasis, and minor venereal diseases, Clin. Obstet. Gynecol. **18**:73, 1975.

Rosen, P. P., and others: *Pneumocystis carinii* pneumonia, Am. J. Med. **58**:794, 1975.

Schultz, M. G.: The surveillance of parasitic diseases in the United States, Am. J. Trop. Med. Hyg. **23**(suppl. 4):744, 1974.

Schultz, M. G.: Giardiasis (editorial), J.A.M.A. **233**:1383, 1975.

Seah, S. K. K., and Flegel, K. M.: African trypanosomiasis in Canada, Can. Med. Assoc. J. **106**:902, 1972.

Smith, H. A., and others: Veterinary pathology, Philadelphia, 1972, Lea & Febiger.

Spencer, H., and others: Tropical pathology, New York, 1973, Springer-Verlag New York, Inc.

The facts about toxoplasmosis, Today's Health **50**:64, July, 1972.

Thompson, J. H., Jr.: Useful laboratory procedures. Examination of blood smears for malarial and other organisms, III, Bull. Pathol. **10**:61, 1969.

Thompson, J. H., Jr.: How to detect intestinal parasites, Lab. Med. **1**:31, Apr., 1970.

Thomspon, J. H., Jr.: How to detect blood and tissue parasites, Lab. Med. **2**:42, Apr., 1971.

Trager, W.: Some aspects of intracellular parasitism, Science **183**:269, 1974.

Walzer, P. D., and others: *Pneumocystis carinii* pneumonia in United States, Ann. Intern. Med. **80**:83, 1974.

Warren, K. S.: Regulation of the prevalence and intensity of schistosomiasis in man: immunology or ecology? J. Infect. Dis. **127**:595, 1973.

Wilcocks, C., and Manson-Bahr, P. E. C., editors: Manson's tropical diseases, Baltimore, 1972, The Williams & Wilkins Co.

Wolfe, M. S.: Giardiasis, J.A.M.A. **233**:1362, 1975.

LABORATORY SURVEY OF UNIT FIVE*

PROJECT
Pyogenic cocci
Part A—Study of morphology of pyogenic cocci

1. Examine microscopically prepared gram-stained smears of the following:
   a. *Staphylococcus aureus*
   b. *Streptococcus* species
   c. *Streptococcus pneumoniae*
   d. *Neisseria gonorrhoeae*
2. Examine microscopically the capsule stain of *Streptococcus pneumoniae*. (A suitable specimen of sputum may be used if available.)
3. Note details of morphologic arrangement, gram-staining reactions. Make comparisons.

Part B—Study of cultural characteristics

1. Make cultures of a hemolytic *Staphylococcus aureus* on the following:
   a. Nutrient agar—note pigment.
   b. Blood agar—note hemolysis.
   c. Mannitol salt agar—note growth on medium with a high concentration of salt and fermentation of mannitol.
2. Make culture of *Streptococcus pneumoniae* on blood agar. Note alpha hemolysis and characteristic colonies.
3. Examine a culture of *Neisseria gonorrhoeae* on chocolate agar.
4. Examine a blood agar plate showing colonies of *Streptococcus pyogenes*. Note beta hemolysis.

Part C—Identification of colonies of cocci

1. Make a culture of a nasal swab from the nose.
   a. Inoculate a blood agar plate and mannitol salt agar. Incubate.
   b. Note characteristics of colonies. Use gram-stained bacterial smears to study morphology of bacteria in colonies present.
      Note: The differential characteristics of the organisms growing on blood agar are discussed by the instructor.
2. Make a culture of a throat swab rubbed over your tonsils (if present) and the back part of your throat.
   a. Inoculate a blood agar plate.
   b. Note colonies. Make gram-stained smears.

---

*Since it would be impossible to investigate thoroughly all of the organisms in this unit, we will present only a few preliminary tests indicating the nature of some of the microbes and the diagnosis of their infections.

Part D—Typing of pneumococci with demonstration by instructor and discussion

1. Discuss methods of typing pneumococci *(Streptococcus pneumoniae)*.
2. Compare pneumococci from cultures and from body fluids as sputum or peritoneal fluid of mouse. What important structure is not so well developed in pneumococci from cultures as in organisms from the body?
3. Demonstrate method of taking cultures to detect meningococcal carriers.
4. Distribute prepared smears.
   a. Pus containing gonococci stained with methylene blue and by Gram's method
   b. Purulent cerebrospinal fluid containing meningococci stained with methylene blue and by Gram's method
   Note the similarity of meningococci and gonococci.

PROJECT
Enteric bacilli
Part A—Study of morphology of gram-negative bacilli

1. Examine microscopically gram-stained smears of the following:
   a. *Escherichia coli*
   b. *Salmonella typhi* (or other *Salmonella* species)
   c. *Shigella flexneri*
   d. *Klebsiella pneumoniae*
   e. *Proteus vulgaris*
   f. *Vibrio cholerae**
2. Note detail and gram-staining reaction.

Part B—Motility in hanging drop preparation with demonstration by instructor and discussion

1. Emphasize technic of making hanging drop and precautions in handling a preparation of living infectious organisms.
2. Demonstrate motility of *Proteus vulgaris* or one of the salmonellae.

Part C—Study of cultural characteristics

1. Make cultures of *Escherichia coli* on the following:
   a. Blood agar
   b. MacConkey agar
   c. *Salmonella-Shigella* agar
   d. Endo medium
   e. Eosin–methylene blue agar
   f. Russell double sugar agar
   g. Bismuth sulfite agar
   h. Lead acetate agar

---

*Not one of the enteric bacilli, this vibrio causes an important disease of the intestine.

2. Select a pathogen such as one of the salmonellae and make corresponding cultures on the media just mentioned for culture of *Escherichia coli.*
3. Compare the growth obtained on these media for the pathogen with that of the nonpathogen.
   Note: The instructor discusses the varied appearance of the colonies obtained and the reasons. The student identifies the lactose and nonlactose fermenters. The importance of this biochemical test is stressed.
4. Make culture of *Pseudomonas aeruginosa** on blood agar. Note pigment produced. Examine smears of *Pseudomonas aeruginosa.*

Part D—Demonstration of a serologic reaction by instructor with discussion

1. The macroscopic slide agglutination test using growth of *Salmonella typhi* (or other *Salmonella*) and *Salmonella* polyvalent diagnostic serum
2. Discussion of bacterial agglutination in type-specific serum
3. Review of the significance of the agglutination test as a serologic reaction and diagnostic role of the Widal test in typhoid fever

PROJECT
Small gram-negative rods
Part A—*Brucella* and *Francisella*

1. Demonstration by the instructor
   a. Cultures of *Brucella*
   Note: Because of danger of infection, cultures of *Francisella* should *not* be handled.
   b. Prepared stained smears of small, gram-negative rods
   c. Slide agglutination test for undulant fever
   d. Preserved liver of a guinea pig that has died of tularemia if available
2. Discussion by instructor: diagnostic tests for tularemia and brucellosis

Part B—The hemophilic bacteria

Demonstration by instructor with brief discussion of the cough plate method of detecting the presence of *Bordetella pertussis*

PROJECT
Gram-positive bacilli
Part A—Study of morphology of the gram-positive bacilli—both aerobic and anaerobic

1. Examine microscopically prepared smears.
   a. *Corynebacterium diphtheriae*—in smears stained with:
   (1) Methylene blue
   (2) Albert stain

---

*Although not discussed as an enteric bacillus, it is found in the intestinal tract.

b. *Clostridium tetani*—in gram-stained smears

c. *Bacillus anthracis*—in gram-stained smears

2. Note morphology of the bacteria, gram-staining reaction, arrangements, presence and position of spores. Make comparisons.

3. Compare prepared slides of the diphtheria bacillus with those of diphtheroid bacilli. What is the value of differential stains?

Part B—Study of cultural characteristics

1. Study prepared cultures of *Corynebacterium diphtheriae*.

   a. Examine a 24-hour culture on Löffler serum medium.

   b. Examine a culture on blood-tellurite agar.

2. Study prepared cultures of diphtheroid bacilli.

3. Note morphology of colonies, especially on the differential media.

4. Compare organisms grown on one medium when examined in stained smears with those grown on the other.

5. Compare the growth of the diphtheria bacilli with that of the diphtheroid bacilli.

Part C—Introduction to anaerobic culture methods

1. Demonstration by instructor with discussion

   a. Selected technics for culture of strict anaerobes

      (1) Inoculation of thioglycollate broth

      (2) Manipulation of stab or solid tube cultures to contain growth within depths of tube

      (3) Use of layering technic to cover surface of medium with airtight substance—petroleum jelly or paraffin

      (4) Use of Brewer anaerobic jar

      (5) Use of GasPak anaerobic jar system

   b. Distinctions between anaerobes and microaerophiles

   c. Technics for culture of capnophiles—use of candle jar

2. Student participation

   a. Study of anaerobic cultures prepared by some of the technics demonstrated

   b. Comparison of simple technics with more elaborate ones

   c. Evaluation of results obtained with various technics

Part D—Study of a biochemical reaction

1. Demonstrate the stormy fermentation of milk.

   a. Boil a tube of skim milk for 10 minutes to drive off oxygen.

   b. Use sterile technic to inoculate heavily the tube of milk from an anaerobic culture of *Clostridium perfringens*. Allow the milk to cool to around 50° C before the inoculation is made.

   c. Obtain a tube of sterile mineral oil.

    d. Pour the oil over the surface of the milk to form a surface layer about ½ inch in width.

    e. Incubate the tube of milk at 37° C.

  2. Observe nature of bacterial growth in the milk under anaerobic conditions. What method for anaerobiosis was used?

Part E—Demonstration of animal inoculation by instructor with discussion

  1. Determination of the toxigenicity of a strain of suspected *Corynebacterium diphtheriae* in the guinea pig by the intracutaneous virulence test

  2. Emphasis on applications of animal virulence tests

  3. Comparison of a positive and a negative test result

Part F—Demonstration of the Schick test by instructor with discussion

Part G—Demonstration of the technic of throat culture by instructor with brief discussion

  1. Emphasis on proper technic, precautions, and reasons for taking throat cultures

  2. Use of differential stains for throat smears
    Students may stain suitable smears with:
    a. Albert stain
    b. Löffler alkaline methylene blue stain
    c. Gram I stain with Gram II stain as a mordant

  3. Discussion of differential features noted microscopically

---

PROJECT
Acid-fast bacteria
Part A—Study of morphology of acid-fast bacteria

  1. Use a specimen of tuberculous sputum that has been autoclaved. Why is this necessary?

  2. Prepare smears of the sputum and make acid-fast stains. Make a drawing in color (red and blue) of the organisms present.

  3. Make gram-stained smears of sputum. Compare the two stained preparations.

  4. Examine microscopically prepared smears of:
    a. *Mycobacterium tuberculosis*
    b. *Mycobacterium leprae*
    c. *Mycobacterium kansasii* (atypical mycobacterium)
    d. *Mycobacterium smegmatis* (atypical mycobacterium)
    e. *Nocardia asteroides* (actinomycete)

Part B—Study of cultural characteristics

  1. Examine prepared cultures of *Mycobacterium tuberculosis* on Lowenstein-Jensen medium.

2. Examine prepared cultures of the following atypical mycobacteria. Note salient features.
   a. *Mycobacterium kansasii* (group I)
   b. *Mycobacterium scrofulaceum* (group II)
   c. *Mycobacterium intracellulare* (or *xenopi*) (group III)
   d. *Mycobacterium smegmatis* (group IV)
3. Compare the cultural characteristics of atypical mycobacteria with those of the mycobacteria of tuberculosis.
4. Why do we not examine cultures of *Mycobacterium leprae?* Can this microbe be cultured?

Part C—Animal inoculation in the laboratory diagnosis of tuberculosis with demonstration by instructor and discussion

1. Study of organs of a guinea pig that has died from tuberculosis
   The instructor can carefully perform an autopsy on a tuberculous guinea pig. The effect of the bacilli on different organs should be demonstrated and discussed. How do the intestinal ulcers of tuberculosis differ from those of typhoid fever?
   Note: Organs must be fixed and noninfectious.
2. Demonstration of fixed tuberculous specimens from a human being, for example, a tuberculous lung with cavity formation
3. Presentation of microscopic tissue sections of tubercles
4. Demonstration of technic of inoculating a guinea pig to detect the presence of *Mycobacterium tuberculosis*

Part D—Skin testing in tuberculosis with demonstration by instructor and discussion

1. Technic of doing a Mantoux test
2. Technic of the tine (multiple-puncture) test
3. Interpretation of test results, where positive and where negative

---

PROJECT
Spirochetes
Part A—Study of morphology of spirochetes

1. Examine microscopically prepared smears of:
   a. *Borrelia recurrentis*—Wright stain of peripheral blood film
   b. *Treponema pallidum*—silver stain of infected tissue
   c. Fusospirochetal organisms of Vincent's angina—stained with Gram stains I and II
2. Compare.

Part B—Study of spirochetes from the mouth

1. Take scrapings from between the teeth with a toothpick.
2. Make a thin smear. Let dry and fix in flame.

3. Stain with carbolfuchsin for 1 minute.
4. Note morphology of organisms present.

PROJECT
Fungi
Part A—Study of unicellular fungi

1. Observe the morphology of yeasts in a wet mount.
   a. Rub up a small portion of yeast cake in about 5 to 10 ml of distilled water.
   b. Place a drop of the suspension on a microslide.
   c. Cover with a cover glass carefully so as to avoid bubbles.
   d. Examine with the low- and high-power objectives of the microscope.
2. Observe the morphology of yeasts in a wet mount admixed with Gram's iodine.
   Note: The starch granules from the yeast cake are stained a dark blue. By contrast the yeast cells appear yellowish.

Part B—Study of the morphology of multicellular fungi

1. Observe the morphology of the common bread mold *Rhizopus*.
   a. Obtain a piece of bread containing a black moldy growth.
      (1) Examine the natural growth macroscopically, that is, with the naked eye.
      (2) Examine the growth using a strong hand lens.
      Note: Observe the color, consistency, texture, and general appearance of the black mold growing on the bread.
   b. Observe the black mold microscopically.
      (1) Carefully tease off some of the mold and transfer to microslide. Place in a drop of water. Carefully position a cover glass.
      (2) Use the low- and high-power objectives of the microscope to examine the mycelial fragments of this mold. Reduce the opening of the iris diaphragm to enhance contrast in the specimen.
      Note: Observe the appearance of the hyphae and note whether they possess partitions. Observe the nuclei and the appearance of the fruiting bodies.
2. Observe the morphology of the common powdery green mold on citrus fruit. Proceed as above.
3. Make drawings. Note the presence of hyphae in test specimens of mold. Are they septate or nonseptate? Are nuclei present?

Part C—Study of morphology and cultural characteristics of *Candida albicans*

1. Observe the growth of *Candida albicans* on:
   a. Blood agar (Incubate at 37° C.)
   b. Sabouraud agar (Incubate one plate at 37° C and the other at room temperature −20° C.)

2. Prepare smears from colonies on blood and Sabouraud agar. Stain by Gram's method.
3. Prepare wet mounts of fungous growth. Examine microscopically.

Part D—Study of a biochemical reaction of yeast

1. Place about 20 ml of 2% lactose in one large test tube (1 by 8 inches).
2. Place about 20 ml of 2% dextrose in another.
3. Rub up a cake of yeast in a few milliliters of water and add one-half the substance to each of the above tubes.
4. Place either test tube at an angle of about 45 degrees in a vessel of water warmed to 42° C. What happens?

Note: This test is used to determine whether sugar found in the urine of a nursing mother is lactose or dextrose.

PROJECT
Protozoa
Part A—Study of morphology and biologic functions of protozoa in hay infusion

Note: Look for a protozoon of the genus *Paramecium*. Note its motility, feeding pattern, and expulsion of waste material from the body of the parasite. Observe cell detail.

1. Procure a few drops of hay infusion containing paramecia; place on a clean slide and add a few grains of sand to hold cover glass away from the organisms.
2. Gently apply cover glass.
3. Observe microscopically. Note movement of paramecia. Do they seem to have a purposeful or aimless movement?
   Note: In the microscopic study of the wet mount, use the low-power objective of the scope first. Then study the organisms under the high-power objective. Reduce the amount of light coming through the microscope by closing the iris diaphragm. What is the effect of this?
4. Sketch a quiescent organism.
5. Place a drop of India ink on a microslide and mix with a drop of hay infusion (containing *Paramecium*).
6. Examine microscopically. Note activities of protozoa in regard to particles of carbon in the ink. Note currents in oral groove. Note ingestion of particles, attempt at digestion (India ink is indigestible), and expulsion of particles.

Part B—Study of morphology of an intestinal ciliate

1. Prepare a wet mount of a specimen of fecal material from a guinea pig.
   a. Mix a small portion of the contents of the cecum of the animal with a drop of isotonic saline solution on a microslide.
   b. Superimpose a cover glass.
2. Examine microscopically using both low-power and high-power objectives.

3. Look especially for the large ciliate *(Balantidium)* that often inhabits the cecum.
4. Note the presence of various flagellates.
5. Observe morphology and movements of protozoa present.
6. Note comparative size of protozoa and bacteria. Make a drawing of each type of protozoon found.

Part C—Study of morphology of important pathogenic protozoa

1. Obtain prepared slides for study of the following:
   a. Malarial parasites in stained blood films
      (1) Thick films
      (2) Thin films
   b. Trypanosomes in stained blood film
   c. *Entamoeba histolytica* in iron hematoxylin–stained fecal smear
   d. *Giardia lamblia* in iron hematoxylin–stained fecal smear
   e. *Trichomonas vaginalis* in a Papanicolaou-stained smear from cervix of uterus
   f. *Toxoplasma gondii* in hematoxylin and eosin–stained tissue section
2. Examine the protozoa microscopically. Note morphologic features and location in the body where found. Make drawings.
3. Compare the vegetative and cyst forms of *Entamoeba histolytica* in stained fecal smears.
   Note: The vegetative forms of *Entamoeba histolytica* may be demonstrated by the instructor in a fresh specimen of feces, if such is available.

EVALUATION FOR UNIT FIVE

Part I

In the following exercises, indicate the one number on the right that completes the statement or answers the question.

1. Which of the following are gram-positive cocci occurring typically in pairs?
   - (a) Pneumococci
   - (b) Gonococci
   - (c) Streptococci
   - (d) Meningococci
   - (e) Staphylococci

   1. all of these
   2. a, c, and e
   3. a, b, and d
   4. a
   5. a and e

2. Which of the following diseases may be caused by *Streptococcus pyogenes?*
   - (a) Puerperal sepsis
   - (b) Osteomyelitis
   - (c) Scarlet fever
   - (d) Erysipelas
   - (e) Septic sore throat

   1. a and b
   2. a, b, and d
   3. c
   4. none of these
   5. all of these

3. *Coxiella burnetii* is different from the rickettsias causing typhus fever and Rocky Mountain spotted fever in that:
   - (a) It can produce a skin rash in the patient.
   - (b) It does not require an insect vector.
   - (c) It is found only in the United States.
   - (d) It is more resistant to heat.

   1. a
   2. b
   3. b and d
   4. none of these
   5. c

4. Which of the following statements apply to the enteric bacilli?
   - (a) They are all gram negative.
   - (b) They are all airborne infections.
   - (c) They are all strictly anaerobic.
   - (d) Their mode of invasion is through the alimentary tract.
   - (e) They all invade the bloodstream during the course of disease.

   1. b
   2. c
   3. d and e
   4. a and d
   5. all but c

5. Which one of the following facts about typhoid fever is used as the basis for a diagnostic laboratory test?
   - (a) Bacilli enter the lymphatics of the intestines.
   - (b) Symptoms are caused by the release of endotoxins when the organisms disintegrate.
   - (c) The bacilli often lodge in the bone marrow.
   - (d) The antibodies formed against the bacilli are agglutinins.
   - (e) Bacilli often lodge in "rose spots" on the skin.

   1. e
   2. d
   3. c
   4. b
   5. a

6. What characteristic do members of the genus *Haemophilus* have in common?
   - (a) They destroy red blood cells by dissolving them.
   - (b) They are found only in the red blood cells of an infected person.
   - (c) They cannot be grown in the laboratory except in living animals.
   - (d) They grow in the laboratory only in a medium that contains hemoglobin.
   - (e) They all cause diseases of the central nervous system only.

     1. a
     2. b
     3. c
     4. d
     5. d and e

7. Which of the following answer this description: anaerobic, sporeforming organisms that produce exotoxins?
   - (a) *Corynebacterium diphtheriae*
   - (b) *Clostridium tetani*
   - (c) *Clostridium perfringens*
   - (d) *Clostridium botulinum*
   - (e) *Streptococcus pyogenes*

     1. a
     2. all but e
     3. all but a
     4. b and d
     5. all but a and e

8. Which of the following statements concerning diphtheria are true?
   - (a) There is a method of testing a person's immunity to the disease.
   - (b) Most children from 1 to 6 years of age are susceptible to the disease.
   - (c) Diphtheria is an airborne disease.
   - (d) A patient is proved to have diphtheria if his throat smears show an acid-fast organism.
   - (e) A susceptible child exposed to the disease should be given diphtheria antitoxin as soon as possible.

     1. all but d
     2. all but b
     3. all but d and e
     4. all but e
     5. all of these

9. The following types of tuberculosis seldom, or never, occur in man:
   - (a) Human
   - (b) Bovine
   - (c) Avian
   - (d) Cold blooded

     1. c
     2. c and d
     3. a
     4. b
     5. b, c, and d

10. In testing sputum from a patient suspected of having tuberculosis, the bacteriologist can do the following for the laboratory identification of the organism:
    - (a) Culture the specimen in the absence of oxygen
    - (b) Use a special medium for the culture of the specimen
    - (c) Attempt to identify the causative organism in smears by means of the Gram stain
    - (d) (Inoculate a guinea pig with the suspected material
    - (e) Attempt to identify the causative organism in smears by means of the acid-fast stain

      1. a
      2. b
      3. c and d
      4. b and d
      5. b, d, and e

630

11. Which of the following diseases confer little, if any, immunity?
    (a) Pneumonia
    (b) Meningococcal meningitis
    (c) Gonorrhea
    (d) Syphilis
    (e) Osteomyelitis

    1. a, b, and d
    2. a, b, and e
    3. c and d
    4. a
    5. all but b

12. We are aware of the following facts concerning pathogenic rickettsias:
    (a) They grow only within living cells.
    (b) They may be grown within the embryonated hen's egg.
    (c) They are gram positive.
    (d) They are always found within the nucleus of the cell.
    (e) They are transmitted by insects.

    1. a
    2. a, b, and e
    3. c
    4. d
    5. none of these

13. We are aware of the following facts concerning viruses:
    (a) They cannot be seen with an ordinary light microscope.
    (b) They grow only within living cells.
    (c) They cannot be separated from media by filtration.
    (d) They cause only one important communicable disease—smallpox.
    (e) They confer no immunity with disease.

    1. a
    2. a and b
    3. a, b, and c
    4. c
    5. none of these

14. The rickettsial diseases are spread primarily by the following route:
    (a) Droplet infection
    (b) Contaminated water
    (c) Contaminated food
    (d) Insects and rodents
    (e) Direct contact

    1. a
    2. b
    3. c
    4. d
    5. e

15. The presence of an infection caused by *Treponema pallidum* can be demonstrated by three laboratory procedures as follows:
    (a) Gram's stain
    (b) Complement fixation test
    (c) Dark-field examination
    (d) Growth on special culture media
    (e) Precipitation tests

    1. b, c, and d
    2. b, c, and e
    3. c, d, and e
    4. a, d, and e
    5. a, b, and c

16. Mr. W. is diagnosed as having malaria. His clinical course is characterized by chills and fever that occur regularly once every 3 days. Which of the following statements apply to this patient?
    (a) His disease is the estivoautumnal type.
    (b) He is infected with *Plasmodium vivax.*
    (c) He is suffering from the tertian type of malaria.
    (d) His malaria is the quartan type.
    (e) He is infected with *Plasmodium malariae.*

    1. a and e
    2. a and b
    3. c and e
    4. b and d
    5. d and e

17. What sort of specimen should be collected from a patient if the doctor wants to ascertain the presence of *Entamoeba histolytica?*
    (a) Early morning sputum placed in refrig-   1. c
        erator until examined   2. b and c
    (b) Catheterized urine specimen kept at   3. d and c
        room temperature until examined   4. b
    (c) Stool specimen kept warm until ex-   5. e
        amined
    (d) Voided urine specimen incubated until examined
    (e) Stool specimen refrigerated until examined

18. In meningococcal meningitis the following are true:
    (a) Credé's technic is useful in prophylaxis.   1. a and b
    (b) An exotoxin is prominent.   2. a and e
    (c) The Waterhouse-Friderichsen syndrome   3. c and d
        may be a feature.   4. c, d, and e
    (d) The causative organism is found in the   5. none of these
        cerebrospinal fluid.
    (e) There are never skin lesions.

19. Rubella (German measles) is an important disease because
    (a) The virus is related to the togaviruses.   1. a
    (b) There is a mild skin rash.   2. c and d
    (c) The virus is teratogenic.   3. b, c, and d
    (d) The virus is oncogenic.   4. c and e
    (e) The virus persists.   5. e

20. Hospital strains of coagulase-positive staphylococci:
    (a) Are a threat to newborn infants   1. a
    (b) Are maintained by carriers among per-   2. a and b
        sonnel   3. b and c
    (c) Are distinguished by phage-typing pat-   4. a, b, and e
        terns   5. d and e
    (d) Never become resistant to antibiotics
    (e) Do not cause infections of surgical wounds

21. The substance elaborated by streptococci that initiates the dissolution of fibrin clots is:
    (a) Spreading factor   1. a
    (b) Leukocidin   2. b
    (c) Streptodornase   3. c
    (d) Streptolysin   4. d
    (e) Streptokinase   5. e

22. Which of the following is not typical for cholera:
    (a) Explosive outbreaks   1. a, b, c, and d
    (b) Severe dehydration   2. all of these
    (c) Transmitted by water   3. a and c
    (d) Caused by a vibrio   4. a, b, and c
    (e) Exo-endotoxin important   5. none of these

23. Pulmonary tuberculosis:
    (a) Invariably develops when tubercle bacilli   1. e
        enter the respiratory tract   2. d
    (b) Is usually caused by the bovine bacillus in   3. c
        the United States   4. b
    (c) Results in immediate-type allergy   5. a
    (d) Is more likely to develop in a tuberculin-
        positive than in a tuberculin-negative
        person
    (e) Involves the lymphatics in the primary
        complex
24. Which of the following is most likely to constitute a source of streptococcal upper
    respiratory tract infection?
    (a) A healthy throat carrier   1. a
    (b) A healthy nasal carrier   2. b
    (c) The skin of a patient with scarlet fever   3. c
        rash   4. d
    (d) Dust contaminated with streptococci   5. e
    (e) The hands of a throat carrier
25. *Mycobacterium leprae:*
    (a) Tends to appear in large numbers in nasal   1. a and b
        mucosa   2. c and d
    (b) Causes a highly contagious disease   3. b, c, and d
    (c) Infections exist in United States   4. a, c, and e
    (d) Grows well on media used for the tuber-   5. e
        cle bacillus
    (e) Causes leprosy
26. Urinary tract infection is rarely, if ever, caused by
    (a) *Proteus vulgaris*   1. a
    (b) *Escherichia coli*   2. b
    (c) *Pseudomonas aeruginosa*   3. c
    (d) *Mycobacterium tuberculosis*   4. d
    (e) *Mycobacterium smegmatis*   5. e
27. Compared to most pathogenic bacteria, most pathogenic fungi require:
    (a) A higher temperature for growth   1. a, b, and c
    (b) More complicated nutrients for growth   2. b
    (c) More time to grow   3. c and d
    (d) Less oxygen for growth   4. c
    (e) A higher moisture content for growth   5. d and e
28. A diagnosis of tuberculous meningitis is confirmed by:
    (a) X-ray examination of lung   1. e
    (b) Subcutaneous injection of tuberculin   2. d and e
    (c) Sputum examination   3. d
    (d) Blood culture   4. b
    (e) Culture of cerebrospinal fluid   5. a, b, c, and d
29. All of the following are notable features of *Mycobacterium tuberculosis except:*
    (a) Grows very slowly   1. a
    (b) Highly resistant to drying   2. b and c
    (c) Sporeformer   3. c
    (d) Strict aerobe   4. d and e
    (e) Rich in lipids   5. e

30. Which one of the following parasites most commonly causes blockage of the lumen of the intestinal tract?
    (a) *Taenia solium*   1. a
    (b) *Taenia saginata*   2. b
    (c) *Giardia lamblia*   3. c
    (d) *Ascaris lumbricoides* 4. d
    (e) *Necator americanus* 5. e
31. A blood sample is often useful in the diagnosis of all of the following *except:*
    (a) Malaria   1. a
    (b) Brucellosis   2. b and d
    (c) Meningococcal meningitis   3. c and d
    (d) Bacillary dysentery   4. d and e
    (e) Cholera   5. e
32. The diagnosis of hookworm disease is based mainly on:
    (a) An increase in antibody titer   1. a, b, and c
    (b) Identification of skin lesions 2. b, c, and d
    (c) Characteristic eggs in stool   3. c
    (d) Inclusion bodies in cells   4. c, d, and e
    (e) Complement fixation test   5. d
33. A parasite seen often in the urine, especially in men is:
    (a) *Trichomonas vaginalis*   1. a
    (b) *Enterobius vermicularis* 2. b
    (c) *Giardia lamblia*   3. c
    (d) *Chilomastix mesnili*   4. d
    (e) *Trichuris trichiura*   5. e
34. A parasite prevalent in the Arctic is:
    (a) *Loa loa*   1. b
    (b) *Fasciolopsis buski* 2. c
    (c) *Onchocerca volvulus* 3. a
    (d) *Trichinella spiralis* 4. e
    (e) *Clonorchis sinensis* 5. d
35. The majority of respiratory tract infections are caused by:
    (a) *Staphylococcus aureus* 1. a and b
    (b) *Staphylococcus albus*   2. c
    (c) Hemolytic streptococcus 3. d and e
    (d) Friedländer's bacillus   4. c, d, and e
    (e) A virus   5. e
36. Pneumonia in a meat packer may be caused by
    (a) Q fever   1. a
    (b) Tularemia   2. b
    (c) Brill's disease 3. c
    (d) Rickettsialpox 4. d
    (e) Psittacosis   5. e
37. In tropical Africa 50% to 70% of the malignant tumors occurring in children are represented by:
    (a) Burkitt's lymphoma 1. a
    (b) Acute leukemia   2. b
    (c) Kaposi sarcoma   3. c
    (d) Hodgkin's disease 4. d
    (e) Neuroblastoma   5. e
38. Echovirus is easily isolated from all of these except:

(a) Cerebrospinal fluid	1. a
(b) Nose	2. a and b
(c) Stools	3. c
(d) Throat	4. d and e
(e) Fingernail scrapings	5. e

39. A 20-year-old man has a hard, indurated, penile ulcer. He believes it is healing with applications of Vaseline and denies that it was ever painful. The most useful diagnostic procedure would be to take scrapings for:

(a) Gram stain	1. a and b
(b) Acid-fast stain	2. c
(c) Tissue culture	3. d
(d) Dark-field microscopy	4. d and e
(e) Fluorescent antibody study	5. e

40. Nosocomial infections:

(a) Must be reported to local health department	1. a, b, c, and d
	2. a, b, c, d, and e
(b) Are acquired in the hospital	3. b
(c) Usually are viral	4. c
(d) Are invariably airborne	5. b and e
(e) Are caused in great number by enteric bacilli	

41. The presence of "sulfur granules" in an exudate from a draining sinus is diagnostic for:

(a) Nocardiosis	1. a
(b) Actinomycosis	2. a and b
(c) Coccidioidomycosis	3. b
(d) Dermatophytosis	4. c and d
(e) Histoplasmosis	5. e

42. Disease in the central nervous system may result from

(a) Epidemic typhus	1. none of these
(b) Rabies immunization	2. all of these
(c) Mumps	3. a
(d) Cryptococcosis	4. a and b
(e) Torulosis	5. a, b, c, and d

43. The agent that causes ornithosis is classified as:

(a) Virus	1. a
(b) Protozoa	2. b
(c) Fungus	3. c
(d) Mycoplasma	4. d
(e) Rickettsia	5. none of these

44. Mycoplasmas differ from other bacteria in that:

(a) Their colonial morphology is different.	1. all of these
(b) They lack cell walls.	2. a, b, c, and d
(c) They do not grow on cell-free media.	3. a, b, and c
(d) They produce disease only in animals other than man.	4. a and b
	5. e
(e) Their colonies are usually quite large.	

45. Diagnostic procedures of value in the laboratory identification of the fungi include:

(a) Gram stain	1. all of these
(b) Slide cultures	2. all of these except a
(c) Phenomenon of fluorescence	3. none of these
(d) Culture on Sabouraud agar	4. a
(e) Fermentation reactions	5. d and e

46. Spontaneous mutation of the virus is now known to be a problem in the epidemiology and control of:
  (a) Influenza   1. all of these
  (b) Rabies   2. none of these
  (c) Rubeola   3. a, b, and c
  (d) Rubella   4. a
  (e) Yellow fever   5. d and e

47. The patient with leukemia or bone marrow failure is vulnerable to infection with which of the following fungi:
  (a) *Histoplasma*   1. a
  (b) *Coccidioides*   2. b
  (c) *Candida*   3. c
  (d) *Blastomyces*   4. d and e
  (e) *Actinomyces*   5. d

48. The following are important fungi and ordinarily benign but they may be serious opportunist pathogens:
  (a) *Candida*   1. all of these
  (b) *Aspergillus*   2. none of these
  (c) *Mucor*   3. all of these except b
  (d) *Rhizopus*   4. all of these except e
  (e) *Cryptococcus*   5. c

49. The three most important genera of dermatophytes are:
  (a) *Taenia*   1. a, b, and c
  (b) *Trichophyton*   2. b, c, and d
  (c) *Epidermophyton*   3. c, d, and e
  (d) *Microsporum*   4. b, d, and e
  (e) *Candida*   5. a, c, and d

50. The following observations have been made in regard to the Australia antigen:
  (a) It does not grow in tissue culture.   1. a, b, d, and e
  (b) It has neither DNA nor RNA backbone.   2. b, c, d, and e
  (c) It is of no consequence in blood banking.   3. a, b, and c
  (d) It is significantly related to viral hepatitis, 4. d and e
      type A.   5. none of these
  (e) It is prevalent in drug addicts.

Part II

Comparisons: Match the item in Column B to the phrase (or word) in Column A best describing it.

COLUMN A	COLUMN B
Staphylococcus aureus with Streptococcus pyogenes:
_____ 1. Can produce hyaluronidase	(a) *Staphylococcus aureus*
_____ 2. Phage typing in epidemiologic study	(b) *Streptococcus pyogenes*
_____ 3. Serologic typing in epidemiologic study	(c) Both
_____ 4. Can produce greenish discoloration on blood agar	(d) Neither
_____ 5. Healthy carrier state possible
_____ 6. Nonpathogenic in man
_____ 7. Septicemia a dangerous event
_____ 8. Produces localized abscesses
_____ 9. Produces cellulitis

Typhoid fever with bacillary dysentery:

_____ 10. Ulceration of bowel

_____ 11. Man only source of infection

_____ 12. Blood culture important

_____ 13. May be transmitted by water

_____ 14. Penicillin effective in infection

_____ 15. Organisms nonmotile

_____ 16. Organisms ferment lactose

(a) Typhoid fever
(b) Bacillary dysentery
(c) Both
(d) Neither

Tetanus with gas gangrene:

_____ 17 Antibiotic therapy highly effective

_____ 18. Polyvalent antiserum useful

_____ 19. Active immunity advised

_____ 20. Organisms found in feces of man and animals

_____ 21. Dirt of wound enhances growth of organisms

_____ 22. Organisms produce potent exotoxin

_____ 23. Disease result of wound infection

_____ 24. Blood culture positive early in disease

(a) Tetanus
(b) Gas gangrene
(c) Both
(d) Neither

Togaviruses with picornaviruses:

_____ 25. Transmitted by arthropods

_____ 26. Includes polioviruses

_____ 27. Includes viruses of equine encephalomyelitis

_____ 28. Several immunologically distinct viruses

_____ 29. Birds are important in cycle in nature

_____ 30. Includes rabies virus

_____ 31. Includes herpes virus

_____ 32. Man accidental host usually

_____ 33. Important diseases in man

_____ 34. Ecologic hodgepodge

_____ 35. Many member viruses in the group

(a) Togaviruses
(b) Picornaviruses
(c) Both
(d) Neither

Coccidioidomycosis with histoplasmosis:

_____ 36. Disease occurs in primary and progressive forms

_____ 37. Organisms reproduce by endosporulation

_____ 38. Found in southwestern part of United States

_____ 39. Found in central Mississippi Valley

_____ 40. Causes paracoccidioidal granuloma

_____ 41. Organisms parasitize cells of reticuloendothelial system

_____ 42. Also known as speleonosis

_____ 43. Disease contracted from inhalation of spore-bearing dust

_____ 44. Causes disease in lungs

_____ 45. Organism grows in the laboratory as a mold

(a) Coccidioidomycosis
(b) Histoplasmosis
(c) Both
(d) Neither

Part III

1. Place the letter indicating the organism from the column on the right in front of the statement regarding it in the column on the left.

_____ 1. The organism infecting cattle and causing undulant fever in man

_____ 2. The organism resembling the enteric group but causing respiratory infection in man

_____ 3. The aerobic, gram-negative organism that caused epidemics of the Black Death during the Middle Ages

_____ 4. An organism attacking persons who handle infected animal products such as hides, wool, or hair

_____ 5. The organism causing rabbit fever

_____ 6. An organism losing its importance as a pathogen in man since the advent of the automobile

(a) *Klebsiella pneumoniae*
(b) *Bacillus anthracis*
(c) *Pseudomonas mallei*
(d) *Vibrio cholerae*
(e) *Brucella abortus*
(f) *Yersinia pestis*
(g) *Francisella tularensis*

2. Match the given disease in the following list with its causative agent and a specific diagnostic test. From Column 1, select the appropriate causative agent and place its corresponding number opposite the disease in the first blank to the right. From Column 2, select the appropriate diagnostic test and place its corresponding number opposite the disease in the second blank to the right.

DISEASE	PATHOGENIC AGENT	DIAGNOSTIC TEST
Malaria	_____	_____
Lobar pneumonia	_____	_____
Diphtheria	_____	_____
Infectious mononucleosis	_____	_____
Typhoid fever	_____	_____
Typhus fever	_____	_____
Enterobiasis	_____	_____
Lymphopathia venereum	_____	_____
Herpes simplex	_____	_____
Hookworm disease	_____	_____
Epidemic cerebrospinal meningitis	_____	_____
Ringworm of scalp	_____	_____
Rabies	_____	_____
Primary atypical pneumonia	_____	_____
Whooping cough	_____	_____
Boils	_____	_____
Syphilis	_____	_____
Tuberculosis	_____	_____
Scarlet fever	_____	_____

COLUMN 1 (Pathogenic agents)
 (1) *Staphylococcus aureus*
 (2) *Streptococcus pyogenes*
 (3) *Salmonella typhi*
 (4) *Bordetella pertussis*
 (5) *Mycobacterium tuberculosis*
 (6) *Treponema pallidum*
 (7) *Neisseria meningitidis*
 (8) *Streptococcus pneumoniae*
 (9) *Plasmodium*
(10) Neurotropic virus
(11) *Rickettsia prowazekii*
(12) *Chlamydia*
(13) *Microsporum*
(14) *Corynebacterium diphtheriae*

(15) Epstein-Barr virus

(16) *Mycoplasma pneumoniae*

(17) *Herpesvirus hominis*, type 2

(18) *Necator americanus*
(19) Pinworms
(20) *Herpesvirus hominis*, type 1.

COLUMN 2 (Diagnostic tests)
 (1) Widal agglutination test
 (2) Weil-Felix reaction
 (3) Phage typing
 (4) Thick blood smears
 (5) Demonstration of Negri bodies
 (6) Neufeld typing method
 (7) Cold agglutination test
 (8) Wassermann test
 (9) Mantoux test
(10) Cough plate culture method
(11) Cerebrospinal fluid culture
(12) Heterophil antibody test
(13) Schultz-Charlton phenomenon
(14) Inoculation of Löffler blood serum culture medium
(15) Use of 10% potassium hydroxide in wet mount
(16) Intranuclear inclusion bodies in skin scrapings
(17) Stool examination for ova and parasites
(18) Anal swabs taken at night
(19) Frei test

3. True-False. Circle either *T* or *F* accordingly.

T F   (1) Madura foot is one manifestation of the dermatomycosis also known as "athlete's foot."

T F   (2) The gonococcus is especially dangerous as it is able to live for a long time outside of the human body.

T F   (3) The tubercle bacillus is able to live but for a brief period after being expelled from the body.

T F   (4) A most important characteristic of yeasts is their ability to produce fermentation.

T F   (5) Abscesses are circumscribed or localized areas of inflammation.

T F   (6) A substance, aflatoxin, produced by a mold is also known to induce cancers in animals.

T F   (7) Cellulitis is a limited type of inflammation.

T F   (8) Pneumococci cause pseudomembranes to be formed in the body.

T F   (9) Leprosy is infectious but not contagious.

T F   (10) Chancre and chancroid are identical terms and mean the same.

T F   (11) Chancres always manifest themselves within 48 hours after contact.

T F   (12) A father cannot give syphilis to his unborn child unless he gives it first to the mother and she in turn transmits it to the fetus.

T F   (13) Enlarged buboes are large ulcers on the skin of patients with tuberculosis.

T F   (14) Mucous patches always occur during the primary stage of syphilis.

T F   (15) Serologic examinations for syphilis are most accurate if run immediately after the chancre occurs.

T F   (16) Congenital syphilis always appears right after birth.

T F   (17) The Wassermann test is very specific in the diagnosis of syphilis; that is, it will be positive only in those instances in which the patient has syphilis and no other disease.

T F   (18) Hansen's disease is tuberculosis.

T F   (19) Leprosy is also known as Hansen's disease.

T F   (20) From the infectious standpoint, it is safer to live with an active case of tuberculosis than with a case of leprosy.

T F   (21) Tuberculin testing is more important in adults than in children.

T F   (22) Tuberculosis usually produces lesions in the lungs, but it may occur in other areas of the body without pulmonary changes.

T F   (23) Only a few fungi are normal inhabitants of the human body.

T F   (24) The main insect vector for yellow fever is *Aedes aegypti*, which can survive and transmit the infection in the southern parts of the United States.

T F   (25) Viruses are commonly grown in the laboratory in semisolid media containing egg.

T F   (26) Latent syphilis is caused by a weak type of spirochete.

T F   (27) *Actinomyces* is often called ray fungus.

T F   (28) Pyogenic bacteria are so called because they produce pus in tissues.

T F   (29) *Candida* occurs and causes disease on mucous membranes.

T F   (30) Coccidioidomycosis produces a chronic lesion in the lung that simulates tuberculosis.

T F   (31) Meningitis may be caused by meningococci, staphylococci, and pneumococci.

T F   (32) Sulfur granules are found in the lesions of blastomycosis.

T F   (33) Sporotrichosis involves the skin and regional lymph vessels.

T F   (34) Actinomycosis is acquired by eating infected pork.

T F   (35) Bronchopneumonia is usually a bilateral inflammation of the lung.

UNIT SIX
MICROBES
PUBLIC WELFARE

Persons in daily contact with diseases caused by bacteria are likely to look on microbes as agents of harm only, but such is not the case. *The majority of bacteria and other microbes are helpful to man, animals, and plants, all of which depend on microbes for their very existence.* In a broad sense microorganisms producing disease form but a small and inconspicuous group. Nothing gives us a better idea of the broad scope of microbial activity than a consideration of how bacteria and other microbes affect our daily lives.

MICROBES IN THE PROCESSES OF NATURE

The *microbiology of nature* defines the role that microbes play in the various processes of nature. Bacteria and other microbes have been said to be nature's garbage disposal system and fertilizer factory for they are the active agents in the decomposition of dead organic matter of animal origin, releasing elements needed for growth of plants and returning them to the soil. Bacteria purify sewage by living on the impurities in it and converting these to inoffensive substances that serve as food material for plants. As water trickles through the soil, bacteria help to filter it. Though contaminated when it enters the earth, water may trickle out pure and clean. In short, microbes represent a major natural resource.

Participation in the cycle of an element. The chief elements entering the bodies of animals and plants are nitrogen, carbon, hydrogen, and oxygen, which are combined with other less common ones to form complex proteins. Animals depend on plants for these elements. Plants receive their hydrogen and oxygen from water and must obtain their carbon and nitrogen from an inorganic compound containing them. The assimilation of an element from inorganic compounds, its conversion into organic compounds to form the bodies of plants and animals, and its subsequent reappearance in the inorganic state to be used again, constitute the *cycle* of the element.

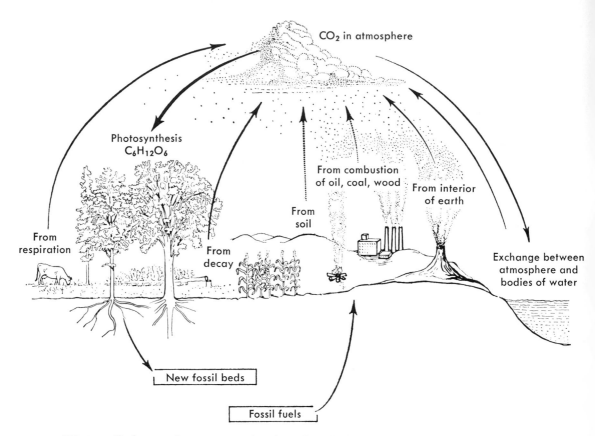

FIG. 33-1. Carbon cycle: sources of carbon dioxide from the earth's surface and within its depths; assimilation of carbon dioxide from atmosphere by plant life through photosynthesis. (From Arnett, R. H., Jr., and Bazinet, G. F., Jr.: Plant biology: a concise introduction, ed. 4, St. Louis, 1977, The C. V. Mosby Co.)

CARBON CYCLE. Plants can assimilate carbon only from carbon dioxide. Carbon dioxide is present in the air in small quantities, but the supply must be constantly renewed. This comes about as carbon dioxide, a waste product of metabolism, is constantly eliminated in the breath of all animals (Fig. 33-1).

If animals were sufficiently numerous to eat all plants, converting them into carbon dioxide and excreting the gas, the supply of plant food for animals and of carbon dioxide for plants would be perfectly balanced. However, tissues of certain plants, such as wood and cellulose that contain much carbon, when eaten by animals, pass through the intestinal canal unchanged. Therefore the carbon in them is not converted into carbon dioxide and released. Other plants die with much carbon bound in their bodies. Carbon is also present in the body tissues of all animals when they die. If there were not some method of recovering this carbon in a form available for plant growth (CO_2), plant life would soon cease and animal life would shortly thereafter. To recover the carbon, fungi and bacteria attack animal excreta, dead plants, and animal bodies, and initiate the process of decay and decomposition. The

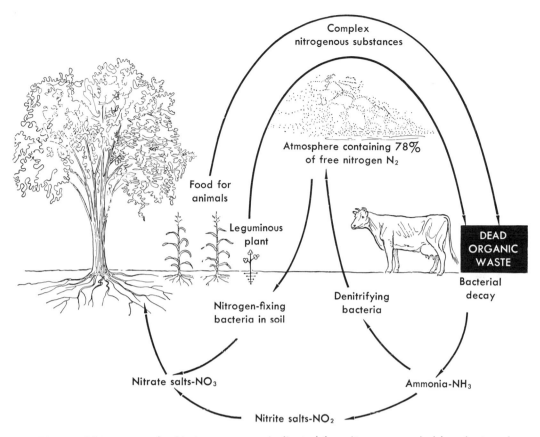

FIG. 33-2. Nitrogen cycle. Various sources indicated for nitrogen needed by plant and animals for protein synthesis. Note that microbes of soil convert some nitrogen stepwise into ammonia, nitrite, and nitrate. Nitrate is also derived from nitrogen of air by bacterial action and through oxidizing action of lightning. Nitrate from these sources is taken up by roots of plants. (From Arnett, R. H., Jr., and Bazinet, G. F., Jr.: Plant biology: a concise introduction, ed. 4, St. Louis, 1977, The C. V. Mosby Co.)

complex carbon compounds are broken down by microbes into carbon dioxide, which floats away in the air to become plant food.

NITROGEN CYCLE. Nitrogen is one of the most important elements in the composition of the plant body. It is abundant in the air but not in a form suitable for plant use. To be so, it must be in the form of nitrates. Nitrates present in the soil in very small amounts must constantly be renewed. Microbes participate by (1) decomposing organic matter and (2) converting the nitrogen of the air into nitrates (Fig. 33-2). Organic matter is broken down by different microorganisms with various actions to form simpler compounds, among which is ammonia. Ammonia is attacked by *nitrifying* bacteria (genus *Nitrosomonas*), which convert it into nitrites. Other nitrifying bacteria (genus *Nitrobacter*) then convert the nitrites into nitrates, which are absorbed by plant roots and built into complex nitrogen compounds like those of the original organic matter.

The term *nitrogen fixation* applies to the recovery of free nitrogen from the air and its synthesis into nitrates for use by plants. The first step in nitrogen fixation is the formation of ammonia. This is accomplished by two groups of microorganisms. One (genus *Azotobacter*) lives in the soil. Another group (genus *Rhizobium*) penetrating the root hairs, develops symbiotically in the cells of certain legumes like clover. It receives its nourishment from the plant and forms nodules filled with the rapidly multiplying bacteria. They bind the nitrogen of the air into a compound incorporated into the bacterial body, which is removed and used by the plant. Animals eat the plants. When the plant dies and decays, the nitrogenous compounds are set free in the soil. Soils in which such products of decay are present are fertile. Within the upper several centimeters of well-tilled soil, there are 300 to 3000 kg of living microorganisms per hectare. An acre of alfalfa well nodulated by nitrogen-fixing rhizobia has the capacity to fix in the ground 200 kg of nitrogen from the air. The microbes in the earth's crust fix annually 10^8 metric tons of nitrogen from the air, six times more than can be accomplished artificially in industry. *Microbiology of the soil* deals with the many aspects of the population of microorganisms, on which fertility depends.

OTHER CYCLES. Other elemental cycles in which microbes participate are those of phosphorus and sulfur in fresh and marine waters.

MICROBES IN ANIMAL NUTRITION

Many years ago Pasteur expressed the idea that animal life would not be possible were it not for microbial inhabitants of the intestinal tract.* Bacteria take an active part in the digestive processes of both lower animals and man, and in the large bowel they form many vitamins essential to animal and human nutrition. Large amounts of vitamins in the B complex group are so produced, and here is the major source of vitamin K, important in the prevention of hemorrhage. Vitamin K is primarily supplied to the body from the intestinal lumen, where about 1000 units of vitamin K per gram weight of dry bacteria are synthesized by *Escherichia coli*.

Administration of the modern broad-spectrum antibiotics (for example, the tetracyclines) may so sharply reduce the microbial population of the intestinal tract as to cause a sizable reduction in the production of vitamin substances there. Depletion of vitamin K is especially serious prior to surgery, and the vitamin may have to be administered to the patient as a drug to prevent hemorrhage. Microbes may have even much greater importance in the physical well-being of lower animals and man than is apparent with our present state of knowledge.

MICROBES IN INDUSTRY

Microbiology of industry involves the roles that microbes play in processes of commerce and industry, such as the following.

Manufacture of dairy products. *Microbiology of agriculture* is a broad subject

---

*"If microscopic beings were to disappear from our globe, the surface of the earth would be encumbered with dead organic matter and corpses of all kinds, animal and vegetable. . . . Without them, life would become impossible because death would be incomplete." (Louis Pasteur, 1861.)

encompassing the relations of microbes to domestic animals and to the soil. One important subdivision is *dairy microbiology*, the manufacture and processing of milk products.

When milk is churned, the fat globules coalesce, taking up a small amount of milk solids and water to form butter, and leaving a residue of buttermilk. When milk is allowed to stand, it sours and curdles because bacteria convert the milk sugar, lactose, into lactic acid, a process of *ripening*. Normal milk bacteria play the key role in ripening, that is, lactic acid bacteria primarily in genera *Streptococcus*, *Leuconostoc*, *Pediococcus*, and *Lactobacillus*. Originally ripening served to preserve milk until enough to churn had accumulated, since the acidity of sour milk inhibits growth of other microbes. Ripening is often hastened by adding cultures of these lactic acid–producers known as *starters*, and today ripening is done so as to give flavor to butter. Butter is usually prepared from cream instead of whole milk, but for cultured buttermilk ordinary butter starters are added to fresh sweet milk.

Cheese consists of the curd separated from the liquid portion of milk either by bacterial fermentation of the lactose to form lactic acid or by the action of the milk-curdling enzyme, *rennin* (from the stomach of a calf). Cheese prepared by the action of lactic acid is known as cottage cheese. Green cheese is the product when rennin is the curdling agent. The water is drained away and the curd pressed into a block and partly dried. When green cheese is set aside for a time, it ripens; that is, it undergoes changes in color, flavor, and odor, which are induced by enzymes in the curd as well as in certain molds, yeasts, and bacteria. The kind of cheese resulting depends, to a great extent, on the kind of organisms causing it to ripen. For instance, Swiss cheese is ripened by *Streptococcus thermophilus* and *Lactobacillus bulgaricus*, as starters, after which a gas-producing colony of *Propionibacterium freudenreichii* subsp. *shermanii* literally blows out the familiar holes. In the preparation of blue cheese (example, Roquefort), the curd is sprinkled with spores of *Penicillium roqueforti* or similar mold to produce the blue-green mottling. The creamy consistency of Camembert is largely due to the growth of *Penicillium camemberti*, and the strong smell of limburger can be traced to growth of certain yeasts and bacteria. Cheese is usually named after the locality in which it is produced, and a certain type of cheese may have several different names, depending on where it is produced. For instance, cheddar (after Cheddar, England), Wisconsin, American, and brick cheeses are basically the same.

In the United States milk used to make cheese must either be pasteurized (heated to 161° F for 15 seconds) or heat treated (150° F for 15 seconds). When cheese is made from heat-treated milk, it must also be aged for at least 60 days at cool temperatures. In pasteurized milk, unlike heat-treated milk, both pathogens and bacteria responsible for cheese flavors are killed. Generally, milder cheeses like processed cheese are made of pasteurized milk; whereas stronger cheeses like cheddar are made from heat-treated milk.

Manufacture of alcohol and alcoholic beverages. Alcoholic beverages are of two classes: those manufactured by fermentation alone and those in which the alcohol is distilled off the fermented mixture. The distillate has a higher alcoholic content than the beverage manufactured by fermentation alone. Among beverages manufactured by fermentation alone are wine and beer. Among those manufactured by distillation are commercial alcohol, whiskey, rum, and brandy. The manufacture of both fermented and distilled beverages depends primarily on the conversion of sugar into alcohol by yeasts. Commercial fermentations rely mainly on enzymes of yeasts of *Saccharomyces* species.

Wines are made by allowing grape juice to ferment. Beer is made from grains, most often barley. Grains do not contain sugar at first, but when the grain is soaked in water, the seeds germinate and the starch-splitting enzyme known as *diastase* (amylase) is produced. It converts the starch of the grain into the sugars maltose and glucose. Grain starch that has been converted into sugar is known as *malt*. The maltose and glucose in the malt are converted into alcohol by yeasts. If foreign organisms grow in beer, they give it an undesirable taste. This is "disease" of beer.

Purification of alcoholic beverages by distillation depends on the fact that alcohol has a much lower boiling point than water. When a fermented mixture is heated to a temperature considerably below the boiling point of water, the alcohol distills over, leaving the water behind. Commercial alcohol and whiskey are usually made by distilling fermented grains. Brandy is made by distilling fermented fruits. Rum is made by distilling fermented cane juice. The mixture of fermenting grains used in the manufacture of alcohol and alcoholic beverages is known as *mash*. The mash from which rye whiskey is made contains 51% or more rye, whereas that used for the manufacture of bourbon contains 51% or more corn. The mash used in the production of alcohol is usually a mixture of corn (88.5%), barley (9.75%), and rye (1.75%), known as *spirit mash*.

Industrial alcohol (ethyl) has been for years manufactured from molasses. Alcohol may be produced by the action of bacteria on the cellulose of plants. In this method by-products of the lumber industry such as sawdust, trimmings, and remains of trees are used. A large amount of alcohol is obtained as a by-product when petroleum is cracked to form gasoline.

Baking. An essential step in baking is the same as that for the manufacture of alcoholic beverages, conversion of sugar into alcohol and carbon dioxide by yeast. It is the carbon dioxide, not the alcohol, that plays the important role in baking. When yeasts are mixed with dough, their enzymes, notably zymase, ferment the sugar present; the carbon dioxide thus produced riddles the dough with small holes, causing it to rise. When the dough is baked, the heat volatilizes the alcohol and causes the carbon dioxide to expand and to be driven off, leaving the bread light and spongy. Some sugar must be present or fermentation will not take place. The dough may contain a small amount of diastase, which converts some of the starch of the dough into sugar. If white flour is used in which the amylase has been destroyed in the refining process, sugar must be added.

Manufacture of vinegar. There are several methods of making vinegar. It may be made from wine or cider or from grains that have undergone alcoholic fermentation. In most methods the first step is the production of alcohol by fermentation, and the second step is the conversion of the alcohol into acetic acid. The first step differs in no manner from the production of alcohol for other purposes. When the alcoholic liquor is exposed to the air, a film known as mother of vinegar and containing acetic acid bacteria (genera *Acetobacter* and *Gluconobacter* [*Acetomonas*]) forms on the surface, and these bacteria convert alcohol into acetic acid. If the process continues too long, the bacteria decompose the acetic acid into carbon dioxide and water, and the vinegar loses some of its strength.

Production of sauerkraut. Sauerkraut is finely shredded cabbage that has been allowed to ferment in brine formed by salt and cabbage juice. The finely shredded cabbage is placed in layers in a cask, and salt is sprinkled over each layer. The layers are packed closely together by a weight placed on them. The salt extracts juice from the cabbage, and bacteria present convert the sugar of the juice into lactic acid, which helps to give sauerkraut its peculiar flavor.

Tanning. The recently removed hide or skin of an animal must be preserved in some way, or its microbial population will destroy it. The leather is soaked and softened in numerous changes of cool water to remove dirt, blood, and nonspecific debris. Cool water retards bacterial growth. The excess flesh is trimmed from the hide or skin and the hair is loosened. One method of loosening the hair is to soak the hide in a solution of lime. It has been found that old solutions, which contain many more bacteria, remove the hair more effectively than fresh ones, which contain few. During the process the bacteria partially decompose and soften the leather.

After the loosened hair is scraped away, minerals are removed by washing or by chemical action. Before they are tanned, hides of some animals are pickled. Pickling is carried out by treatment with sulfuric acid and salt. Tanning is accomplished by the use of vegetable tannins or chemicals such as alum.

Curing of tobacco. When tobacco leaves are cured, they change in texture and flavor and take on a brown color. These changes are accompanied by a decrease in sugar, nicotine, and water. The process is apparently one of oxidation induced by fermentation.

Retting of flax and hemp. In flax and hemp plants the fibers are closely bound to the wood and bark of the plant by a gluelike pectin. When these fibers are separated from the bark and wood, the commercial products linen and hemp are obtained. The dissolution of the pectin to effect separation is accomplished by the action of certain aerobes and the anaerobic butyric acid bacteria (organisms in genus *Clostridium*).

Manufacture of antibiotics. The manufacture of antibiotics is a major pharmaceutical activity. Over 50 are available today for plants, animals, and man. The basic sources are cultures of molds or soil bacteria. Penicillin is produced by certain molds of the genus *Penicillium*, and antibiotics such as streptomycin and the tetracyclines are obtained from bacteria in genus *Streptomyces*. The development of penicillin and streptomycin fermentations into industrial processes was the beginning of the field of bioengineering. (Fig. 33-3). At present, antibiotic production by microbes exceeds 6 million pounds annually, and the output of penicillin alone is easily more than 1 million pounds. Within recent years the sales of internal anti-infectives in the United States have amounted to more than $500 million, and antibiotics form

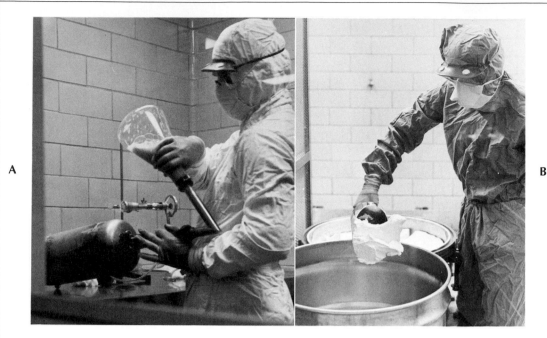

FIG. 33-3. Production of antibiotics. Manufacture of antibiotics is a complex process involving fermentation, filtration, extraction, and other intricate operations, many of which must be performed under sterile conditions. Antibiotic is grown in large fermentation tanks under controlled conditions, filtered to remove impurities, and finally recovered in crystalline form. **A,** Recharging seed tank with microbial culture in fermentation process. **B,** Handling of bulk antibiotic under sterile conditions. (Courtesy Eli Lilly & Co., Indianapolis, Ind.)

a large part of such sales. Prescriptions written for such pharmaceutical compounds have numbered around 150 million.

Oil prospecting. Certain types of bacteria are known to be quite plentiful in the soil overlying deposits of oil, a finding utilized to locate such deposits. The growth of bacteria is supported by gases that seep from the oil to the surface. Microbes have also been used to recover uranium compounds from low-grade ores and slag of mines.

Other industrial processes.* In addition to practical applications of microbes just given, the following are mentioned:

1. Manufacture of acetone (fermentation of sugars—*Clostridium acetobutylicum*)
2. Manufacture of lactic acid (fermentation of molasses—lactic acid bacteria)
3. Manufacture of dextran, a plasma expander (fermentation of cane sugar—*Leuconostoc mesenteroides*)
4. Manufacture of butyl alcohol (fermentation of sugars—butyric acid bacteria)
5. Manufacture of vitamins (examples, cyanocobalamin or vitamin B_{12} [at least 2000 pounds annually], riboflavin or vitamin B_2, and vitamin C)

*The theoretical study of fermentation has been constantly stimulated by industrial values of the many end products obtained.

6. Microbiologic assay—determination of amount of amino acids and vitamins in tissues and body fluids (example, assay of vitamin B_{12})

7. Manufacture of steroid hormones, cortisone, and hydrocortisone (action of certain molds on plant steroids)

8. Manufacture of citric acid used in lemon flavoring (oxidation of sugar—*Aspergillus niger*)

9. Manufacture of enzyme detergents (subtilisins, alkaline proteases from *Bacillus subtilis*, added to final detergent product)

10. Manufacture of insecticides* (microbial insecticides or bioinsecticides—protein of *Bacillus thuringiensis* lethal to larvae [caterpillars] of serious food crop pests in order Lepidoptera; possible application of known insect pathogens as some fungi in genera *Beauvera* and *Coelomomyces*, protozoa in genus *Nosema*, and insect viruses such as baculoviruses or granulosis viruses)

11. Production of food in seas and oceans (photosynthesis carried on by certain marine microbes, role similar to that of plants on land)

12. Production of ensilage for animal feed (fermentation of sugars in shredded green plants by lactic acid bacteria)

13. Manufacture of plant growth factors (example, gibberellins)

14. Manufacture of enzymes (examples, amylases, proteases, pectinases)

15. Manufacture of amino acid (examples, glutamate, lysine; 50 million pounds of glutamate and 1 million pounds of lysine produced annually in the United States, with 100,000 pounds of glutamic acid manufactured by bacterial processes for the pharmaceutical industry)

16. Manufacture of flavor nucleotides (examples, inosinate, guanylate)

17. Manufacture of polysaccharides (example, xanthan polymer)

18. Sewage disposal (see Chapter 35)

---

*Microbes live in symbiosis with insects, furnishing essential vitamins and amino acids needed by the insect and probably nitrogen through nitrogen fixation. The herbivorous (wood-eating) termite is thought to depend on the nitrogen-fixing bacteria of its intestinal flora as a significant source of its nitrogen.

QUESTIONS FOR REVIEW

1. What is meant by the cycle of an element?
2. Briefly outline the nitrogen cycle and the carbon cycle.
3. What is nitrogen fixation? Its importance?
4. What determines fertility of the soil?
5. What causes root nodules on clover plants? What purpose do they serve?
6. What organism is important in the synthesis of vitamin K in the intestine?
7. List 10 practical applications made of microorganisms. Give specific examples of microbes (by name) used in industry.
8. Explain how cheese is made. What is dairy microbiology?
9. What is diastase? Its role in fermentation?
10. Briefly tell what is meant by starters, rennin, hectare, ripening, mother of vinegar, bioinsecticide, nitrifying bacteria, fertility of soil.

REFERENCES. See at end of Chapter 37.

34 Microbiology of water

Practically all waters under natural conditions contain microbes, including protozoa, bacteria, fungi, and viruses. Some contain many, others, only a few. The number and kind of microbes present depend on the source of water, the addition of the excreta from man and animals, and the addition of other contaminated material.

Since sewage holds the pooled excreta from both the sick and well, it necessarily contains pathogenic organisms, especially those that leave the body by feces or urine. Sewage must be properly disposed of to avoid contamination of water supplies and the spread of disease by flies or other vectors.

Sanitary microbiology (the microbiology of sanitation) pertains to the treatment of drinking water supplies and sewage disposal.

Sanitary classification. From a sanitary standpoint waters may be classified as potable, contaminated, and polluted. *Potable* water is free of injurious agents and pleasant to the taste. In other words, it is a satisfactory drinking water. *Contaminated* water contains dangerous microbial or chemical agents. A contaminated water may be of pleasing taste, odor, and appearance. *Polluted* water has an unpleasant appearance, taste, or odor. Because of its content, it is unclear and unfit for use. It may or may not be contaminated with disease-producing agents.

The pollution of water supplies is a major health problem today. Not only sanitary sewage but also complex wastes from industry and agriculture may be discharged into water. In increasing volume such substances as plastics, detergents, pesticides, herbicides, insecticides, oils, animal and vegetable matter, chemical fertilizers, and even radioactive materials are released from factories, canneries, poultry-processing plants, oil fields, and farms. The persistence of synthetic detergents in wastewater and sewage reflects a failure of microbial action. Bacteria do not possess the enzymes to degrade them chemically.

Fish reflect a measure of the purity of a water supply, since they cannot live and thrive in the water of a river or lake that is heavily polluted.

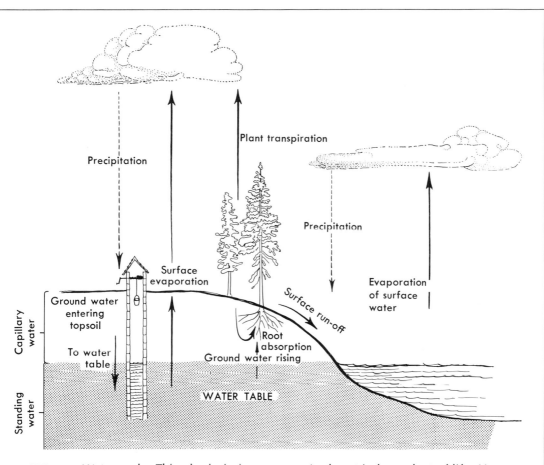

FIG. 34-1. Water cycle. This physical phenomenon is almost independent of life. However, water is necessary for metabolism in all organisms. (From Arnett, R. H., Jr., and Bazinet, G. F., Jr.: Plant biology: a concise introduction, ed. 4, St. Louis, 1977, The C. V. Mosby Co.)

Sources of water. Water is the most abundant of substances on the earth and, singly, an inorganic fluid found simultaneously in nature as gas, liquid, or solid. It is surprisingly difficult to obtain or keep as a pure substance. It is contaminated easily (or recontaminated after purification).

A water supply may come from (1) rain or snow, (2) surface water (shallow wells, rivers, ponds, lakes, and wastewater), and (3) groundwater (deep wells and springs) (Fig. 34-1). Generally, surface water contains more microbes than either groundwater or rainwater. Surface water contains many harmless microbes from the soil, and in the vicinity of cities it is often contaminated with sewage bacteria. (The majority of soil microorganisms are found in the upper 6 inches of the earth's crust.) Surface water in sparsely settled localities may not be dangerous, but *the only safe rule is not to use surface water without purification.*

Unless properly constructed, shallow wells may become contaminated with the drainage from outhouses, stables, and other outbuildings. Improperly constructed shallow wells have been responsible for many outbreaks of typhoid fever and dysentery in rural communities. For a shallow well to be safe, its upper portion must be lined with an impervious material so that water from the surface does not seep into it, and it should be located so that outhouses drain away from it. Of course, it must be kept tightly closed. Shallow springs may be just as dangerous as shallow wells.

Deep-well water and deep-spring water usually contain few microorganisms because these are filtered out as the water trickles through the layers of the earth, but like shallow wells, deep wells must be protected against pollution by surface water. Microbial contamination may occur when a well is situated within 200 feet of the source of contamination, and chemical contamination may occur for a distance of 400 feet. Contamination is more likely during wet weather.

Waterborne diseases. The sources for microbes in water are many—soil, air, the water itself, and the decaying bodies and excreta of man and animals. As a result, many different kinds are found therein, including well-known pathogens as well as the harmless water bacteria. There is increased concern in this country as to the presence of viruses in water, since most human ones multiply in the alimentary canal. Human enteric viruses have been recovered from a third of surface water samples examined. Groundwaters may also transmit virus.

The varied diseases spread by water are notably those whose agents leave the body by way of the alimentary or urinary tracts, and waterborne infections are mostly contracted by drinking contaminated water. When a waterborne epidemic occurs, most cases appear within a few days, indicating that all were infected at about the same time. Most significant of the waterborne diseases are typhoid fever, bacillary dysentery, amebic dysentery, salmonellosis, viral hepatitis, and cholera. Epidemics of acute diarrhea often follow heavy contamination of water with sewage. Diseases that on occasion are transmitted by water are tularemia, anthrax, leptospirosis, schistosomiasis, ascariasis, poliomyelitis, and other enterovirus (coxsackievirus and echovirus) infections. Pathogenic bacteria do not live long in water, and they do not multiply there; but they may persist for a time in water that is cool and contains considerable organic matter.

Certain opportunist pathogens, including many species of *Pseudomonas*, are referred to as water organisms (water bugs), since they survive and propagate in a wet, stagnant environment. They can grow in sterile water on traces of phosphorus and sulfur and are generally resistant to antibiotics and disinfectants. In hospitals they lurk as an unsuspected source of infection in sinks, drains, faucets, and the oxygen therapy apparatus. They can gain a foothold in the air ducts, mist generators, and humidifying pans in incubators of newborn and premature nurseries. Most susceptible to these water organisms are newborn infants, surgical patients, and chronic invalids. Control of this source of infection lies in heat sterilization, absolute cleanliness, and removal of the moist, stagnant areas.

Bacteriologic examination of water. We know from experience that it is almost impossible to isolate from water the organisms responsible for the most important waterborne diseases because few are present to begin with and they do not multiply

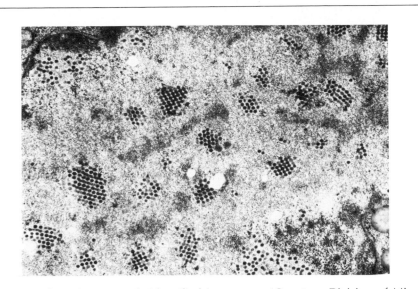

FIG. 34-2. Adenovirus crystals identified in sewage. (Courtesy Division of Microbiology and Infectious Diseases, Southwest Foundation for Research and Education, San Antonio, Tex.)

in water. Sanitarians and public health workers therefore have concluded that the only safe method to prevent waterborne diseases is to condemn any fecally polluted water as being unfit for human use. It *might* contain harmful organisms, since these practically always gain access to water in the feces or urine of man. Whether fecal pollution exists would be determined by examining the water for colon bacilli *(Escherichia coli)*, abundant in feces and *not* found outside the intestinal tract in nature. Certain other bacteria that resemble colon bacilli may or may not be of fecal origin. Also found in water, these bacteria ferment lactose with the formation of gas. In practical water analysis, therefore, testing includes the group spoken of as the *coliform* bacteria (The presence of coliforms increases the likelihood of viruses being present. Note that viruses have been identified in sewage as indicated in Fig. 34-2). The method of procedure recommended by the American Public Health Association and the American Water Works Association consists of three tests: presumptive, confirmed, and completed. The following are the tests as outlined in *Standard Methods for the Examination of Water and Wastewater.*

PRESUMPTIVE TEST. The presumptive test is based on the fact that coliform bacteria ferment lactose with production of gas. Lactose broth fermentation tubes are inoculated with 0.1, 1, and 10 ml portions of the test water and incubated for 24 hours ± 2 hours at 35° ± 0.5° C. If no gas has formed, the tubes are incubated for another 24 ± 3 hours. The presence of gas is *presumptive* evidence that coliform bacteria are present and suggests fecal contamination. The smallest amount in which the fermentation of lactose occurs is an index of the number of organisms present.

CONFIRMED TEST. In the confirmed test, fermentation tubes of brilliant green lactose bile broth or solid differential media such as Endo medium or eosin–methylene blue agar are inoculated with material from the tubes showing gas. If, on incubation at

$35° \pm 0.5°$ C, the brilliant green lactose bile broth shows gas at the end of 48 ± 3 hours or if colonies with the characteristic coliforms appear on the solid medium within 24 ± 2 hours, the *confirmed* test is positive.

COMPLETED TEST. If brilliant green lactose bile broth was used in the confirmed test, the completed test is done by streaking Endo medium or eosin–methylene blue agar plates from the tubes showing gas and incubating them at $35° \pm 0.5°$ C for 24 ± 2 hours. From these plates (or the original plates if Endo medium or eosin–methylene blue agar was used in the confirmed test) colonies typical for or similar to those of the coliform bacteria are transferred to lactose broth fermentation tubes and agar slants. These are incubated at $35° \pm 0.5°$ C and inspected at the end of 24 ± 2 hours and 48 ± 3 hours. After incubation, if microscopic examination of the growth on the agar slants reveals small gram-negative organisms and if the lactose fermentation is positive, the identification of coliform bacteria is said to be *completed*. Further elucidation of bacteria identified in the completed test is not done as sanitarians in the United States believe that a satisfactory drinking water should be free of both fecal and nonfecal organisms.

Included in *Standard Methods for the Examination of Water and Wastewater* is the *membrane filter procedure* of detecting coliform bacteria. In this method, developed in Germany during World War II, the water is filtered through a thin disk of bacteria-retaining material. After filtration, the disk is inverted over an absorbent pad saturated with modified Endo medium and incubated for 20 ± 2 hours at $35° \pm 0.5°$ C. All organisms that produce a dark purplish green colony with a metallic sheen are considered to be coliforms. Colonies of coliforms are counted, and an estimation of the number per 100 ml of water sample is made.

European bacteriologists have devised systems whereby water is also examined for *Streptococcus faecalis* and *Clostridium perfringens*, both of which are just as constantly present in feces as coliforms but not so plentiful. Also, the use of phages as accurate indicators of fecal pollution of water has been suggested.

Drinking water standards. Drinking water must contain no impurity that would offend sight, taste, or smell, and substances with deleterious physiologic effects must be eliminated or not introduced. The United States Public Health Service has set standards for drinking water used in interstate commerce. Originally designed to protect the traveling public, these, to a great extent, have been used throughout the country. Specifications relate to (1) source and protection of the water, (2) physical and chemical properties, (3) bacteriologic content, and (4) limits on concentrations of radioactivity. At the end of 1974, Congress passed the Safe Drinking Water Act, which calls for stringent federal standards to protect consumers to "maximum feasible extent" from contaminants in water. The Environmental Protection Agency (EPA) is charged to set the interim standards for safe drinking water, which will become effective by the end of 1977. Revised standards will follow based on further studies of the health effects of the various drinking water contaminants.

Investigation of the problem of viruses in water is just beginning. Since minimal amounts of virus can produce infection, total removal of viruses from a water supply for human consumption seems indicated.

Water purification. Water may be purified by natural means. As water trickles through the earth, microbes are filtered out. Water standing in lakes and ponds undergoes some degree of purification because of the combined action of sunlight, sedimentation, dilution of impurities, and destruction of bacteria by protozoa. Streams may be purified as they flow along. However, the fact that all the foregoing methods of natural purification are slow and uncertain in their action and the possibility being great that contaminated material will enter the flowing stream along its course mean that one cannot and must not depend on self-purification of water.

FIG. 34-3. East Side water purification plant, Dallas, Texas. Purification process extends from chemical plant in the foreground where ferrous sulfate and lime are added to water to pump house in background from where water is pumped to reservoir. Between these points are rapid mixers, flocculators, settling basins, and sand filters. Water is chlorinated several times at points along way. (Courtesy Dallas Water Utilities, Dallas, Tex.)

Therefore all surface waters must be regarded as dangerous unless subjected to artificial purification.

In emergency or unusual circumstances drinking water can be boiled, or small quantities can be made relatively safe for consumption by the addition of either 2 to 4 drops of 4% to 6% chlorine bleach or 5 to 10 drops of 2% tincture of iodine to one quart of water, provided the mixture is allowed stand for 30 minutes. Halazone, a commercial, iodine-containing tablet may also be used.*

Generally, on a large scale, artificial water purification (Fig. 34-3) is carried out by a combination of physical and chemical methods—sedimentation and filtration combined with the action of chemicals to soften and clarify the water, and chlorination (p. 284). *Sedimentation* is the process whereby solids in the water such as mud and organic matter are removed. Water is held in settling basins for a number of hours to allow the large particles to settle out. A chemical such as alum or ferric sulfate is added to coagulate the suspended organic matter and microbes present, and the precipitated coagulum is removed mechanically. Some clarification of water occurs with sedimentation, but *filtration* is the process that puts the final sparkle in the water.

The average rapid filter that is in use has the following units from floor to surface: (1) watertight floor with grooves and tile for draining the filtered water away, (2) layer of gravel from 12 to 18 inches thick, and (3) layer of sand. The gravel at the bottom is of such a size as to pass through a 2-inch screen mesh. The size of the gravel gradually decreases as the surface of the layer where the gravel will pass through a $1/16$-inch mesh is approached. The sand layer is about 2.5 feet thick. Rapid sand filters can filter over 200 million gallons per day per acre of filter area. By rapid filtration 90% to 95% of all bacteria are removed more or less mechanically by the layer of sand at the top of the filter. Efficient filtration removes all pathogenic and most saprophytic organisms. Filters must be cleaned often. This is done by reversing the flow of water and agitating the upper layer of sand. In some cases air is forced through the filter from the bottom to the top. A rapid filter is made in units so that just a part of it is closed for cleaning at any one time.

After filtration, water is universally chlorinated. There are three methods in common use:

1. *Simple chlorination.* This method is the addition of a standard effective amount of chlorine to water. Chlorine is usually added from cylinders connected with regulating equipment to control the exact amount.
2. *Ammonia-chlorine treatment.* This treatment results in compounds called chloramines, formed when chlorine is added to water containing ammonia. Chloramines are less active as germicidal agents but more stable than free chlorine.
3. *Superchlorination.* This process involves the addition of a much larger dose of chlorine than with simple chlorination and the subsequent removal of the excess.

---

*See section "Health hints for the traveler" in Health information for international travel 1976, Morbid. Mortal. Wk. Rep. **25**(suppl.):9, Oct., 1976, HEW publication No. (CDC) 76-8280.

Certain chemicals can soften hard water by removing dissolved limestone. When mixed with raw, untreated water, granular activated carbon absorbs odors and tastes resulting from decaying vegetation and other organic matter present.

Fluoride content of water and tooth decay. It is well known that the fluoride content of the drinking water of infants and children profoundly influences the development of their teeth. If too much fluoride is present (4 to 5 ppm water), the condition known as *fluorosis* or mottled enamel develops; if too little, the incidence of dental caries is greatly increased. Adult teeth do not seem to be affected. The optimum ratio of fluoride to water for drinking by infants and young children is 1 ppm. When this is present, there is sufficient fluoride to retard tooth decay significantly but not enough to stain or mottle the teeth. Reduction of the incidence of tooth decay is on the order of 60% to 70%. In nature the fluoride content of water varies from much less to more than 1 ppm, but the proper level may be maintained by adding fluoride

FIG. 34-4. Wastewater treatment plant, Dallas, Texas, aerial survey. Note layout of settling basins, microbial digestors, and trickling filters. In primary settling basins, organic wastes are oxidized by bacteria and algae growing on rocks. Sludge from primary settling basins goes to a separate group of heated tanks called digestors, where, by a process of anaerobic decomposition, solids are broken down by microbes. Effluent from primary settling basins goes to 29 circular trickling filters, each with a half-acre rock bed. (Courtesy Dallas Water Utilities, Dallas, Tex.)

to water whose content is low or by chemically removing fluoride from water whose content is high.

Ice. When pathogenic bacteria gain access to ice, the majority die, but a few survive. Perhaps 10% survive for more than a few hours. Ice that has been stored for 6 months is practically bacteria free. The use of water of high sanitary quality in the manufacture of ice has lessened the possibility that ice itself will convey disease. However, the handling of ice in the home, eating places, and fast food stands may be a source of infection.

Purification of sewage. Sewage must be processed in treatment plants to render it harmless before it is discharged into any body of water (river or lake) (Fig. 34-4). Many methods of sewage purification are used. The usual sequence is screening and sedimentation to remove larger particles, chemical treatment to remove small ones, microbial action to digest organic matter, aeration to induce oxidation, filtration through sand, and chlorination. Breakdown of human waste in sewage by microbes is essentially decay of organic material, and the sanitary engineer designs the disposal plant to facilitate and speed up microbial activity and ensure its completion. Puri-

FIG. 34-5. Water reclamation. An "activated sludge" pilot plant in Dallas, Texas, of a type used by many cities over nation; capacity 1 million gallons a day. In activated sludge, microbes are mixed with sewage to consume its organic products. Here is one large tank maintained in an aerobic state to facilitate growth of sewage microbes. It is seeded by sludge (primarily microorganisms) settled from a second large tank, the secondary clarifying tank. Input of raw sewage or sewage already settled into aeration tank provides substrate for microbial action. (Courtesy Dallas Water Utilities, Dallas, Tex.)

fication of sewage so accomplished efficiently and economically makes reclamation of water practical (Fig. 34-5).

There is no known way to remove all viruses from sewage (Fig. 34-2); they are not regularly inactivated by primary and secondary treatment and easily pass into the effluent. Enteroviruses, reoviruses, and adenoviruses have been demonstrated to come through standard treatment. Parasites readily disseminated in sewage include *Entamoeba histolytica, Giardia lamblia, Ascaris,* hookworms, and tapeworms.

Swimming pool sanitation. Swimming pools convey conjunctivitis, ear infections, skin diseases, and intestinal infections unless kept sanitary. The source of infection is a person using the pool. Swimming pool water should be kept as pure as drinking water. This can be done with frequent changes of water and the use of disinfectants. Chlorine is the most widely used one, and the highest possible concentration should be maintained, about 1 ppm (see also p. 284). A higher concentration irritates the eyes. Before entering the pool, each bather should take a shower using soap. Persons with infections of any kind should not be allowed entry to the pool.

Some authorities prefer iodine to chlorine as a disinfectant for swimming pools. Its action is less hindered by organic matter, and there is less eye and skin irritation than with chlorine. Bromine has also been used as a swimming pool disinfectant.

QUESTIONS FOR REVIEW

1. Name the diseases spread by water. Give causal agent.
2. Discuss the bacteriologic examination of water.
3. Describe the purification of water by (a) natural means, (b) artificial means.
4. List the sources of water. From a public health standpoint how do waters from these sources differ? State the sources of the microbes in water.
5. Why is sewage of such sanitary significance?
6. What is the relation of the fluoride content of water to tooth decay?
7. Discuss water pollution today as a public health problem. (Consult references.)
8. Give the sanitary classification of water.
9. How are swimming pools made sanitary?
10. Briefly explain water cycle, water bugs, fluorosis, activated sludge, water reclamation.

REFERENCES. See at end of Chapter 37.

35 Microbiology of food

MILK

If obtained in a pure state and kept pure, milk is our best single food; if improperly handled, our most dangerous one. Milk conveys infection because it is an excellent culture medium and is consumed uncooked. *Microbiology of dairy products* deals with (a) the microbes that make milk and milk products unfit for human consumption, (b) prevention of disease in cattle, and (c) manufacture of milk products.

Bacteria in milk. Bacteria gain access to milk in different ways. Milk as secreted by the mammary glands of perfectly healthy cows is usually sterile, but as the cow is milked, bacteria from the teats and milk ducts get in, and by the time the milk enters the receptacle, contamination has taken place. Bacteria may be present within the udders of diseased cows. Unclean and unsanitary milking utensils, pasteurizing tanks, and milk containers as well as dust and manure are important sources of contamination.

Under the best conditions the bacterial content of milk is relatively high. Some bacteria, even though they come from external sources, are so commonly present as to be regarded as normal milk bacteria. Normal milk bacteria sour milk; they also may destroy its food value.

The two main sources for pathogenic bacteria found in milk are the people who handle it and infected cows. The pathogens most often present are staphylococci, streptococci, lactic acid bacteria, and enteric microorganisms.

Although it is not the only criterion, the number of bacteria in milk is the best index of its sanitary quality. The number present depends on that originally introduced and the temperature at which the milk is kept. Many species of bacteria begin to multiply as soon as they gain access to milk. This can be largely prevented if the milk is rapidly chilled to 10° C (50° F) as soon as it is obtained and kept cold. If the milk is allowed to warm up, there is a rapid rise in the number of bacteria.

662

Milk of high quality may contain only a few hundred bacteria per milliliter; bad milk, millions per milliliter. All authorities agree that a single high bacterial count does not invariably mean poor quality milk and that to be significant a high bacterial count must recur daily. The presence of coliform bacteria points to fecal contamination.

Milkborne diseases*. The diseases spread by milk may be classified as (1) diseases of human origin and (2) diseases of milk-producing animals transmitted to man in their milk. Most notable of those that primarily affect man and may be transmitted by milk are typhoid fever, salmonelloses, bacillary and amebic dysentery, scarlet fever, infantile diarrhea, diphtheria, poliomyelitis, infectious hepatitis, and septic sore throat. The diseases that primarily affect milk-producing animals and that may be transmitted to man by the milk of the infected animal are undulant fever, Q fever, and bovine tuberculosis. Undulant fever may be contracted from the milk of infected cows or goats.

The spread of human diseases by milk usually stems from contamination of the milk by the discharges of a handler who is ill of the disease or a carrier. Milkborne epidemics typically appear among the patrons of a given dairy, and the source of infection usually is traced to a food handler there.

Bovine tuberculosis, especially in children under 5 years of age, usually comes from ingestion of the milk of tuberculous cows. The milk of a single cow excreting *Mycobacterium tuberculosis* may render the mixed milk of the whole herd infectious. The danger of contracting tuberculosis through milk is completely eliminated by pasteurization and tuberculin testing of cows.

Pasteurized milk. *Pasteurization* is the process of heating milk to a temperature high enough to kill all *nonsporebearing pathogenic* bacteria but not so high as to affect the chemical composition. There are two methods of pasteurization. One is known as the holding method; the other as the continuous-flow, high-temperature, short-time, or flash method. In the former the milk is held in tanks or vats, where it is subjected to a comparatively low temperature for a comparatively long period of time while being constantly agitated. In the latter the milk is subjected to a higher temperature for a short period of time as it flows through the pasteurizing equipment (Fig. 35-1). After both methods the milk should be transferred through pipes, where it is chilled, to the packaging machines. All milk should be received into clean sanitary containers (glass, cardboard, plastic).

The *Grade "A" Pasteurized Milk Ordinance—1965 Recommendations of the Public Health Service* are that (1) in the holding method the milk must be subjected to a temperature of *at least* 145° F (63° C) for *at least* 30 minutes and (2) in the continuous-flow method the milk must be subjected to a temperature of *at least* 161° F (72° C) for *at least* 15 seconds. Note carefully that pasteurization does *not* completely

---

*A disease not of bacterial origin but of historic interest is *milk sickness* (called milksick by the pioneers). This disease cost the life of many midwestern pioneers including the mother of Abraham Lincoln. It is caused by the ingestion of milk from cows that have been poisoned by eating certain species of goldenrod and the white snakeroot. The poisonous principles in these plants are excreted in the milk of the affected cow.

FIG. 35-1. High-temperature short-time (HTST) method of pasteurization. Milk is received from dairy farm in tank trucks and placed in holding tanks. Milk is pasteurized at 175° F for 16 seconds through a plate heat-exchanger with holding tube and timing pump. It is also clarified, standardized, homogenized, and steam heated to 190°-194° for 30 seconds, vacuum-cooled back to 175°, and cooled at 33° through a plate heat-exchanger containing cooling medium. All is done automatically from a central control panel, utilizing sanitary air-operated valves for routing the milk. Cleaning and sanitizing is also done by automatic circulation of cleaning and sanitizing compounds. Note the Vac-Heat unit in foreground. This unit ensures the same milk flavor year-round by removing various volatile feed-off flavors, which vary during the year's four seasons. (Courtesy CP Division, St. Regis, Dallas, Tex.)

sterilize milk and that it should never be used to cover gross negligence in sanitation and the sanitary handling of milk.

Pasteurized versus raw milk. Pasteurization is the single most crucial item in the maintenance of a safe milk supply. Objections to pasteurization might be that (1) it destroys vitamins, (2) dairymen may not be so careful when they know that the milk is to be pasteurized, and (3) pasteurization may be carelessly and ineffectively done. The latter two objections can be eliminated by thorough inspection systems. Vitamin C, which is destroyed, can easily be replaced in the diet or given as ascorbic acid.

The temperature of pasteurization is carefully controlled, for heating milk to high temperatures can drastically change it. Boiling decomposes milk proteins, changes the phosphorus content, precipitates calcium and magnesium, drives off carbon dioxide, combusts milk sugar, and destroys enzymes in the milk.

Requirements for a safe milk supply. To ensure a safe milk supply, the following requirements must be met:
1. Cows must be healthy, well fed, and free of diseases such as tuberculosis, brucellosis, and mastitis.
2. All individuals handling milk must be free of infectious organisms, and their hands and person must be clean.
3. The premises must be kept clean.
4. The udders and flanks of the cows must be washed before milking.
5. Milking utensils and machinery that contain the milk must be kept sterile and should be so constructed as to keep out dust and flies. (Practically all milking today in commercial dairies is done mechanically.)
6. Milk must be chilled immediately to 10° C (50° F) and kept cold.
7. It must be pasteurized under carefully controlled conditions and again chilled.
8. Directly after pasteurization, it must be placed into the container reaching the customer.
9. It must be delivered cold (10° C) in refrigerated vehicles and immediately refrigerated at its destination.
10. A uniform statewide sanitary milk code must provide for proper inspection of dairies and pasteurization plants.
11. Local laboratory services must check on the purity, cleanliness, and safety of the milk.

Milk grading. Different states have different systems of grading their milk supply, but all are based generally on the sanitation of the dairy, health of the cows, methods of handling the milk, and bacterial content. In addition to the bacterial count* of milk, testing for coliforms is done to detect contamination after pasteurization. Milk is also checked for possible adulteration.

The phosphatase test is used in grading to check whether milk has been adequately pasteurized. The enzyme phosphatase is mostly destroyed at the temperature of pasteurization. A significant amount detectable in milk indicates incomplete pasteurization or the presence of raw milk mixed in with the sample of pasteurized milk.

Several grades of milk are recognized by the *Milk Ordinance and Code—1953 Recommendations of the Public Health Service*, but only one, Grade A pasteurized milk, is allowed in interstate commerce for retail sale. This is milk that has been

---

*Bacterial count is obtained by either the *plate count* (p. 116) or *direct microscopic clump count*. In the direct microscopic clump count, an accurately measured amount of milk (0.01 ml) is placed on a slide and spread into a thin smear of uniform thickness. The dried smear is stained. With a microscope calibrated so that the area of the field is known, the bacteria in representative fields are counted and the number of bacteria in the whole specimen calculated. In this clump count, both individual bacteria and unseparated groups of bacteria are counted as units.

pasteurized, cooled, and packaged in accord with precise Public Health Service specifications. Its bacterial content has been reduced by pasteurization from no more than 300,000/ml to no more than 20,000/ml, with coliforms no more than 10/ml.

The general improvement in milk sanitation has been such that most authorities feel that there is no longer any practical necessity for special grades of milk. Therefore the present tendency is to stress the production of one standard, high-quality, safe, pure milk readily available to people in all walks of life at reasonable cost.

Milk formulas for hospital nurseries. The preparation of infant milk formulas in hospitals should be in accord with standards set by the hospital medical staff or appropriate health agency. For proper microbiologic control, the American Academy of Pediatrics in its Standards and Recommendations for the Hospital Care of Newborn Infants recommends that technics of formula preparation be checked at least once a week. As part of their surveillance plan, random samples of the ready-to-use formula are sent to the laboratory to be cultured. The limit for the bacterial plate count of a formula sample is 25 organisms per milliliter. In the identification of organisms present there should be none other than sporeformers. Otherwise, a break in technic is indicated, and responsible authorities should be notified at once.

Bacterial action on milk. Milk may be decomposed by the action of bacteria on its carbohydrates (fermentation of milk) or proteins (putrefaction of milk). Souring or normal milk decomposition comes from lactic acid fermentation, the result of the formation of insoluble casein by the action of lactic acid on caseinogen. Lactic acid is formed by bacterial action on lactose (milk sugar). The organisms responsible are *Streptococcus lactis, Lactobacillus acidophilus, Bifidobacterium bifidum (Lactobacillus bifidus), Lactobacillus casei, Lactobacillus bulgaricus, Escherichia coli*, and others.

Putrefaction of milk seldom occurs, but it is dangerous when it does. Putrefactive changes are caused by the action of sporebearing and anaerobic bacilli on the milk proteins. The milk is finally converted into a bitter liquid, having little resemblance to fresh milk.

Slimy or ropy milk usually comes from the action of *Alcaligenes viscolactis* or a similar organism, but it may be due to streptococci and the lactic acid bacteria. Alcoholic fermentation may rarely be spontaneous, but it is a feature of manufacturing processes in which sugar and yeasts (or bacteria) are added to milk. Among the products are kumyss or koumiss, kefir, leben, and yogurt or matzoon.

Yogurt is the milk to which Metchnikoff referred in his book on the prolonged life of the Bulgarian tribes, and is the forerunner of present-day Bulgarian buttermilk. The chief characteristic of yogurt is its high acidity, produced by the growth of *Lactobacillus bulgaricus*. Because of this, it is considered to be one of the safest of all foods. Archaeologists working in underdeveloped areas of the world consider that natural yogurt, a major food item in the diet of many peoples, is always nutritious and safe to eat. Yogurt and buttermilk are at times used in the nursing care of patients with advanced cancerous growths, secondarily infected and malodorous. The alkaline medium of the growth allows bacterial growth and fermentation activities to flourish.

The application of yogurt or buttermilk changes the tissue environment of the cancerous area to an acid one, thus inhibiting the growth of the odor-forming microbes.

Bacterial growth may color milk red, yellow, or blue.

Ice cream. Although ice cream is frozen, not all bacteria are destroyed. In fact, some pathogens live longer in ice cream than in milk. Typhoid bacilli have been known to live as long as 2 years in ice cream. Outbreaks of typhoid fever, septic sore throat, diphtheria, food poisoning, and scarlet fever have been traced to ice cream. To prevent it from spreading disease, it must be made with pasteurized milk and handled properly thereafter. Proper handling means cleanliness of utensils and factory and proper health supervision of employees.

Butter. Although derived from milk, butter only poorly supports microbial growth because it is made up chiefly of fat and water and is deficient in protein and sugar. However, certain bacteria do grow in it and render it rancid. The ordinary saprophytic organisms found in butter are the various types of lactic acid bacteria, especially streptococci. Improperly prepared butters may contain *Escherichia coli* and various yeasts and molds. The presence of an excessive quantity of yeasts in butter indicates that proper sanitary precautions were not exercised in preparation. *Mycobacterium tuberculosis* and the atypical mycobacteria have been found in butter. Typhoid bacilli have been found, but they tend to die therein. Contamination of butter with pathogenic organisms may be eliminated by pasteurization of the cream used in making the butter, proper sterilization of utensils, and strict supervision of personnel.

FOOD

The *microbiology of food* is studied from many standpoints. The soil and agricultural microbiologist thinks of the microbes (free living and symbiotic) in the soil. These are utilized by food plants in growth and development. Microbes aid in the return of waste products of food utilization to the soil, where the waste products are used again by microbes and food plants. The food microbiologist studies the part that microbes play in (1) the manufacture of bread, butter, cheese, and many other foods and (2) the spoilage of foods and how this may be prevented. Public health microbiologists, primarily interested in the health of the community, study food as to what diseases it may convey, the harmful products of its spoilage, and how both foodborne disease and food spoilage may be prevented. The diseases conveyed by food are discussed here.

Food poisoning

When foods are related to disease, they are considered first as carriers of infection, as given on p. 171 (examples, poliomyelitis, hepatitis).

Next comes a group of illnesses spoken of as food poisoning. By definition, food poisoning is an acute illness resulting from the ingestion of some injurious agent in food. It is classified as:

1. Nonmicrobial type
   a. Individual idiosyncrasy

      b. Chemical factor from:
        (1) Foods naturally poisonous (toadstools and the like)
        (2) Poisons accidentally or intentionally added (plant sprays and the like)
    2. Microbial type
      a. Food intoxications—effect of ingested toxin (for example, botulism)
      b. Food infections—multiplication of ingested microorganisms
        (1) Bacterial (for example, salmonellosis)
        (2) Parasitic (for example, trichinosis)

A consistent feature of food infections and intoxications is that many persons eat the offending food and most develop the disease. In food infection, the diagnosis is made if the offending organism is cultured from the feces of the patient and from the food eaten. In food intoxication diagnostic efforts are directed toward finding the offending toxin in the food eaten. Examination of the feces of the patient is of little value.

Staphylococci,* *Clostridium botulinum,* and, less often, certain other organisms produce food intoxication. *Salmonella* species and rarely *Streptococcus faecalis* (enterococcus) causes food infection. Table 35-1 lists bacterial types of food poisoning and the prominent features of each.

Food intoxication. Staphylococcal food intoxication, the most common type of food poisoning, is caused by the action of an enterotoxin liberated by some strains of staphylococci. The staphylococci multiply in the offending food before it is eaten and elaborate their toxin there. (They do not multiply in the intestinal tract.) Manifestations come from the ingested toxin. Staphylococci require a period of not less than 8 hours to elaborate enough toxin in food to cause symptoms. After the food is eaten, the disorder appears within 6 hours. Recovery occurs in 24 to 48 hours. Staphylococcal food intoxication usually follows the ingestion of starchy foods, especially potato salad, custards, and pies. When the offending food is meat, it is usually ham. Many outbreaks have been traced to chicken salad.

Botulism is a very serious type of true food intoxication produced by the ingestion of food containing the potent exotoxin of *Clostridium botulinum* (p. 409). The major source of botulism is improperly processed home-canned vegetables of low acid content. *Before they are eaten,* such foods should always be cooked *at the temperature of boiling water for at least 15 minutes* and thoroughly stirred and mixed. Contaminated foods may have unpleasant odors somewhat characteristic for a given food, but these off-odors are not easily recognized and do not suggest spoiled food. The cans or containers holding the food do not necessarily bulge, a conventional sign of food spoilage.

Bacterial food infections. Food infection is caused by the multiplication of bacteria that have been taken into the intestinal canal. The responsible bacteria are most often salmonellae. Of special consequence in food infection are *Salmonella enteritidis* (Gärtner's bacillus), *Salmonella typhimurium,* and *Salmonella cholerae-*

---

*Food poisoning, especially infection and intoxication caused by staphylococci, was formerly known as "ptomaine" poisoning. This is incorrect because ptomaines play no part in food poisoning.

TABLE 35-1. BACTERIAL TYPES OF FOOD POISONING

Organism	Foods	Onset (after food eaten)	Clinical features	Toxin	Outcome
Food intoxication					
Staphylo-coccus	Potato salad Chicken salad Cream fillings Milk Meats Cheese	1 to 6 hours	Vomiting Diarrhea Abdominal cramps Prostration	Enterotoxin	Severe symptoms of short duration (death rare)
Clostridium botu-linum	Canned string beans, corn, beets, ripe olives	2 hours to 8 days	Vomiting Diarrhea Double vision Difficulty in swallowing, speaking Respiratory paralysis	Neurotoxin	Mortality high
Food infection					
Strepto-coccus	Sausage Poultry dressing Custard	2 to 18 hours	Nausea Pain Diarrhea	None	Recovery 1 to 2 days
Salmonella	Eggs Poultry Meats	7 to 72 hours	Diarrhea Severe abdominal pain Fever Prostration	None	Recovery — few days (death rare)

suis. Once inside the intestine, they multiply rapidly and induce widespread inflammation with severe manifestations—nausea, vomiting, diarrhea, and fever.

The disease usually occurs as an explosive outbreak following a meal attended by a large number of persons. Food cooked in large quantities is more often the source of infection than food cooked in small quantities, since heat is not so likely to penetrate and destroy the organisms in a large quantity. Most outbreaks occur during the warm months. The incubation period for *Salmonella* food infection is 7 to 72 hours, and recovery occurs usually within 3 or 4 days; but in severe infections death may occur within 24 hours.

The insufficiently cooked meat of infected animals may convey the disease to man, but usually salmonellae reach the food (often, but not necessarily, meat) from outside sources. Such sources are the intestinal contents of the slaughtered animals, the intestinal contents of animals (especially rats and mice) that have contacted the food incriminated, and (probably most important) human carriers of the bacteria. *Salmonella* infections are common in rats, and these animals often become carriers. Cockroaches have been shown to convey *Salmonella* food poisoning. Salmonellae are abundant in the intestinal tract of poultry, a very important source, and may be present on the shell of an egg. Contaminated eggs broken commercially may be a source of organisms in egg products. Foods such as the meringue on pies contaminated with *Salmonella* may have no abnormal odor or taste.

Prevention of *Salmonella* food poisoning depends on cleanliness in handling food, proper cooking of food, proper refrigeration of food that has been cooked, and detection of carriers. Frequent and careful handwashing by food handlers is mandatory. Food that remains on their skin for a period of time provides nourishment for microbial contaminants, some foodborne, and thus contributes to their survival in the environment.

Under the provisions of the Egg Products Inspection Act of 1970, all commercial eggs broken out of the shell for manufacturing use must be pasteurized. The egg or egg product is heated to 60° to 61° C (140° to 143° F) and held there for 3½ to 4 minutes. The final product must be free of *Salmonella*.

Rarely streptococci cause food poisoning of the infection type. At times the addition of large numbers to the already resident enterococci in the bowel triggers gastroenteritis. The food usually responsible contains meat that has stood at warm temperatures for several hours, and the organism usually recovered from the feces is *Streptococcus faecalis*. The incubation period is 2 to 18 hours, and the illness lasts a very short time.

Other organisms sometimes infecting food are *Clostridium perfringens, Bacillus cereus, Proteus* bacilli, and *Escherichia coli*. In outbreaks traced to these, it has been thought that the microbes were able to multiply in food, especially meat, that had stood at room temperature overnight or longer.

Parasitic food infection. For a discussion of trichinosis, see p. 613.

Measures to safeguard food

To safeguard a food supply, remember that:
1. Appearance, taste, and smell are not always reliable indicators that a food is safe for consumption.
2. Any unusual change in color, consistency, or odor or the production of gas in food means that the food should be discarded *without being eaten. Note:* If *suspect* food has been tasted or eaten, it should not be discarded but kept for 48 hours in the event that it might have to be examined for the presence of *Clostridium botulinum* exotoxin.
3. Dishes, cutlery, can openers, utensils, equipment of all kinds used to prepare, serve, and store food should be clean and sanitary.
4. Hands that prepare, serve, and store food should be clean. Preferably food is handled as little as possible.
5. Foods served raw should be washed carefully and thoroughly.
6. Bacteria grow fastest in the temperature range 40° to 140° F (Fig. 35-2). Therefore:
   a. Hot food should be kept hot (above 140° F)
   b. Cold food should be kept cold (below 40° F). All dairy foods should be refrigerated.

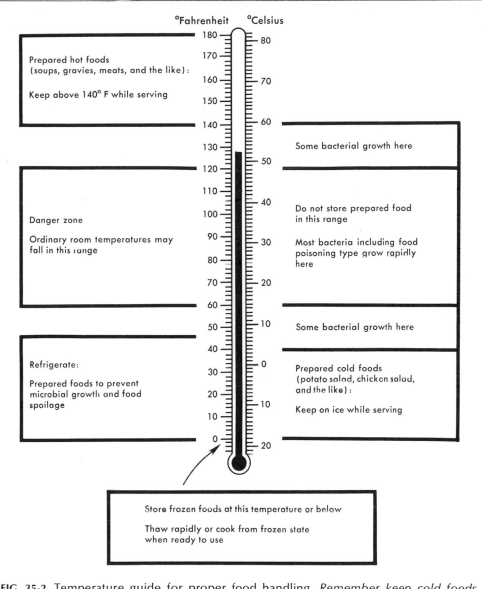

FIG. 35-2. Temperature guide for proper food handling. *Remember keep cold foods cold and hot foods hot!* Temperatures above 45° F and below 140° F provide breeding grounds for many foodborne illnesses.

   c. Cooked food that is to be refrigerated should be cooled quickly and refrigerated promptly. If cooked food is to be kept longer than a few days, it should be frozen.

   d. Perishable foods should be kept chilled if taken on a trip or picnic. Foods may spoil easily if exposed to a warmer temperature for no longer than a half hour before they are eaten.

   e. Extra care should be taken with foods easily contaminated by microbial growth—meats, poultry, stuffing, gravy, salads, eggs, custards, and cream pies.

Food preservation

There are numerous methods of preserving food. The following are a few.

Refrigerating. Refrigeration is one of the best and most universally used methods of food preservation. It has several distinct advantages: (1) there is little change in the composition of food; therefore taste, odor, and appearance are preserved; (2) the nutritive value of the food is not reduced; (3) there is no decrease in digestibility; and (4) there is little effect on vitamin content. During refrigeration pathogenic bacteria are inhibited and many destroyed. When refrigeration is discontinued, those that remain viable begin to multiply. Some saprophytes are able to multiply at refrigeration temperatures.

Freezing. Freezing as a way of preserving food is in widespread use today because of its application in the preparation of "convenience" food items. Quick freezing is carried out at very low temperatures ($-35°$ F with a holding temperature of $-10°$ F). A preliminary step to freezing is blanching. The food is heated quickly to inactivate enzymes that would break down proteins and change the texture, flavor, and appearance of the food. Blanching is followed by instant cooling in ice water. Food that is frozen retains its nutrients and is palatable. Freezing of meat kills encysted larvae therein. Since cold storage foods decompose rapidly when thawed or warmed to room temperature, they should be consumed right away.

Drying. Microbes are unable to multiply for lack of water.

Smoking. The food is dried and preservatives from the smoke added to it.

Pickling. The high acid content of the medium prevents microbial multiplication.

Salting and preserving in brine. The osmotic pressure of the medium is so changed that microbes will not multiply. Sometimes water is extracted from the microbial cell.

Canning. * Heat destroys microorganisms. The container with its content of food is hermetically sealed to keep out contaminants. The sealing also excludes free oxygen, which would aid the growth of molds, yeasts, and most species of pathogenic bacteria.

Commercial canning is a very safe method of food preservation. It is done scientifically and under supervision by the indicated health agency. Home canning on the other hand may be done under rough-and-ready conditions by unskilled hands. Since there is always a possibility for food contamination, home canned foods, especially meats and vegetables, should always be heated prior to consumption.

Preserving. The food is heated and sugar added. A large amount of sugar retards bacterial multiplication in the same manner as does salt. However, it does not retard the multiplication of fungi. Heating and sealing have the same effect as in canning.

---

*In 1810, Nicholas Appart (1752-1841), encouraged by Napoleon Bonaparte, laid the foundation of modern canning and preserving industries by his discovery of the process of heating foods to boiling temperatures and sealing them hermetically in suitable containers. About the same time, the tin can was patented.

Cooking. Microbes are killed by heat, the composition of the food is changed, and water may be removed.

Chemicals. The addition of chemicals such as benzoic acid in the form of the sodium salt, sodium and calcium propionates, sodium and potassium nitrites, and esters of para-hydroxybenzoic acid (parabens) to foods as preservatives is regulated by the Food and Drug Administration and by the U.S. Department of Agriculture.

Ionizing radiation. Ultraviolet irradiation has already been used successfully by food industries to treat the air in storage and processing rooms, to prevent the growth of fungi on shelves and walls of preparation rooms, and to destroy parasites in meat.

Cold sterilization of foods by means of ionizing radiation is under investigation by the United States Armed Forces and the Atomic Energy Commission, and one of the major obstacles encountered has been the terrific resistance of the spores of *Clostridium botulinum* to the effects of radiation (a problem with seafoods).

Food in vending machines. Vending machines today dispense many foods that support the rapid growth of infectious or toxigenic microbes. The food to be safe for consumption must be prepared under sanitary conditions and be transported in properly refrigerated vehicles. In the vending machine they must be maintained at proper temperatures and replaced often.

QUESTIONS FOR REVIEW

1. Name diseases spread by contaminated milk and food.
2. Give differential characteristics of milkborne and waterborne epidemics.
3. What is meant by Grade A pasteurized milk?
4. What is flash pasteurization?
5. Give the requirements for a safe milk supply.
6. Make an outline showing the different kinds of food poisoning. Name causal agents.
7. Differentiate food intoxication from food infection. Give examples.
8. List the different methods of food preservation.
9. List practical measures to safeguard food.
10. Explain briefly lactic acid bacteria, normal milk bacteria, yogurt, buttermilk, holding method of pasteurization, milk sickness, phosphatase test, souring of milk, food idiosyncrasy, low-acid foods, high-acid foods, blanching of foods.
11. Indicate how a bacterial count of a milk sample is done.

REFERENCES. See at end of Chapter 37.

36 Immunizing biologicals*

The two terms *vaccine* and *immune serum* are often confused, any product of this nature being spoken of as a "serum." Vaccines and immune serums differ in fundamental properties, method of production, and resultant type of immunity.

A *vaccine* is the causative agent of a disease (bacterium, toxin, virus, or other microbe) so modified as to be incapable of producing the disease yet at the same time so little changed that it is able, when introduced into the body, to elicit production of specific antibodies against the disease. Vaccines are always antigens; therefore they always induce active immunity. They are most useful in the *prevention* of disease.

An *immune serum* is the serum of an animal (or man) that has been immunized against a given infectious disease. Its salient feature pertains to its contained antibodies. Immune serums confer passive immunity; their antibodies do *not* stimulate further antibody production. Both vaccines and immune serums are specific in their action; that is, they induce immunity to no disease other than the one for which they are prepared. Immune serums are of two types: antitoxic and antibacterial.

IMMUNE SERUMS (PASSIVE IMMUNIZATION)

Antitoxins. Antitoxins are immune serums that neutralize toxins. They may be prepared artificially, and they also develop in the body as a result of repeated slight infections. This is why some adults are immune to diphtheria. Antitoxins have no action on the bacteria that produce the toxins. For example, diphtheria antitoxin neutralizes diphtheria toxin in the tissues and body fluids but has no effect on the

---

*In this chapter are presented the biologic products that have been standardized and are available. The administration of certain of these for the production of passive immunity is given. In the next chapter are recommended schedules for the production of active immunity.

diphtheria bacilli growing in the throat and producing the toxin. Neutralization of the circulating toxin in the body does favor the defense mechanisms operating to eliminate the membrane formed in the throat, however.

PREPARATION. Antitoxins can be successfully prepared only against exotoxins. The antitoxins that have been used longest and saved most lives are those against diphtheria and tetanus toxins. Both are prepared in the same general way. The bacteria are grown in liquid culture until a large amount of exotoxin has been released. The bacilli are then filtered from the medium and its toxin content. It is necessary to determine the strength of each lot of toxin or antitoxin because two lots prepared exactly alike seldom have the same strength.

The unit of toxin is measured as the *minimum lethal dose* (MLD), the amount of toxin that when injected into a test animal will kill it in a prescribed time. For diphtheria toxin it is the amount that will kill a guinea pig weighing 250 g in 4 days, and for tetanus toxin, the amount that will kill a 350 g guinea pig in the same length of time.* To determine the MLD, one begins with a very small dose and gives increasing amounts of toxin to a series of guinea pigs. The animals receiving small doses may not be affected or else become but slightly ill and recover; those receiving larger doses become ill and some may die, but it is more than 4 days before death occurs. Finally, an animal receiving a still larger dose dies on the fourth day. The amount of toxin given this animal contains one MLD.

When the toxin is found to be of suitable strength, horses† are immunized with it. First, a small dose is used; it is increased at each of several successive injections. The first dose of toxin may be preceded by an injection of antitoxin. At the time that experience has shown antitoxin production to be at its height, the horse is bled, and the antitoxin strength of its serum tested. If the serum is found to contain sufficient antitoxin, it is further refined and purified for use. Refining serves to (1) concentrate the antitoxin and (2) eliminate horse serum protein. Horse proteins can sensitize the recipient of the antitoxin. If an immune serum is given the second time to a sensitized person, anaphylactic shock may result. It is *not* the antibodies in antitoxins and other immune serums that lead to allergic manifestations but the *serum protein* of the animal used in preparing the immune serum.

STANDARDIZATION. There are three methods of standardizing antitoxins‡: (1) animal protection tests, (2) skin reactions, and (3) flocculation tests. We shall briefly describe how diphtheria antitoxin is standardized by determining its protective action against diphtheria toxin in a susceptible animal. First, there are certain definitions to be understood.

1. *Standard antitoxin*—an antitoxin of known strength prepared, stored, and distributed according to precise specifications

---

*Tetanus toxin is so potent a poison that 1 ml of a broth culture of tetanus bacilli may contain enough toxin to kill 75,000 guinea pigs.
†In a few instances, cattle instead of horses are used.
‡In 1896, Paul Ehrlich introduced methods of standardizing toxins and antitoxins. To him goes the credit for the concept of MLD.

2. *Unit of antitoxin*—an amount of antitoxin equivalent to 1 unit of standard antitoxin *

3. *L+ dose of diphtheria toxin*—an amount of toxin that, when combined with 1 unit of antitoxin, causes the death of a 250 g guinea pig on the fourth day

Different amounts of diphtheria toxin are mixed with 1 unit of standard diphtheria antitoxin, and the mixtures are tested in 250 g guinea pigs. The L+ dose of toxin as thus determined is mixed with different amounts of the antitoxin being tested. These are injected into 250 g guinea pigs, and the mixture that causes the death of a guinea pig on the fourth day obviously contains 1 unit of antitoxin. Tetanus antitoxin is standardized in much the same manner as diphtheria antitoxin, but different amounts of antitoxin and toxin are used. Botulism and gas gangrene antitoxins are standardized in a similar manner, but the mouse is the test animal.

Standardization of antitoxins by skin reactions is based on the same underlying principles as standardization by animal protection tests. The toxin and antitoxin are injected into the skin of rabbits or guinea pigs. The production of a skin reaction by this method has the same significance as the death of a guinea pig in the guinea pig antitoxin-toxin injection method.

The flocculation method depends on perceptible flocculation in a test tube when antitoxin and toxin are brought together in certain proportions.

Since antitoxin can be successfully prepared only against the few organisms producing extracellular toxins, the number of antitoxins is limited. Generally speaking, all antitoxins should be given *early* in the disease and *in sufficient amount* because they cannot repair injury already done.

LISTING OF ANTITOXINS. *Botulism antitoxin, trivalent (equine)*, types A, B, and E, is a refined and concentrated preparation distributed and stored by the Center for Disease Control in Atlanta, Georgia. † It is given to prevent damage from toxin not already taken up by the central nervous system. Therefore it is of value when given before manifestations have appeared but is of little effect after that time. Antitoxin against one type of *Clostridium botulinum* is ineffective against the toxin of the other types. To be therapeutically expedient, antitoxin must be available against toxins of the A, B, and E types of bacilli, the ones causing disease in man.

Before he receives antitoxin, the individual must be *skin tested*—about 15% of persons receiving the antitoxin show allergic reactions. For the treatment of botulism it is best to give large doses early, with the first one given intravenously and an additional dose intramuscularly. More antitoxin is given in 2 to 4 hours if indicated. For prophylaxis in an individual who has consumed suspect food, antitoxin is given intramuscularly, and the individual watched for signs of botulism.

---

*This definition is used instead of the original definition of Ehrlich—the amount of antitoxin required to neutralize 100 MLD of toxin—since toxin also contains variable quantities of toxoid. Although toxoid has no disease-producing capacity, it can combine with antitoxin. Different lots of antitoxin tested against different lots of toxin therefore have different strengths. The standard antitoxin unit used in this country contains sufficient antitoxin to neutralize 100 MLD of the particular toxin that Ehrlich used originally in establishing his unit of antitoxin.

†See also p. 411.

Diphtheria antitoxin (purified, concentrated globulin, equine) is a therapeutic agent that has given possibly more brilliant results than any other. By its use much suffering has been prevented and many lives saved. In the production of diphtheria antitoxin in horses, toxoid has replaced diphtheria toxin as the immunizing agent. Refinements in manufacture give a diphtheria antitoxin that contains more than 5000 units per milliliter and is less likely to trigger serum reactions than previous antitoxins, since in production a very high proportion of the horse serum protein is removed.

Diphtheria antitoxin is *not* effective against toxin that has combined with the body cells, and no amount of antitoxin given late in the disease can repair injury already incurred. For best results, the disease must be recognized early and sufficient antitoxin given at once. After subcutaneous administration, the antitoxin content of the blood does not reach its maximum for 72 hours. Therefore injections must be made *intravenously* (or *intramuscularly*).

The Committee on Infectious Diseases of the American Academy of Pediatrics *insists* that antitoxin be given as soon as the clinical diagnosis of diphtheria is made, with *no delay* even in waiting for bacteriologic results. The site of the membrane, the degree of toxicity, and the length of illness are much more reliable guidelines for determining the dose of diphtheria antitoxin than the weight and age of the patient. (Children and adults receive the same doses.) The Committee's schedule for the administration of diphtheria antitoxin is given in Table 36-1.

Note: When diphtheria antitoxin is given, *remember a serum reaction is a pos-*

TABLE 36-1. ADMINISTRATION OF ANTITOXIN IN DIPHTHERIA*

	Duration of illness†			
	48 hours	**Over 48 hours**		
Lesions	Throat and larynx	Membrane in nasopharynx	Brawny swelling of neck	Extensive disease (3 + days)
Dose of antitoxin (units)	20,000 to 40,000	40,000 to 60,000	80,000 to 120,000	80,000 to 120,000
Route of administration‡	1. Intravenously or 2. Intravenously up to one-half; rest intramuscularly	Intravenously	Intravenously	Intravenously

*Recommendations of the Committee on Infectious Diseases (1974) of the American Academy of Pediatrics.
†For prophylaxis, 1000 to 5000 units of antitoxin are given to the Schick-positive individual exposed.
‡Preferred route is intravenous (after eye and skin tests for hypersensitivity) to neutralize toxin rapidly. If patient reacts to antitoxin, desensitization is indicated.

sibility, especially when the antitoxin is given *intravenously*. Preliminary testing for serum hypersensitivity is mandatory.

Although penicillin G and erythromycin have no effect on the toxin of *Corynebacterium diphtheriae*, they attack the microbe itself. The administration of these antibiotics is a valuable adjunct to serum therapy, never a substitute. It reduces the number of secondary invaders, decreases the severity and length of the illness, and helps to eliminate carriers.

Tetanus antitoxin (TAT) or *Antitetanus serum* (ATS) is a product usually obtained from horses, but it may also be produced in cattle. Although patients have undoubtedly benefited by its use, to the person allergic to horse serum there is a real hazard. *Tetanus immune globulin (human)* (TIG[H]) is prepared from the blood of persons actively immunized with tetanus toxoid. As a gamma globulin fraction of a hyperimmunized human being, it contains antitoxin without any foreign protein. Table 36-2 compares the human antitoxin with that of animal origin.

If the tetanus-susceptible wound is seen immediately after injury and if it can be adequately cared for by standard surgical technics and antibiotics, most authorities believe that the risk involved in giving equine or bovine tetanus antitoxin is not justified regardless of the immune status of the patient. On the other hand, if the patient delays seeking medical attention for a day or more or if the wound is one in which adequate surgical care is impossible because of its extent and nature, protection must be given to nonimmunized persons and to those patients who have failed to take their tetanus booster in the preceding 5 years. Equine or bovine antitoxin is never recommended if human tetanus immune globulin is available.

Gas gangrene antitoxin is prepared against the organisms important in causing gas gangrene. Its value is questioned. Many surgeons do not give it.

TABLE 36-2. EVALUATION OF BIOLOGIC PRODUCTS FOR PASSIVE IMMUNIZATION IN TETANUS

Point of comparison	Tetanus antitoxin (TAT)	
	Equine or bovine*	Human (tetanus immune globulin [human] — TIG[H])
Comparative effectiveness	Less effective	More effective, 10:1
Dosage	Usual — 3000 to 10,000 units†	Usual adult dose — 250 to 500 units
Duration of protective levels of antitoxin in recipient	5 days	Up to 30 days (detectable levels reported to 14 weeks)
Danger of allergic reaction	Present; estimated 6% to 7% or more	Remote
Cost	Less expensive	More expensive
Evaluation	Not advocated if human product available	Recommended in allergic persons and generally

*The majority of persons sensitive to horse serum are also sensitive to bovine serum.
†Recommendations as to administration of tetanus antitoxin are variable — 500 units may be as effective according to some authorities.

Antivenins. Antivenins have been prepared against the venoms of snakes and the black widow spider. They are prepared in the same general way as antitoxins, that is, by immunization of a horse with serial doses of the venom of the snake or spider. Antivenins may be prepared against a single species of snake or against a group of closely related species. Whether snake antivenin is prepared against a species or a group of species depends on the geographic location in which it is to be used. *North American antisnakebite serum* (also known as Antivenin [*Crotalidae*] Polyvalent and crotaline antitoxin, polyvalent) is effective against rattlesnakes, copperheads, and cottonmouth moccasins, the most common poisonous snakes of North America. It is not effective against the venom of the coral snake. To combat that potent poison, *North American coral snake antivenin* (Antivenin [*Micrurus fulvius*] [Equine origin]) is distributed to state and local health departments through the Center for Disease Control, Atlanta.

The use of antivenin in snakebite or black widow spider bite should in no way replace first aid and supportive measures.

Antibacterial serums. Before the discovery of the antimicrobial compounds, antibacterial serums played a crucial role in the treatment of various infections. Among such were those against meningococci (antimeningococcal serum) and pneumococci (antipneumococcal serum). Many were prepared by injecting horses with the given bacteria. Rabbits were employed in the manufacture of antipneumococcal serum, and a specific serum for each known type of pneumococcus was prepared. A serum against *Haemophilus influenzae*, type b, was prepared by injecting rabbits with the bacilli and was formerly used in the treatment of meningitis caused by *Haemophilus influenzae*, type b. Antipertussis serum was prepared by injecting rabbits with *Bordetella pertussis* and its products and used with success in very young children to prevent the disease in those exposed (when given early enough) and to decrease the severity of the established disease.

Antibacterial serums act to destroy bacteria. It is thought that they combine with a surface antigen, thereby rendering the bacteria more susceptible to leukocytes. A serum that acts on several strains of bacteria is *polyvalent*, one that acts on only one strain, *univalent*. Today the treatment of disease processes with antibacterial serums has almost completely been supplanted by sulfonamide and antibiotic therapy.

Antiviral serums. The best known antiviral serum is *antirabies serum*, prepared by injecting horses with rabies virus. (With any hyperimmune serum of equine origin, remember there is always the danger of an anaphylactic reaction or severe serum sickness.) Its purpose is to establish an immunity directly after exposure and to provide a passive immunity until an active one can be established by vaccination. See Table 37-2, p. 701, for the recommendations of the World Health Organization.

For passive immunization in rabies, there is currently an effective human antiserum, *human rabies immune globulin* (HRIG). It is obtained by immunizing human volunteers.

Convalescent serum. Convalescent serum therapy consists of the injection of the whole blood or serum of a person *recently* recovered from a disease into one ill of the disease (as a therapeutic measure) or into one exposed to the disease (as a pre-

TABLE 36-3.   CLINICAL VALUE OF IMMUNE SERUM GLOBULIN (GAMMA GLOBULIN) (ISG)*

Disease	Indications	Preparation
Measles†	Modification of symptoms after vaccination; use only with Edmonston live virus vaccine Prevention:	Immune serum globulin (human) (also measles immune globulin)
	1. Children with normal immune mechanisms	Immune serum globulin (human)
	2. Children with immune system disorder	Immune serum globulin (human)
	Modification	Immune serum globulin (human)
Rubella‡	Prevention only in female in first trimester of pregnancy	Immune serum globulin (human)
Varicella (chickenpox)	Modification	Immune serum globulin (human)
Immunologic deficiency syndromes (agammaglobulinemia, hypogammaglobulinemia, dysgammaglobulinemia)§	Replacement therapy	Immune serum globulin (human)
Rh hemolytic disease	Prevention	Rh_0(D) immune globulin (human)
Poliomyelitis‡	Prevention	Immune serum globulin (human)
Viral hepatitis type B	Prevention (value questioned)	Immune serum globulin (human) Note: Hepatitis B immune globulin now for investigational use with good results. ISG — low titer; HBIG — high titer antibodies.

*Compiled from the 1976 Physicians' Desk Reference, Oradell, N. J., 1976, Medical Economics, Inc., and 1974 Red book of the American Academy of Pediatrics. Diphtheria immune (human) globulin, zoster immune globulin (ZIG), and human rabies immune globulin (HRIG) have also been prepared.
†See also p. 704 for schedule of measles vaccination.
‡See also Chapter 37.
§See also p. 202 for discussion.
‖Rh_0 (D) immune globulin (human) (RhoGAM) is discussed on p. 212.

Dose	Schedule	Route	Comments
04 ml/kg body weight	One dose	Intramuscular	Inject measles vaccine in one arm, this in other, with separate syringe
25 ml/kg body weight			Inject within 4 days of exposure; give live virus vaccine in 8 or more weeks
to 30 ml			
05 ml/kg body weight			Give within 6 days, if possible (rarely indicated)
25 to 0.44 ml/kg body weight (20 ml to 30 ml total advised by some)	One dose	Intramuscular	Subject controversial
to 1.2 ml/kg body weight (20 to 30 ml total sometimes advised)	One dose	Intramuscular	Give within 3 days of exposure; may decrease severity of illness
6 ml/kg (20 to 30 ml total sometimes advised) uble dose at onset of therapy	One dose every 3 to 4 weeks	Intramuscular	Maximum dose 20 to 30 ml at any one time; gamma globulin deficiency here; adequate levels must be supplied to protect against infections
c vial, usually	One dose	Intramuscular	Give to nonsensitized Rh-negative mothers after delivery of Rh-positive infant or after abortion
ml/kg body weight	One dose (second dose can be given 5 weeks after first)	Intramuscular	Protection for 2 to 6 weeks if given before exposure or development of symptoms
ml (for adult)	One dose within 1 week after transfusion; one dose (10 ml) 1 month later	Intramuscular	For adults—not recommended for children; recommended for high-risk adults only

Continued.

TABLE 36-3. CLINICAL VALUE OF IMMUNE SERUM GLOBULIN (GAMMA GLOBULIN) (ISG)*—cont'd

Disease	Indications	Preparation
Viral hepatitis type A	Prevention or modification in children or adults 1. Short-term, moderate risk	Immune serum globulin (human)
	2. Long-term, intense exposure (also prophylaxis in institutions)	
Pertussis‡	Prevention in infants under 2 years	Pertussis immune serum globulin (human)
	Treatment	
Tetanus‡	Prevention—children and adults	Tetanus immune globulin (human)
	Treatment	

ventive measure). In this way, the blood of the convalescent patient containing antibodies confers a passive immunity on the recipient. This type of therapy has been used with good results in measles, whooping cough, and scarlet fever.

Caution should be observed in giving convalescent serum or any human serum because it may transmit viral hepatitis!

Gamma globulin. Gamma globulin is the main globulin fraction of blood plasma with which antibodies are associated. It is prepared commercially by a series of precipitations in pooled normal adult plasma, in venous blood, or in pooled extracts of human placentas, with varying concentrations of alcohol at a low temperature. The gamma globulin fraction removed is more than 40 times as rich in antibodies as the original plasma from which it is taken. Five hundred milliliters of blood yields an average dose.

Gamma globulin, usually dispensed as *immune serum globulin (human) (ISG)*, is emphasized as indicated in Table 36-3. *Gamma globulin is always given intramuscularly—never intravenously** (Fig. 36-1). As ordinarily given, gamma globulin is one of the benign injectables, being associated with few side reactions.

---

*The consequence can be a severe shocklike state.

Dose	Schedule	Route	Comments
02 to 0.05 ml/kg body weight	One (repeat in 3 to 5 months if necessary)	Intramuscular	Give within 1 week of exposure
06 ml/kg body weight	One (repeat in 5 to 6 months if necessary)		Protection for 1 year
5 ml	One dose (repeat in 5 to 7 days)	Intramuscular	Product from donors hyperimmunized with pertussis vaccine; protection not reliable
25 ml	One dose daily for three to five doses; or 3 to 6.75 ml at once		Use larger doses with severe disease
units/kg body weight; 250 units, usual adult dose	One dose	Intramuscular	Product from person hyperimmunized with tetanus toxoid
00 to 10,000 units	May be repeated 1 month later		Active immunization should be started with tetanus toxoid

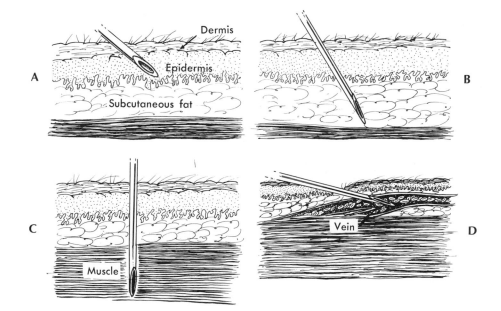

FIG. 36-1. Routes of injection, diagrammatic sketches. Note position and depth of needle. **A,** Intradermal. **B,** Subcutaneous. **C,** Intramuscular. **D,** Intravenous.

VACCINES (ACTIVE IMMUNIZATION)

Bacterial vaccines. Bacterial vaccines are suspensions of killed bacteria in isotonic sodium chloride solution. After culture for 24 to 48 hours, the bacterial growth is emulsified in sterile saline solution and the number of bacteria in each milliliter of the washings determined. Isotonic sodium chloride solution is added until the desired number per milliliter is obtained. The bacteria are then killed at the lowest possible lethal temperature in the shortest possible time. Without heat, they may be killed by formalin, cresol, or Merthiolate. In some vaccines the bacteria are killed by ultraviolet light. The finished product is cultured to verify its sterility. Vaccines are usually prepared so that the number of dead microorganisms given at a single injection ranges from 100 million to 1 billion.

Bacterial vaccines are of two types: *stock*, made from stock cultures maintained in the laboratory, and *autogenous*, made from the organisms of a specific lesion in the patient for whom the vaccine is prepared. A *mixed* vaccine contains bacteria belonging to two or more species. A *polyvalent vaccine* contains several strains of organisms belonging to the same species, for instance, one containing several strains of *Pseudomonas aeruginosa*. Bacterial vaccines show their best results in the prevention of typhoid fever, whooping cough, and plague.

Prompted by the growing resistance of bacteria to standard antibiotics, researchers are presently concentrating on the development of bacterial vaccines. As improvements in laboratory technology have come and as more is learned of the mechanisms of host immunity and how bacteria cause disease, they are accomplishing more. Major bacterial antigens, both polysaccharide and protein, have been isolated, purified, and studied, and of special interest in vaccine production, certain avirulent bacterial hybrids have been formulated.

LISTING OF BACTERIAL VACCINES. *Bacille Calmette-Guérin (BCG) vaccine (Tuberculosis vaccine)* is derived from a strain of bovine tubercle bacilli cultivated on artificial media (containing bile) so long that they have completely lost their virulence for man. There are many BCG vaccines in the world today, all derived from the original strain, but they are quite variable in their properties. In the United States freeze-dried BCG vaccines are available in two concentrations, the higher for use in the multiple-puncture technic and the lower for inoculation into the skin. Administered either way, the vaccine is given *only to persons with a negative tuberculin test*.

Being tested experimentally is a BCG vaccine aerosol. Persons to be vaccinated breathe in vaccine droplets suspended in a man-made mist for some 45 minutes or so.

BCG vaccine has been used extensively in European countries to immunize children. The World Health Organization in recent years has vaccinated over 250 million persons with BCG. This has been an especially worthwhile program in parts of the world where the incidence of tuberculosis is high. In the United States the prevalence rate is low. BCG vaccination, a controversial issue for over 50 years, has not been used widely.

Cholera vaccine given to the military personnel of the United States Armed

Forces sent to duty in cholera-endemic areas contains 8 billion killed vibrios per milliliter. Cholera vaccine is toxic and must be injected at frequent intervals to maintain immunity. A new oral vaccine that contains live, attenuated vibrios in water is being tested. It can be given daily and induces antibody formation in the intestinal tract, the actual site of involvement in the disease. Another new vaccine, a cholera toxoid (inactivated cholera enterotoxin), will soon be available.

Pertussis vaccine is a saline suspension of the killed organisms. Pertussis vaccine is commonly prepared and used in the alum-precipitated and aluminum hydroxide– and aluminum phosphate–adsorbed forms in combination with diphtheria and tetanus toxoids. Severe neurologic reactions rarely complicate administration of pertussis vaccine.

Plague vaccine of the United States Armed Forces contains 2 billion killed plague bacilli per milliliter.

Tularemia vaccine, live, attenuated is a lyophilized, viable, weakened variant of *Francisella tularensis*. Tularemia vaccine has also been prepared by the U.S. Army as an inhalant-type vaccine. Large numbers of persons can be vaccinated against this disease simply by marching through a room containing vaccine droplets in the atmosphere. The interest of the military in this disease stems from its possibilities in bacteriologic warfare. Airborne immunization is being evaluated experimentally in other diseases transmitted by droplet infection. *Foshay's tularemia vaccine* is made of killed organisms and produces an immunity that lasts about 1 year.

Typhoid vaccine for the U.S. Armed Forces is an acetone-inactivated dried strain of *Salmonella typhi* (the AKD vaccine). An oral vaccine containing live attenuated typhoid bacilli is being tested. The U.S. Public Health Service discredits paratyphoid A and B vaccines as immunizing agents and has implicated them in reactions following administration of mixed vaccines.

RICKETTSIAL (BACTERIAL) VACCINES. Rickettsias for vaccine production have been grown in lice, rodent lung, and cell cultures. The vaccines in use at the present time are prepared by growing the rickettsias in the yolk sac of the developing chick embryo. A vaccine so prepared is referred to as a *Cox vaccine*. Cox vaccines containing the respective rickettsias have been prepared against epidemic typhus fever, murine typhus fever, Rocky Mountain spotted fever, and Q fever. In addition, a Rocky Mountain spotted fever vaccine found to be effective has been made by grinding up the bodies of infected ticks.

Vaccination against the rickettsial diseases is advised for persons likely to contact the infectious agent because of living conditions or occupation. Protection against epidemic typhus is strongly urged for those who have been in contact with a patient. Although typhus immunization is not required by any country in the world as a condition of entry, it is nevertheless recommended to travelers when travel plans include a geographic area that is not only infected but one in which living conditions are generally poor.

Toxoids. *Botulinum toxoid*, an effective pentavalent preparation available for active immunization against botulism, incorporates aluminum phosphate–adsorbed toxoids (see p. 686) for the antitoxins of *Clostridium botulinum*, types A, B, C, D,

685

and E. It is given in three spaced injections followed by a booster to laboratory workers at risk.

Diphtheria toxoid is diphtheria toxin detoxified so that it cannot cause disease but can induce formation of specific antitoxin. Formalin, 0.2% to 0.4%, is added to diphtheria toxin and the mixture incubated at 37° C until detoxication is complete (several weeks). This treatment, used with other bacterial toxins as well, reduces toxicity but preserves the antigenic properties of toxin. After purification to remove inert protein, the preparation is available as *plain* or *fluid toxoid.* Fluid toxoids are used in the United Kingdom and Canada.

Alum added to diphtheria toxoid precipitates the antigenic portion. The precipitate, after being washed and suspended in sterile physiologic saline solution, is *diphtheria toxoid, alum precipitated.* Its advantage is that the alum is not absorbed but remains at the injection site. The toxoid slowly separates from it to give a prolonged antigenic stimulation. If aluminum hydroxide or aluminum phosphate is added to the liquid toxoid, the antigenic portion adheres to the particles of the aluminum compound. This process is known as *adsorption,* and diphtheria toxoid so treated is known as *aluminum hydroxide–* or *aluminum phosphate–adsorbed diphtheria toxoid* (depot toxoid or antigen). Its properties are similar to those of the alum-precipitated toxoid. Diphtheria toxoid should not be given to a person more than 10 to 12 years of age without preliminary sensitivity tests.

Tetanus toxoid establishes permanent immunity to tetanus. It is manufactured in the liquid (or fluid), alum-precipitated, aluminum hydroxide–adsorbed, and aluminum phosphate–adsorbed forms.

Most manufacturers market mixtures of diphtheria and tetanus toxoids (DT) or mixtures of diphtheria and tetanus toxoids and pertussis vaccine (DTP). Combinations are available in the unconcentrated, alum-precipitated, aluminum hydroxide–adsorbed and aluminum phosphate–adsorbed forms. They have proved to be very satisfactory.

Viral vaccines. With the exception of the antibiotics, nothing has done more to protect us against infection than the unborn chick. Viral vaccines have been possible largely because of the development of modern technics for cultivating viruses in embryonated hen's eggs and also in cell cultures, thus furnishing the large supply of virus essential to the production of vaccines. In chick embryos or in cell cultures, cultivation of a virus involving multiple transfers from one medium to another (serial passage) can alter the organism's pathogenicity. A normally virulent virus can thus be *attenuated* (weakened) or domesticated. Fortunately, with loss of virulence there is *no* loss of antigenicity. This makes possible the production of very effective vaccines.

In a killed or *inactivated* virus vaccine the infectivity of the virus and its ability to reproduce have been destroyed by physical or chemical means, but its capacity to induce antibody formation has been preserved. Formalin is a standard inactivating agent.

Killed virus vaccines must be injected in several doses, but live virus vaccines can be taken orally or inhaled (as an aerosol). Virologists have long preferred living agents to killed or inactivated viruses for vaccines. A living agent continues to multiply in the body of the animal or man to which it is given and thus exerts a prolonged and increasingly strong stimulation to the host to make antibodies. The immunity so produced is stronger and longer lasting because the presence of the live virus simulates actual infection. The possibility that the virus will revert to its original level of pathogenicity in the vaccinated person is an obvious disadvantage to the use of the living agent. Live viral vaccines are well known in veterinary medicine. The two best known in human medicine are the measles vaccine and the live poliovirus vaccine, Sabin-type strains.

LISTING OF VIRAL VACCINES. *Adenovirus vaccine* has been prepared by the formalin and ultraviolet inactivation of viral types grown on monkey kidney cells. Because of the discovery that some of the *inactivated* adenovirus serotypes produced tumors in experimental animals, active immunization with these agents is suspended. Trials with live adenovirus types 4 and 7 vaccine given orally in an enteric-coated tablet are in operation with good results, since live adenoviruses presumably have no capacity to cause cancers.

Influenza virus vaccine is prepared from virus grown in the fertile hen's egg, inactivated with formalin, and concentrated (Fig. 36-2).

The Bureau of Biologics, Food and Drug Administration, regularly reviews the formulation of influenza vaccines, suggesting changes as needed to take in the strains expected to cause trouble during the next flu season. Usually the strains of the recent

FIG. 36-2. Influenza vaccine, commercial production. Virus is injected into fertile hen's eggs along an assembly line. (Courtesy Eli Lilly & Co., Indianapolis, Ind.)

epidemics are known, and the most practical arrangement from year to year seems to be one wherein two strains recently implicated are placed in a bivalent vaccine. It is thought that a bivalent vaccine with greater amounts of antigen is preferable to a polyvalent one containing smaller amounts of four or five strains. In the event that a large-scale epidemic is anticipated, a monovalent vaccine might be indicated.

The peculiar ability of influenza viruses to change their antigenic structure from time to time poses a problem to the manufacturer of flu vaccines. A strain against which there is no protection in an existing polyvalent vaccine can easily emerge. Great care must be taken that strains containing a wide pattern of antigenic substances are selected for vaccine production.

The production of active immunity by influenza virus vaccine has met with considerable success. However, the immunity is short-lived, the vaccine sometimes gives fairly severe reactions, and the subject can become sensitized to egg protein as well as react to a previously existing sensitivity. Highly purified vaccines with most of the egg protein eliminated are being made available.

New vaccines are being tested. One is an attenuated live virus vaccine, a recombinant, administered intranasally (by nose drops). Another utilizes the subunit proteins, hemagglutinin and neuraminidase, in the viral envelope to induce antibody formation in the recipient. Antibody to neuraminidase reduces the amount of virus replicating in the respiratory tract and lessens the likelihood of transmission to contacts.

Measles virus vaccines recommended for use are either (1) *live, attenuated virus vaccine* (Edmonston B vaccine) prepared from chick embryo or canine renal cell cultures of the Edmonston B strain of the measles virus (developed by John F. Enders and associates) or (2) *live virus vaccines* (Edmonston strains) which have been further attenuated by additional chick embryo passages (the *further attenuated vaccines* of Schwarz and Moraten). Live virus vaccines give permanent protection. Inactivated virus vaccines are in disrepute. *Note:* Measles vaccines must be carefully refrigerated.

Mumps virus vaccine is an attenuated live virus vaccine adapted to the chick embryo and given in a single injection. The vaccine contains egg proteins and a small quantity of neomycin. Routine administration of mumps vaccine is contraindicated before puberty. At the present time, it is thought best to let the younger child develop his own immunity. The duration of protection afforded by the mumps vaccine is at least 4 years.

Salk poliomyelitis vaccine, inactivated poliovirus vaccine (IPV), is a formaldehyde-inactivated viral vaccine, which is highly effective. It must be injected.

Sabin poliomyelitis vaccine, oral poliovirus vaccine (OPV), is prepared with the three types of attenuated live poliovirus grown in human (not monkey) cell culture. After the vaccine has been fed to an individual, the virus of the vaccine multiplies in the lining and in the lymphoid tissue of the alimentary tract. A satisfactory immune response from the vaccinated person occurs usually within 7 to 10 days. The implantation of the virus in the bowel blocks further infection by the same type of a wild poliovirus. This materially cuts down on the number of carriers. The fear that competition from other enteroviruses might check growth of poliovirus in the intestinal tract has been largely obliterated by studies of children in the tropics. Their alimentary tracts literally swarm with viruses and yet they show antibody responses indicating growth of poliovirus. To minimize possible interference from other enteric viruses, it is recommended that the oral vaccine be given during the spring and winter months in temperate climates.

The chief disadvantage of the Sabin vaccine is that it is difficult to preserve. It can be kept frozen for years but only *for 7 days in an ordinary refrigerator* and *3 days at room temperature!* Tap water cannot be used to dilute it, since *the contained chlorine destroys the poliovirus.* Like other vaccines manufactured on kidney cells, the final preparation contains traces of penicillin and streptomycin from the culture medium. The overall effectiveness of the Sabin vaccine is 94% to 100%.

Rabies vaccines are of two kinds, one made from nervous tissues and the other from nonnervous tissues. The first is prepared from the brain and spinal cord of rabbits suffering with rabies induced by the inoculation of fixed virus (p. 534). Just before death the brain and cord are removed under strict aseptic precautions, and the virus is inactivated. The phenol-inactivated vaccine is the *Semple vaccine.* An emulsion is prepared by grinding the virus-containing material in water or isotonic saline solution, either during the process of inactivation or after it is complete. The fact that this vaccine contains brain and spinal cord tissue is thought to be responsible for the severe complication of encephalomyelitis that sometimes follows its administration. Probably on an allergic basis, this disease may lead to paralysis and even death.

The other kind of rabies vaccine is free of nervous tissue. It is prepared by growing the fixed rabies virus in the tissues of the embryonated duck egg (Fig. 36-3). To the suspension of embryonic duck tissue, a viricidal agent is added to inactivate the

FIG. 36-3. Virus-laden duck embryo used for rabies vaccine. Frozen, finely ground embryonic tissue is inactivated, filtered, and dried under sterile conditions. (Courtesy Eli Lilly & Co., Indianapolis, Ind.)

virus. The duck embryo vaccine (DEV) is dried and administered after dilution with distilled water. Because of the absence of central nervous system tissue, this vaccine is less hazardous than the one prepared from the nervous tissues of the rabbit. However, it is not quite as immunogenic. Because of the rarity of serious complications, preexposure prophylaxis with duck embryo vaccine is practical in veterinarians, animal handlers, spelunkers, and research workers at risk. In the United States today, 96% of rabies vaccine given is of this type.

The U.S. Public Health Service recommends that all dogs be immunized against rabies by the age of 3 months. For dogs a vaccine is grown on a series of chick embryos to reduce its virulence. Such a live virus vaccine, given in a single dose, will produce immunity of over 3 years' duration in the dog. In areas where the vampire and other bats exist, it may be necessary to immunize livestock.

Currently a potent, experimental, killed-virus rabies vaccine has been grown in nonneural tissue culture. There is also work being done to minimize the amount of fat in nervous tissue used to grow the virus.

Rubella virus vaccine, live, attenuated, exists as three licensed preparations. One vaccine is a live virus strain originally attenuated by 77 passages in green monkey kidney and then prepared in duck embryo culture. A second is the same strain prepared in dog kidney culture. The third vaccine, the Cendehill strain, was isolated in green monkey kidney and then passed 51 times in rabbit kidney culture. These three are effective ones.

Episodes of arthritic joint pain, found especially in the older age groups, complicate the use of the licensed vaccines, especially the dog kidney one. Although arthritis was a disturbing event when first recognized as a consequence to vaccination, it has usually turned out to be mild and self-limited, without sequelae. Hypersensitivity reactions to egg protein and neomycin may occur.

Smallpox vaccine has long been prepared with cowpox (vaccinia) virus inoculated into female calves 6 months to 1 year old. The skin of the abdomen of the calf is scarified, and virus is rubbed into the scratches. In 6 days the area is thickly covered with vesicles containing the modified virus within the dermal lymph. The sticky exudate is carefully removed from the vesicles, and from it vaccine is processed and purified.

A smallpox vaccine as effective as the calf vaccine is produced by growing the vaccinia virus in the bacteria-free tissues of the chick embryo. Vaccine may also be prepared by growing the virus in cell cultures.

If smallpox vaccine is not shipped and stored at freezing temperatures, it rapidly loses its potency. There is a lyophilized (freeze-dried) preparation that can be stored at temperatures ranging from 35° to 50° F without losing its potency for a period of 18 months. Once this vaccine is reconstituted, it must be refrigerated.

Smallpox vaccination (see also p. 698) means infecting a person with a strain of vaccinia or cowpox virus. The viruses of cowpox and of smallpox are quite similar in makeup and serologic properties. Some authorities believe that cowpox originated by the transfer of smallpox from man to the cow, with an adaptation of the virus to this animal to produce a milder disease.

Yellow fever vaccine is prepared by growing a strain of the virus in chick embryos.

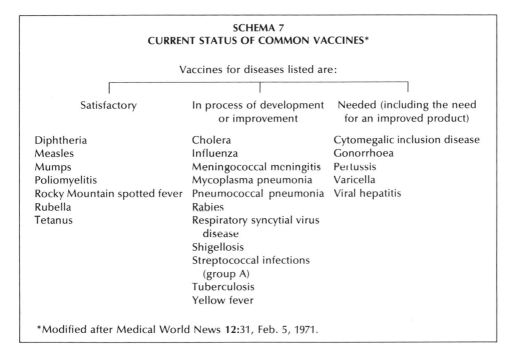

SCHEMA 7
CURRENT STATUS OF COMMON VACCINES*

Vaccines for diseases listed are:

Satisfactory	In process of development or improvement	Needed (including the need for an improved product)
Diphtheria	Cholera	Cytomegalic inclusion disease
Measles	Influenza	Gonorrhoea
Mumps	Meningococcal meningitis	Pertussis
Poliomyelitis	Mycoplasma pneumonia	Varicella
Rocky Mountain spotted fever	Pneumococcal pneumonia	Viral hepatitis
Rubella	Rabies	
Tetanus	Respiratory syncytial virus disease	
	Shigellosis	
	Streptococcal infections (group A)	
	Tuberculosis	
	Yellow fever	

*Modified after Medical World News **12:**31, Feb. 5, 1971.

It was developed at a time when little thought was given to the avian leukosis viruses. A new strain of the virus is being tested with technics that filter out any such contaminants. It will ultimately replace the old one. For purposes of international travel, yellow fever vaccine must be approved by the World Health Organization and be administered at a designated Yellow Fever Vaccination Center.

Summary. Schema 7 is a summary for material just presented and indicates prospects for the future.

PRECAUTIONS FOR ADMINISTRATION OF BIOLOGIC PRODUCTS

To ensure the development of the desired immunity and to prevent insofar as possible certain untoward side effects or complications, the following precautions in the administration of vaccines, serums, and other biologic products are strongly advised:

1. *Read carefully the label on the package and the accompanying leaflet when any biologic product is to be given!*
2. Disinfect the skin properly at the injection site. (Note also proper disinfection of the site of entry into the vial of medication.)
3. Use a sterile syringe and needle for each injection, preferably a sterile *disposable* unit. (Using several needles with the same syringe can be a way to spread viral hepatitis.)
4. When the needle is placed into the subcutaneous or intramuscular area, pull

the plunger of the syringe outward to check that the needle has not inadvertently entered a vein. If it has, blood wells up in the syringe.

5. Immunize only well children. Delay any such procedure for sick children until their recovery.

6. *Anaphylactic and allergic accidents in biotherapy are unpredictable—the next one may be yours!* Carefully question the recipient of the injection as to reactions to previous doses of the same or comparable biologic products. Inquire carefully as to previous injection of horse serum, any known allergy to horse serum protein, and any history of allergic conditions such as asthma. Did the child have reactions to a previous dose, such as fever or sleeplessness? Was there an area of redness and tenderness about the injection site? If the child had only a mild reaction, the injections may be continued. In some instances the amount of the material must be reduced and the overall number of injections increased. If the child had a severe reaction, *do not* repeat the injection. Consult the physician.

7. Ascertain the presence of allergy to egg or chicken dander if certain viral vaccines prepared in chick embryos (influenza, yellow fever, and measles) are to be given. Practically speaking if the recipient can eat eggs without event, it is safe to give the vaccine. Some of the newer vaccines may contain neomycin to which an allergic state may also exist.

8. Note these special situations:

   a. Do *not* immunize persons with altered or deficient immunologic mechanisms.

   b. Do *not* immunize persons receiving steroid hormones.

   c. Do *not* vaccinate children for smallpox who have a generalized dermatitis such as eczema because of the danger of a generalized vaccinia. Children with skin diseases should be segregated from children being vaccinated against smallpox.

   d. Unless there is an emergency, do *not* immunize during a poliomyelitis outbreak.

Tests for hypersensitivity. Preliminary testing to detect hypersensitivity by all means is mandatory.

The following tests for hypersensitivity must be used discreetly, and a syringe containing 1 ml of epinephrine 1:1000 should be handy (see also p. 236).

1. In the *scratch* test, a 1:10 dilution of the biologic product in isotonic saline solution (tetanus antitoxin, for example) is applied to the abraded skin.

2. In the *eye* test, one drop of a 1:10 dilution is placed in the conjunctival sac. If this is negative, the eye test may be repeated with undiluted material.

3. The *intradermal* skin test is usually carried out with 0.02 to 0.03 ml of a 1:10 dilution. *(An intradermal test with 0.1 ml of undiluted TAT may be fatal.)* If indicated, skin tests may be carried out serially with 0.02 to 0.03 ml of 1:10,000, 1:1000, 1:100, and 1:10 dilutions. The positive reaction is a hive-like wheal with redness.

4. In the *intravenous* test, the patient's blood pressure is recorded before the

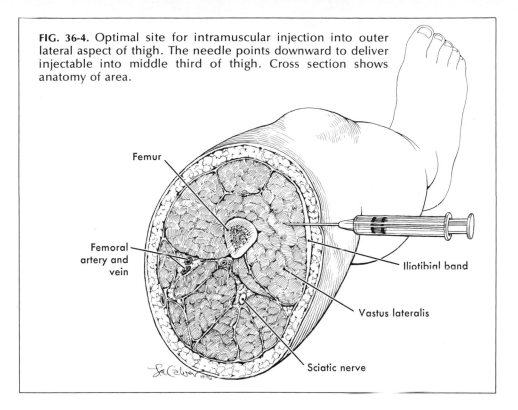

FIG. 36-4. Optimal site for intramuscular injection into outer lateral aspect of thigh. The needle points downward to deliver injectable into middle third of thigh. Cross section shows anatomy of area.

Femur

Femoral artery and vein

Iliotibial band

Vastus lateralis

Sciatic nerve

5 minutes taken to inject 0.1 ml of biologic product in 10 ml of isotonic saline solution and every 5 minutes afterward for 30 minutes. If the blood pressure falls 20 points or more, a hypersensitive state is indicated.

Injection site for biologic products. *The optimal site for injection* is the outer lateral aspect of the thigh (Fig. 36-4). By directing the needle downward as it enters the tissues at the junction of the upper third with the lower two-thirds, one can deliver serum or other biologic product into the middle third of the thigh. This area is preferred because (1) in the event of an impending serum reaction a tourniquet can be placed on the thigh above the injection site, and the absorption of the biologic product greatly delayed, and (2) here there are no important anatomic structures to be injured by pressure of injected material. If speed of absorption is desired, the fascia lata, a tense band stretched across this area, facilitates it.

STANDARDS FOR BIOLOGIC PRODUCTS

Manufacture. It is so important that serums, vaccines, and other biologic products, including human blood and its derivatives, meet acceptable standards of safety, purity, and potency that, when offered for sale, import, or export in interstate commerce, they must be manufactured under the license and regulations of the federal government. Authority for this is delegated to the Food and Drug Administration of the United States Department of Health, Education and Welfare.

Label. Each package must be properly labeled. The name, address, and license number of the manufacturer, the lot number, and expiration date must be shown.

Date of expiration. On the *date of issue* the product is placed on the market. This must be within a certain time after manufacture, depending on the kind of product and the temperature of storage. The *expiration date* means the date beyond which the product cannot be expected to exert its full potency. The expiration date of most biologic products is 1 year after manufacture or issue.

QUESTIONS FOR REVIEW

1. Define vaccine, immune serum, antitoxin, antivenin, antibacterial serum, convalescent serum, MLD, date of issue (of vaccine), expiration date (of vaccine), autogenous vaccine, lyophilized vaccine, immune serum globulin, gamma globulin.
2. Cite at least four differences between vaccines and immune serums.
3. What is the term used to designate the strength of tetanus and diphtheria antitoxins?
4. Briefly discuss the three methods of standardizing diphtheria antitoxin.
5. Name three antitoxins of importance. Indicate the manufacture of one.
6. List seven diseases for which immune serum globulin may be of benefit. Indicate how gamma globulin is used to modify or prevent disease.
7. How is toxoid prepared? Name three important toxoids.
8. What is BCG? Comment on its usefulness.
9. Compare live virus vaccines with inactivated virus vaccines.
10. List and briefly describe the measles vaccines.
11. Name immunizing agents used in combination.
12. How are rabies vaccines prepared?
13. List the rubella virus vaccines.
14. Give the optimal site for injection of biologic products.
15. List the tests for hypersensitivity.
16. Give precautions in the administration of biologic products.

REFERENCES. See at end of Chapter 37.

37 Immunizing schedules

In this chapter methods of immunization for the prevention of the more important infectious diseases are given. Since opinions from public health physicians and authorizing agencies concerning the most efficacious methods of immunization are not uniform, the student will note variations in procedures endorsed in standard references.

Dosage may be indicated in text discussions, but in actual practice, one must always carefully check the manufacturer's recommendations and admonitions concerning the product stated in the package insert and label.

RECOMMENDATIONS OF COMMITTEE ON INFECTIOUS DISEASES OF AMERICAN ACADEMY OF PEDIATRICS*

The Red Book of the American Academy of Pediatrics is widely accepted as a source of the best immunization procedures in both children and adults. Table 37-1 schedules the protection recommended by the Committee on Infectious Diseases for normal infants and children from 2 months of age through the sixteenth year of life (after which the immunizing procedures are those for adults).

Combined active immunization. In the schedule of Table 37-1, diphtheria and tetanus, adsorbed, (depot) toxoids combined with pertussis vaccine (DTP) are preferred as primary immunizing agents over the nonadsorbed (fluid or plain) mixtures. A dose† is injected deep into the deltoid or midlateral muscles of the thigh or given by intradermal jet injection (Fig. 37-1).

The standard preparations of combined diphtheria-tetanus toxoids (DT) used in

---

*A copy of the Red Book is obtained from the American Academy of Pediatrics, Inc., Evanston, Ill. The Academy has developed a personal immunization card, billfold size and plastic, to provide a permanent record.

†The concentration of antigens varies in various products. The package insert supplied with the vaccine should be carefully studied as to volume of dose.

TABLE 37-1. SCHEDULE FOR ACTIVE IMMUNIZATION AND TUBERCULIN TESTING IN NORMAL INFANTS AND CHILDREN*

Age		Administration of immunizing agents†		Tuberculin testing‡
Months	Years	Combined antigens	Others	
2		DTP (diphtheria-tetanus toxoids, adsorbed, and pertussis vaccine)	TOPV (trivalent oral polio-virus vaccine)	
4		DTP	TOPV	
6		DTP	TOPV	
12		Measles-rubella or measles-mumps-rubella§	Live measles vaccine (2 to 3 days after tuberculin test)	X
18		DTP	TOPV	
	4 to 6	DTP	TOPV	
	14 to 16	Td (adult-type combined tetanus-diphtheria toxoids)		
	Thereafter	Td every 10 years		

*See also text.
†Routine smallpox vaccination no longer recommended.
‡Risk of exposure indicates frequency of tuberculin testing.
§Combined viral vaccines (single injection) can be given from ages 1 to 12.

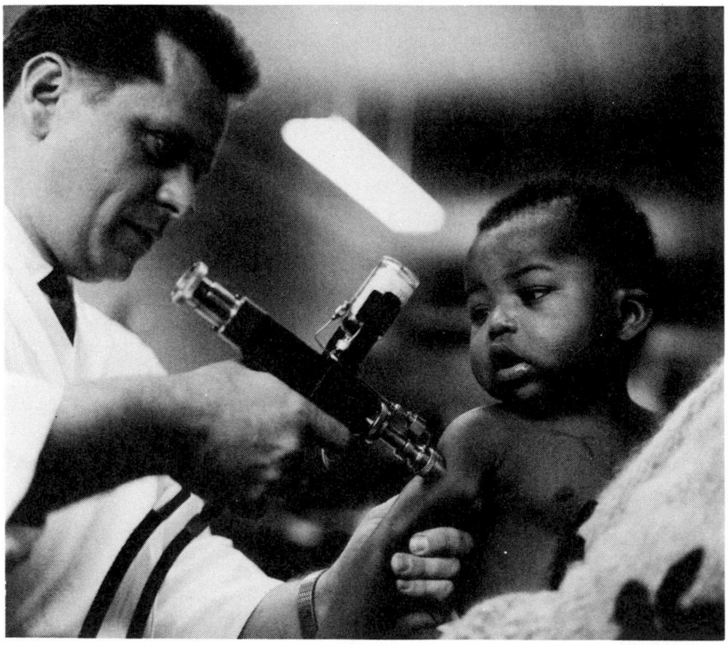

FIG. 37-1. Jet injector (air gun) delivery of measles vaccine to young Chicagoan—instant, pain free. This method is also used for other vaccines mentioned in text. (From Medical News, J.A.M.A. **196**:29, [May 23] 1966; courtesy American Medical Association.)

infants and young children contain 7 to 25 Lf (flocculating units) diphtheria toxoid per dose. The *adult type* of combined tetanus-diphtheria toxoids (Td) with adjuvant used in teenagers and adults contains not more than 2 Lf diphtheria toxoid per dose. Td contains less antigen than DT, since fairly severe reactions in older children and adults may be associated with the increase of diphtheria antigen. Adult-type toxoids are specially purified preparations.

Live oral poliovirus vaccine (OPV) is recommended over the killed vaccine because it is more effective as an immunizing agent, antibodies induced persist longer, and it is easier to give. TOPV (trivalent oral poliovirus vaccine) is the one preferred. (If monovalent virus-type feeding is carried out, the order is type 1 followed by type 3 and then type 2.)

The immunity for each disease produced by the combined method of immunization is as great as would be obtained by separate immunization against the given disease.

Tetanus. For individuals to whom the combined primary immunizations of early childhood do not apply, the Red Book recommends administration of tetanus toxoid, adsorbed. The total recommended dose of the manufacturer may be given in fractional doses of 0.05 or 0.1 ml. A booster (recall) injection of alum tetanus toxoid should be given 1 year later.

When the child or also an older person sustains an injury and the wound remains clean, the Committee on Infectious Diseases feels that in the fully immunized person no booster dose is needed unless more than 10 years have elapsed since the last dose. When the child or older person incurs a contaminated wound likely to be complicated by tetanus, such as a deep puncture wound, dog bite, certain crusted or suppurating lesions, and wounds containing soil or manure, T (tetanus toxoid with adjuvant) or Td should be given regardless of the immunization status of the patient *unless* the patient is known to have received Td or T within the past five years. If he has, the inoculation at this time need not be given.* It has been found that the immunity conferred by tetanus toxoid is extremely long lasting† and that frequent recall doses seem to provoke reactions.

Pertussis. Combined immunization as indicated in Table 37-1 is preferable, utilizing the antigens combined with adjuvant, especially when immunization is begun under the age of 6 months. Pertussis vaccine is not given beyond the age of 6 years.

When the young child has been intimately exposed to whooping cough or there is an epidemic and protective immunity must be developed as soon as possible, pertussis vaccine adsorbed should be given intramuscularly in three doses of 4 protective units each, 8 weeks apart. Routine recall injections are given 12 months later and at school age with adsorbed vaccine. Exposure recall injections indicated for children up to the age of 6 years are given with adsorbed vaccine.

---

*Nonimmunized individuals also require passive immunization. See pp. 678 and 682.
†The U.S. Public Health Service recommends that tetanus boosters of adsorbed (*not* fluid) toxoid be given at 10-year intervals.

Diphtheria. In infants and young children diphtheria toxoid is given alone *only* when there is definite reason for not using the combined immunization schedule, such as the occurrence in the child of a severe reaction following the injection of the multiple vaccines or the exposure of the child to a possible or known case of diphtheria. The toxoid with adjuvant is given in three fractional (0.05 to 0.1 ml) doses.

TOXOID SENSITIVITY TEST (ZOELLER, MOLONEY). Since even small doses of diphtheria toxoid can cause a reaction in a highly sensitive person, when the adult-type toxoids are not available, a toxoid sensitivity test should be done beforehand. One-tenth milliliter of *fluid* diphtheria toxoid diluted 1:50 or 1:100 in saline solution is injected intracutaneously. The test site is read after 18 to 24 hours. A positive reaction, indicating hypersensitivity, is a red indurated area. In that person the positive toxoid sensitivity test is presumptive evidence of immunity from natural exposure, and further toxoid should *not* under any circumstances be given. Persons with no reaction, a negative test, should receive diphtheria toxoid adsorbed as a single antigen, or diphtheria and tetanus toxoids, childhood type (DT).

Smallpox. The preferred method of smallpox vaccination is either the multiple pressure method or intradermal jet injection with calf lymph virus.* In preparation of the arm acetone or ether is suitable; alcohol, especially if medicated, should *not* be used because it destroys the virus. However, no preparation at all is better than vigorous cleaning that might abrade the skin. The vaccination site† should be kept dry, preferably uncovered, until the scab falls off. The site of a primary vaccination should be inspected between the sixth and the eighth days, the revaccination site between the fourth and the seventh days.

After vaccination *primary vaccinia* (Fig. 37-2) occurs in persons with no immunity. On the fourth day a small circumscribed solid elevation of the skin *(papule)* appears; on the eighth day the lesion contains fluid *(vesicle)* and has a pink areola around it that averages several centimeters in diameter. On the ninth day it is filled with pus *(pustule)* and has begun to dry up by the end of 14 days. The formation of a scar at the site is evidence of a successful vaccination.

---

*There are three methods for smallpox vaccination:

1. In the *multiple-pressure* technic, a sterile sharp needle is held tangentially to the skin. Pressure is applied to an area about one-eighth inch in diameter. For primary vaccination in a smallpox-free area, 6 to 10 pressures within 5 or 6 seconds suffice. For revaccination 20 to 30 pressures within 30 seconds are given. A trace of blood welling up within half a minute indicates the right amount of force exerted against the drop of vaccine on the skin.

2. In the *multiple-puncture* method, a forked needle dipped in vaccine is touched to the surface of the skin. The needle is held perpendicular to the skin, and (if the geographic area is free of smallpox) two to three punctures are made in an up-and-down manner within a small area about one-eighth inch in diameter.

3. The *jet injection* method requires a specially constructed injector that delivers about 0.1 ml of a specially purified vaccine into the skin (Fig. 37-1).

†"Primary vaccination and revaccination are best performed on the outer aspects of the upper arm, over the insertion of the deltoid muscle, or behind the midline. Reactions are less likely to be severe on the upper arm than on the lower extremity or other parts of the body. With proper technic, resultant scars are small and unobtrusive." (Kempe, C. H., and Benenson, A. S.: Smallpox immunization in the United States, J.A.M.A. **194**:161, 1965.)

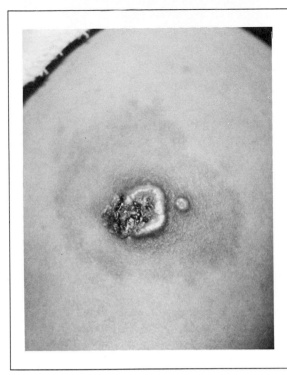

FIG. 37-2. Smallpox vaccination. Primary vaccinia on arm of year-old child. A red areola surrounds umbilicated, partly encrusted pustule with small satellite.

After revaccination, the following may be seen:

1. *Major reaction**—a vesicular or pustular lesion, an area of hardening, or a reddened zone about a central crust or ulcer—indicates virus has multiplied at the site.
2. *Equivocal reaction*— any other reaction—indicates vaccination did not "take." Equivocal reactions may come as a consequence of immunity adequate to suppress the virus or may represent only an allergic reaction to an inactive vaccine.

No reaction means inactive vaccine or faulty technic. In such a case vaccination should be repeated. Vaccine virus is very susceptible to even moderate warmth and should be kept in the freezing compartment of the refrigerator. *The vaccine virus must be kept frozen.*

INDICATIONS. The U.S. Public Health Service currently recommends that primary smallpox vaccination be discontinued on a routine basis in the United States. The risk from the disease is virtually nonexistent, and on occasions severe reactions are present to the vaccine. (Accordingly, the American Academy of Pediatrics no longer advocates this immunization routinely.) Smallpox vaccination is still necessary for travelers going to countries requiring a valid international certificate of vaccination and for travelers that were in Ethiopia the two weeks just prior to their return to the United States. The Public Health Service no longer advises routine smallpox

*The World Health Organization's Expert Committee on Smallpox defines only two responses to smallpox vaccination—"major" and "equivocal."

immunization for hospital and health care personnel but still of course stresses it for laboratory workers in contact with the variola virus.

CONTRAINDICATIONS. Children with eczema and skin lesions such as poison ivy or impetigo, extensive wounds, or burns should not be vaccinated against smallpox because generalized vaccinia, a condition resembling smallpox, may develop; nor should they be allowed to contact a person who has an active vaccination lesion. It would be dangerous to vaccinate a patient with depressed immunologic responses, either because of a disease such as dysgammaglobulinemia, lymphoma, leukemia. or blood dyscrasia, or because of therapy being given with steroids, antimetabolites, alkylating agents, or ionizing radiation. Pregnant women must *not* be vaccinated.

Mumps. Live, attenuated mumps virus vaccine given in a single 0.5 ml sub-cutaneous injection or in combination with measles and rubella vaccines (Table 37-1) protects for 6 years. Febrile reactions have not been noted. The vaccine is advocated for susceptible children approaching puberty, adolescents, and adults, particularly males with no history of the disease. In special circumstances vaccination may be considered for younger children in closed populations such as special schools, camps, institutions, and the like. It should *not* be given to pregnant women or to children less than 12 months of age.

Rubella. Rubella immunization is carried out with a single subcutaneous injection of vaccine (or given in a combined vaccine as indicated in Table 37-1). It is advocated for boys and girls between the ages of 1 year and puberty and especially for those of kindergarten age. Vaccinees will shed virus from their throats for 2 or more weeks afterward, but they are not thought to be infectious. It is thought that vaccination gives long-range protection, and a booster is not now advised.

Immunization is contraindicated for infants less than 1 year of age, for pregnant women, and for individuals with altered immune states (leukemia, lymphoid malignancies, recipients of immunosuppressive therapy). It should not be given routinely to adolescent girls (after the twelfth birthday) or adult women because of the danger that it might complicate a pregnancy. It is not given in severe febrile illness.

Rabies. With rabies we have an exception to the rule that there is not enough time between exposure and onset of disease to establish an *active* immunity to protect the person exposed. Since the incubation period of rabies may be much longer than that of most diseases, an active immunity can usually be established before the disease begins.* The onset of the disease *must* be prevented because there is no treatment; *the disease must be considered to be 100% fatal.*

Table 37-2 outlines the recommendations of the WHO Expert Committee on Rabies for rabies prevention. Active immunization, as indicated in this table, is carried out in a series of 21 daily doses of rabies vaccine (duck embryo or brain tissue types). A booster dose is given 10 days and 20 days after the last dose.

---

*Smallpox and rabies are the only diseases in which the exposed person can be protected against development of the disease by the production of an active immunity. In smallpox an active immunity develops very quickly. In rabies the period of incubation is sometimes so long that an active immunity can be produced before the disease takes effect.

TABLE 37-2.   RECOMMENDATIONS FOR THE PREVENTION OF RABIES*

| Exposure | Condition of vaccinated or unvaccinated animal inflicting injury | | Recommended treatment† |
	At exposure	During 10 days' observation	
1. Indirect contact	Healthy or rabid	Healthy or rabid	None
2. Licks			
a. Intact skin	Rabid	—	None
b. Skin lesions, scratches; mucous membrane intact or abraded	1. Healthy	Clinical signs of rabies or laboratory diagnosis	Serum now; start vaccine at first signs of rabies in the animal
	2. Signs veterinarian judges to suggest rabies	Healthy	Serum now; start vaccine; stop at 5 days if animal is normal
	3. Rabid, escaped, killed, or unknown	Rabid	Serum now; start vaccine
3. Bites			
a. Mild exposure	1. Healthy	Clinical signs of rabies or laboratory diagnosis	Serum immediately; start vaccine at first sign of rabies in animal
	2. Signs veterinarian judges to suggest rabies	Healthy	Serum and vaccine at once; stop at 5 days if animal is normal
	3. Rabid, escaped, killed, or unknown	Rabid or unknown	Serum and vaccine at once
	4. Wild wolf, fox, bat, skunk, others	—	Serum and vaccine at once
b. Severe exposure such as multiple wounds or face, finger, and neck wounds	1. Healthy	Clinical signs of rabies or laboratory diagnosis	Serum at once; start vaccine at first sign of rabies in animal
	2. Signs veterinarian judges to suggest rabies	Healthy	Serum and vaccine at once; stop at 5 days if animal is normal
	3. Rabid, escaped, killed, or unknown	Rabid or unknown	Serum and vaccine at once
	4. Wild wolf, fox, bat, skunk, others	—	Serum and vaccine at once

*Modified from Technical Report series no. 321 (1966), WHO Expert Committee on Rabies, Fifth Report, Geneva, 1966, World Health Organization.
†See text for the dosage and administration of vaccine and serum.

The best prophylaxis against rabies after exposure is the use of hyperimmune serum in combination with vaccine. Human rabies immune globulin (HRIG) is given as soon as possible. The recommended dosage is 20 units per kilogram body weight with at least half the dose being used to infiltrate the wound inflicted by the animal for local antiviral effect. The rest of it is given intramuscularly.

If a nonbite exposure to rabies occurs in an immunized individual, one booster dose (1 ml) of duck embryo vaccine is injected. If reexposure is a bite by a rabid animal, 5 daily doses of vaccine are given with a booster 20 days later. Prophylactic vaccination consists of two subcutaneous inoculations of 1 ml of duck embryo rabies vaccine each into the deltoid area, 1 month apart, followed by a booster dose 6 months later. Boosters are recommended every 1 to 3 years.

The amount of protection needed from hyperimmune serum and vaccine depends on many factors (Table 37-2) such as the type of animal, location of the bite, extent of injury, protection afforded by clothing, whether rabies is present in the community, and whether observation of the animal is possible. *It is very important that the animal inflicting the bite be apprehended and kept for a 10-day period of observation.* (Rats and mice do not transmit rabies.)

Although the method of preparing rabies vaccine differs greatly from the original method of Pasteur, the administration of the vaccine is still referred to as the *Pasteur treatment.* It is befitting that his name be thus remembered.

Tuberculosis. BCG vaccination against tuberculosis is done as superficially as possible into the skin (intracutaneously) of the upper arm over the deltoid or triceps muscle. To the newborn infant 0.05 ml (or half the usual dose) is given; to the older *tuberculin-negative* individual, 0.1 ml. Tuberculin testing—Mantoux test with 5 units of purified protein derivative (PPD) tuberculin—must be done before vaccination except in infants less than 2 months of age. A preliminary radiograph of the chest may be indicated. Two to three months after vaccination tuberculin testing is repeated, and if the test is negative, vaccination is repeated.

BCG vaccination is advised for children or those persons traveling in areas of the world where tuberculosis is a major health problem and for those likely to be exposed to adults with the disease. It should be carried out 2 months before exposure. The Public Health Service stresses vaccination also for the health workers at risk from unrecognized cases of tuberculosis and the members of socially disaffiliated groups or those without a regular source of health care such as alcoholics, drug addicts, and migrants. BCG vaccination is said to be 75% effective with protection lasting for 10 to 15 years. A low incidence of postvaccinal complications is reported.

RECOMMENDATIONS OF U.S. PUBLIC HEALTH SERVICE

Influenza. The U.S. Public Health Service recommends annual influenza vaccination for persons over 65 years of age and for those in whom the onset of flu would represent an added health risk, such as individuals of all ages with chronic debilitating diseases (diabetes, cardiovascular ailments, pulmonary disease, and others). There is some indication for immunization of persons responsible for furnishing essential public services, such as law enforcement officers, firemen, and many others.

TABLE 37-3. ADMINISTRATION OF POLIOVIRUS VACCINES

Group	Doses	Poliovirus type(s)	Time interval	Comments
I. Administration of trivalent oral poliovirus vaccine (TOPV)				
Infants	First	TOPV (I, II, III)		Primary immunization begun at 6 to 12 weeks of age; may be given with DPT
	Second	TOPV	8 weeks after first dose (no less than 6 weeks)	
	Third	TOPV	8 to 12 months after second dose	
	Booster	TOPV		Given at a time child enters school; boosters with OPV otherwise not indicated
Children and adolescents	First	TOPV		
	Second	TOPV	6 to 8 weeks after first dose	
	Third	TOPV	8 to 12 months after second dose	Third dose given as early as 6 weeks after second dose in unusual circumstances
Adults*	First	TOPV		Recommended for adults at high risk; pregnancy not indication or contraindication for immunization
	Second	TOPV	6 to 8 weeks after first dose	
	Third	TOPV	8 to 12 months after second dose	
II. Administration of monovalent oral poliovirus vaccine (MOPV).				
All ages	First	MOPV (I)		
	Second	MOPV (III)	6 to 8 weeks after first dose	
	Third	MOPV (II)	6 to 8 weeks after second dose	
	Fourth	TOPV (I, II, III)	8 to 12 months after third dose of monovalent vaccine	

*Routine immunization is not indicated for most adults residing in continental United States.

The primary series has been traditionally 2 doses given subcutaneously (beneath the skin) 6 to 8 weeks apart. A single dose is now considered reasonable for either the primary or annual booster vaccination. The second dose provides little extra benefit. (In the package insert supplied with the vaccine, the manufacturer gives the recommended dose for adults and children and directions for administration.) Vaccination should be done by the middle of November.

The status of influenza immunization (both vaccine and schedule) can change rapidly. For this reason the recommendations of its Committee on Immunization Practices are regularly published by the U.S. Public Health Service and should be consulted before any procedure is carried out.

Poliomyelitis. The U.S. Public Health Service is urging regular immunization against poliomyelitis for all children from early infancy. Routine immunization of infants should begin at 6 to 12 weeks of age and be completed against all three types of the virus. The schedule for oral vaccination is given in Table 37-3. In the United States inactivated poliovirus vaccine (IPV) is not used, and TOPV has largely replaced the monovalent form in immunization programs.

TABLE 37-4. SCHEDULES FOR MEASLES IMMUNIZATION

Schedule	Type of vaccine*	Age	Dose (subcutaneous)†	Remarks
1.	Live attenuated (Edmonston B strain) measles virus vaccine	12 months and older	One	Although live attenuated viral vaccine may be given safely, many physicians may wish to add immune serum globulin (human) because of fewer clinical reactions
2.	Live attenuated (Edmonston B strain) plus immune serum globulin (human)	12 months and older	One, plus 0.01 ml per pound of body weight of immune serum globulin (human) at different site with different syringe	
3.	Live, further attenuated (Schwarz and Moraten strains)	12 months and older	One	Clinical reactions those of schedule 2; immune serum globulin (human) *not* indicated

Only live attenuated virus vaccines should be used! As soon as possible children who may have received inactivated vaccine should be revaccinated with live virus vaccine. It has been found that children given inactivated vaccine develop a sinister illness when exposed to natural measles several years later.
†No booster given here. Some state health departments are urging a booster for the child immunized before 12 months of age, the booster to be given at 12 to 16 months of age or even later.

Vaccination is contraindicated in patients with severe underlying diseases such as leukemia or generalized malignancy or those with lowered resistance because of therapy.

Measles. The U.S. Public Health Service urges immunization for all susceptible children (those not having had vaccine or natural measles). The prime target groups are children at least 1 year of age and susceptible children entering nursery school, kindergarten, or elementary school (Fig. 37-1). Protection against measles is especially needed for children in the high risk groups, that is, children in institutions and those with chronic heart or pulmonary disease and cystic fibrosis. Vaccination of adults is not stressed since most individuals are serologically immune by the age of 15 years.

Table 37-4 gives the currently acceptable schedules for measles immunization.

Live attenuated vaccine should not be given to pregnant women or to patients being treated with steroids, irradiation, antimetabolites, or those agents that depress an individual's immunologic capacities. Leukemia, lymphoma, and other generalized malignancies are also contraindications, as is untreated tuberculosis.

Typhoid and paratyphoid fevers. The U.S. Public Health Service no longer recommends routine typhoid immunization in the United States even for individuals in flood disaster areas. Circumstances where vaccination would be indicated include an intimate exposure to a known carrier, an outbreak of the disease in a community, or the event of travel to an endemic area. The Public Health Service does *not* advise paratyphoid immunization because there is good evidence that paratyphoid A and B vaccines are worthless. (Schedules for typhoid immunization may be found in the Red Book of the American Academy of Pediatrics.)

REQUIREMENTS FOR U.S. ARMED FORCES

Immunization procedures for the U.S. Armed Forces are regularly evaluated and updated. So that the requirements may be precisely related to the geographic area of military duty, the world is divided into the following areas:

Area I (routine)—United States (includes 50 states, District of Columbia, Virgin Islands, Puerto Rico, Wake Island, Midway Island), Canada, Greenland, Iceland, Marshall Islands, Guam, Pacific islands east of the 180th meridian, North Pole, South Pole, Bermuda, Bahama Islands, Baja California, and a strip of Mexico 50 miles south of the U.S. border

Area II—All areas outside Area I

Area IIC (cholera)—Arabian peninsula, Afghanistan, Burma, Ceylon, Republic of China, Hong Kong, India, Indonesia, Korea, Peoples Republic of China, Macao, Malaysia, Pakistan, Philippines, Thailand, Iran, Iraq, Syria, Turkey, Lebanon, Israel, Jordan, Kuwait, African continent north of the Sahara, and Madagascar

Area IICP (cholera and plague)—Laos, Cambodia, and Vietnam

Area IIY (yellow fever)—Central America southeast of the Isthmus of Tehuantepec, Panama, and South America

Area IIYC (yellow fever and cholera)—Africa south of the Sahara

TABLE 37-5. IMMUNIZATIONS REQUIRED FOR U.S. ARMED FORCES

Immunizing agent	Basic series*	Required Area I (excluding alert forces)	Required Area II	Reimmunization Area I	Reimmunization Area II
Smallpox vaccine, USP (freeze-dried)	Vaccination to be read after 6 to 8 days; successful reaction required	✔	✔	3 years	3 years
Typhoid vaccine (acetone-killed, dried)	Two 0.5 ml injections 4 weeks apart	✔	✔	None	0.5 ml — 3 years
Tetanus and diphtheria toxoids, adsorbed, USP (for adult use)	Two 0.5 ml injections 1 to 2 months apart with third dose of 0.1 ml 12 months later	✔	✔	0.1 ml — 6 years; 0.5 ml after injury or burn	0.1 ml — 6 years; 0.5 ml after injury or burn
Poliovirus vaccine, live oral, trivalent, types I, II, and III	Two oral doses 6 to 8 weeks apart; third dose 8 to 12 months later	✔	✔	None	None
Yellow fever vaccine, USP†	One 0.5 ml injection of concentrated vaccine diluted 1:10		✔ (Areas IIY, IIYC)		0.5 ml of 1:10 dilution — 10 years
Influenza virus vaccine, USP†	One 0.5 ml injection	✔	✔	0.5 ml — 1 year	0.5 ml — 1 year
Cholera vaccine, USP†	0.5 ml injection followed by 1.0 ml in 1+ weeks		✔ (Areas IIC, IIYC, IICP)		0.5 ml — 6 months (while in area)
Plague vaccine, USP†	Two *intramuscular* injections — 1.0 ml followed in 3 months by 0.2 ml		✔ (Area IICP)		0.2 ml (IM) — 6 months (while in area)

*Smallpox vaccination is given either by multiple-puncture technic or by intradermal jet. Plague vaccine must be given intramuscularly. Other injections are subcutaneous or intramuscular. (See Fig. 36-1.)
†The vaccine listed in the United States Pharmacopeia.

Table 37-5 summarizes the immunization requirements for United States military personnel. Military dependents and civilians in the employ of the armed forces or organizations serving the armed forces are not required to take immunizations while within the United States or Canada. Outside the United States or Canada they are required to take essentially the same ones as members of the armed forces.

TABLE 37-6. SPECIAL IMMUNIZATION REQUIREMENTS BY AGE FOR INTERNATIONAL TRAVEL*

Disease	Travel destination	Age	Doses of vaccine	Route of inoculation	Intervals	Recall needed
Cholera	Asian countries, Middle East	6 months to 5 years	(1) 0.1 ml (2) 0.3 ml Booster: 0.1 ml	Subcutaneously	1 month between 1 and 2; 6+ months between 2 and 3	6 months
		5 to 10 years	(1) 0.3 ml (2) 0.5 ml Booster: 0.3 ml			
		10 years and older (adults also)	(1) 0.5 ml (2) 1.0 ml Booster: 0.5 ml			
Yellow fever†	Central and South America, Africa	All ages	0.5 ml of 1:10 dilution	Subcutaneously		10 years
Plague	High-risk area	Less than 1 year	(1) 0.1 ml (2) 0.1 ml (3) 0.04 ml Booster: 0.04 ml	Intramuscularly only	30 days between 1 and 2; 4 to 12 weeks between 2 and 3	6 to 12 months
		1 to 4 years	(1) 0.2 ml (2) 0.2 ml (3) 0.08 ml Booster: 0.08 ml			
		5 to 10 years	(1) 0.3 ml (2) 0.3 ml (3) 0.12 ml Booster: 0.12 ml			
		10 years and older (adults)	(1) 0.5 ml (2) 0.5 ml (3) 0.2 ml Booster: 0.2 ml			
Typhoid fever	Developing countries and those with low standards of sanitation	6 months to 10 years 10 years and older (adults also)	(1) 0.25 ml (2) 0.25 ml (1) 0.5 ml (2) 0.5 ml	Subcutaneously	4+ weeks	1 to 3 years if exposed (0.1 ml intradermally or 0.25 to 0.5 ml subcutaneously)

*Based on data from the Red Book of American Academy of Pediatrics, 1974.
†Note that yellow fever vaccination for international travel must be given at a designated Yellow Fever Vaccination Center, registered by the World Health Organization. Centers are located in 46 states, District of Columbia, Puerto Rico, Canal Zone, and American Samoa. Yellow fever vaccination certificate is not valid until 10 days after primary immunization.

INTERNATIONAL TRAVEL

Immunizations. The Red Book of the American Academy of Pediatrics also contains immunization information and recommendations for international travel.* In Table 37-6, which presents vaccination schedules recommended by this organization, there is a breakdown of special requirements for travel by age groups.

Where to find a doctor. The International Association for Medical Assistance to Travelers, Inc., 745 Fifth Avenue, New York, N.Y. 10022 (also note Intermedic, 777 Third Avenue, New York, N.Y. 10017, and World Medical Association, 10 Columbus Circle, New York, N.Y. 10019) supplies without charge a directory of English-speaking physicians available 24 hours a day in the different cities of the world. A long way from home one may also obtain medical aid from U.S. embassies and consulates, the Red Cross, travel agencies, the police, medical associations, hospitals, clinics, and U.S. Armed Forces bases and installations.

Traveler's diarrhea. Traveler's diarrhea is known by the many colorful synonyms —Aztec two-step, Montezuma's revenge, Cromwell's curse, Delhi belly—that mark out a widespread geographic distribution, and preventive immunization does not exist for this unpleasant complication of the initial phase of the tourist's trip abroad (or out of the country). Traveler's diarrhea is described consistently by nausea and vomiting, abdominal cramps, chills, low-grade fever, and diarrhea productive of loose and watery, foul-smelling stools. Manifestations, though distressing and temporarily incapacitating, disappear within a few days. Relapses or sequelae are rare. Microbes found in the stools and thereby implicated include *Shigella* species, *Salmonella* species, enteropathogenic *Escherichia coli*, and *Giardia lamblia*. However, the cause is uncertain and a cure-all nonexistent.

To minimize the hazard, especially where sanitary practices are substandard, certain precautions are stressed. The tourist can safely eat only foods thoroughly cooked or recently peeled. He can drink with impunity only boiled water, carbonated mineral water, or beverages boiled or carbonated in cans. Very hot water out of the tap is not likely to transmit viable enteric bacteria. After it has cooled, it can be used for oral hygiene and various other purposes.

Center for Disease Control (CDC). The responsibility for assisting the states in controlling communicable and vector-borne diseases and for participation in international programs of disease eradication rests with an agency of the U.S. Public Health Service—the Center for Disease Control (CDC). Its headquarters at Atlanta is a vast complex of specialty laboratories and other disease control facilities, and there are field stations across the United States and Puerto Rico. Valuable advice for the traveler is given in its booklet *Health Information for International Travel*.†

World Health Organization (WHO). The World Health Organization, or WHO as it is usually called, is an agency of the United Nations set up to provide for the

---

*This book also advises suppressive medication for malaria (a major worldwide problem) prior to entry into an infected area, during sojourn there, and even for a number of weeks thereafter.
†Health information for international travel, Morbid. Mortal. Wk. Rep. 25(suppl.):Oct., 1976, Department of Health, Education and Welfare Pub. no. (CDC) 76-8280.

highest level of health for all peoples of the world. It is composed of 135 member states and 2 associate members. It works with the United Nations, governments, and special health groups to devise standards, train personnel, carry on research, and improve public health in every way at the international level. To protect the sightseer of the jet age, WHO is concerned for uniform standards of hygiene in aviation. Guidelines are detailed for aspects of sanitation applied to aircraft and international airports, the design to provide every safety factor for the traveler in rapid transit over the globe.

For the traveler, the World Health Organization (Geneva, Switzerland) publishes the bilingual (English and French) *Vaccination Certificate Requirements for International Travelers, Situation as on 1 January, 1975.*

REFERENCES FOR UNIT SIX

Artenstein, M. S.: The current status of bacterial vaccines, Hosp. Pract. **8:**49, June, 1973.

Balagtas, R. C., and others: Treatment of pertussis with pertussis immune globulin, J. Pediatr. **79:**203, 1971.

Barkin, R. M.: Measles: regaining control (editorial), J.A.M.A. **231:**737, 1975.

Barrett-Connor, E.: Chemoprophylaxis of malaria for travelers, Ann. Intern. Med. **81:**219, 1974.

Bass, J. W., and others: Booster vaccination with further live attenuated measles vaccine, J.A.M.A. **235:**31, 1976.

Bauman, H. E.: Food microbiology, ASM News **38:**312, 1972.

Beardmore, W. B.: A new form of smallpox vaccine, J. Infect. Dis. **127:**718, 1973.

Bell, J. A.: Viruses and water quality (editorial), J.A.M.A. **219:**1628, 1972.

Blake, P. A., and others: Serologic therapy of tetanus in the United States, 1965-1971, J.A.M.A. **235:**42, 1976.

Boffey, P. M.: Color additives: botched experiment leads to banning of red dye no. 2, Science **191:**450, 1976.

Brooks, S. M.: Ptomaine: the story of food poisoning, Cranbury, N.J., 1974, A. S. Barnes & Co., Inc.

Brown, M. S.: What you should know about communicable diseases and their immunizations. Part II Diphtheria, pertussis, tetanus, and polio, Nursing '75 **5:**56, Oct., 1975.

Bulla, L. A., Jr.: Microbial control of insects: a synopsis, ASM News **39:**97, 1973.

Burt, B. A.: Fluoride—the case for hardening teeth, Nursing Times **70:**374, Mar., 1974.

Cairns, J., Jr., and Dickson, K. L., editors: Biological methods for the assessment of water quality, Philadelphia, 1973, American Society for Testing and Materials.

Collected recommendations of Public Health Service Advisory Committee on Immunization Practices, Morbid. Mortal. Wk. Rep. **21**(25, suppl.):1, 1972.

Colwell, R. R., and Morita, R. Y., editors: Effect of the ocean environment on microbial activities, Baltimore, 1974, University Park Press.

Committee on Fetus and Newborn, American Academy of Pediatrics: Hospital care of newborn infants, ed. 5, Evanston, Ill., 1971, American Academy of Pediatrics.

Cooper, L. Z., and Schweitzer, M. D.: Hospital personnel: the immunization gap, Hosp. Pract. **9:**11, Mar., 1974.

Corey, L., and Hattwick, M. A. W.: Treatment of persons exposed to rabies, J.A.M.A. **232:**272, 1975.

Demain, A. L.: Application of the microbe to the benefit of mankind: challenges and opportunities, ASM News **38:**237, 1972.

Diehl, J. F.: Irradiated food, Science **180:**214, 1973.

Doetsch, R. N., and Cook, T. M.: Introduction to bacteria and their ecobiology, Baltimore, 1973, University Park Press.

Dowling, H. F.: Diphtheria as a model, J.A.M.A. **226:**550, 1973.

Dugan, P. R.: Biochemical ecology of water pollution, New York, 1972, Plenum Publishing Corp.

DuPont, H. L.: Recent developments in immunization against diarrheal diseases, South. Med. J. **68:**1027, 1975.

Findlay, W. P. K., editor: Modern brewing technology, Cleveland, Ohio, 1976, CRC Press, Inc.

Francis, B. J.: Current concepts in immunization, Am. J. Nurs. **73**:646, 1973.

Frazier, W. C.: Food microbiology, New York, 1967, McGraw-Hill Book Co.

Fraser, D. W.: Preventing tetanus in patients with wounds (editorial), Ann. Intern. Med. **84**:95, 1976.

Ganguly, R., and others: Rubella virus immunization of preschool children via respiratory tract, Am. J. Dis. Child. **128**:821, 1974.

Garner, W. R., and others: Problems associated with rabies preexposure prophylaxis, J.A.M.A. **235**:1131, 1976.

Grade "A" Pasteurized Milk Ordinance—1965 Recommendations of the United States Public Health Service, U.S.P.H.S. Bull. no. 229 (revision) Washington, D.C., 1967, U.S. Government Printing office.

Grady, G. F., and Lee, V. A.: Hepatitis B immune globulin—prevention of hepatitis from accidental exposure among medical personnel, N. Engl. J. Med. **293**:1067, 1975.

Gregory, P. H.: The microbiology of the atmosphere, ed. 2, New York, 1973, John Wiley & Sons, Inc.

Hardy, R. W. F., and Havelka, U. D.: Nitrogen fixation research: a key to world food? Science **188**:633, 1975.

Hattwick, M. A. W., and others: Postexposure rabies prophylaxis with human rabies immune globulin, J.A.M.A. **227**:407, 1974.

Health information for international travel 1976, Morbid. Mortal. Wk. Rep. 25(suppl.), Oct., 1976, HEW Pub. no. (CDC) 76-8280.

Hersh, E. M., and others: BCG vaccine and its derivatives, J.A.M.A. **235**:646, 1976.

Hill, S. A.: Prophylaxis against tetanus in the wounded (editorial), South. Med. J. **67**:759, 1974.

Hockenhull, D. J. D., editor: Progress in industrial microbiology, vol. 13, Edinburgh, 1974, E. & S. Livingstone, Ltd.

Holder, A. R.: Compulsory immunization statutes, J.A.M.A. **228**:1059, 1974.

Horstmann, D. M.: Controlling rubella: problems and perspectives, Ann. Intern. Med. **83**:412, 1975.

Houran, J. B.: Is your lab water up to standards? Lab. Management **10**:52, Jan., 1972.

Hungate, R. E.: Status of microbiological research in relation to waste disposal, ASM News **38**:364, 1972.

James, G. V.: Water treatment, Cleveland, 1971, CRC Press, Inc.

Jay, J. M.: Modern food microbiology, 1970, New York, Van Nostrand Reinhold Co.

Johnson, R. H., and Ellis, R. J.: Immunobiologic agents and drugs available from the Center for Disease Control, Ann. Intern. Med. **81**:61, 1974.

Karchmer, A. W., and others: Simultaneous administration of live virus vaccines, Am. J. Dis. Child. **121**:382, 1971.

Kermode, G. O.: Food additives, Sci. Am. **226**:15, Mar., 1972.

Kilbourne, E. D.: Influenza: the vaccines, Hosp. Pract. **6**:103, Oct., 1971.

Kilbourne, E. D., and Smillie, W. G., editors: Public health and human ecology, ed. 4, New York, 1969, The Macmillan Co.

Kolata, G. B.: Phage in live virus vaccines: are they harmful to people? Science **187**:522, 1975.

Lawson, R. B.: Current status of immunizations, Postgrad. Med. **49**:111, June, 1971.

Lees, R.: Food analysis: analytical quality control methods for the food manufacturer and buyer, Cleveland, 1975, CRC Press, Inc.

Lerner, A. M.: Committee on Immunization (Infectious Disease Society of America): guide to immunization against mumps, J. Infect. Dis. **122**:116, 1970.

Linnemann, C. C., Jr.: Viral vaccines: the question of revaccination, J.A.M.A. **235**:63, 1976.

Maier, F. J.: Fluoridation, Cleveland, 1972, CRC Press, Inc.

Malina, J. F., Jr., and Sagik, B. P., editors: Virus survival in water and wastewater systems. Water Resources Symposium Number Seven, Austin, 1974, Center for Research in Water Resources.

Margolis, F. J.: Value of fluoride supplements in prevention of dental caries, J.A.M.A. **234**:312, 1975.

Marx, J. L.: Insect control, (II): hormones and viruses, Science **181**:833, 1973.

Marx, J. L.: Nitrogen fixation: research efforts intensify, Science **185**:132, 1974.

McLean, D. M.: Sewage irrigation: health benefit or hazard? Ann. Intern. Med. **82**:112, 1975.

Medical News: Immunization vs complacency: are we ready for the challenge? J.A.M.A. **229**:1557, 1974.

Medical Services, Immunization requirements and procedures; Army Regulation no. 40-562;

BUMED Instruction no. 6230 1G; Air Force Regulation no. 161-13;CG COMDTINST 6230. 4B, Department of the Army, The Navy, The Air Force, and Transportation, Washington, D.C., 22 March, 1974.

Merson, M. H., and Gangarosa, E. J.: Travelers' diarrhea, J.A.M.A. **234:**200, 1975.

Miller, L. W., and others: Diphtheria immunization, Am. J. Dis. Child. **123:**197, 1972.

Mitchell, R.: Introduction to environmental microbiology, Englewood Cliffs, N.J., 1974, Prentice-Hall, Inc.

Mitchell, R., editor: Water pollution microbiology, New York, 1972, John Wiley & Sons, Inc.

Nader, R.: Not *all* of our drinking water meets healthful standards, Today's Health **53:**10, Dec., 1975.

Nickerson, J. T., and Sinskey, A. J.: Microbiology of foods and food processing, New York, 1972, American Elsevier Publishing Co.

Nickerson, W. J.: Microbial degradation and transformation of wastes, ASM News **38:**367, 1972.

Okada, H.: The eradication of infectious diseases (editorial), J. Infect. Dis. **127:**474, 1973.

Petricciani, J. C.: Bacteriophages and vaccines, ASM News **39:**435, 1973.

Porter, J. R.: Microbiology and the disposal of solid wastes, ASM News **40:**826, 1974.

Porter, J. R.: Microbiology and the food and energy crisis, ASM News **40:**813, 1974.

Phillips, C. F.: Immunization against influenza, South. Med. J. **69:**353, 1977.

Physicians' desk reference to pharmaceutical specialties and biologicals, Oradell, N.J., 1976, Medical Economics, Inc.

Plotkin, S. A.: Rubella vaccination (editorial), J.A.M.A. **215:**1492, 1971.

Prevention the best medicine, immunization during pregnancy, Med. Opinion **5:**59, Feb., 1976.

Price, J. F., and Schweigert, B. S., editors: The science of meat and meat products, ed. 2, San Francisco, 1971, W. H. Freeman & Co.

Rauch, P.: Avoiding injuries from injections, Nursing '72 **2:**12, May, 1972.

Recommendation of the Public Health Service Advisory Committee on Immunization Practices: yellow fever vaccine, Ann. Intern. Med. **71:**365, 1969.

Recommendation of the Public Health Service Advisory Committee on Immunization Practices: combination live virus vaccines, measles and rubella, and measles, mumps, and rubella, Morbid. Mortal. Wk. Rep. **20**(16):145, 1971.

Recommendation of the Public Health Service Advisory Committee on Immunization Practices: measles vaccines, Morbid. Mortal. Wk. Rep. **25**(45):359; **25**(46):376, 1976.

Recommendation of the Public Health Service Advisory Committee on Immunization Practices: influenza vaccine, Morbid. Mortal. Wk. Rep. **25**(21):165; **25**(28):221; **25**(45):357, 1976.

Recommendation of the Public Health Service Advisory Committee on Immunization Practices: BCG vaccines, Mortal. Morbid. Wk. Rep. **24:**69, 1975.

Recommendation of the Public Health Service Advisory Committee on Immunization Practices: meningococcal polysaccharide vaccines, Morbid. Mortal. Wk. Rep. **24:**381, 1975.

Recommendation of the Public Health Service Advisory Committee on Immunization Practices: smallpox vaccination of hospital and health personnel, Morbid. Mortal. Wk. Rep. **25:**9, 1976.

Recommendation of the Public Health Service Advisory Committee on Immunization Practices: general recommendations on immunization, Morbid. Mortal. Wk. Rep. **25:**349, 1976.

Recommendation of the Public Health Service Advisory Committee on Immunization Practices: rabies, Morbid. Mortal. Wk. Rep. **25:**403, 1976.

Recommendations of the Committee on Infectious Diseases of the American Academy of Pediatrics: immunization of children at high risk from influenza infection, Morbid. Mortal. Wk. Rep. **25**(36):285, 1976.

Redeker, A. G., and others: Hepatitis B immune globulin as a prophylactic measure for spouses exposed to acute type B hepatitis, N. Engl. J. Med. **293:**1055, 1975.

Report of the Committee on Infectious Diseases, Evanston, Ill., 1974, American Academy of Pediatrics, Inc.

Rosenbaum, M. J., and others: Recent experiences with live adenovirus vaccines in Navy recruits, Milit. Med. **140:**251, 1975.

Rubin, R. J., and Corey, L.: Preventing rabies in humans, South. Med. J. **67:**1472, 1974.

Sanford, J. P.: Personal communication, 1976.

Sartwell, P. E., editor: Maxcy-Rosenau preventive medicine and public health, New York, 1973, Appleton-Century-Crofts.

Schoenbaum, S. C., and others: Epidemiology of congenital rubella syndrome, the role of maternal parity, J.A.M.A. **233:**151, 1975.

Schwartz, R. D.: Microbial production of methane on a commercial scale, ASM News **39:**247, 1973.

Smith, E. H., Jr.: Fluoridation of water supply, J.A.M.A. **230:**1569, 1974.

Sparks, F. C., and others: Complications of BCG immunotherapy in patients with cancer, N. Engl. J. Med. **289:**827, 1973.

Standard methods for the examination of water and wastewater, ed. 13, Washington, D.C., 1971, American Public Health Association, Inc.

Stanfield, J. P., and others: Diphtheria-tetanus-pertussis immunization by intradermal jet injection, Br. Med. J. **2:**197, 1972.

Stevens, R. B., editor: Mycology guidebook, Seattle, 1974, University of Washington Press.

Stokes, J., Jr.: Pediatric immunization and the new Academy schedule, Hosp. Pract. **7:**127, May, 1972.

Stokes, J., Jr., and others: Trivalent combined measles-mumps-rubella vaccine, J.A.M.A. **218:**57, 1971.

Subunit flu vaccine held less toxic, more immunogenic, Hosp. Pract. **11:**25, Feb., 1976.

Surgenor, D. M. N., and others: Clinical trials of hepatitis B immune globulin N. Engl. J. Med. **293:**1060, 1975.

Toxicological evaluation of certain food additives, Eighteenth report of the Joint FAO/WHO Expert Committee on Food Additives, World Health Organization, Technical Report Series, 1974, no. 557, Geneva.

Udall, S. L., and Stansbury, J.: Sewage: our most neglected resource (editorial), Ann. Intern. Med. **81:**849, 1974.

Uduman, S. A., and others: Should patients with zoster receive zoster immune globulin? J.A.M.A. **234:**1049, 1975.

Vaccines: an update. Dept HEW publication no. (FDA) 74-3013. Available from Consumer Information, Public Documents Distribution Center, Pueblo, Colo., 81009. U.S. Government Printing Office, no. 1974-544-190/57 (no charge).

Wade, N.: Insect viruses: a new class of pesticides, Science **181:**925, 1973.

Weibel, R. E., and others: Persistence of immunity following monovalent and combined live measles, mumps, and rubella virus vaccines, Pediatrics **57:**467, 1973

Weinstein, L.: Whatever happened to the "old-time" infections, J.A.M.A. **229:**196, 1974.

Weiser, H. H., and others: Practical food microbiology and technology, Westport, Conn., 1971, Avi Publishing Co.

When is infectious waste not infectious waste? Hospitals, J.A.H.A. **46:**56, May 1, 1972.

White, C. S., III, and others: Repeated immunization: possible adverse effects, Ann. Intern. Med. **81:**594, 1974.

Wiktor, T. J., and others: Human cell culture rabies vaccine, J.A.M.A. **224:**1170, 1973.

Wilson, Sir G. S., and Miles, Sir A. A., editors: Topley and Wilson's principles of bacteriology and immunity, Baltimore, 1975, The Williams & Wilkins Co.

Witte, J. J.: Healthy criticism: we're not immunizing enough of our children, Today's Health **51:**4, Sept., 1973.

Zeikus, J. G., and Ward, J. C.: Methane formation in living trees: a microbial origin, Science **184:**1181, 1974.

LABORATORY SURVEY OF UNIT SIX

PROJECT
Bacteriology of water
Part A—Bacterial plate count of water

1. Obtain from the instructor the following:
   a. Sample of water—may be water of low bacterial count (tap water) or water of high bacterial count (river water)
   b. Sterile pipettes
   c. Three tubes, each of which contains 9 ml of sterile water
   d. Three sterile Petri dishes
   e. Three tubes of nutrient agar
2. Prepare dilutions of water sample.
   a. To one tube of sterile water, add 1 ml of water sample and mix.
   b. Transfer 1 ml from this first tube with a new sterile pipette to a second tube of 9 ml of sterile water and mix.
   c. Repeat procedure by taking a new pipette, transferring 1 ml from the second to third tube and mixing.
3. Calculate dilution of water in each tube. Why is a new pipette used for each procedure?
4. Mix diluted water samples with a suitable culture medium.
   a. With a new pipette, transfer 1 ml from each tube to a Petri dish, beginning with highest dilution and proceeding to lowest. Why do you go from the highest to the lowest dilution?
   b. Melt tubes of agar in water bath, let cool to 42°-45° C, and add one tube to each Petri dish.
   c. Mix agar with water by rotating, let the agar solidify, invert plates, and incubate at 37° C for 24 hours.
5. Select a plate on which colonies are distinctly separated and count the colonies.
6. Calculate from the dilution the bacterial content of the water per milliliter. Record your results below.

Plate no.	Dilution	No. of bacterial colonies present	Estimated no. of bacteria per ml in water sample
1			
2			
3			

Part B—Presumptive test for coliforms in water

1. Obtain from the instructor the following:
   a. Sample of contaminated water
   b. Sterile pipettes
   c. Nine fermentation tubes containing lactose (or lauryl tryptose) broth
2. Inoculate a sample of contaminated water as follows:
   a. To each of two fermentation tubes, add 0.1 ml of water sample.
   b. To two more tubes, add 1 ml of water sample.
   c. To remaining five fermentation tubes, add 10 ml portions of water sample.
3. Mix water sample with contents of each tube by gently rolling the tube between palms of the hands.
   Note: Do not let the contents of a tube come in contact with the cotton plug.
4. Place tubes in incubator at 37° C and examine at end of 24 and 48 hours. For each period, note percentage of gas formed in each tube as indicated by portion of collecting tube that is filled with gas.
5. Record your findings below:

Tube no.	Amount of water sample	Percentage gas formed in fermentation tube in		Presumptive test	
		24 hrs	48 hrs	Positive	Negative
1	0.1 ml				
2	0.1 ml				
3	1.0 ml				
4	1.0 ml				
5	10.0 ml				
6	10.0 ml				
7	10.0 ml				
8	10.0 ml				
9	10.0 ml				

Part C—Demonstration of a positive confirmed test by the instructor with discussion

1. Selection of one of the fermentation tubes from the presumptive test in which gas formation is nicely shown
2. Demonstration of plates (Endo medium or eosin–methylene blue agar) made from tube selected
3. Observation of characteristics that differentiate coliform colonies from those of other organisms on the medium

PROJECT
Bacteriology of milk
Part A—Bacterial plate count of milk

1. Use same method here that was used in obtaining bacterial plate count of water (p. 713). Use milk samples instead of water. Adapt technic so that dilution bottles contain 99 ml instead of 9 ml of sterile water.
2. Plate milk sample dilutions and incubate at 37° C for 48 hours.
3. Count the colonies and calculate the number of bacteria per milliliter of original milk sample. Are colonies on the plates all alike, or are there varieties?
4. Chart results.

Plate no.	Dilution	No. of bacterial colonies present	Estimated no. of bacteria per ml of milk
1			
2			
3			

Part B—Effect of pasteurization on milk

1. Obtain from the instructor the following:
   a. Two tubes of raw milk
   b. Two 99 ml dilution bottles
   c. Two tubes of nutrient agar
   d. Pipettes
2. Pasteurize one raw milk sample as follows:
   a. Place one tube of milk in a water bath and raise temperature of the water to 65° C.
   b. Hold at this temperature for 30 minutes.
3. Prepare dilutions of both milk samples.
   a. Make a 1:100 dilution of this milk and the unheated milk.
   b. With separate sterile pipettes, place 1 ml of each dilution in a Petri dish and mix with melted agar as outlined in steps (b) and (c) in the experiment on counting bacteria in water (p. 713).
   c. Observe. Which plate contains more bacteria? Why?
   d. Chart results.

Sample	Bacterial plate count	No. of bacteria per ml in milk sample
Pasteurized milk (heated to 65° C for 30 min)		
Raw milk (not heated)		

Part C—Phosphatase test*

Note: This test depends on the fact that raw milk contains an enzyme known as phosphatase, which is more resistant to pasteurization than is any known pathogenic bacterium. Its inactivation therefore is a reliable index that the milk has been properly pasteurized and is safe.

1. Test the two samples of milk with which you have previously worked—one a sample of raw milk and one a sample of raw milk that you have pasteurized. Proceed as follows for a given sample.
    a. Pipette 5 ml of buffered substrate into a test tube.
    b. Add 0.5 ml of milk sample and shake. Label tube.
    c. Place the test tube in a 37° C incubator or water bath for 30 minutes. Remove.
    d. Add 6 drops of CQC reagent and 2 drops of catalyst to the tube. Stopper test tube, mix contents, and incubate for 5 minutes.
    Note: If milk sample has not been properly pasteurized, a blue color will appear after about 5 minutes. If milk has been properly pasteurized, the color will be gray or light brown. No attempt is made in this exercise to extract the pigment and quantitate it.
2. Compare results obtained for the two samples of milk—one pasteurized and one not.

Part D—Methylene blue reduction test

1. Work in pairs.
2. Obtain two milk samples from the instructor—one of low bacterial count and one that is heavily contaminated.
3. On each sample, carry out the test as follows:
    a. Place 10 ml of each sample of milk in a test tube. Label tube.
    b. Add 1 ml of methylene blue thiocyanate solution† to each test tube.
    c. Place rubber stoppers in the two tubes and mix thoroughly by inverting.
    d. Place in water bath at 37° C.
    e. Observe tubes occasionally to determine decolorization.
    Note: The more bacteria present, the more rapid the methylene blue is decolorized. On this basis test milk may be classified as follows:
    Class 1. Excellent quality milk, not decolorized in 8 hours.
    Class 2. Milk of good quality, decolorized in less than 8 hours, but not less than 6 hours.
    Class 3. Milk of fair quality, decolorized in less than 6 hours, but not less than 2 hours.

---

*The method of preparing the reagents needed in the test may be found in Standard Methods for the Examination of Dairy Products, published by the American Public Health Association.
†Prepared from methylene blue thiocyanate tablets (about 8.8 g) certified by the Commission on Standardization of Biological Stains. Tablets may be obtained from laboratory supply houses. Dissolve one tablet in 200 ml hot sterile distilled water and allow solution to stand overnight. Store in amber glass bottles away from light.

Class 4. Poor quality milk, decolorized in less than 2 hours.

4. Classify the two milk samples that you and your partner have tested. Record findings.

REFERENCES

Standard methods for the examination of water and wastewater, Washington, D.C., 1976, American Public Health Association, Inc.

Standard methods for the examination of dairy products, microbiological and chemical, Washington, D.C., 1972, American Public Health Association, Inc.

EVALUATION FOR UNIT SIX

Part I

Check the one number from the column on the right that indicates the correct answer to the question or that correctly completes the statement.

1. Why does the bacteriologist measure the safety of a water supply by testing for the comparatively nonpathogenic colon bacillus?

   (a) Although the colon bacillus is normal in the intestines, if ingested, it usually causes disease.

   (b) The colon bacillus indicates that the water supply is probably contaminated with fecal material.

   (c) Pathogenic organisms from infected fecal material are hard to isolate in water.

   (d) Although fecal material from some sources may be safe, that from other sources may carry disease germs.

   (e) Presence of the colon bacillus means that the typhoid bacillus is also present.

   1. a
   2. all of these
   3. b and c
   4. c, d, and e
   5. b, c, and d

2. Measles vaccines are licensed by:

   (a) American Medical Association
   (b) National Institutes of Health
   (c) Children's Bureau
   (d) Food and Drug Administration
   (e) American Public Health Association

   1. a
   2. b
   3. c
   4. d
   5. e

3. Authorities recommend that all children be actively immunized against the following diseases:

   (a) Pertussis
   (b) Diphtheria
   (c) Smallpox
   (d) Rabies
   (e) Scarlet fever

   1. a and c
   2. b and d
   3. c, d, and e
   4. a, b, and c
   5. all of them

4. In which of the following diseases can toxoid be used for active immunization?

   (a) Smallpox
   (b) Typhoid fever
   (c) Diphtheria
   (d) Tetanus
   (e) Whooping cough

   1. c and d
   2. a, c, and d
   3. c and e
   4. a and e
   5. all of these

5. In which of the following is active immunization of doubtful, uncertain, or not universally accepted value?

   (a) Smallpox
   (b) Tuberculosis
   (c) Influenza
   (d) Typhoid fever
   (e) Diphtheria

   1. e
   2. d
   3. c
   4. b
   5. a

6. A woman, 8 weeks pregnant, is exposed for the first time to rubella. What is recommended of the following?

    (a) Nothing   1. a

    (b) Administration of live attenuated rubella   2. b

        vaccine   3. c

    (c) Administration of killed vaccine   4. d

    (d) Administration of gamma globulin from pooled   5. e

        normal adult blood

    (e) A course of antibiotic therapy

7. Which of the following tests for the pasteurization of milk is most practicable to detect raw milk mixed with pasteurized milk or to indicate milk incompletely heated in the pasteurization process?

    (a) Acidity test   1. a and b

    (b) Phosphatase test   2. a, b, and c

    (c) Catalase test   3. b and c

    (d) Test for coliforms   4. c and e

    (e) Methylene blue test   5. b

8. An unprotected family of five—two parents and children aged 2, 4, and 6 years— is exposed to a case of poliomyelitis. (Members of the family have not been immunized against the disease.) Subsequently, the 6 year old becomes ill, and the clinical diagnosis of paralytic poliomyelitis is made. Which of the following is effective in preventing infection in the other members of the family?

    (a) Administration of gamma globulin from pooled   1. a

        normal adult blood   2. b

    (b) Administration of Sabin vaccine   3. c

    (c) Administration of Salk vaccine   4. d and e

    (d) A course of antimicrobial therapy   5. none of these

    (e) Administration of an antiviral compound

9. Active immunization against tetanus confers immunity lasting:

    (a) Six months   1. a

    (b) One year   2. b

    (c) Three years   3. c

    (d) Five years   4. d

    (e) Ten years   5. e

10. The presence of 1 ppm of fluoride in drinking water may lead to:

    (a) Mottled tooth enamel   1. a and b

    (b) Decreased number of dental caries   2. b

    (c) Impaired renal function   3. a and e

    (d) Softening of the bones   4. b and c

    (e) Fluorosis   5. d and e

Part II

Most of the people who ate the potato salad at the picnic became ill with nausea and vomiting within 2 to 6 hours after the meal was served. Of those who did *not* eat the potato salad, none became ill.

1. The most likely organism causing the illness would be: (Circle one)

    (a) Group A streptococcus

    (b) Enterococcus

    (c) Coagulase-positive staphylococcus

    (d) Coagulase-negative staphylococcus

    (e) None of above

2. The symptoms were probably caused by:
   (a) An infection of the alimentary canal
   (b) A toxin produced by the organisms growing in the food before it was eaten
   (c) A toxin produced by organisms growing in the intestinal tract
   (d) A substance resulting from bacterial decomposition of the food
   (e) None of these
3. The organism associated with the illness would most likely be identified by
   (a) Culture of the potato salad
   (b) Culture of the vomitus from a patient
   (c) Culture of stool specimens
   (d) Microscopic examination of the remaining salad
   (e) None of these
4. Epidemiologic investigation of the episode of food poisoning revealed that the food handler who prepared the salad had an infection of the hand. An organism was cultured from the sample of the salad dressing used in preparation of the potato salad. To determine the source of the food contamination, one should compare this organism with any organisms recovered from cultures taken. If organisms are found to be of the same species, they are compared by means of:
   (a) A precipitin test to determine serologic type
   (b) The fluorescent antibody technic
   (c) A test to determine bacteriophage type
   (d) A slide agglutination test
   (e) A complement fixation test

Part III

1. Using the letter in front of the appropriate term from column B, indicate the nature of the immunizing substance in column A.

COLUMN A	COLUMN B
_____ Antivenin (Crotalidae)	(a) Dead bacteria
_____ BCG Vaccine	(b) Attenuated bacteria
_____ Botulinum toxoid	(c) Inactivated virus
_____ Cholera vaccine	(d) Live attenuated virus
_____ Diphtheria toxoid	(e) Modified exotoxin
_____ Gamma globulin	(f) Antibodies (immunoglobulins)
_____ Immune serum globulin (human)	
_____ Influenza vaccine	
_____ Measles immune globulin	
_____ Measles vaccine (Edmonston B strain)	
_____ Mumps immune globulin	
_____ Mumps vaccine	
_____ Pertussis vaccine	
_____ Plague vaccine	
_____ Poliomyelitis immune globulin	
_____ Rabies chick embryo vaccine	
_____ Rabies duck embryo vaccine	
_____ Rabies immune globulin	
_____ Rubella vaccine	
_____ Sabin poliomyelitis vaccine	
_____ Salk poliomyelitis vaccine	
_____ Semple vaccine	
_____ Smallpox vaccine	
_____ Tetanus antitoxin	
_____ Tetanus immune globulin	
_____ Tetanus toxoid	
_____ Yellow fever vaccine	

2. The following is a list of diseases that are spread by ingestion of either contaminated milk or water. If the disease is spread by contaminated water, place a W in the blank space in front of the disease; if it is spread by contaminated milk, then place an M in the blank space. Some of these diseases may be spread by both milk and water, in which case both an M and a W are to be placed in the blank space. If the disease is not spread by either medium, the space is to be left blank.

_____ Typhoid fever
_____ Bacillary dysentery
_____ Septic sore throat
_____ Syphilis
_____ Bovine tuberculosis
_____ Actinomycosis
_____ Scarlet fever
_____ Infantile diarrhea
_____ Rabies
_____ Amebic dysentery
_____ Q fever
_____ Relapsing fever
_____ Infectious jaundice
_____ Salmonelloses
_____ Viral hepatitis
_____ Cholera
_____ Rocky Mountain spotted fever
_____ Brucellosis
_____ Diphtheria
_____ Tetanus

3. True-False. Circle either the T or the F.

T F   1. Bacteria are able to obtain carbon and nitrogen from the air because they contain chlorophyll.

T F   2. Pathogenic bacteria that grow in the soil never form spores.

T F   3. Bacteria are seldom found below 6 feet in the soil.

T F   4. Diseases may be transmitted by means of an intermediate host.

T F   5. Drinking water contains a large number of bacteria always.

T F   6. A silo depends on bacteria for its action.

T F   7. The causative organism of a disease must be isolated before a vaccine may be made.

T F   8. In food poisoning from *Salmonella*, digestive disturbances are prominent.

T F   9. *Salmonella* food poisoning has been chiefly associated with canned vegetables.

T F   10. Toxin preformed in food may cause food poisoning.

T F   11. Toxoids are not immunogenic because the part of the molecule responsible for toxicity has been masked or destroyed.

T F   12. A person with a negative Schick test is always completely immune to diphtheria.

T F   13. Material for active immunization against diphtheria is best produced by boiling the diphtheria toxin.

T F   14. Ultraviolet radiation is bactericidal because the heat that results from the ultraviolet radiation coagulates the bacterial proteins.

T F   15. Bacterial counts of pasteurized milk cannot be expected to yield any information useful in preventing disease because pasteurization kills all pathogenic organisms except spore forms.

T F   16. Most cases of pulmonary tuberculosis are acquired by the drinking of infected milk.

721

T　F　17. Food poisoning is often associated with picnics because the food served has not been properly refrigerated for several hours.

T　F　18. Food is an important vehicle in the transmission of syphilis.

T　F　19. Most microbes in nature are not harmful to man.

T　F　20. Ammonia is converted by nitrifying bacteria into nitrites.

T　F　21. The term *nitrogen fixation* refers to the recovery of free nitrogen from the air by unicellular animals.

T　F　22. Lactic acid bacteria are normal milk inhabitants.

T　F　23. The chief milk-curdling enzyme is diastase.

T　F　24. *Saccharomyces* species are important in commercial fermentations.

T　F　25. *Bacillus subtilis* provides alkaline proteases for the manufacture of enzyme detergents.

GLOSSARY

abscess Circumscribed collection of pus.

acquired immunity Immunity acquired after birth.

active carrier Person or animal who becomes carrier after recovery from given disease.

active immunity Immunity brought about by activity of certain body cells of person becoming immune on direct exposure to antigen.

acuminate Pointed, tapering.

acute disease Disease that runs a rapid course with more or less severe manifestations.

adjuvant Substance mixed with antigen to enhance antigenicity and antibody response.

aerobe Organism whose growth requires the presence of oxygen.

afferent Bringing to or into.

affinity Attraction.

agammaglobulinemia Deficiency or absence of gamma globulin in blood usually associated with increased susceptibility to infection.

agar Gelatinous substance prepared from Japanese seaweed used as a base for solid culture media.

agglutination Visible clumping of cells suspended in a fluid.

agglutinins Antibodies that cause agglutination.

agglutinogen Any substance that, acting as an antigen, stimulates production of agglutinins.

algid Cold; an algid fever is one in which the patient goes into a state of collapse.

alkaloid Basic substance found in plants that is usually the part of the plant with medicinal properties.

allergen Substance that induces allergic state when introduced into body of susceptible person.

allergy (hypersensitivity) State in which the affected person exhibits unusual manifestations on contacting an allergen.

alum toxoid Toxoid treated with an aluminum compound.

amboceptor Substance that combines with cells and complement to dissolve cells (example, hemolysin).

ameba, amoeba (pl., amebas, amoebae) Protozoon that moves by extruding fingerlike processes (pseudopods).

ameboid Resembling an ameba.

amino acids Organic chemical compounds containing an amino (NH_2) group and a carboxyl (COOH) group that form the chief structure of proteins.

anaerobe Organism that grows only or best in the absence of atmospheric oxygen.

anaphylactoid Like anaphylaxis.

anaphylaxis State of hypersusceptibility to a protein resulting from a previous introduction of protein into body.

antagonism Mutual resistance.

antibacterial serum Antiserum that destroys or prevents the growth of bacteria.

antibiotic Agent produced by one organism that destroys or inhibits another organism.

antibody Agent in the body that destroys or inactivates certain foreign substances that gain access to body, particularly microbes and their products.

anticoagulant Agent that prevents coagulation.

antigen Substance that, when introduced into the body, causes production of antibodies.

antiluetic Antisyphilitic.

antiseptic Substance that prevents the growth of bacteria.

antiserum Immune serum.

antitoxin Immune serum that neutralizes the action of a toxin.

antivenin Antitoxic serum for snake venom.

aphthous Characterized by the presence of small ulcers.

ascites Abnormal collection of fluid in the peritoneal cavity.

aseptic Free from living microorganisms.

ataxia Failure of muscular coordination.

atopy Human allergy with hereditary background.

attenuated Weakened.

autoantibody Antibody formed against autoantigens.

autoantigen Antigens present in same individual as antibody-producing cells and not "foreign" to body.

autoclave Apparatus for sterilizing by steam under pressure; pressure steam sterilizer.

autogenous vaccine Vaccine made from culture of bacteria obtained from the patient himself.

autoimmune disease Disease wherein autoimmunization against certain body proteins is pathogenetic mechanism.

autoimmunity Unusual state resulting from production *(autoimmunization)* by body of antibodies against its own proteins.

autoinfection Infection of one part of body by bacteria derived from some other part.

autopsy Examination of internal organs of dead body.

autotrophic Organisms that can form their proteins and carbohydrates out of inorganic salts and carbon dioxide.

B lymphocyte (B cell) Thymus-independent lymphocyte.

bacteremia Condition in which bacteria are in the bloodstream but do not multiply there.

bactericide (bactericidal) Agent lethal to bacteria.

723

bacteriology Science that treats of bacteria.

bacteriolysins Antibodies that lyse bacteria in presence of complement.

bacteriophages (phages) Bacterial viruses.

bacteriostasis Inhibition of bacterial growth; bacteria, however, not directly killed.

Bang's disease Contagious abortion of cattle.

BCG (bacillus of Calmette-Guérin) Vaccine against tuberculosis made from bovine strain of tubercle bacilli attenuated through long culturing; name derived from two French scientists developing the strain.

biologic transfer of infection Mode of transfer of infection from host to host by an animal or insect in which agent causing disease undergoes a cycle of development.

biotherapy Treatment of disease with a living agent or its products.

blood serum see serum.

boil Abscess of the skin and subcutaneous tissue.

broad-spectrum Term used to indicate that an antibiotic is effective against large array of microorganisms.

bronchiolitis Inflammation of bronchiole, finer subdivision of respiratory tract.

bronchopneumonia Small focal areas of inflammatory consolidation in the lungs.

bubo Inflammatory enlargement of a lymph node often with formation of pus.

buffer Substance or system of substances in body fluids that lessens effect of addition of acids or alkalies.

cancer Lay term for neoplasm.

capnophilic Growing best in presence of carbon dioxide.

capsule Envelope that surrounds certain bacteria.

carbohydrates Class of organic chemical compounds composed of carbon, hydrogen, and oxygen, latter two in proportion to form water; to this class belong sugars, starches, and cellulose.

carditis Inflammation of the heart.

caries Caries in teeth (dental caries) are layman's cavities.

carrier Person in apparent health who harbors pathogenic agent in his body.

caseous Cheeselike.

catabolism Metabolic breakdown of complex substances into products of simpler makeup.

catarrhal Characterized by an outpouring of mucus.

cell Minute protoplasmic structure, anatomic and physiologic unit of all animals and plants; that is, all animals and plants are made up of one or more cells, and their activities depend on combined activities of those cells.

cell (tissue) culture Cultivation of tissue cells away from human or animal body.

cellulitis Diffuse inflammation of connective tissues.

Celsius Inventor of the temperature scale setting 100° between the freezing point at zero and the boiling point of water.

CDC Center for Disease Control, Atlanta, Georgia.

centigrade Thermometer temperature scale with 100° between the melting point of ice at zero and the boiling point of water at 100.

centrioles Two or more granules contained in centrosome and prominent during cell division.

centrosome (cell center) Condensed portion of cytoplasm of certain cells containing centrioles; it plays an important part in cell division.

chain reaction Series of successive reactions in which each succeeding reaction depends on preceding one and which, when once begun, continues until one or more of chemicals taking part in reaction are exhausted.

chancre Initial lesion of syphilis.

chemotaxis Reaction to a chemical whereby cells are attracted (positive chemotaxis) or repelled (negative chemotaxis) by chemical.

chemotherapy Treatment of disease by administration of drugs that destroy causative organism of disease but do not injure patient.

cholesterol Fatlike alcohol ($C_{27}H_{45}OH$) occurring as crystals with notched corners; found in blood, bile, egg yolk, seeds of plants and animal fats.

chromatin Stainable part of the nucleus of a cell forming a network of fibrils; it is deoxyribonucleic acid attached to protein base and is carrier of the genes in inheritance.

chromatin granules Karyosomes.

chromatography Method of chemical analysis whereby certain compounds of a mixture are separated by use of their solubility and absorptive properties.

chromogenic Pigment producing.

chromosomes Rod-shaped masses of chromatin that appear in the cell nucleus during mitosis; they play an important part in cell division and transmit the hereditary characteristics of the cell.

chronic Long, continued.

cilia (sing., cilium) Hairlike processes that spring from certain cells and by their action create currents in lipids; if the cells are fixed, the lipid is made to flow, but if the cells are unicellular organisms suspended in liquid, the cells move.

ciliates Unicellular organisms that move by means of cilia.

clinical Founded on actual observation.

clinical case Person ill and showing signs of a disease.

clone Progeny of a single cell.

coagulase Enzyme that hastens coagulation.

coagulation Formation of a blood clot.

coliform bacteria Group of bacteria consisting of

Escherichia coli and related intestinal inhabitants.

colon bacillus *Escherichia coli.*

colony Visible growth of bacteria on culture medium; all progeny of single preexisting bacterium.

commensalism Symbiosis in which one individual is benefited and other is not affected.

communicable Capable of being transmitted from one person to another.

complement fixation Destruction or inactivation of complement by combination of antigen, antibody, and complement; the basis of complement fixation tests for syphilis and other diseases.

complication Disease state concurrent with another disease.

condyloma Wartlike growth.

congenital Existing at the time of birth or shortly thereafter.

consolidation Solidification, as of the lung in pneumonia.

consumption Wasting away; a lay term meaning pulmonary tuberculosis.

contagious Highly communicable; in common parlance, a disease easily "caught."

contamination Soiling with infectious material.

convalescent carrier Carrier who harbors organisms of a disease during recovery from disease.

convalescent serum Serum of person recently recovered from a disease; in a few cases injection of convalescent serum seems to be of value in treatment or in prevention of the disease in others.

Coombs' test Test to detect presence of globulin antibodies on surface of red blood cells; used to detect sensitized red cells in hemolytic diseases.

coproantibodies Antibodies formed in the colon.

coprophilic Affinity for feces and filth (bacteria).

coprozoic Living in or found in feces.

counterstain Second stain of different color applied to a smear to make effects of first stain more distinct.

Credé's method Instillation of 2% silver nitrate solution into each eye of newborn infant to prevent ophthalmia neonatorum.

crisis Sudden change in course of a disease; diseases that terminate by sudden change for better are said to end by crisis.

cryobiology Science that treats of effects of low temperatures on biologic systems.

culture Growth of microorganisms on nutrient medium; to grow microorganisms on such a medium.

culture media *(sing.,* **medium)** Artificial food material on which microbes are grown (cultured).

cutaneous Pertaining to the skin; cutaneous inoculation is done by rubbing the infectious material on the abraded skin.

cycle Series of changes considered to lead back to starting point.

cyst Abnormal saccular structure containing liquid, air, or solid material; stage in history of certain protozoa during which time encysted organism is protected by surrounding wall.

cysticercus Larval form of tapeworm in tissues of intermediate host; develops into adult worm on entry into intestinal canal of definitive host.

cytology Science treating of study of cells, their origin, structure, and function.

cytolysin Antibody that dissolves cells.

cytopathogenic (cytopathic) effect Pathologic changes in cells of given cell (tissue) culture referable to action of some injurious agent, especially those from viruses grown in the cell culture.

Dakin's solution Neutral solution of sodium hypochlorite once used in disinfection of wounds.

defibrinate To remove the fibrin of blood to prevent clotting.

definitive Final, ending.

degenerate To undergo progressive deterioration.

degeneration Deterioration; change from higher to lower form.

degerm To remove bacteria from skin by mechanical cleaning or application of antiseptics.

deoxyribonucleic acid (DNA) One of two nucleic acids that have been identified; essential for biologic inheritance.

dermatitis Inflammation of skin.

dermatomycoses Superficial fungous infections of skin and appendages.

dermatophyte Fungus parasitic for skin.

dermotropic Affinity for the skin.

desensitization Condition wherein an organism does not react to a specific antigen to which it previously did; process of bringing about this state.

desquamation Shedding of the superficial layer of the skin in scales or shreds.

diarrhea Abnormal frequency or fluidity of bowel movements.

diatomaceous earth Earth made up of petrified bodies of diatoms or unicellular algae.

Dick test Skin test to determine susceptibility to scarlet fever.

differential stain Stain distinguishing between different groups of organisms.

direct contact Spread of a disease more or less directly from person to person.

disinfectant Substance that disinfects.

disinfection Destruction of all disease-producing organisms and their products.

dissociation Separation; dissolution of relations.

DNA *(see deoxyribonucleic acid).*

droplet infection Infection conveyed by spray thrown off from mouth and nose during talking, coughing, etc.

drug fast Resistance to the action of drugs; microbes able to withstand action of given drug.

DTP Diphtheria-tetanus-pertussis vaccine.

dysentery Diarrhea plus blood and mucus in stool; associated with inflammation of alimentary tract.

EBV Epstein-Barr virus.

ecology Science of organisms as affected by factors of their environment.

ecosystem Basic unit in ecology derived from interaction of living and nonliving elements in given area.

ectoenzyme Enzyme excreted into surrounding medium through plasma membrane of cell forming it.

ectoplasm Outer clear zone of the cytoplasm of unicellular organism.

efferent Bearing away.

electrophoresis Application of an electric current to separate substances that move faster in an electric field from those that move more slowly.

elementary bodies Virus particles.

empyema Collection of pus in a cavity; when used without qualification, collection of pus in pleural cavity.

encephalitis Inflammation of brain.

encephalomyelitis Inflammation of brain and spinal cord.

endemic More or less continuously present in a community.

endocarditis Inflammation of the endocardium or lining membrane of heart including that of the valves.

endoenzyme Enzyme liberated only when cell that produces it disintegrates.

endogenous Originating within the organism.

endoplasm Zone of granular cytoplasm found near nucleus of many unicellular organisms.

endoplasmic reticulum Membrane-limited, canalicular organelle in cytoplasm of cells.

endothelium Flattened cell lining of heart, blood vessels, and lymph channels.

endotoxin Toxin liberated only when cell producing it disintegrates.

enteric bacteria Bacteria isolated from gastrointestinal tract.

enterotoxin Toxin bringing about diarrhea and vomiting usually succeeding ingestion of contaminated food.

enzyme Catalytic substance secreted by living cell capable of changing other substances without undergoing any change itself.

eosinophil Granular leukocyte whose granules stain with eosin.

epidemic Disease that attacks large number of persons in a community at same time.

epidemiology Science that treats of epidemics.

essential (disease) Of unknown cause; idiopathic.

etiology Cause.

eucaryote Protist with true nucleus.

exacerbation Increase in severity of a disease.

exanthem Febrile disease accompanied by skin eruption.

exfoliate To come off in strips or sheets, particularly stripping of skin after certain exanthematous diseases.

exfoliative cytology Study of cells cast off from a body surface.

exogenous Coming from the outside of the body.

exotoxin Toxin secreted by microorganism into surrounding medium.

exudate Fluid and formed elements of blood extravasated into tissues or cavities of body as part of inflammatory reaction.

facultative Able to do a thing although not ordinarily doing it; for example, a facultative anaerobe is an organism that can live in absence of oxygen but does not ordinarily do so.

familial Affecting several members of the same family.

FDA Food and Drug Administration of Department of Health, Education and Welfare.

feedback Return of part of output of a system as input.

ferment Enzyme.

fever Abnormally high body temperature usually related to a disease process.

fibrinolysin Substance that dissolves or destroys fibrin.

filaria Long, threadlike round worm that lives in the circulatory or lymphatic system.

filariform Resembling filariae.

fixed virus (rabies) Street or wild virus adapted to rabbit and less virulent for man and dog.

flagella (sing., flagellum) Long, hairlike processes that by their lashing activity cause organism to move; one or more flagella may be attached to one or both ends of organism or completely around it.

flagellates Organisms that move by means of flagella.

flocculation test Test dependent on coalescence of finely divided or colloidal particles into larger visible particles.

fluid toxoid Toxoid not further treated with aluminum compounds.

fluorescence Property of emitting light after exposure to light.

fluorescent antibodies Antibodies that fluoresce.

focal infection Localized site of more or less chronic infection from which bacteria or their products spread to other parts of body.

fomites Substances other than food that may transmit infectious organisms.

fulminating Sudden, severe, and overwhelming.

fumigation Exposure to fumes of a gas that destroys bacteria, vermin, etc.

fungicide Agent destructive to fungi.

furuncle A boil.

furunculosis Presence of a number of boils.

gamete Cell, either male or female, that undergoes sexual reproduction.

gamma globulin Fraction of globulin of blood with which antibodies are associated.

gene Biologic unit of heredity, self reproducing and located in definite position (locus) on a particular chromosome.

genome Complete set of hereditary factors.

germ Pathogenic microbe.

germ cell Cell specialized for reproduction.

germicide Agent that destroys germs.

Giemsa stain Stain (azure and eosin dyes) used to demonstrate protozoa, viral inclusion bodies, and rickettsias.

globulin Class of proteins characterized by being insoluble in water but soluble in weak solutions of various salts; one of important proteins of the plasma of the blood.

gnotobiosis Science of rearing and keeping animals either born germ free or with limited known microbial flora.

Golgi apparatus Organelle of cell cytoplasm made up of irregular network of canals or solid strands.

gram negative Bacteria decolorized by Gram's method but stain with counterstain (red or brown).

gram positive Bacteria not decolorized by Gram's method; they retain original violet color of Gram stain and are not stained by the counterstain.

Gram stain, Gram's method Method of differential staining devised by Hans Christian Gram, Danish bacteriologist.

granulocyte Cell containing granules within its cytoplasm; granular leukocyte (polymorphonuclear neutrophil, eosinophil, or basophil).

granuloma Circumscribed collection of reticuloendothelial cells surrounding point of irritation.

granulomatous Inflammation with pronounced response of reticuloendothelial system.

grouping Classification *(see also **typing**)*.

gumma Granuloma of late stages of syphilis.

habitat Place where plant or animal found in nature.

halophile Salt loving.

hectare Ten thousand square meters.

hectic fever Daily recurring fever characterized by chills, sweating, and flushed countenance.

hematogenous Originating in blood or borne by blood.

hemoglobin Oxygen-carrying red pigment of red blood cell.

hemoglobinophilic Pertaining to organisms that grow especially well in culture media containing hemoglobin.

hemolysin Antibody dissolving red blood cells in presence of complement.

hemolysis Lysis of red blood cells.

hemolytic Causing hemolysis.

hemophilic Blood loving; hemoglobinophilic.

hepatitis Inflammation of the liver.

hepatogenous Originating in the liver.

hereditary Transmitted through members of family from generation to generation.

herd immunity Resistance to disease related to immunity of high proportion of members of group.

heterologous Derived from an animal of another species.

heterophil Having affinity for antigens or antibodies other than one for which it is specific.

heterotrophic Organisms requiring simple form of carbon for metabolism.

Hinton test Precipitation test for syphilis.

homeostasis Tendency to stability within internal environment or fluid matrix of an organism.

host Animal or plant on which a parasite lives.

hydrolysis Decomposition from incorporation and splitting of water.

hyperimmune Quality of possessing degree of immunity greater than that found under similar circumstances.

hyperplasia Increase in size related to increase in number of component units.

hyperpyrexia Very high fever, usually more than 106° F.

*hypersensitivity (see **allergy**)*.

hypertonic Solution with higher osmotic pressure than that of reference one.

hypha (pl., hyphae) One of filaments composing a fungus.

hypotonic Solution with lower osmotic pressure than that of reference one.

icterus Jaundice.

idiopathic Of unknown cause.

idiosyncrasy Individual and peculiar susceptibility or sensitivity to a drug, protein, food, or other agent.

immune Exempt from a given infection.

immune bodies Antibodies.

immune globulin Sterile preparation of globulin from blood; antibodies normally associated with gamma globulin fraction of blood proteins.

immune serum Serum containing immune bodies.

immunity Natural or acquired resistance to a disease.

727

Glossary

immunohematology Branch of hematology treating of immune bodies in the blood.

immunologist One versed in immunity.

immunology Science that deals with immunity.

impetigo contagiosa Infectious vesicular and pustular eruption most often seen on face and other body areas of children.

inclusion bodies Round, oval, or irregularly shaped particles in cytoplasm or nucleus of cells parasitized by viruses; colonies of viruses.

incompatible Not capable of being mixed without undergoing destructive chemical changes or acting antagonistically.

incubate To promote growth of microorganisms by placing them in an incubator.

incubation period Period intervening between time of infection and appearance of manifestations.

incubator Cabinet in which constant temperature is maintained for purpose of growing cultures of bacteria.

indicator Something that renders visible completion of a reaction.

indirect contact Transfer of infection by means of inanimate objects, contaminated fingers, water, food, and the like.

infection Invasion of body by pathogenic agents with their subsequent multiplication and production of disease; inflammation from living agents.

infectious Having qualities that may transmit disease.

infectious granuloma Granuloma from a specific microbe.

infestation Invasion of body by macroscopic parasites such as insects; refers particularly to parasites on surface of body.

inflammation Protective reaction on part of tissues brought about by presence of an irritant—reaction to injury.

inhibition Diminution or arrest of function.

inoculate To implant microbes or infectious material onto culture media; to introduce artificially biologic product or disease-producing agent into body.

insecticide Substance that destroys insects.

intracellular Between cells.

intercurrent infection Infection that attacks a person already ill of another disease.

intermittent Periods of activity separated by periods of quietude.

intoxication Poisoning.

intracellular Within a cell.

intracutaneous (see intradermal).

intradermal Within substance of skin.

intraperitoneal Within the peritoneal cavity.

intraspinal Within vertebral canal.

intravenous Within a vein.

intrinsic From within.

involution forms Abnormal forms assumed by microorganisms growing under unfavorable conditions.

iodophors Disinfectants with germ-killing iodine carried by surface-active solvent.

isoantigens Antigens from individual of same species.

isolate To close all avenues by which a person may spread infection to others; to separate from others.

isologous Pertaining to same species.

isotonic Solution having the same osmotic pressure as that of standard reference one.

isotopes Atoms of same elements having atomic weights but generally same chemical behavior.

jaundice Deposition of bilirubin in skin and tissues with resultant yellowish discoloration in a patient with hyperbilirubinemia.

Kahn test Precipitation test for syphilis.

karyosome Chromatin mass or knot on linin network of nucleus.

karyotype Chromosomal makeup of a cell, individual, or species.

keratitis Inflammation of cornea.

keratoconjunctivitis Inflammation of the cornea and conjunctiva.

kernicterus Cerebral manifestations of severe jaundice in newborn infant; degeneration of nerve cells caused by accumulation of bilirubin in brain.

Kline test Precipitation test for syphilis.

Koch's postulates Certain requirements that must be met before a given microorganism can be considered the cause of a certain disease.

lag phase Period of time between stimulus and resultant reaction.

larva (pl., larvae) Young of any animal differing in form from its parent.

latent Seemingly inactive, potential, concealed.

lesion Specific pathologic structural or functional (or both) change brought about by disease.

leukocidin Substance that destroys leukocytes.

leukocyte White blood cell.

leukocytosis Transient protective increase in leukocytes in blood in response to injury.

leukopenia Abnormal decrease in leukocytes in blood.

local infection One that is confined to a restricted area.

locus Position.

lues venerea Syphilis.

lumbar Pertaining to that part of the back between the ribs and pelvis.

lumen Space inside a tubular or hollow structure.

lumpy jaw Actinomycosis in cattle.

lymphadenitis Inflammation of a lymph node.

lymphadenopathy Disease of a lymph node (or nodes).

lymphocyte Nongranular white blood cell of lymphoid origin.

lymphocytosis Increase in lymphocytes in the blood.

lymphoid tissue Delicate connective tissue lattice with lymphocytes and related cells in its meshes.

lyophilization Creation of stable product by rapid freezing and drying of frozen product under high vacuum.

macromolecule Very large molecule having polymeric chain structure.

macrophage Large mononuclear wandering phagocytic cell originating in reticuloendothelial system.

macroscopic Visible to naked eye.

macular Consisting of small flat reddish spots in skin.

malignant Virulent; going from bad to worse.

Mantoux test Tuberculin skin test.

marker Something that identifies or is used to identify.

Mazzini test Precipitation test for syphilis.

meatus Opening.

mechanical transfer (of infection) Transfer of infection by insects in which infectious agent is spread mechanically and undergoes no cycle of development in body of particular insect.

medical technologist Specialist in technics of tests vitally related to medicine.

medulla Center of a part or organ.

meiosis Special type of cell division during maturation of sex cells by which normal number of chromosomes is halved.

membranous croup Lay term for diphtheria.

metabolism Sum total of chemical changes whereby nutrition and functional activities of body maintained.

metachromatic granules Granules of deeply staining material found in certain bacteria.

metazoa Multicellular animals.

microaerophilic Applied to microorganisms that require free oxygen for their growth but in an amount less than that of oxygen of atmosphere.

microbe Microscopic unicellular organism.

microgram One thousandth of a milligram or 1/1,000,000 of a gram.

micrometer One thousandth of a millimeter or 1/25,000 of an inch; a micron.

micron (see micrometer).

microorganism Organism of microscopic size.

microscopy Study of objects by means of the microscope.

miliary Small, resembling a millet seed in size.

minimum lethal dose (MLD) Smallest dose that will cause death.

mitochondrion (pl., mitochondria) Small spherical or rod-shaped cytoplasmic organelle(s).

mitosis Indirect cell division.

mixed culture Culture containing two or more kinds of organisms.

mixed infection Infection with two or more kinds of organisms.

mixed vaccine Vaccine containing two or more kinds of organisms.

molds Multicellular fungi.

molt Act of shedding outer body covering (skin, cuticle, feathers).

morbid Pertaining to or affected with disease.

morbidity Departure (subjective or objective) from state of physiologic well-being (WHO definition).

mordant Chemical added to a dye to make it stain more intensely.

morphologic Pertaining to shape or form.

mutation Change or alteration in form or qualities; permanent transmissible change in characters of an offspring from those of its parents.

mycelium Vegetative part of a fungus, consisting of many hyphae.

mycobacteriosis Disease produced by certain unclassified members of the genus *Mycobacterium.*

mycohemia Presence of fungi in bloodstream.

mycology Science that deals with fungi.

mycophage Fungal virus.

mycosis Disease caused by fungi.

mycotoxin Fungal toxin.

nanometer (millimicron) One millionth of a millimeter.

nasopharynx Portion of the pharynx above the palate.

natural immunity Immunity with which a person or animal is born.

necrosis Death of tissue while yet part of living body.

negative staining Staining the background, not the organism.

Negri bodies Diagnostic inclusion bodies found in certain brain cells of animal with rabies.

neoplasm New growth of abnormal cells of autonomous nature (scientific term for layman's "cancer").

neuritis Inflammation of a nerve.

neurotropic Affinity for central nervous system or nervous tissue.

NIH National Institutes of Health.

normal flora Bacterial content of given area during health; reasonably constant as to quantity and proportions.

nosocomial Pertaining to hospital or infirmary.

nosology Science of classification of disease.

Glossary

nuclear medicine Use of radioisotopes in medicine.

nucleic acids Complex chemical substances closely associated with transmission of genetic characteristics of cells; the two identified are ribonucleic acid and deoxyribonucleic acid.

nucleolus (pl., nucleoli) Body within nucleus of cell that takes part in metabolic processes of the cell and plays part in its multiplication.

nucleoprotein Simple basic protein combined with nucleic acid.

nucleus (pl., nuclei) Central, compact portion of cell, the functional center.

old tuberculin (OT) Special type of tuberculin.

oncogenic Tumor producing.

oncology Study of neoplasms.

opportunists Microbes that produce infection only under especially favorable conditions.

opsonins Substances in blood that render microorganisms more susceptible to phagocytosis.

organelle Specialized part of protozoon that performs special function; specific particles of organized living substance present in almost all cells.

osmosis Passage of fluids or other substances through a membrane.

osteomyelitis Inflammation of the bone marrow.

otitis media Inflammation of the middle ear.

pandemic Very widespread epidemic, even of worldwide extent.

parasite Animal or plant organism that lives on another organism.

parasitology Science that treats of parasites and their effects on other living organisms.

parenchyma Specialized functioning tissue or cells of an organ.

parenterally In some manner other than by intestinal tract.

parietal Relating to wall of a cavity.

paroxysm Sudden attack of disease or acceleration of manifestations of existing disease.

passive carrier Carrier who harbors the causative agent of disease without having had the disease.

passive immunity Immunity produced without body of person or animal becoming immune taking any part in its production.

pasteurization Heating of milk for a short time to a temperature that will destroy pathogenic bacteria but not affect its food properties and flavor.

pathogenicity Disease-producing quality.

pathogenesis Sequence of events in development of given disease state.

pathognomonic Specifically characteristic or diagnostic of a disease.

Paul-Bunnell test Heterophil antibody test on serum; important in diagnosis of infectious mononucleosis.

peritonitis Inflammation of the peritoneum.

permanent carrier Carrier who harbors disease-producing agent for months or years.

petechia (pl., petechiae) Pin-point hemorrhage.

Petri dish Round glass dish with cover used for growing bacterial cultures.

Peyer's patches Collection of lymphoid nodules packed together to form oblong elevations of mucous membrane of the small intestine, their long axis corresponding to that of intestine.

phage typing Use of bacteriophages and their lytic properties to classify bacteria.

phagocyte Cell capable of ingesting bacteria or other foreign particles.

phagocytic Related to phagocytes or phagocytosis.

phagocytosis Process of ingestion by phagocytes.

phenol coefficient Measure (ratio) of disinfecting property of a chemical compared with that of phenol (carbolic acid).

phlegmon Acute diffuse inflammation of subcutaneous connective tissue.

photosynthesis Elaboration of glucose from carbon dioxide and water by sunlight in presence of chlorophyll.

pinocytosis Cell-drinking, imbibition of liquids by cells; minute invaginations on surface of cells close to form fluid-filled vacuoles.

Pirquet's test (von Pirquet) Tuberculin skin test.

plaques Visible areas of cellular damage caused by virus inoculated into susceptible cell culture, analogous to colonies of bacteria on agar plate.

plasma Fluid portion of circulating blood; fluid portion of *clotted* blood is *serum*.

plasma membrane Outer membrane encasing the protoplasm of a cell.

plasmids Generic term for intracellular inclusions considered to have genetic functions; extrachromosomal genetic elements.

plasmolysis Shrinking of cell suspended in hypertonic solution.

plasmoptysis Swelling and bursting of a cell suspended in hypotonic solution.

plastids Small bodies found in cytoplasm of cells; they have to do with cell nutrition and contain the chlorophyll of green plants.

pleomorphism Existence of different forms in the same species.

pleurisy Inflammation of the pleura.

pleuropneumonia Infectious pneumonia and pleurisy of cattle.

pneumonia Inflammatory consolidation or solidification of lung tissue; presence of exudate blots out air-containing spaces.

pneumonitis Inflammation of supporting framework of the lung.

pneumotropic Affinity for the lungs.

polar bodies Deeply staining bodies found in one or both ends of certain species of bacteria.

pollution State of being unclean; as used in bacteriology, containing harmful substances other than bacteria.

precipitation Clumping of proteins in solution by addition of specific precipitin.

precipitins Antibodies that cause precipitation.

prehension Act of taking hold; grasping.

preservative Substance added to product to prevent bacterial growth and consequent spoiling.

primary First (first focus of disease).

primary infection First of two infections, one occurring during course of other.

procaryote Protist without true nucleus.

prognosis Forecast of outcome of a disease.

properdin Protein component of globulin fraction of blood playing role in immunity.

prophylaxis Prevention of disease.

proprietary Referring to fact that a given item is a commercial one.

protein One of group of complex organic nitrogenous compounds, widely distributed in plants and animals, forming the principal constituents of protoplasm; they are essentially combinations of amino acids and their derivatives.

proteolytic Bringing about digestion or liquefaction of proteins.

protist Member of kingdom Protista that includes all single-celled organisms (bacteria, algae, slime molds, fungi, protozoa); some plantlike, some animal-like, some neither.

protoplasm Living material of which cells are composed.

protozoology Science that treats of protozoa.

protozoon (pl., protozoa; adj., protozoan) Unicellular animal organism.

provocative dose Dose stimulating appearance of given effect.

pseudomembrane Fibrinous exudate forming a tough membranous structure on surface of skin or mucous membrane.

pseudopod Temporary protoplasmic process put forth by protozoon for purposes of locomotion or obtaining food.

psychrophile Cold-loving organism, growing best at low temperatures.

ptomaines Basic substances resembling alkaloids formed during decomposition of dead organic matter.

pure culture Culture containing only one species of organism.

purulent Containing pus.

pus Fluid product of inflammation consisting of leukocytes, bacteria, dead tissue cells, foreign elements, and fluid from the blood.

pustule Circumscribed elevation on skin containing pus.

putrefaction Decomposition of proteins.

pyelitis Inflammation of renal pelvis.

pyelonephritis Inflammation of kidney parenchyma and pelvis.

pyemia Form of septicemia in which organisms in bloodstream lodge in organs and tissues and set up secondary abscesses.

pyogenic Pus forming.

quiescent Not active.

quinsy Peritonsillar abscess.

racial immunity Immunity peculiar to a race.

radiant energy Energy from radioactive source.

radioisotope Isotope form of element that is radioactive.

raw milk Unpasteurized milk.

receptors Precise chemical groupings on surface of target cell; those on immunologically competent cell can combine specifically with antigen; term used by Ehrlich in his side-chain theory of immunity to denote specialized portions of cell that combine with foreign substances.

recrudescent Recurrence of symptoms of disease after period of days or weeks.

recurrent Reappearance of symptoms after an intermission.

reduction Removal of oxygen from or addition of hydrogen to a compound.

remission Temporary cessation of manifestations of a disease.

remittent Characterized by remissions.

rennin Milk-curdling enzyme.

replication Process by which genetic determinants are duplicated during cell multiplication so that identical genetic characters are passed to next generation.

resident bacteria Bacteria normally occurring at given anatomic site.

resistance Inherent power of body to ward off disease.

reticuloendothelial system System of phagocytic cells scattered through various organs and tissues, particularly spleen, liver, bone marrow, and lymph nodes, playing an important part in immunity.

Rh factor Blood factor, agglutinogen, found on red blood cells so-named because of its occurrence on red blood cells of rhesus monkeys.

rhinitis Inflammation of the nose.

ribonucleic acid (RNA) One of two nucleic acids (*see also* **DNA**).

ribosomes Ribonucleoprotein granules in cell cytoplasm.

ringworm Fungous disease of the skin.

rodent Gnawing mammal (rats, mice, etc.)

rose spots Characteristic spots in skin over lower portion of trunk and abdomen in typhoid fever.

Glossary

sanitary Conducive to health.

sanitize To reduce number of bacteria to safe level as judged by public health standards.

sapremia Condition in which products of action of saprophytic bacteria on dead tissues are absorbed into body and produce disease.

saprophyte Organism that normally grows on dead matter.

Schick test Skin test to detect susceptibility to diphtheria.

scrofula Tuberculosis of the lymph nodes, particularly those of neck.

secondary infection Infection occurring in host suffering from primary infection.

sedimentation Settling of solid matter to bottom of a liquid.

selective action Tendency on part of disease-producing agents to attack certain parts of body.

Semple vaccine Rabies vaccine prepared from rabbit brain treated with phenol.

sepsis Poisoning by microbes or their products.

septic Relating to presence of pathogenic organisms or their poisonous products.

septicemia Systemic disease caused by invasion of bloodstream by pathogenic organisms, with their subsequent multiplication therein.

serology Branch of science that deals with serums, especially immune serums.

serum Fluid that exudes when blood coagulates; portion of plasma left after plasma protein fibrinogen is removed.

signs Objective disturbances produced by disease; observed by physician, nurse, or medical attendant.

simian viruses Viral contaminants in cell cultures of normal monkey cells.

simple stain Stain using only one dye.

sinus Tract leading from an area of disease to body surface.

skin test dose (STD) Unit of measurement of scarlet fever toxin, amount required to produce a positive reaction on skin of person susceptible to scarlet fever.

smear Very thin layer of material spread on glass microslide.

species immunity Immunity peculiar to a species.

sporadic disease Disease that occurs in neither an endemic nor epidemic.

spore Highly resistant form assumed by certain species of bacteria when grown under adverse influences; reproductive cells of certain types of organisms.

sporulation Production of spores or division into spores.

sterile Free of living microorganisms and their products.

sterilization Process of making sterile.

stock vaccine Vaccine made from cultures other than those from patient who is to receive the vaccine.

strain Subdivision of species.

streptolysin O and S Hemolysins produced by streptococci.

stroma Supporting framework of an organ or gland.

subacute Between acute and chronic in time.

subcutaneous Under the skin.

substrate Substance on which an enzyme acts.

sulfur granules Small yellow granules present in pus from lesions of actinomycosis.

superinfection New (superimposed) infection with drug-resistant microbes as complication of antimicrobial therapy for preexisting infection.

suppuration Formation of pus.

sycosis barbae Folliculitis of beard.

symbiosis Mutually advantageous association of two or more organisms.

symptoms Subjective disturbances of disease, felt or experienced by patient but not directly measurable; for example, pain—the patient feels it definitely but it cannot be seen, heard, or touched.

syndrome Set of symptoms and signs occurring together in a complex; etiologic agents variable.

taxon Particular group into which related organisms are classified.

taxonomy Branch of biology treating of arrangement and classification of biologic organisms.

template Pattern, guide, mold, or blueprint.

terminal disinfection Disinfection of room after it has been vacated by patient.

terminal infection Infection that occurs during course of chronic disease and causes death.

Thallophyta Division of plant kingdom to which fungi belong.

thermal death point Degree of heat necessary to kill liquid culture of given species of bacteria in 10 minutes.

T lymphocyte (T cell) Thymus-dependent lymphocyte.

tinea A dermatomycosis.

tissue culture see *cell culture.*

titer Measure of minimal amount of given substance needed for precise result in titration; standard strength of a solution established by titration.

toxemia Presence of toxins in the blood.

toxin Poisonous substance elaborated during growth of pathogenic bacteria.

toxoid Toxin treated in such a manner that its toxic properties are destroyed without affecting its antibody-producing properties.

transduction Transmission of genetic factor from one bacterial cell to another by viral agent.

transformation Artificial conversion of bacterial

732

types within a species by transfer of DNA from one bacterium to another.

trophozoite Active, motile, feeding stage of a protozoan organism.

tubercle Granuloma that forms the unit specific lesion of tuberculosis.

tuberculin Toxic protein extract obtained from tubercle bacilli.

tuberculous Affected with tuberculosis.

tumor Mass.

typing (classification) Determination of category to which an individual, object, microbe, and the like, belongs with respect to given standard of reference.

ulcer Circumscribed area of inflammatory necrosis of epithelial lining of a surface.

ulceration Process of ulcer formation.

unit Standard of measurement.

vaccination Introduction of a vaccine into body.

vaccine Causative agent of disease so modified that it is incapable of producing disease while retaining its power to cause antibody formation.

vaccinia Cowpox.

variation Deviation from parent form.

VDRL (Venereal Disease Research Laboratory) test Precipitation test for syphilis.

vector Carrier of disease-producing agents from one host to another, especially an arthropod (fly, mosquito, flea, louse, or other insect).

vegetative bacteria Nonsporeforming bacteria or sporeforming bacteria in their nonsporulating state.

venereal Transmission of disease by intimate sexual contact.

vesicle (blister) Small circumscribed elevation of skin containing thin nonpurulent fluid.

viremia Presence of virus in bloodstream.

virology Science that treats of viruses and viral diseases.

virucide Agent destroying or inactivating viruses.

virulence Ability of an organism to produce disease.

virus-neutralizing antibodies Antibodies that inactivate viruses.

viscerotropic Affinity for internal organs of chest or abdomen.

viscus (pl., viscera) Internal organ, especially one of abdominal organs.

vital functions Functions necessary for maintenance of life.

vitamins Certain little-understood food substances whose presence in very small amounts is necessary for normal functioning of body cells.

Wassermann test Complement fixation test for syphilis devised by August von Wassermann.

Weil-Felix reaction Nonspecific but highly valuable agglutination test for typhus fever; uses a member of *Proteus* group as organism agglutinated.

wheal Circumscribed elevation of skin caused by edema of underlying connective tissue.

Widal test Agglutination test for typhoid fever.

wild virus Virus found in nature.

Wright stain Mixture of eosin and methylene blue used to demonstrate blood cells and malarial parasites.

X rays (roentgen rays) Highly penetrating form of ionizing radiation produced in special high-voltage equipment.

yeasts Unicellular fungi.

zoonosis (pl., zoonoses) Disease of animals that may be secondarily transmitted to man.

INDEX

A

Å; *see* Angstrom
ABO blood group system, 205, 209-211, 212
Abortion, contagious, 385
Abscesses
  actinomycosis and, 433
  amebiasis and, 582, 583, 584
  anaerobes and, 402
  Bacteroides and, 412
  balantidiasis and, 592
  blastomycosis and, 563
  chancroid and, 392
  coliform bacilli and, 377
  listeriosis and, 462
  mycobacteriosis and, 450
  nocardiosis and, 434, 435
  Pseudomonas and, 453
  pus from, 135, 139
  staphylococci and, 332, 333, 334
  streptococci and, 340
Acetic acid, 648
Acetic acid bacteria, 649
Acetobacter, 649
Acetomonas, 649
Acetone, 73, 650
Acid-fast mycobacteria, 436-451, 624-625
  leprosy and, 446-449
  mycobacteriosis and, 449-450
  stain for, 73, 74-75
  tuberculosis and, 436-446; *see also*
    Tuberculosis
Acidity, bacteria and, 79-80, 146
Acids
  in disinfection, 285-286
  fermentation and, 123-124
Acme, 168
Acriflavine, 286
ACTH; *see* Adrenocorticotropic hormone
Actinobacillus mallei, 455
Actinomyces bovis, 433
Actinomyces israelii, 433
Actinomyces muris ratti, 425
Actinomycetes, 426, 432-435
  as sources of antibiotics, 291
Actinomycin D, 292
Actinomycosis, 433-434
Activators, molecular, 83
Active transport, 86
Adaptation, 62
Adenine, 46, 47
Adenine arabinoside, 290, 301
Adenoidal-pharyngeal-conjunctival viruses, 530;
  see also Adenoviruses
Adenosine diphosphate, 86
Adenosine triphosphate, 86
Adenoviruses, 494
  cancer and, 508-509
  diagnosis and, 501
  hepatitis and, 543
  inclusion bodies and, 492

Adenoviruses—cont'd
  respiratory illness and, 528, 529, 530-531, 532
  sewage and, 655, 661
  vaccines and, 687
Adjuvants, 193
ADP; *see* Adenosine diphosphate
Adrenocorticotropic hormone, 204
Aedes mosquitoes, 172, 540, 541, 542
Aerial mycelium, 555
Aerobacter aerogenes, 379; *see also Enterobacter*
Aerobes, 80
Aerosols for disinfection, 315
Aflatoxins, 562-563
African sleeping sickness, 584-585
African trypanosomiasis, 584-585
Agammaglobulinemia, 200, 203, 680-681
Agar, 108, 131, 149
Agglutination, 219-222, 250-251
  blood cross matching and, 212
  discovery of, 27
  nonspecific, 221
Agglutinins, 216, 217, 219-222
  of blood, 209-212
  mycoplasmas and, 464
  typhoid fever and, 372
Agglutinogens, 219
Agramonte, A., 541
Agranular reticulum, 44
Agriculture, microbiology of, 646-647
Air
  bacteria in, 249
  disinfection of, 315
  infection and, 171
AKD vaccine, 685
Alastrim, 516
Albert's stain, 75
Alcaligenes, 125, 666
Alcohol, 279, 281-282
  butyl, manufacture of, 650
  clostridia and, 408
  ethyl, 279, 281, 312
    manufacture of, 648
  hand disinfection and, 310
  and iodine, 279, 307, 312
  isopropyl, 279, 281, 306, 307, 312
  manufacture of, 647-648, 650
  thermometers and, 314
  vinegar and, 648
Alcoholic beverages, 647-648
Aldehydes, 281-282; *see also* Formalin
Aleppo button, 586
Alexin, 200
ALG; *see* Antilymphocyte globulin
Alimentary tract; *see also* Intestines
  as portal of entry, 165
  resident population of, 163-164
Alkaline glutaraldehyde, 503
Alkaline proteases, 651
Alkalinity, bacteria and, 79-80, 146

734

Index

Index

Index

Index

Index

Index

Index

Gastric juice, 178
Gastroenteritis
  Salmonella and, 370
  staphylococci and, 334
  Vibrio and, 457
Gastrointestinal tract; *see* Digestive system;
    Intestines
Gel diffusion, 225
Gelatin, 108, 124
Gelmo, P., 31
Genera, 4
General paresis, 419
Genes, 48
  mutations and, 62-63
Genetic code, 47
Genetic factors, bacterial variations and, 62-64
Genetic immunity, 189
Gengou, O., 29
Genitourinary tract; *see* Urinary tract
Gentamicin, 291, 292, 298
Gentian violet, 72, 73, 108, 286
Genus, 4
Geopen; *see* Disodium carbenicillin
Geotrichosis, 573
German measles; *see* Rubella
Germicides, 275
Giant cell, 492, 493
Giardia lamblia, 580, 586, 587
  disinfection and, 285
  drugs and, 289
  sewage and, 661
  traveler's diarrhea and, 708
Giardiasis, 586-587
Gibberellins, 651
Gilchrist's disease, 563
Gingivitis, 424, 520
Gingivostomatitis, 520
Glanders, 455-456
Glandular cell, 48
Glandular fever, 547
Globulin, 193-199
Glomerulonephritis, 200, 340
Glottis, 177
Gloves, sterilization of, 309
Gluconobacter, 649
Glucose
  degradation of, 87
  fermentation of, 122-124, 146-147, 560
Glutamate, 651
Glutamic acid, 651
Glutaraldehyde, 279, 282, 306, 307
  viruses and, 503
Glycolytic pathway, Embden-Meyerhof, 87
Golgi apparatus, 44
Gonococcemia, 354
Gonococci; *see* Gonorrhea; *Neisseria gonorrhoeae*
Gonorrhea, 351-360; *see also Neisseria
    gonorrhoeae*
  immunity and, 358
  laboratory diagnosis of, 135, 354-358
  prevention of, 358-359
  social importance of, 358
  sources and modes of infection of, 354
  VD pandemic and, 359-360

Goodpasture, E. W., 30
Gorgas, W. C., 541
Gracastoro, G., 21
Graft-versus-host reaction, 207
Grafts, 205, 206
Gram, H. C., 73
Gram-negative bacteria, 73; *see also*
    Gram-negative rods
  anaerobes and, 412
  Bartonella bacilliformis as, 474
  cell wall of, 54
  chlamydiae as, 475
  Neisseriae as; *see Neisseriae*
  of intestinal tract, 109-112
  rickettsias as, 466
  Vincent's angina and, 424
Gram-negative rods; *see also*
    Gram-negative bacteria
  Bacteroides as, 411
  enteric bacilli as; *see* Enteric bacilli
  Pseudomonas as, 452-456; *see also
    Pseudomonas*
  small, 385-399, 622
    biochemical identity of, 394
    Brucella in, 385-389
    Calymmatobacterium granulomatis in, 399
    Francisella tularensis in, 394, 396-397
    hemophilic bacteria in, 389-393
    Pasteurella in, 394, 398-399
    Vibrio in, 456-458
    Yersinia in, 393-396
Gram-negative shock or sepsis, 382, 453
Gram-positive bacteria, 73, 622-624; *see also*
    Gram-positive cocci
  actinomycetes and, 432-435
  anaerobes and, 412
  anthrax bacillus as, 458-459
  cell wall of, 54
  Corynebacteria as, 426-432
Gram-positive cocci; *see also* Gram-positive
    bacteria
  anaerobes and, 412
  clostridia and, 407-409
  Lactobacilli and, 459-461
  Listeria monocytogenes as, 461-462
  pyogenic, 329-350
    staphylococci in, 329-336; *see also*
      Staphylococci
    streptococci in, 336-345; *see also* Streptococci
    Streptococcus pneumoniae in, 345-350
  as sources of antibiotics, 290
Gram stain, 73-75
  gonococci and, 355
Gram-variable bacteria, 73
Gramicidin, 294
Granular reticulum, 44
Granuloma inguinale, 399, 492
Granulomas
  coccidioidal, 568-569
  leprosy and, 446
  mycobacteriosis and, 449
  nocardiosis and, 435
  paracoccidioidal, 565
Granulomatosis infantiseptica, 462

748

Index

Index

Infectious mononucleosis, 222, 547-548
Inflammation, 167
  granulomatous, 181-183
  immune complex disease and, 200
  as protective mechanism, 179
  viruses and, 493
Influenza
  vaccines for, 687-688, 702-704, 706
  viral, 501, 523-528, 529, 532
Influenza bacillus, 389, 391-392
Influenza-like diseases, coxsackieviruses and, 540
Influenza viruses, 501, 528, 529, 532
INH; *see* Isoniazid
Injections
  routes of, 683
  site for, 693
Injury, inflammation and, 179
Inoculation
  of animals; *see* Animals, inoculation of
  of culture, 112, 113
  infection and, 170
Inosinate, 651
Insecticides, 651
Insects
  bacteria on, 249
  bites of, 237
  in spread of disease, 171
  viral diseases in, 550
Instruments, surgical, 306-307, 308-309
Interference
  bacterial, 336
  phenomenon of, 503
Interferon, 178, 202, 301-302, 503
Interferon inducers, 302, 503
Intermediate host, 577
Intermedic, 708
Intermicrobial transfer, 63-64
Intermittent sterilization, 270-271
International Association for Medical Assistance
    to Travelers, Inc., 708
International travel, 707, 708-709
Interstitial plasma cell pneumonia, 596
Intestinal flagellates, 586-587
Intestinal worms, 617-618
Intestines; *see also* Digestive system
  bacteria of
    culture media and, 109-112
    fermentation of lactose and, 124
  infections of, swimming pools and, 661
  perforation of, 371
  sulfonamides and, 288
  tuberculosis of, 442
Intradermal tests; *see* Skin tests
Intravenous test, 692-693
Invasin, 340; *see also* Hyaluronidase
Invasion of body by microbes, 161-165
Invasion period, 168
Involution, 61
Iodamoeba butschlii (williamsi), 579
Iodides, 238, 241
Iodine, 279, 284, 307, 312, 314
  Gram stain and, 73
  swimming pools and, 661
  viruses and, 503

Iodine—cont'd
  water purification and, 658
5-Iodo-2'-deoxyuridine; *see* Idoxuridine
Iodophors, 279, 284, 307, 310, 312, 313, 314
Ionizing radiations, 273
  foods and, 673
  vaccines and, 700, 705
Iosan, 284
IPV; *see* Inactivated poliovirus vaccine
Iridescent phenomenon, 452
Isoantibodies, 193
Isoantigens, 193
Isoenzyme, 84
Isogeneric transplant, 205
Isograft, 205
Isolation, 312
Isomerases, 85
Isoniazid, 288, 292, 437, 444
Isonicotinylhydrazine; *see* Isoniazid
Isopropyl alcohol, 279, 281, 306, 307, 312
Isotonic solution, 81

J

Jail fever, 468
Japanese B encephalitis, 536
Japanese disease, 473
Jaundice, 423, 543
Jenner, E., 23
Johne's disease, 436

K

Kafocin; *see* Cephaloglycin
Kahn test, 219, 420
Kala-azar, 586
Kalafungin, 290
Kanamycin, 290, 292, 297, 301
Kantrex; *see* Kanamycin
KCN test; *see* Potassium cyanide test
Keflex; *see* Cephalexin
Keflin; *see* Cephalothin
Kendrick, P. L., 30
Kenny, M., 31
Keratitis
  gonorrheal, 353
  interstitial, syphilis and, 422
Keratoconjunctivitis, epidemic, 530
Killed virus vaccines, 686
Kinases, 90
  bacterial, 166
Kingdoms, 4
Kircher, A., 22
Kissinger, J. R., 541
Kitasato, S., 26, 27
Klarer, 31
Klebs, E., 426
Klebs-Löffler bacillus, 426
Klebsiella, 368, 369, 378-379
  endotoxin shock and, 382
  meningitis and, 364
  resistance factors and, 294
Klebsiella pneumoniae, 378
  capsules and, 55
  Schistosoma mansoni and, 602
Kline test, 219, 420

Index

Index

Index

Index

Index

Schultz-Dale reaction, 237
Schwann, T., 23
Schwarz, 688, 704
Sclerosing panencephalitis, subacute, 537
Scolex, 602, 604
Scotochromogens, 449-450
Scrapie, 537
Scratch test, 443, 692
Scrofula, 445
Scrub typhus, 468-469, 473
Scrubbing, 265
Scutula, 561
Seatworm, 599, 610-612
Sedimentation, 266
   water purification and, 658
Selective toxicity, 293
Semipermeable membrane, 80-81
Semisynthetic penicillins, 296
Semlike Forest disease, 549
Semmelweis, I. P., 24
Semple vaccine, 534, 689
Sensitivity, 420; *see also* Hypersensitivity
   to drugs, 241
Sensitizing dose, 236
Septa, 555
Septic sore throat, 340, 344
Septicemia, 170
   anaerobes and, 402
   coliform bacilli and, 377
   Friedländer's bacillus and, 379
   hemorrhagic, 398-399
   listeriosis and, 462
   lymphatic circulation and, 186
   meningococcal, 360, 361, 362
   Pseudomonas and, 455
   Salmonella and, 370
   staphylococci and, 332, 333, 334
   streptococci and, 340, 344
Septra; *see* Co-trimoxazole
Serology, 216, 229, 250
   blood for, 133, 137
   fungi and, 559
   metazoa and, 600
   protozoa and, 581
   syphilis and, 419-421
   viruses and, 500, 501
Serratia, 368, 369, 378-379
   endotoxin shock and, 382
   resistance factors and, 294
Serratia marcescens, 91, 379, 380
Serum, 193
   bactericidal effect of, 26, 27
   collection of, 216
Serum hepatitis, 543
Serum proteins, 675
Serum sickness, 239-240
Sewage, 652, 660-661
   adenoviruses in, 655
Sexual spore, 556
Shigella, 368, 369, 375-377
   biochemical reactions of, 375
   colonies of, 111
   fermentation and, 124
   resistance factors and, 294

Shigella—cont'd
   traveler's diarrhea and, 708
Shigella boydii, 375, 376
Shigella dysenteriae, 375, 376
Shigella flexneri, 89, 375, 376
Shigella sonnei, 375, 376
Shingles, 518-520
Shipyard eye, 530
Shock
   anaphylactic, 235-237, 238
   endotoxin, 382
   gram-negative, *Pseudomonas* and, 453
Shock tissue, 235
Shoes, disinfection of, 313
Silver nitrate, 280-281, 358
Silver stains, 595, 596
Simulium, 617
Single gel diffusion, 225
Sinus infections, 453, 562
Sisomicin, 291
Skin
   abscesses of, 334
   allergy and, 239
   diseases of
      anaerobes and, 402
      swimming pools and, 661
      vaccines and, 692, 700
      viruses and, 501, 513-523
   disinfection of, 312
   normal flora of, 162
   as portal of entry, 165
   preoperative preparation of, 311-312
   as protective mechanism, 177
   sporotrichosis and, 572-573
Skin rash
   in chickenpox, 517
   in measles, 514
   in rubella, 514
   in scarlet fever, 514
   in smallpox, 517
Skin tests, 228, 229, 343, 692
   allergy and, 243, 692
   antitoxins and, 675-676
   for blastomycosis, 564-565
   botulism antitoxin and, 676
   Brucella and, 388
   candidiasis and, 568
   cat-scratch disease and, 480
   chancroid and, 392
   coccidioidomycosis and, 569
   histoplasmosis and, 572
   immunity and, 201
   for leprosy, 448
   lymphopathia venereum and, 478
   mumps and, 547
   Pseudomonas and, 455-456
   tuberculin, 26, 228, 241, 438, 443-444, 702
   viruses and, 502
Skunks, 532-533
Slaked lime, 287
SLE; *see* St. Louis encephalitis
Sleeping sickness, African, 584-585
Slide cultures, 119
Slipping, 52

Index

Index

Index

Index